Flora of Alberta

SECOND EDITION

This thoroughly revised, completely updated, and considerably expanded edition of *Flora of Alberta* retains the basic format and purpose of the first edition published in 1959, and gratefully acknowledged in the intervening years by a wide audience.

This volume provides an inventory of the 1775 known vascular plant species – ferns, conifers, and flowering plants – that are native to Alberta, or if non-native, that have become established and grow wild in the province. All are grouped according to genus and family, and descriptions of species, genera, and families are included.

Keys are provided to allow for the identification of all the known species. Many keys have been revised to improve their effectiveness. There is also information on habitat or habitats in which species generally occur. The text for each species now includes the chromosome number and indicates the geographical distribution in North America.

An important innovation in this edition is the inclusion of 1158 provincial distribution maps, showing the locations of native species, and of two specially prepared maps of the province – one indicating the distribution of major soil groups, the other the distribution of major vegetation types.

Like its predecessor, this revised edition will be invaluable to botanists, agriculturalists, foresters, wild-life specialists, teachers, students, and lovers of nature.

E.H. MOSS (1892–1963) was Professor of Botany at The University of Alberta.

JOHN G. PACKER is a member of the Department of Botany at The University of Alberta.

Flora of Alberta

A Manual of Flowering Plants, Conifers, Ferns and Fern Allies found growing without Cultivation in the Province of Alberta, Canada

E.H. MOSS

Second Edition
Revised by

JOHN G. PACKER

UNIVERSITY OF TORONTO PRESS
Toronto Buffalo London

Printed in Canada

Reprinted 1992, 1996, 2000, 2008

ISBN 0-8020-2508-0

Printed on acid-free paper

Canadian Cataloguing in Publication Data

Moss, E.H. (Ezra Henry), 1892–1963
Flora of Alberta

Bibliography: p.
Includes index.
ISBN 0-8020-2508-0
1. Botany – Alberta. I. Packer, John G. II. Title.
QK203.A4M67 1983 581.97123 C83-098145-4

University of Toronto Press acknowledges the financial assistance to its publishing program of the Canada Council for the Arts and the Ontario Arts Council.

This book has been published with the help of a grant from the Humanities and Social Sciences Federation of Canada, using funds provided by the Social Sciences and Humanities Research Council of Canada.

University of Toronto Press acknowledges the financial support for its publishing activities of the Government of Canada through the Book Publishing Industry Development Program (BPIDP).

Contents

Preface to the Second Edition vii

Preface to the First Edition xi

Introduction 1

Key to the Major Groups and Families of Vascular Plants 8

Number of Taxa 23

Descriptive Flora 24

Geographical Abbreviations 601

Glossary 603

Bibliography 613

Distribution Maps 615

Index 665

Preface to the Second Edition

The first edition of *Flora of Alberta* was described by Moss in his preface as a 'pioneer effort'; this it was and a most successful one. His *Flora* was the first attempt to catalogue the species of Alberta, either with or without the descriptions and keys that he also provided. Given the state of our knowledge of the flora at that time, it was a most adequate account. However, over the past twenty-five years interest and research in the Canadian flora have seen gratifying advances, and now it can be fairly said that his *Flora* no longer reflects the present knowledge of our flora. This revised edition attempts to bring the *Flora* up to date and at the same time incorporates some material not included in the original publication.

The basic format of this edition is essentially unchanged from that of the first. The families are again arranged following the 'Englerian system,' and the genera and species are treated alphabetically within the family.

The revised *Flora* includes all the additional species, both native and introduced, that are now known to occur in the province of Alberta. Twelve species hybrids of *Populus* and *Betula* and one of *Dryopteris* are noted but not keyed, described, or mapped. As in the first edition, cultivated species that have escaped and appear to have become established are included, but casual escapes from cultivation are excluded. Numerous nomenclatural changes have been incorporated, though restraint has been exercised regarding the introduction of new generic names, as for example in *Polygonum*, *Oenothera*, and *Senecio*. Where a name has changed, the name of the taxon used in the first edition is given in the description, with or without additional synonymy. The names of infraspecific taxa, varieties or subspecies (which in our usage are equivalent), are given in the text and when necessary their diagnostic features described.

The keys have been revised to accommodate the additional species and changed taxonomic concepts. Many have been rewritten, even in the absence of significant changes, to improve their utility. They are now all dichotomous, and a significant departure from the first edition is in the change to bracketed couplets. These are preferred because the alternative leads are immediately juxtaposed, reducing the chances of keying errors by the user.

Included in the species descriptions of the revised *Flora* is some information not provided in the original edition. Somatic chromosome numbers (2*n*) have been included, where this information is available, irrespective of the source of the material from which they were obtained. Where a chromosome number has been determined from Alberta material it is marked with an asterisk (*). Chromosome numbers are attributes of species but caution is necessary in applying information obtained from non-Alberta material to Alberta representatives. In some taxa such extrapolation would be reasonable, but for others it is not to be supposed that the chromosome number or numbers prevailing elsewhere are necessarily those to be found in our material.

Also included in the revised edition is the overall distribution of all native species. The North American distribution is generally given in two tiers from west to east across the continent. The first tier represents the northern limit, the second tier the southern limit. A comma indicates a reasonably continuous distribution and a semicolon indicates a disjunction. Where a species extends beyond North America its general distribution is briefly indicated prior to the North American, separated from it by a colon. See page 601 for key to abbreviations.

Dot maps, showing the Alberta distribution of all the native species treated in the revision, are included. They are based on collections from the National Museum, Ottawa, the Department of Agriculture Herbarium, Ottawa, and the major herbaria in Alberta. A few well-authenticated literature and sight records have also been employed. Every dot is referenced and can be documented to herbarium, locality, and collector. All material mapped from the Alberta herbaria has been verified. In the case of the Ottawa herbaria, determinations proved to be most reliable but selected material was examined, namely for critical groups, when collectors were unknown or inexpert, and when distributions were questionable. In a few instances, where verification of Ottawa material was deemed essential but for one reason or another impractical, the maps are based solely on material in provincial herbaria.

In regard to species limits I have taken a pragmatic approach, which in reality is the only possible one to take. Consequently, there is not a 'broad' or 'narrow' concept, a 'conservative' or 'radical' one. In this edition, species had to be 'mappable' and 'keyable,' which required that there be constant, clear-cut, morphological differences between them, allowing identification. What this means in effect is that the majority of species are the same as in the first edition. However, where recent monographic work has established the existence of species meeting the above criteria, in what were poorly understood, broad species in old taxonomies, these modern narrower species have naturally been adopted. Thus it is that in such genera as *Chenopodium*, *Draba*, and *Epilobium*, there are significant differences in the accounts of the first and second editions. In contrast, where species limits are very uncertain and our knowledge of the species meagre some consolidation has been effected, as for example in the genus *Poa*, where only 20 of the 24 species recognized by Moss are included. Similarly, some reduction has been effected with regard to the native species of *Taraxacum*, where the number is down to 1 from 3. It is probable that there is more than one native species of *Taraxacum* in the province, but until the taxonomy of the species in northwestern North America is better understood there is no point in a user of this or any other *Flora* agonizing over which of a possible half-dozen poorly defined and possibly meaningless species a particular *Taraxacum* specimen might be.

There are of course errors and shortcomings in the book, as the author knows better than anyone. It is hoped, however, that they are not too glaring and that none will significantly reduce its utility.

The author wishes to acknowledge the contribution of a number of colleagues who have assisted him in one way or another in the writing of this revision. First are those recognized experts in certain taxonomically difficult genera who prepared accounts specifically for this revision. In taxonomic order, *Salix* was contributed by G. Argus of the National Museum of Canada, Ottawa, and *Betula* by T.C. Brayshaw of the Provincial Museum of British Columbia, Victoria; *Atriplex* is largely the work of I.J. Bassett and C.W. Crompton of the Biosystematics Research Institute, Department of Agriculture, Ottawa, as is the treatment of *Chenopodium*; *Draba* is largely the work of G. Mulligan of the Biosystematics Research Institute, Department of Agriculture, Ottawa; *Oxytropis* was contributed by W. Elisens while a graduate student in the Department of Botany, The University of Alberta, Edmonton; and *Epilobium* was contributed by P. Hoch of the Missouri Botanical Garden, St Louis. The effort of these colleagues in providing modern, revised treatments of these difficult genera has greatly enhanced the overall value of this revision, and I am much indebted to them. These accounts were edited so as to conform with the format of the text, and I take full responsibility for any ensuing errors or loss of clarity. It should be noted that in each of the above genera the species distribution maps were prepared solely from material determined by the contributing authors.

Thanks are also due to Kenton Chambers, F.J. Hermann, J.H. Hudson, J. McNeill, and D.F. Murray for checking certain determinations in the Cyperaceae, Juncaceae, Caryophyllaceae, and Compositae.

I am grateful to B. Hallworth and J. McNeill who brought to my attention a number of species recently found introduced into Alberta, and to P. Achuff, C. Bradley, A. Ceska, I. Corns, P. Lee, C. Wallis, D. Vitt, and others who furnished material of native species newly found in the province.

C.C. Chinnappa, Department of Biology, University of Calgary, Calgary, kindly allowed me to include unpublished chromosome numbers that he has determined from Alberta material.

I am infinitely obliged to W.J. Cody, Curator of the Herbarium, Biosystematics Research Institute, Department of Agriculture, Ottawa, and to J. Gillett, Curator of the National Herbarium, Ottawa, for their invariably prompt attention to my many requests for the loan of specimens, for their assistance in other ways, and for their co-operation with regard to the mapping of their Alberta collections. Likewise to C.D. Bird and B. Hallworth, formerly Curator and Assistant Curator of the Herbarium, University of Calgary, Calgary, for their generous assistance and cooperation on my frequent visits.

Over several years a number of people have assisted in mapping the Alberta distributions of our native species. The University of Alberta collections were largely mapped by M.G. Dumais, and those of the University of Calgary by V. Kullyk. The holdings in the Ottawa herbaria were mapped by B. McMullen, S. Lupu-Krehm, and K. Snarr. In addition P. Bickell, K. Busby, W. Elisens, B. Godwin, R. Kempinsky, and L. Kershaw assisted significantly with the mapping and compilation.

Other people who have assisted me at some time or another, by providing information or services, include K. Denford, the late F.W. Gould, V. Harms, J. Kuijt, G. La Roi, A. & D. Löve, G. Ringius, G. Scotter, R. Whitkus, D. Fabijan, and J. Harris.

In the course of revising the *Flora* I have consulted innumerable books, monographs, and taxonomic papers. For the two latter it is hardly possible to cite references, but those of the former that were used most freely are listed in the bibliography. I wish to acknowledge here specifically three works that have been especially valuable and made my task a good deal easier than it might otherwise have been. They are *Flora of the Prairie Provinces* by B. Boivin; *Vascular Plants of the Pacific Northwest* by C.L. Hitchcock, A. Cronquist, M. Ownbey, and J.W. Thompson; and *Flora of Alaska and Neighboring Territories* by E. Hultén.

Throughout the preparation of this book I have been the recipient of a grant (No. A-2582) from the Natural Sciences and Engineering Research Council of Canada (NSERC), considerable portions of which have at times been allotted to aspects of the revision without which it could never have been completed in its present form. The University of Toronto Press again has provided generous funding towards the cost of publication. Even so, publication would not have been possible had it not been for a substantial publication subsidy made available by the Forest Development Research Trust Fund of Alberta Energy and Natural Resources.

Figures 1 and 2 have been adapted from maps in *Atlas of Alberta* (published in 1969 by University of Alberta Press in association with University of Toronto Press), with permission.

To all those who have assisted me in writing this book and to those agencies that have funded its production and publication I am most truly grateful. Finally, to my wife, Anne, my deepest thanks are due for her expert editorial assistance and unfailing encouragement.

J.G.P.
Edmonton, July 1982

Preface to the First Edition

The Province of Alberta, Canada, affords features of exceptional botanical interest. Here forest and steppe merge, with north–south and east–west transition zones; and the Rocky Mountains present various aspects of altitudinal zonation, from foothills forest to high alpine. Moreover, the region is a meeting ground of Boreal and Cordilleran floristic elements, while in the extreme northeast a transition from Boreal to Arctic begins. Ecological and phytogeographical features of the vegetation of Alberta are reviewed in a recent (1955) article by the present author. Taxonomically, the flora is at once intriguing and beset with difficulties. Botanists in the region soon encounter taxonomic problems and have been handicapped through lack of a suitable guide to the species.

The present volume represents a pioneer effort to record and describe the vascular plants of Alberta with keys to their identification. It is a compilation based on available records and on my field experience over some thirty-five years in the region. Most of the species and varieties described in the manual are represented in the herbarium of the University of Alberta. Included are a few species of which I have not seen Alberta material but which will almost certainly be found in the region. Naturally, extensive use has been made of manuals designed for contiguous areas, and of available catalogues, lists of plants, and recent monographs. Publications most freely consulted (other than the numerous taxonomic papers) are listed in the Bibliography.

The arrangement of families is that found in most of the standard American manuals and is based, in the main, on the Englerian system. However, genera are placed in alphabetical order within each family and the species in the same order under each genus. Subspecies, varieties, and forms are variously noted and described. Specific and subspecific names derived from proper nouns are not capitalized, in accordance with a recent recommendation attached to the 'International Code of Botanical Nomenclature' (1952). Synonyms have been included where they are likely to be helpful to those who consult Rydberg's manuals and other standard works. Common names are given where these are known to have a considerable degree of currency. The keys, with few exceptions, are dichotomous. The metric system of measurement is used throughout. Technical terms have been largely employed, for these are essential to concise and accurate description.

Departure from the customary arrangement of genera and species demands some explanation. Though fairly familiar with the traditional order, I have come to prefer the alphabetical arrangement, mainly because of facility in locating genera within families and species within genera. Moreover, the alphabetical order obviates the necessity of using numbers to designate genera and species, thus simplifying keys and giving cleaner pages throughout. Admittedly, the traditional order has certain desirable features, notably the practical advantage of close proximity of descriptions of similar entities in the text. However, closely related groups do appear together in the keys; there the critical diagnostic features are set forth for comparison.

The taxonomy of the book is on the whole conservative. Generally a broad species concept is favoured but exceptionally a treatment involving quite narrow species limits has been adopted. Where different authoritative treatments of a particular group are available, I have tended to use the one which in my experience is the most satisfactory for the local flora.

This manual is designed for serious students of plant life. The hope is that it will prove useful not only to professional botanists, but to foresters, agriculturists, ranchers, weed inspectors, school teachers, and lovers of plants in general. I am well aware that errors and omissions are inevitable in a pioneer work of this kind. Criticisms and corrections will be welcomed. Specimens of Alberta species not included in the book will be gladly accepted.

Grateful acknowledgment is made to the many persons who have helped in making possible this publication. Especially useful have been the catalogues of Dr Hugh M. Raup for northern Alberta, the lists and collections of Mr August Breitung for the Cypress Hills and Waterton Lakes Park, the splendid collections and list for the Edmonton district by Dr G.H. Turner, the superb herbarium of Dr W.C. McCalla, and the collections of the late A.H. Brinkman. Others who have provided important material for study include: F. J. Hermann, J.J. Sexsmith, J.A. Campbell, J.L. Bolton. E. Woollven, A.C. Budd, Keith Horton, J.S. Rowe, George Pegg, and the writer's present colleagues, Dr J.H. Whyte, Dr R.G.H. Cormack and Dr Lorene L. Kennedy, also earlier colleagues, Dr E. Silver (Dowding) Keeping and the late Professor Francis J. Lewis. Special thanks is extended to my friends Dr W.C. McCalla and Dr G.H. Turner, whose expert advice on local flora has been frequently sought and freely given. Gratitude is expressed to the authors and publishers of the various manuals from which descriptions and keys have been adapted. At appropriate places in the text, acknowledgment is made to authors whose taxonomic keys have been extensively utilized. Of the many taxonomists who have assisted through the years, the following are particularly remembered in connection with determinations of difficult species: Carleton R. Ball, M.O. Malte, Hugh M. Raup, C.O. Rosendahl, Reed. C. Rollins, Harold A. Senn, A.E. Porsild, K.K. Mackenzie, I.M. Johnston, J.R. Swallen, F. J. Hermann, LeRoy E. Detling, F.W. Pennell, Lincoln Constance, Mildred E. Mathias, Lloyd H. Shinners, Donovan Correll, Marion Ownbey, E.C. Ogden, R.C. Barneby, W.G. Dore, C. Frankton, Bernard Boivin, and Marcel Raymond. My wife assisted on numerous field trips and has given much time to checking of manuscript. To her loyalty and encouragement is due a large share of the credit for the completion of this volume.

Gratitude is expressed to the National Research Council of Canada for two

Summer Research Associateships awarded in connection with basic studies for this book, and to the Friends of the University of Alberta and to the Publications Fund of the University of Toronto Press for substantial grants in aid of publication.

E.H.M.
January 1958

Introduction

THE AREA

The province of Alberta is situated in northwestern North America between latitudes 49° and 60° N and longitudes 110° and 120° W. It is 1,200 km (756 miles) from north to south and varies from 291 to 646 km – average 551 (182–404 miles, average 326) – from east to west, and covers an area of 661,185 km^2 (255,285 sq. miles).

The southern half of Alberta has an altitude of 600–900 metres (2,000–3,000 ft) in the east, gradually rising westwards to 900–1500 metres (3,000–5,000 ft) in the foothills of the Rocky Mountains. The mountains, which form the southwestern boundary of the province, rise to a height of over 3,700 metres (12,000 ft). In the southeast, straddling the boundary with the province of Saskatchewan, occur the Cypress Hills, rising to a height of 1,390 metres (4,500 ft). In the northern half the prevailing altitudes are below 600 metres (2,000 ft) and even less than 300 metres (1,000 ft) in the plain of the Athabasca River; the upland areas very rarely exceed 1,200 metres (4,000 ft).

Alberta was extensively glaciated during the last ice age, with the Laurentide ice-sheet advancing from the northeast and the Cordilleran ice-sheet spreading out from the mountains. That the two ice-sheets made contact at some points is generally accepted. However, it has been reliably established that some areas escaped glaciation; for example, parts of the front range of the Rocky Mountains, the Cypress Hills, and the extreme southwestern parts of the province.

Ice had largely receded from our area by 8000–9000 BP, leaving behind surficial deposits of glacial moraine and till, and lake deposits of silt, clay, and sand that largely cover the province today. Apart from the mountains, bedrock is significantly exposed only in the extreme northeast, north of Lake Athabasca, where part of the Canadian Shield enters the province.

The climate of Alberta, as would be expected, is continental, with long, cold winters and and hot summers. Mean temperatures for January vary from −8° to −12° C (10–18°F) in the extreme south and southwest to −19° to −24° C (−2° to −12°F) north of 57°N. Mean July temperatures vary from 18–20°C (64–68°F) in the south to 16°C (60°F) in the north. The average number of frost-free days decreases from 120 in the south to 60 in the north. There is also a significant falling-off westwards, with increasing altitude. Precipitation over most of Alberta averages 50 cm per annum (20 inches) or less. In the southeast it is 30 cm (12 inches) and it increases to 40–50 cm (16–20 inches) in the north. Considerably higher amounts are recorded from the mountain area.

FIGURE 1. Distribution of the major soil groups in Alberta (adapted from *Atlas of Alberta*, 1969)

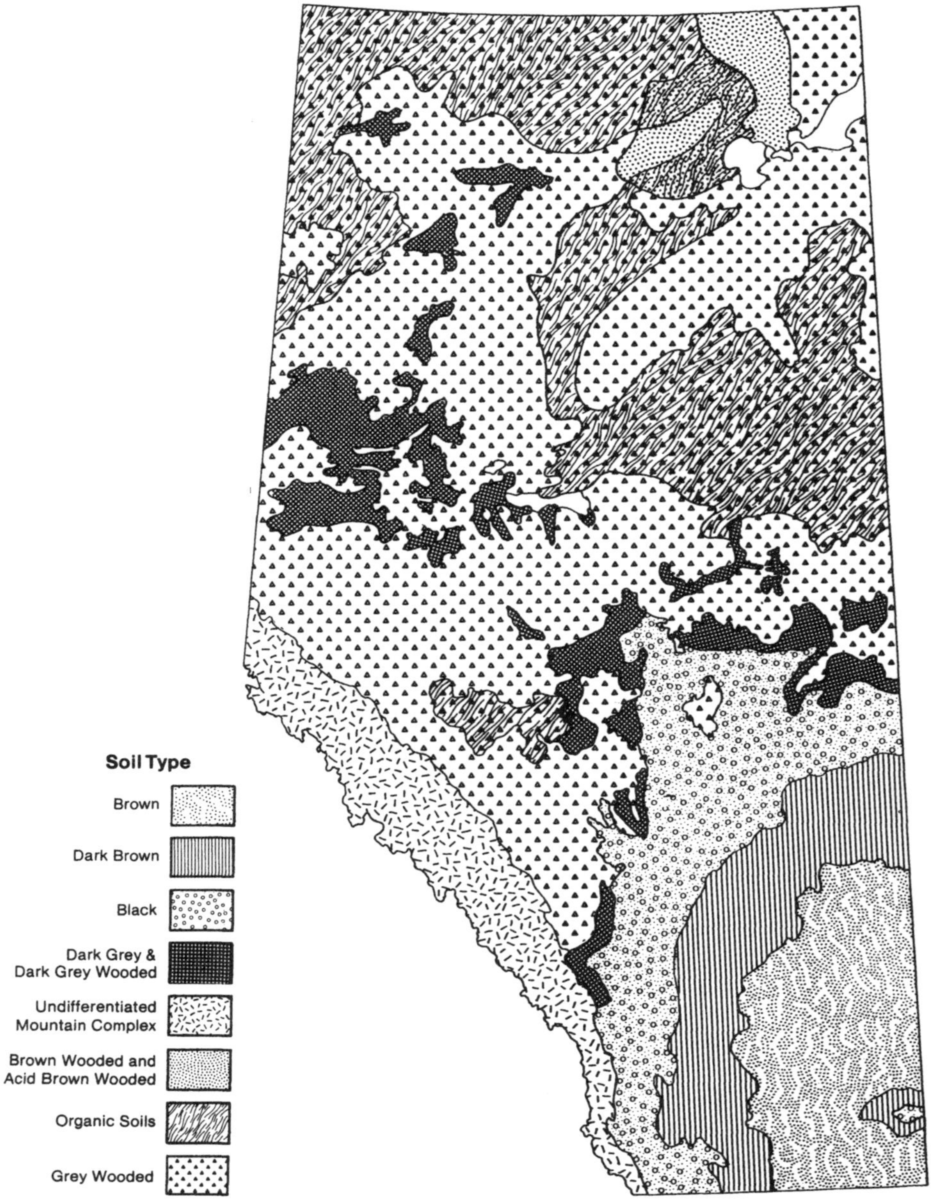

An effective way to summarize the physical environment of our area is by reference to the distribution of the major soil groups within the province (see Figure 1). The semi-arid southeastern part is characterized by Brown soils, the average annual precipitation being about 30 cm (12 inches) with yearly amounts fluctuating between 15 and 65 cm (6–26 inches), and the mean July temperature varying from 18° to 20° C (64–68°F).The Dark Brown soils occupy an area of slightly higher precipitation averaging about 35 cm per annum (14 inches) though fluctuating considerably from year to year, and since the summer temperatures are not markedly different from in the Brown soil zone this zone is somewhat less arid. The Black soils occur in an area where precipitation averages from 38 to 46 cm per annum (15–18 inches), and since the mean July temperature ranges from 17° to 18° C (62–64°F) moisture deficits become less critical. On the northern and western periphery of the Black soil area, and also in northwestern Alberta, islands of Dark Grey and Dark Grey Wooded soils occur, where average annual precipitation varies from 40 to 48 cm (16–19 inches) and mean July temperatures of about 16°C (60°F) are recorded. Grey Wooded soils predominate in the province, occurring in areas with a subhumid climate. Precipitation varies from 48 to 55 cm (19–22 inches) in the western parts to 40 to 50 cm (16–20 inches) in central and northern areas. July mean temperatures vary from 12° to 17° C (50–62°F). Associated with the Grey Wooded soils over much of northern Alberta are Organic soils, which have a peat surface more than 30 cm (12 inches) deep. In the northeast occur Brown Wooded and Acid Brown Wooded soils. With the exception of these, and the Grey Wooded soils, extensive areas of Solonetzic soils are found in association with all soil groups in the province. The mountain area is categorized as the 'Undifferentiated Mountain Complex' because of the limited soil survey data available.

VEGETATION

The distribution of the major vegetation types in Alberta is shown in Figure 2. Occupying the southeastern region is prairie grassland, with a peripheral band in which an occasional aspen poplar stand (*Populus tremuloides*) might be found in a favourable location. This is transitional to the aspen parkland where aspen woodlands, interspersed with quite extensive grassland associations, prevail. In the southwest this gives way to an altitudinally zoned mountain forest in which lodgepole pine (*Pinus contorta*), white spruce (*Picea glauca*), and Engelmann's spruce (*Picea engelmannii*) are dominants. In west central Alberta between the aspen parkland and the mountain forest a fairly extensive area dominated by aspen and spruce, both white and black (*Picea mariana*), is interposed, which extends northwards to Lesser Slave Lake and westwards, as a narrow band, to the provincial boundary. Between these latter extensions and the mountain forest a quite extensive coniferous forest of lodgepole pine and white spruce predominates.

North of the aspen parkland and the aspen-spruce belt exists a mixed-forest complex. In the eastern part, to about 55°N, there is a rather homogeneous aspen woodland, while in the west considerable, disjunct, areas of aspen parkland occur. However, in the north central part of the province, covering a very large

FIGURE 2. Distribution of the major vegetation types in Alberta (adapted from *Atlas of Alberta*, 1969)

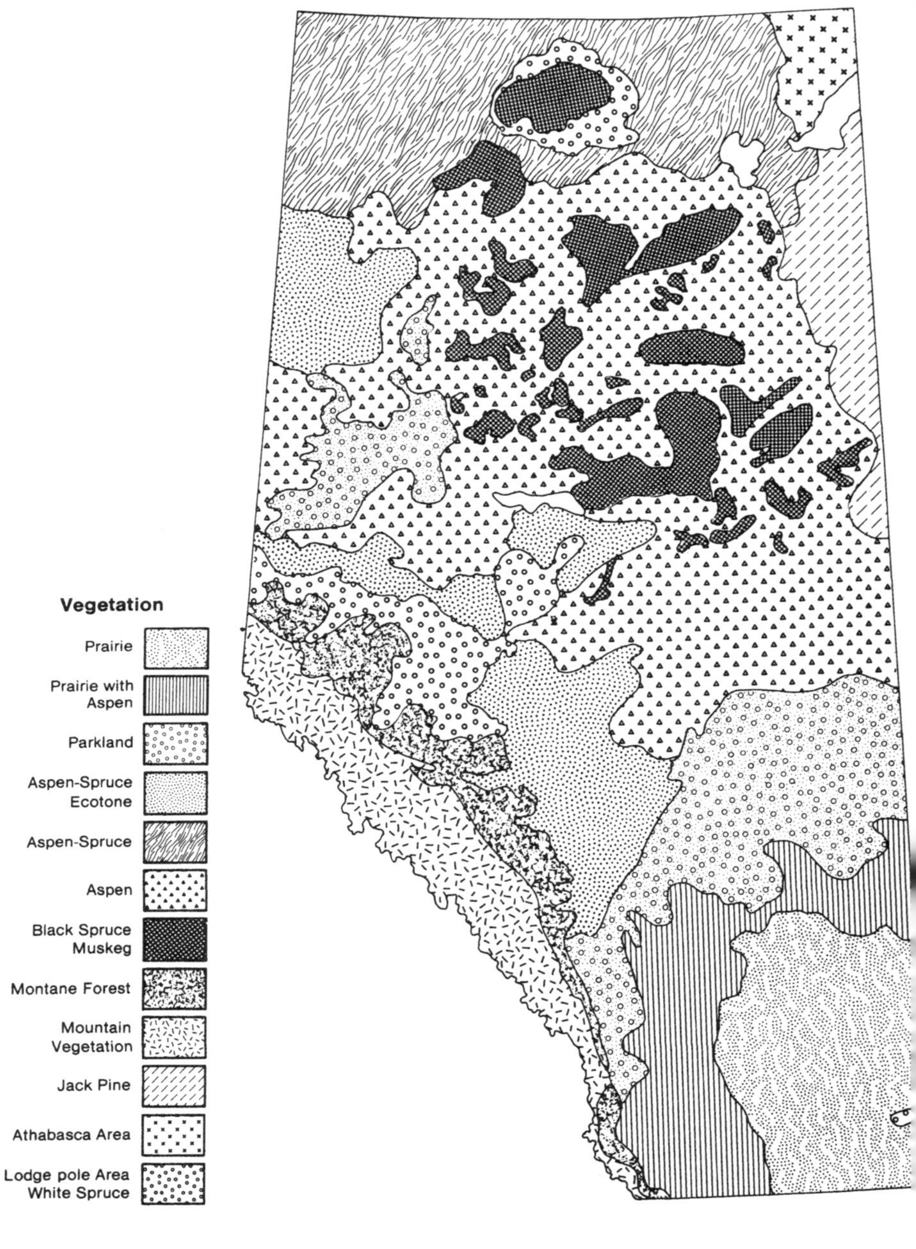

area, there is a mosaic of aspen woods and black spruce muskeg, the latter often very extensive and at least equalling in overall area the aspen woodland.

To the north of this mixed-forest complex there is, in the west, a recurrence of the aspen-spruce forest, often interspersed with lodgepole pine and black spruce muskeg. On the eastern side, and extending north to Lake Athabasca, is an area of coniferous forest, with jack pine (*Pinus banksiana*) dominating in the north, and jack pine associated with white and black spruce prevailing in the south.

Except for the centrally located Caribou Hills and the Precambrian outcrop north of Lake Athabasca, the northern extremity of the province is dominated by an aspen-spruce forest interspersed with some black spruce muskeg. The Precambrian area is characterized by areas of jack pine and black spruce and extensive areas with only a lichen cover. The Caribou Hills have a peripheral forest of spruce with some lodgepole pine, and on the central plateau a cold, subarctic-type black spruce muskeg occurs.

SOME NOTABLE COLLECTORS AND AUTHORITIES FOR ALBERTA SPECIES

The name (generally abbreviated) following a generic or species name in this as in any book or flora is that of the person who first named and described the taxon or, where there is a combination of names, includes the name of that person. The names (authorities) are technically important but they can also tell us something about the history of botany and the history of exploration of a given area.

The most frequently cited authority in this book, by far, is Linnaeus (L.). There are several reasons for this. Most significant is that he was responsible (in 1753) for the binomial system of naming plants (and animals) that we use today, and although many species had been described earlier he is taken as the starting-point for their names. The second reason is that many plant species have wide distributions and occur not only in Alberta but in other parts of North America and often in Asia and Europe as well. Many of these so-called circumpolar plants were named and described from specimens collected in Europe and Asia, by Linnaeus and others, long before they were discovered in North America. Also adding to the number of Linnaean species here is the fact that with the colonization of North America came many European weeds named by him. Finally, being the foremost botanist of his day, he received collections from his students and others from all parts of the globe, including North America, as for example from Peter Kalm and John Clayton.

Carolus Linnaeus (1707–78) was born in Rashult, Sweden. He attended a number of universities, and eventually obtained a medical degree from the University of Haderwijk in the Netherlands. He travelled widely in Europe before and after graduating and was appointed Professor of Medicine at Uppsala in 1741, a position he held until 1775. He published many works, but none more important than *Species Plantarum* in 1753.

David Douglas (1798–1834) – Dougl. – was born in Scone, Perthshire, Scotland, and hired in 1823 as a plant collector by the Horticultural Society of London, having previously been employed by the Glasgow Botanical Garden. In 1823–4 he collected in the northeastern United States and adjacent Canada. He came again to North America in 1824 and collected extensively in the northwestern United States and British Columbia. He crossed the Rocky Mountains and prairies in 1827 en route to Hudson Bay for the return to England. His last expedition to western North America began in 1829, and he intended to return overland, by way of Siberia, but was killed in Hawaii. Douglas published the names and descriptions of some of the new species in his collections, but more often his observations were in manuscript only and subsequently used by Hooker in his publications; this is indicated in the text in the following manner: Dougl. *ex* Hook.

Thomas Drummond (1790–1835) was born in Perthshire, Scotland, and in 1825 appointed Assistant Naturalist to Sir John Franklin's Second Overland Expedition in Search of a Northwest Passage. He collected in Saskatchewan and the Rocky Mountains of Alberta in the summers of 1825–7, spending the winter of 1825–6 in Edmonton. He returned with the expedition in 1827. He came again to America in 1830 and for four years collected along the Mississippi River and in Texas. He died in Havana, Cuba, while pursuing more collections.

Asa Gray (1810–88) was born in Sauquoit, New York; he attended Fairfield Academy and in 1831 graduated with a medical degree. Over the next ten years there were a number of teaching appointments and a variety of travels, some sponsored by John Torrey (1796–1873) with whom he sometimes published (T. & G.). In 1842 he became Professor of Natural History at Harvard, a position he retired from in 1873. During the 1840s and 1850s he was the often overwhelmed recipient of collections from many of the official and unofficial expeditions and surveys then being made in the western United States. The published reports included many new taxa. His best-known work is *Manual of the Botany of the Northern United States*, which has run through eight editions, the last (revised) having been published in 1950.

Sir William J. Hooker (1786–1865) – Hook. – was born in Norwich, England, and became early an able and influential botanist. He was appointed Professor of Botany at Glasgow University in 1820 and the first Director of the Royal Botanic Garden at Kew in 1841. He had access to the important collections of his students and protégés, and interested persons sent him plants from all over the world. From North America were included the collections of D. Douglas, T. Drummond, J. Goldie, W. Tolmie, and M. Gairdner, whose various appointments had been largely secured on his recommendation. Hooker's *Flora Boreali-Americana*, published from 1829 to 1840, is a comprehensive account of the flora of northern North America, including the whole of Canada. Among the many species in it that are named and described for the first time, those from western Canada are especially abundant, based largely on the collections of Drummond and Douglas.

Eric Hultén (1894–1981) – Hult. – was born in Halla, Sweden, attended Stockholm University, and eventually became Director of the Stockholm Riksmuseum. Hultén was a great and influential student of the arctic flora; his *Outline of the History of Arctic and Boreal Biota during the Quaternary Period* published in 1937 is crucial in understanding the evolution of the arctic flora. His most notable contributions to American botany were *Flora of Alaska and Yukon* (1941–9) and *Flora of Alaska and Neighboring Territories* (1968). His works *The Amphi-Atlantic Plants* (1958) and *The Circumpolar Plants* (1962, 1971) are also of inestimable value.

John Macoun (1831–1920) was born in Maralin, County Down, Ireland. The family emigrated to Canada in 1850 and settled in the Belleville area of Ontario. A largely self-taught botanist Macoun was appointed Professor of Natural History at Albert University in Belleville in 1868. He resigned in 1879 on becoming Botanist to the Geological and Natural History Survey of Canada. Between 1872 and 1881 he embarked on five official explorations of the Canadian west, two of them transcontinental. His *Catalogue of Canadian Plants,* giving synonymy, habitat, and collectors of all known Canadian species, was published from 1833 to 1902. He died in Sydney, British Columbia.

Thomas Nuttall (1786–1859) – Nutt. – was born in Long Preston, Yorkshire, England, and emigrated to the United States in 1808. Between 1810 and 1836 he embarked on six major collecting expeditions across the length and breadth of the country. His *Genera of North American Plants,* published in 1818, included many new genera and species based on his collections. He was appointed Curator of the Botanical Garden at Harvard in 1822, where he stayed until 1834. He returned to England in 1842.

Frederick Pursh (1774–1820) was born in Grossenhain, Saxony, and prior to his emigration to the United States in 1799 was employed by the Botanic Garden in Dresden. Under the patronage of Benjamin Smith Barton he collected widely in the eastern United States. His *Flora Americae Septentrionalis,* published in 1814, is a description of the plants of North America and includes many new and rare species. He was subsequently variously employed in Canada, by, for example, Lord Selkirk of the Red River Settlement. His Canadian collections of this time were largely destroyed by fire in the winter of 1818–19. He died, destitute, in Montreal.

Per Axel Rydberg (1860–1931) – Rydb. – was born in Westergoethland, Sweden, and emigrated to the United States in 1882. He held an appointment at Lutheran Academy, Wahoo, Nebraska, from 1884 to 1893, and then was employed by the United States Department of Agriculture. From 1899 until his death he was associated with the New York Botanical Garden. He made extensive collections in the Rocky Mountains in the United States over many years. His *Flora of the Rocky Mountains and Adjacent Plains,* first published in 1917, includes Alberta. His *Flora of the Prairies and Plains of Central North America* was published posthumously in 1932.

Key to the Major Groups and Families of Vascular Plants

NOTE: The key allows at some points for more than one interpretation of floral morphology. If a family description appears not to coincide with the interpretation employed in using the key check the specimen carefully to ascertain if it might be consistent with the description given a revised interpretation of the floral morphology.

Details of ovary structure are often most easily obtained using maturing fruits.

1 Plants herbaceous, reproducing by spores Division **Pteridophyta** p. 8

Plants, herbs, shrubs, or trees, reproducing by seeds Division **Spermatophyta** **2**

2 Ovules not enclosed in an ovary; seeds in a dry cone (rarely in a berry-like cone); pollen-sacs on scales arranged in cones; trees or shrubs; leaves needle-like or scale-like, mostly evergreen Subdivision **Gymnospermae**; Class **Coniferae** p. 9

Ovules enclosed in an ovary, which at maturity becomes the fruit; pollen-sacs terminating stamens borne in flowers; trees, shrubs, or herbs; leaves of various forms, mostly deciduous Subdivision **Angiospermae** **3**

3 Leaves usually parallel-veined; vascular bundles of stem irregularly arranged; cambium absent; parts of flowers usually in 3's or 6's, never in 5's; embryo with 1 cotyledon; herbs (rarely shrubby) Class **Monocotyledoneae** p. 10

Leaves usually net-veined; vascular bundles of stem commonly in a single ring; cambium usually present; parts of flowers usually in 5's or 4's; embryo with 2 cotyledons; herbs, shrubs, or trees Class **Dicotyledoneae** p. 11

Division **PTERIDOPHYTA** / Ferns and Fern Allies

1 Leaves slender, 1–2 mm wide, simple, sessile, often scale-like, mostly small **2**

Leaves broader, usually 5 cm long or more, often large, variously incised or dissected **5**

2 Stems conspicuously jointed and hollow; leaves scale-like, in sheath-like whorls at the nodes; sporangia on peltate scales in a terminal cone **Equisetaceae** p. 28

Stems not as above; leaves mostly imbricated; sporangia at the base or in the axils of leaf-like or scale-like sporophylls that often form a terminal cone **3**

3 Leaves slender, grass-like, 5–15 cm long, dilated and sheathing at base; stem short, corm-like, lobed; plant rooting in mud, commonly covered by water **Isoetaceae** p. 27

Leaves small, less than 1.5 cm long; stems slender, elongated, branched, the main ones usually horizontal and creeping; sporangia in club-like cones; plants resembling coarse mosses **4**

4 Leafy shoots mostly 7–15 mm wide; homosporous **Lycopodiaceae** p. 24

Leafy shoots less than 6 mm wide; heterosporous **Selaginellaceae** p. 26

5 Leaves (fronds) long-stalked, palmately divided into 4 leaflets (like a 4-leaf clover); plant rooting in mud, the leaves sometimes floating in water; spores borne in hard, bean-like structures **Marsileaceae** p. 44

Leaves (fronds) not divided into 4 leaflets; spores in sporangia on green leaves or on modified leaf-segments **6**

6 Sporangia relatively large, borne in a terminal (grape-like) cluster, the sterile blade appearing lateral on a common stalk with it **Ophioglossaceae** p. 32

Sporangia minute, borne in clusters (sori) on the back or near the margins of green blades, or on separate modified fronds **Polypodiaceae** p. 34

Division **SPERMATOPHYTA** / Seed Plants

Subdivision **Gymnospermae**

Class CONIFERAE / Conifers

1 Leaves needle-like, borne singly (and alternate) or in clusters; ovulate cones absent, or woody with many spirally arranged scales **2**

Leaves scale-like, if needle-like opposite or whorled; ovulate cones woody or fleshy, the scales 12 or less, opposite or whorled **Cupressaceae** p. 44

2 Plants dioecious; ovulate cones lacking, the seeds solitary, surrounded by a red fleshy cup-like aril **Taxaceae** p. 51

Plants monoecious; ovulate cones woody, several–many-seeded **Pinaceae** p. 46

Subdivision **Angiospermae** / Flowering Plants

Class MONOCOTYLEDONEAE / Monocotyledons

1 Plants submersed or free floating aquatics, only the flowers sometimes elevated a little above the water **2**

Plants terrestrial or emergent aquatics with lower parts in water but the upper stem, leaves, and flowers elevated much above it **8**

2 Plants thallus-like, not differentiated into stem and leaves **Lemnaceae** p. 167

Plants differentiated into stem and leaves **3**

3 Flowers in conspicuous, globose, unisexual heads **Sparganiaceae** p. 51

Flowers not as above **4**

4 Perianth present **5**

Perianth absent **6**

5 Flowers biseriate, 3-merous, imperfect (dioecious); ovary inferior; leaves linear, 1–2 cm long **Hydrocharitaceae** p. 62

Flowers uniseriate, 4-merous, perfect; ovary superior; leaves generally longer or wider **Potamogetonaceae** p. 53

6 Leaves alternate; flowers perfect, in terminal spikes **Ruppiaceae** p. 58

Leaves opposite or whorled; flowers imperfect, axillary **7**

7 Leaf-margins minutely spinulose; ovary solitary **Najadaceae** p. 53

Leaf-margins not spinulose; ovaries 2 or more **Zannichelliaceae** p. 58

8 Perianth with at least some coloured, showy, petal-like parts **9**

Perianth entirely lacking petal-like parts **12**

9 Ovaries several–many; leaves mostly with a distinct blade and petiole **Alismataceae** p. 60

Ovary solitary; leaves usually lacking a distinct blade and petiole **10**

10 Ovary superior; if partly inferior stamens 6 **Liliaceae** p. 176

Ovary inferior; stamens 1–3 **11**

11 Flowers regular; stamens 3 **Iridaceae** p. 186

Flowers irregular; stamens 1 or 2 **Orchidaceae** p. 188

12 Flowers in a fleshy spadix 2–10 cm long, subtended by a conspicuous white spathe or a leaf-like spathe, the spadix then appearing to be lateral; emergent aquatics **Araceae** p. 166

Flowers not in a spadix subtended by a spathe **13**

13 Ovary solitary with 1 locule and 1 ovule **14**

Ovary not as above (either the ovaries 2 or more, more or less united at least towards the base, or if the ovary solitary then with 2 locules or more, or if the ovary solitary and 1-loculed then with 3 or more ovules) **18**

14 Style and stigma 1 **15**

Styles or stigmas 2 or 3 **17**

15 Leaves all basal, terete **Lilaeaceae** p. 59

Leaves not all basal, not terete **16**

16 Flowers in dense cylindrical spikes; achenes stipitate, hairy **Typhaceae** p. 51

Flowers in dense globose heads; achenes not stipitate or hairy **Sparganiaceae** p. 51

17 Stem usually hollow and terete, never triangular; leaves 2-ranked, the sheaths usually open; flowers mostly perfect, distichous on the axis of the spikelet, or only one flower in the spikelet; each flower subtended by 2 bracts (lemma and palea); perianth bristles absent; stigmas 2 **Gramineae** p. 62

Stem usually solid, often triangular; leaves generally 3-ranked, the sheaths usually closed; flowers spirally arranged on the axis, seldom distichous; each flower subtended by a single bract; perianth bristles mostly present or the flowers imperfect; stigmas 2 or 3 **Cyperaceae** p. 120

18 Flowers imperfect (monoecious), in dense globose heads **Sparganiaceae** p. 51

Flowers perfect **19**

19 Perianth dry, often scarious; fruit a capsule; flowers in chaffy heads or in clusters at the end of branches or laterally on the stem **Juncaceae** p. 168

Perianth herbaceous; fruit of follicles; flowers in elongate spike-like racemes or the flowers about 5 in a short raceme **20**

20 Flowers about 5; carpels united only at the base; fruit globose, 5–8 mm long **Scheuchzeriaceae** p. 60

Flowers many; carpels more or less united; fruit narrower or shorter **Juncaginaceae** p. 59

Class DICOTYLEDONEAE / Dicotyledons

1 Perianth uniseriate (i.e. of only one series of appendages all more or less the same) or entirely absent **2**

Perianth biseriate (i.e. of two different series of appendages, the inner generally larger, coloured, the outer generally smaller, greenish) **3**

2 Flowers imperfect, the staminate, and generally the pistillate too, in catkins; trees or shrubs **Group I** p. 12

Flowers perfect, or, if imperfect, the plants herbaceous, or, if woody, the flowers not in catkins **Group II** p. 12

3 Corolla of separate petals **4**

Corolla of united petals; or only 2 or 3 petals connate or united and the flowers markedly irregular **5**

4 Stamens more than twice as many as the petals, usually numerous **Group III** p. 15

Stamens twice as many as the petals or fewer **Group IV** p. 16

5 Stamens more numerous than the corolla-lobes **Group V** p. 19

Stamens the same number as the corolla-lobes or fewer **Group VI** p. 19

GROUP I

Perianth uniseriate or absent; flowers imperfect, both staminate and pistillate normally in catkins; trees or shrubs

1 Plants monoecious **Betulaceae** p. 215

Plants dioecious **2**

2 Ovules many; fruit a many-seeded capsule; seeds hairy **Salicaceae** p. 196

Ovules 1 or 1; fruit a 1-seeded nutlet; seeds not hairy **Myricaceae** p. 215

GROUP II

Flowers perfect, or, if imperfect, the plants herbaceous, or, if woody, the flowers not in catkins; perianth uniseriate or appearing so, rarely absent

1 Plants parasitic on the trunks and branches of coniferous trees (pines) **Loranthaceae** p. 222

Plants not parasitic on coniferous trees **2**

2 Trees or shrubs **3**

Herbs **10**

3 Ovary superior **4**

Ovary inferior or partly so, or closely enclosed by the calyx and appearing inferior **6**

4 Leaves pinnately compound or palmately lobed; fruit winged **Aceraceae** p. 399

Leaves simple, not lobed; fruit berry-like, black **5**

5 Plants 1–2 (5) dm tall; leaves needle-like, 0.5–1.0 cm long **Empetraceae** p. 397

Plants 5–20 dm tall; leaves ovate to elliptic, 3–10 cm long **Rhamnaceae** p. 400

6	Leaves densely spiny	**Araliaceae** p. 424
	Leaves not densely spiny	**7**
7	Leaves alternate	**Elaeagnaceae** p. 410
	Leaves opposite	**8**
8	Stamens 4	**Cornaceae** p. 434
	Stamens more than 4	**9**
9	Stamens 5	**Caprifoliaceae** p. 511
	Stamens 8	**Elaeagnaceae** p. 410
10	Plants aquatic (if leaves alternate see *Potamogetonaceae* p. 53)	**11**
	Plants terrestrial	**14**
11	Leaves opposite	**Callitrichaceae** p. 397
	Leaves whorled	**12**
12	Leaves entire	**Hippuridaceae** p. 423
	Leaves finely dissected	**13**
13	Flowers epigynous; stamens 8; leaves pinnately branched	**Haloragaceae** p. 423
	Flowers hypogynous; stamens 10–16; leaves dichotomously branched	**Ceratophyllaceae** p. 272
14	Ovary inferior or partly so, or closely enclosed by the calyx and appearing inferior	**15**
	Ovary(ies) superior	**21**
15	Ovary or its locules with many ovules	**Saxifragaceae** p. 328
	Ovary or its locules with only 1 or 2 ovules	**16**
16	Inflorescence a capitulum, subtended by an involucre of separate bracts	**Compositae** p. 519
	Inflorescence not a capitulum subtended by an involucre	**17**
17	Leaves alternate or basal	**18**
	Leaves opposite or whorled	**20**
18	Flowers in umbels; leaves generally pinnate or much dissected	**19**
	Flowers not in umbels; leaves entire	**Santalaceae** p. 221
19	Styles 5	**Araliaceae** p. 424
	Styles 2	**Umbelliferae** p. 425
20	Leaves opposite	**Nyctaginaceae** p. 249
	Leaves whorled	**Rubiaceae** p. 509
21	Ovary or its locules with many ovules	**22**
	Ovary or its locules with only 1 or 2 ovules	**26**

22 Ovary solitary **23**
Ovaries 2 or more, separate from one another or united only at the base **Ranunculaceae** p. 272

23 Leaves compound **Ranunculaceae** p. 272
Leaves simple **24**

24 Stamens 2 **Scrophulariaceae** p. 487
Stamens more than 2 **25**

25 Sepals free **Caryophyllaceae** p. 253
Sepals united basally forming a tube **Primulaceae** p. 449

26 Ovary solitary **27**
Ovaries 2 or more, separate from one another **Ranunculaceae** p. 272

27 Style (if any) and stigma 1 **28**
Styles or style branches and stigmas 2 or 3 **30**

28 Leaves opposite; stinging hairs present **Urticaceae** p. 220
Leaves alternate; stinging hairs absent **29**

29 Flowers imperfect, in axillary clusters; ovary 1-loculed; 1-seeded **Urticaceae** p. 220
Flowers perfect, in terminal racemes, ovary 2-loculed; 2-seeded **Cruciferae** p. 294

30 Stipules (ocreae) sheathing the stem at the nodes; or stamens 9 **Polygonaceae** p. 223
Stipules not sheathing; stamens not 9 **31**

31 Plants with ovary and fruit 3-loculed; juice milky; flowers in clusters resembling individual flowers **Euphorbiaceae** p. 395
Plants not as above **32**

32 Leaves alternate **33**
Leaves opposite, at least below **34**

33 Flowers subtended by spiny, scarious bracts; sepals scarious **Amaranthaceae** p. 248
Flowers not subtended by scarious bracts; sepals herbaceous or fleshy **Chenopodiaceae** p. 234

34 Plants 10–20 dm tall, or vine-like; leaves dissected or lobed **Cannabinaceae** p. 219
Plants 1–2 dm tall, more or less prostrate; leaves entire **35**

35 Plants of marshy ground; ovary 2-loculed, 4-lobed, 4-seeded **Callitrichaceae** p. 397
Plants of dry ground; ovary 1-loculed, 1-seeded **Caryophyllaceae** p. 253

GROUP III

Calyx and corolla both present; petals separate; stamens more than twice as many the petals

1 Plants aquatic; leaves floating, peltate, entire **Nymphaeaceae** p. 271
Plants terrestrial; if aquatic, leaves not as above **2**

2 Leaves basal, the margins fused to form winged, obconical pitchers; stigma umbrella shaped, 3–4 cm wide **Sarraceniaceae** p. 326
Leaves and stigma not as above **3**

3 Plants with fleshy stems, beset with long slender spines; petals numerous **Cactaceae** p. 409
Plants not as above **4**

4 Ovary inferior **5**
Ovary superior **7**

5 Herbs, rough-hairy **Loasaceae** p. 408
Shrubs or trees, not rough-hairy **6**

6 Leaves alternate **Rosaceae** p. 346
Leaves opposite **Hydrangeaceae** p. 342

7 Ovaries 2 or more, separate from one another or united only at the base **8**
Ovary solitary **10**

8 Stamen filaments united; ovaries cohering in a ring around a central column **Malvaceae** p. 401
Stamen filaments free; ovaries not as above **9**

9 Stamens inserted on the receptacle, at the base of the ovaries; stipules absent **Ranunculaceae** p. 272
Stamens inserted on a disc or hypanthium; stipules usually present **Rosaceae** p. 346

10 Leaves punctate with translucent dots or black-dotted **Hypericaceae** p. 403
Leaves not punctate or black-dotted **11**

11 Shrubs **12**
Herbs **13**

12 Plants 1–2 dm tall; flowers yellow; fruit a capsule **Cistaceae** p. 404
Plants taller; flowers white; fruit a fleshy drupe **Rosaceae** p. 346

13 Stamen filaments united; ovary 2-loculed or more **14**
Stamen filaments free; ovary 1-loculed **15**

14 Flowers irregular; ovary 2-loculed **Polygalaceae** p. 394
Flowers regular; ovary several-loculed **Malvaceae** p. 401

15 Leaves trifoliolate **Capparidaceae** p. 293
Leaves not trifoliolate **16**

16 Ovary flattish apically, conspicuously radially ridged; fruit a poricidal capsule **Papaveraceae** p. 291
Ovary not as above; fruit a berry **Ranunculaceae** p. 272

GROUP IV

Calyx and corolla both present, the petals separate; stamens not more than twice as many as the petals

1 Plants white, pink or reddish brown; leaves reduced to scales; saprophytes **Monotropaceae** p. 438
Plants green; leaves well developed **2**

2 Plants aquatic; leaves whorled **Haloragaceae** p. 423
Plants not as above **3**

3 Shrubs **4**
Herbs **18**

4 Plants armed with prickles, spines or thorns **5**
Plants unarmed **10**

5 Leaves and branches both armed **Araliaceae** p. 424
Leaves or branches armed, not both **6**

6 Leaves spiny-margined **Berberidaceae** p. 290
Leaves not spiny-margined **7**

7 Leaves pinnate **Leguminosae** p. 369
Leaves not pinnate **8**

8 Leaves opposite **Rhamnaceae** p. 400
Leaves alternate **9**

9 Branches with stout thorns (1) 2–7 cm long; stamens 10 **Rosaceae** p. 346
Branches with needle-like spines; stamens 5 **Grossulariaceae** p. 342

10 Plants 1–2 dm tall, with small appressed leaves; pubescent; flowers yellow; of sandy habitats **Cistaceae** p. 404
Plants not as above **11**

11 Stamens 3; leaves needle-like 5–10 mm long **Empetraceae** p. 397
Stamens 4, 5 or more; leaves not needle-like **12**

12 Leaves opposite or whorled **13**
Leaves alternate **15**

13 Stamens 8 or 10; leaves palmately lobed; fruit winged **Aceraceae** p. 399
Stamens 4; leaves not palmate; fruit not winged **14**

14 Leaves opposite, serrate, 1–2 (3) cm long; branches brown, ridged **Celestraceae** p. 399

Leaves whorled at stem tip; if opposite 4–10 cm long, entire and the branches reddish **Cornaceae** p. 434

15 Stamens 5 **16**

Stamens more than 5 **Ericaceae** p. 439

16 Stamens opposite the petals **Rhamnaceae** p. 400

Stamens alternate with the petals **17**

17 Ovary superior; fruit white or orange-red, 1-seeded **Anacardiaceae** p. 398

Ovary inferior; fruit yellow, red, or black, several-seeded **Grossulariaceae** p. 342

18 Flowers markedly irregular **19**

Flowers more or less regular (one sepal or petal, or more, may be somewhat larger than the others) **23**

19 Ovary 1-loculed **20**

Ovary 2-loculed or more **22**

20 Stamens 10 **Leguminosae** p. 369

Stamens less than 10 **21**

21 Petals 4; stamens 6, in two bundles of 3 **Fumariaceae** p. 292

Petals 5; stamens 5 **Violaceae** p. 405

22 Flowers white or pink; plants 1–5 dm tall **Polygalaceae** p. 394

Flowers yellow or orange; plants generally 5 dm or more tall **Balsaminaceae** p. 400

23 Plants semi-aquatic, matted; leaves opposite; flowers minute, axillary **Elatinaceae** p. 404

Plants not as above **24**

24 Ovary inferior **25**

Ovary(ies) superior **30**

25 Flowers with 2 sepals, 5 petals, 6–10 stamens, unilocular ovary, 4–6 styles **Portulacaceae** p. 250

Flowers not as above **26**

26 Stamens 4 or 8 **27**

Stamens 5 or 10 **28**

27 Leaves whorled at stem tip; inflorescence subtended by 4 conspicuous white bracts **Cornaceae** p. 434

Leaves and inflorescence not as above **Onagraceae** p. 412

28 Ovary 5-loculed; styles 5; fruit a blackish berry **Araliaceae** p. 424
Ovary 1- or 2-loculed; styles 2 or 3; fruit dry **29**

29 Inflorescence umbellate, rarely of dense globose heads subtended by spiny bracts **Umbelliferae** p. 425
Inflorescence a raceme or cyme **Saxifragaceae** p. 328

30 Ovary solitary **31**
Ovaries 2 or more, separate from one another or united only at the base **44**

31 Leaves basal, with conspicuous gland-tipped hairs; small insectivorous plants of bogs **Droseraceae** p. 326
Leaves not as above **32**

32 Stamens the same number as and opposite the petals; sepals 2 **Portulacaceae** p. 250
Stamens not the same number as the petals, or, if so, alternate with them; sepals 4 or 5 **33**

33 Ovary 1-loculed **34**
Ovary 2-loculed or more **38**

34 Style 1 **35**
Styles 2 or more **36**

35 Stamens 5; leaves pinnate **Leguminosae** p. 369
Stamens more than 5; leaves trifoliolate **Capparidaceae** p. 293

36 Stamens alternating with 5 branched, gland-tipped staminodes **Parnassiaceae** p. 341
Stamens not alternating with branched, gland-tipped staminodes **37**

37 Leaves opposite; placentation free-central; styles 2–5 **Caryophyllaceae** p. 253
Leaves basal or alternate; placentation parietal; styles 2 **Saxifragaceae** p. 328

38 Styles 5; leaves entire; flowers blue or yellow **Linaceae** p. 393
Styles 1–3; leaves and flowers various **39**

39 Leaves trifoliolate; stem woody below; flowers small, yellowish white; fruit a berry **Anacardiaceae** p. 398
Leaves not trifoliolate; stem generally herbaceous below; flowers mostly showy; fruit dry **40**

40 Stamens 6, tetradynamous (rarely 2 or 4) **Cruciferae** p. 294
Stamens 10, rarely 8 **41**

41 Leaves deeply, palmately divided **Geraniaceae** p. 391
Leaves more or less entire or merely toothed **42**

42 Plants more than 3 dm tall; leaves opposite or in whorls of 3 **Lythraceae** p. 412
Plants less than 3 dm tall; leaves generally basal **43**

43 Style 1 **Pyrolaceae** p. 435
Styles 2 **Saxifragaceae** p. 328

44 Ovaries entirely separate; 1 ovule in each ovary **45**
Ovaries united somewhat at the base; several ovules in each ovary **46**

45 Flowers perigynous, the stamens inserted on a disc; stipules present **Rosaceae** p. 346
Flowers hypogynous, the stamens inserted on the receptacle; stipules absent **Ranunculaceae** p. 272

46 Leaves fleshy; ovaries 5 **Crassulaceae** p. 327
Leaves not fleshy; ovaries 2 **Saxifragaceae** p. 328

GROUP V

Calyx and corolla both present; petals all united (or only 2 or 3 petals connate or united and the flowers markedly irregular); stamens more numerous than the petals or corolla-lobes

1 Plants reddish brown; leaves reduced to scales; saprophytes **Monotropaceae** p. 438
Plants green; leaves well developed **2**

2 Shrubs **Ericaceae** p. 439
Herbs **3**

3 Flowers regular **Adoxaceae** p. 515
Flowers markedly irregular **4**

4 Stamens 10 **Leguminosae** p. 369
Stamens less than 10 **5**

5 Leaves finely divided **Fumariaceae** p. 292
Leaves entire **6**

6 Flowers yellow or orange **Balsaminaceae** p. 400
Flowers white or pink **Polygalaceae** p. 394

GROUP VI

Calyx and corolla both present; petals all united; stamens the same number as the corolla-lobes or fewer

1 Plants pinkish white or brown; leaves reduced to papery scales; parasites **Orobanchaceae** p. 504
Plants green; leaves well developed **2**

2 Stamens opposite the corolla-lobes and the same number **Primulaceae** p. 449
Stamens alternate with the corolla-lobes, the same number or fewer **3**

3 Ovary superior **4**
Ovary inferior **26**

4 Corolla markedly irregular **5**
Corolla more or less regular, one or two lobes possibly somewhat larger than the others **7**

5 Plants with square stems; leaves opposite; ovary 4-lobed **Labiatae** p. 478
Plants not as above **6**

6 Ovary 1-loculed; aquatic plants with dissected leaves or scapose plants of moist habitats **Lentibulariaceae** p. 505
Ovary 2-loculed **Scrophulariaceae** p. 487

7 Stems twining **Convolvulaceae** p. 461
Stems not twining **8**

8 Stamens 2 (3) **9**
Stamens 4 or 5 **11**

9 Stems square; ovary 4-lobed **Labiatae** p. 478
Stems round; ovary not 4-lobed **10**

10 Leaves basal; corolla scarious **Plantaginaceae** p. 507
Leaves opposite; corolla not scarious **Scrophulariaceae** p. 487

11 Stamens 4 **12**
Stamens 5 **15**

12 Leaves opposite **13**
Leaves basal **14**

13 Leaves deeply divided; more or less prostrate plants of dry habitats **Verbenaceae** p. 477
Leaves merely serrate; erect plants of moist habitats **Labiatae** p. 478

14 Flowers solitary **Scrophulariaceae** p. 487
Flowers many, in long spikes **Plantaginaceae** p. 507

15 Shrubs **16**
Herbs **17**

16 Plants less than 5 dm tall, generally prostrate **Ericaceae** p. 439
Plants more than 5 dm tall, erect **Solanaceae** p. 485

17 Leaves trifoliolate; corolla-lobes fimbriate **Menyanthaceae** p. 459
Leaves not trifoliolate; corolla-lobes not fimbriate **18**

18 Plants with milky juice; ovaries 2, separate except sometimes at the apex; stamens united or the anthers adnate to the stigma **19**
Plants lacking milky juice; ovary and stamens not as above **20**

19 Stamens united **Asclepiadaceae** p. 460
Stamens free, the anthers merely adnate to the stigma **Apocynaceae** p. 459

20 Ovary deeply 4-lobed; style basal **Boraginaceae** p. 468
Ovary not deeply 4-lobed **21**

21 Plants prostrate, glabrous; leaves alternate, lanceolate-obovate, entire; flowers white, stigma sessile, as broad as the ovary; of saline mud-flats **Boraginaceae** p. 468
Plants not as above **22**

22 Ovary 1-loculed **23**
Ovary 2-loculed or more **24**

23 Leaves opposite, entire **Gentianaceae** p. 455
Leaves alternate or basal, usually deeply dissected or lobed **Hydrophyllaceae** p. 465

24 Ovary 3-loculed; stigma 3-cleft; fruit a loculicidal capsule **Polemoniaceae** p. 462
Ovary 2-loculed; stigma capitate or bilobed; fruit a berry, or an operculate or septicidal capsule **25**

25 Stamen filaments or some of them densely hairy **Scrophulariaceae** p. 487
Stamen filaments not densely hairy **Solanaceae** p. 485

26 Plants with tendrils; stamens united; fruit fleshy, spiny **Cucurbitaceae** p. 517
Plants not as above **27**

27 Flowers (often dimorphic) in heads subtended by an involucre of bracts **28**
Flowers not in heads subtended by an involucre of bracts **30**

28 Stamens 5, the anthers united in a tube around the bifid style **Compositae** p. 519
Stamens 4, the anthers free **29**

29 Cauline leaves whorled, entire **Rubiaceae** p. 509
Cauline leaves opposite, pinnate **Dipsacaceae** p. 516

30 Shrubs or woody climbers **Caprifoliaceae** p. 511
Herbs **31**

31 Corolla irregular; stamens united **Lobeliaceae** p. 518
Corolla regular; stamens free **32**

32 Leaves alternate; flowers blue or purplish; ovary 3–(5)-loculed **Campanulaceae** p. 517
Leaves opposite or whorled; flowers white, pinkish, lilac or yellow **33**

33 Flowers white or yellow **34**
Flowers pinkish or lilac **35**

34 Leaves whorled, entire **Rubiaceae** p. 509
Leaves opposite, pinnately lobed **Valerianaceae** p. 516

35 Plants trailing; leaves crenate towards the tip **Caprifoliaceae** p. 511
Plants erect; leaves entire **Rubiaceae** p. 509

Number of Taxa

	Families	Genera	Species
Pteridophyta			
Native	7	21	61
Introduced	0	0	0
TOTAL	7	21	61
Spermatophyta			
Gymnospermae			
Native	3	9	20
Introduced	0	0	0
TOTAL	3	9	20
Angiospermae			
Monocotyledoneae			
Native	19	100	413
Introduced	0	14	43
TOTAL	19	114	456
Dicotyledoneae			
Native	80	312	981
Introduced	4	91	237
TOTAL	84	403	1,218
Grand total	113	547	1,755

Descriptive Flora

LYCOPODIACEAE / Club-moss Family

Lycopodium L. / Club-moss

Low evergreen herbs with leafy stems, more or less moss-like in aspect; spreading by prostrate stems trailing on the ground or by subterranean rootstocks; roots adventitious, from the under side of creeping stems; erect or ascending branched shoots arising usually from the creeping stems, the branching commonly dichotomous (forked); leaves, small, crowded, often overlapping, in 4–16 ranks, usually linear or lanceolate, 1-nerved; sporangia (spore-cases) solitary in the axils of sporophylls, which are like or unlike ordinary leaves and often organized in club-like spikes (strobili); spores resembling sulphur-yellow powder, shed through a transverse opening of the sporangium; prothallus (gametophyte) rarely seen, commonly subterranean and mycorrhizal, sometimes terrestrial, thalloid and green.

1 Sporangia borne in distinct terminal cones (strobili); gemmae absent — 2
Sporangia scattered along the length of the stem, in leaf-axils; gemmae often present among the upper leaves — **L. selago**

2 Sporophylls leaf-like; sterile branches prostrate — **L. inundatum**
Sporophylls different from leaves; sterile branches erect or ascending — 3

3 Cones borne on long peduncles — 4
Cones sessile or nearly so — 5

4 Leaves in 8–10 rows with an attenuate hair-like tip; cones 1–6 — **L. clavatum**
Leaves in 4 rows, acuminate, lacking hair-like apex; cones 1–2 — **L. complanatum**

5 Leaves spreading or reflexed, in 6 or more rows, the free portion of the leaf 3.0 mm or more long — 6
Leaves appressed, generally in 4 rows, the free portion of the leaf less than 3.0 mm long — 7

6 Ascending branches simple or few-forked; prostrate stems trailing on ground — **L. annotinum**
Ascending branches much forked, the plant tree-like; prostrate stems below ground — **L. obscurum**

7 Leaves all more or less the same shape; stem rounded **L. sitchense**
Leaves of one row distinguished from the others, being shaped like a bricklayer's trowel; stem appearing winged **L. alpinum**

L. alpinum L. / Alpine Club-moss
Creeping stems slightly underground, with few distant bract-like leaves; ascending branches repeatedly forked, 4–12 cm high, glaucous; sterile branchlets flattened, the leaves 4-ranked, the upper ones ovate-lanceolate, 4–5 mm long, the lower (ventral) ones concave at base, 2–3 mm long; fertile branchlets nearly cylindrical, the leaves uniform, 6-ranked, appressed; strobili 1–3, sessile; sporophylls about 3.5 mm long, acuminate, the margins slightly erose. $2n = 46$. *Diphasiastrum alpinum* (L.) Holub. Alpine meadows and rocky slopes. Circumpolar, arctic-alpine: Alas, Yuk s to BC; nQue, Lab. Map 1 ○

L. annotinum L. / Stiff Club-moss
Prostrate stems elongate, trailing on the ground, often with a shallow covering of humus; erect branches simple or once or twice forked, 0.5–3 dm high; leaves often spreading or reflexed, firm, mostly linear, acuminate, tipped by a sharp spinule, 3–10 mm long; strobili sessile, solitary, terminating leafy stems; sporophylls ovate, sharply acuminate at apex, the margins erose-crenate. $2n = 68$. Moist woods. Circumpolar; transcont s of Arct Arch s to Wash, Colo, Gt Lakes, and Apal. Map 2 ●

L. clavatum L. / Common or Running Club-moss
Prostrate stems much elongated, forked, creeping on the ground, rooting at intervals, the leaves curved upwards; ascending stems at first simple, becoming forked; leaves mostly in 10 ranks, uniform, spreading or ascending, often incurved, linear-subulate, usually attenuate to a soft hair-like tip; peduncle of strobili 3–12 cm long, with remote bract-like leaves; strobili often in pairs; sporophylls abruptly acuminate, the margins scarious and erose. $2n = 68$. Dry woodlands. Circumpolar, boreal: similar to *L. annotinum*, s to nCalif, nMont. Map 3

L. complanatum L. / Ground Cedar
Horizontal creeping stems mostly at or just beneath the surface of the ground; ascending shoots 0.5–3 dm high, much branched, with compact or open habit, the branchlets flattened, and often with marked annual constrictions; leaves 4-ranked, sharply acuminate, adnate over half their length, the dorsal ones appressed, those of the lateral ranks with short free tips, the ventral ones 1 mm or less long; peduncles forked, with small distant scale-like uniform leaves. $2n = 46$. *Diphasiastrum complanatum* (L.) Holub. Dry woodlands. Circumpolar, boreal: similar to *L. annotinum*, s to nMont. Map 4

L. inundatum L.
Stem prostrate, 5–20 cm long; erect, sterile, branches lacking; leaves crowded, 4–6 (8) mm long, linear-subulate, twisted towards the upper side of the stem; fertile branches ascending to erect, 3–10 cm long; cones 1–3 cm long, sessile,

solitary, the sporophylls expanded at base but otherwise similar to leaves. $2n = 156$. *Lycopodiella inundata* (L.) Holub. Bogs. wEurope, Japan: seAlas to Calif, nMont; Gt Lakes to Nfld s to Gulf states. Map 2 ○

L. obscurum L. / Tree Club-moss; Ground Pine
Creeping stem deep in the ground; erect aerial stems, 1–3 dm high, tree-like, mostly unbranched at base, densely bushy-branched upwards; leaves crowded on the branchlets, in about 6 ranks, linear, 4–6 mm long, sharp-pointed; strobili sessile at the tips of leafy branchlets, mostly 4 or 5 mm thick. $2n = 68$. Moist woods and boggy clearings. eAsia: cAlas, Yuk e to Lab, Nfld s to Wash, Mont, Man, Gt Lakes, Apal. Map 1 ●

L. selago L.
Creeping and rooting stem short, mostly 2–5 cm long; ascending stems tufted, 5–20 cm long, simple below, repeatedly forked upwards; leaves 8-ranked, densely crowded on the branches, ascending or erect, lanceolate-subulate, usually hollow at base, glossy, not adnate, 3–8 mm long; gemmae often borne in the upper axils; strobili absent, the sporangia formed in the axils of ordinary leaves, each season's growth having a lower sterile zone and an upper fertile zone. $2n = 264$. *Huperzia selago* (L.) Bernh. *ex* Schrank & Mart. Damp mossy ledges, alpine slopes, and subalpine woods. Circumpolar: Alas, Yuk to Ellesmere Is s to Wash, Colo, Gt Lakes, Apal. Map 5

L. sitchense Rupr.
Creeping stems mostly beneath the surface of the ground; erect stems tufted, several times forked, 5–15 cm high; sterile branchlets not at all flattened; leaves uniform, subulate, 3–4 mm long, spirally arranged, the free tips longer than the adnate bases; strobili mostly solitary at the ends of peduncle-like branches (these with reduced appressed leaves in 6 ranks); sporophylls 2.5–3.5 mm long, ovate-deltoid, abruptly narrowed to a slender spreading tip, the margins scarious and erose. $2n = 46$. *L. sabinaefolium* Willd. var. *sitchense* (Rupr.) Fern., *Diphasiastrum sitchense* (Rupr.) Holub. Open woods and barrens. Japan, Kamchatka: sAlas to Ore, nMont, nSask. Map 6

SELAGINELLACEAE

Selaginella Beauv. / Little Club-moss

Small perennial plants, resembling mosses, forming mats or dense tufts; stems slender, branching and leafy, erect or prostrate, the roots fibrous; leaves small, 1-nerved, spiral, sometimes 4-ranked, often appressed and overlapping, obscurely ligulate; sporangia (spore-cases) solitary in axils of sporophylls, these leaf-like and organized in terminal strobili; spores of 2 kinds, large ones (megaspores) and small ones (microspores), borne in separate sporangia but usually in the same strobilus; prothalli (gametophytes) very small, largely contained within the spore-walls.

1 Plants forming small mats; leaves lanceolate, spinulose, not bristle-tipped; strobili conspicuous, more or less subcylindric **S. selaginoides**

Plants tufted or trailing; leaves of sterile branches linear-lanceolate, rigid, densely crowded, overlapping, minutely ciliate on the margins, bristle-tipped; strobili inconspicuous, sharply 4-angled 2

2 Leaves abruptly adnate to stem which is easily seen when leaves detached **S. wallacei**

Leaves strongly decurrent, the stem clothed with the decurrent bases when leaves detached 3

3 Leaf-bristle scarcely broader at base, c. 0.5 mm long **S. rupestris**

Leaf-bristle broader at base, c. 1.0 mm long **S. densa**

S. densa Rydb.
A low densely matted plant; stems short, sterile branches ascending, forming dense flexuous tufts; leaves linear-subulate, with a long white bristle at tip, pale green to greyish green, 2–3 mm long, with 5–12 cilia on each side; strobili 4-sided, 1–3 cm long, the sporophylls ovate-triangular, bristle-tipped. $2n = 18$. Includes *S. scopulorum* Maxon and *S. standleyi* Maxon. Dry prairie, open sandhills, and other exposed places. sAlas to swMan s to nCalif, in Rocky Mts s to NM. Map 7

S. rupestris (L.) Spring
Similar to *S. densa,* but differing in the characters of the key; generally more trailing and somewhat less tufted. $2n = 18$. Dry, open areas. nSask to Gt Lakes e to Que s to Okla, Ga. Map 8 ●

S. selaginoides (L.) Link
Delicate plant; sterile stems prostrate, commonly 1–4 cm long, with spreading lanceolate leaves; fertile branches erect, usually 4–8 cm high, the leaves ascending, lanceolate-acuminate, with spinulose margins; strobili 1–3 cm long, subcylindric, the sporophylls larger than the leaves, conspicuously spinulose. Wet mossy banks, mounds, and rocks. $2n = 18$. Circumpolar: Alas, Yuk, Mack, nSask, nMan to Nfld s to BC, Colo, Gt Lakes. Map 9

S. wallacei Hieron.
Similar to *S. densa,* but differing in the characters of the key; generally more trailing, the ascending branches not flexuous or densely tufted; leaf bristle shorter, c. 0.5 mm long. Dry, rocky slopes. sBC to Calif. Map 8 ○

ISOETACEAE / Quillwort Family

Isoetes L. / Quillwort

Small perennial herbs, mostly aquatic, growing in ponds or shallow lakes; general appearance that of a rush or tufted grass; stem short, fleshy, corm-like, 2–3-

lobed, bearing a dense tuft of onion-like leaves; roots numerous, fibrous, forked; leaves slender, linear, somewhat 2–4 angled, the apex narrowed and terete, the base expanded and clasping; sporangia solitary, borne in the hollow leaf-base, somewhat covered by a thin membrane (velum); spores of two kinds, usually in separate sporangia on the same plant; microsporangia with numerous minute microspores and megasporangia with fewer but much larger megaspores arise in alternating cycles of leaf-bases.

Megaspores 0.3–0.4 mm thick, the surface with low tubercles or short crests; microspores finely spinulose **I. bolanderi**

Megaspores mostly 0.4–0.5 mm thick, densely spinulose; microspores smooth or finely papillose **I. echinospora**

I. bolanderi Engelm.
Leaves mostly 5–10 cm long, erect or spreading, rather soft, tapering to a fine point; sporangia 3–6 mm long; megaspores 0.3–0.4 mm thick, white or bluish; microspores fawn-coloured, ovoid, 23–30 microns long. Mountain ponds and lakes. sBC to Calif e to Colo. Map 10 ○

I. echinospora Dur.
Leaves mostly 7–15 cm long, erect or somewhat recurved, stiff or flaccid, tapering to a blunt tip; sporangia 4–7 mm long; megaspores 0.3–0.6 mm thick, white; microspores fawn-coloured, 25–35 microns long. $2n = 22$. *I. muricata* Dur. of Ed. 1, *I. braunii* Dur. Ponds and lakes. Circumpolar: Alas, Yuk, Mack, nSask, nMan to Nfld s to BC, Colo, Gt Lakes. Map 10 ●

EQUISETACEAE / Horsetail Family

Equisetum L. / Horsetail; Scouring Rush

Rush-like plants with jointed, branched, creeping rootstocks, often tuber-bearing; roots felted, in whorls from the nodes; aerial stems jointed, with scale-like leaves at the nodes, the internodes usually hollow, the surface ridged and overlaid with siliceous tubercles or bands; stomata in the grooves in regular rows or broad bands; each internode commonly with a large central air-cavity, smaller cavities (vallecular) under the grooves and still smaller ones under the ridges; leaves small, whorled, fused laterally into a sheath, only their tips free, these sometimes deciduous; branches, when present, mostly in whorls from the nodes; sporangia borne in terminal cones (strobili); cone with whorled peltate sporangiophores each bearing 5–10 sporangia; spores green, spherical, each provided with 4 elastic hygroscopic elaters; prothallus (gametophyte) small, thalloid, variously lobed; found in damp places.

1 Sheaths green below, reddish brown above, the teeth cohering in broad papery lobes; branches often arcuate and themselves branched **E. sylvaticum**

Sheaths and branches not as above 2

2 Stems green, unbranched (occasionally a few random branches occur, but regular whorled branches at successive nodes lacking); cones mostly apiculate (except in *E. fluviatile* and sometimes in *E. laevigatum*) 3

Stems not as above; *either* green and branched *or* unbranched with a greyish white to a light yellow-brown colour; cones not apiculate 7

3 Stems 2–10 dm high, more than 3.0 mm wide, with 10–40 ridges; sheaths generally more than 4.0 mm long 4

Stems 1–3 dm high, less than 3.0 mm wide, with 3–12 ridges; sheaths generally less than 4.0 mm long 6

4 Teeth present on all sheaths, brownish black or with a very narrow hyaline margin, stem papery, readily flattened; cones not apiculate **E. fluviatile**

Teeth missing from many or all sheaths, where present, scarious or with a broad scarious margin; stem rigid, not easily flattened; cones often apiculate 5

5 Cones sharply apiculate; sheaths generally with two continuous black bands, some teeth generally persistent **E. hyemale**

Cones rounded at apex or only slightly apiculate; sheaths with only one black band (often appearing as a row of dots), teeth invariably deciduous, except for sheath subtending cones **E. laevigatum**

6 Stem slender, wiry, 0.5–1.0 mm thick, much bent or zig-zagged; sheath teeth 3 **E. scirpoides**

Stem thicker, 1.0–2.0 (3) mm thick, straight or curved (no angular bends); sheath teeth more than 3 **E. variegatum**

7 Stems green, branched 8

Stems not green, unbranched (bud-like outgrowths sometimes found at base of upper sheaths) 11

8 Lowest internode of each branch generally much longer than the corresponding sheath of the stem **E. arvense**

Lowest internode of each branch generally only as long as or shorter than the corresponding sheath of the stem 9

9 Sheath of the branches 3-toothed **E. pratense**

Sheath of the branches more than 3-toothed (4–6) 10

10 Stem sheaths with 6–8 teeth **E. palustre**

Stem sheaths with 10–30 teeth **E. fluviatile**

11 Sheaths (2) 3.0–6.0 mm across at mouth, teeth separate, with a narrow brown midvein and a broad hyaline margin; bud-like outgrowths sometimes occur at base of upper sheaths **E. pratense**

Sheaths (3) 5.0–10.0 mm across at mouth, teeth often coherent, with a broad dark brown midvein and scarcely paler margin; outgrowths from the upper sheaths absent **E. arvense**

E. arvense L. / Common or Field Horsetail

Shoots of 2 kinds, from a rootstock which is extensively creeping and branching, dark-felted and tuber-bearing; fertile stems appearing in early spring, erect, unbranched, usually thick and succulent, 1–3 dm tall, brownish to whitish, soon withering, the sheaths with 8–12 large, brown, acuminate teeth; cone long-peduncled, obtuse, not apiculate; sterile shoots appearing as the fertile ones wither, solitary or clustered, 1–5 dm high, erect or decumbent, green, slender, much branched, usually with 10 or 12 ridges, the sheaths with about 12 sharp brownish teeth; branches numerous in dense whorls, ascending or spreading, mostly unbranched, 3–5-angled, the sheaths with lanceolate-attenuate teeth. Variable in appearance, depending partly on environmental conditions. $2n = 216$. Moist woods, meadows, banks, roadsides, and railway embankments; often persisting in cultivated fields. Circumpolar: throughout NAm. Map 11

E. fluviatile L.
Rootstock wide-creeping, not felted, shining, often reddish; stems erect, 5–10 dm high, simple or branching, 3–8 mm thick, with 10–30 shallow grooves, smooth, soft and brittle; central cavity large, four-fifths the stem diameter; sheaths 6–10 mm long, tightly appressed, the teeth persistent, acuminate, dark brown, scarcely scarious-margined; branches none or sparse and irregular or many and regularly whorled, without branchlets; cones at the tips of main stems, long-peduncled, deciduous. $2n = 216$. Includes *E. limosum* L. Marshes, swamps, and bogs. Circumpolar: transcont s of Arct Arch, s to Wash, nMont, Va. Map 12

E. hyemale L. / Scouring Rush
Rootstock slender, blackish; stems solitary or clustered, mostly unbranched, evergreen, rigid, 0.5–10 dm tall, very rough, 18–40-ridged, the ridges beset with siliceous cross-bands or a row of tubercles; sheaths appressed, at first green, often becoming ashy and black-banded at base and apex, the teeth brownish; cones short-peduncled, apiculate. $2n = 216$. *Hippochaete hyemalis* (L.) Bruhin. In our area var. **affine** (Engelm.) A.A. Eat. (*E. prealtum* Raf.). Produces a sterile hybrid with *E. laevigatum* (*E.* × *ferrissii* Clute), *E. hyemale* var. *intermedium* A.A. Eat. of Ed. 1. Sandy shores, open woods, embankments. Circumpolar, boreal: throughout NAm s of 65 in west, 50 in east. Map 13

E. laevigatum A.Br.
Rootstock brown or black; stems usually annual, erect, solitary or clustered, normally unbranched, 3–8 dm tall, 2–7 mm thick, rather soft, 15–25-ridged, more or less smooth with a single row of low tubercles or cross-bands; sheaths longer than broad, slightly flaring upward, with a terminal dark band and rarely with one at the base of the sheath, the teeth blackish and early deciduous; cones blunt or apiculate. $2n = 216$. *Hippochaete laevigata* (A.Br.) Farw. Includes *E. kansanum* Schaffner. Produces a sterile hybrid with *E. variegatum* (*E.* × *nelsonii* (Eat.) Schaffn.). Meadows, banks, and low prairie grassland. sBC to sQue s to Calif, Tex. Map 14

E. palustre L.
Rootstock black, creeping, branched; stems erect, solitary or clustered, simple or extensively branched, 2–6 dm tall, deeply 5–10-ridged, the ridges narrow and prominent, the central cavity very small, the vallecular cavities nearly as large; sheaths of main stem widened upward, the teeth persistent, lanceolate-acu-

minate, brownish or blackish with broad scarious borders; branches few and irregular, or numerous and regularly whorled, slender; cones borne on the main stems, long-peduncled, deciduous. $2n = 216$. Banks of streams, meadows, wet woods, often in shallow water. Circumpolar, boreal: Alas, Yuk, wMack, Huds Bay to Nfld s to Wash, Mont, sSask, sMan, Gt Lakes. Map 15

E. **pratense** Ehrh.
Rootstock black, creeping; shoots of 2 kinds, annual, erect, 2–5 dm high; fertile stems, appearing in early spring but apparently uncommon in our region, normally unbranched at first, later developing whorls of branches like those of the sterile stems; sheaths pale, the teeth slender, brown, white-margined; cone long-peduncled, slender, deciduous; sterile stems mostly solitary, slender, whitish green, with 10–18 ridges, the sheaths pale with slender, brown, white-margined teeth; branches numerous, regularly whorled, very slender, 3-angled, unbranched, the teeth of the sheaths deltoid. $2n = 216$. Stream banks, thickets, and moist woods. Circumpolar, boreal: similar to *E. palustre.* Map 16

E. **scirpoides** Michx.
Rootstock widely branched, slender, mostly 1–1.5 mm thick; stems evergreen, densely clustered, 6–16 cm long, ascending or prostrate, often arched-recurved, very slender, 0.5–1 mm thick, usually appearing about 6-ridged; central cavity lacking, the vallecular cavities commonly 3; sheaths 3–4 mm long, loose, the 3 or 4 teeth triangular-acuminate, white-margined; cones 2–5 mm long, apiculate. $2n = 216$. *Hippochaete scirpoides* (Michx.) Farw. Damp woods, thickets, and mossy banks. Circumpolar: similar to *E. palustre* but n to Arct coast. Map 17

E. **sylvaticum** L. / Woodland Horsetail
Rootstock creeping, deep in the ground; shoots of 2 kinds, annual, erect, 3–6 dm high; fertile stems green or pink-brown, at first unbranched, later producing whorls of compound branches; cones deciduous, not apiculate; sterile stems mostly solitary; all primary stems 10–18-ridged, the ridges each with 2 rows of sharp spinules, the sheaths 1–2 cm long, loosely flaring and reddish brown upward, the teeth united into 3 or 4 lobes; branches numerous, regularly whorled, simple or again branched, the branchlets often curved downward. $2n = 216$. Open woodlands and thickets. Circumboreal: similar to *E. palustre* but not Wash. Map 18

E. **variegatum** Schleich.
Rootstock much branched and creeping; stems evergreen, clustered, ascending to erect, 1–4 dm long, 1–3 mm thick, branched near the base, 5–10-ridged, the ridges grooved, with 2 rows of siliceous tubercles; central cavity about one-third the stem diameter; vallecular cavities large; sheaths with a broad black band at the summit, the teeth abruptly acuminate, broadly white-margined; cones 5–8 mm long, strongly apiculate. $2n = 216$. *Hippochaete variegata* (Schleich. *ex* Weber & Mohr) Bruhin. Produces a sterile hybrid with *E. hyemale* (*E.* × *trachydon* A.Br.). See also under *E. laevigatum.* Wet thickets, bogs, and sandy shores. Circumpolar: similar to *E. palustre* but s to Colo in Rocky Mts and n to Ellesmere Is. Map 19

OPHIOGLOSSACEAE / Adder's-tongue Family

Botrychium Sw. / Grape Fern

Plants with short erect soft rootstocks and a cluster of fleshy roots; frond consisting of a single sterile blade and a sporophyll both on a common stalk (stipe) arising from the rootstock; base of stipe containing the bud for next year's frond; sterile blade (leaf) sessile or stalked, usually pinnately or ternately dissected or compound; fertile blade (sporophyll) spike-like or panicle-like with clusters of large globose sporangia arranged in double rows; sporangium without an annulus, opening into 2 valves; spores numerous, yellow in colour; gametophytes mostly subterranean.

1 Sterile blade inserted near base of stipe 2
Sterile blade inserted about or above middle of the stipe 3

2 Plants large, 10–50 cm high; sterile blade ternately compound with 2–4 orders of divisions, about as long as wide or wider; bud at base of stipe hairy **B. multifidum**
Plants smaller, 2–15 (20) cm high; sterile blade slightly longer than wide; mostly pinnate, pinnae often lobed; bud at base of stipe glabrous **B. simplex**

3 Sterile blade generally much longer than broad, pinnate, pinnae broadly clavate to flabellate, often remote 4
Sterile blade not much longer than broad, often shorter, mostly bipinnate or ternately divided, the divisions not clavate or flabellate, pinnae normally not remote. 5

4 Base of stipe swollen; sterile blade inserted at apex; pinnae broadly clavate **B. dusenii**
Base of stipe not swollen; sterile blade inserted about stipe middle; pinnae flabellate **B. lunaria**

5 Sterile blade deltoid, wider than long, sessile 6
Sterile blade ovate to lanceolate, as long as broad or somewhat longer, distinctly short-petioled **B. boreale**

6 Bud at base of stipe hairy; stalk of fertile blade generally 4–17 cm long; sterile blade large, delicate, generally more than 6 cm long, 10 cm wide **B. virginianum**
Bud at base of stipe glabrous; stalk of fertile blade 1–3 cm long; sterile blade smaller, firm, less than 6 cm long, 10 cm wide **B. lanceolatum**

B. boreale Milde
Plant 2–25 cm tall; bud glabrous; sterile blade sessile, ovate, mostly inserted above the middle of the plant, pinnate or bipinnate, pinnae palmately lobed to crenate, conspicuously veiny; fertile blade 2–10 cm long, erect. $2n = 180$. Open, grassy subalpine and alpine areas. Greenland, nEurope, eAsia: Alas, Yuk, nwMack, s to neOre, Nev. Map 20 •

B. dusenii (Christ) Alston
Similar to *B. lunaria;* differing in the characters of the key. $2n = 90$. Dry open areas. Alta s to Calif and Ariz. Map 21

B. lanceolatum (Gmel.) Angstr.
Plant 5–25 cm tall, rather stout and fleshy; sterile blade deltoid, sessile or nearly so, once or twice ternate or pinnately divided, the pinnae lobed; fertile panicle mostly extending above the sterile blade, the sporangia crowded on ascending branches; sterile and fertile blades both reflexed at emergence from the bud. $2n = 90^*$. Mountain slopes. More or less circumpolar, subarctic-alpine: sAlas, Yuk s to Colo, Utah; Gt Lakes to Nfld, s to NC. Map 22 ●

B. lunaria (L.) Sw. / Moonwort
Plant 3–25 cm tall, slender or stout and fleshy; sterile blade sessile or nearly so, with a sheathing base, inserted about the middle of the plant, once-pinnate, the 3–7 pairs of segments broadly fan-shaped, obovate or semicircular, often overlapping, entire or sometimes toothed; fertile panicle raised well above the sterile blade. $2n = 90, 180$. In our area var. **lunaria** and with less flabellate pinnae var. **onondagense** (Underw.) House. Meadows, banks, slopes, and open woods. Circumpolar: Alas, Yuk, Mack s to Calif, Colo, sSask, sMan e to Lab s to Gt Lakes. Map 23

B. multifidum (Gmel.) Rupr. / Leather Grape Fern
Plant 10–50 cm tall, stout and very fleshy; sterile blade on a stout stalk, inserted near the base of the stipe, evergreen, coriaceous, deltoid in outline, ternate to 2–4-pinnately dissected, the ultimate pinnules toothed or nearly entire. $2n = 90$. In our area var. **intermedium** (D.C. Eat.) Farw. (*B. silaifolium* Presl.) with the stalk of the fertile blade 3–15 cm long, the blade normally 5–20 cm broad and about as long, and the pinnules not crowded. Moist, sandy areas. Circumpolar, boreal: transcont 60 to 40. Map 20 ○

B. simplex E. Hitchc.
Plant 3–15 (25) cm tall; sterile blade inserted at base, sessile or petiolate (–3 cm), blade 0.5–4 (6) cm long, simple, lobed or pinnate, the pinnae flabellate, lowest pair at times enlarged and pinnately lobed; fertile blade 3–15 (25) cm long, sporangia in a simple spike or pinnate on short branches. $2n = 90$. Moist meadows and shores. nEurope, Japan: BC to Sask s of 54, s to Calif, NM; Gt Lakes to Nfld. Map 22 ○

B. virginianum (L.) Sw. / Grape Fern
Plant tall and erect, commonly 20–50 cm high; bud hairy, partially exposed at a slit in the sheathing base of the stipe; sterile blade sessile or nearly so, broadly triangular in outline, more or less ternate, decompound, the lower pair of pinnae much larger than those above, the ultimate segments oblong-lanceolate, variously incised or toothed; fertile panicle bipinnate or tripinnate, elevated on a stalk 5–20 cm long. $2n = 184$. *Botrypus virginianus* (L.) Holub. Moist woodlands and thickets. Circumpolar, boreal: transcont n to 54, s to sUSA. Map 24

POLYPODIACEAE / Fern Family

Plants of diverse habit with leaves (fronds) arising from underground stems (rhizomes), these horizontal and creeping or short and more or less erect; fronds coiled circinately in the bud, often scaly, commonly with large blades, pinnate to decompound, and elevated on stiff stalks (stipes), fertile and sterile ones alike or dissimilar; sporangia aggregated in brownish strips or clusters (sori) on the back or margins of the frond or its pinnae; sori naked or covered by an indusium or by the recurved leaf-margin; sporangium long-stalked, opening and discharging the spores through the movements of a special structure, the annulus; spores microscopic, all alike; prothallus (gametophyte) minute, flat, green, formed on damp ground.

1	Fronds strongly dimorphic, sterile green, fertile turning brown	2
	Fronds not strongly dimorphic	3
2	Plants up to 20 cm high; fertile fronds longer than sterile	**Cryptogramma**
	Plants more than 40 cm high; fertile fronds shorter than sterile	**Matteuccia**
3	Sori set on leaf-margin, not round	4
	Sori set in from leaf-margin (if marginal, sori round)	7
4	Sori continuous along leaf-margin	5
	Sori not continuous, kidney-shaped	**Adiantum**
5	Plants more than 30 cm high; fronds bipinnate	**Pteridium**
	Plants mostly less than 30 cm high; fronds pinnate to bipinnate	6
6	Blades glabrous, distal pinnae 5 mm or longer	**Pellaea**
	Blades densely woolly below, distal pinnae less than 5 mm long	**Cheilanthes**
7	Fronds pinnate, pinnately lobed or entire	8
	Fronds compound pinnate, 2 or more divisions	10
8	Pinnae lobed, about as long as broad, frond usually less than 10 cm long	**Asplenium**
	Pinnae simple, much longer than broad, frond usually more than 10 cm long	9
9	Pinnae rounded at end, marginal spines lacking	**Polypodium**
	Pinnae acute at apex, with conspicuous marginal spines	**Polystichum**
10	Stipe generally 2 or 3 times longer than the blade; lowest pair of pinnae much larger than those above, the blade almost appearing to be trifoliolate	11
	Stipe generally much less than twice as long as blade; lowest pair of pinnae not much larger than those above	12
11	Indusium present	**Cystopteris**
	Indusium absent	**Gymnocarpium**
12	Small tufted plants, mostly less than 20 cm high; lower pinnae somewhat remote	13
	Larger plants, generally more than 20 cm high, tufted or with scattered or single fronds; lower pinnae not remote	14

13 Indusium evident, hood-like, attached laterally; stipe bases not persistent **Cystopteris**
Indusium not obvious, below sorus; stipe bases persistent **Woodsia**

14 Indusium absent 15
Indusium present 16

15 Lowest pair of pinnae reflexed, conspicuous hyaline hairs on midrib and veins of lower surface; woodland species **Phegopteris**
Lowest pair of pinnae not reflexed, conspicuous hyaline hairs absent; alpine species **Athyrium**

16 Indusium elongate, flap-like, attached along one margin to the vein subtending the sorus, margin often fimbriate **Athyrium**
Indusium rounded or reniform, attachment localized at the sinus, margin not fimbriate **Dryopteris**

Adiantum L.

A. pedatum L. / Maidenhair Fern
Rhizome 2–5 mm thick, short-creeping or oblique; stipes deep reddish-brown to black, lustrous, stiff, 1–4 dm high, forking above into 2 frond-branches, with 3–5 pinnae on each branch; pinnules numerous, very thin, glaucous, short-stalked, mostly oblong, spreading at right angles, the outer margins often deeply incised; sori oblong-lunate, appearing marginal, each covered by a false indusium (reflexed marginal lobe of the pinna). $2n = 60$. On cliffs and among boulders; at high altitudes. Asia: sAlas to Calif; Gt Lakes to Nfld. Map 31 ○

Asplenium L. / Spleenwort

A. viride Huds. / Green Spleenwort
Fronds numerous in crowded crowns from the rhizome, the latter also with many stipe-bases; stipe reddish brown below, green above, rachis green, flattened but not winged; blades linear, pinnate, 5–15 cm long; pinnae 10–20 pairs, alternate upwards, rounded to rhombic ovate, sessile, mostly crenate; sori 2–4 pairs on a pinna; indusium delicate. $2n = 72$. Calcareous rocks and crevices of cliffs. Circumpolar: Alas, Yuk to Calif, Colo; Gt Lakes to Nfld. Map 25

Athyrium Roth

Large ferns with short-creeping, stout, densely scaly rhizomes; fronds in clusters; stipes scaly, with 2 vascular bundles at base and a single U-shaped bundle upwards; blades 2–3 pinnate, mostly thin and delicate, up to 10 dm long, the segments mostly pinnatifid; sori elongated and curved across the veins, sometimes horseshoe-shaped, occasionally nearly straight or almost circular; indusium attached along one side, delicate, sometimes lacking.

Indusium present **A. filix-femina**
Indusium absent **A. distentifolium**

A. distentifolium Tausch *ex* Opiz
Similar to *A. filix-femina* but smaller, 2–8 dm high; fronds narrower, larger pinnae mostly 2–8 cm long, the lower distant. $2n = 80$. *A. alpestre* (Hoppe) Rylands of Ed. I. Rocky slopes and alpine meadows. Circumpolar, scattered: seAlas to Calif, Colo; Gt Lakes to Nfld. Map 35 ●

A. filix-femina (L.) Roth / Lady Fern
Fronds 4–10 (20) dm long; stipes about a third as long as the blade, with light brown, fibrous scales at base; blades acuminate, somewhat narrowed at base, the rachis pale; pinnae 20–30 pairs below the pinnatifid tip, the larger mostly to 15 cm long, 4 cm wide; pinnules pinnatifid or merely serrate; indusia inconspicuous, crescent-shaped to reniform, ciliate, mostly crossing the vein. $2n = 80$. In our area var. **cyclosporum** Rupr. Moist places, woods, and thickets. Circumpolar: transcont s of 65 in west, 55 in east, s to Calif and Gulf states. Map 26

Cheilanthes Sw.

Rhizomes short, scaly; fronds erect, rigid, 1–4-pinnate; petioles wiry, bundle solitary; sori marginal, solitary or more often confluent; indusium often formed by the revolute margin of pinnule.

Leaf-blade with conspicuous scales on rachis and costae below	**C. gracillima**
Leaf-blade lacking scales (dehisced sporangia are not scales)	**C. feei**

C. feei Moore / Slender Lip Fern
Fronds tufted, 8–25 cm tall, from a short rhizome; scales of rhizome orange to brown with a distinct median stripe; fertile and sterile fronds alike; stipes slender, 4–10 cm long, brown or purplish, soft-hairy or glabrate; blades linear-oblong to ovate, 1–3.5 cm wide, usually 3-pinnate; segments small, rounded, thinly soft-hairy above or glabrous, densely brown-hairy on the lower surface; indusia obscure, mature sporangia often appear to cover the lower surface of pinnae. $2n = 87$. Dry, calcareous, rocky outcrops. sBC, Alta s to Calif, Tex, Wisc. Map 27 ●

C. gracillima D.C. Eat.
Similar to *C. feei* but fronds narrower, 1–2 cm wide, generally bipinnate, the pinnae crowded but not overlapping, the lower about as long as the middle, the margins strongly revolute. Ledges and crevices of igneous rocks. sBC to Calif, Nev, Mont. Map 27 ○

Cryptogramma R.Br. / Rock Brake

Small ferns, strongly dimorphic, the fertile fronds taller and with narrower segments than the sterile; stipes weak, with one bundle, greenish above, yellowish and scaly at base; sterile fronds spreading, bipinnate with broad flat divisions, dentate or crenate; fertile fronds much longer than the sterile ones, 2–3 pinnate, the pinnules oblong to linear, entire, the revolute margin covering the more or less confluent sori; true indusium lacking; spores tetrahedral.

Rhizome short, ascending; fronds clustered, the sterile ones rather firm **C. acrostichoides**

Rhizome slender, long-creeping; fronds scattered, the sterile ones delicate **C. stelleri**

C. acrostichoides R.Br.
Fronds densely clustered, the stipes mostly straw-coloured, the rachis green; blades of sterile fronds ovate-deltoid, about 5 cm long, with 4–6 pinnae on each side of the winged rachis; lower pinnae with 2–4 pairs of pinnules, these ovate-oblong, obtuse, mostly crenate; fertile fronds stiffly erect, up to 2 dm tall, the stipes much longer than the blades, the ultimate segments linear, mostly 9–12 mm long, entire, the revolute margins thin and yellowish. $2n = 60^*$. *C. crispa* (L.) R.Br. var. *acrostichoides* (R.Br.) C. Clarke of Ed. 1. Rocky slopes and crevices. eAsia: Alas, Yuk, wMack to nSask, Man, sOnt s to Calif, NM. Map 28

C. stelleri (S.G. Gmel.) Prantl
Differs from the last species in the key characters and the following: blades of sterile fronds ovate; pinnae deltoid-ovate, the lower of these stalked; pinnules mostly obovate and often crenately lobed; fertile fronds with stipes somewhat longer than the blades, the ultimate segments linear-lanceolate, up to 2 cm long. $2n = 60$. Cool, shaded, calcareous rock or springy places. More or less circumpolar: Alas, Yuk, wMack, nSask s to Wash, Colo; Gt Lakes to Lab, Nfld, and NJ. Map 29

Cystopteris Bernh. / Bladder Fern

Rhizome creeping, slender, scaly; fronds delicate, clustered or scattered; stipes slender, straw-coloured, scaly at base only, with two bundles extending to the base of the blade; blades 2–3-pinnate, the veins reaching the margin; sori roundish, not confluent; indusium delicate, hood-like or convex, attached at one side and partly under the sorus, soon withering.

Fronds tufted; rhizome rather thick, unbranched; blades of fronds broadly lanceolate, 2–3-pinnate; pinnae about equal at base **C. fragilis**

Fronds scattered; rhizome slender, branched; blades of fronds deltoid-ovate, somewhat ternate and 2–3-pinnate; pinnae very unequal at base **C. montana**

C. fragilis (L.) Bernh.
Fronds clustered near the tip of a creeping, light brown, scaly rhizome; fronds variable, mostly 10–25 cm long, the stipe shorter than the blade; stipe and rachis brittle; pinnae lanceolate to ovate, the segments variable, often rhombic-ovate, sometimes pinnatifid, otherwise cleft or toothed; sori small; indusium often toothed. $2n = 168, 252$. In our area ssp. **fragilis** with spiny spores and ssp. **dickieana** (Sim.) Hyl. with rugose spores. Moist ledges and slopes. Circumpolar: throughout NAm. Map 30

C. montana (Lam.) Desv.
Differing from the last species in the key characters and the following: fronds 20–35 cm long, the stipe much longer than the blade and sparingly glandular; lowermost pinnae large, unequally deltoid-ovate and the pinnules deeply divided; indusium whitish, inconspicuous. $2n = 168$. Springy or damp calcareous places. Circumpolar, scattered records: Alas, Yuk, wMack s to BC, nMont; Colo; Ont; Que to Nfld. Map 31 •

Dryopteris Adans. / Shield Fern

Mostly large ferns of woodlands; rhizome stout, closely covered with old stipe bases; fronds 1–3 pinnate, rather firm, often green through the winter; stipes rather stout, with abundant basal scales, not jointed to the rhizome; pinnae mostly obliquely inserted on the rachis; sori rounded, on the veins; indusia round-reniform, persistent, attached by a deep notch (sinus) on one side.

1 Fronds 0.2–2 dm long, aromatic; blades glandular on margins, veins and rachis; indusia often overlapping; xerophytic **D. fragrans**
Fronds 1–10 dm long, not especially aromatic; blades rarely glandular; indusia not overlapping; chiefly in woodlands 2

2 Blades mostly bipinnate to tripinnate, the segments with spinulose teeth; indusia small, dot-like 3
Blades mostly bipinnate, the segments not spinulose or rather obscurely so; indusia about 1 mm wide 4

3 Basal pinnule on the lowest pinna not much larger than the one opposite; scales of stipe uniformly coloured **D. carthusiana**
Basal pinnule on the lowest pinna much larger than the one opposite; scales of stipe darker in the middle than near the margin **D. assimilis**

4 Pinnae lanceolate or linear-lanceolate, uniform; the longest 8–18 cm **D. filix-mas**
Pinnae oblong-lanceolate and (the lower) triangular-ovate; the longest 6–8 cm **D. cristata**

D. assimilis S. Walker / Broad Spinulose Shield Fern
Fronds clustered, up to 1 m long; stipe and rachis with light brown scales; stipe shorter than the tripinnate-pinnatifid blade; blades 1–4 dm broad, triangular-ovate, acuminate; lower pinnules only slightly decurrent on a narrowly winged midrib and the basal pinnule on the lowest pinna much larger than the one opposite; pinnules and lobes serrate, the teeth usually incurved and spinulose; sori small, the indusia soon shrivelling. $2n = 82$. *D. dilatata* (Hoffm.) A. Gray of Ed. 1. Moist woods. More or less circumpolar, boreal: Alas, Yuk, swMack s to Calif, Colo; Gt Lakes to James Bay, Nfld. Map 32 •

D. carthusiana (Vill.) H.P. Fuchs / Narrow Spinulose Shield Fern
Resembling *D. assimilis* but with narrower, often oblong-lanceolate fronds, usually bipinnate-pinnatifid and more uniform at the base; lower pinnules decurrent to form a distinctly winged midrib. $2n = 164$. Hybridizes with *D. cristata.*

D. spinulosa (Muell.) Watt of Ed. 1. Moist woods, slopes, and meadows. More or less circumpolar: Alta, Mont to Nfld s to Gt Lakes and Apal. Map 33 ●

D. cristata (L.) A. Gray
Fronds 3–8 dm long, the stipes 1–4 dm long; fertile fronds longer than the sterile ones; blades ovate-lanceolate, acuminate, bipinnate-pinnatifid, the pinnae mostly oblong-lanceolate, the lower ones triangular-ovate with their basal pinnules or segments incised to deeply divided; segments incurved-serrate or biserrate, the teeth short-spinulose; sori midway between margin and midvein; indusium 0.7–2 mm wide. $2n = 164$. The hybrid **D. × uliginosa** (A. Braun *ex* Doell) Druce ($2n = 164$*), formed between this and the previous species, has been reported from our area. Moist woods and marshes. Europe, wAsia: BC to Nfld s to Mont, Ark, Apal. Map 32 ○

D. filix-mas (L.) Schott / Male Fern
Fronds 4–10 dm long; stipes 0.5–1 dm long, stout and with long-attenuate scales; blades lanceolate to oblong-lanceolate, acuminate, narrowed at base, bipinnate-pinnatifid; pinnae 20–30 pairs, their segments oblong, obtuse, incurved-serrate, the teeth not spinulose; sori large, often confined to the lower half or two-thirds of the pinnule; indusium persistent, 0.8–2 mm wide. $2n = 164$. Wooded slopes. Circumpolar, boreal, scattered records: BC to Wash, Mont; Gt Lakes to Nfld; swUSA. Map 33 ○

D. fragrans (L.) Schott / Fragrant Shield Fern
Fronds 0.2–2 dm long, spicy-aromatic; stipes short, glandular, densely covered with reddish or brownish scales; blades linear-lanceolate, tapering at both ends, coriaceous, capitate-glandular; pinnae pinnately incised or merely crenate, overlapping and often inrolled; rachis and veins with reddish or cinnamon-brown scales; sori crowded, medial; indusium 1–2 mm wide, the margin often toothed. $2n = 82$. Siliceous rocks. More or less circumpolar: transcont s to cBC, nSask, nMan, Gt Lakes to Nfld. Map 32 ▲

Gymnocarpium Newman

Delicate ferns with slender creeping blackish rhizomes and scattered fronds; stipe slender, straw-coloured with a dark base, longer than the blade, with 2 bundles throughout; blades broadly triangular or 5-sided in outline, bipinnate-pinnatifid to tripinnate-pinnatifid; lower pinnae articulate to the rachis, long-stalked; veins forked, reaching the margin; sori rounded; indusium lacking.

Blade quite delicate, glabrous or nearly so, the two lower divisions nearly as long as the terminal one; pinnules acutish; stipe and rachis lustrous, nearly glabrous **G. dryopteris**

Blade fairly firm, glandular on both sides, the two lower divisions about half as long as the terminal one; pinnules obtuse; stipe and rachis dull, glandular **G. jessoense**

G. dryopteris (L.) Newm. / Oak Fern
Rhizome slender, branched, blackish with appressed brownish scales; stipe filiform, 0.5–5 dm high, sparsely scaly at base, glabrous, shiny; blades deltoid-pentagonal, 5–25 cm broad, 5–18 cm long, with 3 stalked primary divisions, these pinnate or bipinnate, the ultimate segments oblong, acutish, toothed or entire; sori small, rounded, without indusia. $2n = 80, 160$. *Dryopteris disjuncta* (Rupr.) Morton, *Phegopteris dryopteris* (L.) Fée and *Thelypteris dryopteris* (L.) Slosson. In our area the western var. **disjuncta** (Rupr.) Ching, $2n = 80$, with a larger, more dissected leaf and smaller spores and the not readily distinguishable eastern var. **dryopteris,** $2n = 160$. Moist woods and thickets. Circumpolar, boreal: Alas to Lab, Nfld s to Ore, Mont, Sask, Man, Gt Lakes; Colo, NM, Ariz. Map 34

G. jessoense (Koidz.) Koidz.
Similar to *G. dryopteris;* mainly distinguished by the characters of the key and the narrower fronds. $2n = 80, 160$. In our area ssp. *parvulum* Sarvela. Produces a sterile hybrid, *G.* × *intermedium* Sarvela, with the previous species. *G. robertianum* (Hoffm.) Newm. of Ed. 1. Rock crevices. Circumpolar: Alas, Yuk, Mack, nSask to Gt Lakes, NB s to BC, Alta. Map 35 ○

Matteuccia Todaro

M. struthiopteris (L.) Todaro / Ostrich Fern
Rhizome coarse, scaly, forked, producing an erect caudex covered with persistent stipe-bases, and sending out black underground stolons; fronds dimorphic; sterile fronds in whorls, 0.5–1.5 m tall; stipes much shorter than the blades, deeply channelled on upper side, scaly when young; blades gradually narrowed at base, abruptly acuminate at top, pinnate-pinnatifid; pinnae very many, long-acuminate, deeply pinnatifid, mostly spreading; fertile fronds relatively short, 2–6 dm tall, erect, rigid, coarse, at first olive-green, finally blackish, pinnate; pinnae crowded, ascending-spreading, less than 5 cm long, undulate and segmented; sori several to a segment, covered by the revolute margin of the pinna; indusia hyaline, hood-like, lacerate. $2n = 78, 80$. *M. nodulosa* (Michx.) Nieuwl. *Pteretis nodulosa* (Michx.) Nieuwl. *P. pensylvanica* (Willd.) Fern. Damp woods. Circumpolar boreal: BC to Nfld s to Gt Lakes; Alas. Map 36 ●

Pellaea Link

Rhizome short, scaly; fronds stiffly erect, often dimorphic, 1–4-pinnate; stipe wiry, bundle solitary; pinnae coriaceous, short-petiolate, often glaucous; sori marginal, usually confluent, the common indusium formed by the revolute pinna margin.

Fronds dimorphic, fertile much longer; stipe purplish black, densely pubescent; stalks of pinnae spreading **P. atropurpurea**

Fronds similar, fertile and sterile subequal; stipe brownish, glabrous or sparsely pubescent; stalks of pinnae ascending **P. glabella**

P. atropurpurea (L.) Link
Fronds caespitose, dimorphic, 10–50 cm long; stipe slender, 5–20 cm long, purplish-black, densely pubescent; blades to 30 cm long, bipinnate below, pinnate above, lower pinnae 1–6 cm long, longer than the upper; pinna stalks divergent at a wide angle (+45°) from the rachis. $2n = 87$. An apogamous triploid species. Dry limestone rocks. sBC to NM, Ariz; nSask; eOnt, wQue s to Fla, Tex. Map 37 ▲

P. glabella Mett. *ex* Kuhn
Fronds caespitose, monomorphic, 0.5–20 cm long; stipe slender, 1–10 cm long, reddish brown to brown, glabrous or sparsely pubescent; blades 3–10 cm long, glaucous, pinnate, the lower pinnae sometimes lobed, or bipinnate, pinnae 0.5–3.0 cm long, hardly reduced above, pinnae stalks ascending, the angle to the rachis less than 45°. $2n = 58, 116$. *P. atropurpurea* (L.) Link of Ed. 1. In our area var. **occidentalis** (Nels.) Butters (*P. occidentalis* (Nels.) Rydb.), $2n = 58$, reproducing sexually and with the lower pinnae entire or merely lobed (Banff area), and var. **simplex** (Suks.) Butters (*P. sukdorfiana* Butters), $2n = 116$, with apogamous reproduction and the lower pinnae with 3–5 basal pinnules. Dry limestone rocks. wMack, BC to Que s to Wash, Utah, Ariz, Tenn. Map 36 ○

Phegopteris Fée

P. connectilis (Michx.) Fée
Rhizome slender, creeping, scaly when young; fronds 10–50 cm long, erect, solitary, arising at intervals from rhizome; stipe 5–30 cm long, as long or up to twice as long as blade, sparsely pubescent and with occasional brown scales; blade triangular-ovate; 5–15 (25) cm long, 5–15 (20) cm wide, pinnae in 10–25 opposite to subopposite pairs, pinnate to pinnatifid, the lowest pair sharply deflexed; sori marginal to submarginal, indusium absent. $2n = 60, 90$*. Moist woodlands. Circumpolar: Alas, Yuk, wMack, nSask to Nfld, s to Ore, Mont, La, and Apal. Map 39 ○

Polypodium L.

Rhizome creeping, scaly, slender or rather thick; fronds somewhat scattered, the stipe articulate near the base; blades pinnate or pinnatifid, the segments linear-oblong, spreading nearly at right angles, the margin entire or inconspicuously serrulate; sori round or oval, rather large; indusium none.

Sori oval, set midway between costa and margin, paraphyses absent, rhizome scales uniformly coloured **P. hesperium**

Sori round, marginal, paraphyses present, rhizome scales with darker centre **P. virginianum**

P. hesperium Maxon
Rhizome thick, licorice-flavoured, scales of uniform colour; fronds 5–40 cm long;

stipe 1–10 (15) cm long, glabrous; blade 4–20 (25) cm long, elliptic-oblong, 1–5 (10) cm wide, pinnae in 5–10 (15) subopposite pairs; sori oval, set midway between costa and margin, soral paraphyses mostly absent. $2n = 148$. *P. vulgare* L. var. *columbianum* Gilbert of Ed. 1. Moist, rocky outcrops. Alas, Yuk s to Calif, Ariz, NM. Map 37 ○

P. virginianum L.
Rhizome thick, acrid, scales with a darker centre stripe; fronds mostly 10–35 (40) cm long; stipe 5–15 cm long, glabrous; blade oblong to lanceolate, 5–20 (25) cm long, 2–7 cm wide; pinnae in 10–20 alternate to subopposite pairs; sori round, marginal, sporangia interspersed with paraphyses. $2n = 74, 148$. *P. vulgare* L. var. *virginianum* (L.) Eat. of Ed. 1. Moist, rocky outcrops. wYuk, nBC, nSask, Man e to Nfld s to Ala, Ga. Map 37 ●

Polystichum Roth

P. lonchitis (L.) Roth / Northern Holly Fern
Rhizome stout, scaly, erect or ascending; fronds tufted, coriaceous, evergreen, pinnate, 1–4 dm long; stipes very short, with rusty-brown scales; blades linear-lanceolate, gradually narrowed to the base; pinnae alternate, 25–40 on each side of rachis, placed at right angles, lanceolate (the basal ones short-triangular), falcate, the base sharply auriculate on upper side, the tip acuminate and spinose, the margin serrate-dentate with conspicuously spinulose teeth; sori usually in 2 rows; indusium rounded, peltate, attached in the centre of the sorus, persistent. $2n = 82$. Shaded rocks and slopes. Circumpolar: Alas to Calif, Colo; Gt Lakes, eQue to Nfld. Map 38

Pteridium Gled. / Bracken

P. aquilinum (L.) Kuhn
Coarse fern with elongated, branched, creeping rhizomes and large firm fronds; stipes stiff, with several vascular bundles; blades 2–3-pinnate, somewhat triangular in outline, 2–6 dm wide; pinnae opposite or nearly so, the pinnules alternate; ultimate segments narrowly oblong, entire or lobed, with revolute margins, nearly glabrous above; sori marginal, linear, continuous, protected by the recurved margin and a delicate, obscure inner indusium, nearly concealed by the sporangia; indusia ciliate; spores tetrahedral. $2n = 104$. Woods and thickets. In our area var. **pubescens** Underw. of ssp. **aquilinum.** Circumpolar: sAlas to Calif; sMan to Nfld s to Gulf states. Map 39 ●

Woodsia R.Br.

Low ferns with densely tufted fronds; stipes shorter than the blades, rather stiff, with 2 bundles at the base, these united upwards; blades 1–2-pinnate, linear to ovate-oblong in outline, the ultimate segments lobed to pinnatifid; veins forked, free, not reaching the margin; sori rounded, separate but often confluent with age; indusium attached by its base, under the sporangia, appearing like a basal disc, open at the top, often with a lobed or fringed edge; spores large, bilateral, winged.

1 Stipes jointed near the base, the joint appearing as a small dark enlargement; blades not glandular; indusium a minute disc with large dark hairs, curved above or around the sporangia — 2

Stipes not jointed; blades sometimes glandular; indusium concealed by the mature sporangia or appearing as a basal rosette with a few thread-like segments — 3

2 Stipes straw-yellow or green, delicate, not scaly or hairy above the joint; blades delicate, 6–14 mm wide, glabrous; indusial hairs 5–8, only slightly longer than the sporangia — **W. glabella**

Stipes brown, firm, scaly and hairy; blades 10–35 mm wide, hairy; rachis and midribs scaly; indusial hairs 10–20, conspicuous — **W. ilvensis**

3 Fronds without jointed hairs on the midribs and veins but sometimes minutely glandular; indusium with very slender moniliform segments, often hair-like — **W. oregana**

Fronds with long whitish jointed hairs, also glandular; indusium with broad lobes lacerate into slender segments — **W. scopulina**

W. glabella R.Br. / Smooth Woodsia
Forming small tufts; scales of rhizome and stipe-bases brown, concolorous, 3–4 mm long, denticulate, not ciliate; stipes delicate; fronds 4–16 cm long, the blades linear, pale green, glabrous and without scales, pinnate; pinnae mostly 8–14 pairs, divergent, suborbicular or ovate, trilobed or pinnatifid with 2 or 3 pairs of segments, these crenate or entire. $2n = 78$. Moist places on calcareous rocks and shaded cliffs. Circumpolar, arctic-alpine: transcont n, s to BC, nSask, cMan, Lab, Nfld, NY; Lk Superior. Map 40 ○

W. ilvensis (L.) R.Br. / Rusty Woodsia
Forming dense tussocks; scales of rhizome abundant, brown, concolorous, 4–6 mm long, denticulate, sparingly ciliate; stipes firm; fronds 5–20 cm long, the blades oblong-lanceolate, firm, dark green, hairy and usually scaly beneath, commonly aromatic, pinnate-pinnatifid; pinnae mostly 10–20 pairs, deeply pinnatifid, the segments oblong, obtuse, often revolute and lobed. $2n = 78, 82$. Rock crevices. Circumpolar: Alas, Yuk to cBC, sMack, nSask, Man to Nfld s to Gt Lakes and Apal. Map 41 ○

W. oregana D.C. Eat. / Oregon Woodsia
Scales of rhizome light brown, sometimes with a dark median stripe, nearly entire; fronds 5–25 cm long, the blades mostly narrowly oblong-lanceolate, pinnate-pinnatifid or bipinnatifid, bright green, glabrous or sometimes glandular-puberulent, the rachis pale; pinnae 10–15 pairs, the larger ones pinnately divided with 5–7 pairs of segments, the margins crenulate-serrate, often revolute. $2n = 76$. Produces a sterile hybrid with *W. scopulina* (*W.* × *maxonii* Tyron). Rocky cliffs and ledges. BC to Man s to Calif, NM; Gt Lakes, Que. Map 40 ●

W. scopulina D.C. Eat. / Mountain Woodsia
Forming large tufts; scales of rhizome lanceolate, hair-tipped, somewhat toothed and ciliate, pale brown with an irregular black median stripe; fronds numerous, 10–40 cm long, the stipes stout, the blades mostly lanceolate, 1–2-pinnate, glandular-pubescent especially beneath, the rachis sparingly hairy, not scaly; pinnae 9–20 pairs, the pinnules or segments 5–9 pairs, obtuse, crenate to incised. $2n = 76$. On rocks and in crevices of cliffs. Alas to Calif, NM; nSask; cMan, Gt Lakes, eQue. Map 41 •

MARSILEACEAE / Marsilea Family

Marsilea L.

M. vestita Hook. & Grev.
Plants rooting in mud, and forming dense colonies; rhizomes slender and creeping; leaves resembling a 4-leaf-clover, aerial or floating in water, assuming aerial or aquatic forms according to habitat; leaf-stalks 2–10 cm long, terminated by 4 obovate or spatulate leaflets, 5–15 mm long, folded upwards or spreading, entire or undulate at the apex, sometimes sparsely pubescent, the venation dichotomous; spores of 2 kinds, borne in hard bony bean-like structures (sporocarps), these 4–6 mm long, solitary on short peduncles attached near the leaf-bases, greenish when young, becoming brown, coarsely strigose with long hairs, mostly with 2 teeth near the base. *M. mucronata* A.Br. of Ed. 1. Ditches, ponds, and lakes, sometimes on dry ground where a temporary pool has dried. sBC to Sask, Minn s to Calif, Tex, Ark. Map 42

CUPRESSACEAE

Trees or shrubs, evergreen; leaves opposite or whorled, scale-like or needle-like; plants monoecious or dioecious; reproductive structures produced in cones (strobili) of 2 kinds; male cones small, mostly terminal on branches, sporangia 2–5 on peltate scales; seed-cones small, dry or somewhat fleshy; seeds 1 or more per scale, often winged.

Seeds in a small dry cone; large trees — **Thuja**
Seeds in a berry-like cone; small trees or shrubs — **Juniperus**

Juniperus L. / Juniper

Shrubs or small trees; leaves scale-like or awl-shaped, opposite or in whorls of 3; pollen and seed cones borne on same or separate plants; pollen cones very small, catkin-like; seed-cones berry-like, greenish at first, becoming bluish, with 1–6 seeds.

1 Leaves awl-shaped, in whorls of 3, spreading, 5–15 mm long — **J. communis**
Leaves mostly scale-like, opposite, appressed, 1–1.5 mm long — 2

2 Shrub, branches prostrate; scale leaves strongly apiculate **J. horizontalis**
Shrub, branches erect, or small tree; scale-leaves not strongly apiculate **J. scopulorum**

J. communis L. / Ground Juniper
Prostrate or spreading shrub, usually forming carpets or broad clumps, sometimes 1 m high; leaves in whorls of 3, crowded, 5–15 mm long, all needle-like, sharp-pointed, jointed at the base, deeply grooved and glaucous white on upper surface; cones axillary; 'berry' pale blue, sessile, ovoid, angled, 1–3-seeded. *J. sibirica* Burgsd. In our area, var. **saxatilis** Pall., entirely prostrate or decumbent with the leaves incurved, imbricate and abruptly short-pointed, and var. **depressa** Pursh, somewhat bushy but wide-spreading and often flat-topped, up to about 1 m high, the leaves spreading or ascending, nearly straight, about 10 mm long and acuminate-pointed. $2n = 22$. Woods and open slopes. Circumpolar: Alas to Lab and Nfld s to Calif, Colo, sSask, sMan, Gt Lakes, Apal. Map 43

J. horizontalis Moench / Creeping Juniper
Prostrate shrub, often forming large mats, the main branches greatly elongate and producing numerous ascending branchlets, 1–3 dm high; foliage green or bluish; leaves of 2 kinds; juvenile leaves awl-shaped, 5–7 mm long, often on young vigorous branches; adult leaves scale-like, appressed, overlapping, glandular on back, tips apiculate; cones terminal; 'berry' dark blue, usually on a recurved stalk; seeds 1–6. *Sabina horizontalis* (Moench) Rydb. $2n = 22$. Sandy and rocky areas. Alas to Nfld s to Colo, Sask, sMan, Gt Lakes, NY, Me. Map 44

J. scopulorum Sarg. / Rocky Mountain Juniper
Small tree or erect shrub with fissured brown bark; resembling *J. horizontalis* in its leaves, but the tips of the scale-leaves subacute; 'berry' bright blue beneath a bloom, its stalk straight or somewhat curved; seeds 1–4. *Sabina scopulorum* (Sarg.) Rydb. $2n = 22$. At Banff and in the Crow's-nest Pass occur semi-erect individuals which may have been produced by browsing, or they may possibly be hybrids between *J. scopulorum* and *J. horizontalis*, *J.* × *fassettii* Boivin (*J. scopulorum* var. *patens* Fassett). Open rocky areas. cBC to sSask s to Ore, Ariz, NM. Map 45

Thuja L.

T. plicata D. Don / Western Red Cedar
A common forest tree species in BC but known in Alta from only a few records; bark of older trees thin, reddish brown, shallowly grooved and separating into long loose shreds; leaves small, scale-like, overlapping, about 3 mm long, opposite, 4-ranked, falling with the branchlets in about 3 years; branchlets generally arranged in flat sprays; seed-cone narrowly ovoid, about 12 mm long; seeds small, each with 2 membranous wings. $2n = 22$. Cool, moist mountain slopes. Alas to Calif e to nwMont. Map 46

PINACEAE / Pine Family

Trees or shrubs, resinous, mostly evergreen; leaves needle-shaped, linear, spirally arranged, single or in groups; plants monoecious; reproductive structures produced in cones (strobili) of 2 kinds; male cones small, formed singly or in dense clusters, scales flattened, sporangia 2 per scale on the lower surface; seed-cones relatively large and woody at maturity, producing seeds at the base of the scales, usually 2 seeds on a scale; seeds winged.

1 Leaves borne singly on the branches 2
Leaves in groups of 2 or more 5

2 Branches (from which the needles have fallen) rough with peg-like projections 3
Branches not rough with peg-like projections 4

3 Leaves 4-sided, not narrowed at the base **Picea**
Leaves flattened, narrowed at the base **Tsuga**

4 Cones erect, their scales falling away at maturity; bracts shorter than the scales; winter buds blunt, rounded; leaf-scars circular **Abies**
Cones drooping, their scale persistent on the axis; bracts longer than the scales and 3-lobed; winter buds sharp-pointed; leaf-scars oval **Pseudotsuga**

5 Leaves evergreen, 2–5 in a cluster, surrounded by a papery sheath at the base **Pinus**
Leaves deciduous, 10–40 in a cluster on a short spur-shoot, the cluster without a sheath **Larix**

Abies Mill. / Fir

Trees with fairly smooth bark and many resin blisters; needles linear, flattened, sessile, shiny green above and usually whitish beneath, often appearing 2-ranked, mostly blunt at apex; winter buds rounded and blunt; pollen-cones ovoid to short-cylindric, yellowish to red; seed-cones large, cylindric, erect, dark purple, the scales broadly rounded, the bracts usually much shorter than the scales, the central axis of the cone persisting after the scales have fallen.

Bark thin, scaly on old trees; wood odour not unpleasant; leaves of lower branches wide-spreading and distinctly 2-ranked, generally lacking rows of stomata above; pollen-cones yellow-red or tinged with purple; scales of seed-cones about 1.5 cm wide, usually twice as long as the short-awned bracts **A. balsamea**
Bark thick, fissured on old trees; wood odour unpleasant; leaves of lower branches directed forward and upward, imperfectly 2-ranked, rows of stomata conspicuous above with ×10 magnification; pollen-cones blue-violet; scales of seed-cones 2–2.5 cm wide, about 3 times as long as the long-awned bracts **A. lasiocarpa**

A. balsamea (L.) Mill. / Balsam Fir
Tree up to 18 m tall in our region; leaves 12–25 mm long, blunt or minutely notched at apex; seed-cones 5–10 cm long, obtuse, Closely related to *A. lasiocarpa*, from which it may be distinguished by the characters of the key. $2n = 24$. Moist

woods, commonly intermixed with white spruce and poplar. Alta to Nfld and NS s to Gt Lakes. Map 47

A. lasiocarpa (Hook.) Nutt. / Alpine Fir
A somewhat larger species than *A. balsamea,* sometimes 25 m tall, but often a prostrate shrub at timber-line. A common tree of subalpine forests, also occurring at lower altitudes. Probably intergrading with *A. balsamea* in the region of Lesser Slave Lk and the Athabasca R. Alas to Ore e to wMont, NM, and Ariz. Map 48

Larix Adans. / Larch

Trees with rough bark and heavy wood; branches of 2 kinds; long shoots of the current year, with scattered leaves, and dwarf spur-like branches of slow growth producing clusters of 10–40 leaves; all leaves slender, soft, needle-like, pale green, turning yellow in autumn and dropping off; staminate catkins quite small, nearly globose; seed-cones rather small, the scales thin, concave, often reddish, the bracts mucronate.

1 Seed-cones 1–2.5 cm long, their bracts shorter than the scales; leaves 1–2.5 cm long, 3-angled **L. laricina**
Seed-cones 2–5 cm long, their bracts extending beyond the scales; leaves 3–4 cm long 2

2 Leaves 4-angled; twigs tomentose; bracts much exceeding the scales and with reflexed tips **L. lyallii**
Leaves 3-angled; twigs soon becoming glabrate; bracts not greatly exceeding the scales in mature cones **L. occidentalis**

L. laricina (Du Roi) K. Koch / Tamarack
Tree to 20 m tall; bark thin, roughened by numerous reddish brown scales; twigs slender and pliable; leaves 12–20 in a cluster, soft, very slender, with 2 resin ducts near outside edge (as seen in cross-section); seed-cones globose-ovoid, stalked. $2n = 24$. A common tree of muskegs in central and northern areas. Alas, Yuk to cBC e to Lab, Nfld s to Gt Lakes, NJ. Map 49 ●

L. lyallii Parl. / Alpine Larch
A small tree, rarely over 10 m tall; bark thin, reddish brown, furrowed, with loose scales; branches hanging down, their tips curved upward; twigs very tough and tomentose; leaves 30–40 in a cluster, bluish green, with 2 resin ducts, one in each of 2 corners of the 4-sided leaf; seed-cones ellipsoid-oblong, the scales hairy on the lower side, the bracts with long reflexed tips. A tree of high altitudes, usually at the upper timber-line. sBC, Wash, nwMont. Map 50

L. occidentalis Nutt. / Western Larch
Large tree, reaching 50 m or more in British Columbia; bark deeply furrowed near the base of older trees and forming large reddish brown plates; leaves

15–30 in a cluster, yellowish green, flatly 3-angled, without resin ducts; scales of seed-cones tomentose below when young; bracts of cone with long pointed tips. $2n = 24$. Moist mountain slopes. sBC to Ore, nwMont. Map 49 ○

Picea A. Dietr. / Spruce

Mostly large trees with narrow crowns and dark scaly thin bark; leaves 4-sided, often pointed, spirally arranged and extending from all sides of the twigs, jointed at the base to a persisting peg-like sterigma; seed-cones maturing the first season, often terminal, pendent at maturity; seed-scales thin, much longer than the small bracts; seeds with thin wings.

1 Seed-cones persisting for several years, usually less than 2.5 cm long, often purplish; twigs hairy and dull brown; leaves usually blunt at apex **P. mariana**

Seed-cones deciduous, 2.5–8 cm long, brown when mature; twigs glabrous or minutely hairy; leaves more pointed 2

2 Seed-scales stiff, broadly rounded at apex and entire; cones 2.5–5 cm long; leaves mostly less than 1.5 cm long and sharp-pointed; twigs glabrous **P. glauca**

Seed-scales flexible, truncate to pointed at apex and erose; cones 3–8 cm long; leaves mostly over 1.5 cm long, the apex flattened and short-pointed; twigs minutely hairy **P. engelmannii**

P. engelmannii Parry *ex* Engelm. / Engelmann Spruce
Large tree, often 25–30 m tall, sometimes much taller, the lower branches drooping; a depressed shrub at high elevations; bark of older trees often cinnamon-red. $2n = 24$. A leading species in the subalpine forest on the eastern slopes of the Rocky Mts. seBC to Calif e to Mont, NM, Ariz, eNev. Map 51

P. glauca (Moench) Voss / White Spruce
A common forest tree, reaching a height of 40 m in favourable situations; bark ashy brown, sometimes reddish. $2n = 24$. *P. canadensis* (Mill.) BSP. In our area var. **albertiana** (S. Brown) Sarg., characterized by cones c. 3.5 cm long, and var. **porsildii** Raup, found sporadically from the Banff region northward, distinguished by a broad crown and smooth balsam-like bark beset with resin blisters. At middle altitudes in the Rocky Mts this species intergrades with *P. engelmannii.* Mountains and northern forests. Alas to nMont e to Lab, Nfld s to Gt Lakes. Map 52

P. mariana (Mill.) BSP. / Black Spruce
A small tree, commonly reaching a height of 10 m; bark thin, scaly, greyish brown, the inner bark with a greenish tinge; twigs pubescent; leaves short and thickish, usually obtuse or emarginate, dark green, bluish green or glaucous; seed-cones ovoid or nearly globose, 2–3 cm long, greyish or purplish brown, generally persistent for several years, their scales usually with erose margins. $2n = 24$. The most common tree of muskegs in Alta, also found on drier soils northward and in the mountains. Alas, BC to Lab, Nfld s to Gt Lakes. Map 53

Pinus L. / Pine

Trees, sometimes shrubby, usually with rough bark; needle-like leaves long, in groups of 2–5, enclosed at the base in chaffy sheaths; pollen-cones in clusters at the base of the current year's growth; seed-cones maturing in the second year, becoming very woody and conspicuous, often long persistent on the tree; seed-scales often greatly thickened at apex.

1 Needles 2 in a bundle 2
 Needles 5 in a bundle 3

2 Cones directed towards the apex of the shoot, strongly incurved or divergent, the scales without prickles at maturity **P. banksiana**
 Cones spreading at right angles, or more commonly reflexed, the scales each armed with a persistent prickle **P. contorta**

3 Cones cylindrical, long-stalked; scales thin, needles with finely serrated margins **P. monticola**
 Cones ovoid, short-stalked or sessile; scales thick; needles with smooth margins 4

4 Cones 8–20 cm long, opening at maturity; scales light brown, very thick at tip **P. flexilis**
 Cones 3–7 cm long, remaining closed for some years; scales purplish, very thick in middle and towards the tip **P. albicaulis**

P. albicaulis Engelm. / White-bark Pine
Small alpine tree or often reduced to a shrub, the branches sometimes prostrate on the ground; trunk often crooked and the crown matted; bark smooth, whitish, little broken except at the base of the trunk; twigs yellowish and hairy; leaves 4–8 cm long, stiff; seed-cones 3–7 cm long, purplish, the scales much thickened, with a stout pointed umbo, not prickly. Sparingly in the timber-line belt of the Rocky Mts. sBC to cCalif e to Mont and Wyo. Map 54

P. banksiana Lamb. / Jack Pine
A small shrubby tree, 5–10 m high on open or poor sites, but straight and tall in dense stands on good sites, often reaching a height of 20 m; bark reddish brown, splitting into irregular scaly ridges and furrows; lower branches of wide-spreading crowns generally drooping; leaves 2–4 cm long, divergent and often twisted, yellowish green; seed-cones 3–5 cm long, conical, directed towards the apex of the branch, strongly incurved or spreading, the scales variously thickened and without prickles at maturity; cones remain on the branches for many years. $2n = 24$. Common on sandy and gravelly places in central and northern regions. wMack to Nfld s to Gt Lakes. Map 55 ●

P. contorta Loudon / Lodgepole Pine
Reaching a height of 20–30 m or more, usually in dense stands and then with tall straight clean trunks, occasionally growing in open stands and forming broad crowns; branches generally curving upward; leaves generally longer and less twisted than those of jack pine and forming more dense clusters towards the

ends of the twigs; seed-cones 2–5 cm long, ovoid or conical, spreading or reflexed, the scales umbonate, each scale armed with a minute recurved prickle; cones persisting on the branches for many years. $2n = 24$. The most common tree species at lower and middle altitudes on the eastern slopes of the Rocky Mts in Alta. In our area var. **latifolia** Engelm. Produces hybrids of intermediate morphology with *P. banksiana* where the ranges of the two species overlap in central Alta. seAlas, Yuk to Calif e to Mont, Utah, Colo. Map 56

P. flexilis James / Limber Pine
A small tree, with an irregular crown, a short stout trunk, and whorls of thick limbs; bark light grey, becoming dark brown and much cracked in age; leaves 3–7 cm long, rigid, crowded at the ends of the branches; seed-cones 8–20 cm long, light brown tinged with purple, the scales greatly thickened at the tip. $2n = 24$. Exposed rocky slopes and hilltops to subalpine elevations. seBC to Calif, Nev, NM; wDaks. Map 57

P. monticola Dougl. *ex* D. Don / Western White Pine
A common and large forest tree of southern BC but very rare in Alta; bark of young trees is silvery grey and smooth, on older trees becoming darker and checked into small squarish plates; leaves 5–10 cm long, bluish green, whitened in lines; cones 1–2.5 dm long, the scales thin and without prickles. $2n = 24$. Open rocky slopes. sBC to cCalif e to wMont. Map 55 ○

Pseudotsuga Carr.

P. menziesii (Mirb.) Franco / Douglas Fir
The largest tree in Canada but relatively small on our side of the Rocky Mts; trunks of old trees massive, the bark dark brown, very thick and deeply furrowed; bark of young trees thin and smooth; lower branches often drooping, with long pendent branchlets; leaves 2–3 cm long, flat, spirally arranged but often appearing 2-ranked by a twist at their base, narrowed to a very short stalk, blunt to acute at apex, pale beneath; winter buds sharp-pointed; seed-cones pendent, 5–10 cm long, maturing in one season, the bracts tridentate, projecting beyond the broad rounded scales. $2n = 26$. *P. taxifolia* (Poir.) Britton. Mountain forests. sBC to Calif e to Mont, NM. Map 58

Tsuga Carr. / Hemlock

Tsuga heterophylla (Raf.) Sarg. / Western Hemlock
Large tree, to 50 m or more in BC; bark brown, strongly furrowed; branches drooping, the younger with short pubescence and long multicellular hairs; leaves 5–20 (25) mm long, flattened, yellow-green above, white below, rounded at apex, spreading at right angles to the branches; seed-cones 1.5–2.5 cm long, oblong-ovoid. $2n = 24$. Moist montane forest. Alas to nCalif e to nwMont. Map 59 ●

TAXACEAE / Yew Family

Taxus L. / Yew

T. brevifolia Nutt. / Western Yew
Evergreen shrub or small tree, normally dioecious; leaves linear, 1–2 cm long, flat, spinulose-tipped, yellowish green, slender-petioled; staminate cones axillary, head-like, appearing in early spring; sporangia pendent; ovule solitary, axillary, with scales at the base; seed ovoid, 5–6 mm long, surrounded by a red, fleshy, cup-like aril, 4–5 mm broad. Moist woodland. sAlas to Calif e to nwMont. Map 59 ○

TYPHACEAE / Cattail Family

Typha L.

T. latifolia L. / Common Cattail
Marsh or aquatic herbs with coarse creeping rootstocks, long leaves and conspicuous mace-like flower-spikes; stems erect, 1–2 m tall; leaves linear, upright, flat, sheathing, 1–2 cm wide; flowers very small, unisexual, without perianth, borne in a dense cylindrical spike, the staminate above; staminate flowers of 1–7 (usually 3) stamens and mingled with hairs; filaments very short, united; pollen grains usually in 4's; pistillate part of spike about 2.5 cm in diameter, and including fertile and sterile flowers; pistillate flower of a 1-celled ovary, borne on a hairy stipe; style short and stigma spatulate; fruits are minute achenes on slender stalks and scattered in fluffy masses in late summer. $2n = 30$. Marshes or shallow water. Circumpolar: Alas, Yuk, James Bay to Nfld, generally distributed s. Map 60 ●

SPARGANIACEAE / Bur-reed Family

Sparganium / Bur-reed

Marsh or aquatic plants, perennial from rootstocks, with erect or floating stems, slender leaves, and unisexual flowers in dense globose heads; leaves sessile, linear, 2-ranked, sheathing at the base; staminate and pistillate flowers in separate heads, these scattered along the upper part of the leafy stem; upper heads bearing naked staminate flowers interspersed with minute scales; lower heads composed of pistillate flowers, each with 3–6 sepals and a 1–2-locular ovary; styles 1 or 2; pistillate heads becoming bur-like, composed of achenes; the latter usually beaked, 1–2-seeded.

1 Stigmas 2; achenes broadly obpyramidal, somewhat truncate at the top when mature **S. eurycarpum**
Stigmas 1; mature achenes ellipsoid, tapering at each end 2

2 Beak of the achene less than 1.5 mm long; staminate head 1 3
Beak of the achene 1.5–5 mm long; staminate heads 1–several 4

3 Achene beakless; staminate head adjacent to upper pistillate head; leaves 1–5 mm wide, opaque, often yellow **S. hyperboreum**
Achene beak 0.5–1.5 mm long; staminate head distant from upper pistillate head; leaves 2–8 mm wide, thin, translucent, usually dark green **S. minimum**

4 Beak of achene more or less straight; staminate head solitary **S. glomeratum**
Beak of achene conspicuously curved; staminate heads 2 or more 5

5 Perianth-segments attached near base of stipe; anthers less than 0.8 mm long **S. fluctuans**
Perianth-segments attached near top of stipe; anthers at least 0.8 mm long **S. angustifolium**

S. angustifolium Michx.
Erect or floating, stems 2–10 dm long, longest in floating plants; leaves mostly 2–10 dm long, 2–15 mm wide, base usually dilated and sheathing, often with a conspicuous scarious margin, blade flat, rounded or keeled; staminate heads 2–5 (8); pistillate heads 2–5, axillary and sessile or pedunculate, 1.0–2.5 mm wide at maturity; achenes smooth or ribbed, often contracted in the middle, body 3–6 mm long, beak 1.5–4.0 mm long. $2n = 30$. A highly variable species in which different morphological forms are often treated as species. Includes *S. chlorocarpum* Rydb. and *S. emersum* Rehm. (*S. multipedunculatum* (Morong) Rydb.) of Ed. 1. Shallow lakes, ponds, and streams. Circumpolar, boreal: Alas to Nfld s to Calif, Colo, NM, sSask, sMan, Gt Lakes. Map 61 ●

S. eurycarpum Engelm. / Giant Bur-reed
Stem stout, erect, 5–15 dm high; leaves stiffish, flat or slightly keeled, 6–12 mm wide; infloresence branched, with 3–20 heads of which 1–4 are pistillate, the latter 2–3 cm in diam. at maturity; achenes sessile, obpyramidal, 6–9 mm long, 4–8 mm broad at the truncate and abruptly beaked summit, the beak stout and about 3 mm long. $2n = 30$. Shallow water. sBC to Nfld s to Calif, Fla. Map 62 ●

S. fluctuans (Morong) Robins.
Aquatic, with slender stems up to 15 dm long; leaves all floating, ribbon-like, thin, cross-reticulate beneath, 3–11 mm broad; inflorescence usually branched, each branch with a small number of staminate and pistillate heads; mature pistillate heads 1.5–2 cm in diam.; achenes dull brown, 3–4 mm long, with a stout, flattened, curved beak. Lakes and streams. BC to nSask; Ont to Nfld s to Gt Lakes. Map 61 ○

S. glomeratum Laest.
Erect or floating, stem 2.6 dm long; leaves as long as stem or longer, 3.0–10.0 mm wide, flat or keeled towards the base, the lower sheathing; staminate head 1, rarely 2, contiguous with upper pistillate head; pistillate heads 3–5, crowded, 1.0–1.5 cm wide at maturity; achene fusiform, body 4.0–5.0 mm long, beak

1.0–1.5 mm long. $2n = 30$. Ponds. Possibly introduced. Eurasia: a few scattered localities in NAm, Yuk, Minn, and eQue. Map 60 ○

S. hyperboreum Laest.
Erect or floating, stems 1–3 dm long; leaves as long or longer, mostly 2–3 mm wide, flat; staminate head contiguous with upper pistillate head, sessile; pistillate heads 2–4, axillary or supra-axillary, about 1.0 cm across at maturity; achenes fusiform to clavate, ribbed, body 4–5 mm long, beak absent or short-apiculate, less than 0.5 mm long. $2n = 30$. Shallow alpine lakes. Circumpolar, arctic-alpine: Alas, Yuk, nBC to nMan, Huds Bay, nOnt, nQue, Lab, Nfld; wOnt, seMan. Map 62 ○

S. minimum (Hartm.) Fries / Slender Bur-reed
Stems floating or somewhat erect, very slender, 1–8 dm long; leaves thin, flat, 1–7 mm wide, often floating; inflorescence with 1–4 axillary pistillate heads and a remote staminate head; mature pistillate heads 6–12 mm in diam.; achene body 3–4 mm long, beak 1–1.5 mm long. $2n = 30$. Shallow water. Circumpolar: Alas, Yuk, BC to Nfld s to Ore, Mont, Utah, sSask, ND, Gt Lakes. Map 63

NAJADACEAE

Najas L. / Naiad

N. flexilis (Willd.) Rostk. & Schmidt
Delicate, densely tufted aquatic annual with leaves crowded at the tips of the long slender stems; leaves 1–3 cm long, flaccid, pale green, enlarged at base and tapering to a fine point, the margins beset with minute spinules; flowers imperfect, solitary, the staminate consisting of a single stamen enclosed in a membranous sheath, the pistillate of a single pistil not enclosed; achene 2–3 mm long, tipped with a persistent style; seed pale brown, shining, faintly sculptured with 30–40 rows of hexagonal areolae. $2n = 12, 24$. Ponds and streams. Europe: BC to Calif e to Ida; seMan to Nfld s to SD, Md. Map 64 ●

POTAMOGETONACEAE / Pondweed Family

Potamogeton L. / Pondweed

Water plants with jointed, mostly rooting stems; leaves alternate (or the uppermost opposite), floating and submersed, the latter delicate and narrower; stipules membranous, more or less united and sheathing; spike-like flower clusters mostly raised on a peduncle to the surface of the water; small greenish flowers in whorls; stamens 4, their anthers bearing sepal-like appendages; ovaries 4, separate; stigma sessile or on a short style; fruit 1–5 mm long, drupe-like with an enclosed nutlet; many of the species reproducing more freely by tubers or detached winter-buds than by seeds.

1 Submersed leaves with stipules adnate to the base, forming a sheath around the stem 2

Submersed leaves with free stipules 5

2 Submersed leaves 3–8 mm wide, margins serrulate, closely crowded on stem, distichous **P. robbinsii**

Submersed leaves 1–2 (3) mm wide, not serrulate at the apex, not closely crowded in an obviously distichous arrangement 3

3 Stipular sheaths conspicuous, loose, much wider than the stem, 2–5 cm long **P. vaginatus**

Stipular sheaths inconspicuous, little if any wider than the stem, 0.5–2 cm long 4

4 Fruit 3–4 mm long, beaked; leaves with a long tapering acute tip **P. pectinatus**

Fruit 2–3 mm long, almost beakless; leaves mostly obtuse or rounded at apex, sometimes minutely apiculate **P. filiformis**

5 Submersed leaves mostly less than 5 mm wide 6

Submersed leaves more than 5 mm wide 12

6 Floating leaves present, 2–6 cm wide, base cordate to rounded; submersed leaves 1–2 mm wide, 10–20 cm long **P. natans**

Floating leaves absent 7

7 Leaves with 15–35 nerves; stem flattened; fruits 4–5 mm long **P. zosteriformis**

Leaves with 3–7 nerves; stem terete or only slightly flattened; fruits smaller 8

8 Stipules fibrous, strongly ribbed, becoming shredded 9

Stipules membranaceous, not shredding 10

9 Leaves flat with 5–7 nerves, abruptly narrowed to a blunt tip **P. friesii**

Leaves revolute with 3–5 nerves, tapered to a slender tip **P. strictifolius**

10 Leaves generally 2–3 mm wide; fruits 3–4 mm long **P. obtusifolius**

Leaves generally less than 1.5 mm wide; fruits 2–2.5 mm long 11

11 Nodes usually without glands; peduncle clavate; fruits with sharp, wavy, dorsal keel **P. foliosus**

Nodes usually with glands, peduncle not markedly clavate, fruits with a low dorsal keel **P. pusillus**

12 Leaves minutely toothed; fruit with beak 2–3 mm long and with a prominent basal appendage **P. crispus**

Leaves usually entire; fruit with beak 1.5 mm long or less 13

13 Floating leaves absent; submersed leaves with blades rounded-cordate at the clasping base 14

Floating and submersed leaves generally present; submersed leaves petioled or tapering to a sessile base, scarcely clasping 15

14 Stipules conspicuous, white, persistent; leaves linear-lanceolate to oblong, usually more than 10 cm long **P. praelongus**

Stipules inconspicuous, soon disintegrating; leaves ovate-lanceolate to narrowly lanceolate, usually less than 10 cm long **P. richardsonii**

15 Submersed leaves not markedly different from the floating leaves, plants generally reddish **P. alpinus**

Submersed leaves and floating leaves markedly dissimilar; plants generally greenish **P. gramineus**

P. alpinus Balbis
Whole plant with a reddish tinge; stems slender, simple or branched; submersed leaves sessile, thin, translucent, usually with 7 prominent nerves, narrowly lanceolate to elliptic-oblong, 5–20 cm long; floating leaves often lacking, delicate, 1–2 cm wide, oblanceolate to obovate, tapering into a petiole; stipules broad, 2–3 cm long; peduncles 6–12 cm long; fruits 3–3.5 mm long, with a curved beak 1–1.3 mm long. $2n = 52$. Produces hybrids of intermediate morphology with *P. gramineus* (*P.* × *nericus* Hagstr.) and *P. praelongus* (*P.* × *griffithii* A. Benn). Circumpolar: Alas, Yuk, sMack to Huds Bay, Nfld s to Calif, Utah, Colo, Gt Lakes. Map 65

P. crispus L.
Stems branched, flattened with the broad sides channelled; leaves all submersed, linear to oblong, 3–8 cm long, the margins serrulate and undulate, 3–5-nerved, the midrib prominent; peduncles often curved; fruits 3–6 mm long, with an erect conic beak and a basal appendage; winter-buds bur-like, with short hardened leaves. $2n = 36, 42, 50, 52, 72, 78$. Introduced. Rare.

P. filiformis Pers.
Stems slender, commonly bushy branched; leaves all submersed, erect or ascending, narrowly linear, 1–12 cm long, 0.2–2 mm wide, blunt or apiculate at apex, 1-nerved with remote cross-veins; stipules joined to blade for 5–10 mm; peduncles slender; spikes 1–5 cm long; fruits 2–3 mm long, the beak nearly central and wart-like. $2n = 78$. In our area var. **alpinus** (Blytt) Aschers. & Graebn. (var. *borealis* (Raf.) St. John). Produces a hybrid with *P. pectinatus* (*P.* × *suecicus* K. Richt.). Circumpolar, boreal: Alas to Nfld s to Calif, Ariz, Colo, sSask, SD, Gt Lakes. Map 66

P. foliosus Raf.
Stems very slender, variously branched; leaves all submersed, linear, 2–8 cm long, 1–3 mm wide, 3–5-nerved, the lateral nerves indistinct; usually without nodal glands; stipules soon disappearing; peduncles slightly clavate; spike short; fruits flattened, 2–2.5 mm long, the dorsal keel thin and undulate or dentate, the beak erect. $2n = 28$. In our area var. **foliosus** (var. *macellus* Fern.). sBC to Nfld s to Mex; Alas; Gt Slave–Athabasca Lks area. Map 67

P. friesii Rupr.
Stem compressed, sparingly branched except near the summit; leaves all submersed, linear, 3–8 cm long, 1.5–3.5 mm wide, the apex blunt and mucronate, 5–7-nerved, the lateral nerves delicate; stipules strongly fibrous, whitish, in age lacerate, usually with 2 basal glands; peduncles flattened and slightly clavate, 1.5–5 cm long; spike slender with 3–4 remote whorls; fruits 2–3 mm long, with

obscure keels. $2n = 26$. More or less circumpolar: Alas to Nfld s to Wash, sSask, Daks, Gt Lakes, Mass. Map 68

P. gramineus L.
Stems slender, 0.5–1 mm thick, usually much branched; submersed leaves sessile, linear, narrowly elliptic or oblanceolate, 1–12 cm long, 3–9-nerved; floating leaves coriaceous, ovate to elliptic, 2–6 cm long, 13–19-nerved; petioles slender, often longer than the blades; stipules persistent, about 2 cm long; peduncles stout, usually about 4 cm long; spikes compact, 1–2.5 cm long; fruits about 2.5 mm long, more or less keeled, the beak short and recurved. $2n = 52$. In our area var. **gramineus** and at the upper end of submersed leaf size, var. **maximum** Morong *ex* Benn. Produces hybrids with *P. natans* (*P.* × *sparganifolius* Laest. *ex* Fries) and *P. alpinus*. Circumpolar, boreal: Alas to Nfld s to sUSA Map 69

P. natans L.
Stems simple or sparingly forking; submersed leaves often lacking, narrowly linear, firm, 1–2 dm long, 0.8–2 mm wide, with 3–5 obscure nerves; floating leaves usually well developed, the petioles longer than the blades and connected by a flexible brownish joint; blades coriaceous, lustrous, elliptic to ovate, 4–9 cm long, rounded or cordate at base, 13–37-nerved; stipules acute, strongly keeled, 4–10 cm long; peduncles stout, 4–10 cm long; spikes 2–5 cm long, about 1 cm thick; fruits 3–5 mm long, rounded on back or slightly keeled, beak short. $2n = 52$. Produces a hybrid with *P. gramineus*. Circumpolar, boreal: Alas to Nfld s of 55, s to Calif, Ariz, c and ne USA. Map 70

P. obtusifolius Mert. and Koch
Stem slender, much branched; leaves all submersed, often reddish, 3–9 cm long, 1–4 mm wide, apex rounded or apiculate, nerves 3–5, central nerve bordered by 2–4 rows of whitish, septate, lacunae, base with two conspicuous glands; stipules 1–2 cm long, scarious, open; peduncles 1–4 cm long, stout; spikes 8–17 mm long, ovate to cylindrical; achene 2.5–4 mm long, mostly keeled, beak short, straight. $2n = 26$. More or less circumpolar, boreal: sBC, nWash; cYuk, wMack, nSask, Man to Nfld s to Gt Lakes, NJ. Map 71

P. pectinatus L. / Sago Pondweed
Rhizomes slender, tuber-bearing; stems filiform, repeatedly forked above and very leafy; leaves all submersed, narrowly linear, 1-nerved or nerveless, 3–10 cm long, tapering to an acute apex; stipules joined to blade for 1–3 cm; peduncles 3–10 cm long; spikes 2–4 cm long, with remote whorls; fruits 3–4 mm long, often with a short, curved beak. $2n = 78$. Usually in shallow water. Produces an abundance of tubers, which are said to be an important duck-food. Circumpolar, boreal: Alas to Lab and Nfld, widely distributed s. Map 72

P. praelongus Wulf. / White-stem Pondweed
Rhizomes with rusty spots; stems whitish to olive-green, zig-zag, freely branched; leaves all submersed, linear-lanceolate to oblong, boat-shaped at apex, 10–30 cm long, rounded and slightly clasping at base, with 3 or 5 strong nerves and several

weaker ones; stipules conspicuous, 2–8 cm long, white, persistent; peduncles up to 4 dm long; spikes 3–4 cm long; fruits 4–5 mm long, with a prominent dorsal keel, beak short and thick. $2n = 52$. Produces a hybrid with *P. alpinus*. In deep water. Alas to Nfld s to Calif, NM, sSask, Neb, Gt Lakes, NY. Map 73

P. pusillus L.
Stems capillary, freely branched above; leaves all submersed, linear, 1–8 cm long, 3-nerved, the lateral nerves often obscure; nodal glands usually present; stipules delicate, scarious, soon deciduous, their margins free or somewhat united; peduncles very slender, 1–6 cm long; spikes subcapitate to short-cylindric, with 1–5 whorls; fruits 2–2.5 mm long, rounded, with a very low dorsal keel. $2n = 26$. Includes *P. panormitanus* Biv. In our area var. **pusillus** and var. **tenuissimus** Mert. and Koch (*P. berchtoldii* Fieb.) distinguished by the non-connate stipules and shorter peduncles, 0.5–3 cm long. Circumpolar, boreal: Alas to Nfld s to sUSA. Map 74

P. richardsonii (Benn.) Rydb. / Clasping-leaf Pondweed
Rhizomes unspotted (cf. *P. praelongus*); stems usually much branched, densely leafy; leaves all submersed, sessile, cordate-clasping at base, mostly ovate-lanceolate, 3–12 cm long, with 3 (or 5–7) strong nerves and several weaker ones; stipules coarse, soon disintegrating to whitish fibres; peduncles often clavate; spikes dense, 1.5–3 cm long; fruits about 3 mm long, with 3 obscure keels, short-beaked. $2n = 52$. *P. perfoliatus* L. var. *richardsonii* Benn. Alas to Nfld, transcont s to 35. Map 75 ●

P. robbinsii Oakes
Stems stout, much branched, terete; leaves all submersed, crowded, distichous, linear-lanceolate, 2–8 (12) cm long, 3–4 (8) mm wide, margin finely serrulate, apex acute, nerves 20–35 or more; stipules pale, adnate to leaf base below, free above, eventually shredding; peduncles slender, 3–5 (7) cm long; spikes 0.7–2.0 mm long; achenes obovoid, 3.5–5.0 mm long, sharply keeled, beak inconspicuous. $2n = 52$. BC to Calif, Ida, Mont, Wyo; cMan, Gt Lakes to Lab. Map 75 ○

P. strictifolius Benn.
Stems very slender, branching above; leaves all submersed, linear, firm, 2–6 cm long, tapered to a slender tip, usually 3-nerved, the lateral nerves indistinct; stipules whitish, becoming fibrous; usually without nodal glands; peduncles very slender; spikes slender, with 3 or 4 remote whorls; fruits 2–3 mm long, with a very low dorsal keel or none. $2n = 52$. Yuk, Mack to Que, Gt Lakes to Conn s to Alta, Sask, Neb, Pa; Utah. Map 76

P. vaginatus Turcz. / Large-sheath Pondweed
Resembling *P. filiformis* and *P. pectinatus*, but coarser and the stipular sheaths broader; sheaths of lower primary leaves loose and much thicker than the stem, 2–5 cm long, the leaf-blades blunt, 1–4 dm long, 1–2 mm wide; upper leaves more slender, with closer sheaths; spikes 3–8 cm long; fruits 3–3.5 mm long,

with a wart-like beak. $2n = 78$. In deep water. Eurasia: Alas to Lab and Nfld, s to Ore, Mont, Wyo, sSask, sMan, Gt Lakes to NY. Map 77

P. zosteriformis Fern.
Stems wing-flattened, 1–3 mm wide, much branched; leaves all submersed, linear, 1–2 dm long, 2–5 mm wide, acute or cuspidate, scarcely narrowed at base, with 1 or 3 strong nerves and numerous very fine ones; stipules firm, fibrous, 1–3 cm long; peduncles 2–5 cm long, often curved, spike 1.5–3 cm long; fruits 4–5 mm long, the dorsal keel sharp and somewhat toothed, the beak marginal. *P. zosterifolius* Schum. var. *americanus* Benn. BC to Que s to 40; Alas; nwMack. Map 78

RUPPIACEAE

Ruppia L.

R. maritima L. / Widgeon Grass
Submersed aquatic herb with slender forking stems; leaves alternate, delicate, thread-like, 5–20 cm long, often in fan-like groups; stipules adnate and sheathing, their margins scarious; flowers minute, perfect, without perianth; stamens 2; ovaries 4; fruit a small ovoid drupe, 3–4 mm long; the flowers, usually 2 in number, are borne on a spadix which elongates into a peduncle up to 5 dm long, and often spirally coiled; each drupe is also elevated on a stalk, resulting in an umbelliform cluster of fruits. $2n = 20, 40$. *R. occidentalis* S. Wats. of Ed. 1. In saline water. Circumpolar: Alas to Calif, transcont s of 53. Map 64 ○

ZANNICHELLIACEAE

Zannichellia L.

Z. palustris L. / Horned Pondweed
Slender branching aquatic herb, with fragile stems; leaves opposite, filiform, 3–8 cm long, 1-nerved; stipules sheathing, membranous; flowers unisexual, both kinds in the same axil, each with a minute, delicate sheath; perianth none; staminate flower of a single anther on a long filament; pistillate of 1–8 (usually 4) separate carpels, each with a short style and peltate stigma; achenes, usually 2–4, narrowly oblong, flattish, 2–4 mm long, often toothed along one side and with a persistent beak-like style, about 1 mm long. $2n = 12, 24, 28, 32, 36$. Ponds, streams, and ditches. Circumpolar: transcont s of 53. Map 79

JUNCAGINACEAE / Arrow-grass Family

Triglochin L. / Arrow-grass

Perennial herbs of brackish marshes; leaves all basal, linear, somewhat fleshy, sheathing at the base; scape long, erect, terminating in a bractless, spike-like raceme of small greenish flowers; perianth-segments 6, in two whorls, deciduous; stamens 6, attached at the bases of the perianth-segments; anthers oval, nearly sessile; carpels 3 or 6, attached along their inner edges to an axis, each ovary 1-celled and 1-ovuled; style very short or none; stigmas 3–6, plumose; fruit of 3–6 follicles, at maturity separating from each other and the axis, dehiscent on the inner side; seeds elongated.

Carpels usually 6; stigmas 6, spreading; fruit ovoid-oblong, rounded at base and with a wingless axis **T. maritima**

Carpels 3; stigmas 3, very short; fruit linear-clavate, tapering at base and with a winged axis **T. palustris**

T. maritima L. / Arrow-grass
Rhizome stout, covered with persistent leaf-bases, without stolons; leaves up to 5 dm long, linear, shorter than the scapes; raceme 1–4 dm long, the curved, erect pedicels decurrent as ridges on the axis; pistil usually of 6 carpels, ovoid-oblong, about 5 mm long at maturity; fruit rounded at base, winged, splitting into 3–6 follicles and revealing a slender wingless axis; stigmas radiating, persistent on the fruit. $2n$ = 12, 24, 30, 36, 48*, 60, 90, 120, 144. A species complex variously treated. Brackish marshes and boggy areas. Circumpolar: throughout NAm except Arct Arch. Map 80

T. palustris L. / Slender Arrow-grass
Rhizome short, with very slender stolons; leaves narrowly linear, shorter than the scapes; raceme 1–2 dm long, the very slender pedicels erect; pistil of 3 carpels, becoming linear-clavate, with a tapering base; fruit 5–7 mm long, splitting from the base upward, the segments awl-pointed at base and the axis broadly 3-winged. $2n$ = 24, 36, 48. Brackish marshes, shores, and bogs. Circumpolar: throughout Canada, except Arct Arch, s to Calif, NM, Ia, NY, Me. Map 81

LILAEACEAE / Flowering-quillwort Family

Lilaea Humb. & Bonpl.

L. scilloides (Poir.) Haum. / Flowering-quillwort
Small annual marsh herb with rush-like leaves and fibrous roots from a very short basal stem; leaves 8–35 cm long, densely tufted, linear-subulate, soft, with scarious sheathing bases; flowers minute, in inflorescences of 2 types; basal flowers, enclosed by the sheathing leaf bases, all pistillate, each of one carpel and a very long filiform style, up to 10 cm long, terminated by a capitate stigma;

other flowers borne in short spikes on slender scapes, these flowers being pistillate below, staminate above and often perfect ones in the middle of the spike; flowers all without perianth; scapose flowers of a single sessile stamen, a single carpel, or one of each; fruit dry and 1-seeded, that of the upper flowers being ribbed and with a pair of wings, that of the basal flowers ribbed and wingless. *L. subulata* Humb. & Bonpl. $2n = 12$. Sloughs and mud-flats. BC to Calif, Nev, nMont. Map 82 ○

SCHEUCHZERIACEAE

Scheuchzeria L.

S. palustris L.
A low bog herb with zig-zag stem, 2–4 dm tall; leaves alternate, grass-like, with membranous sheaths at the base; blades erect, somewhat tubular and with a round pore at the apex; raceme short and few-flowered; sepals and petals spreading, the latter narrower, persistent; stamens 6, anthers linear; carpels 3 (rarely more), divergent, 2-ovuled, with sessile stigmas; follicles ovoid, divergent, 5–8 mm long, 1–2-seeded, with a short, curved beak; seeds 4–5 mm long, black. $2n = 22$. In our area var. **americana** Fern. Cold, peaty bogs. Circumpolar: Alas, James Bay to Nfld s to Wash, Mont, cSask, sMan, Gt Lakes; Calif. Map 82 ●

ALISMATACEAE / Water-plantain Family

Aquatic or marsh plants with scapose stems and fibrous roots; leaves basal, long-stalked, erect or floating, usually broad, strongly nerved with transverse veinlets; flowers verticellate, usually in large clusters; sepals 3, green, persistent; petals 3, usually white, deciduous; stamens 6 or many; carpels numerous, becoming achenes in fruit.

Leaves usually ovate or elliptic-lanceolate; stamens usually 6; carpels in a single whorl on a small flat receptacle **Alisma**

Leaves sagittate or, when submersed, without blades; stamens many; carpels crowded on a large convex receptacle **Sagittaria**

Alisma L. / Water-plantain

Marsh plants, often in shallow water; the stout corm-like base with fibrous roots, a cluster of long-petioled leaves and erect flowering stalks; leaves with ovate or elliptic-lanceolate blades or, when submersed, often ribbon-like; inflorescence paniculate, with the numerous small flowers in whorls; flowers perfect; petals scarcely longer than the sepals; stamens 6–9; carpels many, in a single ring on a flat receptacle; achenes flattened, ribbed or grooved on the back.

Leaves elliptic-lanceolate or linear; inflorescence usually shorter than the leaves, its branches often recurved; stamens barely as long as the ovaries **A. gramineum**

Leaves ovate to elliptic; inflorescence extending above the leaves, its branches ascending; stamens twice as long as the ovaries **A. plantago-aquatica**

A. gramineum Lej. / Narrow-leaved Water-plantain
Leaves usually erect, with long narrow petioles and lanceolate to narrowly elliptic blades, 4–8 cm long; submersed leaves long, ribbon-like, usually 5–10 mm wide; petals white or purplish. $2n = 14, 16^*$. *A. geyeri* Torr. Marshes, ponds, streams. Eurasia: sBC to Ont, transcont s to 40. Map 83

A. plantago-aquatica L. / Broad-leaved Water-plantain
Leaves ascending, with long petioles; blades narrowly ovate to elliptic-lanceolate, 5–15 cm long, rounded to cordate at base; petals white, or sometimes purple-tipped. $2n = 10, 12, 14, 16, 24, 28$. Includes *A. brevipes* Greene and *A. triviale* Pursh. In our area var. **americanum** Schultes & Schultes. Marshes, ponds, ditches. Circumpolar: BC to Nfld s to sUSA. Map 84 ●

Sagittaria L. / Arrowhead

Aquatic or marsh plants with milky juice; leaves long-petioled, usually with sagittate or hastate blades, or the submersed leaves narrow lanceolate or linear (phyllodes); rhizomes slender, producing tubers; flowers conspicuous, mostly unisexual, often in whorls of 3 on the branches of the inflorescence; the lower flower-heads mostly pistillate, the upper staminate; petals white, larger than the sepals; stamens numerous; carpels numerous, in a dense subglobose head; achenes flattened, distinctly winged.

Beak of achene erect, less than 0.5 mm long; head of achenes at maturity seldom over 1.5 cm wide; monoecious **S. cuneata**

Beak of achene at the inner margin of the top and pointing inward, 0.5–2 mm long; head of achenes at maturity usually more than 1.5 cm wide; often dioecious **S. latifolia**

S. cuneata Sheld. / Arrowhead; Wapato
Plants 2–4 dm tall; leaf-blades mostly sagittate, the basal lobes usually markedly shorter than the terminal part; bracts of inflorescence 0.5–4 cm long, lanceolate; stamens 10–18; achenes 2–2.5 mm long. $2n = 22^*$. In mud and shallow water. Alas, Yuk, Huds Bay to Lab s to Calif, NM, Tex, Gt Lakes. Map 85

S. latifolia Willd. / Arrowhead; Wapato
Plants 3–6 dm tall, more robust than the last species; leaf-blades mostly sagittate, the basal lobes equalling or shorter than the upper part; bracts of inflorescence ovate, 1 cm long or less; stamens 25–40; achenes 2.3–3.5 mm long, with broad marginal wings. $2n = 22$. *S. sagittifolia* L. var. *latifolia* Muhl. Ponds and lakes. sBC to NS, transcont s. Map 84 ○

HYDROCHARITACEAE / Waterweed Family

Elodea Michx.

E. longivaginata St. John
Submersed perennial plants rooting in mud; stems slender, sparingly branched; leaves opposite or occasionally in whorls of 3, generally distant, sometimes crowded and imbricate, linear mostly 1–2 cm long, 1–2 mm wide, finely serrulate, apex rounded, rarely acute; plants dioecious; staminate flowers with spathe 2–15 cm long, dilated at the end, perianth-tube thread-like, 4–30 cm long, sepals 3, 3.5–5 mm long, 2–2.5 mm wide, petals 3, white, 5 mm long, 0.5 mm wide, stamens 9; pistillate flowers with spathe 3–7 cm long, perianth-tube up to several dm long, carrying flowers to water surface, sepals about 3.0 mm long, 1.3 mm wide, petals white, spatulate, 4 mm long, 1.3 mm wide, staminodia 3, ovary 1, inferior, 3–4 mm long, stigmas 3, 1–2 mm long, entire; fruit a capsule; seeds 6, 5–6 mm long. $2n = 24$*. *E. canadensis* Michx. of Ed. 1. Ponds and lakes. Alta, Sask s to Mont, Wyo, Utah, Colo, NM, wNev, ND. Map 86

GRAMINEAE / Grass Family

Annual or perennial herbs; stems (culms) usually hollow except at the solid nodes, frequently rounded; leaves 2-ranked, usually linear, parallel-veined, sheathing at the base, the sheath enveloping the culm like a split tube; a minute appendage (ligule) is at the junction of sheath and blade; flowers minute, arranged in spikelets, these aggregated in spikes, racemes or panicles; spikelet consists of a series of 2-ranked bracts, the lowest two (glumes) empty, the others with flowers, each flower with a subtending lemma and an embracing palea; flowers perfect, staminate or pistillate; perianth none, or reduced to 2 or 3 minute scales (lodicules); stamens commonly 3, the filaments slender and the anthers usually exsert from the spikelet at maturity; ovary 1-loculed, commonly with 2 styles and plumose stigmas; fruit a seed-like grain (caryopsis), sometimes permanently enclosed in the lemma and palea; lemma, palea and flower parts collectively called the floret; axis of the spikelet is the rachilla; the main axis of an inflorescence is called the rachis.

1 Spikelets 2-flowered, with one perfect terminal floret and a sterile or staminate floret below; articulation usually below the glumes (these falling off with the florets at maturity); spikelets, or at least the fruits, more or less dorsally compressed; a few species, mostly introduced 2

Spikelets not as above, usually 1–many-flowered, the sterile florets, if any, above the perfect florets; articulation usually above the glumes (these remaining behind following disarticulation of spikelet), exceptions are *Alopecurus, Beckmannia, Cinna, Polypogon, Spartina, Sphenopholis*; spikelets usually laterally compressed; *or* the spikelets 3-flowered, the upper floret perfect, the lower (lateral) two staminate or sterile; articulation above the glumes; spikelets laterally compressed; most of our species 3

2 Spikelets in pairs, one sessile and perfect, the other pedicellate and sterile; glumes coriaceous; palea membranous **Schizachyrium**

Spikelets not in pairs; glumes membranous; palea thick, indurate **Group 1**

3 Spikelets with 2 staminate or 2 sterile rudimentary florets below the single perfect floret; no sterile or rudimentary floret above 4

Spikelets without sterile florets below the perfect florets 5

4 Lower (lateral) florets staminate, as long as or longer than the upper (central) perfect floret; florets brown **Hierochloe**

Lower (lateral) florets reduced to sterile lemmas not more than two-thirds the length of the upper (central) perfect floret; florets green or pale **Phalaris**

5 Aquatic annuals; culms 10–30 dm tall; spikelets unisexual, 1-flowered; stamens 6 **Zizania**

Plants and spikelets not as above (the majority of our species) 6

6 Spikelets sessile or subsessile in spikes or spike-like racemes **Group 2**

Spikelets pedicellate in open or contracted panicles, sometimes spike-like, but then the spikelets arising all around the rachis, not obviously from one side or in two opposite rows 7

7 Spikelets 1-flowered **Group 3**

Spikelets 2–many-flowered 8

8 Glumes as long as the lowest floret, usually as long as the spikelet **Group 4**

Glumes shorter than the first floret **Group 5**

GROUP 1

1 Spikelet surrounded by 1–many distinct or more or less connate bristles, these forming an involucre **Setaria**

Spikelet not subtended by bristles 2

2 Glumes and sterile lemma awned or awn-pointed **Echinochloa**

Glumes and sterile lemma awnless 3

3 Spikelets in slender spike-like racemes; ligule membranous **Digitaria**

Spikelets in an open panicle, each on a long pedicel; ligule a ring of hairs 4

4 Basal leaves different from the culm leaves, forming a winter rosette; second glume and sterile lemma pilose **Dicanthelium**

Basal leaves similar to culm leaves, not forming a winter rosette; second glume and sterile lemma scaberulous **Panicum**

GROUP 2

1 Spikelets on one side of the rachis (secund); spikes usually more than one 2

Spikelets or spikelet clusters on opposite sides of the rachis (distichous); spikes terminal, solitary 5

2 Spikes comb-like; spikelets more than 4 mm long 3

Spikes not comb-like; spikelets 4 mm long or less 4

3 Culms 2–4 (5) dm tall; spikes 1–2 (3); leaf-blades 1–2 mm wide **Bouteloua**
Culms 3–15 dm tall; spikes 4 or more; leaf-blades 2–15 mm wide **Spartina**

4 Spikelets orbicular, about 3 mm long; leaf-blades flat, 5–10 mm wide; marshy places **Beckmannia**
Spikelets lanceolate, about 4 mm long; leaf-blades twisted, 1–2 mm wide; dry plains **Schedonnardus**

5 Spikelets solitary at each node of the rachis (sometimes 2 in *Agropyron* hybrids, but never throughout the spike; see key to *Elymus*) 6
Spikelets more than one at each node of the rachis (sometimes solitary in part of the spike in *Elymus* hybrids) 10

6 Spikelets placed edgewise to the rachis; first glume wanting except in the terminal spikelet **Lolium**
Spikelets placed flatwise to the rachis 7

7 Plants perennial **Agropyron**
Plants annuals 8

8 Spikes short, 1–2 cm long **Eremopyrum**
Spikes 3–17 cm long 9

9 Glumes ovate, 3-nerved **Triticum**
Glumes awl-shaped, 1-nerved **Secale**

10 Spikelets 3 at each node of the rachis, 1-flowered, the lateral pair stalked, usually reduced to awns **Hordeum**
Spikelets usually 2 at each node, alike, 2–6-flowered 11

11 Culms mostly 1–5 dm tall; rachis disarticulating at maturity; glumes and lemmas awned, 2–10 cm long **Sitanion**
Culms mostly more than 5 dm tall; rachis not disarticulating at maturity (where it does the glumes and lemmas not as above) **Elymus**

GROUP 3

1 Lemma indurate, tightly enfolding the palea and caryopsis 2
Lemma not indurate and closely enfolding the palea and caryopsis 4

2 Awn trifid **Aristida**
Awn simple 3

3 Florets plump; awns deciduous, 1.5 cm long or less **Oryzopsis**
Florets slender; awns persistent, 2.0 cm long or more **Stipa**

4 Panicle open; florets stipitate; stamens 1 **Cinna**
Plants not as above 5

5 Disarticulation below the glumes; panicle dense, spike-like 6
Disarticulation above the glumes; panicle various (including spike-like) 7

6 Glumes long-awned **Polypogon**
Glumes awnless **Alopecurus**

7	Ligule a ring of hairs	8
	Ligule membranous	9
8	Spikelets less than 3 mm long	**Sporobolos**
	Spikelets more than 3 mm long	**Calamovilfa**
9	Lemmas awn-tipped, short-pointed, or mucronate, less than 4 mm long	**Muhlenbergia**
	Lemmas not as above, awnless or with a dorsal awn	10
10	Glumes shorter than lemma	11
	Glumes, at least one of the pair, longer than the lemma	12
11	Spikelets 4–5 mm long	**Arctagrostis**
	Spikelets shorter	**Agrostis**
12	Panicle dense, spike-like; glumes keeled, awn-tipped	**Phleum**
	Panicle open to somewhat congested, but not spike-like; glumes rounded on back, awnless	13
13	Floret evidently bearded at base; palea well-developed	**Calamagrostis**
	Floret glabrous or minutely hairy at the base; palea small or wanting	**Agrostis**

GROUP 4

1	Spikelets 2-flowered, disarticulating below the glumes; glumes markedly dissimilar, the first narrow-lanceolate, the second wedge-shaped; lemmas generally scabrous	**Sphenopholis**
	Spikelets not as above	2
2	Lemmas bifid at apex, with a conspicuous geniculate awn arising between the lobes; ligule a ring of hairs	**Danthonia**
	Lemmas and ligules not as above	3
3	Glumes 2–3 cm long; spikelets 2-flowered, or with a rudimentary third floret, pendulous	**Avena**
	Glumes shorter; spikelets not pendulous	4
4	Spikelets 3–6-flowered, 1–1.5 cm long	**Helictotrichon**
	Spikelets mostly 2-flowered, less than 1 cm long	5
5	Lemmas convex, awned from below the middle, truncate, erose-dentate at apex	6
	Lemmas keeled, awn, when present, arising from above the middle	7
6	Leaf-blades flat, 3–6 mm wide, spikelets purple; awn about 3 mm long	**Vahlodea**
	Leaf-blades involute or folded, less than 3 mm wide; spikelets light brown to green, awn over 3 mm long	**Deschampsia**
7	Lemmas with a dorsal geniculate awn	**Trisetum**
	Lemmas awnless or short-awned	8
8	Glumes as long as or longer than the spikelet	**Trisetum**
	Glumes about as long as the first floret	**Koeleria**

GROUP 5

1 Tall reed-like grass with plume-like panicle; rachilla with silky hairs as long as the lemmas **Phragmites**

Smaller grasses, rarely over 12 dm tall 2

2 Plants dioecious; culms erect from creeping rhizomes; in saline or alkaline soils **Distichlis**

Plants not dioecious (except in a few species of *Poa* with villose lemmas) 3

3 Plants low, matted annuals, mostly 3–15 cm tall, the leaves and spikelets congested at the ends of naked culms; ligule a ring of hairs; blades white-margined, spiny-tipped **Munroa**

Plants not as above 4

4 Spikelets 2-flowered, disarticulating below the glumes; glumes markedly dissimilar, the first narrow-lanceolate, the second wedge-shaped; lemmas generally scabrous **Sphenopholis**

Spikelets not as above 5

5 Lemmas 3-nerved, the nerves prominent 6

Lemmas 5–many-nerved, the nerves sometimes obscure 7

6 Spikelets many-flowered; glumes and lemmas obtuse to acute, with wart-like projections covering these and other parts **Eragrostis**

Spikelets 2-flowered; glumes and lemmas erose-truncate; wart-like projections lacking; plants growing in water or very wet soil **Catabrosa**

7 Callus of florets bearded 8

Callus not bearded (lemmas cobwebby at base in some species of *Poa*) 10

8 Lemmas erose at summit, awnless; tall grass of wet places **Scolochloa**

Lemmas bifid at summit, awned 9

9 Lemmas about 10 mm long, excluding awn; sheaths closed **Schizachne**

Lemmas 5–7 mm long, excluding awn; sheaths open **Trisetum**

10 Lemmas awned, the awn 1 mm or more long 11

Lemmas awnless or awn-pointed, the point less than 1 mm long 14

11 Spikelets over 12 mm long; lemma awned from a bifid apex; sheaths closed for most of their length 12

Spikelets less than 12 mm long; lemmas with a terminal awn; sheaths open 13

12 First glume 1-nerved; second glume 3-nerved **Bromus**

First glume obscurely 3-nerved; second glume 5-nerved **Melica**

13 Plants annual **Vulpia**

Plants perennial **Festuca**

14 Panicle congested; spikelets in dense clusters towards the ends of the panicle branches, forming a solitary spike or several spike-like clusters 15

Panicle open; spikelets not in dense clusters 16

15 Panicle of one-sided spikelet clusters; primary panicle branches 1–10 cm long; leaf-blades flat, mostly 3–10 mm wide; ligules mostly 3–9 mm long **Dactylis**
Panicle a solitary spike, branches short; leaf-blades generally folded or involute, mostly 1–2 mm wide; ligules 1–2 mm long **Koeleria**

16 Glumes rather papery; lemmas of 2 or more upper florets often folded together as a unit; culms commonly bulbous at base; sheaths closed **Melica**
Glumes not papery; lemmas not as above; culms not bulbous at base; sheaths open or closed 17

17 Lemmas 8 mm long or more 18
Lemmas less than 8 mm long 19

18 Sheaths closed; lemma-nerves prominent **Bromus**
Sheaths open; lemma-nerves often obscure **Festuca**

19 Lemmas with parallel nerves, these not converging towards the tip, the apex broadly obtuse or truncate 20
Lemmas with nerves converging towards the tip 21

20 Sheaths closed; nerves prominent; plants of ponds, streams, moist woods **Glyceria**
Sheaths open; nerves generally faint; plants mostly of saline soil **Puccinellia**

21 Lemmas compressed-keeled **Poa**
Lemmas rounded on the back 22

22 Lemmas 5 mm or more long; glumes not compressed-keeled **Festuca**
Lemmas usually less than 5 mm long; glumes usually compressed-keeled **Poa**

X Agroelymus G. Camus

Hybrids between species of *Agropyron* and *Elymus;* of intermediate morphology.

X A. bowdenii Boivin
Rhizomatous; culms 6–8 dm tall; blades flat, 3–10 mm wide, scabrous below, smooth to scabrous above; spike 7–12 cm long, open, the lower spikelets somewhat remote; spikelets 1–2 per node, 1.4–2.2 cm long; glumes subulate, 3–9 mm long, glabrous to pubescent, green to purplish; lemmas 6–10 mm long, glabrous to pubescent; awnless or short-awned, 1–3 mm long. *Agropyron smithii* × *Elymus innovatus* ssp. *innovatus.* BC, Alta. Map 91 ○

X A. hirtiflorus (A.S. Hitchc.) Bowden
Culms tufted, 4–9 dm tall, with slender creeping rhizomes; leaves firm, 1–4 mm wide, flat or involute; spike erect, 5–15 cm long; glumes narrow, firm, hirsute, with an awn about the length of the body; lemmas hirsute, mostly about 8 mm long, with an awn 5–10 mm long. $2n = 28$*. *Agropyron trachycaulum* × *Elymus innovatus* ssp. *innovatus. Elymus hirtiflorus* A.S. Hitchc. of Ed. 1. Dry grassland. BC to Sask; Ont; Wyo. Map 87 ●

X A. mossii Lepage
Perennial; culms to 8 dm tall; blades flat, 3–7 mm wide, scabrous above and along margin; spikes 8–12 cm long, compact, spikelets 1–2 per node, 1.2–1.7 cm long, mostly 5-flowered; glumes subequal, 8–15 mm long, lanceolate, aristate, 1–3-nerved, mostly pubescent at base, purplish; lemmas 8–10 mm long, the backs pubescent, purplish; awn 4–15 mm long. *Agropyron trachycaulum* × *Elymus canadensis*. Open woods. Alta. Map 87 ○

X A. turneri Lepage
Rhizomatous; culms 5–15 dm tall; blades flat, 3–6 mm wide, scabrous above; spikes 10–22 cm long, open, the lower spikelets often distantly remote; spikelets 1–2 per node, 1.5–2.5 cm long, 5–8-flowered; glumes generally unequal, subulate, 3–5-nerved, pilose; lemmas 6–13 mm long, pilose; awns mostly 2–10 mm long. $2n = 28$*. *Agropyron dasystachyum* × *Elymus innovatus* ssp. *innovatus*. Moist banks. Alta, Sask. Map 87 ▲

X Agrohordeum G. Camus

Hybrids between species of *Agropyron* and *Hordeum;* of intermediate morphology.

X A. macounii (Vasey) Lepage / Macoun's Wild Rye
Culms tufted, erect, slender, 5–10 dm tall; blades firm flat or involute, 2–5 mm wide, usually scabrous on both sides; spike slender, 4–12 cm long, usually about 5 mm thick excluding the awns; rachis tending to break up; spikelets appressed, mostly 2-flowered; glumes setaceous, scabrous, awned; lemmas with slender straight awns 1–2 cm long. $2n = 28$. *Agropyron trachycaulum* × *Hordeum jubatum* ssp. *jubatum*. *Elymus macounii* Vasey of Ed. 1. Moist meadows and saline flats. Alas, Yuk, sMack to Ont, Que s to BC, Alta, Sask, Man, and wUSA. Map 88

Agropyron Gaertn. / Wheat Grass

Plants perennial, many with creeping rhizomes; spikelets in a spike, several-flowered, solitary, sessile, usually compressed, with the flat side towards the rachis of the spike, the rachilla disarticulating above the glumes and between the florets; glumes subequal, usually several-nerved and shorter than the first lemma, awnless or awned; lemma rather firm, 5–7-nerved, acute or awned from the apex; palea about as long as the lemma or shorter, ciliate on the two keels.

Species of this genus so intergrade that recognition of taxonomic units is somewhat arbitrary. Hybrids between species of *Agropyron* and those of other genera also occur.

1 Spike more or less pectinate, oval-oblong, flattened, mostly less than 10 cm long; lemmas awnless or awns less than 5 mm long — 2
  Plants not as above — 3

2 Spikelets strongly spreading; spike distinctly pectinate — **A. pectiniforme**
  Spikelets spreading to ascending; spike not so distinctly pectinate — **A. sibiricum**

3 Rhizome short or absent; spike open, the spikelets generally shorter than the internodes; glumes strongly 4–6-nerved; lemmas glabrous, the awns, if present, divergent, 1–2 cm long; anthers over 5.0 mm long **A. spicatum**
Plants not as above 4

4 Rhizome present; anthers (2.5) 3–7 mm long 5
Rhizome absent, plants caespitose; anthers 1.5–2.0 mm long 8

5 Lemma awns divergent, 0.5–2.0 cm long; glumes 2–3-nerved **A. albicans**
Lemma awns, if present, straight; glumes 1–5 (7)-nerved 6

6 Leaves flat, 4–8 (10) mm wide, green **A. repens**
Leaves mostly involute, if flat 2–3 (4) mm wide, strongly glaucous 7

7 Glumes usually keeled, attenuate, broadest below the middle, often as long as or just shorter than the mostly glabrous, occasionally hirsute, lemma; lemma usually at least short-awned; anthers 3–4 mm long **A. smithii**
Glumes usually flat or rounded on the back, broadest above the middle, acute at the apex, mostly about half the length of the lemma or slightly longer; lemmas generally awnless, usually densely hirsute, occasionally glabrous; anthers 4.0 mm long or longer **A. dasystachyum**

8 Culms often decumbent; awns strongly divergent; spike disarticulating at maturity **A. scribneri**
Culms erect; awns, if present, straight, spike not disarticulating at maturity 9

9 Spikes mostly less than 10 cm long, the spikelets imbricate; glumes 3–4-nerved, the hyaline margin 0.5–1.0 mm wide; lemmas usually pubescent **A. violaceum**
Spikes mostly over 10 cm long, the spikelets scarcely imbricate; glumes mostly 6–7-nerved, the hyaline margins 0.1–0.5 mm wide; lemmas usually glabrous **A. trachycaulum**

A. albicans Scribn. & Smith
Rhizomatous; culms 4–8 (10) dm tall; blades mostly involute, 1–3 mm wide; spikes 4–12 (15) cm long; spikelets 3–7-flowered; glumes 3–5-nerved, oblong-lanceolate; acute to acuminate, usually short-awned, scabrous or glabrous; lemmas 7–10 mm long, glabrous or pubescent, the awn arcuate, 5–15 mm long. $2n = 28^*$. In our area var. **albicans** with pubescent lemmas, and var. **griffithsii** (Scribn. & Smith) Beetle (*A. griffithsii* of Ed. 1) with glabrous lemmas. Dry prairie grassland. BC to Sask s to Wash, Ida, Utah, Colo, SD. Map 89

A. dasystachyum (Hook.) Scribn. / Northern Wheat Grass
Plants with long creeping rhizomes; culms loosely tufted, 4–13 dm tall; leaves flat to involute, often glaucous; spikes 5–20 cm long; spikelets loosely to closely imbricated, mostly 4–8-flowered; glumes mostly 3–5-nerved, broad in middle, scarious-margined, acute or awn-pointed, scabrous or pubescent; lemmas about 1 cm long, densely to sparsely pubescent, awnless or short-awned. $2n = 28^*$, 42^*. In our area var. **dasystachyum** with lemmas densely pubescent and var. **riparium** (Scribn. & Smith) Bowden (*A. riparium* of Ed. 1) with lemmas glabrous to scabrous, sometimes with a few long hairs at the base. Produces hybrids with *A. trachycaulum* and *A. spicatum*. The former (**A. × pseudorepens** Scribn & Smith,

$2n = 28$*, Map 91 ●) is similar to *A. dasystachyum* and not readily distinguished from it, but generally the blades flat, lemmas short-awned, anthers about 2 mm long. The latter ($2n = 21$*) is rhizomatous, has glabrous spikelets and lemmas awnless or short-awned. Also hybridizes with *Elymus innovatus* Beal. Prairie grassland, sand-hills, dry open woodland. BC to Man s to Calif, Ariz, Neb, Ill. Map 90

A. pectiniforme R. & S.
Culms densely tufted, 3–4 dm tall; leaf-blades 2–4 mm wide; spike oblong-oval, pectinate, flattened, 4–7 cm long; 1.5–2.5 cm broad; spikelets compressed, 0.7–1.5 cm long; glumes firm, hyaline-margined, gradually tapered into an awn; lemma lanceolate-subulate, with an awn about 2 mm long. $2n = 14, 28, 42$. *A. cristatum* (L.) Gaertn. of Ed. 1. Widespread along roadsides and on waste ground. Introduced. Common.

A repens (L.) Beauv. / Couch Grass; Quack Grass
Plants with long creeping yellowish rhizomes; culms 4–8 dm tall; leaf-blades flat, thin, mostly 4–8 mm wide; spikelets 4–8-flowered; glumes 3–7-nerved, 8–12 mm long; lemmas similar to the glumes in size and shape, acuminate or with an awn 1–10 mm long. $2n = 42$*. Roadsides, waste ground, gardens. Introduced. Common.

A. scribneri Vasey
Densely tufted; leaves mostly basal, the blades short, rather rigid, puberulent; culms 1–4 dm tall, decumbent to ascending and flexuous; spikes dense, often flexuous or nodding, the rachis usually disarticulating at maturity; glumes narrow, with divergent awns; lemmas nerved towards the tip, tapering to a divergent awn 1.5–3 cm long. $2n = 28$. Produces a hybrid with *A. violaceum* (Hornem.) Lange (**A. × brevifolium** Scribn., Map 87 △), which has a disarticulating rachis, awned glumes, lemmas with backs short pubescent and awns 2–10 mm long. Dry alpine slopes. BC s to Calif, Ariz, NM. Map 92

A. sibiricum (Willd.) Beauv.
Similar to *A. pectiniforme*, but with longer spikes and awns and with the upper surface of leaf-blades glabrous. $2n = 28$. *A. desertorum* (Fisch.) Schult. An escape from cultivation, waste ground. Introduced. Rare.

A. smithii Rydb. / Western Wheat Grass
Plants often blue-glaucous; rhizomes creeping; leaves firm, strongly nerved, tapering to a sharp point, flat or involute; culms 3–6 dm tall; spikelets rather closely imbricate, occasionally 2 at a node, 6–10-flowered; glumes mostly 10–12 mm long, rigid, attenuate from below the middle with nearly straight sides; lemmas similar to the glumes in size and shape, obscurely nerved, acuminate to short-awned. $2n = 56$*. In our area var. **smithii** with lemma back glabrous and var. **molle** (Scribn. & Smith) Jones with the lemma backs pubescent. Produces a hybrid with *Elymus innovatus* Beal (× *Agroelymus bowdenii* Boivin). Prairie grassland. BC to Ont s to nCalif, nNev, Ariz, Tex, Ark, Tenn, introduced elsewhere. Map 93

A. spicatum (Pursh) Scribn. & Smith / Bluebunch Wheat Grass
Caespitose, rarely with rhizomes, often forming large bunches; culms slender, 3–10 dm tall; leaf-blades slender, flat to involute, glabrous or pubescent above; spike 8–15 cm long; spikelets rather distant, each extending barely beyond the base of the next one in the spike; glumes 5–10 mm long, 3–4-nerved, obtuse to short-awned; lemmas 8–10 mm long, glabrous or scabrous, awnless or with a curved-divergent awn (usually at right angles) 1–2 cm long. $2n = 14^*$, 21, 28. In our area var. **spicatum** with lemmas awned, and var. **inerme** (Scribn. & Smith) Heller (*A. inerme* of Ed. 1) with lemmas awnless or short-awned. Dry open areas. BC to Sask s to nCalif, Ariz, Tex; Alas, Yuk, Gt Lakes. Map 94

A. trachycaulum (Link) Malte
Tufted plants, lacking rhizomes; culms 5–15 dm tall; leaf-blades long, mostly 3–6 mm wide, flat; spikes erect or slightly curved, 5–25 cm long, sometimes appearing partially secund; spikelets somewhat imbricate to remote, few-flowered; glumes 3–7-nerved, nearly as long as the spikelet, awn-tipped; lemmas scabrous, obscurely 5-nerved, tapering into a straight awn, 1–3 cm long. $2n = 28$. In our area var. **trachycaulum** (*A. trachycaulum* of Ed. 1) with lemmas awnless or short-awned, var. **unilaterale** (Cassidy) Malte (*A. subsecundum* (Link) A.S. Hitchc. of Ed. 1) with lemma awns 15–40 mm long, glumes 10–16 mm long, and var. **glaucum** (Pease & Moore) Malte, with lemma awns 5–15 (20) mm long, glumes 6–10 mm long. Produces hybrids with *Hordeum jubatum* L. (× *Agrohordeum macounii* (Vasey) Lepage), *Elymus innovatus* Beal (× *Agroelymus hirtiflorus* (A.S. Hitchc.) Bowden), and possibly *Elymus canadensis* L. (× *Agroelymus mossii* Lepage). Prairie, parkland, open woods. Alas, Yuk, sMack to Huds Bay, Nfld s to Calif, Ariz, NM, Ia, Gt Lakes. Map 95

A. violaceum (Hornem.) Lange
Loosely tufted; culms 2–5 dm tall; leaf-blades short, flat, 3–5 mm wide, pubescent to glabrous; spikes 3–7 cm long; spikelets usually closely imbricate; glumes broad, flat, scarious-margined, unsymmetrical and short-awned at tip; lemmas pubescent, acute or short-awned. $2n = 28^*$. *A. latiglume* (Scribn. & Smith) Rydb. of Ed. 1. Produces a hybrid with *A. scribneri*. Montane to alpine slopes and meadows. Greenland: Alas, Yuk to Ellesmere Is s to Wash, Colo, sMack, Huds Bay, Lab. Map 96

Agrostis L. / Bent Grass

Perennials, mostly tufted, some with rhizomes or stolons; inflorescence an open or contracted panicle; spikelets numerous, small, 1-flowered, disarticulating above the glumes, the rachilla usually not prolonged; glumes equal or nearly so, 1-nerved, somewhat keeled, persistent after the fall of the floret; lemma shorter than or nearly equalling the glumes, thin, obtuse, awned or awnless, sometimes minutely hairy at the base; palea lacking in some species, when present delicate in texture and often quite small.

1 Palea evident, at least half as long as the lemma 2
Palea lacking, or minute and nerveless, not over one-third as long as the lemma; rhizomes absent 4

2 Rhizomes or stolons present; culms over 3 dm tall; widespread species **A. stolonifera**
Rhizomes or stolons absent; culms mostly 1–3 dm tall; mountain species 3

3 Rachilla produced as a distinct bristle behind the palea **A. thurberiana**
Rachilla not produced beyond the palea **A. humilis**

4 Panicle narrow, at least some of the lower branches bearing spikelets to near their base 5
Panicle open, the branches spreading or reflexed at maturity, naked at their base 6

5 Culms over 2 dm tall; leaf-blades over 1.5 mm wide; ligules 2–6 mm long **A. exarata**
Culms mostly 1–2 dm tall; leaf-blades less than 1.5 mm wide; ligules 1–2 mm long **A. variabilis**

6 Lemmas awned **A. mertensii**
Lemmas awnless **A scabra**

A. exarata Trin. / Spike Redtop
Variable in size and habit, mostly tufted; culms 2–10 dm tall; leaf-blades flat, 2–6 mm wide; ligule prominent, often 6 mm long; panicle narrow, 5–30 cm long, somewhat open and interrupted to spike-like and dense, the branches spikelet-bearing to the base; spikelets mostly 3–4 mm long; glumes subequal, acuminate to awn-tipped, very scabrous on the keel; lemma about 2 mm long, sometimes with a prickle or short awn; palea minute. $2n$ = 28, 42, 56. In our area ssp. **minor** (Hook.) C.L. Hitchc. Moist open areas. sAlas, Yuk, wMack s to Calif, Nev, NM, SD, Neb. Map 97 •

A. humilis Vasey
Plants tufted, glabrous; culms 5–15 cm tall; leaves basal, blades less than 1 mm wide; ligule 0.5–1.0 mm long; panicle narrow, 1–3 cm long, branches appressed; glumes subequal, about 2 mm long; lemma about as long as glumes, awnless; palea two-thirds to three-quarters as long as lemma; anthers 0.6–0.7 mm long. $2n$ = 14. *Podagrostis humilis* (Vasey) Björkm. Moist alpine areas. BC to Ore, Nev, Utah, Colo. Map 98

A. mertensii Trin.
Tufted, dwarf plant of alpine and far northern situations; culms 4–40 cm tall; leaves mostly basal, the blades commonly involute, 0.2–2 mm broad; panicle pyramidal, 1–14 cm long, the divergent branches smooth, forking near the middle; spikelets purplish, 2–3 mm long; glumes subequal; lemma nearly as long as the glumes, usually bearing a bent, exserted awn, 3–6 mm long; palea short or obsolete. $2n$ = 42, 56. *A. borealis* Hartm. of Ed. 1. Moist alpine slopes. Cir-

cumpolar: Alas, Yuk, Mack to sBaffin Is s to BC, Utah, Colo, nSask, James Bay, NY, Nfld; NC. Map 97 ○

A. scabra Willd. / Hair Grass; Tickle Grass
Forming small dense tufts; culms mostly 3–7 dm high, slender; leaves mostly basal and filiform, the broader blades 1–3 mm wide; ligule hyaline, 2–5 mm long; panicle 1–2.5 dm long and nearly as wide at maturity, very diffuse, the branches elongate, very slender, scabrous, brittle ascending or spreading, sometimes drooping, spikelet-bearing towards the ends; spikelets 2–2.7 mm long; glumes unequal, acuminate, scabrous on the keels; lemma 1.5–1.7 mm long; palea wanting; at maturity the broad panicle often breaks away and is carried by the wind like a tumble-weed. $2n = 42$. Includes *A. geminata* Trin. Common in disturbed areas and open woods to subalpine elevations. eAsia: Alas, Yuk, Mack to Huds Bay, Nfld, widely distributed s. Map 99

A. stolonifera L.
Rhizomatous or stoloniferous; culms 2–10 dm tall; leaf-blades 1–10 mm wide, scabrous; ligule 3–5 mm long, ciliate or lacerate; panicle 5–30 cm long, slender to pyramidal, branches whorled, generally reddish; spikelets 2–3 mm long; glumes subequal, acute; lemma shorter than glumes, awnless or rarely short-awned; palea bifid, half to two-thirds as long as lemma. $2n = 28, 30, 32, 35, 42, 44, 46, 56$. *A. alba* L. and *A. palustris* Huds. of Ed. 1. Moist ground. Introduced. Common.

A. thurberiana A.S. Hitchc.
Small tufted grass; culms slender, 1–4 dm tall; leaves somewhat crowded at base, the blades about 2 mm wide; panicle rather narrow, 5–7 cm long, lax and often drooping; spikelets 2 mm long; glumes 2 mm long and lemma slightly shorter; palea about two-thirds as long; rachilla prolonged, hairy. $2n = 14, 28$. *Podagrostis thurberiana* (A.S. Hitchc.) Hult. Moist alpine areas. BC to Calif, Ida, Utah, Colo; sAlas. Map 100

A. variabilis Rydb. / Alpine Redtop
Small and densely tufted; culms slender, 1–2.5 dm tall; leaves basal, soft and fine, the blades usually less than 1 mm wide, flat or conduplicate; panicle narrow, 2–6 cm long, rather dense, usually purple; spikelets about 2 mm long, purplish or bronze; lemma 1.5 mm long, awnless; palea minute. $2n = 28$. Alpine ridges. BC to Calif, Mont, Colo. Map 101

Alopecurus L. / Foxtail

Perennials with flat leaf-blades and soft dense spike-like panicles; spikelets 1-flowered, disarticulating below the glumes, flattened; glumes equal, usually united at base, ciliate on the keel; lemma about as long as the glumes, obtuse, 5-nerved, the margins united at base, bearing from below the middle a slender awn, this included or exserted; palea none.

1	Glumes about 5 mm long	**A. pratensis**
	Glumes 2–4 mm long	2

2 Spikelets densely woolly; panicle oblong or ovoid, about 1 cm thick **A. occidentalis**
Spikelets not woolly; panicle linear or oblong-linear, less than 1 cm thick **A. aequalis**

A. aequalis Sobol. / Water Foxtail
Culms solitary or in small tufts, 2–5 dm tall, erect or decumbent at base, sometimes rooting at nodes; leaf-blades 1–4 mm wide; panicle slender, spike-like, 2–7 cm long, 3–5 mm thick; glumes 2–2.5 mm long, blunt and scarious at apex, villose on keel and at base; awn 0.5–3.0 mm long, inserted about the middle of the lemma; anthers 0.5–1 mm long. $2n = 14^*$. Shallow water and muddy shores. Circumpolar: Alas, Yuk, sMack, Huds Bay, nQue, Nfld, generally distributed s. Map 102

A. occidentalis Scribn. & Tweedy / Alpine Foxtail
Forming slender, creeping rhizomes; culms 1–8 dm tall, erect, or often decumbent at base, glabrous, rather stiff, often coarse; leaves few, the sheaths often inflated, the blades scabrous, 3–5 mm wide, usually flat; panicle oblong or ovoid, about 1 cm wide, woolly; glumes 3–4 mm long, woolly, 3-nerved; lemmas about as long as the glumes, awned, the awn attached near base and more or less exserted. $2n = 112^*$. Often included in *A. alpinus* J.E. Smith. Shores and open woodland. Alta to Colo. Map 103

A. pratensis L. / Meadow Foxtail
Culms erect, 3–6 dm tall; leaf-blades 2–6 mm wide; spike-like panicle 3–7 cm long, 5–8 mm thick; glumes about 5 mm long, acute or acuminate, conspicuously villose-ciliate on the keel; lemma with awn exserted 2–5 mm. $2n = 28, 42$. Fields. Introduced. Rare.

Arctagrostis Griseb.

A. arundinacea (Trin.) Beal
Perennial with culms 4–10 dm tall; leaves with prominent ligule and blades 0.5–2 dm long, 4–10 mm wide; panicle narrow but somewhat open, mostly 1–2 dm long; spikelets often purplish, 1-flowered, 3–4 mm long; glumes unequal, acute; lemma longer than glumes, densely hispidulous; palea similar to the lemma. $2n = 28$. *A. latifolia* (R.Br.) Griseb. var. *arundinacea* (Trin.) Griseb. Marshy ground. Alas, Yuk, wMack to BC, Alta. Map 104 ●

Aristida L.

A. longiseta Steud. / Red Three-awn
Stiff perennial, densely tufted, often in large bunches; culms 2–3 dm high; leaves often curved, not over 1 mm wide, involute; panicle (excluding the awns) 4–8 cm long, narrow, with few spikelets on short ascending branches; spikelets 1-flowered; glumes unequal, the first 8–10 mm long, the second about twice as long; lemma 10–15 mm long, slender, indurate, with 3 awns, 4–8 cm long, the awns similar and equally divergent. $2n = 22, 44$. Dry plains. sBC to swMan, s to Ore, Ariz, Tex, SD. Map 104 ○

Avena L. / Oat

Annual grasses with coarse leaves and lax panicles of large spikelets; panicle 1–3 dm long, widely branched; spikelets 2–3-flowered and drooping; glumes nearly equal, 7–11-nerved, 2–3 cm long; lemma shorter than the glumes, indurate, tapering to a 2-toothed apex, rounded on the back, awned or awnless.

Lemma usually hairy on the back, the base much enlarged and with a tuft of stiff hairs; awn 2–5 cm long, geniculate **A. fatua**

Lemma glabrous, the base not enlarged or hairy; awnless or with a short straight awn only in the first floret **A. sativa**

A. fatua L. / Wild Oat
Culms stout, smooth, 3–10 dm tall; rachilla of spikelet soon disarticulating, the grains dropping as soon as mature; awn 2–5 cm long, stout, twisted. $2n = 42$. A noxious weed of arable land. Introduced. Common.

A. sativa L. / Cultivated Oat
Differs from the previous species as shown in the key; the grain is not shed at maturity. $2n = 42$. Roadsides. An escape from cultivation. Occasional.

Beckmannia Host

B. syzigachne (Steud.) Fern. / Slough Grass
Coarse annual of low wet ground; culms stout, 3–10 dm high, light green; leaf-blades flat, the larger 3–10 mm wide, firm, scabrous; sheaths loose, overlapping; ligules 5–8 mm long; inflorescence a narrow, crowded or more or less interrupted panicle, composed of numerous appressed or ascending spikes; spikelets 1-flowered, 3 mm long, orbicular, flat, nearly sessile in 2 rows along one side of a slender rachis; glumes about length of spikelet, cross-wrinkled, keeled, the apex rounded but apiculate; lemma narrow, acuminate, 5-nerved, its tip barely protruding beyond the glumes; palea nearly as long as lemma. $2n = 14^*, 28$. In our area ssp. **baicalensis** (Kusnez) Koyama & Kawano. Asia: Alas, Yuk, Mack, Huds Bay, NS s to Calif, Nev, Ariz, NM, Ia, Gt Lakes. Map 105

Bouteloua Lag. / Grama Grass

B. gracilis (HBK) Lag. / Blue Grama
Low perennial of dry prairie; plants densely tufted, often with rhizomes and forming a sod; culms 1–4 dm tall, leafy at base; leaves flexuous or curled, 1–2 mm wide, flat or involute, the sheaths glabrous but collar pilose on margins; ligules mostly of hairs; fertile culm slender, bearing 1–3, usually 2, spikes near the summit; spikes comb-like, 1.5–4 cm long, straight or arcuate, often bluish purple; spikelets with 1 perfect flower and 1 or more sterile rudiments, sessile, inserted in 2 rows on one side of the rachis; first glume 2–3 mm long; second glume 3–5 mm long, often hispid on keel; fertile lemma pilose, 3-nerved and 3-awned; palea about as long as lemma; sterile rudiment densely bearded at

summit of rachilla, long-awned. $2n = 20, 21, 28, 35, 42, 61, 77, 84$. Common in prairie grassland. BC to sMan s to Calif, Tex, Mo, Ill. Map 106

Bromus L. / Brome Grass

Annuals or perennials with flat blades and closed sheaths; spikelets several–many-flowered, relatively large, usually lax or drooping in an open or contracted panicle; glumes shorter than the lemmas, unequal; lemmas convex on back or keeled, 5–9-nerved, 2-toothed at apex, awnless or awned from between the teeth; palea usually shorter than the lemma, 2-keeled, the grain adherent to it; ovary crowned with a hairy appendage, at the base of which are the lateral plumose stigmas.

1 Weedy annuals; mostly rather small slender plants with pubescent sheaths and blades 2
Perennials; relatively large grasses, pubescent or glabrous 7

2 Glumes slender, hairy, the first 1-nerved; lemmas slender, acuminate, the teeth 2–5 mm long **B. tectorum**
Glumes broad, glabrous, the first 3–5-nerved; lemmas broad, rounded at apex, the teeth very short 3

3 Pedicels all or mostly shorter than spikelets; panicle congested; lemmas prominently nerved **B. mollis**
Pedicels mostly longer than spikelets; panicle open; lemmas less prominently nerved 4

4 Lemma margins inrolled at maturity; palea almost equalling lemma; rachilla visible **B. secalinus**
Lemma margins not inrolled; palea much shorter than lemma; rachilla concealed 5

5 Awns straight, of more or less equal length; anthers 1–2 mm long **B. commutatus**
Awns divergent, the upper longer; anthers about 1 mm long 6

6 Panicle somewhat secund; spikelets 5–8 mm wide, inflated; awns strongly divergent; first glume 5-nerved, the second obscurely 7–9-nerved **B. squarrosus**
Panicle not secund; spikelets less than 5 mm wide, not much inflated; awns flexuous, often somewhat divergent; first glume 3-nerved, the second 5-nerved **B. japonicus**

7 Spikelets strongly flattened, the lemmas compressed-keeled; first glume 3–5-nerved, second 5–9-nerved **B. carinatus**
Spikelets rounded or somewhat flattened; lemmas not compressed-keeled; first glume 1-nerved, second 3-nerved (except in *B. anomalus*) 8

8 Plants rhizomatous; lemmas awnless or short-awned **B. inermis**
Plants not rhizomatous; lemma awns longer than 3 mm 9

9 Lemmas hairy along the margin and sometimes on lower part of the back, the upper part glabrous 10
Lemmas hairy rather evenly over the back, usually more densely so along the lower part of the margin 11

10 Lower lemmas about 3 mm wide; awn 3–5 mm long; ligule about 1 mm long **B. ciliatus**

Lower lemmas about 2 mm wide; awn usually over 5 mm long; ligule 3–5 mm long **B. vulgaris**

11 Panicle up to 20 cm long; leaf-blades mostly 2–4 mm wide, auricles absent; first glume 3-nerved **B. anomalus**

Panicle up to 10 cm long; leaf-blades mostly over 5 mm wide, auricles present; first glume 1-nerved **B. altissimus**

B. altissimus Pursh
Perennial; culms 8–15 dm tall, glabrous, rarely pubescent at nodes; blades flat, 5–15 mm wide, mostly glabrous, with conspicuous auricles at base; panicle 10–20 (24) cm long, branches spreading; spikelets 1.5–3 cm long; first glume 4–7 mm long, 1–rarely 3-nerved, second glume 6–10 mm long, 3–5-nerved, lemma 9–14 mm long, pubescent; awn 3–7 mm long. $2n = 14$. *B. purgans* L. of Ed. 1, *B. latiglumis* (Shear) A.S. Hitchc. Moist banks. BC to sMan, Minn, Wisc, Mich, NY, Me, sQue s to Neb, Mo, Ind, Pa. Map 107

B. anomalus Rupr. *ex* Fourn. / Nodding Brome
Culms 3–6 dm high, the nodes pubescent; sheaths often sparsely short-hairy; blades scabrous, mostly 2–4 mm wide; ligule about 1 mm long; panicle about 1 dm long, drooping; first glume 3-nerved, the second 5-nerved; lemmas about 12 mm long, evenly and coarsely hairy over the back; awn 2–4 mm long. $2n = 14^*$, 28. *B. porteri* (Coult.) Nash. Moist prairie grassland and open woods. BC to Sask, s to Calif, Ariz, NM, Tex. Map 108

B. carinatus Hook. & Arn.
Culms usually robust, 6–12 dm high; sheaths often retrorsely hairy; blades commonly 5–10 mm wide, pubescent or glabrate; panicle rather stiff, 1–2 dm long; spikelets large, 2–4 cm long, 5–7 mm wide, strongly flattened; lemmas 10–15 mm long, strongly flattened and keeled, usually pubescent; awns mostly 2–5 mm long. $2n = 28, 42, 56, 70$. *B. marginatus* Nees of Ed. 1. Open woods and meadows. BC to swSask s to Calif, NM; introduced Alas, Yuk, Ont. Map 109

B. ciliatus L. / Fringed Brome
Culms 6–10 dm high, slender, the nodes glabrous or pubescent; sheaths glabrous or the lower short-hairy; blades mostly 4–8 mm wide; panicle loose, open, drooping; spikelets greenish; first glume 1-nerved, the second 3-nerved; lemmas 10–15 mm long, long-hairy near the margin, glabrous or sparingly hairy on the back, rarely glabrous throughout; awn 3–5 mm long. $2n = 14^*$, 28. Includes *B. richardsonii* Link. Open woods and meadows. Alas, Yuk, swMack to James Bay, Nfld s to Calif, Ariz, NM, Tex, Neb, Gt Lakes, Apal. Map 110

B. commutatus Schrad.
Annual; culms 2–8 (10) dm tall, pubescent; sheaths pilose; ligule 0.5–1.5 mm long; blades flat, 2–4 (6) mm wide; panicle 0.5–2.0 dm long, branches ascending;

spikelets 1.5–2 (3) cm long; first glume 5–7 mm long, 3–5-nerved, second glume 6–9 mm long, 5–9-nerved; lemmas 8–10 mm long, nerves obscure; awn about as long as lemma. $2n$ = 14, 28, 56. Roadsides and waste ground. Introduced. Rare.

B. inermis Leyss. / Awnless Brome
Rhizomatous perennial; culms 2–15 dm tall; blades 5–15 mm wide; panicle mostly 5–20 cm long, open, branches erect or arcuate; spikelets mostly 1.5–3 cm long; first glume 4–8 mm long, 1(3)-nerved, second glume 6–10 mm long, 3(5)-nerved; lemmas 7–16 mm long, 5–9-nerved, awnless or with a short awn generally less than 3 mm long. In NAm, including our area, both native and introduced elements occur. The former, ssp. **pumpellianus** (Scribn.) Wagnon (*B. pumpellianus* Scribn. of Ed. 1), $2n$ = 56*, 84, with pubescent lemmas, leaves and culms (at least the nodes), ligules 1–2 mm long, and auricles generally present; the latter, ssp. **inermis,** $2n$ = 28, 42, 49, 54–56–58, 70, generally glabrous (nodes seldom hairy), ligules about 1 mm long and auricles often absent. Open woods and grassland to subalpine elevations, roadsides, fields, and waste ground. Circumpolar: Alas, Yuk, sMack to James Bay, Nfld s to Calif, Ariz, Colo, Okla, Ia, Ind, NJ. Map 111

B. japonicus Thunb. / Japanese Chess
Annual, with slender culms 2–6 dm tall; sheaths and blades softly villose; panicle loose, open, somewhat nodding, with very slender branches; first glume 3-nerved, the second 5-nerved; lemmas broad, obtuse, 2-toothed at apex, 7–9 mm long, glabrous; awns 8–10 mm long, strongly flexuous at maturity. $2n$ = 14, 28. Roadsides and waste ground. Introduced. Occasional.

B. mollis L.
Annual; culms 2–9 dm tall; mostly soft-pubescent; blade flat, 1.5–4.0 mm wide; ligules mostly 1 mm; auricles lacking; panicle erect, 3–10 cm long, somewhat congested; spikelets 1–2 cm long; first glume 3–5-nerved, 4–6 mm long, second glume 5–7-nerved, 6–8 mm long; lemmas 6–9 mm long, nerves prominent; awn 6–10 mm long. $2n$ = 14, 28. Waste ground. Introduced. Rare.

B. secalinus L.
Annual; culms 3–8 dm tall, glabrous, sometimes pubescent at nodes; blades flat, 2–8 mm wide, generally pubescent; panicle 5–20 cm long, open, nodding, somewhat secund; spikelets 1–2 cm long, the florets spreading at maturity to expose rachilla; first glume 4–6 mm long, 3–5-nerved, second glume 5–8 mm long, 5–7-nerved; lemmas 7–9 mm long, margins curved around fruit, the back strongly convex, nerves obscure; awn shorter than lemma. $2n$ = 14, 28. Roadsides, fields, and waste ground. Introduced. Rare.

B. squarrosus L.
Similar to *B. japonicus;* distinguished by the characters of the key. $2n$ = 14. Roadsides, fields, and waste ground. Introduced. Rare.

B. tectorum L. / Downy Chess
Annual, with slender culms 2–5 dm tall; sheaths and blades softly pubescent; panicle repeatedly branched, with rather crowded, drooping spikelets on slender pedicels; glumes narrow, villose, the first one 1-nerved; lemmas lanceolate, villose or puberulent, acuminate into slender teeth; awns 12–15 mm long, straight. $2n = 14, 28$. Locally abundant on roadsides and waste ground, especially in the southwest. Introduced. Common.

B. vulgaris (Hook.) Shear
Slender woodland grass; culms mostly 8–10 dm tall, the nodes pubescent; sheaths and blades soft-hairy, the blades commonly 5–8 mm wide; ligule prominent, 3–5 mm long; panicle slender, with few spikelets, these narrow and drooping; first glume 1-nerved, the second 3-nerved; lemmas 8–10 mm long, narrow, sparsely hairy over the back, more densely near the margins, or nearly glabrous; awn 6–8 mm long. $2n = 14$. Open woods. BC to Calif e to Mont, Wyo. Map 112

Calamagrostis Adans. / Reed Grass

Perennials, mostly with creeping rhizomes; mainly tall grasses, with narrow elongate leaves; panicle open or congested and narrow; spikelets numerous, 1-flowered, disarticulating above the glumes; glumes acute or acuminate, keeled, the first usually slightly larger; lemma equalling or shorter than the glumes, awned from below the middle of the back, surrounded at the base (callus) with a ring of white hairs; palea shorter than the lemma; rachilla usually prolonged into a hairy bristle.

1 Awn of lemma geniculate, more or less exserted; callus-hairs usually much shorter than the lemma — 2
Awn of lemma straight or nearly so, included; callus-hairs equalling the lemma or nearly so — 4

2 Awn of lemma much exceeding the glumes; plant tufted, sometimes with short rhizomes — **C. purpurascens**
Awn of lemma about equalling the glumes — 3

3 Plant rhizomatous, not tufted; culms scabrous below panicle; glumes sharply keeled; spikelets strongly flattened; leaves glabrous at junction of sheath and blade — **C. montanensis**
Plant tufted, but often with rhizomes; culms smooth; glumes not strongly keeled; spikelets not strongly flattened; leaves often densely hairy at the junction of sheath and blade — **C. rubescens**

4 Panicle rather loose and open; callus-hairs fully as long as the lemma — 5
Panicle more or less contracted; callus-hairs mostly shorter than the lemma, of unequal lengths — 6

5 Panicle 2 cm wide or more; glumes scabrous on keel; callus-hairs of uniform length except for an outer shorter ring — **C. canadensis**
Panicle less than 2 cm wide; glumes scarcely scabrous on keel; callus-hairs of varying length — **C. lapponica**

6 Culms and leaves harsh and scabrous, blades firm; ligules mostly over 3 mm long, often lacerate; panicle firm **C. inexpansa**

Culms and leaves smooth, or the culms scabrous only below panicle, the blades at the tip and along margins, blades lax; ligules 1–3 mm long, entire; panicle rather soft 7

7 Glumes 2–3 mm long; callus-hairs distinctly shorter than lemma **C. stricta**

Glumes 3–4 mm long; callus-hairs not much shorter than lemma **C. lapponica**

C. canadensis (Michx.) Beauv. / Bluejoint; Marsh Reed Grass
Large grass with creeping rhizomes; culms tufted, 6–12 dm tall; leaves numerous, rather lax, flat, 4–8 mm wide; ligule 3–6 mm long; panicle large, 1–2 dm long, rather broad, open or dense, somewhat nodding; glumes usually 3–4 mm long; lemma with straight awn attached just below the middle; callus-hairs abundant, about as long as the lemma. $2n = 42–66$. A species complex variously treated. In our area ssp. **canadensis** (var. *macouniana* (Vasey) Stebbins of Ed. 1) with the glumes 3–4 mm long, acute, rounded on the back, scarious to scabrous, and ssp. **langsdorfii** (Link) Hult. (var. *scabra* (Kunth) A.S. Hitchc. of Ed. 1) with the glumes 4–6 mm long, acuminate, keeled, and strongly scabrous. Marshes and moist woodland. Circumpolar: Alas, Yuk to sBaffin Is s to Calif, NM, Neb, Ohio, Del. Map 113

C. inexpansa A. Gray / Northern Reed Grass
Tufted, 4–10 dm tall, with slender rhizomes; leaves firm, flat or involute, 2–4 mm wide, very scabrous; ligules 4–7 mm long; panicle narrow, 5–15 cm long, dense, more or less lobed; glumes mostly 3–4 mm long; lemma scabrous, the awn straight or nearly so, about as long as the glumes; callus-hairs half to three-fourths as long as lemma. Intergrading varieties have been described, differing in shape of panicle and size of spikelets. $2n = 28, 56, 70, 84–105$. Marshy places and low meadows. eAsia, Greenland: Alas, Yuk, sMack to Huds Bay, nQue, Nfld s to Calif, Ariz, NM, Neb, Mo, Ill, Va, Me. Map 114

C. lapponica (Wahl.) Hartm.
Plants tufted; culms 3–6 (8) dm tall, smooth or scaberulous below panicle; sheaths smooth; ligules 0.5–3 (4) mm long; blades flat or involute, 2–3 (4) mm wide, margins smooth or scabrous; panicle open or contracted, 5–15 cm long, 1–2 cm wide, often purple; glumes mostly 3–4 mm long, smooth or sparsely scabrous on back and keel; lemmas somewhat shorter than glumes; awn mostly straight, shorter than lemma; callus-hairs of varying length, the longest about equalling lemma or slightly longer. $2n = 28, 42, 49–140$. Alpine ridges. Circumpolar: Alas, Yuk to sBaffin Is s to BC, Alta, nSask, Huds Bay, Lab. Map 115 ○

C. montanensis Scribn. / Plains Reed Grass
Culms stiffly erect, usually 2–4 dm tall, usually arising singly from slender, deep-seated rhizomes; leaves mostly 1–2 mm wide, involute, scabrous; ligules 1–3 mm long; panicle erect, dense, 5–10 cm long, usually pale; glumes 4–5 mm long, acuminate, sharply keeled; lemma with geniculate awn attached about 1 mm above base, about as long as the lemma and often exserted from side of

glumes; callus-hairs about half as long as lemma. $2n = 28$. Prairie grassland, open pine woods, and dry hillsides. BC to Man s to Mont, Colo, SD, Minn. Map 115 ●

C. purpurascens R.Br. / Purple Reed Grass
Tufted, sometimes with short rhizomes; culms 3–7 dm tall; sheaths rough; ligule 2–5 mm long; blades scabrous, 2–4 mm wide, flat or involute; panicle narrow, dense, 4–12 cm long, usually pinkish or purplish; glumes 5–8 mm long, scabrous; lemma with 4 short bristle-like teeth at the apex, the awn inserted near the base, geniculate and exserted about 2 mm; callus-hairs about one-third as long as lemma. $2n = 28$, 40–(42*)–58, 84. In our area spp. **purpurascens.** Open slopes and woodlands. eAsia, Greenland: Alas, Yuk to Baffin Is n to Ellesmere Is s to Calif, Utah, Colo, nSask, Huds Bay, nQue, Lab; SD; Lk Superior; Gaspé. Map 117

C. rubescens Buckl. / Pine Grass
Tufted, with creeping rhizomes; culms slender, 6–9 dm tall; sheaths smooth but usually hairy near the collar; ligule 1–4 mm long; blades 2–4 mm wide, flat or involute; panicle narrow, dense or somewhat loose, sometimes interrupted, 6–14 cm long, pale or purplish; glumes 4–5 mm long, acuminate; lemma with awn inserted near the base, geniculate, often slightly exserted from side of glumes; callus-hairs scant, about one-third as long as lemma. $2n = 28$, 42, 56. Open woods and banks. BC to swMan s to Calif, Nev, Colo. Map 118

C. stricta (Timm) Koeler
Similar to *C. inexpansa* and intergrading with it, but generally a smaller grass, less scabrous, with a shorter, less dense, somewhat more interrupted panicle. $2n = 28$, 56, 70, 84. *C. neglecta* (Ehrh.) Gaertn., Mey. & Scherb. of Ed. 1. Wet meadows, banks, and shores. Circumpolar: Alas, Yuk to Devon Is s to Ore, Utah, Colo, sSask, ND, Gt Lakes. Map 116

Calamovilfa (Gray) Hack.

C. longifolia (Hook.) Scribn. / Sand Grass
Coarse perennial with long stout scaly rhizomes; culms mostly solitary, 8–12 dm tall; sheaths much overlapping at base, usually glabrous except at the throat, where often long-villose; ligule 0.5–2 mm long; blades firm, flat and 4–8 mm wide at base, usually involute above, tapering to a long slender apex; panicle 1–3 dm long, narrow, the branches ascending or somewhat spreading; spikelets pale, 6–7 mm long, 1-flowered, awnless, the rachilla not prolonged; glumes rigid, 1-nerved, 4–7 mm long; lemma 4–7 mm long, glabrous; callus-hairs copious, more than half as long as lemma; palea about as long as lemma. $2n = 40$, 60. Sandy prairie and open woods; common on sand-dunes where an effective stabilizer. sBC to wOnt s to nIda, Colo, Mo. Map 119

Catabrosa Beauv.

C. aquatica (L.) Beauv. / Brook Grass
Aquatic perennial grass with stems prostrate and rooting at the lower nodes; ascending culms 1–4 dm long, striate and glabrous; sheaths loose and overlapping; leaves flat, soft, short, 3–8 mm wide; panicle 1–2 dm long, erect, pyramidal or rounded-oblong, the capillary branches spreading in somewhat distant whorls; spikelets mostly 2-flowered, disarticulating above the glumes and between the lemmas; glumes unequal, shorter than the lemmas, truncate, scarious and often erose at apex; lemmas strongly 3-nerved, 2.5–3 mm long, yellow or brown, broadly truncate and more or less erose at the scarious apex; palea about as long as the lemma. $2n = 20$. Streams and in wet ground near springs. Greenland, Iceland, Eurasia: BC to cMan, James Bay, Lab, Nfld s to Ore, Nev, Ariz, Colo, Ia, Wisc. Map 120

Cinna L.

C. latifolia (Trev.) Griseb. / Drooping Wood Reed
Tall perennial without creeping rhizomes; culms slender, 6–12 dm high; leaves thin, lax, flat, scabrous, 5–12 mm wide; panicle green, 1–3 dm long, usually very loose, the clusters of slender branches ascending or drooping; ligule 5–10 mm long; spikelets 3–4 mm long, 1-flowered, disarticulating below the glumes; rachilla prolonged as a minute bristle; glumes narrow, acuminate, about 3 mm long; lemma similar to the glumes, 3-nerved, with a minute apical awn; palea appearing 1-keeled. $2n = 28$. Moist woods. Circumpolar, boreal: Alas, BC, sMack, to James Bay, Lab, Nfld s to Calif, Nev, NM, ND, Gt Lakes, Apal. Map 121

Dactylis L.

D. glomerata L. / Orchard Grass
Coarse, tufted perennial, glaucous and scabrous; culms 6–10 dm tall; blades flat, 2–8 mm wide; panicle 5–20 cm long, densely crowded or interrupted, with the spikelets nearly sessile in one-sided clusters at the ends of the branches; spikelets 3–6-flowered, flat, disarticulating between the florets; glumes unequal, acuminate, keeled, more or less hispid; lemmas 5–8 mm long, 5-nerved, often ciliate-keeled and mucronate; palea a little shorter than the lemma. $2n = 14, 21, 28, 42$. Roadsides and waste ground. Introduced. Rare.

Danthonia Lam. & DC. / Oat Grass

Tufted perennials, lacking rhizomes; leaves narrow; often involute; ligule reduced to a tuft of hairs; inflorescence a small racemiform panicle of relatively large spikelets (or of 1 spikelet); spikelets 3–8-flowered, disarticulating above the glumes and between the florets; cleistogamous spikelets are often produced within the lower sheaths; glumes about equal, broad, papery, usually extending beyond the florets; lemmas rounded on the back, obscurely nerved, hairy at least on the margins, 2-toothed at apex, with a twisted awn between the teeth;

awn stout, flat, geniculate, the base tightly twisted when dry; palea somewhat shorter than the lemma, papery.

1	Panicle usually reduced to a single spikelet; sometimes 2, rarely 3	**D. unispicata**
	Panicle of 2–several spikelets	2
2	Lemmas glabrous on the back, hairy on the margins only	**D. californica**
	Lemmas hairy on the back, sometimes sparsely so	3
3	Glumes mostly about 20 mm long; lemmas over 9 mm long	**D. parryi**
	Glumes 9–12 mm long; lemmas less than 9 mm long	**D. spicata**

D. californica Boland
Tufted; culms 9 dm tall, often disarticulating at the nodes; leaves flat or involute, sparsely to conspicuously spreading-hairy; panicle 3–9 cm long, branches erect or spreading; glumes usually 10–18 mm long; lemmas (excluding awns) 8–12 mm long, pilose on callus and on lower part of margin, otherwise glabrous, the teeth long-aristate, the awn 5–10 mm long; anthers 2–4 mm long or minute. $2n = 36^*$, c. 96^*. Includes *D. intermedia* Vasey of Ed. 1, which appears to be merely a cleistogamous variant. Dry to moist open areas and open woodland to subalpine elevations. Yuk, nSask, nOnt, Lab, Nfld s to Calif, NM, sSask, Lk Superior, eQue (Gaspé); sAlas. Map 122

D. parryi Scribn. / Parry Oat Grass
Often forming tough clumps, with many persistent sheaths at the base; culms 3–6 dm tall; leaves 1–4 mm wide, 1–2.5 dm long, slender, pubescent near throat and on collar, otherwise glabrous; panicle pale, becoming straw-coloured, 3–7 cm long, usually with 3–8 spikelets, the branches more or less pubescent, ascending or appressed; glumes 20–22 mm long; lemmas about 1 cm long, densely to sparsely hairy over the back, strongly pilose on the callus at the sides, the teeth aristate, the awn 10–15 mm long. $2n = 36^*$. Foothills grassland. BC to NM in Rocky Mts. Map 123

D. spicata (L.) Beauv. / Poverty Oat Grass
Tufted; culms slender, 2–7 dm tall; leaves mostly involute, filiform, glabrous or sparsely hairy, a tuft of long hairs at the throat; panicle 2–5 cm long, the branches short, erect or ascending, each with 1 or 2 spikelets; glumes 10–12 mm long; lemmas 3.5–5 mm long, hairy on the back, the teeth acuminate-pointed, the awn 5–8 mm long. $2n = 36$. Dry to moist open areas and open woodland. sAlas, BC to Nfld s to Ore, NM, generally distributed e. Map 124 ○

D. unispicata (Thurb.) Munro / One-spike Oat Grass
Densely tufted; culms 1.5–2.5 dm tall; leaves with pilose blades and sheaths, the hairs on the sheaths spreading or reflexed; panicle reduced to a single spikelet or sometimes 2, rarely 3; spikelets 1–2 cm long; lemmas about 1 cm long, pilose below on margins and on callus, otherwise glabrous; awn 6–8 mm long. $2n = 36$. Open, rocky, or sandy ground. BC to wSask s to Calif, Utah, Colo. Map 124 ●

Deschampsia Beauv. / Hair Grass

Mostly tufted perennials, with shining, pale or purplish spikelets; leaves slender; panicles loose or narrow; spikelets 2-flowered, disarticulating above the glumes, the hairy rachilla prolonged and sometimes with a reduced floret; glumes about equal, thin, scarious, keeled, acute; lemmas thin, somewhat truncate at apex, 2–4-toothed, with a slender awn from or below the middle, bearded at base; palea shorter than lemma.

Panicle narrow, spike-like; leaves mostly filiform, 1–1.5 mm wide **D. elongata**
Panicle open, the branches spreading or drooping; leaves wider, flat or folded **D. cespitosa**

D. cespitosa (L.) Beauv. / Tufted Hair Grass
Culms in dense tufts, 2–12 dm tall, leafy at base; leaves often elongate, mostly 1.5–4 mm wide, rather firm, flat or folded; panicle loose, open, nodding, 10–25 cm long, the capillary branchlets spikelet-bearing toward the tips; spikelets 3–5 mm long, pale or purplish, shining; glumes 3–4.5 mm long; lemmas smooth, 2.5–3.5 mm long, awned from near the base, the callus-hairs short. $2n = 26^*$, (27, 28). In our area ssp. **cespitosa.** Moist open areas. Circumpolar: Alas to Ellesmere Is s to Calif, Ariz, NM, SD, Gt Lakes, Apal. Map 125

D. elongata (Hook.) Munro / Slender Hair Grass
Culms densely tufted, slender 3–10 dm tall; leaves soft, slender, 1–1.5 mm wide, flat or folded, those of the basal tuft filiform; panicle very narrow, 1–3 dm long, the fine branches ascending; spikelets on short appressed pedicels; glumes 4–6 mm long, 3-nerved, equalling or exceeding the florets; lemmas 2–3 mm long, shining, pilose at base; awns 3–5 mm long, straight. $2n = 26$. Meadows and open slopes. sAlas, Yuk, to Calif, NM. Map 126

Dicanthelium (A.S. Hitchc. & Chase) Gould

Perennials, with distinct seasonal phases, the winter phase a rosette of leaves, the spring phase of simple flowering culms, the autumnal phase with axillary fascicles and/or branches bearing axillary panicles; leaves flat; ligules mostly of hairs; panicles open or contracted, the autumnal panicles mostly enveloped by the leaf-sheaths; spikelets of 2 florets; glumes unequal, the first small, the second larger, membranous, about equal to the lemma of the first floret; this floret sterile or staminate; lemma membranous, several-nerved, palea reduced or absent; second floret fertile, lemma shorter than first lemma, indurate, rounded on back, shiny, enveloping the palea. *Panicum,* in part, of Ed. 1.

1 Spikelets 3–4 mm long **D. leibergii**
Spikelets shorter 2

2 Plants velvety-villose, greyish green; ligule 3 mm long; spikelets about 2 mm long **D. acuminatum**
Plants copiously papillose-hispid; ligule 1 mm long; spikelets 2.7–3.0 mm long **D. oligosanthes**

D. acuminatum (Swartz) Gould & Clarke
Densely tufted or matted, velvety-villose; culms 1–3 (10) dm tall, ascending or spreading, the nodes densely hairy; blades thick, 5–12 mm wide; ligule of conspicuous hairs, 3 (5) mm long; panicle 3–6 (16) cm long; spikelets about 2 mm long, pilose; culms of the autumn are repeatedly branched and widely spreading, the whole forming a dense cushion. $2n = 18$. *Panicum thermale* Boland of Ed. 1. (*P. ferventicola* Schmoll var. *papillosum* Schmoll). In our area var. **acuminatum**. Marshy places, in our area around hot springs. BC to NS, generally distributed s. Map 127 ●

D. leibergii (Vasey) Freckman
Plants tufted, erect or ascending, geniculate below; culms solitary or clustered, 3–6 dm tall; leaves 7–12 mm wide, 7–11 cm long, papillose-hirsute above and below, to nearly glabrous above; sheaths shorter than internodes; panicle 5–10 cm long, often not fully exserted from leaf-sheath, branches ascending; spikelets 3–4 mm long, soft villose; autumnal phase of a few erect branches from middle and lower nodes. Prairies and dry open woods. Alta to Ont s to ND, Kans, Mo, Ind, Pa, NY. Map 127 ▲

D. oligosanthes (Schultes) Gould
Tufted perennial, 1–3.5 dm high; culms of the spring and summer are erect, long-hairy; sheaths mostly overlapping, copiously hirsute; blades firm, erect or strongly ascending, 5–8 cm long, 3–6 mm wide, hairy; panicles more or less included, becoming short-exserted; spikelets few, ellipsoid, 2.7–3 mm long, pubescent; autumnal phase bushy branched from the nodes, with tufts of overlapping reduced leaves and mostly hidden panicles. *Panicum wilcoxianum* Vasey of Ed. 1. Dry open areas. Alta to sMan s to NM, Kans, Ind, Wisc. Map 127 ○

Digitaria Heist. / Crab Grass

Annual or perennial; ligule membranous; inflorescence of spike-like, mostly unbranched terminal racemes, digitate or approximate; glumes unequal, the first minute or absent, the second well developed, from about one-third to about as long as the spikelet, 3-nerved; sterile and fertile lemmas the same size, the former conspicuously 5-nerved, the latter induarate, with hyaline margins, indistinctly 3-nerved.

Second glume about as long as spikelet; leaves glabrous	**D. ischaemum**
Second glume about half as long as spikelet; leaves pilose	**D. sanguinalis**

D. ischaemum Schreb. *ex* Muhl.
Annual; culms 2–5 dm long, decumbent, rooting at the nodes; leaf-blades mostly 2–6 mm wide, essentially glabrous; inflorescence of 2–6 slender, spike-like racemes, 3–10 cm long; first glume lacking or minute and hyaline; second glume 1.5–2.2 mm long, 3-nerved, about as long as the sterile lemma, both pubescent; fertile lemma 2.0–2.3 mm long, indurate, purple-black. $2n = 36$. Lawns and waste ground. Introduced. Rare.

D. sanguinalis (L.) Scop. / Crab Grass
Tufted annual, branching and prostrate, often rooting at the nodes; culms usually 3–6 dm long; leaf-blades 4–10 mm wide, hairy; sheaths loose, often papillose-hirsute; inflorescence of 3–12 slender spike-like racemes from the summit of the culm; spikelets about 3 mm long, 1-flowered, in 2's on one side of a broadly winged rachis; first glume minute; second glume about half as long as spikelet; sterile lemma strongly 5-nerved; fertile lemma pale or brownish, hard. $2n = 18$, 36–48, 54, 76. A weed of gardens and waste ground. Introduced. Rare.

Distichlis Raf.

D. stricta (Torr.) Rydb. / Salt Grass
Low perennial; rhizomes extensively creeping, covered with whitish scales; leaves 5–12 cm long, flat to involute, with sharp tips; plants dioecious; culms 1–4 dm long, erect or decumbent at base; inflorescence a small crowded panicle with large flattish spikelets; staminate spikelets straw-coloured, mostly 1–1.5 cm long; pistillate spikelets greenish, not so flat, 0.5–1 cm long; spikelets several-flowered, the rachilla disarticulating above the glumes and between the florets; glumes unequal, acute; lemmas closely imbricate, 3–6 mm long, firm, the pistillate coriaceous; palea with winged keels; grain 3–5 mm long, beaked. $2n = 40$. Saline areas in prairie. BC, sMack to Man s to Calif, Ariz, Tex, Mo. Map 128

Echinochloa Beauv.

E. crusgalli (L.) Beauv. / Barnyard Grass; Cockspur Grass
Coarse annual; culms thick, simple or branching at base, erect or decumbent, usually 3–8 dm tall, glabrous; leaves long, mostly 5–15 mm broad, glabrous, somewhat scabrous; sheaths compressed; ligule lacking; inflorescence a panicle of stout racemes; spikelets crowded, each with a fertile floret and a sterile floret below; glumes unequal, the first one less than half the length of the spikelet, 3-nerved, pointed; second glume and sterile lemma as long as the spikelet, prominently nerved, papillose-hispid on the nerves, the sterile lemma usually long-awned; fertile lemma planoconvex, smooth and shining (finely striated, under magnification), acuminate-pointed; fertile palea about as long as the lemma. $2n = 36$, 42, 48, 54, 56, 72. *E. pungens* (Poir.) Rydb. of Ed. 1. Fields, ditches, and roadsides. Introduced. Occasional.

Elymus L. / Wild Rye

Mostly coarse, tufted perennials, with or without rhizomes; leaves flat or sometimes involute; inflorescence a spike, often bristly, the spikelets in pairs (sometimes 3 or 4, or only 1, at some nodes); spikelets 2–6-flowered; rachilla disarticulating above the glumes and between the florets; glumes equal, twisted in front of the spikelet, somewhat asymmetrical, narrow to subulate, usually rigid, acute to aristate; lemmas rounded on back, obscurely 5-nerved, acute or awned from the tip; paleas somewhat shorter than the lemmas.

Elymus species hybridize with those of several other genera. The *Elymus* hybrids in our area (× *Agroelymus*) are included in the following key, also × *Agrohordeum*.

1 Creeping rhizomes present; spikelets mostly hairy 2
Creeping rhizomes absent; spikelets mostly glabrous 7

2 Glumes lanceolate, awnless or acuminate-pointed **E. mollis**
Glumes subulate or very narrow 3

3 Spikes 5–10 cm long; glumes often purplish; lemmas densely pubescent with long silky hairs **E. innovatus**
Spikes and spikelets not as above 4

4 Lemmas awnless or short-awned, 1–2 mm long 5
Lemmas awned 6

5 Spikes mostly 10–20 cm long, dense; glumes 8–16 mm long **E. piperi**
Spikes mostly 7–12 cm long, more open; glumes 3–9 mm long **X Agroelymus bowdenii**

6 Spikelets ascending, mostly overlapping; glumes subequal, generally purplish **X Agroelymus hirtiflorus**
Spikelets often spreading, the lower remote; glumes generally markedly unequal, seldom purplish **X Agroelymus turneri**

7 Glumes subulate, not broadened above the base 8
Glumes narrow-lanceolate, somewhat broadened above the base, often awned 9

8 Leaf-blades 2–4 (5) mm wide; ligules about 1 mm long; glumes 3–7 mm long **E. junceus**
Leaf-blades 4–15 mm wide; ligules mostly 2–7 mm long; glumes 7–16 mm long **E. piperi**

9 Spike slender, about 1 cm wide including awns, disarticulating at maturity or on drying; glumes 1–3-nerved, narrow, lacking a thinner margin **X Agrohordeum macounii**
Spike thicker, more than 1 cm wide including awns, not disarticulating; glumes 3–5-nerved, with an evident thinner margin 10

10 Glumes thin, bases not indurate; awns straight 11
Glumes firm, bases indurate, ivory-like; awns straight or divergent 12

11 Lemmas glabrous to scabrous **E. glaucus**
Lemmas densely pubescent **X Agroelymus mossii**

12 Awns straight; base of glume terete and more or less bowed out **E. virginicus**
Awns outwardly curved; base of glume flattened, scarcely bowed out **E. canadensis**

E. canadensis L. / Canada Wild Rye
Culms tufted, stout, 7–12 dm tall; blades flat, or involute when dry, 4–20 mm wide, scabrous; spikes 1–1.5 dm long, thick and bristly, nodding; spikelets commonly in threes or fours, slightly spreading; glumes 2–4-nerved, the awn about as long as the body; lemmas strongly nerved, 8–14 mm long, usually scabrous-

hairy, the awns divergently curved, 2–4 cm long. $2n = 28^*, 42$. Possibly hybridizes with *Agropyron trachycaulum*. Riverbanks, sandy areas. BC, sMack, to Que s to nCalif, Ariz, NM, Tex, Mo, SC. Map 129 ●

E. glaucus Buckl. / Smooth Wild Rye
Culms tufted, slender, 5–10 dm tall, often short-decumbent at base; leaves flat, usually 5–10 mm wide, soft and lax but scabrous; ligule truncate, lacerate, 1–2 mm long; spike erect or somewhat nodding, usually dense, 5–15 cm long; glumes flat, 3–5-nerved, acuminate to short-awned; lemmas glabrous or nearly so, long-awned. $2n = 28$. In our area var. **glaucus.** Moist to dry open woods and hillsides. Alas, Yuk, nSask to nwOnt s to Calif, NM, Ia, Ind, Mich. Map 130

E. innovatus Beal / Hairy Wild Rye
Culms 4–10 dm tall, erect, from slender, creeping rhizomes with brown scales; leaves firm, flat or involute on drying, 2–5 mm wide, glabrous beneath, scabrous above, often glaucous; ligule truncate, 0.5 mm long or less; spike 4–10 cm long, rather dense, purplish or greyish villose; glumes very narrow, densely villose; lemmas broader, coarsely villose, the awns mostly 1–4 mm long. $2n = 28^*, 56$. Hybridizes with *Agropyron dasystachyum*, *A. smithii*, and *A. trachycaulum*. In our area ssp. **innovatus.** Open woodlands, especially pine. Alas, Yuk, sMack to James Bay s to BC, Mont, Wyo, SD. Map 131

E. junceus Fisch.
Caespitose; culms 4–10 dm tall; ligule 0.2–1.0 mm long, truncate; blades flat or involute, 1.5–4 mm wide, scabrous; spikes mostly 5–10 cm long, about 1 cm wide, erect, disarticulating at maturity; glumes 3–7 mm long, subulate; lemmas 6–10 mm long, generally scabrous; awns 0.5–2 (3) mm long. $2n = 14$. Waste ground. Introduced. Rare.

E. mollis Trin. / American Dune Grass
Forming extensively creeping rhizomes; culms stout, 6–12 dm tall, glaucous, pubescent below the spike; sheaths crowded and overlapping at base; blades firm, 6–12 mm wide, rough above; spike erect, dense, thick, 1–2 dm long; spikelets scabrous to pubescent; glumes lanceolate, flat, 3–5-nerved, 12–25 mm long; lemmas soft-hirsute, acuminate or mucronate. $2n = 28$. Sand-dunes. eAsia, Greenland: Alas, nYuk to nBaffin Is s to Calif, nAlta, nSask, Lk Superior, James Bay, nQue, Lab, Nfld, Mass. Map 129 ○

E. piperi Bowden / Giant Wild Rye
Plants robust, forming large clumps, the culms usually 2–3 m tall; blades elongate, 6–18 mm wide, firm, strongly nerved; spike erect, dense, 1–2.5 dm long, 1–1.5 cm thick; spikelets often 3 to 5 at a node; glumes subulate, awn-pointed; lemmas acuminate or mucronate, thinly pubescent to glabrous. $2n = 28, 56$. *E. cinereus* Scribn. & Merr. of Ed. 1. Riverbanks, ravines, moist slopes. BC to Sask s to Calif, Ariz, Colo. Map 132

E. virginicus L. / Virginia Wild Rye
Culms stout, erect, tufted, 5–10 dm tall; blades mostly 3–12 mm wide; spike usually erect, dense, 5–12 cm long; glumes firm, strongly nerved, terete, indurate and outwardly curved at base, often yellowish, acuminate and usually awned; lemmas glabrous, scabrous, or sparsely hirsute, tapering into a straight awn. $2n = 28$. Open woods and thickets. BC to Nfld s to neWash, Ida, Colo, Tex, Fla, SC. Map 133

Eragrostis Beauv.

E. megastachya (Koel.) Link / Skunk-grass
Annual, with wart-like projections throughout; culms 1–5 dm tall, decumbent to erect; ligule a ring of hairs; leaf-blades flat, 2–7 mm wide; panicle open to somewhat congested, 5–20 cm long, 1–6 cm wide; spikelets mostly 5–10 mm long, many-flowered, strongly compressed; glumes subequal, about 2 mm long, shorter than the first floret; lemmas 1.5–2.5 mm long, conspicuously 3-nerved. $2n = 20, 40$. *E. cilianensis* (All.) Mosher. Roadsides and waste ground. Introduced. Rare.

Eremopyrum (Ledeb.) Juab. & Spach

E. triticeum (Gaertn.) Nevski / Annual Wheat Grass
Annual plant, branching at base; culms slender, usually decumbent, 1–3 dm tall; leaf-blades 2–3 mm wide, flat, less than 1 dm long; spike oval or ovate, 1–1.5 cm long; spikelets crowded, about 7 mm long; glumes and lemmas acuminate, scarcely awned. $2n = 14$*. *Agropyron triticeum* Gaertn. of Ed. 1. Waste ground. Introduced. Rare.

Festuca L. / Fescue

Perennials, often densely tufted, with spikelets in narrow or spreading panicles; leaves mostly narrow; spikelets few–several-flowered, the rachilla disarticulating above the glumes and between the lemmas; glumes narrow, unequal, usually shorter than the lemmas, the first sometimes very small; lemmas rounded on the back, 5-nerved, the nerves often obscure, usually awned from the apex, sometimes awnless; palea about as long as the lemma.

1 Blades flat, soft and lax, mostly more than 3 mm wide — 2
Blades mostly involute; if flat, less than 3 mm wide — 3

2 Panicle usually erect; awn of lemma none or very short — **F. pratensis**
Panicle drooping; awn of lemma 5–20 mm long — **F. subulata**

3 Lower sheaths friable, becoming fibrillose; new basal offshoots chiefly divergent, decumbent below; blades smooth — **F. rubra**
Lower sheaths chartaceous, persistent; new basal offshoots strongly ascending — 4

4 Glumes nearly as long as spikelet; lemmas 7–10 mm long, scabrous; culms stout; forming large tussocks — **F. scabrella**
Glumes much shorter than spikelet; lemmas mostly not more than 7 mm long — 5

5 Awn as long as or longer than the body of the lemma; blades soft, sulcate, glabrous **F. occidentalis**

Awn shorter than the body of the lemma; blades firm, slender, involute, usually scabrous 6

6 Culms usually over 3 dm high; panicle 1–2 dm long, somewhat open; leaf-blades elongate **F. idahoensis**

Culms usually less than 3 dm high; panicle less than 1 dm long, narrow; leaf-blades relatively short 7

7 Panicle loosely open, with divergent branches during anthesis; anthers 2.5–3 mm long **F. ovina**

Panicle spike-like, the branches little-spreading; anthers less than 2 mm long 8

8 Panicle 2–10 cm long, interrupted, linear to lanceolate; anthers 1.0–1.7 mm long; lemmas greenish, strongly involute **F. saximontana**

Panicle usually 1–3 cm long, dense at least above, lanceolate to ovoid; anthers 0.3–1 mm long; lemmas purplish, membranaceous, not strongly involute 9

9 Culms glabrous or somewhat scabrous-hirsute; anthers 0.6–1 mm long; panicle narrow **F. brachyphylla**

Culms inconspicuously but densely tomentose; anthers 0.3–0.5 mm long; panicle almost ovoid, often somewhat secund **F. baffinensis**

F. baffinensis Polunin
Densely tufted, forming low tussocks with wiry culms and numerous short leaves; culms slender, 0.5–2 dm tall, densely puberulent or tomentose with minute, curved hairs (seen only with a lens); leaves mostly basal, 2–6 cm long, rather soft, flattish to involute; panicles lance-ovoid, compact, 1–3 cm long, purplish; lemmas elliptic to ovate-lanceolate, not strongly involute; anthers usually 0.3–0.5 mm long. $2n = 28$. Dry alpine slopes. Greenland, Spitzbergen: Alas to Ellesmere Is s in Rocky Mts to Colo, Huds Bay, Baffin Is. Map 134

F. brachyphylla Schultes
Tufted; culms 0.5–2.0 dm tall, generally glabrous; leaf-blades 0.5–1.0 mm wide, involute, 1–9 cm long, smooth, glabrous; panicle compact, mostly 1–3 cm long, green to dark purple, bronze-edged; spikelets 2–4-flowered; glumes lanceolate, 2–3 mm long; lemmas ovate-lanceolate, 2–3 (4) mm long, scaberulous on back, awn 1–3 mm long; anthers 0.6–1.0 mm long. $2n = 42$. *F. ovina* L. ssp. *brachyphylla* (Schultes) Piper. Alpine slopes. Eurasia, Greenland: Alas to Ellesmere Is s to Calif, NM, nSask, James Bay; eQue (Gaspé). Map 135 ●

F. idahoensis Elmer / Bluebunch Fescue
Densely tufted; culms 3–8 dm tall; leaves numerous, filiform, usually elongate, involute, very scabrous or rarely smooth, often blue-glaucous; panicle narrow, 0.5–2 dm long, the branches somewhat spreading or ascending; spikelets mostly 5–7-flowered; glumes narrow, acute; lemmas about 7 mm long, nearly terete, the awn usually 2–4 mm long. $2n = 28$. *F. ovina* L. var. *ingrata* Hack. *ex* Beal. Prairie grassland. BC to Calif, Colo. Map 136 ●

F. occidentalis Hook.
Closely tufted; culms slender, 4–8 dm tall, glabrous, shining; leaves filiform, soft, smooth; panicle narrow, 5–20 cm long, lax, more or less secund; spikelets 6–10 mm long, loosely 3–5-flowered; glumes broadly lanceolate, soft, glabrous; lemmas rather thin, scabrous, 5–6.5 mm long, with awns 4–7 mm long. $2n = 28$. Dry wooded slopes. BC to Calif e to Ida, Mont, Wyo; Gt Lakes. Map 137

F. ovina L. / Sheep Fescue
Densely tufted; culms very slender, 1.5–6 dm high, glabrous; basal sheaths long persistent; blades firm, slender, involute; panicle stiff, 2–10 cm long, loosely open, its branches divergent at anthesis; glumes firm and attenuate; lemmas coriaceous, strongly involute, mostly 4–5.5 mm long, short-awned; anthers 2.5–3 mm long. $2n = 14, 21, 28, 35, 42, 49, 56, 70$. Lawns and roadsides. Introduced. Occasional.

F. pratensis Huds. / Meadow Fescue
Loosely tufted, often with short creeping rhizomes; culms 5–10 dm tall, smooth; blades elongate, 3–8 mm wide, with conspicuous auricles at base; panicle erect, or nodding at tip, 1–2 dm long, the branches spikelet-bearing nearly to base; spikelets 8–12 mm long; glumes 3–5 mm long; lemmas 5–7 mm long, acute, awnless or occasionally short-awned. $2n = 14, 28$ (42, 70). *F. elatior* L. of Ed. 1. Grown for forage, an escape on roadsides and waste ground. Introduced. Occasional.

F. rubra L. / Red Fescue
Loosely tufted, with matted rootstocks, sometimes short-creeping; culms erect, often bent at base, 1–8 dm tall, the basal sheaths brown or reddish, soon becoming shredded; basal offshoots often emerging laterally, sometimes from the summits of the sheaths; blades soft, smooth, 0.5–3 mm wide, usually folded or involute; panicle 3–20 cm long, the branches mostly ascending; spikelets often purple-tinged; glumes unequal; lemmas lanceolate, 4–8 mm long, narrowly scarious-margined, usually with short awns; anthers 2–4 mm long. $2n = 14, 21, 28, 42, 49, 56, 63, 70$. Native and introduced. Dry open areas. Circumpolar: Alas, Yuk, Huds Bay to Lab, Nfld s to Calif, NM, Tex, SC. Map 138

F. saximontana Rydb.
Tufted, culms 1–5 (6) dm tall, glabrous; leaf-blades involute, scaberulous; panicle 2–8 (12) cm long, interrupted, yellowish green, rarely purple; spikelets 2–4 (5)-flowered; glumes lanceolate, 2–3 mm long; lemmas lanceolate to lance-attenuate, 2–4 (5) mm long, scaberulous on back, awn 1–3 mm long; anthers 1–1.5 (1.7) mm long. $2n = 42$. *F. ovina* L. ssp. *saximontana* (Rydb.) St.-Yves, *F. brachyphylla* (Schultes) ssp. *saximontana* (Rydb.) Hult. Resembling *F. brachyphylla* and perhaps a lowland variant of it, but generally distinct in our area. Grassland and open woods. Alas, Yuk, sMack, to James Bay, nQue, Nfld s to BC, Colo, sSask, sMan, Gt Lakes, sQue. Map 139

F. scabrella Torr. / Rough Fescue
Densely tufted, often as large tussocks, these enlarging by short rhizomes; culms coarse, 3–9 dm high; basal leaves firm, scabrous, involute, erect; old basal sheaths long persistent, their blades early disarticulating; panicle 5–15 cm long, the branches appressed or ascending, the spikelets toward the tips; spikelets elliptic or ovate; glumes scarious, lustrous, somewhat unequal, 5–9 mm long; lemmas firm, 7–10 mm long, scabrous, 5-nerved, the midnerve forming a keel upwards. $2n = 56$*. *F. altaica* Trin. ssp. *scabrella* (Torr.) Hult. Our few alpine records are no doubt referrable to **F. altaica** ($2n = 28$), Map 136 ○, but at present the two cannot be distinguished morphologically. Dry grassland, where a dominant component. sBC to sMan s to Ore, Ida, Mont, Colo, ND. Map 140

F. subulata Trin.
Culms slender, mostly 5–10 dm tall; blades soft, flat, lax, 3–10 mm wide; panicle open, drooping, 1.5–4 dm long, the branches spreading or reflexed, the lower as much as 15 cm long; spikelets about 1 cm long, excluding the slender awns; glumes narrow, acuminate; lemmas somewhat keeled, the tips with awns 5–20 mm long. Moist woods. sAlas to Calif e to Mont, Wyo, Utah. Map 135 ○

Glyceria R.Br. / Manna Grass

Aquatic or marsh perennials, usually tall, often rooting at the decumbent base or rhizomatous; sheaths mostly closed or connate; blades flat; inflorescence an open or contracted panicle, often large; spikelets few–many-flowered, rounded or somewhat flattened, the articulation above the glumes and between the florets; glumes unequal, shorter than the lemmas, usually scarious; lemmas broad, firm, convex on back, 5–9-nerved, the nerves parallel, rib-like and usually prominent; paleas elliptic or obovate.

1 Panicles narrow, erect; spikelets linear, nearly terete, over 7 mm long **G. borealis**
Panicles usually broad, open, nodding; spikelets ovate or oblong, somewhat flattened, 2–7 mm long 2

2 First glume usually not more than 1 mm long 3
First glume usually about 1.5 mm long 4

3 Leaf-blades mostly 2–4 mm wide; lower ligules 1.5–3.0 mm long; first glume 0.5–0.8 mm long; lemmas barely 2 mm long **G. striata**
Leaf-blades mostly 4–10 mm wide; lower ligules 3–6 mm long; first glume about 1 mm long; lemmas 2.0–2.2 mm long **G. elata**

4 Lemmas with broad scarious margins; glumes commonly obtuse **G. pulchella**
Lemmas gradually narrowed to a firm blunt apex; glumes acute **G. grandis**

G. borealis (Nash) Batchelder
Culms 5–10 dm tall, slender; blades flat or folded, usually 2–4 mm wide; panicle 1–4 dm long, narrow, the branches and pedicels appressed; spikelets mostly 6–12-flowered, 7–14 mm long, linear, nearly terete; glumes about 1.5 and 3 mm long; lemmas 3–4 mm long, obtuse and scarious at apex, strongly 7-nerved, the nerves hispidulous or scaberulous. $2n = 20$. Shallow water and

wet meadows. Alas, sYuk, wMack to James Bay, Lab, Nfld s to Calif, NM, SD, Gt Lakes. Map 141

G. elata (Nash) A.S. Hitchc.
Tufted, often in large bunches, dark green; culms 6–15 dm tall, rather succulent; blades flat, lax, 2–4 dm long, 4–10 mm wide; panicle 1.5–3 dm long, oblong, spreading, the lower branches often reflexed; spikelets 3–5 mm long; glumes about 1 and 2 mm long; lemmas 2–2.5 mm long, 7-nerved, with a hyaline apex. $2n = 20$. Stream sides and wet meadows. BC to Calif, Ariz, NM. Map 143 ○

G. grandis S. Wats. *ex* A. Gray
Culms tufted, usually stout, 10–15 dm tall; blades flat, 1.5–4 dm long, 6–12 mm wide; panicle large, 2–4 dm long, open, often nodding at summit; spikelets 5–6 mm long; glumes whitish, 1.5 and 2 mm long, acute; lemmas purplish, about 2.5 mm long, 7-nerved, narrowly ovate with a firm blunt apex. $2n = 20$. *G. maxima* (Hartm.) Holmb. ssp. *grandis* Hult. Common in wet areas. Alas, Yuk, swMack to James Bay, Nfld, s to Ore, Nev, Ariz, NM, Ia, Tenn, Va. Map 142

G. pulchella (Nash) K. Schum.
Resembling *G. grandis,* but generally smaller, being 4–6 dm tall; blades 2–6 mm wide; panicle less than 2 dm long; glumes mostly obtuse; lemmas with broad scarious margins. Ponds and ditches. Alas, Yuk, Mack, nSask to cMan s to sBC, cSask, sMan. Map 143 ●

G. striata (Lam.) A.S. Hitchc. / Fowl Manna Grass
Plants pale green, rhizomatous and often in large clusters; culms slender, 3–8 dm tall; blades flat or folded, mostly 2–4 mm wide; panicle ovoid, open, 1–2 dm long, erect or drooping at the tip; spikelets 3–4 mm long, often purplish; glumes obtuse, the first 0.5–0.8 mm long, the second 1–1.3 mm long, 1-nerved; lemmas about 2 mm long, oblong, prominently 7-nerved, obtuse and more or less scarious at the tip. $2n = 20^*$. In our area ssp. **stricta** (Scribn.) Hult. Common in shallow water and boggy meadows. sYuk, Mack to Huds Bay, Lab, Nfld s to Calif, Tex, SD, Ia, Ala, Fla. Map 144

Helictotrichon Besser

H. hookeri (Scribn.) Henr. / Hooker's Oat Grass
Prairie perennial; culms densely tufted, 2–4 dm high; leaves firm, flat or folded, mostly basal, 2–4 mm wide, the margins thickened with a whitish nerve; sheaths glabrous, keeled; ligules 2–3 mm long; panicle narrow, 4–10 cm long, the branches erect, each usually with a single spikelet; spikelets shining, brownish, with 3–6 florets, these extending beyond the glumes; rachilla villose, disarticulating above the glumes and between the florets; glumes thin, sub-equal; lemmas 10–12 mm long, firm, brown, scaberulous, the callus short-bearded, the apex with narrow teeth, awned from a little below the middle; awn 1–1.5 cm long, twisted below, geniculate, sometimes bent twice, much exceeding the spikelet; palea somewhat shorter than the lemma. $2n = 14$. *Avena hookeri* Scribn.,

Avenochloa hookeri (Scribn.) Holub. Parkland prairies and foothills. BC, swMack to sMan s to Mont, NM, ND, Minn; swYuk. Map 145

Hierochloe R.Br.

Perennial, sweet-smelling grasses, with flat leaves and small panicles of broad, bronze-coloured spikelets; spikelets articulated above the glumes; 3-flowered, the terminal floret perfect, the others staminate or neutral; glumes nearly equal, broad, very thin, glabrous, shining, obscurely 3-nerved; staminate lemmas firm, brown, boat-shaped, hairy, each with a 2-nerved palea, awned or awnless; fertile lemma similar but awnless, smooth or nearly so, and with a 1-nerved palea.

Staminate lemmas awnless or nearly so	**H. odorata**
Staminate lemmas bearing exserted awns 5–8 mm long	**H. alpina**

H. alpina (Sw.) R. & S.
Culms tufted, 1–4 dm tall, with leafy shoots at base and short rhizomes; blades of basal shoots elongate, 1–2 mm wide, those of the culms shorter and wider; panicle 2–4 cm long; spikelets 6–8 mm long, olivaceous; staminate lemmas ciliate on margins, awned below the apex, the lower lemma with a short straight awn, the second with a longer bent awn; fertile lemma mucronate, pubescent near apex. $2n = 56, 58, 63, 64–78$. Dry alpine slopes. Circumpolar: Alas to Ellesmere Is s to BC, swAlta, sMack, Huds Bay, nQue, NY, Nfld. Map 146 ○

H. odorata (L.) Beauv. / Sweet Grass
Culms 3–6 dm tall, along with leafy shoots, from creeping rhizomes; cauline leaves 2 or 3, short, lanceolate, with ligules 1–2 mm long; leaves of sterile shoots elongate, mostly 2–5 mm wide; panicle pyramidal, 4–10 cm long, somewhat compact or loose with spreading to reflexed branches; spikelets 4–7 mm long, brownish, bronze or purple; glumes ovate, 4–6 mm long; staminate lemmas about equalling the glumes, often hairy on the back, awnless or nearly so; fertile lemma hairy at apex. $2n = 28, 42, 56$. Moist to dry open areas. Circumpolar: Alas, Yuk, Mack to Huds Bay, nQue, Nfld s to Ore, Nev, Ariz, SD, Gt Lakes. Map 146 ●

Hordeum L. / Barley

Tufted annual or perennial grasses with flat blades and dense bristly spikes; spike with 3 spikelets at a node, the middle one sessile, the lateral pair pedicelled or sessile and often reduced to awns; spikelets 1-flowered (rarely 2); rachilla disarticulating above the glumes, prolonged beyond the palea as a bristle or minute floret; glumes rigid, narrow, often awl-shaped and long-awned, placed in front of the spikelet; lemma rounded on back, obscurely 5-nerved, tapering into a long awn (except in a variety of cultivated barley); palea slightly shorter than the lemma.

1 Leaf-blades with conspicuous basal auricles 2
Leaf-blades lacking conspicuous basal auricles 3

2 Glume margins, of at least the central spikelet, ciliate-pectinate; spike disarticulating **H. glaucum**
Glume margins not ciliate-pectinate, spike not disarticulating **H. vulgare**

3 Plants annual; glumes not all alike, some capillary, others 0.8–1.8 mm broad above the base **H. pusillum**
Plants perennial, glumes all capillary **H. jubatum**

H. glaucum Steud.
Annual; culms 2–5 dm tall; leaf-blades flat, 3–7 mm wide; spike 4–9 cm long (including awns), sometimes not fully exserted from the upper sheath, disarticulating; all spikelets well developed; glumes only slightly broadened below, the margins ciliate-pectinate, awn 1.5–2 cm. $2n = 14$. Roadside near Jasper the only Canadian record. Introduced. Rare.

H. jubatum L. / Foxtail Barley
Tufted perennial; culms erect or decumbent at base, 3–10 dm tall; blades 2–9 mm wide, scabrous; spikes nodding, 5–12 cm long, about as wide (including awns) at maturity, pale green to purple; rachis disarticulating at maturity, the joints falling with the spikelets attached; lateral spikelets on pedicels and reduced to 1–3 spreading awns; glumes of perfect spikelet 0.6–9.0 cm long; lemma 5–7 (10) mm long, with an awn as long as the glumes. $2n = 28$ (14, 42). In our area spp. **breviaristatum** Bowden (*H. brachyantherum* Nevski of Ed. 1) with glumes 0.5–1.8 cm long and lemma awns of the central spikelets 0.4–1.2 cm long, ssp. **jubatum** with glumes 3.5–9.0 cm long, and ssp. × **intermedium** Bowden (*H. jubatum* var. *caespitosum* (Scribn.) A.S. Hitchc. of Ed. 1), a hybrid of the foregoing ssp. with glumes 1.0–3.6 cm long and lemma awns of central spikelets 1.0–3.5 cm long. Fields, waste ground, and roadsides. Asia: Alas, Yuk, Mack to Huds Bay, Lab, Nfld, generally distributed s. Map 147

H. pusillum Nutt. / Little Barley
Annual; culms 1–4 dm tall; leaves flat, 2–4 (5) mm wide; spike 3–7 cm long (including awns); glume bases of central spikelet and inner glumes of lateral spikelets broadened (0.8–1.8 mm) for about the length of the florets, narrowed above to capillary awns 10–15 mm long, outer glumes entirely capillary; lemma of central spikelet 10–15 mm long, including awn, of the lateral spikelets awn-tipped. $2n = 14$ (28). Grassland and waste ground. Wash to sAlta, generally distributed s. Map 148

H. vulgare L. / Barley
Annual; culms erect, 4–12 dm tall; blades mostly 5–15 mm wide; spike 2–10 cm long, excluding awns; lateral spikelets sessile, well developed and usually fertile; glumes flat, linear, tapering into an awn; lemma about 1 cm long, tapering into a straight erect awn 6–15 mm long. $2n = 14$. Cultivated. Occasional in fields and along roadsides but not persistent. Beardless barley, var. **trifurcatum** (Schlecht.) Alefeld, has much reduced awns, commonly 3-cleft, with the middle division

in the form of a hooded lobe. Two-row barley, **H. distichon** L., with only the central spikelet fertile is also an occasional escape from cultivation.

Koeleria Pers.

K. macrantha (Ledeb.) J.A. Schultes f. / June Grass
Tufted perennial; culms mostly 2–5 dm tall, puberulent below the panicle, leafy at base; sheaths often retrorsely pubescent or scabrous; blades flat or involute, 1–3 mm wide, glabrous or pubescent; ligule 0.5–1 mm long; inflorescence a spike-like panicle, 2–8 cm long, shining, pale green or purplish, dense, often interrupted below; spikelets mostly 4–5 mm long, 2–4-flowered, compressed, the rachilla disarticulating above the glumes and between the floret, prolonged beyond the perfect florets; first glume 2–3 mm long, 1-nerved; second glume 3–4 mm long, 3–5-nerved, about as long as the spikelet; lemmas rounded on the back, somewhat scarious, obscurely 5-nerved, acute or short-awned, the awn borne just below the apex; palea hyaline. $2n = 14, 28$. *K. cristata* (L.) Pers. of Ed. 1, *K. nitida* Nutt. Common in prairie grassland throughout the region. Circumpolar, boreal: BC, sMack to sOnt, sQue, Me s to Calif, Tex, La; sYuk. Map 149

Lolium L. / Ryegrass; Darnel

Annual or perennial; inflorescence a simple spike, the spikelets sessile and solitary at each node, set edgewise on the rachis; spikelets several-flowered; the first glume absent in lateral spikelets, the second equalling or exceeding first floret, 5–9-nerved; lemmas 5-nerved, awned or awnless; palea about equalling lemma.

1	Glume shorter than spikelet	2
	Glume (of at least some spikelets) as long as or longer than spikelet	3
2	Lemma awnless; leaves seldom more than 3 mm wide	**L. perenne**
	Lemma awned; leaves 3–10 mm wide	**L. multiflorum**
3	Lemma 9–10 mm long; leaves 2–4 mm wide	**L. persicum**
	Lemma 6–8 mm long; leaves 3–10 mm wide	**L. temulentum**

L. multiflorum Lam. / Italian Ryegrass
Similar to *L. perenne*, but annual or biennial and generally somewhat larger; distinguished by the characters of the key. $2n = 14$. Lawns, roadsides, and waste places. Introduced. Occasional.

L. perenne L. / Ryegrass
Tufted perennial; culms 2–8 dm tall; leaves 2–4 mm wide, flat, folded or involute; inflorescence 5–25 cm long; spikelets mostly 10–15 mm long, (3)5–10-flowered; glume 7–10 mm long, lanceolate, longer than first floret, shorter than spikelet; lemmas acute, scabrous below, awnless. $2n = 14$. Lawns, roadsides, waste places. Introduced. Rare.

L. persicum Boiss. & Hohen. / Persian Darnel
Annual; culms 1.5–5 dm tall; leaves 2–4 mm wide, scabrous above, smooth below; inflorescence 3–10 (15) cm long; spikelets somewhat dorsally compressed, 10–20 mm long, 5–7-flowered; glumes mostly 10–20 mm long, almost equalling to exceeding the spikelet; lemma 9–10 mm long, awn about equalling length of lemma. $2n = 14$. Fields, roadsides, and waste places. Introduced. Occasional.

L. temulentum L. / Darnel
Annual; culms 3–10 dm tall, scabrous; leaves flat, 3–10 mm wide; inflorescence 10–25 cm long; spikelets 15–25 mm long, mostly 5–8-flowered; glumes equalling or exceeding the spikelet; lemma 6–8 mm long, usually awned; awn 6–12 mm long. $2n = 14$. Roadsides and waste places. Introduced. Rare.

Melica L. / Melic Grass

Tall perennials, with rhizomes; culm often with a bulbous corm at the base; leaves usually flat, the sheaths closed for most of their length; inflorescence a narrow panicle; spikelets relatively large, 2–several-flowered; rachilla disarticulating above the glumes and between the florets (in our species); terminal florets usually reduced and several folded together; glumes unequal, thin, with scarious margins, 3–5-nerved; lemmas rounded on back, several-nerved, scarious-margined, awnless or sometimes awned from between the teeth of the bifid apex; palea shorter than the lemma.

1	Lemmas awned; plants not bulbous at base	**M. smithii**
	Lemmas not awned; plants mostly bulbous at base	2
2	Lemmas narrow, long-acuminate, hairy on the nerves; glumes narrow	**M. subulata**
	Lemmas broad, acute or obtuse, glabrous; glumes broad, papery	**M. spectabilis**

M. smithii (Porter) Vasey / Melic Grass
Culms 5–10 dm tall, slender; sheaths retrorsely scabrous; blades lax, scabrous, 6–12 mm wide; panicle 1–2.5 dm long, the branches solitary, distant, spreading or reflexed; spikelets 16–20 mm long, often purplish; first glume 4–6 mm long, obscurely 3-nerved; second glume 5–8 mm long, 5-nerved; lemmas about 10 mm long, strongly nerved, with an awn 3–5 mm long from the bifid apex. Woodlands. sBC to Ore, Mont, Wyo; Gt Lakes. Map 150

M. spectabilis Scribn. / Onion Grass
Culms 3–6 dm tall, bulbous at base, its bulb-like corm attached to the rhizome by a short, slender stalk; sheaths and blades smooth or scabrous, the blades 2–4 mm wide, usually flat; panicle 6–20 cm long, narrow, the branches short, rather stiff, appressed; spikelets 7–15 mm long, rather papery in age, purple-tinged; glumes scarious, the first 4–6 mm long, the second 6–8 mm long; lemmas 6–8 mm long, 7–11-nerved, scabridulous on back, the apex obtuse or slightly emarginate, scarious. $2n = 18$. Woodlands. sBC to Calif e to Mont, Wyo, Colo. Map 151

M. subulata (Griseb.) Scribn. / Alaska Onion Grass
Culms 5–10 dm high, mostly bulbous at base; sheaths retrorsely scabrous, often long-hairy; blades thin, mostly 2–5 mm wide, scabrous; panicle 1–2 dm long, usually narrow; spikelets slender, 15–20 mm long; glumes narrow, obscurely nerved; lemmas about 12 mm long, tapering-acuminate, pilose on the nerves. $2n = 18, 27$. Woodlands. sAlas to Calif e to Mont, Wyo. Map 152

Muhlenbergia Schreb. / Muhly Grass

Wiry perennials of small to medium size, mostly rhizomatous; culms simple or branched, the inflorescence a slender (sometimes spike-like) or open panicle; spikelets small, 1-flowered (occasionally 2-flowered in *M. asperifolia*); rachilla disarticulating above the glumes; glumes awl-shaped to ovate, varying from very small to equalling the lemma, usually 1-nerved; lemmas rather thin, rounded on the back, usually 3-nerved, acute to awl-shaped, mucronate or awned.

1 Panicle open, 4–12 cm wide; spikelets on long slender pedicels **M. asperifolia**
Panicle narrow, spike-like; spikelets on short pedicels 2

2 Glumes awned, the awn much exceeding the lemma; panicle usually dense, 0.5–1.5 cm wide 3
Glumes shorter than to almost equalling the lemma; panicle very slender with short appressed branches 4

3 Internodes puberulent, dull; ligules 0.2–0.6 mm long; anthers 0.8–1.5 mm long **M. glomerata**
Internodes smooth, shiny; ligules 0.6–1.5 mm long; anthers 0.4–0.8 mm long **M. racemosa**

4 Glumes lance-subulate, mostly 2–2.5 mm long; ligule 0.5 mm long or less; rhizomes lacking **M. cuspidata**
Glumes ovate-lanceolate, 1–1.5 mm long; ligule 1–3 mm long; rhizomes present **M. richardsonis**

M. asperifolia (Nees & Mey.) Parodi / Scratch Grass
Rhizomatous and forming patches; culms caespitose, branching at base and spreading, slender, 1–4 dm long; sheaths somewhat keeled; ligule erose-toothed; blades flat, short, crowded, harsh; panicle diffuse, 5–15 cm long and about as wide, with slender branches, at maturity breaking away; spikelets 1.5–2 mm long, sometimes 2-flowered; glumes acute; lemma thin, broad, exceeding the glumes. $2n = 20, 22, 28$. Moist alkaline soil. sBC to sMan s to Calif, Ariz, NM, Tex, Mo, Ill. Map 153

M. cuspidata (Torr.) Rydb. / Plains Muhly
Culms 2–4 dm tall, tufted, wiry, erect, from a knotted or bulbiform caudex; ligule erose-truncate, 0.5 mm long or less; blades 1–2 mm wide, flat, or involute on drying; panicle slender, loose, 5–10 cm long, the branches appressed; spikelets about 3 mm long; glumes about two-thirds as long as the spikelet; lemma acuminate-cuspidate, scaberulent or minutely pubescent. The grain is often parasitized by an insect which causes it to develop into a hard yellow rounded body 1 mm thick. Dry slopes and banks. Alta to Man s to NM, Okla, Mo, Ky. Map 154

M. glomerata (Willd.) Trin. / Bog Muhly
Rhizomes hard, scaly, creeping; culms erect, slender, 3–6 dm tall, simple or with a few erect branches at the base, the nodes and internodes puberulent, or the internodes glabrous but roughish to the fingernail; blades erect, firm, flat, scabrous; sheaths scarcely keeled; ligule minute; panicle greenish or purplish, mostly 4–7 cm long and 0.5–1 cm wide, usually interrupted at base, the lower branches somewhat remote; spikelets 5–6 mm long; glumes narrowly lanceolate, tapering to long scabrous awns; lemma much exceeded by the awns of the glumes, pilose below the middle. $2n = 20$. Bogs, peaty meadows, and wet shores. BC, sYuk, sMack to Nfld s to Nev, Utah, Colo, Neb, Ia to WVa. Map 155

M. racemosa (Michx.) BSP.
Differing from *M. glomerata* in the characters of the key, and in its somewhat stouter culms, thicker panicle, keeled sheaths and the lemma pilose only basally. $2n = 40$. Usually on sand-hills, dry slopes, and eroded banks. BC to wOnt s to Wash, Ida, Utah, NM, Tex, Mo, Gt Lakes. Map 156

M. richardsonis (Trin.) Rydb. / Mat Muhly
Forming a dense mat from numerous creeping rhizomes; culms wiry, nodulose-roughened, erect or spreading at base, 1–4 dm high; ligule 1–3 mm long, acute; blades 0.5–1 mm wide, involute; panicle slender, interrupted or sometimes rather close and spikelike, 1–6 cm long; spikelets 2–3 mm long; glumes ovate or ovate-lanceolate, about half as long; lemma tightly rolled; lanceolate, acute, mucronate. $2n = 40$. Prairie grassland, thickets, and shores. BC, Yuk, sMack to NB s to Calif, Ariz, NM, Neb, Mich, Me. Map 157

Munroa Torr.

M. squarrosa (Nutt.) Torr. / False Buffalo Grass
Low tufted annual, prostrate and mat-forming; culms spreading, branching freely, often rooting at the nodes, the leaves fascicled and the internodes scabrous; leaf-blades stiff, 1–3 cm long, 1–3 mm wide, the margin scabrous with a whitish band; sheaths crowded, the collar pilose on the margins; spikelets clustered among the leaves, crowded in 2's or 3's, the lower 3–4-flowered, the upper 2–3-flowered; glumes of lower spikelets equal, 1-nerved, narrow, acute, those of the upper spikelet unequal, the first short or obsolete; lemmas 3-nerved, those of the lower spikelet coriaceous, acuminate-mucronate, those of the upper spikelet membranaceous; palea narrow. $2n = 16$. Dry plains, slopes, and disturbed areas. sAlta, sSask s to Calif, Nev, Utah, Tex. Map 158

Oryzopsis Michx. / Rice Grass

Tufted perennials; leaf-blades flat or involute; inflorescence a narrow or open panicle; spikelets 1-flowered, disarticulating above the glumes; glumes equal or nearly so, rather broad, about as long as the spikelet; lemma convolute, indurated, usually pubescent, with a terminal deciduous awn; palea enclosed by edges of lemma.

1 Panicle diffuse, dichotomously angular-branched; glumes stiff-pointed; lemma long silky pubescent **O. hymenoides**
Panicle narrow; glumes blunt to soft apiculate; lemma short-pubescent to nearly glabrous 2

2 Leaves flat; spikelets, excluding awn, 6–8 mm long **O. asperifolia**
Leaves involute; spikelets, excluding awn, 3–5 mm long 3

3 Lemma glabrous or nearly so **O. micrantha**
Lemma distinctly pubescent 4

4 Awn 0.5–2 mm long, straight or nearly so; lemmas obtuse **O. pungens**
Awn longer, flexuous; lemmas acute 5

5 Panicle 3–6 cm long, branches appressed; awn 4–6 mm long, geniculate **O. exigua**
Panicle 5–15 cm long, branches ascending or spreading; awn 6–12 mm long, twisted **O. canadensis**

O. asperifolia Michx.
Culms loosely tufted, erect or wide-spreading, 2–7 dm long, nearly naked, the 2 or 3 sheaths bearing reduced or obsolete blades; basal blades erect, firm, flat or somewhat rolled, scabrous, especially on the glaucous lower side; panicle 5–8 cm long, narrow, nearly simple, the branches erect; spikelets 6–8 mm long, excluding awn; glumes elliptic to obovate, about 7-nerved, green with scarious margins and tips, abruptly pointed or apiculate; lemma about as long as the glumes, sparingly pubescent, densely hairy on callus, the awn 5–10 mm long; lodicules large, about 4 mm long. $2n = 46, 48$. Banks and wooded slopes. BC, sMack to Nfld s to Ida, Utah, NM, SD, Gt Lakes. Map 159

O. canadensis (Poir.) Torr.
Culms loosely tufted, 3–7 dm tall, erect, slender; ligule about 2 mm long; blades flat to involute, scabrous; panicle open and lax, 5–10 cm long, with flexuous branches, ascending or spreading; glumes 4–5 mm long, very thin, slightly exceeding the dark, pubescent lemma; awn of lemma 6–12 mm long, twisted and weakly geniculate. $2n = 22$. Open woods and hillsides. BC to Nfld s to Gt Lakes, WVa. Map 160 ●

O. exigua Thurb.
Culms densely tufted, 1–3 dm tall, stiffly erect, scabrous; blades 5–10 cm long, erect, filiform, involute; panicle 3–6 cm long, slender, the branches appressed; glumes 4 mm long, acute; lemma about as long as the glumes, appressed-hairy, the awn about 5 mm long, geniculate. $2n = 22$. Dry open ground or thin woods to subalpine elevations. sBC to Ore, Nev, Utah, Colo. Map 160 ○

O. hymenoides (R. & S.) Ricker / Indian Rice Grass
Culms tufted, stout, 3–7 dm high; leaves nearly as long as the culms, slender, stiff; ligules 3–7 mm long, pointed; panicle 1–3 dm long, diffuse, the many slender branches in pairs, the branchlets forked, wide-spreading and flexuous at ma-

turity; spikelets solitary at the tips of long pedicels; glumes 6–7 mm long, papery, abruptly pointed, hairy to glabrous; lemma about 3 mm long, brown or blackish when mature, densely long-pilose with white hairs, the awn 3–6 mm long, soon deciduous. $2n = 28, 48, 65, 130$. Produces a sterile hybrid with *Stipa viridula*, × *Stiporyzopsis caduca* (Beal) Johnson & Rogler. Sand-dunes, dry banks, rocky slopes. sBC to swMan s to Calif, Ariz, NM, Tex. Map 161 ●

O. micrantha (Trin & Rupr.) Thurber
Densely tufted; culms erect, 3–8 dm tall; leaf-blades 1–2 mm wide, strongly involute; ligules about 1 mm long; panicle 5–15 (20) cm long, erect or nodding; spikelets towards the branch ends; glumes subequal, 2–3 mm long, glabrous or scabridulous; lemmas 1.8–2.8 mm long, glabrous, becoming indurate; awn 4–11 mm long, straight or weakly geniculate. $2n = 22$. Dry open areas and rocky slopes. BC to sMan s to Calif, Nev, Ariz, NM, Tex. Map 162

O. pungens (Torr.) A.S. Hitchc.
Similar to *O. exigua*, differing as shown in the key, and being generally taller (up to about 5 dm), with leaves not quite so slender. $2n = 22, 24$. Open sandy and gravelly places. sYuk, sMack to Huds Bay, nQue, NS s to BC, Wyo, Colo, SD, Gt Lakes. Map 161 ○

Panicum L. / Panic Grass

Annual or perennial of various habit, the spikelets in open or compact panicles; leaf-blades flat; ligules mostly of hairs; spikelets compressed dorsiventrally, with 2 florets; glumes usually very unequal, the first often minute, the second about as long as the sterile lemma; the latter simulating a third glume but enclosing a palea and sometimes a staminate flower; fertile lemma indurate, usually white or pale and smooth or shining, typically obtuse, nerveless, the margins inrolled over the palea.

Spikelets less than 4 mm long	**P. capillare**
Spikelets more than 4 mm long	**P. miliaceum**

P. capillare L. / Witch Grass
Annual; culms 1–6 dm high, erect or spreading at base, copiously papillose-hispid to nearly glabrous; blades 1–2.5 dm long, 5–15 mm wide, hispid on both surfaces; ligule almost entirely of hairs; panicle often half the length of the plant, very diffuse, included at base until maturity, the branches finally wide-spreading, eventually breaking off as a 'tumble weed'; spikelets 2–3 mm long, glabrous; first glume about half as long as spikelet; second glume and sterile lemma about the spikelet length, 7-nerved, fertile lemma smooth, shining, acute. $2n = 18, 36, 54$. A weed of fields and waste places. Introduced. Occasional.

P. miliaceum L. / Broomcorn Millet
Coarse annual; culms stout, 2–10 dm tall, erect or decumbent at base; blades 7–30 cm long, 5–20 mm wide, pubescent or sometimes glabrous; panicle 1–3 dm

long, rather compact, usually somewhat included at base, nodding at maturity; spikelets ovoid, acuminate, 4.5–5 mm long, glabrous; glumes and sterile lemma strongly nerved; fertile lemma elliptic, about 3 mm long, smooth and shining. $2n = 36, 40, 42, 49, 54, 72$. An escape from cultivation. Introduced. Occasional.

Phalaris L. / Canary Grass

Annual or perennial; tall plants with numerous flat blades, large ligules and dense or spike-like panicles; spikelets large, laterally compressed, with 1 perfect terminal floret and 2 sterile lemmas below; rachilla disarticulating above the glumes, all the florets falling attached; glumes about equal, keeled, often winged on the keel, longer than the lemmas; sterile lemmas small, appearing like scales attached to the fertile floret; fertile lemma firm, shining in fruit, enclosing the palea.

Annual; mostly 3–8 dm tall; panicle ovoid, dense, 1.5–4 cm long **P. canariensis**
Perennial; mostly 6–15 dm tall; panicle narrow, 6–18 cm long **P. arundinacea**

P. arundinacea L. / Reed Canary Grass
Robust perennial with creeping rhizomes; culms glaucous, 6–15 dm tall; leaf-blades 1–3 dm long, 6–15 mm wide; ligules 2–5 mm long; panicle 6–18 cm long, narrow, dense to somewhat open and lobed; spikelets 5–6 mm long, pale or suffused with purple; glumes about 5 mm long, narrow, the keel scabrous, very narrowly winged; fertile lemma 3–4 mm long; sterile lemmas minute hairy scales. $2n = 14, 27–35, 42, 56$. A form with white-striped leaves, var. **picta** L., is often grown as an ornamental. Shores and marshes. More or less circumpolar: Alas, sYuk, sMack to James Bay, Nfld, generally distributed s. Map 163

P. canariensis L. / Canary Grass
Annual, 3–8 dm high; blades short, 4–7 mm wide; panicle 1.5–4 cm long, ovate to oblong-ovate, dense; spikelets broad, 5–6 mm long, imbricated, pale with green stripes; glumes 7–8 mm long, broad, whitish, the green keel with a prominent pale wing; fertile lemma about 5 mm long, appressed-hairy; sterile lemmas about 2.5 mm long. $2n = 12, 48$. Roadsides and waste ground. Introduced. Occasional.

Phleum L. / Timothy

Perennials, more or less tufted, with flat leaves and dense spike-like panicles; spikelets 1-flowered, flattened, usually disarticulating above the glumes; glumes equal, keeled, ciliate on keel, short-awned; lemma shorter than the glumes, broadly truncate, hyaline, 3–5-nerved; palea nearly as long as the lemma, narrow, hyaline.

Panicle ovoid or short-cylindric, usually 1.5–3 cm long; awns of glumes often over 2 mm long; base of culms not bulbous; upper sheaths inflated **P. commutatum**
Panicle cylindric, usually 5–10 cm long; awns of glumes rarely over 2 mm long; base of culms enlarged or bulbous; sheaths not inflated **P. pratense**

P. commutatum Gaudin / Mountain Timothy
Culms solitary or in small tufts, 1–5 dm tall, from a decumbent, somewhat creeping base; sheaths glabrous, the upper and middle sheaths dilated; blades 2–10 cm long; ligules 1.5–3 mm long; panicle and spikelets as described above. $2n = 14, 28^*, 56$. *P. alpinum* L. of Ed. 1. Alpine meadows, forest borders, and open slopes. Circumpolar: Alas, Yuk, swMack s to Calif, Ariz, NM; James Bay to nQue, Nfld s to NH; Lk Superior. Map 164

P. pratense L. / Timothy
Culms tufted, 4–9 dm tall, erect from a corm-like base, minutely scabrous above; upper sheaths terete, close, not dilated; principal blades 5–20 cm long; ligules 2–4 mm long; panicle and spikelets as described above. $2n = 14, 21, 35, 42, 56, 63, 70, 84$. Grown for hay and pasture and escaping along roadsides and trails. Introduced. Common.

Phragmites Trin.

P. australis (Cav.) Trin. *ex* Steud. / Reed
Very large perennial with stout creeping rhizomes, and often forming long stolons; culms 1.5–3 m tall, stout, leafy, the sheaths overlapping; blades flat, 1–3 cm wide, glabrous; inflorescence a large panicle 1–4 dm long, dense, the branches ascending to nodding, tawny or purplish; spikelets large, several-flowered, 12–15 mm long; rachilla pubescent with long silky hairs, disarticulating above the glumes and between the lemmas; glumes unequal, lanceolate; lemmas narrow, long-acuminate, 8–12 mm long, 3-nerved, glabrous; palea much shorter than the lemma. $2n = 36, 48, 54, 72, 84, 96$. *P. communis* Trin. of Ed. 1. Marshes and lakes. Circumpolar, boreal: BC to NS, generally distributed s. Map 165

Poa L. / Bluegrass

Low or rather tall slender grasses, mostly perennials; leaf-blades relatively narrow, flat, folded, or involute, ending in a boat-shaped tip; inflorescence an open or narrow panicle; spikelets 2–several-flowered, the uppermost floret reduced or rudimentary; rachilla disarticulating above the glumes and between the florets; glumes keeled, somewhat unequal, 1–3-nerved; lemmas definitely keeled to rounded on the back, acute or acutish, rarely obtuse, awnless, often scarious-tipped, 3–5-nerved, glabrous to pubescent, sometimes with a tuft of cobwebby hairs at the base; palea as long as the lemma or shorter.

1	Creeping rhizomes present	2
	Creeping rhizomes lacking	9
2	Culms conspicuously flattened	**P. compressa**
	Culms terete or slightly flattened	3
3	Lemmas glabrous or scabrous, not pubescent	4
	Lemmas usually with crisp-puberulent or villose hairs at least on the nerves below	5

4 Lemmas keeled and flattened; rhizomes long; panicle open **P. nervosa**
Lemmas rounded on back, the keel obscure; rhizomes, when present, short; panicle narrow **P. juncifolia**

5 Base of lemma webbed (a cluster of crinkly hairs) 6
Base of lemma not webbed 7

6 Lemmas less than 4 mm long, not pubescent on the internerves near base **P. pratensis**
Lemmas 4 mm or more long, pubescent on the internerves below, as well as on the nerves; alpine **P. arctica**

7 Panicle narrow, the branches ascending; leaves folded or involute, rather stiff **P. arida**
Panicle open, the branches usually spreading at maturity; leaves flat, or sometimes folded but not stiff 8

8 Internerves of lemmas glabrous, the keel and marginal nerves with or without pubescence **P. nervosa**
Internerves of lemmas pubescent near base (as well as the keel and marginal nerves) **P. arctica**

9 Annual species; panicle open; culms usually less than 2 dm tall; leaf-blades wrinkled **P. annua**
Perennials 10

10 Lemmas webbed at base (a cluster of crinkly hairs) 11
Lemmas not webbed at base 15

11 Lemmas less than 3 mm long; spikelets less than 4.5 mm long 12
Lemmas more than 3 mm long; spikelets over 4.5 mm long 14

12 Panicle usually more than 1 dm long; leaf-blades usually more than 8 cm long; culms often decumbent at base; ligule 1–5 mm long **P. palustris**
Panicle usually less than 1 dm long; leaf-blades usually less than 8 cm long; culms erect at base; ligule mostly about 1 mm long 13

13 Glumes acuminate, about equalling the first lemma; ligule 0.3–0.6 mm long **P. nemoralis**
Glumes acute, shorter than the first lemma; ligule 0.3–1 (2.0) mm long **P. interior**

14 Panicle narrow and condensed, 1–4 cm long; culms 1–2 dm tall, not much longer than the basal leaves **P. pattersonii**
Panicle open, the branches spreading, 5–10 cm long, nodding; culms 2–5 dm tall, much longer than the basal leaves **P. leptocoma**

15 Lemmas glabrous or scabrous, not pubescent 16
Lemmas with crisp-pubescent, silky or villose hairs, at least near the base 20

16 Spikelets flattened, the lemmas definitely keeled; panicle less than 8 cm long 17
Spikelets little flattened, the lemmas rounded on the back; panicle over 8 cm long 19

17 Plants 5–10 cm tall; lemmas smooth **P. lettermanii**
Plants taller; lemmas generally scabrous 18

18 Leaf-blades filiform, scabrous **P. cusickii**
Leaf-blades flat, 1–3 mm wide, smooth; alpine **P. epilis**

19 Ligule decurrent, 1.5–6.5 mm long, acuminate to acute **P. nevadensis**
Ligule not decurrent, 1–2 mm long, round to truncate **P. juncifolia**

20 Panicle open, the lower branches spreading and naked below 21
Panicle narrow, contracted, the branches appressed or ascending 23

21 Panicle about as broad as long; spikelets broad, the bases cordate or subcordate; leaf-blades seldom over 5 cm long **P. alpina**
Panicle longer than broad; spikelets not so broad, the bases not at all cordate; leaf-blades longer 22

22 Spikelets little flattened, the lemmas rounded on back, not keeled **P. gracillima**
Spikelets flattened, the lemmas V-shaped and keeled **P. stenantha**

23 Spikelets little flattened, the lemmas rounded on back, not keeled 24
Spikelets flattened, the lemmas keeled 25

24 Panicles less than 1 dm long; culms seldom over 3 dm tall; leaves less than 1 dm long; lemmas 3.5–4 mm long **P. sandbergii**
Panicles longer; culms over 3 dm tall; leaves over 1 dm long; lemmas 4–5 mm long **P. canbyi**

25 Panicle about as wide as long; spikelets broad, cordate or subcordate at base **P. alpina**
Panicle longer than broad; spikelets not at all cordate at base 26

26 Culms not much longer than basal leaves; panicle dense, branches ascending; lemmas 3.5–4.5 mm long **P. pattersonii**
Culms clearly longer than basal leaves, panicle open, branches spreading; lemmas 2.5–3.5 mm long **P. glauca**

P. alpina L. / Alpine Bluegrass
Culms erect, 1–3 dm tall, from a rather thick crown; leaves mostly on short basal shoots, the blades short, 2–4 mm wide; cauline leaves 1 or rarely 2, erect; panicle pyramidal or ovoid, rather compact, the lower branches spreading or reflexed; spikelets broad, purplish or bronze; glumes broad, ovate, abruptly acute; lemmas 3–4 mm long, strongly hairy on the keel and marginal nerves. $2n = 14$, 21–(39*)–74. Subalpine to alpine meadows and rocky slopes. Circumpolar: Alas, Yuk to Baffin Is s to Ore, Utah, Colo, sMack, nSask, sMan, Gt Lakes, Nfld. Map 166

P. annua L. / Annual Bluegrass
Tufted annual, often forming mats, bright green; culms 5–20 cm long, decumbent or ascending, sometimes rooting at the lower nodes; blades soft, 1–3 mm wide; sheaths loose; panicle narrow, 2–7 cm long; spikelets 3–5 mm long; glumes broadly lanceolate; lemmas elliptic, thin, distinctly 5-nerved, obtuse, more or

less hairy on the lower parts of the nerves, the long hairs on the lower part of the keel sometimes simulating a web. $2n = 28$. A weed of lawns and gardens. Introduced. Common.

P. arctica R.Br.
Forming rhizomes; culms in small tufts, 1–3 dm tall, decumbent at base; blades mostly basal, flat or folded; ligules 1–4 mm long, pointed or erose-truncate; panicle 5–10 cm long, open, pyramidal, the lower branches often in pairs, slender, wide-spreading; spikelets 4–6 mm long; lemmas about 4 mm long, usually purplish and suffused with orange below the scarious apex, densely hairy on keel and nerves, often webbed at the base. $2n = 28$–106. Includes *P. longpila* Nash of Ed. 1. Alpine meadows. Circumpolar: Alas, Yuk to Ellesmere Is s to Calif, Utah, NM, sMack, James Bay, Nfld. Map 167

P. arida Vasey / Plains Bluegrass
Rhizomatous; culms tufted, 2–5 dm tall, usually with a single short leaf below the middle; leaves mostly basal, firm, folded; panicle narrow, compact, 2–10 cm long, the branches ascending; spikelets 5–7 mm long; lemmas 3–4 mm long, densely hairy on keel and marginal nerves and on the lower parts of the intermediate nerves. $2n = 63$–(c.76*)–103. Dry plains and saline meadows. BC to sMan s to Utah, NM, Tex, Ia. Map 168

P. canbyi (Scribn.) Piper
Culms tufted, 5–10 dm tall; blades flat to involute; ligules 2–5 mm long; panicle narrow, dense or rather loose, 0.5–1.5 dm long, the branches mostly short and appressed; spikelets little flattened, much longer than wide, sometimes pale and shining, 3–5-flowered, glumes acuminate or acute, 3–5 mm long; lemmas scarcely keeled, obscurely nerved, minutely hairy on lower part of back. $2n = 56$*, 70*, c.72, 82–(c.98*)–106. Flowers in summer. Dry to moist grassland. BC to Sask s to nCalif, Nev, Colo, Neb, Minn, Mich; Alas, Yuk; Lk Superior; Gaspé. Map 169

P. compressa L. / Canada Bluegrass
Culms from long rhizomes, erect, decumbent at base, 2–7 dm tall, wiry, strongly flattened, often blue-green; blades mostly short, 1–4 mm wide; panicle narrow, 2–8 cm long, the branches usually in pairs, short and bearing spikelets nearly to the base; spikelets 4–6 mm long; glumes 2–3 mm long; lemmas firm, 2–3 mm long, slightly hairy towards the base, with or without a web at the base. $2n = 14$, 35, 39, 42, 45–56. Meadows and waste ground. Introduced. Occasional.

P. cusickii Vasey
Densely tufted, the culms 2–6 dm tall; often dioecious; blades filiform, scabrous; ligules very short; panicle narrow, oblong, contracted or somewhat open at anthesis, purplish, pale or tawny; spikelets 7–9 mm long; lemmas 4–6 mm long, flattened and keeled, glabrous or scabrous. $2n = 28$, 42, 56*, 59*. Dry grassland and sand-hills. BC to swMan s to Calif, Nev, Colo, ND, introduced Alas, Yuk. Map 170

P. epilis Scribn.
Similar to *P. cusickii*; differing as shown in the key and having longer ligules (about 3 mm); also said to be pistillate only. $2n$ = 56, 84. Subalpine to alpine meadows. BC to Calif, Nev, Utah, Colo. Map 171

P. glauca Vahl
Densely tufted, often glaucous; culms stiff, flattened, 1–6 dm tall, usually with 3 nodes, the uppermost leaf borne below the middle of the culm; blades mostly basal; ligule about 2 mm long; panicle 3–7 cm long, narrow, lax and open to rather dense; spikelets bluish green to purplish, 2–3-flowered, lemmas 3–4 mm long, pubescent on the lower half of the keel and on the marginal nerves, with no web of hairs at base. $2n$ = 42–78. A complex species and often not readily separated from *P. interior*. Includes *P. rupicola* Nash of Ed. 1. Rocky slopes at subalpine to alpine elevations. Circumpolar: Alas, Yuk to Ellesmere Is s to Calif, Nev, Utah, NM, sMack, nSask, Gt Lakes, Nfld. Map 172

P. gracillima Vasey
Rather loosely tufted, the culms 3–6 dm tall, often decumbent at base; blades lax, flat or folded, the basal ones mostly filiform; ligules 2–5 mm long on culm leaves; panicles pyramidal, 4–10 cm long, the lower branches spreading, sometimes reflexed, naked below; spikelets 4–6 mm long; lemmas about 4 mm long, rounded on back, minutely scabrous, crisp-pubescent near base, especially on the nerves, not webbed. $2n$ = 81, 84, 86. Moist montane woods and meadows. BC, Mont s to Calif, Utah, Colo. Map 176 ○

P. interior Rydb.
Culms 2–5 dm tall, usually tufted, stiffly erect, slender; blades 1–2 mm wide; ligules usually less than 1 mm long; panicle narrow, 4–10 cm long, the branches ascending or occasionally spreading; spikelets somewhat crowded, 2–4-flowered glumes ovate-lanceolate; lemmas 2–3 mm long, pubescent on the lower part, webbed at the base or sometimes apparently webless. $2n$ = 28*, 42*, 56*. *P. nemoralis* L. var. *interior* (Rydb.) Butters & Abbe. Plains and open woods. BC, sMack to sMan s to Wash, Ariz, NM, Tex, SD, Gt Lakes. Map 173

P. juncifolia Scribn. / Alkali Bluegrass
Culms tufted, 5–10 dm tall; blades rather stiff, involute, flat or folded, 1–3 mm wide, smooth; panicle dense, 1–2 dm long, narrow, the branches appressed; spikelets 3–7-flowered, 7–10 mm long; lemmas about 4 mm long, rounded on the back, glabrous or scabrous near base, not webbed. $2n$ = 62*–65, 70, 78, 84, 96, c.100. Includes *P. ampla* Merr. of Ed. 1. Moist open ground. BC to Sask s to Calif, Nev, Ariz, NM, SD; Yuk. Map 174

P. leptocoma Trin. / Bog Bluegrass
Culms slender, 2–5 dm tall, solitary or few in a tuft, often decumbent at base; blades flat, lax, 2–4 mm wide; ligules 3–4 mm long on upper leaves; panicle 5–10 cm long, nodding, the branches spreading; spikelets narrow, pointed, about 6 mm long; lemmas about 4 mm long, acuminate, compressed-keeled, pubescent

on keel and marginal nerves or sometimes nearly glabrous, webbed at the base. $2n = 42$. Includes *P. paucispicula* Scribn. & Merr. Bogs and wet meadows. eAsia: Alas, Yuk, swMack s to Calif, Mont, NM. Map 175

P. lettermanii Vasey
Densely tufted perennial; culms 3–12 cm tall, not much longer than basal leaves; blades mostly flat, 1–2 mm wide; ligules 0.5–2.0 mm long, truncate; panicle narrow, 1–2 (3) cm long; spikelets compressed, purple-tinged, 2.5–4 (4.5) mm long, 2–5-flowered; glumes subequal, 2–4 mm long, almost equalling spikelet, somewhat scabrous on keel; lemmas 2–3 (3.5) mm long, glabrous, smooth; anthers about 0.5 mm long. Exposed alpine ridges and summits. BC to Calif, Nev, Utah, Colo. Map 176 ●

P. nemoralis L.
Similar to *P. interior* Rydb.; differing by the characters of the key. $2n = 28–49$, 70. Moist meadows. Introduced. Rare.

P. nervosa (Hook.) Vasey
Rhizomatous; culms 3–6 dm tall; blades 2–4 mm wide, flat or folded, mostly basal; sheaths glabrous or the lower retrorsely pubescent; ligules 1–2 mm long; panicle 5–10 cm long, open, the tip nodding, the branches mostly in 2's or 3's, naked below; spikelets 4–6 mm long; lemmas rather strongly nerved, compressed keeled, with or without pubescence on the lower part of the nerves, the base not webbed. $2n = 28$, 56, 61–(c.74*)–91. Open woods. BC, Alta s to Calif, Nev, Utah, NM. Map 177 ●

P. nevadensis Vasey *ex* Scribn.
Densely tufted perennial; culms 3–10 dm tall; basal leaves to 50 cm long; leaf-blades involute to flat, 1–3 mm wide; ligule decurrent, 1.5–6.5 mm long, acute to acuminate; panicle narrow, mostly 10–20 cm long; spikelets rounded, 5–10 mm long, 2–6-flowered; glumes 2.7–5.7 mm long, scabrous; lemmas mostly 3–5 mm long, rounded on back, glabrous to scabrous; anthers 1.5–3.0 mm long. $2n = 62–70$. Moist meadows. BC, Alta s to Calif, Nev, Ariz, Colo; Yuk. Map 177 ○

P. palustris L. / Fowl Bluegrass
Culms loosely tufted, 3–10 dm tall, often decumbent at the purplish base; blades 1–2 mm wide; ligules on primary culms 3–5 mm long; panicle 1–3 dm long, pyramidal or oblong, often nodding, the branches in whorls of about 5; spikelets 2–4-flowered, about 4 mm long; glumes ovate-lanceolate, acute; lemmas 2–3 mm long, usually bronze at tip, pubescent on the keel and marginal nerves, webbed at base. $2n = 28^*$, 30, 32, 42. Moist meadows and open woods. Circumpolar: sAlas, Yuk, sMack to Huds Bay, Nfld s to Calif, Nev, Utah, NM, Mo, Gt Lakes, Apal. Map 178

P. pattersonii Vasey
Culms densely tufted, mostly 5–20 cm tall; blades 1–2 mm wide, usually folded, rather lax; ligules 1–4 mm long; panicle 1–4 cm long, narrow, dense, purplish;

spikelets 5–6 mm long, 2–4-flowered; lemmas about 4 mm long, strongly pubescent on keel and marginal nerves, sometimes sparsely webbed at base. Stony alpine slopes. BC, Alta s to Ida, Nev, Ariz, Colo. Map 179

P. pratensis L. / Kentucky Bluegrass
Spreading by rhizomes and sod-forming; culms often tufted, 3–8 dm tall; sheaths somewhat keeled; blades mostly 2–4 mm wide, soft, flat or folded; panicle pyramidal or ovoid, dense or rather open, the lowermost branches usually in a whorl of 5, these of different lengths; spikelets crowded, 3–5-flowered, 3–6 mm long; lemmas distinctly 5-nerved, pubescent on the nerves, copiously webbed at base. $2n$ = 21–(56*, c.66*, c.70*, 86*)–147. Includes *P. alpigena* (Fr.) Lindm., *P. agassizensis* Boivin & D. Löve. Introduced and native elements, which are indistinguishable, occur in our area. Prairies and woodland. Circumpolar: Alas, Yuk to Ellesmere Is, generally distributed s. Map 180

P. sandbergii Vasey / Sandberg Bluegrass
Forming small dense tufts; culms mostly 1.5–3 dm tall; blades short, crowded at the base, soft, often curled; ligules rather prominent, pointed; panicle narrow, 4–8 cm long, the branches short and appressed, or somewhat spreading in anthesis; spikelets little flattened; lemmas scarcely keeled, 3.5–4 mm long, pubescent on the lower half, especially on the keel and near the margins. $2n$ = 42*, 56*, 70*, c.78*, 80*, 82–(c.90*)–106. Flowers in spring. *P. secunda* Presl. of Ed. 1. Dry plains. BC to Sask s to Calif, Nev, Ariz, NM, Neb. Map 181

P. stenantha Trin.
Culms tufted, 3–5 dm tall; blades mostly basal, flat or loosely involute, rather lax; ligule as much as 5 mm long; panicle 5–15 cm long, nodding, longer than broad, the branches in 2's or 3's, drooping; spikelets 6–8 mm long, 3–5-flowered; lemmas about 5 mm long, keeled, not webbed, but pubescent on lower part of keel and marginal nerves, sparsely hairy on the internerves. $2n$ = 42. Open woods at montane elevations. Asia: Alas, Yuk s to Ore, Ida, Mont; Colo. Map 182

Polypogon Desf.

P. monspeliensis (L.) Desf. / Rabbitfoot Grass
Annual; culms erect from a decumbent base, 1–5 dm tall, usually tufted; leaf-blades flat, 3–5 mm wide, scabrous; ligules 5–6 mm long; inflorescence a dense spike-like panicle, 3–7 cm long, silky-hairy, tawny yellow when mature; spikelets 1-flowered, disarticulating below the glumes; glumes about 2 mm long, short-hairy, awned from the bifid apex, the awn 3–8 mm long, rough; lemma about 1 mm long, smooth and shining, with a fragile awn; palea delicate. $2n$ = 28, 35. Dry stream banks. Introduced. Rare.

Puccinellia Parl. / Alkali Grass

Low, tufted, smooth perennials; leaf-blades flat or involute; panicles narrow or spreading; spikelets several-flowered, terete or subterete, disarticulating above the glumes and between the lemmas; glumes unequal, shorter than the first lemma, the first 1-nerved, the second 3-nerved; lemmas ovate to oblong, rounded on the back, awnless, generally obscurely 5-nerved, the tip broad, thin, hyaline and often erose; palea about as long as lemma.

1 Nerves on lemma conspicuous **P. pauciflora**
Nerves on lemma obscure 2

2 Glumes rounded on back; lemmas 1.5–2.0 mm long, apex rounded to truncate, anthers less than 1 mm long; lower branches often reflexed 3
Glumes more or less keeled; lemmas 2–3 mm long, apex abruptly triangular; anthers more than 1 mm long; lower branches seldom reflexed 4

3 Leaves 2–5 mm wide; anthers 0.6–0.8 mm long **P. distans**
Leaves 1–2 mm wide; anthers 0.3–0.6 mm long **P. hauptiana**

4 Second glume 2.0–2.8 mm long; anthers mostly 1.5–2.0 mm long **P. cusickii**
Second glume 1.5–2.2 (2.5) mm long; anthers 0.7–1.2 mm long **P. nuttalliana**

P. cusickii Weatherby
Similar to *P. nuttalliana;* distinguished by the characters of the key. $2n = 28$*. Moist, generally alkaline areas. Wash, eOre, Ida, Mont, Sask, ND. Map 183 ●

P. distans (L.) Parl.
Culms tufted, decumbent at base, mostly 2–4 dm tall; leaf-blades mostly 2–4 mm wide, flat or becoming involute, the sheaths scabrous; panicle ovoid or pyramidal, 5–15 cm long, loosely branched, the lower branches spreading or reflexed and bearing spikelets chiefly above the middle; spikelets 4–6-flowered, slender, 4–5 mm long; first glume about 1 mm long; second glume 1.2–1.7 mm long, erose at summit; lemmas broadly ovate, 1.5–2.4 mm long, the summit broad, obtuse or truncate, hyaline, erose-ciliate. $2n = 28, 42$*. Possibly introduced. Moist, alkaline areas. More or less circumpolar: BC, sYuk to NS and NEng states. Map 184 ●

P. hauptiana (Krecz.) Kitagawa
Erect or sometimes prostrate; culms 1–4 dm tall; leaf-blades 1–1.5 mm wide, 3–8 cm long, flat or loosely involute; sheaths generally exceeding the internodes; ligules 1.5–2.5 mm long, obtuse; panicle 3–15 cm long, with slender, scabrous, spreading to reflexed branches; spikelets 3–4 mm long, 3–5-flowered, closely appressed; glumes firm, acute to subobtuse, the first 1–1.5 mm long, second 1.2–2 mm long; lemmas obtuse, 1.6–2 mm long, glabrous to sparsely puberulent at base; anthers 0.3–0.5 mm long. $2n = 28$*. Marshy areas. Asia: Alas, Yuk, Mack; Huds Bay. Map 184 ○

P. nuttalliana (Schult.) A.S. Hitchc.
Culms mostly 3–6 dm tall; panicle slender to broadly ovoid or pyramidal, mostly 1–2 dm long, the branches erect to wide-spreading; spikelets 4–7 mm long; first glume 1–1.4 mm long; second glume 1.4–2.5 mm long; lemmas broadly oblong,

2–3 mm long, somewhat narrowed to a broadly triangular summit. $2n = 42$, 56*. *P. airoides* (Nutt.) Wats. & Coult. Moist alkaline areas. BC to Man, Wisc s to Calif, Tex, Kans. Map 185

P. pauciflora (Presl) Munz
With creeping rhizomes; culms 4–10 dm tall; blades thin, flat, mostly 8–15 cm long, 4–12 mm wide; sheaths open; ligules 5–6 mm long; panicle open or rather dense, nodding, 1–2 dm long; spikelets 4–5 mm long, often purplish; glumes 1–2 mm long, the second one 3-nerved, the margins erose-scarious; lemmas 2–2.5 mm long, with 5 prominent nerves and usually 2 faint marginal ones, the tip rounded or truncate, broadly scarious and sometimes erose. $2n = 14$. *Glyceria pauciflora* Presl of Ed. 1, *Torreyochloa pauciflora* (Presl) Church. In our area var. **holmii** (Beal) C.L. Hitchc. Wet places. sAlas, sYuk to Calif, NM, SD. Map 183 ○

Schedonnardus Steud.

S. paniculatus (Nutt.) Trel. / Tumble Grass
Small grass, densely tufted or diffusely branching; culms 2–4 dm high, erect or decumbent at base, leafy below; blades smooth, flat, mostly 2–5 cm long, 1 mm wide, often wavy or twisted; ligules prominent; inflorescence large, often about half the entire height of the plant, composed of very slender spikes arranged on an axis, appressed or wide-spreading, the whole simulating a panicle, breaking away at maturity and behaving as a tumble-weed; spikelets 3–6 mm long, slender, 1-flowered, sessile, appressed along 2 sides of a 3-angled rachis; disarticulating above the glumes, these lanceolate-subulate; lemma a little longer than the glumes, narrow, acuminate, 3-nerved, scabrous; palea shorter than the lemma. $2n = 20$, 30. Plains and waste places. sAlta to sMan s to Utah, Ariz, Tex, La, Wisc. Map 186

Schizachne Hack.

S. purpurascens (Torr.) Swallen / False Melic
Slender perennial, loosely tufted; culms erect from a short-decumbent base, 4–8 dm tall, very slender; sheaths closed; ligules 0.5–1.5 mm long; blades flat, 1–4 mm wide; inflorescence an open lax panicle, the branches single or in pairs, more or less drooping, bearing 1 or 2 awned spikelets about 2 cm long; spikelets purplish, bronze or pale, 3–5-flowered, disarticulating above the glumes and between the florets; glumes unequal, 3-nerved and 5-nerved, much shorter than the adjacent lemma, purple at base; lemmas lanceolate, 8–10 mm long, 7-nerved, long-hairy on the callus, awned from just below the teeth of the bifid apex, the awns as long as the lemma or longer, slightly divergent at maturity; palea with hairy submarginal keels. $2n = 20$. In our area ssp. **purpurascens**. Open woods and northern prairie. Asia: BC, Mack to James Bay, nQue, Nfld s to NM, SD, Gt Lakes; sAlas, sYuk. Map 187

Schizachyrium Nees

S. scoparium (Michx.) Nees / Little Bluestem
Shoots tufted, from an erect or ascending base; culms 2–6 dm high, wiry, often purplish, branching above; leaf-blades flat or folded, 2–5 mm wide; inflorescence a spike-like raceme, 2–5 cm long, the peduncle wholly or partly included in a sheath; rachis white-hairy and each internode widened upward to a cup-shaped apex; each node with a sessile spikelet which is fertile and also 2 pedicellate, sterile, rudimentary spikelets, the pedicels white-hairy; fertile spikelet with glumes and lemmas 5–10 mm long; fertile lemma hyaline, 2-toothed at apex and with a twisted bent awn 7–15 mm long; palea shorter than the lemma. $2n = 40$, c.60. *Andropogon scoparius* Michx. of Ed. 1. In our area var. **scoparium**. Prairie grassland. seBC to swQue s to Ida, Utah, generally distributed e. Map 188

Scolochloa Link

S. festucacea (Willd.) Link
Tall perennial with a thick creeping rhizome; culms erect, stout, 8–12 dm tall; leaf-blades 5–10 mm wide, elongate, extending into a fine point; ligules firm, 3–5 mm long; panicle 1–2 dm long, loose, widely spreading, the distant branches fascicled, ascending, the lowermost much elongate; spikelets 7–11 mm long, 3–4-flowered, disarticulating above the glumes and between the lemmas; glumes unequal, thin, the first 3-nerved, the second 5-nerved and about as long as the lemma above it; lemmas lanceolate, 7-nerved, usually lacerate at apex, densely villose on the callus; palea 2-toothed. $2n = 28$, 42. *Fluminia festucacea* (Willd.) A.S. Hitchc. Shallow water and marshes. Eurasia: BC, sMack to sMan s to Ore, Wyo, Neb, Ia. Map 189

Secale L.

S. cereale L. / Rye
Annual or winter annual, branched from the base; blades flat, 6–12 mm wide, scabrous; inflorescence a spike, 8–15 cm long, or smaller in wild plants, dense, nodding, the rachis internodes pubescent on the edges; spikelets usually 2-flowered, solitary, placed flatwise against the rachis, the rachilla eventually disarticulating above the glumes and produced beyond the upper floret as a minute stipe; glumes linear-subulate, scabrous on the keel; lemmas longer than the glumes and broader, 5-nerved, sharply keeled, ciliate on keel and exposed margins, tapering into a long awn. $2n = 14$, 28. An escape from cultivation. Introduced. Occasional.

Setaria Beauv. / Bristly Foxtail

Annuals with flat blades, hairy ligules and dense, spike-like, bristly panicles; spikelets all alike, with 1 perfect flower, subtended by an involucre of 1–several bristles (sterile branchlets), disarticulating below the glumes but above the bristles; first glume half as long as the spikelet or less; second glume longer, sometimes equalling the spikelet; sterile lemma as long as the spikelet, several-nerved; fertile lemma hard, smooth or finely rugose; palea similar to the fertile lemma in texture.

Bristles below each spikelet 5–20, yellowish brown; spikelets 2–3 mm long; second glume half as long as the spikelet; fertile lemma conspicuously rugose **S. glauca**
Bristles below each spikelet 1–3, green, purplish or tawny; spikelets 1.5–2 mm long; second glume about as long as the spikelet; fertile lemma minutely rugose **S. viridis**

S. glauca (L.) Beauv. / Yellow Foxtail
Annual; culms usually erect, solitary or tufted, 3–6 dm tall; leaf-blades flat, often twisted, with occasional scattered hairs; panicle spike-like, mostly 5–10 cm long, about 1 cm thick, yellow at maturity; bristles 5–20 in a cluster, the longer 2 to 3 times as long as the spikelet; spikelets planoconvex; glumes and lemmas as described above. $2n = 18, 36, 72$. Waste ground, fields, and gardens. Introduced. Occasional.

S. viridis (L.) Beauv. / Green Foxtail
Similar in general appearance to *S. glauca*, differing as shown in the key and as follows: culms usually tufted, sometimes geniculate-spreading, 2–4 dm high; panicle 3–5 cm long, green or pale; bristles mostly 3 to 4 times as long as the spikelet. $2n = 18, 36$. A weed of cultivated ground and waste places. Introduced. Common.

Sitanion Raf.

S. hystrix (Nutt.) J.G. Smith / Squirreltail
Tufted perennial; culms 1–5 dm high, erect or spreading, rather stiff; leaf-blades flat or involute, 1–4 mm wide, glabrous or pubescent; sheaths glabrous to closely pubescent, the upper often inflated; ligules less than 1 mm long; inflorescence a spike, 2–10 cm long, as broad as long (including awns), partly included, green or purplish, later pale, readily breaking up into segments; spikelets 2–few-flowered, 2 at each node of the rachis, this disarticulating at the base of each joint which remains as a pointed stipe to the spikelets above it; glumes very narrow, 1–2-nerved, each nerve extending into a divergent awn 5–9 cm long, sometimes bifid or bearing a bristle on one margin; lemmas firm, convex on back, 1–3-awned at apex, the awns widely spreading, 2–10 cm long; palea 2-keeled. $2n = 28$. Dry plains and dry open woods. BC to Sask s to Calif, Ariz, NM, Tex. Map 190

Spartina Schreb. / Cord Grass

Stout perennials, rhizomatous; inflorescence of ascending, racemose spikes; spikelets 1-flowered, disarticulating below the glumes, strongly compressed laterally, in two rows secund on the rachis; glumes unequal, keeled, 1–3-nerved; lemmas keeled, 1-nerved, the lateral nerves obscure, mostly shorter than second glume; palea thin, hyaline, 2-nerved.

Glumes scabrous, awned; ligules usually 2–3 mm long **S. pectinata**
Glumes long-ciliate, awnless; ligules usually about 1 mm long **S. gracilis**

S. gracilis Trin. / Alkali Cord Grass
Coarse, fibrous plants; culms 4–10 dm tall, usually solitary, glabrous; leaf-blades long, tough, flat, becoming involute, very scabrous above; inflorescence of 4–8 appressed spikes, 2–4 cm long; spikelets much flattened, sessile, closely crowded in 2 rows but 1-sided on the rachis; glumes keeled, very unequal, the first 5–6 mm long, acuminate to awn-tipped, the second 7–9 mm, ciliate on the keel; lemma nearly as long as the second glume, flat, ciliate on the keel, the apex rather obtuse; palea as long as lemma, obtuse. $2n = 40$ (42). Saline meadows. BC, wMack, Sask, sMan, wOnt s to Calif, Ariz, NM, Kans. Map 191

S. pectinata Link
Coarse fibrous plants; culms 5–20 dm tall; leaves flat to involute, the margins scabrous; spikes mostly 10–20, 2–10 cm long, erect or ascending; spikelets flattened, crowded, glumes scabrous, the first 5–10 mm including awn tip, the second mostly 8–20 mm long including awn; lemma 7–9 mm long, keel scabrous, shorter than palea. $2n = 40$ (28, 42, 56, 84). Saline shores and marshes. BC, sMack to Nfld s to Ore, Utah, NM, Tex, Mo, NC. Map 192

Sphenopholis Scribn. / Wedge Grass

Slender perennials with flat blades, tufted culms and shining, slender or spike-like panicles; spikelets 2-flowered, sometimes with a rudimentary third flower, disarticulating below the glumes and below the upper lemma; rachilla prolonged behind the upper palea; glumes unlike, the first narrow, the second broadly ovate to obovate, usually exceeded by the upper floret; lemmas firm, slender, obscurely nerved, awnless or rarely short-awned below the apex; palea hyaline, narrowed toward the base.

Panicle open, lax, nodding; second glume about 3 times as long as wide, pointed or rounded, not hooded **S. intermedia**
Panicle dense, usually spike-like, erect or nearly so; second glume nearly as broad as long, rounded and somewhat hooded at summit **S. obtusata**

S. intermedia (Rydb.) Rydb. / Slender Wedge Grass
Culms 3–10 dm tall, erect, in small tufts; blades flat, soft, 2–5 mm wide; panicle 5–15 cm long, 1–3 cm wide, more or less irregular in outline, lax, nodding, the branches spikelet-bearing from the base; spikelets 3–4 mm long; glumes and lemmas as described above. $2n = 14$. Moist meadows and shores. BC to Nfld s to Ore, Nev, Ariz, generally distributed e; Alas; sMack. Map 193 ●

S. obtusata (Michx.) Scribn. / Prairie Wedge Grass
Culms 2–8 dm tall, densely tufted; blades flat, mostly 2–5 mm wide, glabrous or pubescent; panicle 3–10 cm long, 0.5–1.5 cm wide, dense, usually spike-like but often more or less lobed, erect or nearly so; spikelets 2.5–3 mm long; glumes

and lemmas as described above; lemmas minutely papillose, rarely mucronate or with a short awn. $2n = 14$. Moist meadows, shores, and open woods. BC to Man s to Calif, generally distributed e. Map 193 ○

Sporobolus R.Br. / Dropseed

Annual or perennial grasses, with narrow blades and open or contracted panicles; ligules minute or none; spikelets 1-flowered, the rachilla disarticulating above the glumes; glumes lanceolate to ovate, 1-nerved, the second often as long as the spikelet; lemma membranaceous, rounded on the back, 1-nerved, awnless; palea as long as the lemma or longer, usually prominent; grain soon falling free from the lemma and palea; pericarp loosely enclosing the plump seed, often slipping off, at least when moist.

Perennial; culms 3–8 dm tall; sheaths with a conspicuous tuft of hairs at summit; mature panicle usually open and spreading **S. cryptandrus**

Annual; culms 0.5–3 dm tall; sheaths glabrous or sparsely pilose near collar, upper often inflated; panicle spikelike, mostly included in the sheath **S. neglectus**

S. cryptandrus (Torr.) A. Gray / Sand Dropseed
Culms in small tufts, 3–8 dm tall, erect or spreading, sometimes prostrate; leaf-blades 2–5 mm wide, tapering to a fine point, scabrous on margins; sheaths with a conspicuous tuft of white hairs at the summit; ligules mostly of hairs; panicles as much as 2.5 dm long, usually included at base, the branches finally spreading or reflexed, rather distant, the spikelets crowded along the upper parts of main branches; spikelets 1.8–2.5 mm long, pale to leaden; first glume one-third to half as long as the spikelet, the second about the same length as the acute lemma. $2n = 18, 36^*, 38, 72$. Dry, sandy soil. BC to swQue, generally distributed s. Map 194

S. neglectus Nash
Culms tufted, freely branching, 0.5–3 dm high, erect or spreading, very wiry; leaf-blades slender; upper sheaths often inflated; ligule mostly of hairs; panicles 2–6 cm long, spike-like, usually more or less included; spikelets 1.8–3 mm long, pale, plump; glumes nearly as long as the lemma, subequal, white-scarious; lemma whitish, glabrous, acute. $2n = 36$. Dry open ground. Alta to Que s to Mont, Ariz, NM, widespread c and e USA. Map 195 ○

Stipa L. / Needle Grass

Perennial bunchgrasses with narrow, elongate blades and open or contracted panicles; spikelets often large, 1-flowered, disarticulating above the glumes and in such a manner as to leave a bearded sharp-pointed callus at the base of the floret; glumes about equal, usually long and narrow, often papery, acute to bristle-tipped; lemma narrow, terete, indurate, strongly convolute, terminating in a prominent awn, this usually twice-geniculate and twisted below; palea enclosed in the convolute lemma.

1 Glumes 15–40 mm long; lemma 8–20 mm long 2
Glumes 8–10 mm long; lemma 4–6 mm long 4

2 Lemma evenly pubescent above callus, 8–15 mm long, its awn very slender and flexuous, indistinctly geniculate; glumes 15–28 mm long **S. comata**
Lemma pubescent only at base and along margins, 12–25 mm long, its awn stiff and strongly geniculate; glumes 20–40 mm long 3

3 Lemma-awn 10–20 cm long; glumes 30–40 mm long **S. spartea**
Lemma-awn 5–11 cm long; glumes 20–30 mm long **S. curtiseta**

4 Panicle open, the branches spreading and drooping **S. richardsonii**
Panicle narrow, the branches appressed 5

5 Sheaths villose at the throat; lower nodes of panicle villose; fruit rather turgid, the callus broad and short **S. viridula**
Sheaths not villose at throat or only slightly so; lower nodes of panicle glabrous; fruit slender, the callus narrow, sharp-pointed **S. columbiana**

S. columbiana Macoun / Columbia Needle Grass
Culms mostly 4–6 dm tall, sometimes up to 12 dm; blades 1–4 mm wide, usually involute; ligule 1–2 mm long; sheaths naked at the throat; panicle 9–30 cm long, narrow, the branches short and appressed, compact to rather loose, often purplish; glumes about 1 cm long; lemma 4–6 mm long, appressed-hairy, the awn 2–3.1 cm long, usually twice-geniculate. $2n = 36, 44$. *S. occidentalis* Thurb. var. *minor* (Vasey) C.L. Hitchc., *S. nelsonii* Scribn. *sensu* Barkworth et al. In our area var. **columbiana** as described above (*S. nelsonii* ssp. *dorei* Barkworth & Maze). Prairie grassland and open woods. Yuk, BC to Sask s to Calif, Colo. Map 195 •

S. comata Trin. & Rupr. / Needle and Thread; Spear Grass
Culms 3–7 dm tall; blades 1–3 dm long, 1–2 mm wide, flat or involute; ligules mostly 2–4 mm long; panicle 1–2 dm long, often partly included in the inflated sheath; glumes subequal, 15–20 mm long, 5-nerved, the apex soft and attenuate; lemma 8–12 mm long, pale or finally brownish, awn 10–15 cm long, indistinctly twice-geniculate, very slender, loosely twisted below, coiled or sinuous above, terminal segment usually more than 5.5 cm long. $2n = 44–46$. In our area var. **comata** as described above (common) and var. **intermedia** Scribn. & Tweedy (rare), with the panicle usually fully exserted; the glumes mostly 25–28 mm long; the lemmas 12–15 mm long; the terminal segment of awn straight, less than 5.5 cm long. Dry plains and hills. sYuk, cSask, sMan, Wisc, Ia, Ind s to Calif, Ariz, NM, nTex. Map 196

S. curtiseta (A.S. Hitchc.) Barkworth
Similar to *S. spartea* but smaller, 4–6 dm tall; lower nodes glabrous; ligules mostly 0.2–0.6 mm long; lemmas 7–11 mm long; awns 5–11 cm long. *S. spartea* var. *curtiseta* A.S. Hitchc. of Ed. 1. Prairie grassland. BC, sMack to sMan s to Mont, SD. Map 197

S. richardsonii Link / Richardson Needle Grass
Culms 5–10 dm tall; leaves mostly basal, the blades involute, very slender; ligules 1 mm long or less; panicle 1–2 dm long, open, the branches slender, distant, spreading or drooping, naked below; glumes 8–9 mm long; lemma about 5 mm long, brown at maturity, villose or glabrate near apex, the awn 2–3 cm long. Foothills grassland and open woods. BC to Man s to Wash, Colo, SD; sYuk. Map 198

S. spartea Trin. / Porcupine Grass
Culms mostly 5–10 dm tall, lower nodes pubescent; blades 2–3 dm long, 2–5 mm wide, flat or involute on drying; ligules mostly 0.7–3.0 mm long; panicle 1–2 dm long, the branches few, slender, each with 1 or 2 spikelets; glumes 3–4 cm long, 5-nerved, the apex soft and attenuate; lemma 10–18 mm long, brown, the awn stout, 10–20 cm long, twice-geniculate, the last segment straight. Prairie grassland. sBC to sMan, Minn, Wisc, Mich, Pa s to NM, Okla, Mo, Ind. Map 199

S. viridula Trin. / Green Needle Grass
Culms mostly 4–8 dm tall; blades usually 1–3 mm wide and often involute; ligule 0.5–1 mm long; sheaths villose at the throat, sometimes sparingly so, more or less hairy across the collar; panicle 1–2 dm long, narrow with appressed branches, greenish, pale at maturity; glumes 7–10 mm long, hyaline-attenuate; lemma 5–6 mm long, appressed-hairy, brownish at maturity, the awn 2–3 cm long, finally twice bent. $2n = 82$. Hybridizes with *Oryzopsis hymenoides*. Dry prairie grassland. BC, sMack to sMan s to Ariz, NM, Kans, Minn. Map 200

Trisetum Pers.

Tufted perennials with slender, open or contracted, shining panicles; leaf-blades flat or somewhat involute; spikelets usually 2-flowered, sometimes 3–5-flowered, disarticulating above the glumes and between the florets, the rachilla prolonged behind the upper floret; glumes somewhat unequal, acute, the second longer than the first floret; lemmas usually short-bearded at base, keeled, 2-toothed at apex, usually awned from the back just below the cleft apex, the awn commonly bent and exserted; palea equalling the lemma or shorter.

1 Awnless or with an awn less than 2 mm long and hidden by the glumes **T. wolfii**
Awned, the awn exserted and bent 2

2 Panicle dense, spike-like, sometimes slightly interrupted below; first glume lanceolate, abruptly acuminate; plants densely tufted **T. spicatum**
Panicle loose and open or contracted but not spike-like; first glume attenuate-acuminate; culms in small tufts or solitary 3

3 Second glume 3.5–4.5 mm long, erose near the apex, about twice as long as the first **T. cernuum**
Second glume 4–7 mm long, not erose at the apex, less than twice as long as the first 4

4 Spikelets yellowish; ligules 0.5–1.0 mm long; ovary usually glabrous **T. flavescens**
Spikelets green or purple tinged; ligules 1.0–4.0 mm long; ovary pubescent 5

5 Lemma 4.0–5.0 mm long; awn 5–8 mm **T. montanum**
Lemma 5.0–6.0 mm long; awn about 10 mm long **T. canescens**

T. canescens Buckl.
Culms 5–10 dm tall; leaf-blades flat, 2–7 mm wide; sheaths, at least the lower, sparsely to densely hairy; panicle 10–25 cm long, narrow, usually loose, sometimes interrupted and spike-like; spikelets 2–3-flowered; lemmas 5–6 mm long, the awn usually about 10 mm long, geniculate, attached one-third below the tip; callus-hairs rather scant; rachilla hairs copious. $2n = 42$. *T. cernuum* ssp. *canescens* Calder & Taylor. Moist woods. sBC to Calif e to Ida, Mont. Map 201

T. cernuum Trin. / Nodding Trisetum
Culms tufted, rather lax, 5–10 dm tall; leaf-blades mostly 4–10 mm wide, flat, lax, scabrous; sheaths glabrous to sparsely hairy; panicle 15–25 cm long, open, lax, drooping, the branches filiform, flexuous, spikelet-bearing toward the tips; spikelets usually with 3 florets; lemmas 5–6 mm long, the awn 5–10 mm long, slender, curved; callus-hairs 0.5–1 mm long; rachilla hairs up to 2 mm long. $2n = 42$. Moist woods. seAlas, BC to Calif e to Ida, Mont. Map 202

T. flavescens (L.) Beauv.
Culms 2–5 dm tall; leaf-blades 2–4 mm wide, pubescent above; lower sheaths pubescent; panicle 5–10 (15) cm long, compact, yellowish; spikelets 2–4-flowered; first glume 2.5–3.5 mm long, second glume 5–6 mm long, about equalling spikelets; lemmas 5–6 mm long. $2n = 12, 24, 28^*$. Roadsides. Introduced. Rare.

T. montanum Vasey
Similar to *T. canescens* and possibly only a subalpine variant of it, but smaller; leaves narrower; panicle shorter and narrower; and less interrupted than *T. canescens*. Montane slopes and ledges. Utah, Ariz, Colo, NM. Map 203

T. spicatum (L.) Richt. / Spike Trisetum
Culms densely tufted, erect, 10–50 cm tall, glabrous or puberulent; leaf-blades and sheaths glabrous to densely pubescent; ligules 1–2 mm long; panicle 2–15 cm long, variable but usually dense and spike-like, often interrupted at base, pale or purplish; glumes somewhat unequal; lemmas 5 mm long, the awn 5–6 mm long, geniculate, attached about one-third below the tip of the lemma. $2n = 28, 42$. A complex of almost world-wide distribution. In our area ssp. **spicatum**, with the florets clearly exceeding both glumes, and culms densely long-pilose; ssp. **molle** (Michx.) Hult. with the florets slightly exceeding the glumes, and culms densely short pubescent above, glabrous below; and ssp. **majus** (Vasey) Hult., similar to ssp. **molle** but the culms glabrous. Open woods and mountain slopes. Alas to Ellesmere Is s to Calif, Ariz, NM, sMack, sMan, Gt Lakes, Apal. Map 204

T. wolfii Vasey
Culms loosely tufted, 5–10 dm tall, sometimes with short rhizomes; leaf-blades 2–4 mm wide, flat, scabrous, rarely hairy; panicle 8–15 cm long, erect, rather dense but scarcely spike-like, green or pale, sometimes purplish; spikelets 2-flowered, sometimes 3-flowered; glumes nearly equal, about 5 mm long; lemmas 4–5 mm long, obtusish, scaberulous, awnless or with a minute awn below the tip; callus-hairs scant; rachilla long-villose. Moist woodland. Wash to Calif e to Mont, Alta. Map 205

Triticum L.

T. aestivum L. / Common Wheat
Annual, branching at the base, with erect culms and thick dense spikes; culms 5–10 dm tall; blades flat, 1–2 cm wide; spike mostly 4–10 cm long; spikelets 2–5-flowered, solitary, placed flatwise at each joint of the rachis, the rachilla disarticulating above the glumes and between the florets; glumes 6–8 mm long, rigid, broad, usually keeled towards one side, the other side often obtusely angled at apex, the keel extending into a point or awn; lemmas longer than the glumes, broad, keeled, asymmetric, many-nerved, abruptly pointed or awned, the awn in some varieties up to 8 cm long; palea about as long as the lemma; mature grain about 2–3 times as long as wide. $2n = 42$. An escape from cultivation. Introduced. Occasional.

Vahlodea Fries

V. atropurpurea (Wahl.) Fries
Perennial; culms loosely tufted, purplish at base, 4–8 dm tall; leaves 4–6 mm wide, flat; panicle loose, open, 5–10 cm long, the few slender branches drooping; spikelets 2–3-flowered; mostly purplish; glumes about 5 mm long, exceeding the florets; lemmas scabrous, about 2.5 mm long, the callus-hairs one-third to half as long, the awn straight or geniculate, about 2.5 mm long. $2n = 14$. *Deschampsia atropurpurea* (Wahl.) Scheele of Ed. 1. In our area ssp. **latifolia** (Hook.) Pors. Woods and wet meadows. More or less circumpolar: sAlas, Yuk to Calif, and in Rocky Mts to Colo; wHuds Bay to nQue, Nfld s to NH. Map 206 ●

Vulpia C.C. Gmel.

V. octoflora (Walt.) Rydb. / Six-weeks Fescue
Small annual grass; culms solitary or in small tufts, usually 1.5–3 dm tall; blades narrow, 2–10 cm long, involute; panicle slender, the branches short, usually ascending; spikelets flattened, 6–8 mm long, 6–10-flowered; glumes subulate or lanceolate; lemmas firm, involute, 3–5 mm long, with awns 2–5 mm long. $2n = 14$. *Festuca octoflora* Walt. of Ed. 1. Sterile ground or depleted rangeland. BC to sOnt s to Calif, NM, Fla. Map 206 ○

Zizania L.

Z. aquatica L. / Wild Rice
Tall aquatic annual plant; culms 1–2 m tall, slender or rather stout; leaf-blades flat, 4–15 mm wide; sheaths long; ligules 8–15 mm long, acute; inflorescence a large erect panicle, 3–6 dm long, the lower branches ascending or pendulous and bearing staminate florets, the upper branches erect or ascending and bearing pistillate florets; spikelets unisexual, 1-flowered, disarticulating at the base; glumes none; lemma of staminate spikelet thin, acuminate or short-awned; stamens 6; pistillate borne on clavate pedicels, the lemma firm at maturity, 3-ribbed and awned, the awn 1–6 cm long; grain long-cylindrical, blackish, closely enveloped in the lemma and palea. $2n = 30$. Ponds and lakes. Introduced. Occasional.

CYPERACEAE / Sedge Family

Grass-like or rush-like herbs, mostly perennial; roots fibrous, from a very short or elongated creeping rhizome; stems (culms) mostly solid, often 3-sided; leaves commonly 3-ranked, with closed sheaths, the blades in some species or genera greatly reduced or absent; flowers very small, hypogynous, often unisexual, in the axils of scales (glumes) and arranged in spikes or spikelets; perianth of bristles or scales, or lacking; stamens usually 3; pistil 1, the ovary with a single ovule and sometimes (in *Carex*) enveloped by a sac (perigynium); style, 1, 2–3-cleft; fruit an achene, flattened or 3-sided.

1 Flowers imperfect; plants monoecious or dioecious 2
Flowers all perfect, or at least one in each spikelet perfect 3

2 Achenes enclosed in a perigynium **Carex**
Achenes not enclosed in a perigynium **Kobresia**

3 Scales of the spikelets 2-ranked; perianth lacking; spikelets in terminal heads or spikes **Cyperus**
Scales of the spikelets spirally imbricate; perianth usually present as bristles 4

4 Achenes crowned with a tubercle 5
Achenes without a persistent tubercle 6

5 Spikelets solitary; stem leafless **Eleocharis**
Spikelets several; stem with bristle-like leaves **Rhynchospora**

6 Bristles 1–6, rarely lacking (elongating only in *S. hudsonianus*) **Scirpus**
Bristles numerous, much elongating in fruit, silky **Eriophorum**

Carex L. / Sedge

Perennial herbs; stems (culms) mostly triangular and solid; leaves 3-ranked, long and narrow, with closed sheaths and commonly with inconspicuous ligules; spikes 1–many, usually subtended by a bract, which may be leaf-life or much reduced; flowers monoecious or sometimes dioecious, solitary in the axils of scales (glumes); staminate and pistillate flowers either in different spikes or in different parts of the same spike; an *androgynous* spike has staminate flowers at the apex, with pistillate flowers below; a *gynaecandrous* spike has pistillate flowers above the staminate; perianth none; staminate flowers of 3 stamens, the filaments filiform; pistillate flowers of a single pistil, the ovary one-ovuled and surrounded by a sac (the perigynium), this with a terminal pore through which the style protrudes; style with 2 or 3 branches (stigmas); fruit a 3-angled, lens-shaped or rounded achene enveloped by the perigynium.

In the following key, adapted from K.K. Mackenzie 'Cariceae. N. Am. Flora,' mature fruits are essential for identification.

1 Spikes normally one **Group 1**
Spikes normally more than one (but sometimes aggregated into a head) 2

2 Stigmas 2; achenes lens-shaped to planoconvex **Group 2**
Stigmas 3; achenes triangular **Group 3**

GROUP 1

1 Stigmas 2; achenes lens-shaped 2
Stigmas 3; achenes triangular 4

2 Spike entirely staminate or pistillate (or very nearly so); plants with long slender rhizomes **C. gynocrates**
Spike androgynous; plants caespitose or with short rootstocks 3

3 Perigynia 3 mm long, striate, stipitate **C. nardina**
Perigynia 2–2.5 mm long, essentially nerveless, sessile **C. capitata**

4 Pistillate scales deciduous; perigynia stipitate, at least the lower reflexed at maturity 5
Pistillate scales persistent; perigynia not reflexed at maturity 8

5 Perigynia dark-tinged; style jointed with the achene 6
Perigynia straw-coloured; style continuous with the achene 7

6 Leaves flat, 1.5 mm wide or wider; staminate flowers conspicuous; perigynia deflexed **C. nigricans**
Leaves involute, 1 mm wide; staminate flowers few, inconspicuous; perigynia erect or spreading **C. pyrenaica**

7 Perigynia 4–5 mm long; culms with 4–8 leaves; rachilla present, conspicuously exserted **C. microglochin**
Perigynia 6–7 mm long; culms with 2–3 leaves; rachilla absent **C. pauciflora**

8 Perigynia rounded and beakless at apex, many-nerved **C. leptalea**
Perigynia not rounded at apex, beakless, or beaked 9

9 Spike entirely staminate or pistillate (or nearly so); perigynia pubescent or puberulent **C. scirpoidea**
Spike androgynous 10

10 Perigynia coriaceous, shining; rhizome elongated; common prairie species **C. obtusata**
Perigynia not both coriaceous and shining 11

11 Perigynia finely striate **C. nardina**
Perigynia 2-keeled, otherwise nerveless 12

12 Perigynia pubescent or puberulent at least on beak, 2.5–4 mm long; densely caespitose plants of dry grassland **C. filifolia**
Perigynia glabrous, 3–7 mm long; loosely caespitose plants 13

13 Perigynia 3–4 mm long; high alpine species **C. rupestris**
Perigynia 5–7 mm long; woods and mountain slopes **C. geyeri**

GROUP 2

1 Lateral spikes sessile, short (rarely somewhat elongate); terminal spike very rarely staminate 2
Lateral spikes peduncled, or if sessile elongate; terminal spike staminate, or if gynaecandrous the lateral spikes peduncled 20

2 Perigynia whitish-puncticulate; upper spikes gynaecandrous (except in *C. disperma*); perigynia spongy-thickened at base, not wing-margined; culms in clumps but rootstocks often slender-spreading **Group 4**
Perigynia not whitish-puncticulate 3

3 Culms arising singly or sometimes few together from long-creeping rootstocks; terminal spikes usually androgynous or heads strongly tending to be dioecious 4
Culms densely or loosely caespitose, the rootstocks either none or very short 11

4 Culms becoming decumbent and producing new culms from the axils of old dried-up leaves; sphagnum-bog species **C. chordorrhiza**
Culms not branching; mostly not sphagnum-bog species 5

5 Perigynia wing-margined at least above, flattened, planoconvex, the beak deeply bidentate; spikes closely aggregated in the head, the uppermost gynaecandrous, the others mostly staminate or pistillate **C. siccata**
Perigynia not wing-margined, planoconvex, or unequally biconvex, the beak obliquely cut dorsally, becoming bidenticulate or even bidentate in age 6

6 Spikes densely aggregated into a globose or ovoid head appearing like one spike; spikes androgynous, not tending to be dioecious **C. maritima**
Spikes not all densely aggregated, the lower ones distinct; spikes mostly androgynous, but often some spikes in a head or all spikes on some culms nearly or entirely unisexual 7

7 Upper sheaths green-striate ventrally; perigynia with body thin-margined above **C. sartwellii**
Upper sheaths hyaline or brown ventrally; perigynia not thin-margined 8

8 Culms obtusely triangular, generally smooth above, from slender rhizomes; inflorescence up to twice as long as broad; plants of dry habitats 9
Culms sharply triangular, scabrous above, from slender or stout rhizomes; inflorescence generally more than twice as long as broad; plants of moist habitats 10

9 Plants normally dioecious, perigynia 3.5–4.0 mm long, shorter than scales; anthers 3–4 mm long **C. douglasii**

Plants monoecious; spikes androgynous; perigynia 2.5–3.0 mm long, as long as or longer than scales; anthers about 2 mm long **C. stenophylla**

10 Rhizomes black, 2–3 mm thick; perigynia 3–4 mm long, beak prominent **C. praegracilis**

Rhizomes brown, 1–2 mm thick; perigynia 2.0–2.5 mm long, beak short **C. simulata**

11 Spikes androgynous 12

Spikes not androgynous, the upper gynaecandrous, rarely staminate, the lateral gynaecandrous or pistillate 17

12 Perigynia with body tapering into a long beak, yellowish; culms stout but weak and flattened in drying **C. stipitata**

Perigynia with body abruptly contracted into the beak 13

13 Spikes few (4–10); sheaths not red-dotted ventrally 14

Spikes numerous, often very numerous; sheaths often red-dotted ventrally, at least at mouth 15

14 Spikes closely aggregated in an orbicular or ovoid-oblong head, 1–2 cm long; bracts absent or occasionally a small awl-shaped one; scales acute to short-cuspidate; perigynia 3.5–5 mm long **C. hoodii**

Spikes in a linear head, the lower 3–6 separate, the upper aggregated; bracts strongly developed, the lower at times exceeding the head; scales strongly awned; perigynia about 3 mm long **C. hookerana**

15 Perigynia yellowish or straw-coloured, or in age black, planoconvex; leaf-sheaths cross-wrinkled ventrally **C. vulpinoidea**

Perigynia brownish, thick-planoconvex or unequally biconvex; leaf-sheaths not as above 16

16 Leaf-sheaths copper-tinged at summit; head of spikes lax and open, pale brown; perigynia 2.5–3.5 mm long, dull, with inner flat face, appressed and covered by scales **C. prairea**

Leaf-sheaths not copper-tinged at summit; head straight, dense or slightly open, dark brown; perigynia 2–2.7 mm long, shining, with convex inner face, soon wide-spreading, not wholly covered by scales **C. diandra**

17 Perigynia at most thin-edged, the lower part of the body spongy-thickened 18

Perigynia narrowly or broadly wing-margined, the lower part of the body not spongy-thickened **Group 5**

18 Perigynia appressed **C. deweyana**

Perigynia spreading or ascending at maturity 19

19 Spikes aggregated to form a single head; perigynia entire-margined **C. illota**

Spikes generally distinct; perigynium serrulate distally and on beak **C. interior**

20 Perigynia turgid, obovoid, elliptic in cross-section, beakless **C. aurea**

Perigynia flat, planoconvex, or biconvex, with a beak 21

21 Style continuous with the achene, indurate and persistent; perigynia lustrous, 4–10 mm long **C. saxatilis**

Style jointed with the achene, in time withering; perigynia not lustrous, 2–4 mm long **Group 6**

GROUP 3

1 Perigynia glabrous; style continuous or jointed with the achene 2

Perigynia pubescent, puberulent, hispid, granular-roughened or scabrous; style jointed with achene, finally withering and deciduous 23

2 Style continuous with the achene, indurate and persistent; perigynia strongly nerved or ribbed 3

Style jointed with the achene, finally withering and deciduous; perigynia nerved or otherwise 10

3 Perigynia finely and closely ribbed; scales over 4 times as long as broad 4

Perigynia coarsely ribbed; scales not more than twice as long as broad 5

4 Perigynia relatively thin in texture, somewhat inflated, nearly circular in cross-section, ascending or spreading at maturity; forming long stolon-like rootstocks; ligule short, not longer than wide **C. hystricina**

Perigynia coriaceous, rigid, not inflated, flattened-triangular in cross-section, reflexed at maturity; not stoloniferous; ligule very conspicuous, much longer than wide **C. pseudo-cyperus**

5 Perigynia somewhat leathery and firm, 12–25-nerved 6

Perigynia membranaceous, 7–10-nerved 7

6 Teeth of perigynium-beak short, 0.5 mm long, erect or nearly so; leaf-blades and sheaths glabrous **C. lacustris**

Teeth of perigynium-beak prominent, 1.5–3.0 mm long, ascending to divergent; leaf-blades and sheaths pubescent **C. atherodes**

7 Lower perigynia reflexed; lower bracts several times longer than the inflorescence **C. retrorsa**

Lower perigynia not reflexed (except rarely in *C. rostrata*); lower bracts not more than a few times exceeding the inflorescence 8

8 Pistillate spikes globose to short oblong, 5–15-flowered **C. oligosperma**

Pistillate spikes oblong to cylindric, more than 15-flowered 9

9 Perigynia spreading at maturity; elliptic to subglobose, abruptly contracted to an evident beak **C. rostrata**

Perigynia ascending at maturity; lanceolate-ovate, tapering gradually to an ill-defined beak **C. vesicaria**

10 Achenes slightly depressed at summit, constricted at base, the sides convex above; staminate scales few, their margins united at base; perigynia mostly concealed by the leaf-life scales **C. backii**

Achenes more or less strongly apiculate-tipped, not constricted at base; staminate scales with margins separate to base 11

11 Lower bracts sheathless or very short-sheathing 12

Lower bracts (at least) long-sheathing 16

12 Roots closely covered with a yellowish felt; perigynia glaucous-green or straw-coloured; terminal spike staminate 13

Roots not covered with a yellowish felt; perigynia dark-tinged to glaucous-green; terminal spike gynaecandrous or staminate 14

13 Culms borne singly or in small tufts; plant spreading by long forking stolon-like rootstocks; leaves somewhat involute, often glaucous; pistillate scales persistent, about as wide and long as the perigynia and covering them; perigynia minutely beaked; staminate spike 10–30 mm long **C. limosa**

Culms densely tufted from deep-seated rootstocks; leaves flat, light green; pistillate scales soon deciduous, narrower than the perigynia, long-acuminate; perigynia scarcely beaked; staminate spike 4–12 mm long **C. paupercula**

14 Culms less than 15 cm tall; leaves shorter, often arcuate or flexuous; inflorescence 10–15 mm long, 2–3 mm wide; terminal spike 4–7 mm long, 1.0–1.5 mm wide **C. glacialis**

Plants not as above 15

15 Leaves pubescent; perigynia glabrous but appearing felty, conspicuously 15–25-nerved **C. torreyi**

Leaves glabrous; perigynia not as above **Group 7**

16 Bracts bladeless, white-hyaline above; leaf-blades filiform **C. eburnea**

Bracts with well-developed blades 17

17 Perigynia with beak bidentate 18

Perigynia with beak not bidentate, at most emarginate 20

18 Spikes peduncled; pistillate scales lanceolate, acuminate, greenish white or brownish, often nearly as long as the perigynia; beak of perigynium about equal to the body length **C. sprengelii**

Spikes sessile; pistillate scales ovate, acute or short-cuspidate, reddish or yellowish, much shorter than the perigynia; beak of perigynium one-third the body length or shorter 19

19 Perigynia 3.5–6 mm long, yellowish, some or all of them curved or deflexed, the beak half to three-fourths as long as the body **C. flava**

Perigynia 2–3.5 mm long, greenish, spreading or ascending, not curved, the beak scarcely half as long as the body **C. viridula**

20 Pistillate spikes elongate, linear, drooping on capillary peduncles; plants tufted, the rootstocks short **C. capillaris**

Pistillate spikes short, oblong or linear, erect; plants with long-creeping rootstocks 21

21 Perigynia rounded at base, suborbicular in cross-section; achenes loosely enveloped **C. crawei**

Perigynia tapering at base, triangular; achenes closely enveloped 22

22 Perigynia beakless; pistillate spikes usually not widely separated; lower peduncles rarely long **C. livida**

Perigynia with a beak about 1 mm long; pistillate spikes usually widely separated; lower peduncles long **C. vaginata**

23 Achenes with sides convex above, closely enveloped in upper part of perigynium-body **Group 8**
Achenes with flat or concave sides 24

24 Blades of bracts of non-basal pistillate spikes absent or rudimentary, their sheaths often more or less purple-tinged; achenes closely enveloped, filling perigynium 25
Blades of bracts well developed 27

25 Staminate spike 3–6 mm long; pistillate scales obtuse, ciliate, half as long as the perigynia; stigmas very short **C. concinna**
Staminate spike 8–25 mm long; pistillate scales acutish to acuminate, from somewhat shorter than to exceeding the perigynia; stigmas long, slender 26

26 Perigynia loosely pubescent, wider and longer than the ciliate scales; staminate spike sessile or nearly so; pistillate spikes 5–10-flowered; bracts very short-sheathing; culms smooth or nearly so; stigmas normally 4 **C. concinnoides**
Perigynia appressed-pubescent, narrower and shorter than the smooth scales; staminate spike noticeably peduncled; pistillate spikes 10–25-flowered; lowest bract long-sheathing, spathaceous; culms strongly roughened above; stigmas 3 **C. richardsonii**

27 Perigynia with strongly bidentate beaks; scales not strongly dark-tinged; lowland species 28
Perigynia with minute beaks obliquely cleft, becoming bidentulate; scales reddish brown to purplish black; alpine or mountain species 30

28 Perigynia 5–7 mm long, conspicuously nerved, finely pubescent; species of dry habitats **C. houghtoniana**
Perigynia 3–5 mm long, nerves obscured by the dense pubescence; species of damp or aquatic habitats 29

29 Leaves involute, 2 mm wide or less **C. lasiocarpa**
Leaves flat, 2–5 mm wide **C. lanuginosa**

30 Terminal spike gynaecandrous; lateral spikes pistillate **C. misandra**
Terminal spike usually androgynous; upper lateral spikes staminate or androgynous 31

31 Lateral spikes nearly as broad as long; perigynia 1.5–1.7 mm wide, dark brown; plants relatively small, 1–3 dm high **C. petricosa**
Lateral spikes much longer than broad; perigynia 2–2.5 mm wide, yellowish brown; plants usually tall, 3–9 dm high **C. franklinii**

GROUP 4 (Section **Heleonastes** Kunth)

1 Spikes androgynous; perigynia unequally biconvex **C. disperma**
Spikes (at least uppermost) gynaecandrous; perigynia planoconvex 2

2 Lowest bract bristle-form, much prolonged, many times exceeding its 1–5-flowered spike; spikes widely separate **C. trisperma**
Lowest bract much shorter or none; spikes several–many-flowered, the upper approximate 3

3 Perigynia beakless or nearly so; scales white-hyaline; rootstocks loosely caespitose and with slender stolons 4

Perigynia very short-beaked to strongly beaked; scales often darker-tinged; rootstocks densely caespitose 5

4 Spikes closely aggregated, forming an ovoid or suborbicular head; perigynia with slender nerves **C. tenuiflora**

Spikes remote; perigynia many-ribbed **C. loliacea**

5 Perigynia ovate, broadest near base, the beak conspicuous, strongly serrulate; spikes 5–15, closely aggregated **G. arcta**

Perigynia broadest near middle; beak short, smooth or moderately serrulate; spikes 2–9 6

6 Spikes 2–4, closely approximate; scales chestnut or reddish-tinged 7

Spikes 3–9, the lower at least remote; scales usually hyaline, sometimes brownish-tinged 8

7 Culms smooth or nearly so; perigynia brownish yellow, the beak prominent, hyaline at apex **C. bipartita**

Culms very rough above; perigynia cinereous or at length brown, the beak short, ferruginous at apex **C. heleonastes**

8 Perigynia 5–10 in each spike, distinctly beaked, the beak bidentate and rough-margined; leaf-blades 1–2 mm wide, not glaucous **C. brunnescens**

Perigynia 10–30 in each spike, the beak very short; leaf-blades mostly 2–4 mm wide, glaucous **C. curta**

GROUP 5 (Section **Ovales** Kunth)

1 Bracts either leaf-like or conspicuously exceeding the head 2

Bracts not leaf-like nor conspicuously exceeding the head 3

2 Bracts leaf-like, many times exceeding the head; perigynia subulate **C. sychnocephala**

Bracts conspicuously exceeding the head but not leaf-like; perigynia lance-ovate **C. athrostachya**

3 Perigynia with beak slender and nearly terete, scarcely margined at the tip, the upper 1–2 mm little if at all serrulate, sometimes becoming bidentulate, rarely bidentate 4

Perigynia with beak flattened and serrulate to the tip, often strongly bidentate 13

4 Scales about length of perigynia and nearly the same width above, concealing perigynia above or nearly so, often reddish brown 5

Scales shorter than the perigynia and noticeably narrower above, largely exposing the perigynia above, often dark chestnut, dull brown, or brown-black 8

5 Perigynia 6–8 mm long, oblong-lanceolate **C. petasata**

Perigynia 3.5–6.5 mm long 6

6 Culms and heads stiff; culms 1–3 dm high, in large clumps; spikes usually 3 or 4; perigynia oblong-ovate or obovate **C. phaeocephala**

Culms slender, 2–8 dm high; heads not stiff; spikes 2–8 7

7 Spikes 5–8, closely approximate except sometimes the lowest; perigynia 4–4.5 mm long, oblong-lanceolate **C. platylepis**

Spikes 2–7, in a flexuous or moniliform inflorescence; perigynia 4.5–6.5 mm long, ovate-lanceolate **C. praticola**

8 Perigynia 2.5–3.2 mm long, the body with 1 or 2 transverse folds; heads small, blackish **C. microptera**

Perigynia 3.5–6 mm long, the body without a conspicuous ventral fold; heads generally larger and not so dark 9

9 Perigynia plump, almost biconvex, the beak margined almost to tip, upper margin dark-green; achenes at least 1.75 mm long, generally quadrate **C. preslii**

Perigynia and achenes not as above 10

10 Distance from the top of achene to perigynium tip one-half or less than the total length of the perigynium; perigynia reddish to dark coppery-brown 11

Distance from top of achene to perigynium tip one-half or more than the total length of the perigynium; perigynia stramineous or light green to dark brown 12

11 Scale margins, perigynia tips, and dorsal suture margins noticeably white hyaline; perigynia wings darkened, contrasting with the body of the perigynia **C. macloviana**

Scale margins and perigynia tips not differentiated or narrow hyaline margined; perigynia wings concolorous with the body of the perigynia, wings at most dark edged **C. pachystachya**

12 Perigynia (4.0) 4.5–6.5 mm long; achenes 1.4–1.7 mm long; achene stipes (0.4) 0.5–0.7 mm long **C. haydeniana**

Perigynia 2.9–4.3 (4.7) mm long; achenes 1.1–1.5 mm long; achene stipes 0.3–0.5 mm long **C. microptera**

13 Scales about length of the perigynia and nearly the same width above, concealing the perigynia or very nearly so 14

Scales shorter than the perigynia and noticeably narrower above, largely exposing the perigynia above 16

14 Inflorescence moniliform, not stiff **C. aenea**

Inflorescence stiff; spikes aggregated or approximate 15

15 Lateral spikes ovoid-elliptic, round-tapering at base; perigynia closely appressed; bracts not prominent **C. xerantica**

Lateral spikes suborbicular, rounded at base; perigynia ascending in age; lower 1–2 bracts prominent, stiff **C. adusta**

16 Perigynia subulate to narrowly ovate-lanceolate, 3–4 times as long as wide, thin and scale-like except where distended by the achene 17

Perigynia ovate-lanceolate or broader, at most twice as long as wide 18

17 Perigynia subulate, the margin almost obsolete at base **C. crawfordii**

Perigynia lanceolate, margin conspicuous to base **C. scoparia**

18 Perigynia 4–5.5 mm long, broadly ovate to suborbicular, firm, coriaceous **C. brevior**

Perigynia 3–4 mm long, narrowly to broadly ovate, membranaceous 19

19 Perigynia 3–3.5 mm long, brownish at maturity; spikes closely aggregated, truncate to rounded at base **C. bebbii**

Perigynia 3.5–4 mm long, straw-coloured at maturity; spikes aggregated to quite separate, often short-clavate at base 20

20 Spikes closely aggregated; scales light reddish-brown or deep brown **C. tincta**

Spikes usually in a moniliform inflorescence; scales greenish-hyaline or tawny-tinged **C. tenera**

GROUP 6 (Section **Acutae** Fries)

1 Perigynia nerveless ventrally or with obscure impressed nerves 2

Perigynia distinctly nerved ventrally, the nerves raised 4

2 Scales clearly longer than the perigynia, divergent; perigynia inflated at maturity **C. aperta**

Scales about as long as the perigynia or slightly longer, appressed; perigynia not inflated at maturity 3

3 Lowest bract shorter than the inflorescence **C. scopulorum**

Lowest bract longer than the inflorescence **C. aquatilis**

4 Perigynia leathery, strongly ribbed, the beak 0.4–1 mm long, bidentate **C. nebraskensis**

Perigynia membranaceous, the beak apiculate, entire or nearly so 5

5 Terminal spike gynaecandrous; pistillate scales with a narrow or obsolete midrib and without hyaline margins **C. enanderi**

Terminal spike staminate; pistillate scales with a light centre and narrow hyaline margins 6

6 Upper lateral spike gynaecandrous; northern species **C. lenticularis**

Upper lateral spike pistillate; mountain species **C. kelloggii**

GROUP 7 (Section **Atratae** Kunth)

1 Culms lateral, arising outside tufts of leaves of the preceding year; not clothed at base with the dried-up fragments of the previous year, the lower sheaths conspicuously filamentose **C. buxbaumii**

Culms central, arising from the midst of the persistent leaves of the preceding year or their remains 2

2 Pistillate scales 1.5–2.5 mm long, obtuse or acutish, or the lower acute, broadly ovate, broadly obovate or orbicular; perigynia 2–3.5 mm long 3

Pistillate scales more than 2.5 mm long or sharp-pointed or both; perigynia 3 mm long or more 4

3 Culms stiff, obtusely triangular; scales about covering the perigynia, reddish-brown with prominent midvein; perigynia 2–2.3 mm long, slightly ciliate at mouth, scarcely beaked **C. parryana**

Culms not stiff, sharply triangular above; scales purple-black; perigynia 2–3.5 mm long, not ciliate at mouth of beak **C. norvegica**

4 Terminal spike staminate 5

Terminal spike gynaecandrous 8

5 Perigynia nearly circular in cross-section, little or not at all flattened **C. raynoldsii**
Perigynia strongly flattened 6

6 Plants phyllopodic; leaves crowded near base of culm; culms not purplish at base, clothed with dried-up leaves of previous year **C. paysonis**
Plants aphyllopodic; leaves of fertile culm widely separate, only the upper 2–5 with blades; culms purplish at base; leaves of previous year not prominent 7

7 Midvein of scales inconspicuous; lower spikes drooping **C. podocarpa**
Midvein of scales conspicuous, often excurrent as a short cusp; lower spikes erect **C. spectabilis**

8 Perigynia subinflated triangular, scarcely flattened **C. atrosquama**
Perigynia strongly flattened 9

9 Spikes 6–10; perigynia 5 mm long, lightly 3-nerved, light green, the walls papery; scales much shorter than the perigynia **C. mertensii**
Spikes 3–6; perigynia 2.5–4 mm long, nerveless or obscurely nerved on face, dull green, brownish or black, the walls not papery 10

10 Perigynia not granular-roughened, yellowish green or brownish **C. epapillosa**
Perigynia granular-roughened (under a lens) especially on the upper margins, green-brown or purplish black 11

11 Spikes usually not crowded, the lower nodding on the slender peduncles; perigynia green-brown **C. raymondii**
Spikes contiguous, sessile or short-peduncled; culms stiff, erect; perigynia purplish black **C. albo-nigra**

GROUP 8 (Section **Montanae** Fries)

1 Culms all elongate; the inflorescences all normally with the terminal spike staminate, none of them included within basal leaf-sheaths and wholly pistillate 2
Culms of various lengths, the longer ones usually with both staminate and pistillate spikes, the shorter more or less hidden in the basal leaf-sheaths and often wholly pistillate 4

2 Scales about as long as the perigynia; body of perigynium subglobose; plants densely caespitose **C. pensylvanica**
Scales much shorter than the perigynia; body of perigynium ellipsoid to fusiform obovoid; plants loosely caespitose 3

3 Perigynia 3–4 mm long, narrowed to a thick base; scales with white sides **C. peckii**
Perigynia 2.5–3 mm long, abruptly narrowed to a slender stipe; scales with purple sides **C. deflexa**

4 Staminate spike either separate or associated with a pistillate spike, in the latter case the subtending bract not exceeding the staminate spike; remnants of old leaves stiff and shredded **C. umbellata**
Staminate spike closely associated with a pistillate spike, the bract of the latter exceeding the staminate spike; remnants of old leaves soft, only slightly if at all shredded 5

5 Perigynia 2.5–3 mm long, with thick beak about a fourth as long as the body; staminate spike 2–5 mm long; leaves soft, often spreading; longer culms capillary, flexuous **C. deflexa**

Perigynia 3–4.5 mm long, with lanceolate beak two-thirds to fully as long as the body; staminate spike up to 1.5 cm long; leaves stiff, erect; longer culms stiffly erect **C. rossii**

C. adusta Boott
Caespitose from short rootstock; culms 2–8 dm tall, stiffly erect, rough just below the head, otherwise smooth; leaves 2–4 mm wide, firm; spikes 4–15, aggregated in stiff, erect, ovoid to linear-oblong head, 2–3 cm long; lowest bract dilated, cuspidate-prolonged, the upper scale-like; scales concealing the perigynia, broadly ovate, mostly reddish brown with narrow white-hyaline margins; perigynia 4–5 mm long, thick planoconvex, narrowly wing-margined, stipitate, serrulate above, the beak flat, bidentate. Dry soil. BC to Nfld s to Gt Lakes. Map 207 ●

C. aenea Fern.
Caespitose from short rootstock; culms 4–9 dm tall, wiry but often nodding, smooth; leaves flat, 2–4 mm wide, rather soft; spikes 3–10, in a flexuous moniliform inflorescence, 3–6 cm long, all separate or the upper 2 or 3 approximate; lowest bracts cuspidate; scales covering the perigynia, acute or acuminate, hyaline-margined; perigynia 4–5 mm long, ovate, concavoconvex, several-nerved dorsally, nerveless or more or less nerved ventrally, tapering into a beak half the length of the body. Dry places and open woods. Alas, Yuk, James Bay to Nfld s to BC, nIda, nMont, SD, Gt Lakes. Map 208

C. albo-nigra Mack.
Caespitose from short rootstocks; culms 1–3 dm high, rigid, erect, purplish red and fibrillose at base; leaves clustered at base, the blades 2.5–4 mm wide; spikes usually 3, the terminal gynaecandrous, the lateral pistillate, sessile or nearly so; scales mostly wider than and nearly as long as the perigynia, purplish black, conspicuously white-hyaline at apex and on the margins, the midvein nearly obsolete; perigynia 3–3.5 mm long, flat, purplish black, granular, the beak 0.4 mm long, bidentate. $2n = 52$. Exposed alpine tundra. Alas, Yuk, wMack s to Wash, Utah, Ariz, Colo. Map 209

C. aperta Boott
Loosely tufted from stout woody rootstocks; culms 3–10 dm tall, sharply triangular, rough above; leaf-blades 2–5 mm wide; terminal spike staminate; lateral spikes 2 or 3, mostly pistillate, sometimes staminate at apex; lower bracts leaf-like; pistillate scales longer than the perigynia, narrow, obtuse to acuminate, purplish or blackish with light midvein; perigynia about 3 mm long, obovoid-orbicular, flattish, inflated at maturity, nerveless, abruptly short-beaked. Low, wet ground. sBC to Ore, nwMont. Map 207 ○

C. aquatilis Wahlenb.
Densely tufted and forming long stolon-like rootstocks; culms 2–8 dm tall, reddish at base, sharply angled at summit; leaves often glaucous-green, the blades 2–5

mm wide; staminate spikes 1 or 2, the upper peduncled, the scales purplish black to brown; pistillate spikes 2–4, linear to oblong, 1–4 cm long, the upper occasionally staminate at apex; lower bracts leaf-like; pistillate scales purplish black, shorter than or exceeding the perigynia, obtuse to cuspidate; perigynia 2.5–3 mm long, erect-appressed, elliptic or elliptic-obovate, flat, nerveless or nearly so, often red-dotted, abruptly short-beaked. $2n = 76$. Several subspecies recognized but not readily distinguished; in our area ssp. **aquatilis**, ssp. **stans** (Drej.) Hult. (*C. stans* Drej.), and ssp. **altior** (Rydb.) Hult. (*C. altior* Rydb.). Marshy places. Circumpolar: throughout Can s to Calif, Ariz, NM, Neb, Ind, NJ. Map 210

C. arcta Boott
Culms soft, 1.5–6 dm tall, densely caespitose from short rhizomes; leaves soft, flat, 2–4 mm wide, often overtopping the culms; spikes 5–15, closely aggregated in an oblong head 2–4 cm long; perigynia 2–3 mm long, ovate, nerved on both sides, densely white-punctate, tapered into a flat serrulate beak; scales ovate, hyaline with green centre, often brownish. $2n = 60$. Moist woods. Alas, Yuk, Huds Bay to NB s to Calif, Ida, cSask, cMan, Gt Lakes. Map 211 ○

C. atherodes Spreng.
Loosely tufted and forming long stolon-like rootstocks; culms 3–12 dm high, stout, the basal sheaths purplish, bladeless and becoming fibrillose; leaves mostly with pubescent sheaths, the blades strongly septate-nodulose, 3–12 mm wide, glabrous above, often sparsely hairy below; staminate spikes 2–6, erect, the lowest rarely with a few perigynia; pistillate spikes 2–4, erect, widely separate, 2–12 cm long; bracts leaf-like, the lowest strongly sheathing; pistillate scales ovate, aristate, exceeding or shorter than the perigynia, reddish brown; perigynia 7–12 mm long, glabrous, strongly many-ribbed, tapering to a slender beak, the firm teeth 1.5–3 mm long. *C. trichocarpa* Muhl. var. *aristata* Bailey. Common in wet meadows. Circumpolar, boreal: Alas, Yuk, swMack to James Bay, sOnt, sQue s to Ore, Utah, Colo, Mo, NY. Map 212

C. athrostachya Olney
Forming small clumps, the rootstocks very short; culms 1–5 dm high, roughened above; leaf-blades flat, firm, roughened towards the apex; spikes 4–20, densely aggregated in an ovoid head 1–2 cm long; one to three of the lower bracts strongly developed, 1–5 cm long and exceeding the head, usually brownish at the base; scales lance-ovate, somewhat shorter than the perigynia, brownish with white-hyaline margins; perigynia 3–4 mm long, lanceolate-ovate, flat, tapering into a brownish beak. $2n = 68$. Marshy ground. Alas to Sask s to Calif, Colo, ND. Map 211 ●

C. atrosquama Mack.
Caespitose from short rootstocks; culms 2–5 dm high, slender, purplish red at base, the dried-up leaves of the previous year conspicuous; leaves with well-developed blades 7–10 to a fertile culm, on lower fourth; spikes 3 or 4, the terminal gynaecandrous, the lateral pistillate, the lower 1 or 2 on short erect

peduncles; scales somewhat shorter than the perigynia, purplish black with obsolete midvein; perigynia 3.2 mm long, olive-green, becoming yellowish green, sometimes purple-spotted, somewhat inflated, suborbicular, the beak 0.5 mm long, purplish black. C. *atrata* L. ssp. *atrosquama* (Mack.) Hult. Subalpine meadows. Alas, Yuk, wMack s to neOre, Ida, Mont; Colo. Map 213

C. aurea Nutt.
Loosely caespitose, stoloniferous; culms mostly 1 dm tall; leaf-blades 1–3 mm wide; bracts erect, the blades overtopping the inflorescence; terminal spike commonly staminate throughout, sometimes with a few perigynia at the top; lateral spikes usually 3–5, pistillate, 5–15 mm long; pistillate scales broadly ovate, often cuspidate, pale to reddish brown with a broad green midrib, wide-spreading at maturity; perigynia 2–3 mm long, elliptic-obovoid, plump, light glaucous green to orange-yellow or brownish, the latter fleshy and translucent when fresh, becoming white-pulverulent and coarsely ribbed when dried, beakless. $2n = 52$. Includes C. *garberi* Fern. of Ed. 1. Moist banks and meadows. Alas, Yuk, sMack to James Bay, Lab, Nfld s to Calif, Nev, Utah, NM, Neb, Gt Lakes. Map 214

C. backii Boott
Caespitose, forming dense mats; leaves abundant, usually dark green, the blades 3–6 mm broad; culms slender, weak, narrowly winged, from very short to 2.5 dm high, well exceeded by the leaves; sheaths thin and hyaline ventrally; spikes 1–3, androgynous; staminate part inconspicuous, with about 3 flowers, the scales somewhat tubular, their margins united below; pistillate flowers 2–5, on a zigzag rachis, the scales leaf-like, overtopping the perigynia and partly concealing them; perigynia ellipsoid-ovoid, 4–6 mm long, beak 1–2 mm long. Includes C. *saximontana* Mack. Dry, shady woods. sBC to NB s to Ore, Utah, Colo, Neb, Gt Lakes. Map 215

C. bebbii Olney *ex* Fern.
Caespitose; rootstocks very short; culms 2–8 dm high, slender, roughened beneath the head; leaves light green to yellowish green, flat, 2–4 mm wide; spikes gynaecandrous, truncate to rounded at base, 3–12, closely aggregated into an oblong to linear-oblong head 1–2.5 cm long, becoming brownish; one or two of the lower spikes with inconspicuous setaceous bracts; scales brown, their narrow tips exceeded by the perigynia-beaks; perigynia 3–3.5 mm long, narrowly ovate, nerveless or obscurely nerved, narrowly wing-margined, tapering into a flat serrulate beak. $2n = 68$. Marshes and low meadows. nBC, sMack, James Bay e to Nfld s to Ore, Mont, SD, Gt Lakes; Alas, Colo. Map 216

C. bipartita All.
Caespitose; culms slender, scabrous at summit, 0.5–3 dm tall; spikes 2–4, brownish, the terminal gynaecandrous, approximate in a head 0.5–2 cm long; perigynia planoconvex, pale brown to reddish, densely punctate, 2–3.5 mm long, beaked; scales ovate, obtuse, brownish with white-hyaline margins. $2n = 58$, c.60, 62, 64. C. *lachenalii* Schk., C. *lagopina* Wahl. Moist alpine areas. Circumpolar, arctic-

alpine: Alas e to sBaffin Is s to Mont, sMack, nMan, James Bay, nQue, Lab, Nfld; Utah, Colo; eQue (Gaspé). Map 217

C. brevior (Dewey) Mack.
Caespitose; rootstock short, somewhat woody; culms 3–10 dm high, rough beneath the head; leaves 1.5–4 mm wide, thickish; spikes gynaecandrous, usually 4–8, aggregated in a short stiff head or separated in a moniliform inflorescence 2–5 cm long; bracts scale-like, the lowest often prolonged; scales ovate, acuminate, yellowish brown with narrow hyaline margins; perigynia slightly exceeding the scales, broadly ovate to suborbicular, 4–5.5 mm long, brown, firm, strongly winged, striate dorsally, nerveless or faintly nerved ventrally, the beak flat, serrulate, 1 mm long. Fairly moist, open places. sBC to wQue s to Ore, NM, Ark, Tenn, DC. Map 218 ○

C. brunnescens (Pers.) Poir.
Densely caespitose; culms filiform but erect, mostly 1–4 dm tall, overtopping the leaves; leaf-blades 1–2 mm wide, not glaucous; spikes 5–10, short-ovoid or globose, separate or the uppermost approximate; perigynia planoconvex, ovate-elliptic, white-punctate, lightly nerved, 2–2.5 mm long, the distinct beak serrulate at base; scales ovate, silvery to brownish with a green midnerve. $2n = 56$. Wet woods and bogs. Circumpolar: Alas, Yuk, swMack to James Bay, nQue, Nfld, s to Ore, Utah, Colo, sSask, sMan, Gt Lakes, Apal. Map 218 ●

C. buxbaumii Wahlenb.
Loosely caespitose from long scaly rootstocks; culms 2–8 dm high, stiff, reddish purple at base; leaves with well-developed blades 2–4 to a fertile culm, on lower half, not clustered; spikes 2–5, erect or ascending, approximate or the lower more or less separate, the terminal gynaecandrous, usually 2–3 cm long, the lateral sessile or nearly so, pistillate; scales often awned and exceeding the perigynia, usually brownish; perigynia 3–4 mm long, glaucous-green, 2-ribbed and finely many-nerved, densely papillose, the minute beak 0.2 mm long, purplish. $2n = 106$. Boggy, often calcareous areas. More or less circumpolar: Alas, Yuk, Mack, James Bay, nQue, Nfld s to Calif, Utah, Colo, Ark, Ky, NC. Map 219

C. capillaris L.
Densely tufted; culms filiform, erect, 0.2–6 dm tall; leaves mostly basal, the old ones conspicuous, the blades 0.5–2 mm wide; terminal spike staminate or sometimes with a few pistillate flowers; pistillate spikes 2–4, on slender drooping peduncles, 6–20-flowered, slender, 4–17 mm long; bracts sheathing the lower leaf-like; pistillate scales shorter than the perigynia, broadly hyaline-margined; perigynia 2–4 mm long, greenish brown, lustrous, 3-angled, 2-ribbed, the beak 1 mm long. $2n = 52, 54$. In our area ssp. **capillaris**, 2–20 cm tall, spikes 4–10 mm long, 6–8-flowered, perigynia 2–3 mm long, and ssp. **chlorostachys** (Steven) Löve, Löve & Raymond, 20–60 cm tall, spikes 7–17 mm long, up to 20-flowered, perigynia 2.5–4.0 mm long. Springy and boggy places. Circumpolar, arctic-alpine: Alas to sBaffin Is n to Ellesmere Is s to Ore, Nev, Utah, NM, sSask, sMan, Gt Lakes, Nfld. Map 220

C. capitata L.
Caespitose plant with rootstocks ascending obliquely; culms wiry, 1–3 dm tall; leaf-blades filiform; spike solitary, ovoid, 5–10 mm long, androgynous, with 5–25 crowded perigynia; pistillate scales ovate-orbicular, obtuse, with broad hyaline margins; perigynia much longer than the scales, spreading, ovate, planoconvex, sharply margined, smooth, abruptly tapering into a slender beak, 0.5–0.9 mm long; stigmas 2; achenes lens-shaped. $2n = 50$. Boggy, often calcareous areas. Circumpolar: Alas, Yuk, sMack to Huds Bay, nQue, Nfld s to Calif, Nev, Utah, Wyo, cSask, cMan, James Bay, NH. Map 221 ○

C. chordorrhiza L.f.
Culms elongate, leaning or reclining, bearing a few tufts of leaves, and producing new culms from the axils of the old leaves; fertile culms 1–3 dm tall; leaf-blades 1–5 cm long; sterile culms with several much longer leaves; spikes 3–5, androgynous, in a crowded head 5–15 mm long; pistillate scales rounded, deep brown, acuminate; perigynia 2–3.5 mm long, compressed-ovoid, coriaceous, many-nerved, obscurely margined, the short beak emarginate. $2n = 60, 70$. Wet sphagnum bogs. Circumpolar: Alas to sBaffin Is to Nfld s to cBC, cSask, sMan, Gt Lakes. Map 221 ●

C. concinna R.Br.
Loosely caespitose, often with long slender rootstocks; culms 5–20 cm high, very slender, curved or ascending, much exceeding the leaves; leaves clustered near the base, recurved-spreading, the blades soft, 1–3 mm wide; terminal spike staminate, 3–6 mm long, often sessile; pistillate spikes 2 or 3, aggregated or the lowest remote, erect; bracts reduced to bladeless sheaths, green except for reddish-tinged tips; pistillate scales half the length of the perigynia, reddish brown with white-hyaline margins, obtuse, ciliate; perigynia about 3 mm long, greenish to yellowish white, densely hairy, blunt, the beak very short. Woods. Alas, Yuk, Huds Bay to Nfld transcont sCan; Ore; Wyo; Colo; SD; Mich. Map 222

C. concinnoides Mack.
Loosely caespitose, with long slender rootstocks; culms 15–35 cm high, slender, erect or somewhat curved, exceeding the leaves, smooth or nearly so, dark purplish brown at base; leaves clustered near the base, erect to recurved-spreading, 2–4 mm wide; terminal spike staminate, 8–20 mm long, sessile or nearly so; pistillate spikes 1 or 2, approximate, 5–10 mm long; bracts bladeless, short-sheathing, strongly purple-tinged and hyaline-margined; pistillate scales shorter than the perigynia, acutish or acuminate, ciliate, dark purple with hyaline margins; perigynia 2.5–3 mm long, light green, pubescent, rounded at apex with a beak about 0.5 mm long; stigmas usually 4, sometimes 3. Dry woods. BC to Calif e to neOre, wMont. Map 223

C. crawei Dewey
Forming small tufts from long, creeping rootstocks; culms 0.5–3 dm tall, stiff; leaf-blades stiff, pale green, 1–3 mm wide; terminal spike staminate, 1–2 cm

long; lateral spikes 2–4, pistillate, distant, short-peduncled or sessile, 1–2 cm long; bracts leaf-like, the upper reduced; pistillate scales much shorter than the perigynia, reddish brown with hyaline margins; perigynia about 3 mm long, often resinous-dotted and obscurely nerved, the beak very short. Calcareous meadows. BC to Nfld s to Wash, Utah, Wyo, Kans, Mo, Ala, NJ. Map 224 ○

C. crawfordii Fern.
Very densely caespitose; rootstock short; culms 1–6 dm tall, stiff, rough beneath the head; leaf-blades firm, yellowish green, mostly 1–3 mm wide; spikes gynaecandrous, usually 5–10, rather closely aggregated in a linear-oblong or oblong head 1–2.5 cm long; lower 1 or 2 bracts often setaceous, shorter than the head; scales much shorter than the perigynia, ovate or lanceolate, dull, light brown; perigynia lanceolate-subulate, 4 mm long, brownish, narrowly winged, obscurely nerved, the beak flat, serrulate, reddish brown tipped. $2n$ = c.66, 68*, c.70. Wet, non-boggy areas. BC, sMack to Nfld s to Wash, nIda, cSask, sMan, Gt Lakes, NJ; Alas, Yuk. Map 224 ●

C. curta Good.
Densely caespitose and glaucous; culms erect, usually 2–6 dm tall, mostly overtopping the leaves; leaf-blades 2–4 mm wide, glaucous; spikes 4–8, ovoid or short-oblong, greyish or silvery green, distant to approximate in an erect or flexuous head; perigynia ovoid-oblong, 2–3 mm long, densely punctate, becoming pale brown, the beak inconspicuous; scales hyaline or brownish with a green midvein. $2n$ = 54, 56, 62. *C. canescens* L. of Ed. 1. Bogs and swamps. Circumpolar: Alas, Yuk, to Huds Bay, nQue, Nfld s to Calif, Ariz, Colo, sSask, sMan, Gt Lakes. Map 225

C. deflexa Hornem.
Loosely tufted from ascending or horizontal rhizomes, tinged with purple-red at base, not fibrillose; culms 1–2 dm tall, slender; leaf-blades soft, 1–2 mm wide, often spreading; culms often of various lengths, the longer ones capillary; staminate spike 2–5 mm long, inconspicuous, often overtopped by the pistillate; lateral spikes commonly 2 or 3, pistillate, approximate, sessile or short-peduncled; lowest spike subtended by a leaf-like bract exceeding the inflorescence; scales shorter than the perigynia, acute to cuspidate, mostly purplish brown or reddish brown with very narrow hyaline margins; perigynia stipitate, 2.5–3 mm long, ellipsoid-obovoid, short-pubescent, 2-ridged, the short beak 0.5 mm long. Dry, open woodland. Greenland: Alas, Yuk, Mack to Huds Bay, nQue, Nfld s to cBC, cSask, cMan, Gt Lakes. Map 229 ○

C. deweyana Schwein.
Caespitose, lax and glaucous; rootstocks short; culms 2–8 dm long, much longer than the soft flat leaves; old leaves persistent; spikes 2–7, the terminal largest, gynaecandrous, in a flexuous inflorescence 2–6 cm long; lowest spikes quite remote (1–3 cm apart), usually subtended by a long slender bract; scales thin, whitish or brownish tinged; perigynia 3–12 to a spike, 4–5 mm long, ovate-lanceolate, the beak about 2 mm long, serrulate. $2n$ = 54, 56. Includes *C. leptopoda*

Mack. Moist shady woods. BC, sMack to Nfld s to Calif, Ariz, NM, Ia, Pa; Alas, Yuk. Map 226

C. diandra Schrank
In large tussocks from a short rootstock; culms densely caespitose, 3–7 dm tall, slender, stiff, harshly roughened on the angles; leaf-blades 1–2.5 mm wide, erect-ascending, the sheaths not copper-tinged at summit but often dotted with reddish brown ventrally; spikes numerous in a somewhat compound, dark brown, stiff, dense or slightly open head, 1–5 cm long; staminate flowers apical, inconspicuous; bracts awl-shaped, 1 cm long or less, often rudimentary or absent; scales brownish with hyaline margins; perigynia soon widely divergent, narrower than the scales, deltoid-ovate, biconvex, 2–2.7 mm long, dark brown and shining at maturity, with prominent nerves on the back, the beak long and strongly serrulate. $2n = 54, 60$. Bogs and peaty marshes. Circumpolar: Alas, Yuk, Mack, Huds Bay to Nfld s to Wash, Wyo, Neb, Ia, Pa; Utah, Colo. Map 227

C. disperma Dewey
Loosely caespitose, with very slender stolons; culms very slender and weak, 1–6 dm long; leaves flat, soft, the larger blades 1–2 mm wide; spikes 2–5, androgynous, scattered or the upper approximate, each with 1–6 perigynia and 1 or 2 apical staminate flowers; bract filiform or obsolete; perigynia ellipsoid-ovoid, finely nerved, 2–3 mm long, the minute beak 0.2 mm long; scales shorter than the perigynia, ovate-triangular, acuminate or mucronate, white-hyaline with green midvein. $2n = 70$. Damp woods and marshes. Circumpolar: Alas, Yuk to Huds Bay, nQue, Nfld s to Ore, Nev, Utah, NM, sSask, SD, Gt Lakes. Map 228

C. douglasii Boott
Rootstocks long-creeping, slender, brown, with fibrillose scales; culms scape-like, stiff, 0.5–3 dm high; leaves clustered near the base of the culms, 5–20 cm long; spikes in dense, usually dioecious, heads; staminate heads 1.5–4 cm long, the scales acute or acuminate, straw-coloured or brownish; pistillate heads 1.5–5 cm long and 1–2.5 cm broad, the scales acuminate to cuspidate; perigynia ovate-lanceolate, 3.5–4 mm long, the slender beak more than half as long as the body, serrulate. Dry soil, often alkaline, in open places. sBC to sMan s to Calif, NM, Neb, Ia. Map 229 •

C. eburnea Boott
Densely caespitose from elongate rootstocks; culms capillary, wiry, smooth, 1–3 dm tall; leaf-blades filiform, involute, 0.5 mm wide or less; spikes subtended by short bladeless sheaths, white-hyaline and tinged with yellow or brown ventrally; pistillate spikes 2 or 3, on long erect peduncles, overtopping the sessile, linear, staminate spike; scales shorter than the perigynia, broadly ovate, whitish or pale brown, caducous; perigynia only 2–6 to a spike, obovoid, 1.5–2 mm long, olive-green, becoming brown, abruptly narrowed to a minute beak. Moist calcareous areas. Alas, Yuk, wMack to James Bay, Nfld s to BC, sSask, Tex, Neb, Ala, Va. Map 230 •

C. enanderi Hult.
Rootstocks slender, vertical; culms 1.5–3 dm tall, slender, somewhat longer than the leaves, these 2–4 to a culm, mostly clustered near the base; blades 0.5–1.5 mm wide; inflorescence congested, with short, thick, sessile or subsessile spikes; terminal spike gynaecandrous; lateral spikes 2–3, pistillate, 7–10 mm long; bracts usually 2, the lowermost exceeding the inflorescence, with short, coloured open sheaths; pistillate scales oblong-lanceolate, much shorter and narrower than the perigynia, blunt, dark purple; perigynia 3–3.5 mm long, ovate-lanceolate, biconvex, heavily 12–20-nerved, strongly stipitate (stipe 0.5 mm long), the body smooth and white, the apex coloured and tapering to an apiculate beak; achene filling only the lower half of the perigynium. *C. eurystachya* F.J. Hermann of Ed. 1. Clay-gravel shores of mountain lakes. sAlas, BC, Alta. Map 231

C. epapillosa Mack.
Caespitose, the rootstocks short; culms 1.5–6 dm high, purplish red and brown-fibrillose at base, the old leaves conspicuous; well-developed leaves 5–8 to a fertile culm, on lower fourth, the blades 3–7 mm wide, stiff, erect; spikes usually 4 or 5, the terminal gynaecandrous, the lateral pistillate, the upper closely crowded, the lowermost on short erect peduncles; scales about the length of the perigynia but much narrower, blackish with light midvein, mostly acuminate or cuspidate; perigynia flat, not inflated, 3–4 mm long, not at all papillose, yellowish green, puncticulate, the beak 0.5 mm long, purple, bidentate. Mountain meadows. sBC, Alta s to Calif, Mont, Colo. Map 230 ○

C. filifolia Nutt.
Very densely caespitose; rootstocks very short, stout; old basal sheaths conspicuous, striate, brownish, fibrillose; culms 8–30 cm tall, slender but wiry; leaf-blades filiform, 3–20 cm long, involute; spike solitary, narrow, 1–2 cm long, androgynous, densely flowered; pistillate scales obtuse, light reddish brown with very broad hyaline margins; perigynia 3 mm long, obovoid, more or less triangular, puberulent above, abruptly and minutely stout-beaked; stigmas 3; achenes triangular. Dry plains and ridges. Alas, Yuk, wMack to cSask, sMan, s to Ore, Nev, Utah, NM, Tex, Neb. Map 232

C. flava L.
Culms tufted, firm, 1–6 dm high; leaves flat, 2–5 mm wide, soft but stiffish, often yellowish green; staminate spike usually 1–2 cm long, sometimes partly pistillate above; pistillate spikes 2–5, globose to short-cylindric, 1–2 cm long, the upper crowded, the lower often remote; bracts leaf-like, divergent or reflexed, the lowest sheathed; pistillate scales shorter than the perigynia, tinged with copper-brown, acuminate; perigynia yellowish, often becoming brown, spreading or reflexed, 4–6 mm long, strongly ribbed, the beak 2–3 mm long, bidentate. $2n = 60, 70$. Marshy places. More or less circumpolar: sAlas to sBC, Mont; sMan, James Bay to Nfld s to Gt Lakes. Map 233

C. franklinii Boott
Loosely caespitose, the rootstocks slender and elongate; culms mostly 3–9 dm high, exceeding the leaves which are clustered above the base; leaf-blades stiffish, 1–3 mm wide; spikes 3–8, the uppermost (3 or 4) aggregated, androgynous or staminate, the lateral ones (2 or 3) not approximate, pistillate or androgynous; bracts sheathing, the lowest with prolonged blade; scales shorter than the perigynia, mostly reddish brown; perigynia 4.5–6 mm long, 2–2.5 mm broad, strongly flattened, ciliate on margins, yellowish brown, finely nerved, abruptly tapered to a minute beak, 0.2 mm long; stigmas 3 (rarely 2). Dry to moist open areas to alpine elevations. Alas, Yuk, wMack s to BC, Alta. Map 234 ●

C. geyeri Boott
Loosely caespitose; rootstocks long, thick, woody, scaly; culms 1–4 dm high, stiff, very rough; leaves leathery, very rough on the margins, 1–3 dm long, 2–3.5 mm wide, those of the culms very short at flowering time, later elongating, sterile shoots numerous; spike solitary, androgynous, the staminate part in age short-peduncled, the pistillate part loosely flowered; perigynia 1–3, oblong, ovoid, 2-keeled, 6 mm long, glabrous, tapering at the base, very minutely beaked; scales with broad hyaline margins; stigmas 3; achenes triangular. Open woods and dry mountain slopes. BC to nCalif, Utah, Colo. Map 234 ○

C. glacialis Mack.
Caespitose, from short rootstocks; culms 4–15 cm tall, reddish purple at base; leaves basal, 1.0–1.5 mm wide, flat or channelled, shorter than culms, those of former seasons persistent; spikes 2–4, 2–7 mm long, terminal one staminate, lateral pistillate, aggregated to form a loose head; scales as long as or shorter than perigynia, acute to obtuse entire or somewhat lacerate, reddish brown with a broad hyaline margin; perigynia 2.0–2.5 mm long, obglobose, faintly ribbed, greenish below, chestnut to brown above, beak 0.3–0.5 mm long, terete, apex hyaline; stigmas 3. $2n = 34$. Alpine slopes. Circumpolar: Alas, Victoria Is to sBaffin Is to Ellesmere Is s to Alta, nSask, James Bay, nQue, Nfld. Map 234 ▲

C. gynocrates Wormsk.
Plants with long rootstocks; culms slender but stiff, 4–30 cm tall; leaf-blades filiform; spikes solitary, entirely staminate or entirely pistillate or occasionally androgynous; scales sharp-pointed, usually reddish brown; perigynia 4–10, erect, spreading or reflexed, about 3 mm long, glabrous, nerved, abruptly narrowed into a beak about 0.5 mm long; stigmas 2; achene lens-shaped. $2n = 48, 50$. *C. dioica* L. ssp. *gynocrates* Hult. Sphagnum bogs and marshy places. eAsia, Greenland: Alas, Yuk, Mack to Huds Bay, sBaffin Is, Nfld s to sBC, Mont, sSask, sMan, Gt Lakes; Ore, Nev, Utah, Colo. Map 235

C. haydeniana Olney
Densely caespitose; rootstocks short; culms 1–3.5 dm long, erect, curving or decumbent; leaf-blades 2–3 mm wide; spikes 4–7, gynaecandrous, in a dense head 1–1.8 cm long; bracts scale-like or the lowest more or less prolonged but shorter than the head; scales ovate, acute, blackish with lighter centre and hyaline

margins; perigynia usually ovate, 4.5–6 mm long, very flat and strongly winged, dark-tinged, contracted into a serrulate beak half the length of the body. $2n = 82^*$. *C. macloviana* D'Urv. ssp. *haydeniana* Taylor & MacBryde. A quite distinctive species in our area. Moist open places at subalpine to alpine elevations. sBC to Calif, Utah, Nev, Colo. Map 236 ○

C. heleonastes Ehrh.
Loosely caespitose with slender rootstocks; culms 1.5–3.5 dm high, slender but stiffly erect, very rough at least above; leaves 4–8 to a fertile culm, the blades 1–2 mm wide, more or less involute; spikes 2–4, gynaecandrous, distinct but closely aggregated, the head 8–18 mm long; bracts scale-like, occasionally short-cuspidate; scales obtuse, reddish brown with yellow-brown centre and broad white-hyaline margins; perigynia elliptic-obovate, planoconvex, 2.5–3 mm long, their tips exceeding the scales, densely punctate, finely nerved on both sides, the beak short, reddish brown. $2n = 56$. Bogs and marshes, often calcareous. More or less circumpolar: BC to nwMan, James Bay; Alas, Yuk; Lk Superior; nQue. Map 236 ●

C. hoodii Boott
Densely caespitose; rootstocks, short, stout, black; culms 2.5–8 dm high, brownish at base; leaf-blades 1.5–3.5 mm broad; spikes 4–8, androgynous, in a dense head 1–2 cm long and 8–15 mm thick, the lowermost rarely slightly separate; scales ovate-triangular, acute to cuspidate, brownish with hyaline margins; perigynia 5–10 to a spike, ovate-elliptic, 3.5–5 mm long, the beak about 1 mm long. $2n = 58, 60$. Dry to moist open slopes. BC, Alta to Calif, Utah, Colo, SD; swSask. Map 237

C. hookerana Dewey
Caespitose; rootstocks stout, black, fibrillose; culms very slender, 1.5–3 dm high, brownish at base; leaves flat, thin, 1–2.5 mm broad; spikes 5–10, androgynous, the lower 3–6 separate, the upper ones aggregated, the whole head slender, often rather flexuous, 2–5 cm long; bracts well-developed, enlarged at base and prolonged as a cusp, sometimes exceeding their spikes; scales prominently awned; perigynia ovate-lanceolate, 2.7–3.5 mm long, narrowed into a serrulate beak scarcely half the length of the body. Plains, dry banks, and open woods. sAlta to sMan and ND. Map 238

C. houghtoniana Torr.
Loosely tufted, forming long scaly stolon-like rhizomes; culms 2–6 dm tall, stoutish, purplish at base; leaf-blades flat, 2–5 mm wide; staminate spike usually solitary, sometimes with small secondary ones; pistillate spikes 1–3, sessile or short-peduncled, erect, widely separated; lowest bract sheathless or with a very short sheath, prolonged; scales shorter than the perigynia, reddish brown , acute or awned; perigynia ovoid, 5–7 mm long, strongly nerved, short-hirsute, the bidentate beak about 2 mm long. Dry, sandy, or gravelly places. Alta to Nfld s to Gt Lakes. Map 239 ●

C. hystricina Muhl. *ex* Willd.
Tufted and forming a few long horizontal rootstocks; culms 2–10 dm tall, slender, sharply triangular; leaves flat, mostly 3–8 mm wide; staminate spike solitary; pistillate spikes 1–4, oblong-cylindric, 1–6 cm long, the lower nodding on long peduncles; bracts leaf-like; scales shorter than the perigynia, with small hyaline bodies and long rough awns; perigynia narrowly ovoid, 5–7 mm long, membranous, glabrous, shining, strongly ribbed, tapering into a deeply bidentate beak about 2 mm long. $2n = 58$. Shady marshes. sBC to Nfld s to Calif, NM, Tex, Tenn, Va. Map 239 ○

C. illota Bailey
Caespitose, from short rootstocks; culms 1–4 dm tall; leaves flat, 1–3 mm wide, shorter than culms; spikes several, sessile, gynaecandrous, 4–7 mm long, closely aggregated to form a head in which the individual spikes are not easily distinguished; scales obtuse, shorter and generally narrower than the perigynia, dark brown to greenish black, shiny; perigynia spreading, 2.5–3.2 mm long, ovate-lanceolate, planoconvex, sharp-edged but not winged, nor serrulate, green to brown; stigmas 2. $2n = 64$. Moist places in mountains. sBC to Calif, Mont, Wyo, Colo. Map 240 ○

C. interior Bailey
Densely caespitose, the rootstocks short; culms 1.5–5 dm tall, slender but wiry; leaves 1–3 mm wide; spikes 2–4, rarely 6, distant or approximate, the terminal spike usually gynaecandrous and often with a prolonged staminate base; bracts scale-like or the lowest prolonged; scales obtuse, brownish with scarious margins; perigynia 1–10 to a spike, wide-spreading or recurved at maturity and brownish, 2.2–3.2 mm long, oblong-ovoid to ovoid, plump, the beak 0.5–0.8 mm long, sparingly serrulate. Bogs and swamps. sAlas, sYuk, sMack to James Bay, Nfld s to Calif, NM, Kan, Gt Lakes. Map 241

C. kelloggii Boott
Tufted, forming short or long slender rootstocks; culms 1–6 dm tall, often shorter than the leaves; leaf-blades about 2 mm wide, long-attenuate; terminal spike staminate; lateral spikes 3–5, pistillate, sessile or short-peduncled; lowest bract leaf-like; scales usually shorter than the perigynia, purplish brown with light centre and narrow hyaline margins; perigynia soon deciduous, ovate, 2–3 mm long, lightly several-nerved on each side, stipitate, abruptly apiculate-beaked. $2n = 88$. Often included in *C. lenticularis* Michx., perhaps rightly so. Moist banks and lake shores. Alas, swYuk to Calif, Nev, Utah, Colo. Map 240 ●

C. lacustris Willd.
Caespitose and with long creeping rootstocks; culms 6–12 dm high, stout, sharply angled; basal sheaths purplish, bladeless, becoming fibrillose; leaf-blades glaucous, 5–15 mm wide; staminate spikes 2 or 3; pistillate spikes 2 or 3, erect or the lower sometimes drooping, 2–10 cm long; lowest bract leaf-like; scales ovate, about half as long as the perigynia, purplish; perigynia oblong-ovoid, 5.5–7 mm long, olive-green, glabrous, lightly many-nerved, tapering to a thick bidentate

beak, the teeth about 0.5 mm long. $2n = 74$. Marshes and swampy woods. BC to Nfld s to Ida, SD, Mo, DC. Map 242 ○

C. lanuginosa Michx.
Similar to *C. lasiocarpa* but readily distinguished from it by the characters of the key. *C. lasiocarpa* var. *latifolia* (Böck.) Gleason of Ed. 1. Marshy places. BC to Nfld, generally distributed s. Map 243

C. lasiocarpa Ehrh.
Tufted and forming horizontal scaly rhizomes; culms 4–10 dm tall, often dark purplish red at base; leaves very long and attenuate; staminate spikes 1–3, on a scabrous peduncle; pistillate spikes 1–3, widely separate, erect, sessile or nearly so, mostly 1–3 cm long; lowest bract sheathless or short-sheathing, usually overtopping the inflorescence; pistillate scales, from shorter than to exceeding the perigynia, reddish brown or purplish, acute or acuminate-mucronate; perigynia oblong-ovoid, 3–5 mm long, densely hairy, the nerves obscured by pubescence; the bidentate beak 1 mm long. $2n = 56$. In our area ssp. **americana** D. Löve & Bernard. Bogs and lake shores. Circumpolar: BC, sMack to James Bay, Nfld s to Wash, nIda, cSask, Ia, Ohio, NJ; Alas, Yuk. Map 244

C. lenticularis Michx.
Densely tufted from a short rootstock; culms 1–6 dm tall, erect, slender; leaves 1–2 mm wide, long-attenuate; terminal spike staminate, sometimes with a few perigynia; lateral spikes 3–5, pistillate or gynaecandrous, erect, the upper sessile, the lower peduncled; lowest bract leaf-like, erect, short-sheathing; pistillate scales obtuse, purplish or reddish brown with broad 3-nerved centre and hyaline margins and apex; perigynia soon deciduous, 2–3 mm long, lightly few-nerved on each side, short-stipitate, abruptly apiculate-beaked. Moist shores. BC, Mack to Nfld s to nAlta, cSask, sMan, Gt Lakes. Map 245

C. leptalea Wahlenb.
Densely tufted but with slender elongated rootstocks; culms slender, 2–6 dm high; leaf-blades 0.5–1.3 mm wide; spike solitary, 4–15 mm long, androgynous; scales much shorter than the perigynia, yellowish green with green midrib and hyaline margins; perigynia 1–10, 2.5–5 mm long, oblong-elliptic, 2-edged, many-nerved, rounded at apex, beakless; stigmas 3, short; achenes triangular. $2n = 52$. In our area ssp. **leptalea**. Shady bogs and moist woodland. Alas, Yuk, Mack to James Bay, nQue, Nfld, generally distributed s. Map 242 ●

C. limosa L.
Spreading by slender elongated rootstocks; roots long, felt-covered; culms solitary or in small clusters from old rootstocks, usually 2–4 dm tall; leaves 1–3 mm wide; terminal spike staminate, 1–3 cm long; pistillate spikes 1–3, short-cylindric, 1–2 cm long, drooping; lowest bract 2–6 cm long; scales ovate to elliptic, almost concealing the perigynia, brown or purple or straw-coloured; perigynia elliptic-ovoid, 2.5–4 mm long, glaucous-green, 2-ribbed and obscurely few-nerved, minutely beaked. $2n = 62, 64$. Bogs and marshes. Circumpolar: Alas, Yuk, Mack

to Huds Bay, nQue, Nfld s to Calif, Mont, sSask, sMan, Gt Lakes; Nev, Utah. Map 246

C. livida (Wahlenb.) Willd.
Spreading by long slender rootstocks and forming small clumps; culms 1–5 dm tall, erect, smooth; leaf-blades glaucous, stiff, thickish, deeply channelled, 1–3 mm wide; terminal spike staminate or rarely gynaecandrous, 1–2 cm long; pistillate spikes 1–3, approximate or somewhat separate, nearly sessile or the lowest long-peduncled; scales ovate, shorter than the perigynia, purple-brown with broad lighter centre, perigynia glaucous-green, 2.5–4.5 mm long, fusiform to obscurely triangular, tapering at both ends, nerved, beakless. $2n = 52, 54$. Wet, peaty calcareous bogs. More or less circumpolar: sAlas, sYuk, nSask to James Bay, Nfld s to Wash, sSask, sMan, Gt Lakes; nMack; Calif. Map 247 ○

C. loliacea L.
Loosely caespitose with slender stolons; culms weak, 2–4 dm tall; leaf-blades 0.5–2 mm wide; spikes 2–5, gynaecandrous, the upper contiguous, the lower separate, forming a head 1–2.5 cm long; lowest bract 2–8 mm long; scales ovate, keeled, white-hyaline with green midrib; perigynia 3–8 to a spike, 2.5–3 cm long, thick planoconvex, finely ribbed, densely white-punctate, beakless. $2n = 54$. Marshes and moist banks. More or less circumpolar: Alas, Yuk, Mack, nMan, cOnt s to cBC, cSask. Map 247 ●

C. macloviana D'Urv.
Densely caespitose from short rootstock; culms 1–5 dm high, stiff, striate, smooth below, scabrous and sharply angled above; leaves short, firm, mostly 2–4 mm wide; spikes 3–8, mostly divergent and crowded in an ovoid head 1–2 cm long; bracts scale-like; scales ovate, obtuse, brownish-copper-coloured or purple-black, with hyaline margins; perigynia ovate, 3–4 mm long, greenish or copper-brown to dark brown, finely nerved dorsally, the beak slender. $2n = 86^*$. Includes *C. incondita* F.J. Hermann of Ed. 1. Moist grassy slopes. Greenland, Norway: Mack to nQue, Lab s to Wyo in Rocky Mts. Map 248

C. maritima Gunn.
Low plant, extensively creeping, with slender but tough, conspicuously scaly rootstocks and stolons; culms solitary, usually 2–10 cm high, sometimes curved; leaves in ascending tufts, often longer than the culms; spikes 3–5, crowded in a dense head, 5–15 mm long, androgynous; bracts below lowest spike papery, brown, obtuse or mucronate; scales ovate-orbicular, brown, scarious; perigynia divergent at maturity, 3.5–5 mm long, usually ovate, faintly to conspicuously nerved, stipitate, beak short. $2n = 60$. Includes *C. incurviformis* Mack. of Ed. 1. In our area var. **incurviformis** Boivin. Gravelly alpine areas. Circumpolar: Alas to nBaffin Is to Ellesmere Is s to nBC, sMack, Huds Bay, nQue, Nfld; swAlta (adventive in Edmonton), Colo. Map 249

C. mertensii Prescott
A handsome sedge; densely caespitose, the rootstocks stout and very short; culms 3–10 dm tall, very sharply triangular and winged, purplish red at base; well-developed leaves 3–6 to a fertile culm, regularly disposed on lower half, the blades 4–7 mm wide, the sheaths often tinged with cinnamon-brown; spikes 5–10, the terminal gynaecandrous or staminate, the lateral pistillate with a few staminate flowers at base, drooping on slender peduncles; scales much shorter than the perigynia, dark purplish-brown with light-coloured midvein and hyaline margins; perigynia 4.5–5 mm long, very thin and flattened, light green, becoming whitish or brownish, sometimes purple-spotted; beak 0.2–0.5 mm long, purplish. $2n = 62$. In our area ssp. **mertensii**. Moist woods and stream banks. Japan: sAlas, BC to Calif, Ida, wMont. Map 250 ○

C. microglochin Wahlenb.
Rootstocks long and slender; culms slender, stiff, 8–25 cm tall; leaves 4–8, clustered near base of culm, the blades stiff, involute; spike solitary, narrow, androgynous; pistillate scales obtuse, soon falling; perigynia 3–12, slender, short-stipitate, 4–5 mm long, straw-coloured, smooth, subulate-beaked, spreading or reflexed; achenes linear-oblong, triangular. $2n = 48, 58$. Peaty marshes, often calcareous. More or less circumpolar: Alas to Baffin Is s to swAlta, Mack, nMan, James Bay, Nfld; Colo. Map 250 ●

C. microptera Mack.
Caespitose from a short rootstock; sterile shoots conspicuous, the leaves near the top; culms 2–10 dm tall, smooth, biennial; leaves (1.5) 2–4.5 mm wide; spikes (4) 5–10, gynaecandrous, crowded in a truncate head 1–2 cm long; lowest bracts awned, the upper scale-like; scales narrower and slightly shorter than the perigynia, ovate-lanceolate, brown; (2–5) 3.0–4.5 mm long, lanceolate-ovate, pale green to brown or black, narrowly winged, tapering to a beak 1.2–1.5 mm long, somewhat serrulate and bidentate. $2n = 80^*, 82, 90$. Includes C. *festivella* Mack. and C. *limnophila* F.J. Hermann of Ed. 1. Moist woods. sBC to swMan s to Calif, Ariz, NM; SD, swYuk, sMack. Map 251 ●

C. misandra R.Br.
Caespitose, from short rootstocks; culms 1.0–3.5 dm tall; leaves basal, 1–3 mm wide, flat or channelled, shorter than culms, those of former seasons largely persistent; bracts conspicuous, sheathing; spikes 2–4, 5–15 mm long, on slender peduncles, often nodding, terminal one gynaecandrous, lateral pistillate; scales lance-ovate, 2.5–3.5 mm long, brownish black with hyaline margin; perigynia lanceolate, 3.5–5.0 mm long, dark brown to black, wingless, beak flattened, strongly serrulate; stigmas 3. $2n = 40$. Dry alpine slopes. Circumpolar: Alas to Ellesmere Is s to swAlta, sMack, Huds Bay, Lab; Utah, Colo. Map 251 ○

C. nardina Fries
Densely tufted; culms slender, 3–10 cm tall, usually surpassed by the filiform blades; basal sheaths persistent, conspicuous; spike solitary, 5–10 mm long, ovoid-orbicular, with 2–5 staminate flowers at apex and several pistillate ones

below; pistillate scales dark brown, broadly ovate, obtuse; perigynia 3–3.5 mm long, short-stipitate, short-beaked; stigmas 2; achenes lens-shaped. $2n$ = 68, 70. Dry alpine slopes. Greenland, Norway: Alas to nBaffin Is n to Ellesmere Is s to Wash, Colo, sMack, nMan, James Bay, nQue; Nev. Map 252 ●

C. nebraskensis Dewey
Caespitose and spreading by stolon-like rootstocks; culms 3–10 dm tall, stout, sharply triangular; dried-up leaves of the previous year very conspicuous at the base; leaf-blades 4–8 mm wide, flat; spikes 3–7, the upper 1–2 staminate, the lower pistillate; lowest bract leaf-life, sheathless, the upper reduced; pistillate scales lanceolate, purplish or brownish black with lighter centre; perigynia 3–3.5 mm long, oblong-ovate, flattened, straw-coloured, red-dotted, strongly many-ribbed; beak 0.4–1 mm long, bidentate. Marshy ground. sBC, Alta s to Calif, Ariz, NM, SD, Neb, Kan. Map 252 ○

C. nigricans C.A. Meyer
Loosely caespitose, with stout creeping rootstocks; culms 5–20 cm tall, stiff, firm, smooth; leaves 4–9 to a fertile culm, the blades flat, 1.5 mm wide or more; spike solitary, 8–15 mm long, androgynous, the pistillate scales dark brown; perigynia 4 mm long, stipitate, glabrous, shining, nerveless, long-beaked; stigmas 3, elongate; achene triangular. $2n$ = c.72. Moist alpine areas. sAlas, swYuk to Calif, Utah, Colo. Map 253

C. norvegica Retz.
Loosely tufted, with short stolon-like rootstocks; culms 2–6 dm tall, slender, erect or arched; leaves bunched near the base, the blades soft, 2–3 mm wide; spikes usually 3, the terminal gynaecandrous, the lateral pistillate; scales much shorter than the perigynia, about 2 mm long, broadly ovate, purple-black, the margins white-hyaline; perigynia 2–3.5 mm long, inflated, 2-ribbed, becoming pale brown, glabrous, somewhat granulose; beak 0.5 mm long. $2n$ = 54, 56. *C. media* R.Br. of Ed. 1. Moist woods and stream banks. Circumpolar: Alas, Yuk, Mack to Huds Bay, nQue, Lab s to sBC, Utah, NM, cSask, sMan, Gt Lakes. Map 254

C. obtusata Lilj.
Rhizomes long-creeping, slender but tough, scaly, purplish black; culms low, 6–15 cm high, sharply triangular, rough above, noticeably exceeding the short leaves; leaf-blades 1–1.5 mm wide, involute; spike solitary, 5–12 mm long, androgynous, the upper two-thirds staminate; perigynia 1–6, ovoid, glabrous, often blackish brown, coriaceous, shining, impressed-nerved, 3–4 mm long, exceeding the pale scales; beak obliquely cut; stigmas 3; achenes triangular. $2n$ = 52. Common in prairie grassland and on dry ridges and hills. Eurasia: Alas, Yuk, nwMack to cSask, sMan s to BC, NM, SD, Minn. Map 255

C. oligosperma Michx.
Rootstocks elongate; culms very slender, stiff, 4–10 dm tall; leaf-blades and bracts filiform-involute, stiff; staminate spike solitary; pistillate spikes 1–3, remote,

globose to short-oblong, 1–2 cm long, with 3–15 flowers; scales broadly ovate, about half as long as the perigynia, chestnut-brown; perigynia ovoid, 4–7 mm long, the beak 1–2 mm long, short-bidentate. Wet meadows and bogs. swMack, nAlta, nSask to James Bay, Lab, Nfld s to sMan, Wisc, Ind, Pa, Conn. Map 256 ○

C. **pachystachya** Cham.
Densely caespitose from short rootstocks; culms 3–10 dm high, much exceeding the leaves, conspicuously striate; leaves well developed, flat, 2–4 mm wide; spikes gynaecandrous, 4–12 crowded in a dense head 1–2.5 cm long; bracts scale-like or the lowest awned; scales about the width of the perigynia, ovate, acute, mostly brownish, scarcely hyaline-margined; perigynia ovate, 3.5–5 mm long, serrulate to the middle, copper-coloured at maturity, several-nerved dorsally, the beak about half the length of the body. $2n = 74, 76, 78$. Moist woods. sAlas, Yuk to Calif, Colo. Map 257

C. **parryana** Dewey
Loosely tufted, putting out scaly stolon-like rootstocks; culms 1.5–4 dm high, slender, stiff, usually well exceeding the leaves which are clustered near the base; leaf-blades thin but stiff; spikes 3–5, the terminal gynaecandrous or sometimes nearly staminate, the lateral pistillate, oblong to linear-oblong, erect, approximate; scales concealing the perigynia, dark reddish brown with white-hyaline margins and prominent midvein; perigynia 2–2.3 mm long, 2-ribbed, scarcely beaked, bidentulate and slightly ciliate at mouth. In our area ssp. **parryana**. Moist habitats. sBC to sMan s to Utah, Colo, Neb; sAlas, swYuk. Map 258

C. **pauciflora** Lightf.
Rootstocks long and slender; culms slender, stiff, 8–25 cm tall; leaves 2–3, near the base of the culm, the blades stiff, involute; spike solitary, narrow, androgynous; pistillate scales obtuse, soon falling; perigynia 1–6, slender, short-stipitate, smooth, obscurely nerved, straw-coloured, 6–7 mm long, subulate-beaked, spreading or reflexed; achenes linear-oblong, triangular. $2n = 38, 44, c.74, 76$. Sphagnum bogs. Circumpolar: sAlas, BC, Wash to Nfld s to cSask, sMan, Gt Lakes. Map 259 ●

C. **paupercula** Michx.
Resembling *C. limosa;* distinguished by the characters of the key. $2n = 58$. *C. magellanica* Lam. ssp. *irrigua* (Wahl.) Hult. Sphagnum bogs. Circumpolar: Alas, Yuk, Mack to James Bay, nQue, Nfld s to BC, Utah, Colo, cSask, sMan, Gt Lakes. Map 260

C. **paysonis** Clokey
Loosely caespitose from tough, branching, densely matted rootstocks; culms 1.5–5 dm high, stiff, erect, brownish and fibrillose at base; well-developed leaves 8–15 to a fertile culm, clustered near the base; terminal 1–2 spikes staminate; lateral 2–6 spikes pistillate, erect or spreading, the upper sessile, the lower on short peduncles; scales narrower than the perigynia and nearly the same length, purplish black, with a more or less prominent midvein; perigynia 2–4 mm long,

flat, purple-blotched, purple-tipped, the beak 0.2 mm long; achene short-stipitate. C. *tolmiei* Boott of Ed. 1. Mountain meadows. sBC to Ore, Utah, Wyo. Map 259 ○

C. peckii Howe
Loosely caespitose from short spreading rhizomes; culms 1–5 dm tall, well exceeding the leaves, reddish purple at base; leaf-blades soft, pale, 1.5–3 mm wide; staminate spike and the 2 or 3 pistillate spikes closely aggregated or approximate, the staminate spike often small and exceeded by the pistillate ones; lowest bract sheathless, leaf-like, shorter than or exceeding the inflorescence; scales much shorter than the perigynia, light reddish brown with broad white-hyaline margins; perigynia stipitate, narrowly ellipsoid-obovoid, 3–4 mm long, 2-ridged, grey-pubescent, the slender beak 0.5 mm long. C. *nigromarginata* Schw. var. *elliptica* (Boott) Gl. Open woods. Alta to NS s to Gt Lakes. Map 261

C. pensylvanica Lam.
Densely caespitose, reddish brown and fibrillose at base, and forming slender stolon-like rhizomes; culms slender, 0.5–3 dm tall; leaves ascending, mostly 1–3 mm wide; terminal spike staminate, clavate; pistillate spikes 1–2 (rarely 3), usually contiguous, about 5 mm long, with few perigynia; lowest bract not sheathing, very short, tinged with reddish brown at the base; scales about as long as the perigynia but not concealing them, obtuse to cuspidate, with white-hyaline sides; perigynia 3–3.5 mm long, short-stipitate, the body subglobose, finely pubescent, abruptly narrowed into a bidentate beak about 1 mm long. $2n = 36$. C. *heliophila* Mack. of Ed. 1. In our area var. **digyna** Boeckl. Dry, sandy prairie. eAsia: BC to NB s to Calif, NM, Kans, Tenn, SC. Map 262

C. petasata Dewey
Tufted, the rootstocks short, fibrillose; the brownish base conspicuously clothed with the dried-up leaves of the previous year; culms 3–8 dm tall, slender but stiff, nearly smooth; spikes 3–6, gynaecandrous, aggregated in an erect head 2–4 cm long; bracts scale-like or the lowest occasionally short-prolonged; scales ovate, tinged with reddish brown, the margins broad and hyaline; perigynia largely concealed by the scales; oblong-lanceolate, 6–8 mm long, narrowly wing-margined, striate, the beak about 2 mm long. Dry grassland and open woods. Yuk to Ore, Nev, Ariz, Colo; swSask. Map 263

C. petricosa Dewey
Similar to C. *franklinii* but differing as follows: plants usually much shorter, 1–3 dm high; spikes commonly fewer, 3–5, the lateral ones nearly as broad as long; pistillate scales purplish black; perigynia lanceolate, 1.5–1.7 mm wide, dark brown, tapering at apex. Dry alpine slopes. Alas, Yuk, Mack; swAlta. Map 264

C. phaeocephala Piper
Forming large clumps, the rootstocks short and densely matted; culms 1–3 dm high, slender but stiff; leaves bunched near the base, those of the previous year conspicuous; spikes 2–5, usually 3 or 4, aggregated in an erect head 1–2.5 cm

long, the lower 1 or 2 spikes occasionally a little separate; lower bracts much shorter than the head; scales ovate, acute, brownish, strongly hyaline-margined; perigynia concealed by the scales until overripe, oblong-ovate or obovate, 4–6 mm long, narrowed into a beak about 1 mm long. $2n$ = c.84. Dry alpine slopes. swYuk, sMack to Calif, Colo; Alas. Map 265

C. platylepis Mack.
Caespitose, the rootstocks short; culms 4–7 dm high, slender, weak; leaf-blades 2.5–3.5 mm wide, prominently wrinkled and shallowly pitted; spikes 5–8 in a head 1.5–3.5 cm long, closely approximate or the lower ones slightly separate; bracts usually scale-like; scales ovate, closely appressed, dull reddish brown with white-hyaline margins and green centre; perigynia nearly concealed by the scales, 4–4.5 mm long, oblong-lanceolate, thin, tapering to a beak about 1 mm long. Dry, open coniferous woods. seBC to Wyo. Map 266

C. podocarpa R.Br.
Loosely caespitose from long slender rootstocks; culms 1–5 dm high, slender, more or less nodding above, reddish purple at base; leaves with well-developed blades usually 2–4 to a fertile culm, widely separate on lower third, not at all clustered; terminal 1–2 spikes staminate; lateral spikes 2–4, pistillate or occasionally androgynous, drooping or erect on slender peduncles; scales narrower and usually shorter than the perigynia, brownish black, the midvein nearly obsolete; perigynia 4 mm long, flat, nerveless, the beak black, 0.2 mm long; achene long-stipitate. $2n$ = 56, 60, c.64, 70. *C. montanensis* Bailey. Alpine meadows. eAsia: Alas, Yuk, Mack s to eOre, Ida, Mont. Map 267

C. praegracilis W. Boott
Rootstocks long, stout, black, fibrillose; culms arising singly or in small clumps, slender but stiff, often arching, 2–7 dm tall, dark brown or black at the base; leaf-blades 1–3 dm long, 2–3 mm wide; spikes 5–15, androgynous, mostly crowded in a head 1–5 cm long; perigynia about 10 in each spike, dark brown at maturity, ovate-lanceolate, 3–4 mm long, the beak about half as long as the body; scales cuspidate to acuminate, brownish with hyaline margins, concealing the perigynia. Common in moist open places. BC to Gt Lakes s to Calif, Tex, Ia; Alas, Yuk. Map 268

C. prairea Dewey
Differing from *C. diandra* as follows: leaf-sheaths strongly tinged with yellow-brown or bronze at the mouth; heads 3–8 cm long, flexuous, open at least toward the base, pale brown; scales reddish brown; perigynia appressed in the spike, straw-coloured to brown, only slightly lustrous, ovate-lanceolate, 2.5–3.5 mm long. $2n$ = 66. *C. diandra* var. *ramosa* (Boott) Fern. Wet meadows and moist woodland. sBC to NS s to Neb, Ohio, NJ. Map 269 •

C. praticola Rydb.
Caespitose, forming rather small clumps from short black rootstocks; culms 2–6 dm tall, slender, more or less nodding, much exceeding the leaves, brownish

at base; leaf-blades light green, flat, mostly 1–3 mm wide; spikes 2–7, gynaecandrous, moniliform or barely crowded in a flexuous head 1.5–5 cm long; bracts scale-like or the lowest cuspidate; scales somewhat brownish-tinged, shining; perigynia concealed by the scales, 4.5–6.5 mm long, ovate-lanceolate, short-stipitate, the beak 1–2 mm long. $2n$ = c.70, 76*, 78*. Open woods and grassland. Greenland: Alas, Yuk, sMack to James Bay, Nfld s to Calif, seIda, Utah, Colo, ND, Gt Lakes. Map 270

C. preslii Steud.
Caespitose; rootstocks short, thick, almost corm-like; culms 2–7 dm tall, slender, rough beneath the head; the crown with conspicuous scale-like sheaths; spikes 3–8, gynaecandrous, well-defined, aggregate or the lower separate in a head 1–2 cm long; lower bracts usually but little prolonged, the upper scale-like; scales ovate, reddish brown with green midvein and hyaline margins; perigynia ovate, thick, planoconvex, 3.5–4 mm long, strongly winged, obscurely nerved, abruptly narrowed to a beak about 1 mm long. $2n$ = c.78, 80, 82. Dry open slopes. BC to Calif e to Mont; Alas. Map 269 ○

C. pseudo-cyperus L.
Densely tufted, in large clumps, from short stout rootstocks; culms 3–10 dm tall, strongly triangular, very rough; leaves mostly 4–10 mm wide, strongly septate-nodulose, the ligules very conspicuous; staminate spike solitary; pistillate spikes 3–6, cylindric, 3–7 cm long, the lower more or less nodding on long peduncles; bracts leaf-like, much exceeding the inflorescence; pistillate scales with small reddish brown bodies and very long rough awns; perigynia slenderly ovoid, 3.5–5 mm long, coriaceous, smooth, shining, strongly ribbed, reflexed at maturity, gradually tapering into a deeply bidentate beak about 1 mm long. $2n$ = 66. Swamps and marshes. Eurasia, Japan: Alta to sMan, ND; Gt Lakes to Nfld. Map 271

C. pyrenaica Wahlenb.
Very densely caespitose; culms 5–20 cm tall, slender, wiry; leaves 2–3 to a fertile culm, the blades about 1 mm wide, involute, channelled; spike solitary, 5–20 mm long, androgynous; pistillate scales mostly chestnut-brown; perigynia 3–4 mm long, stipitate, glabrous, shining, nerveless, the beak 0.5 mm long; stigmas 3; achene triangular. Calcareous alpine slopes. sEurope, eAsia: Alas, Yuk, wMack to Calif, Nev, Utah, Colo. Map 272

C. raymondii Calder
Densely caespitose, the rootstocks short, stout, ascending; culms 3–7 dm high, slender, purplish at base; leaves with well-developed blades 6–15 to a fertile culm, clustered above the base; lowermost leaves reduced, purplish black to light brownish-purple; terminal spike usually gynaecandrous; lateral spikes usually 3, wholly pistillate or with a few basal staminate flowers, the lower somewhat curved on slender peduncles; scales narrowly ovate, 2.5–3 mm long, acute or subacuminate, usually shorter than the perigynia, light to dark purplish-red, with obsolete to broad hyaline margins; perigynia 2.5–3.5 mm long, weakly

inflated, 2-ribbed, nerveless, granular, greenish to chestnut-brown, the beak 0.4 mm long, bidentate. *C. atratiformis* Britt. ssp. *raymondii* (Calder) Pors. Moist places and shady woodland. Alas, Yuk, sMack to nMan s to BC, Alta, cSask. Map 273

C. raynoldsii Dewey
Loosely caespitose from a stout rootstock; culms stout, erect, stiff, 2–7 dm high, purplish red at base; leaves with well-developed blades usually 5–10 to a fertile culm, clustered towards the base; terminal spike staminate; lateral spikes 2–4, mostly close together, pistillate, on stiff slender peduncles; scales shorter than the perigynia, purplish black with light midvein; perigynia 3.5–4.5 mm long, somewhat inflated, suborbicular in cross-section, several-nerved, yellowish green, becoming yellowish brown, abruptly narrowed into a minute purplish bidentate beak. $2n = 70$. Moist, open or wooded slopes. sBC to Calif, Nev, Mont to Utah, Colo; swSask. Map 274

C. retrorsa Schwein.
Densely tufted from short rootstocks; culms coarse, 3–10 dm tall; leaf-blades 4–10 mm wide, soft, flat; staminate spikes 1–3, sessile or short-peduncled; pistillate spikes 3–8, closely crowded or the lower ones somewhat remote, thick-cylindric, 2–6 cm long; bracts leaf-like, more or less sheathing; pistillate scales about half as long as the perigynia, narrow, acuminate; perigynia ovoid, 7–10 mm long, much inflated, prominently nerved, soon spreading or reflexed, tapering into a slender beak about half as long as the body. $2n = 70$. Swampy woods and wet meadows. BC, Mack to Nfld s to Ore, Utah, Colo, SD, NJ. Map 275

C. richardsonii R.Br.
Loosely caespitose, forming small clumps from long ascending rootstocks; culms 1–3 dm high, brownish-fibrillose at base, strongly roughened above; leaves clustered near the base, thick, 2–2.5 mm wide; staminate spike 10–25 mm long, noticeably peduncled; pistillate spikes usually 2, erect, approximate, linear, 1–2 cm long; bracts bladeless, the lowest long-sheathing, purplish and hyaline-margined; scales somewhat longer and wider than the perigynia, acutish, dark purplish; perigynia 2.5–3 mm long, straw-coloured or brownish above, appressed-pubescent, rounded at apex, the beak about 0.5 mm long. Dry, sunny places. BC, sMack to Que s to SD, Gt Lakes. Map 276

C. rossii Boott
Caespitose, the rootstocks ascending, stout, somewhat woody, branching, purplish; leaves firm, erect or stiffly ascending; culms of various lengths, 0.5–3 dm tall, wiry; terminal spike staminate, usually conspicuous, up to 1.5 cm long; lateral spikes usually 2 or 3, pistillate, aggregate or somewhat separate, the lowest one often remote; lowest bract leaf-like, 0.5–5 cm long, usually exceeding the inflorescence; scales shorter than the perigynia, greenish to purplish brown with hyaline margins, acute or long-pointed; perigynia 3–4.5 mm long, short-pubescent, 2-ridged, with a conic bidentate beak about 1 mm long. Dry soil, often in open woodlands. Alas, Yuk, sMack to cMan, Lk Superior s to Calif, Ariz, Colo, SD. Map 277

C. rostrata Stokes
Caespitose from long creeping rhizomes; culms stout and spongy at base, usually 5–10 dm tall; leaves 2–10 mm wide, pale green or glaucous, prominently septate-nodulose when dry; staminate spikes 2–4, peduncled; pistillate spikes 2–4, cylindric, 4–10 cm long, sessile or the lower peduncled, the upper sometimes partly staminate; lower bracts leaf-like; pistillate scales narrow, acuminate, shorter than or slightly exceeding the perigynia, usually purplish brown; perigynia ovoid, 4–8 mm long, widely spreading at maturity, shining, nerved, the beak 1–2 mm long, bidentate. $2n = 76, 82$. Common in marshy places. Circumpolar: Alas, Yuk, Mack to Huds Bay, nQue, Nfld s to Calif, Ariz, NM, Neb, Ind, Ia, Apal. Map 278

C. rupestris All.
Loosely caespitose and with long rootstocks; dried-up leaves of previous year conspicuous; culms 6–12 cm high, stiffly erect, the sharp angles serrulate above; leaves usually exceeding the culms, the blades 2–3 mm wide, the tips attenuate and often curled; spikes solitary, androgynous, 1–1.5 cm long; perigynia 6–15, orbicular-ovate, shining, smooth, 4 mm long, 2-keeled, obscurely nerved, rounded at apex, minutely beaked; scales longer than and largely concealing the perigynia, brownish with broad white-hyaline margins; stigmas 3; achenes triangular, apiculate. $2n = 50, 52$. C. *drummondiana* Dewey of Ed. 1. Dry alpine areas. Circumpolar: Alas to Ellesmere Is s to BC, Alta, sMack, Huds Bay, Nfld; Utah, Wyo, Colo. Map 279

C. sartwellii Dewey
Rootstocks long and tough, fibrillose; culms 4–8 dm high, rather stiff, rough, sharply triangular above, brown at base, much longer than the leaves; upper sheaths green and striate ventrally; leaf-blades 2–4 mm wide, flat, about 3 to each culm; head oblong, 3–5 cm long, about 1 cm wide, the numerous spikes densely aggregated, the upper scarcely distinguishable; perigynia 2.5–3 mm long, finely nerved on both sides, the sharp margins serrate above the middle, abruptly contracted into a short minutely bidentate beak. Wet meadows and open swamps. BC, sMack, nSask to James Bay s to Mont, Colo, Mo, Ind, NY. Map 280 •

C. saxatilis L.
Culms 2–8 dm tall, purplish red at base, loosely tufted or arising singly from a creeping rootstock; leaf-blades flat, 1–4 mm wide; staminate spikes 1 or 2; pistillate spikes 1–3, well separated, oblong-cylindric, 1.5–3.5 cm long, slender-peduncled and spreading or drooping; lowest bract leaf-like; pistillate scales ovate or ovate-lanceolate, acute or acuminate, purplish black with hyaline margins and apex; perigynia ovoid, 4–5 mm long, usually exceeding the scales, smooth, shining, usually purplish or blackish at maturity, abruptly short-beaked; style abruptly bent downward against achene at maturity; achene lens-shaped or rarely triangular; stigmas 2, rarely 3. $2n = 80$. C. *physocarpa* Presl. of Ed. 1. In our area ssp. **laxa** (Trautv.) Kalela. Wet places. Circumpolar: Alas to nBaffin Is to Ellesmere Is s to Wash, Nev, Utah, Colo, sMack, Huds Bay, nQue. Map 281

C. scirpoidea Michx.
Rootstocks stout, woody, dark reddish-purple, scaly; culms 1–4 dm tall, stiff, leafy below, purplish red at base; plants dioecious; spikes single, sometimes compound at base, linear, many-flowered, staminate or pistillate, the latter 1.5–3 cm long; scales ovate, chocolate brown, puberulent, ciliate, hyaline-margined; perigynia 2.0–4.5 mm long, ovate to lanceolate, marginally ribbed, hirsute; stigmas 3; achenes sharply triangular. $2n = 62, 64$. Includes *C. athabascensis* F.J. Herm., *C. scirpiformis* Mack., *C. stenochlaena* (Holm) Mack. of Ed. 1. In our area var. **scirpoidea** and var. **stenochlaena** Holm, with longer (3–4.5 mm), narrower perigynia. Moist open areas to alpine elevations. eAsia, Greenland, Norway: Alas to Devon Is s to neOre, Mont, sSask, ND, Mich, NY, Nfld. Map 282

C. scoparia Schk.
Densely caespitose; rootstock short; culms 2–8 dm high, slender, rough beneath the head; leaves firm, 1–3 mm wide; spikes 3–10, ovoid or obovoid, shining, often straw-coloured, aggregated or scattered, clearly defined; scales ovate, acute, brownish-tinged with pale margins; perigynia longer than the scales, ovate-lanceolate, thin, straw-coloured to brown, 4–7 mm long, with thin wing extending to base, nerved on both faces, the beak flat, serrulate. $2n = 60, 64$. Moist open woodland. sBC, Alta; Man to Nfld s to Ore, NM, Ark, Fla. Map 280 ○

C. scopulorum Holm
Rootstocks stout and elongated, forming culms singly or in small clumps; culms 2–6 dm tall, sharply triangular, very rough above, with bladeless sheaths at base; blades 2.5–5 mm wide, scabrous near apex; terminal spike staminate, 1–3 cm long; lateral spikes 1–3, pistillate or androgynous, 0.5–1.5 cm long, contiguous or somewhat separated; lowest bract sheathless, much shorter than inflorescence; pistillate scales ovate or oblong-ovate, narrower than the perigynia, black with somewhat hyaline margins, the midrib often whitish; perigynia 2.2–3.5 mm long, broadly obovate to suborbicular, flattened, granular, pale below, purplish black above, marginally ribbed, otherwise nerveless, rounded at apex, minutely beaked. *C. gymnoclada* Holm of Ed. 1. In our area var. **bracteosa** (Bailey) F.J. Herm. Moist open areas. sBC to Calif e to Wyo, Colo; Alas, Yuk. Map 283 ○

C. siccata Dewey
Rootstocks long and tough; culms arising at intervals, 2–8 dm tall, roughened on the sharp angles above; sterile culms very leafy; leaf-blades flat, 1–3 mm wide, long-attenuate; spikes 4–12, closely aggregated in an oblong head 2–3 cm long; terminal spike gynaecandrous, the middle staminate and the lowest ones small and pistillate; scales brownish with silvery-hyaline margins and 3-nerved green centre; bracts resembling the pistillate scales, the lowest short-cuspidate; perigynia oblong-lanceolate, 4–6 mm long, winged above, abruptly narrowed into a bidentate beak 2–3 mm long; stigmas 2. $2n = 80$. *C. foenea* Willd. of Ed. 1. Dry open areas. sYuk, sMack to Man, Gt Lakes s to Wash, Ariz, Wyo. Map 284

C. simulata Mack.
Rootstocks slender, brown, with fibrillose scales; culms 3–5 dm tall, solitary or few together, light brownish at base; leaves 2–5 to a culm, the blade 2–4 mm wide; spikes pistillate, staminate or androgynous, 5–15 densely aggregated in a narrow head 1–2.5 cm long; perigynia broadly ovate, unequally biconvex, 1.7–2.3 mm long, ascending, chestnut-coloured and shining when mature, serrulate above, abruptly narrowed to a very short beak; scales ovate-triangular, concealing the perigynia, brown with narrow hyaline margins. Swamps and wet swales. Wash to Sask s to Calif, Utah, NM. Map 285

C. spectabilis Dewey
Loosely caespitose, from short-branching matted rootstocks; culms 3–9 dm tall, purplish at base; leaves with well-developed blades 2–5 to a fertile culm, on the lower half or third, quite separate; terminal 1–2 spikes staminate; lateral spikes 2–4, pistillate, erect; scales usually narrower and slightly shorter than the perigynia, purplish black, the margins slightly hyaline and the midrib thick, whitish, and usually excurrent as a minute cusp; perigynia 4–5 mm long, flat, obscurely nerved, granular-roughened, purplish-blotched, the beak 0.5 mm long; achene short-stipitate. $2n$ = c.84. Mountain meadows. Kamchatka: sAlas, Yuk to Calif, nIda, Wyo. Map 286 ○

C. sprengelii Dewey
Caespitose, with coarse, brownish, strongly fibrillose bases; culms 3–8 dm tall, scabrous at summit; leaves soft, flat, 2–4 mm wide; staminate spikes 1–3, peduncled; pistillate spikes 2–4, widely separate, 1–5 cm long, the slender peduncles drooping or ascending; pistillate scales about as long as the perigynia, acute or long-acuminate; perigynia 5–6 mm long with a globose-ovoid body 2.5–3 mm long, green to straw-coloured, 2-ribbed, shining, and abruptly contracted to a slender beak about as long as the body, the apex obliquely cut and at length deeply bidentate. $2n$ = 42. Moist open woods and meadows. BC to Nfld s to Mont, Colo, Neb, Gt Lakes, NJ. Map 283 ●

C. stenophylla Wahl.
Rootstocks slender, creeping, brownish, scaly; culms often tufted, brownish at base, smooth, 3–20 cm high; leaves clustered near the base of culm, 3–15 cm long, slender and involute; spikes androgynous, few, aggregated into a dense head 7–15 mm long; bracts ovate, cuspidate; scales rounded, acute or obtuse, brownish with white-hyaline margins; perigynia few, 1–8 in each spike, almost black at maturity, ovate, 2.5–3 mm long, the beak about one-third as long as the body. $2n$ = 60, 62. *C. eleocharis* Bailey of Ed. 1. In our area ssp. **eleocharis** (Bailey) Hult. Dry plains. Circumpolar: BC, sMack to sMan s to Ore, Nev, Ariz, Neb, La; Alas, Yuk. Map 287

C. stipata Muhl. *ex* Willd.
Tufted, with a stout, brown, fibrillose base; culms and foliage yellowish green; culms thick, soft, 4–10 dm high, the angles slightly winged and strongly serrulate above; leaf-blades flat, flaccid, 4–8 mm broad, usually cross-puckered on the

thin ventral band; head compound, dense, becoming pale brown, 3–10 cm long, 1–4 cm thick, the branches all crowded or the lower somewhat separated; lowest bract bristle-form, 5 cm long or less, often wanting; scales ovate-triangular, pale; perigynia much longer than the scales, planoconvex, thin, spongy at base, 4–6 mm long, brown-nerved, the beak serrulate, about as long as the body. Wet meadows and thickets. eAsia: seAlas, BC to Nfld, generally distributed s. Map 288

C. **sychnocephala** Carey
Densely caespitose, the rootstock very short; culm 2–6 dm high, smooth; leaves soft, 2–4 mm broad; bracts leaf-like, dilated at base, the larger usually 1–2 dm long; spikes 4–10, gynaecandrous, forming dense ovoid head 2–3 cm long; scales lanceolate, about half as long as the perigynia, acuminate; perigynia lance-subulate, thin, 5–6 mm long, with a long slender beak. Wet meadows and shores. BC, sMack to swQue s to Wash, Mont, SD, NY; Alas, Yuk. Map 286 •

C. **tenera** Dewey
Caespitose, from a very short rootstock; culms slender, erect, often nodding, roughened above, much exceeding the leaves; leaf-blades flat, 1.5–2.5 mm wide; spikes gynaecandrous, ovoid, often clavate at base, 3–8 in a moniliform inflorescence, 2.5–5 cm long, becoming straw-coloured; lowest bract short, setaceous, the upper scale-like; scales ovate, greenish-hyaline to tawny-tinged, narrower and shorter than the perigynia; perigynia 3–4 mm long, straw-coloured at maturity, thick, strongly nerved dorsally, contracted into a beak half the length of the body. $2n = 52, 54, 56$. Moist meadows and open woodland. BC to NS s to Mont, Wyo, Okla, Mo, Ohio, NC. Map 289

C. **tenuiflora** Wahlenb.
Loosely caespitose, often forming large clumps, with slender elongated rootstocks; culms very slender, 2–6 dm tall; leaf-blades 1–2 mm wide; spikes 2–4, gynaecandrous, whitish, closely aggregated in a head 6–12 mm long; perigynia ellipsoid or obovoid, planoconvex, 3–3.5 mm long, pale, obscurely few-nerved, almost beakless; scales obtuse, whitish with green centre, almost concealing the perigynia. $2n = 58$. Sphagnum bogs. Circumpolar: Alas, Yuk, Mack to Huds Bay, nQue, Nfld s to BC, cSask, sMan, Gt Lakes. Map 290

C. **tincta** Fern.
Resembling *C. tenera*, but the inflorescence not moniliform, the spikes being closely aggregated, the scales light reddish brown or deep brown, and the perigynia clearly 5–7-nerved on the ventral face. Resembling *C. bebbii*, but the heads narrower and the perigynia larger, 3.5–4 mm long, and distinctly nerved on the ventral face. Meadows and open woodlands. sBC to sAlta; Mich, Que, NB s to NY. Map 291

C. **torreyi** Tuckerm.
Tufted, with short-extended rootstocks; culms slender, erect, 2–4 dm tall, short-pubescent, rough above; leaves soft-pubescent, the blades deep green, 1.5–3

mm wide, the sheaths brown-tinged; terminal spike staminate; lateral spikes 2–4, pistillate, densely flowered, short-cylindric, approximate or the lower separated, sessile or short-peduncled; lowest bract about as long as the inflorescence, sheathless or nearly so; scales about half as long as the perigynia, ovate-orbicular, brownish; perigynia 2.5–3 mm long, broadly ovoid to obovoid, sometimes appearing pubescent at first glance but actually glabrous and puncticulate, yellowish green, many-nerved, abruptly rounded at apex; beak short, cylindric, truncate at mouth. $2n = 70^*$. *C. abbreviata* Prescott. Moist meadows and thickets. BC to Man s to Colo, SD, Minn. Map 292

C. trisperma Dewey
Loosely caespitose and stoloniferous; culms very slender, weak, 2–7 dm long; leaves soft, 1–2 mm wide; spikes 1–3, gynaecandrous, 1–4 cm apart, the lowest with a slender bract 2–4 cm long; perigynia 2–5 in each spike, 3–4 mm long, elliptic-oblong, planoconvex, finely nerved, the beak minute and emarginate; scales ovate, acute, hyaline with green centre. $2n = 60$. Bogs and wet woods. BC, sMack to nQue, Nfld s to Minn. Ill, Va. Map 293 ○

C. umbellata Schk.
Low, densely caespitose, with short, stout rootstocks; strongly fibrillose at the base; culms of various lengths, up to 2 dm high, mostly much shorter and often concealed by the leaves; leaf-blades firm, 1.5–4 mm broad, very rough above; longer culms usually with a terminal staminate spike, often a pistillate spike near it and 1–3 other pistillate spikes on peduncles below; shorter culms with spikes mostly pistillate; scales about as long as the perigynia and partly concealing them, acute to cuspidate; perigynia commonly 2.5–3.5 mm long, obovoid, 2-keeled above, short-pubescent or nearly glabrous, stipitate, contracted into a 2-edged beak usually 1–1.5 mm long. $2n = 30$. Dry open areas, often sandy. sBC to Nfld s to Mo, Tenn, Va. Map 293 ●

C. vaginata Tausch
Culms few together, from creeping horizontal rootstocks, 1–6 dm tall, smooth and flexuous; leaves soft, the blades mostly 1–4 mm wide; terminal spike staminate, long-peduncled, 1–2 cm long; lateral spikes 2–3, pistillate or sometimes staminate at the summit, rather loosely spreading, widely separated, the lower peduncles long; bracts with sheaths 1–2 cm long and enlarged upwards, the blades short; scales shorter than the perigynia, purplish brown; perigynia usually appearing in 2 rows, 3–5 mm long, yellowish green or brownish, fusiform-obovoid, 2-ribbed and obscurely nerved, glabrous; beak about 1 mm long, somewhat outwardly curved, obliquely cleft. $2n = 32$. Bogs and wet woods. Circumpolar: Alas, sBaffin Is to Nfld s to sBC, sSask, sMan, Gt Lakes. Map 294 ●

C. vesicaria L.
Culms 3–10 dm tall, tufted; rootstocks short-creeping and stout; leaf-blades 2–7 mm wide; staminate spikes 2–4, well elevated above the pistillate; lowest bract leaf-like, sheathless, usually overtopping the inflorescence; pistillate spikes

2 or 3, rarely solitary, cylindric, 1.5–8 cm long; scales narrower and shorter than the perigynia, acute to awned, yellowish to purplish brown; perigynia 4–8 mm long, ascending, shining, nerved, the conic beak 2 mm long, bidentate. $2n$ = 74, 82, 86. In our area var. **vesicaria**. Swamps and marshes. Circumpolar: sBC to Nfld s to Calif, NM, Mo, Ind, Del. Map 294 ○

C. viridula Michx.
Culms tufted, 1–3 dm high, stiff; leaves pale green, 1–3 mm wide; terminal spike staminate or sometimes partly pistillate, 5–15 mm long; lateral spikes 2–6, pistillate, crowded, pale green to drab, the upper sessile or nearly so, the lower often remote and short-peduncled; bracts leaf-like, erect or ascending; scales much shorter than the perigynia, brownish with hyaline margins; perigynia yellowish green or brown at maturity, 2–3 mm long, spreading, ribbed, the beak one-third the body-length. $2n$ = 70, 72. Shores, calcareous bogs, and springy places. Circumpolar: Alas, Yuk, nMack to James Bay, Nfld, s to Calif, Nev, Utah, NM, sSask, ND, Gt Lakes. Map 256 ●

C. vulpinoidea Michx.
Densely caespitose from short, tough rootstocks; culms 2–9 dm high, firm, very scabrous above; leaves usually longer than the culms, mostly flat, 2–5 mm wide, long-attenuate, the sheaths with a cross-wrinkled band; spikes numerous, in a compound head 3–12 cm long, the lower branches somewhat apart, the upper densely aggregated, greenish to yellowish or dull brown; staminate flowers apical, inconspicuous; bracts setaceous, usually several, up to 5 cm long; scales ovate, long-awned; perigynia ovate, planoconvex, 2–2.5 mm long, greenish to straw-coloured, thick-edged, the beak nearly as long as the body. $2n$ = 52, 54. Swampy ground. sBC to Nfld, generally distributed s. Map 295 ○

C. xerantica Bailey
Caespitose from short-creeping rootstocks; culms 3–6 dm high, erect, stiff, much longer than the leaves, brownish-tinged at base; leaves light green, firm, 2–3 mm wide; spikes gynaecandrous, 3–6, ellipsoid, approximate but very distinct or somewhat separated, in a rigid head 3–5 cm long; lowest bracts short-prolonged, not prominent; scales ovate, thin, silvery-hyaline tinged with reddish brown, covering the perigynia until mature; perigynia ovate, flat, 4–5.5 mm long, light coloured, strongly winged at base, tapering into a flat, serrulate beak. Prairies and plains. sBC to sMan s to Colo, SD, Minn. Map 295 ●

Cyperus L.

Perennial or annual herbs with narrow, 3-ranked, mostly basal leaves; culms triangular, leafy at base, and with one or more leaves at the summit beneath the inflorescence; flowers perfect, in the axils of small scales, 2-ranked in spikelets which are mostly flat; spikelets disposed in a simple or compound, head-like or umbellate inflorescence; perianth none; stamens 1–3; style 3-cleft and achenes 3-angled in our species.

Low annual with fibrous roots; scales with an outwardly curved awn; stamen 1 **C. squarrosus**

Tall perennial with corm-like thickenings at base; scales with a short erect awn or mucro; stamens 3 **C. schweinitzii**

C. schweinitzii Torr.
Perennial with a short hard rhizome, forming corm-like branches; culms 1–7 dm tall, firm, slender, rough, thickened at the base; leaves 2–4 at base of culm, firm, 2–4 mm wide; bracts 3–6, the longer exceeding the inflorescence; rays of inflorescence 2–9, unequal, ascending, with spikes bearing several 6–12-flowered spikelets; scales ovate, acuminate, about 3.5 mm long, many-nerved, the uppermost with a mucro less than 1 mm long; achene 2.5–3 mm long, light brown. Rather dry sandy soil. sAlta to swQue s to Utah, NM, Tex, Ind, NY. Map 296 ●

C. squarrosus L.
Low tufted annual, fragrant when bruised or dried; culms 1–15 cm high, slender; leaves 2–10 cm long, usually 1–2 mm wide; bracts 2–4, leaf-like, much longer than the inflorescence; spikes crowded in a head-like cluster or in 3-rayed sessile umbels; spikelets 2–10 mm long, flat, linear to oblong, 5–20-flowered, greenish to pale brown; scales oblong-lanceolate, several-nerved, the body 1–2 mm long, narrowed to a slender recurved tip 0.5–1.2 mm long; stamen 1; rachillas persistent after fall of the scales; achenes about 1 mm long. *C. inflexus* Muhl. of Ed. 1. Moist soil, usually sandy alluvium. sBC to NB, generally distributed s. Map 296 ○

Eleocharis R.Br. / Spike Rush

Herbs of wet places, solitary or tufted, often with long rhizomes; annual or perennial; culms simple, erect or reclining, terete or angled, scape-like, the leaves reduced to bladeless sheaths; spikelets solitary, terminal, erect, ovoid to linear; scales spirally arranged, imbricate; flowers perfect; perianth of 1–12 (commonly 6) bristles, or wanting, the bristles usually retrorsely barbed; stamens 1–3; style 2–3-cleft, its base usually much enlarged as a tubercle and persistent on the summit of the lens-shaped or 3-sided achene.

1 Achenes lens-shaped; style 2-cleft (occasionally 3-cleft in *E. ovata*) 2

Achenes 3-sided to nearly terete; style 3-cleft 3

2 Annual; tubercle very flat, short, and as wide as the truncate achene **E. ovata**

Perennial; tubercle ovoid-conic, constricted at base and not more than half as wide as the achene **E. palustris**

3 Tubercle delimited from the achene by a distinct constriction 4

Tubercle confluent with the achene, not forming a distinct cap **E. quinqueflora**

4 Achene nearly circular in cross-section, the surface with several longitudinal ridges and many very fine transverse lines; culms generally 1 dm tall or less, arising from slender rhizomes, often forming dense mats **E. acicularis**

Achene somewhat 3-sided, the surface granular-roughened; culms 2–6 dm tall, arising from stout rhizomes **E. compressa**

E. acicularis (L.) R. & S.
Perennial with slender creeping rhizomes, forming carpets; culms loosely tufted, 3–12 cm high, filiform, angled and grooved; spikelets 2–7 mm long, ovate to lanceolate, somewhat flattened; scales often reddish brown with green centres and scarious margins; bristles 3 or 4, as long as the achene, or lacking; achenes 0.7–1 mm long, slenderly obovoid, with several longitudinal ridges and numerous very fine transverse lines; tubercle conic. $2n = 20$. Wet places. Circumpolar: Alas, arct coast to Baffin Is, generally distributed s. Map 297 ●

E. compressa Sulliv.
Perennial with stout hard rhizome; culms 2–6 dm tall, wiry, flattened; spikelets ovate to oblong, 5–12 mm long; scales ovate to lanceolate, reddish purple or brown, the scarious tips often toothed or lacerate; bristles few and fugacious, or lacking; achenes golden-yellow to brown, 1–1.3 mm long, obovoid, bluntly 3-angled, granular-roughened; tubercle short, depressed-pyramidal. $2n = 24$, 36. Marshy places and sandy shores. Doubtfully distinct from *E. elliptica* Kunth. Marshy places and sandy shores. sBC to Que, nNY s to Colo, Tex, Ga. Map 297 ○

E. ovata (Roth) R. & S.
Tufted annual; culms usually several, ascending, 1–4 dm high; upper sheaths pale brown, 1-toothed; spikelets 5–15 mm long, ovoid-cylindric; scales brownish with hyaline margins; bristles usually 5–7; achenes 1–2 mm long, broadly obovate, biconvex, smooth, shining, subtruncate, with a minute ridge at the base of the tubercle, the latter covering the top of the achene and only about 0.3 mm high. $2n = 10$. *E. engelmannii* Steud. of Ed. 1. Wet places. Eurasia: sBC to Nfld, generally distributed s. Map 298 ○

E. palustris (L.) R. & S.
Perennial with stout creeping rootstocks; culms tufted, 1–10 dm tall, terete or somewhat flattened; sheaths firm, obliquely truncate at summit, often reddish below; spikelet 5–20 mm long, narrowly ovoid to lanceolate, acute; sterile scales at base of spikelet 1–3, firm, obtuse; fertile scales lanceolate, acuminate, acute or somewhat obtuse, usually with a firm midrib to the apex; bristles usually 4; achene obovoid, lens-shaped, yellow to brown, about 1.5 mm long, smooth; tubercle ovoid-conic, about half as wide as the achene, with constricted neck. $2n = 10–96$. Includes *E. uniglumis* (Link) Schult. A polymorphic complex variously treated taxonomically. Wet places. Circumpolar: Alas, Yuk to Huds Bay, nQue, Nfld, generally distributed s. Map 299

E. quinqueflora (F.X. Hartm.) O. Schwarz
Scattered or tufted from horizontal rootstocks; culms 1–3 dm tall, capillary, strongly grooved; spikelets 3–6 mm long, with about 5 flowers; scales ovate, obtuse, deep brown with hyaline margins; achenes 2–2.5 mm long, obovoid, 3-angled or planoconvex, grey-brown, reticulate with rectangular cells; tubercle beak-like, conic, barely differentiated; bristles longer or shorter than the achene. $2n = 20$, 132, 134, 136. *E. pauciflora* (Lightf.) Link of Ed. 1. Calcareous boggy

areas. Circumpolar: Alas, Yuk to Huds Bay, nQue, Nfld s to Calif, NM, sSask, ND, Ia, Gt Lakes. Map 298 ●

Eriophorum L. / Cotton Grass

Perennial sedge-like plants; culms triangular or terete; leaves with linear blades or some reduced to bladeless sheaths; spikelets solitary and terminal, or few–many in head-like clusters or umbels; leaf-like bracts present in species with more than one spikelet; flowers perfect, each in the axil of a scale; perianth of 6 parts, each 4–6-cleft to the base into silky bristles, commonly 2–3 cm long at maturity; stamens 1–3; style 3-cleft; achene 3-angled, surrounded by the long persistent bristles.

1 Spikelet solitary; leafy bracts none, the lowest scale somewhat enlarged; culm-leaves reduced, usually to bladeless sheaths 2
Spikelets 2 or more, clustered or spreading on drooping peduncles; leafy bracts present 6

2 Stoloniferous; culms solitary or few together; empty lower scales 7 or fewer 3
Not stoloniferous; culms tufted; empty lower scales mostly 10–15 4

3 Scales lead-colour to black with narrow pale margins; anthers 0.5–1 mm long; bristles white; achene-beak 0.1–0.2 mm long **E. scheuchzeri**
Scales brownish to black with wide hyaline margins; anthers 1.5–3 mm long; bristles pale cinnamon-brown to white; achene-beak 0.3–0.5 mm long **E. chamissonis**

4 Lower scales white-margined, spreading or reflexed **E. vaginatum**
Lower scales not white-margined; ascending 5

5 Plants 0.6–2.5 dm high; uppermost sheath conspicuously inflated, inserted below middle of culm **E. callitrix**
Plants 3–6 dm high; uppermost sheath not conspicuously inflated, inserted above middle of culm **E. brachyantherum**

6 Leaf-like bract only one, the other bracts reduced to bladeless sheaths; leaf-blades 1–2 mm wide, channelled **E. gracile**
Leaf-like bracts 2 or 3; leaf-blades 3–6 mm wide, flat below the middle 7

7 Midrib of scales prominent to tip; upper sheaths not dark-girdled at summit; anthers 1–1.5 mm long **E. viridi-carinatum**
Midrib of scales not extending into the scarious tip; upper sheaths dark-girdled at summit; anthers mostly 3–4 mm long **E. polystachion**

E. brachyantherum Trautv.
Culms loosely tufted, very slender, terete, 3–6 dm high; basal leaves elongate, filiform, channelled; stem-leaves reduced to bladeless sheaths; spikelet solitary; scales blackish or lead-colour, scarcely paler at the margins, appressed-ascending; anthers 1–2 mm long; achenes 2–2.5 mm long, the beaks 0.1 mm; bristles cream-coloured. $2n = 58$. *E. opacum* (Bjornstr.) Fern. Boggy areas. Circumpolar: Alas, arct coast to nQue, Nfld s to nBC, transcont s almost to US border. Map 300

E. callitrix Cham.
Caespitose, with persistent leaf bases; culms 0.8–2.5 dm high; basal leaves linear, 1–2 mm wide, cauline leaves generally 1 (0-2) inserted below middle of culm, sheath strongly inflated, blade short or lacking; spikelet solitary, obovoid, 1.5–2.5 cm long; scales erect, black or greenish black; anthers 0.6–1.0 mm long; achenes 1.8–2.0 mm long; bristles shiny, white. $2n = 60$. Alpine bogs. neAsia, Greenland: Alas to nBaffin Is s to BC, Alta, sMack, James Bay, nQue, Nfld; Wyo. Map 301 ○

E. chamissonis C.A. Mey.
Producing stolons; culms solitary or few together, 2–5 dm tall, the upper sheaths bladeless; lower leaves with blades 1–2 mm wide; spikelet solitary, in fruit often 2–4 cm high; scales brownish to blackish with broad pale margins; anthers 1.5–3 mm long; achenes 2–2.7 mm long, the beaks 0.3–0.5 mm; bristles commonly pale cinnamon or white. $2n = 58, 60, 62$. Bogs and marshes. Circumpolar: Alas, Yuk, swMack to Huds Bay s to Ore, cSask; Gt Lakes, Nfld. Map 301 ●

E. gracile Koch
Culms very slender, weak, sometimes reclining, 2–6 dm long, often with no young basal leaves at flowering time; leaf-blades channelled, the uppermost round-tipped; bract of inflorescence solitary, erect, about 1 cm long; spikelets 2–5, lateral on minutely pubescent peduncles; scales lead-colour or blackish, with prominent midrib; bristles bright white or sordid. $2n = 60, 76$. Bogs and peaty marshes. BC, sMack, Huds Bay to Nfld s to Calif, Mont, cSask, sMan, Ia, Gt Lakes; Alas; Colo. Map 302 ●

E. polystachion L.
Culms solitary or tufted, 2–8 dm tall, obtusely angled; leaf-blades mostly 3–6 mm wide, flat below the middle and conduplicate upwards, scabrous-margined; bracts usually 2 or 3, often black at base; spikelets 2–10, often on drooping peduncles; scales ovate-lanceolate, dark with broad scarious margins, the very slender midvein not extending into the tip; anthers 2.5–5 mm long; bristles white. $2n = 58, 60$. *E. angustifolium* Honck. of Ed. 1. Muskegs and boggy marshes. Circumpolar: Alas to nQue, Nfld s of Arct Arch s to Ore, NM, Ia, Gt Lakes. Map 303

E. scheuchzeri Hoppe
Producing stolons; culms solitary, 0.5–4 dm tall, with a few blade-bearing sheaths near the base and often one bladeless sheath above; leaves of basal shoots filiform, channelled; spikelet solitary, in fruit 2–2.5 cm high; scales lead-colour or blackish with narrow pale margins; anthers 0.5–1 mm long; achenes with beaks 0.1–0.2 mm long; bristles bright white, occasionally pinkish in flower. $2n = 58$. Marshy ground. Circumpolar: Alas, Yuk to Ellesmere Is s to sMack, Huds Bay, nQue, Nfld; sBC, Alta; Colo, Utah. Map 302 ○

E. vaginatum L.
Densely caespitose, with many culms and persistent basal stumps forming broad tussocks; culms 2–6 dm high, stiff, 3-sided, with 1 or 2 inflated bladeless sheaths; basal leaves with filiform blades; spikelet solitary; scales blackish with pale margins, acuminate, spreading or reflexed at maturity; anthers 1–2 mm long; achenes broadly obovoid, 2.5–3.5 mm long, the beaks 0.1–0.3 mm; bristles usually bright white. $2n = 58$. In our area ssp. **vaginatum**. Relatively dry bogs, usually in sphagnum peat. Circumpolar: Alas to nBaffin Is s to BC, cSask, sMan, Gt Lakes, Nfld. Map 304

E. viridi-carinatum (Engelm.) Fern.
Culms solitary or in small tufts, 3–9 dm tall, 3-sided; leaf-blades 2–6 mm wide, flat except at the very tip; bracts usually 2 or 3, green throughout or brownish at base; spikelets 3–many, mostly on drooping peduncles; scales greenish-drab to blackish with a prominent midrib extending to the tip and often scabrous; anthers 1–1.5 mm long; bristles white to very pale brown. Boggy woods and meadows. sAlas, sYuk, sMack to Nfld s to sBC, Ida, Colo, sSask, sMan, Ia, Gt Lakes. Map 305

Kobresia Willd.

Tufted, sedge-like perennials, forming small or large tussocks; leaves cinnamon-brown at the base of the plant, the filiform, green blades 2–15 cm long; culms 5–30 cm tall, slender, stiff, obtusely triangular, smooth; spike terminal, simple or compound; flowers unisexual and usually monoecious, arranged in small spikelets subtended by scales; staminate flowers of 3 stamens with filiform filaments and linear anthers; pistillate flowers partly enveloped by a spathiform glume (often regarded as a perigynium), usually accompanied by a staminate flower or by a rudimentary rachilla; perianth none; achenes triangular or rarely lens-shaped, jointed with the slender style; stigmas 3 or rarely 2, exserted.

Spike seemingly simple, undivided, linear; leaves of previous year mostly reduced to bladeless sheaths at flowering time **K. myosuroides**
Spike compound, ovoid; leaves of previous year's growth with conspicuous dried blades at flowering time **K. simpliciuscula**

K. myosuroides (Vill.) Fiori & Paol.
Very densely tufted, the old sheaths brown, fibrillose and very conspicuous; culms 1–3 dm tall, equalling or slightly exceeding the erect leaves; spikes linear, 1–3 cm long, undivided; upper spikelets staminate, the lower ones androgynous with one pistillate and one staminate flower, or rarely with pistillate flower only; scales 2–3 mm long; glumes (perigynia) closely enveloping the achene, but the margins free to the base or nearly so, 3–3.5 mm long, light brown, shining; achenes about 2.5 mm long, short-stipitate, brownish at maturity and short-beaked or nearly beakless. $2n = 58$. *K. bellardii* (All.) Degl. of Ed. 1. Turfy, often exposed, alpine areas. Circumpolar, arctic-alpine: Alas to Ellesmere Is s to sBC, sMack, wHuds Bay (c.60) to Lab; Ore, Calif, Wyo to NM. Map 306

K. simpliciuscula (Wahlenb.) Mack.
Densely tufted, marked by conspicuous dried blades of the previous year and cinnamon-brown bases; leaves erect to strongly curved; culms 1–3.5 dm tall, often much exceeding the leaves; spikes ovoid, compound, 1–3.5 cm long; terminal spikelets staminate, the lower ones androgynous or pistillate; scales small, brownish; glumes (perigynia) 2.5–2.7 mm long; achenes about 3 mm long, short-stipitate, brownish at maturity, short-apiculate. $2n = 72–76$. Boggy areas at montane elevations. Circumpolar: Alas to Ellesmere Is s to BC, sMack, James Bay, nQue; Ore; Nfld, eQue. Map 307 •

Rhynchospora Vahl

R. capillacea Torr.
Caespitose perennial; culms slender, 1–4 dm tall; leaves involute, filiform, equalling or slightly longer than the culms; heads 1–2, terminal or terminal and lateral, the latter often remote; spikelets 1–10 per head, 5–7 mm long, ellipsoid, reddish brown; scales spirally imbricate, the lower sterile, lanceolate, midrib prominent and extended beyond the apex; flowers 2–3 per spikelet, perfect or upper staminate; perianth of 6, rarely 12, retrosely barbed bristles; stamens 3; achene obovoid, abruptly narrowed at the base, 2.5–4.2 mm long, surface granular, tubercle lanceolate, almost as long as achene body. Calcareous bogs. Alta to Nfld s to SD, Ill, Ind, Ohio, Tenn, Va. Map 307 ○

Scirpus L. / Bulrush

Tufted or stoloniferous perennials of wet soil; culms triangular or terete, sheathed at the base, leafy or the blade reduced to a mucro; inflorescence subtended by one or more (involucral) bracts and umbellate, capitate or spicate; spikelets 1–many, terminal or lateral, 3–many-flowered; scales spiral, imbricate, nearly all flower-bearing; flowers perfect; perianth of 1–6 bristles, or sometimes absent; stamens 2 or 3; style 2–3-cleft; achenes obovate, lens-shaped or triangular, usually apiculate.

1 Involucral bract solitary or absent, not leaf-like; cauline leaves usually much reduced 2
Involucral bracts 2 or more, leaf-like; cauline leaves usually well developed 9

2 Spikelets single, terminal; bract, when present, small, often shorter than the spikelet 3
Spikelets 1–many, apparently lateral, the involucral bract continuing the culm 6

3 Culms triangular, scabrous 4
Culms terete, smooth 5

4 Bristles 20–25 mm long, white **S. hudsonianus**
Bristles about 2 mm long **S. clintonii**

5 Spikelet ovate, outermost scale with midrib not exserted or barely so; plants tufted from rhizomes **S. pumilus**

Spikelet lanceolate, outermost scale with midrib prolonged as a blunt awn 1.0 mm or more long; plants in dense tussocks, rhizomes absent **S. cespitosus**

6 Culms slender, triangular; spikelets generally 1–5, sessile 7

Culms stout, terete; spikelets many, usually in umbellate clusters 8

7 Leaves stiffly erect; culm not sharply angled; achene reticulate, not apiculate **S. nevadensis**

Leaves often spreading; culm sharply angled; achene apiculate, not reticulate **S. pungens**

8 Culms firm; inflorescence erect, often compact; scales laciniate-fimbriate, pale with conspicuous reddish brown striae; achenes 1.8–3.0 mm long **S. acutus**

Culms soft; inflorescence often drooping; scales ciliolate, a rich chestnut brown; achenes 1.4–2.2 mm long **S. validus**

9 Spikelets more than 10 mm long, 2–12 in head-like or umbellate clusters; rhizomes with swollen nodes 10

Spikelets less than 8 mm long, numerous, in compound umbels or dense umbellate heads; rhizomes lacking swollen nodes 11

10 Style 2-branched; achene lens-shaped; culms mostly 3–7 dm high **S. paludosus**

Style 3-branched; achene 3-angled; culms mostly 10–15 dm high **S. fluviatilis**

11 Bristles much longer than the scales, persistent; plants caespitose **S. cyperinus**

Bristles as long as or shorter than the scales; plants stoloniferous 12

12 Style 3-branched; achene 3-angled; sheaths green, drying brown **S. pallidus**

Style 2-branched; achene lens-shaped; sheaths usually reddish **S. microcarpus**

S. acutus Muhl. *ex.* Bigel. / Great Bulrush
Rhizome thick, spongy, with swollen roots; culms 0.5–3 m high, terete, firm, olive-green; involucral bract erect, usually 1–4 cm long; primary branches of the compound inflorescence rather stiffly ascending or divergent, often 1–8 cm long, each bearing 1–5 spikelets, sometimes shortened to form a dense glomerule; spikelets ovoid to cylindric, acutish, 7–20 mm long; scales broadly oblong-ovate, about 4 mm long, conspicuously exceeding the achene, often flecked with red or brown, usually short-pubescent, the margin hyaline and often fimbriate, the apex notched and mucronate; bristles variable in length; achenes 1.8–3.0 mm long, pale brown, becoming black and lustrous. $2n = 40, 42$. *S. lacustris* L. var. *occidentalis* S. Wats., *Schoenoplectus acutus* (Muhl. *ex* Bigel.) A. & D. Löve. Marshes and lake shores. cBC to Nfld, generally distributed s. Map 308

S. cespitosus L.
Densely caespitose, forming tough tussocks, clothed at the base with conspicuous, drab or straw-coloured sheaths; culms very slender, wiry, terete, smooth, 1–4 dm tall; basal sheaths tough, mucronate, imbricate and persistent, the uppermost sheath green and with a blade 5–8 mm long; spikelet solitary, terminal, ovoid-oblong, 4–5 mm long, 2–4-flowered; scales chestnut-brown, ovate, acute or awn-tipped, the lowest one (bract) with a blunt awn shorter than to exceeding the spikelet; bristles 6, usually longer than the achene, fragile; achenes 3-sided,

obovoid, 1.5–2 mm long, smooth, short-pointed. $2n = 104$. *Trichophorum cespitosum* (L.) Hartm. Bogs. Circumpolar: Alas e to sBaffin Is s to Wash, Colo, cSask, sMan, Gt Lakes, Apal, Nfld. Map 309 ●

S. clintonii A. Gray
Densely tufted from short rootstocks, clothed at the base with dark sheaths and membranous shreds of old culms and leaves; culms very slender, 3-sided; scabrous on the angles, 1–3 dm high; lower leaves as mere sheaths or with very short blades, the uppermost usually prolonged into a long, slender blade sometimes equalling the culm; spikelet solitary, terminal, 4–5 mm long, 4–7-flowered; outermost scale (bract) prolonged into a blunt awn about as long as the spikelet; inner scales acute or obtuse; bristles 3–6; achenes 3-sided, obovoid, 1.5–2 mm long, smooth, scarcely beaked. Open woodland and turfy shores. Alta, Sask; Ont to NB s to Minn, Mich, NY. Map 309 ○

S. cyperinus (L.) Kunth / Wool Grass
Densely tufted and fibrous-rooted; culms 0.5–1 m high, slender, 3-sided; leaves mostly basal, very scabrous, 3–5 mm broad; bracts leaf-like, unequal, to 2 dm long, black at base, green above; inflorescence a compound umbel, spreading or contracted; spikelets mostly pedicelled, 3–5 mm long; scales 1.5 mm long, hyaline, greenish black; bristles 6, much exceeding the achene, white or light brown, persistent; style reddish; achenes 1 mm long, white or yellowish, sharply beaked. $2n = 66$. *S. atrocinctus* Fern. of Ed. 1. Marshy areas. sBC to nSask, nMan, James Bay to Nfld s to cSask, sMan, Minn, Mo, Tex, Fla. Map 310

S. fluviatilis (Torr.) A. Gray / River Bulrush
Rhizome elongate, with thick nodal enlargements; culms stout, sharply angled, 10–15 dm high, leafy almost to summit; leaves 5–15 mm broad, the sheaths coriaceous at the nodes; involucral bracts 3–5, leaf-like, unequal, the outer often about 2 dm long; inflorescence a compound umbel with 5–10 flat peduncles, each usually terminated by a cluster of 2–5 spikelets, these ovoid-cylindric, 1–3 cm long; scales with an apical awn; bristles 6, stiff, barbed; achenes obovoid, sharply triangular, 4–5 mm long, beaked. $2n = 110$. Margins of ponds and lakes. eAsia: Alta to Que, NB s to Calif, NM, Kans, Va. Map 311 ○

S. hudsonianus (Michx.) Fern.
Loosely tufted from slender rhizomes; culms 1–4 dm high, 3-sided, scabrous on the angles; lower sheaths almost bladeless, the upper ones with blades 1–2 cm long or occasionally much longer; spikelet solitary, terminal, ovoid, 5–7 mm long; outer scale (bract) with a blunt tip, shorter than to equalling the spikelet; inner scales awnless; bristles 6, exserted, 2–2.5 cm long, white, thin and flattened; achenes brownish, 1.5 mm long, 3-sided, apiculate. $2n = 58$. *Eriophorum alpinum* L., *Trichophorum alpinum* (L.) Pers. Boggy and springy places. Circumpolar: sAlas, Yuk, sMack, Huds Bay to Nfld s to sBC, cSask, seMan, Gt Lakes. Map 311 ●

S. microcarpus Presl / Small-fruited Bulrush
Culms solitary, stout, triangular, leafy, 3–10 dm high, from a thick stoloniferous caudex; leaves 5–12 mm broad, the sheaths often tinged with red; inflorescence large, compound-umbellate, the rays spreading or ascending; involucral bracts mostly 3, leaf-like, the longest often exceeding the inflorescence; spikelets few to many in glomerules, narrowly to broadly ovoid, acute, mostly 3–4 mm long; scales ovate, marked with green and black; bristles whitish; achenes whitish, lens-shaped, obovate, about 1 mm long, apiculate. $2n = 66$. Includes *S. rubrotinctus* Fern. Marshy places. Kamchatka: sAlas, Yuk, sMack, Huds Bay to Nfld s to Calif, Ariz, NM, Neb, Mich, WVa. Map 312 ●

S. nevadensis S. Wats.
Rhizome elongated, tough, deep in the soil; culms slender, 1–4 dm tall, 3-sided, conspicuously grooved, smooth, leafy below; leaves narrow, stiff, shorter than the culms; involucral bract erect, 1–7 cm long; spikelets ovoid, sessile, 1–10 in the inflorescence; scales walnut-brown, smooth and shining; bristles 1–3, minute; achenes planoconvex, 2–3 mm long, minutely reticulate, scarcely beaked. Wet alkaline soil. sBC to sMan s to Calif, Utah, Wyo, ND. Map 312 ○

S. pallidus (Britt.) Fern.
Robust plant with a leafy crown; stoloniferous; culms 1–few, mostly 3–10 dm tall, 3-sided, pale green, leafy on the lower half; leaf-blades 5–15 mm broad; involucral bracts 3 or 4, leaf-like but short; inflorescence umbelliform, with 4–8 primary rays terminated by crowded heads (glomerules) of spikelets; scales 2–3 mm long, with pale midrib and conspicuous awn; bristles shorter than the achene, whitish; achenes about 1 mm long, white, apiculate. $2n = 56$. *S. atrovirens* Willd. of Ed. 1. Marshy areas. Alta, sMan to James Bay s to neOre, Utah, NM, Tex, Mo. Map 313 ○

S. paludosus A. Nels. / Prairie Bulrush
Rhizome elongated, often with hard tuber-like bodies at the nodes; culms 2–7 dm tall, sharply 3-angled, smooth, leafy; leaves mostly 3–8 mm wide; involucral bracts 2–5, leaf-like, unequal, the outer up to 2 dm long; inflorescence of 3–many spikelets, sessile or subumbellate; spikelets large, 1–2.5 cm long, ovate to cylindric; scales about 6 mm long, pale brown, often puberulent, the apex notched and short awned; bristles 2–6, short; achenes lens-shaped, obovate, 3–4 mm long, apiculate, the surface minutely reticulate. *S. maritimus* L. var. *paludosus* (A. Nels.) Kük. Low wet ground, especially saline flats. BC, sMack, cSask, cMan s to Calif, Ariz, Tex, Ia; Alas; James Bay; St Lawrence R estuary, Me. Map 313 ●

S. pumilus Vahl
Small plant; loosely tufted, with slender rhizomes, clothed at the base with dead culms and brown persistent sheaths; culms 5–17 cm high, very slender, terete, grooved, smooth; upper sheaths bearing slender green blades 0.5–1.5 cm long; spikelet solitary, terminal, 2–3 mm long, 3–5-flowered; outermost scale (bract) sometimes prolonged to more than half the length of the spikelet and blunt-tipped, usually shorter; inner scales obtuse or acutish, the sides often lustrous-

brown and hyaline-margined; achenes about 1.5 mm long, dark brown or blackish, smooth, planoconvex, scarcely beaked. $2n = 78$. *Tricophorum pumilum* (Vahl) Schinz and Thell. In our area ssp. **rollandii** (Fern.) Raymond (*S. rollandii* Fern.). Calcareous bogs. More or less circumpolar: sYuk, sMack, nBC, wcSask; seQue. Map 314

S. pungens Vahl / Three-square Rush
Rhizome hard, elongated; culms scattered, 1–8 dm tall, 3-sided, sharply angular, occasionally twisted, often somewhat lax; leaves near the base, the blades 0.5–3 dm long; involucral bract erect, slender, pointed; spikelets 1–7, sessile, glomerulate, oblong-ovoid, 5–12 mm long; scales ovate, dark brown, notched and awned at apex; bristles 4; achenes 3 mm long, smooth, apiculate. $2n = 76$, 78, 120–128. *S. americanus* Pers. of Ed. 1. In our area ssp. **pungens**. Marshes and shores. Alas; BC to Nfld, generally distributed s. Map 315

S. validus Vahl / Common Great Bulrush
Rhizome stout, horizontal, with fibrous roots; culms 0.5–2.5 m tall, terete, pale green, usually soft and easily compressed; involucral bract erect, 1–7 cm long; primary branches of the compound inflorescence ascending or spreading-drooping, 1–6 cm long, each bearing 1–several sessile or short-peduncled spikelets or spikelet-clusters; spikelet ovoid, 5–10 cm long; scales rounded-ovate, 2.5–3 mm long, scarcely longer than the achene, variously marked with red, more or less pubescent, the apex rounded or shallowly notched and short-mucronate; bristles usually 6, about equalling the achene, reddish, retrorsely white-scabrous; achenes 1.4–2.2 mm long, brownish or olivaceous, obovate, planoconvex, apiculate. $2n = 42$. *S. lacustris* L. var. *validus* (Vahl) Kük., *Schoenoplectus validus* (Vahl) A. & D. Löve. Lakes, ponds, marshes. Alas, James Bay to Nfld, generally distributed s. Map 316

ARACEAE / Arum Family

Perennial herbs with stout horizontal rootstocks; leaves basal, either petioled and with broad blades or narrowly linear with sheathing bases; flowers borne in a dense fleshy spike (spadix) subtended by or enclosed in a large bract (spathe); flowers usually perfect; perianth of 6 scale-like divisions or wanting; stamens 6; pistil solitary, the ovary superior, 1-loculed or 2–3-loculed; fruit berry-like.

Leaves narrowly linear; spadix cylindric or tapering, borne on the side of the 2-edged stem; spathe leaf-like, continuous to the stem **Acorus**

Leaves broad, cordate or broadly rounded at base; spadix short-cylindric, borne on a slender stem; spathe white, broad, concave to flat, behind the spadix **Calla**

Acorus L.

A. americanus (Raf.) Raf. / Sweet Flag; Calamus
Plant aromatic, especially the thick creeping rootstock; leaves crowded at the base, linear, erect, mostly 4–8 dm long, 8–20 mm wide; stem (scape) resembling

the leaves, bearing a lateral spadix and prolonged into an erect green spathe; spadix 3–8 cm long, about 1 cm thick, covered with yellow brown flowers; flowers perfect; perianth of 6 scale-like sepals; ovary 2–3 loculed; stigma depressed-capitate; fruit hard and dry, gelatinous within. $2n = 24$*. *A. calamus* L. of Ed. 1. Swamps and streams. BC to NS s to Mont, Tex, Fla. Map 317

Calla L.

C. palustris L. / Water Arum; Wild Calla
Low herb, with long creeping rhizome; leaves basal, long-petioled, the blades ovate to rounded-cordate; stem (scape) 1–2 dm long, bearing a spadix and subtending spathe at the summit; spathe ovate, abruptly pointed, the upper surface white; spadix short-cylindric, 1.5–2.5 cm long; flowers all perfect, or the upper staminate; perianth none; ovary 1-loculed; stigma circular; fruits berry-like, red, few-seeded, in dense heads. $2n = 36, 72$. Wet boggy areas. Circumpolar, boreal: Alas, BC e to Nfld s to Gt Lakes. Map 318

LEMNACEAE / Duckweed Family

Minute aquatic plants, floating on the water or submersed; not differentiated into stem and leaf, being merely a flat thallus (frond) bearing 0–few short rootlets; flowers very small, rarely seen, unisexual; perianth none; staminate flower of a single stamen; pistillate flower of a flask-shaped ovary with 1–few ovules; fruit a 1–7-seeded utricle. The plants multiply by budding, producing new fronds from the edge or base of the parent frond; these may separate at once from the parent or may remain attached for some time, producing a colony. Overwintering is chiefly by minute bulblets, formed by fronds in the autumn; they sink to the bottom of the pond or stream, then rise and vegetate in spring.

Plant with a single root; thallus oval or elongated, usually green beneath, 1–5-nerved **Lemna**
Plant with several roots; thallus nearly circular, 3–8 mm broad, reddish beneath, 4–15-nerved **Spirodela**

Lemna L.

Thallus 1–5-nerved, with a single root; flowers 3 (2 staminate, 1 pistillate).

Thalli 6–10 mm long, stalked, remaining together to appear 3-lobed; mostly submersed, forming large colonies **L. trisulca**
Thalli 2–5 mm long, stalkless, forming small rosettes or freely separating, rather thick; forming dense floating masses on shallow water **L. minor**

L. minor L. / Common Duckweed
Thalli oval, solitary or associated in colonies of 2–5, flat or slightly convex above and below, obscurely 3-nerved; root 1–12 cm. $2n = 20, 30, 40, 42, 50, 80$. In

quiet water, forming a floating green carpet. More or less circumpolar: Alas to Nfld, transcont s. Map 319

L. trisulca L. / Ivy Duckweed
Thalli oblong-elliptic to oblanceolate, connected, often forming large submersed mats, obscurely 3-nerved; $2n = 20, 40, 60, 80$. In quiet water. More or less circumpolar: Alas to Que, transcont s. Map 320

Spirodela Schleiden

S. polyrhiza (L.) Schleiden / Larger Duckweed
Distinguished as shown in the key. $2n = 40$. Floating on quiet waters. More or less circumpolar: sBC to Que, transcont s. Map 321

JUNCACEAE / Rush Family

Herbs, mostly perennial, glabrous or sparsely hairy, frequently tufted or with creeping rhizomes; leaves slender, terete or grass-like, with sheathing base, sometimes mere scales; flowers few–many, usually in crowded cymes, sometimes condensed into heads, regular, perfect; perianth segments 6, small, green or brown, more or less scale-like, in 2 whorls (calyx and corolla) but essentially alike; paired bracteoles sometimes present below the perianth; stamens 6 or 3; pollen in tetrads; ovary superior, with 3 parietal placentae; stigmas 3; fruit a loculicidal capsule; seeds 3 or many, often with appendages.

Leaves glabrous, often terete, seldom flat and grass-like; capsule many-seeded **Juncus**
Leaves more or less hairy, at least when young, flat or channelled and grass-like; capsule 3-seeded **Luzula**

Juncus L. / Rush

Smooth perennial herbs mostly 2–8 dm tall, or dwarf annuals; leaves channelled, more or less terete, or flattened, sometimes reduced to bladeless sheaths; stem leaves usually with open sheaths; inflorescence a loosely branched or compact cyme of few–many flowers, the latter solitary, paired or in glomerules; flowers as described above; capsule many-seeded; seeds minute, often with tail-like appendages at the ends; testa often becoming mucilaginous.

1 Inflorescence appearing to be on the side of the stem, the latter extending beyond as a cylindrical bract; leaves all basal or mostly so, or reduced to sheaths 2
Inflorescence appearing terminal or lateral; if lateral, the bract flat or channelled on the upper side 5

2 Flowers few, usually 2 or 3; seeds appendaged; mountain plants 3
Flowers several or numerous; seeds lacking appendages; plants usually at lower elevations 4

3 Upper leaf-sheaths with distinct blades; capsule acute **J. parryi**
Upper leaf-sheaths merely bristle-pointed; capsule retuse **J. drummondii**

4 Bract nearly as long as the stem or longer; flowers usually green **J. filiformis**
Bract usually much shorter than the stem; flowers usually brown **J. balticus**

5 Annual 5–20 (30) cm high; inflorescence diffuse, generally more than two-thirds the height of the plant **J. bufonius**
Perennial; inflorescence not as above 6

6 Leaves all basal; flowering stems 3–15 cm tall; heads solitary (rarely 2) with 1–3 (5) flowers; mostly alpine 7
Leaves basal and cauline; flowering stems mostly more than 15 cm tall; heads more than one and / or more than 5-flowered (occasionally the heads single and 2–4-flowered in *J. stygius*, but the capsule 6.0–8.5 mm long); mostly non-alpine 8

7 Plants 3–10 cm high; lower involucral bract much longer than the head; capsule retuse **J. biglumis**
Plants 10–15 cm high; lower involucral bract about as long as head; capsule acute **J. albescens**

8 Capsule 6–9 mm long, much longer than the perianth; seeds 2.5–4.0 mm long, conspicuously appendaged 9
Capsule shorter, at most not much longer than perianth; seeds mostly lacking appendages (short-appendaged in *J. vaseyi* and *J. brevicaudatus* but seeds much shorter) 10

9 Plants rhizomatous; leaves about 2 mm wide **J. castaneus**
Plants tufted, lacking rhizomes, leaves filiform, less than 1 mm wide **J. stygius**

10 Leaf-blades flattened laterally, positioned edgewise to stem (equitant), 3–6 mm wide 11
Leaf-blades not laterally flattened, nor equitant, either terete or flattened dorsiventrally, mostly less than 3 mm wide 12

11 Style mostly more than 1 mm long, exserted from the perianth; stamens 6; seeds 0.8–1.0 mm long **J. tracyi**
Style rarely 1 mm long; stamens 3–6; seeds about 0.5 mm long **J. ensifolius**

12 Leaves septate, terete 13
Leaves not septate, either dorsiventrally flattened or channelled 19

13 Heads solitary (very rarely 2), blackish brown; montane to alpine species **J. mertensianus**
Heads not as above; lowland species 14

14 Capsule tapered at the apex; the inflorescence with short branches, the heads appearing prickly with spreading, often reflexed flowers 15
Capsule abruptly narrowed at the apex; the inflorescence with elongate branches; flowers erect or ascending 16

15 Flowers 3–4 mm long; heads less than 10 mm wide **J. nodosus**
Flowers 4–5 mm long; heads 10–15 mm wide **J. torreyi**

16 Flowers solitary (each with 2 bracteoles) **J. vaseyi**
Flowers in heads of 2 or more 17

17 Seeds about 1 mm long, appendaged **J. brevicaudatus**
Seeds about 0.4–0.5 mm long, appendages lacking 18

18 Perianth-segments acute; anthers longer than the filaments, capsule as long as or shorter than perianth **J. nevadensis**
Perianth-segments obtuse, anthers shorter than the filaments, capsule as long as or longer than perianth **J. alpinoarticulatus**

19 Flowers in heads; perianth segments minutely roughened dorsally; leaves grass-like **J. longistylis**
Flowers not in heads (borne singly in a panicle or cyme); perianth-segments not roughened dorsally; leaves not markedly grass-like 20

20 Capsule cylindric; longer than the perianth-segments; seeds 0.8–1.2 mm long **J. vaseyi**
Capsule cylindric to subglobose, rarely longer than the perianth-segments; seeds less than 0.5 mm long 21

21 Capsule 3-loculed, perianth-segments 3.5–4.0 mm long **J. confusus**
Capsule 1-loculed, perianth-segments generally more than 4.0 mm long **J. tenuis**

J. albescens (Lange) Fern.
Stems 5–15 cm high, very slender, loosely tufted; leaves 1–5 cm long, slender; head solitary, 1–5-flowered; lower bract long-acuminate or long-awned, equalling or exceeding the lowest flower; perianth whitish or pink-tinged; sepals 3–4 mm long; capsule about as long as the sepals, rounded to nearly truncate at apex; seeds 1.3–2 mm long, smooth, brown, with white tails shorter than the body. $2n$ = c.130, c.134. *J. triglumis* L. ssp. *albescens* (Lange) Hult. Bogs. eAsia, arctic-alpine: Alas, Yuk e to nBaffin s to neBC, Colo, Utah, swMack, Huds Bay, nQue, Lab, Nfld. Map 322

J. alpinoarticulatus Chaix
Stems 1.5–5 dm tall, in small tufts from a creeping rhizome; stem leaves 1 or 2; heads few–several, ellipsoid to hemispherical, 2–10-flowered, terminal or lateral on erect to ascending branches; perianth-segments 2–3 mm long, mostly obtuse; capsule slightly longer than the perianth, pale brown; often some of the flowers are elevated on short pedicels above others of the head. $2n = 40$. *J. alpinus* Vill. of Ed. 1. Wet meadows and shores. Circumpolar: sAlas e to sHuds Bay, nQue, Nfld s to Wash, Colo, sSask, sMan, Gt Lakes. Map 323

J. balticus Willd. / Wire Rush
Stems 2–6 dm tall, slender, terete, rising at intervals from a stout creeping rootstock; basal leaves reduced to loose brownish sheaths; bract of inflorescence continuous with stem and often a fifth to a third as long as the stem; inflorescence appearing lateral, densely head-like to diffusely branched, commonly 2–4 cm long; perianth 4–5 mm long, purplish brown, the segments equal, lanceolate; anthers much longer than the filaments; capsule mucronate, equalling or ex-

ceeding the perianth; seeds 0.8–1 mm long. $2n$ = 40, 80. *J. arcticus* Willd. ssp. *balticus* (Willd.) Hyl. A polymorphic species variously treated, represented in our area chiefly by var. **montanus** Engelm. (*J. arcticus* ssp. *ater* Hult.) as described above. Also occurring var. **littoralis** Engelm. (*J. litorum* Rydb.), more robust, with inflorescence 4–8 cm long, and var. **vallicola** Rydb. (*J. vallicola* Rydb.) with acuminate outer perianth-segments much longer than the acute inner series. Wet meadows and bogs. Circumpolar: Alas, Yuk, Huds Bay, nQue to Nfld, generally distributed s. Map 325

J. biglumis L.
Stems 0.2–3.0 dm tall, tufted from a short rhizome; leaves basal, blades terete; inflorescence terminal, exceeded by the lowest bract; heads solitary, 2(3–4)-flowered; perianth-segments 2.5–4.5 mm long, purplish brown; stamens 6; capsule, obovoid, retuse apically, 3.0–5.5 mm long, much longer than perianth; seeds minutely appendaged. $2n$ = 120. Moist alpine areas. Circumpolar, arctic-alpine: Alas to nEllesmere Is s to Alta, sMack, Huds Bay, nQue; Wyo, Colo. Map 324

J. brevicaudatus (Engelm.) Fern.
Stems 1–5 dm tall, densely tufted; leaves slender, erect; inflorescence narrow, 3–12 cm long, the branches few–many, erect or ascending, with few–many heads, each 2–7-flowered; perianth-segments narrowly lanceolate, 3-nerved, 2.5–3 mm long; capsule 3.5–4.5 mm long; seeds fusiform, with a very short appendage at each end. $2n$ = 80. Shores and marshes. neAlta, nSask, Man e to Nfld s to Gt Lakes, Apal. Map 326

J. bufonius L. / Toad Rush
Annual with fibrous roots; stems 0.5–2 dm high, tufted, filiform, often divergent; leaves very slender, short; inflorescence spreading, freely branched; flowers single on the branches, or in 2's or 3's; perianth narrowly lanceolate, with slender tips and scarious margins; capsule shorter than the perianth, obtuse, mucronate; seeds 0.3–0.5 mm long. $2n$ = 30, 34, 36, c.60, 80, 100–110, 120. Includes *J. sphaerocarpus* Nees *ex* Funk. Moist areas. Circumpolar: Alas, Yuk to Huds Bay, Nfld, widely distributed s. Map 327

J. castaneus J.E. Smith
Producing stolon-like rhizomes; stems 1–4 dm tall, stiffly erect; leaves mostly basal, erect; sheaths with scarious margins; blades 1–2 mm wide; bract subtending the inflorescence 2–8 cm long, erect; heads 1–3, the lowest sessile or nearly so, the upper peduncled; perianth-segments lanceolate, chestnut-brown, 4–7 mm long; capsule exceeding the perianth, 6–9 mm long, narrowed to a conspicuous beak; seeds 2.5–4 mm long, very slender, with an appendage at each end. $2n$ = 60. Marshes and bogs. Circumpolar, arctic-alpine: Alas to Ellesmere Is s to neBC, Alta, sMack, Huds Bay, nQue; Mont, Wyo, Colo, NM. Map 328 ●

J. confusus Coville
Stems 3–5 dm tall, slender, in small tufts; leaves almost filiform; auricles scarious, extending beyond the insertion of the blade; inflorescence 0.5–2 cm long, compact, pale; bracteoles large, scarious; perianth-segments 3.5–4 mm long, yellowish with a dark stripe on each side and scarious margins; capsule slightly shorter than the perianth, completely 3-loculed; seeds oblong, apiculate. Low grassland, thickets, open woods. seBC e to sSask s to Calif, in mts to Colo. Map 328 ○

J. drummondii E. Meyer
Stems 1–3 dm tall, densely tufted from a matted persisting base; leaf-sheaths short, brown, close-fitting, bladeless or with a bristle-like blade; inflorescence 1–5-flowered, the bract appearing like the continuation of the stem; perianth-segments 5–7 mm long, lanceolate, brown with a lighter median stripe; capsule oblong, triangular, obtuse-retuse, equalling or exceeding the perianth; seeds appendaged at both ends. $2n = 120$. Moist montane slopes. Alas, Yuk, wMack s to Calif, NM. Map 329

J. ensifolius Wikstr.
Stems 3–6 dm tall, flattened and 2-edged; leaves equitant, the blades 2–6 mm wide, the sheaths with or without auricles; heads 2–many, about 1 cm thick, many–flowered, usually brownish; perianth-segments about 3 m long, lanceolate, acuminate, dark brown; stamens 3 or 6, occasionally some aborted; capsule oblong, obtuse, mucronate, often slightly longer than the perianth. $2n = 40$. In our area var. **ensifolius**, with mostly 3 stamens, auricles generally lacking, and var. **montanus** (Engelm.) C.L. Hitchc., with mostly 6 stamens and auricles generally present (*J. saximontanus* A. Nels. of Ed. 1). Marshy areas. eAsia: sAlas, Yuk to Calif, Utah, NM, Tex; James Bay. Map 330 ●

J. filiformis L.
Stems 1–5 dm tall, very slender, finely striate, from slender rootstocks; leaves basal, reduced to sheaths or blades represented by short bristles; inflorescence 5–10-flowered, open, 1–3 cm high; bract appearing like the continuation of the stem, usually longer than the stem proper; perianth-segments 2.5–3.5 mm long, lanceolate, green with hyaline margins; capsule obovoid, short-mucronate, about as long as the perianth. $2n = 40, 80$. Bogs and marshes. Circumpolar: Alas, Yuk, Huds Bay, nQue s to Wash, Colo, nSask, sMan, Gt Lakes. Map 330 ○

J. longistylis Torr.
Stems slender, 2–6 dm tall, from slender rootstocks; basal leaves flat, 1.5–4 mm wide, pale green; cauline leaves flat or involute; sheaths with scarious margins; inflorescence of 2–8 heads, each with 3–8 flowers; bracts conspicuous, white-scarious; perianth-segments broadly lanceolate, scarious-margined, 5–6 mm long; style elongate; capsule oblong, abruptly long-beaked; seeds ellipsoid, 0.5 mm long, often apiculate. Moist meadows and shores. sBC to Nfld s to Calif, NM, Neb, Gt Lakes. Map 331

J. mertensianus Bong.
Stems densely tufted, 1–4 dm tall, from matted rootstocks; leaves 1–2 mm wide, somewhat flattened, the septa often obscure; inflorescence usually a solitary head, rarely of 2 or 3 heads, dark brown, almost spherical, densely flowered; perianth-segments lanceolate, brownish black; capsule obtuse, about as long as the perianth; seeds with or without tails. $2n = 40, 80$. Wet ground, banks, and slopes. Japan: sAlas, Yuk to Calif, NM. Map 332

J. nevadensis S. Wats.
Stems 1–7 dm tall, from slender rhizomes; leaf-blades septate, terete; sheaths with conspicuous auricles 1–3 mm long; inflorescence 2–12 cm long, open; heads mostly 5–30; perianth-segments 3.0–4.5 (5.5) mm long, lanceolate-acuminate, subequal, outer series longer; stamens 6, anthers 1–2 mm, longer than the filaments; capsule about as long as perianth or shorter; seeds 0.4 mm, lacking appendages. Wet areas. sBC to Calif, Ariz, NM. Map 334 ○

J. nodosus L.
Stems 1–4 dm tall, arising singly from slender rhizomes with tuber-like thickenings; leaves terete, the upper cauline and the bract exceeding the inflorescence; heads spherical, few–several, 8–20-flowered; perianth-segments 3–4 mm long, narrowly lanceolate, long-acuminate; capsule narrowly lanceolate to awl-shaped, 3-angled, exceeding the perianth; seeds 0.5 mm long, apiculate. Common on moist ground. BC, sMack, to Nfld s to Calif, NM, Tex, Neb, Gt Lakes; Alas. Map 333

J. parryi Engelm.
Stems densely tufted, 1–3 dm tall, slender; sheaths brown, mostly bladeless, only the uppermost with an elongate blade; floral bract usually 2–3 cm long, terete, appearing to be the continuation of the stem; inflorescence 1–3-flowered; perianth-segments lanceolate, 6–7 mm long, acute or acuminate, scarious-margined; capsule slightly longer than the perianth; seeds appendaged at both ends. Mountain slopes and meadows. BC to Calif, Colo. Map 334 ●

J. stygius L.
Stems 0.6-3.5 dm tall, arising singly or in tufts from a slender rhizome; leaves (1) 2–3, basal and cauline, filiform; inflorescence terminal, bract as long or slightly longer; heads mostly 1 (2), 1–4-flowered; perianth-segments lanceolate to lance-subulate, 3–6 mm long, midrib reddish; stamens 6; capsule 6.0–8.5 mm long, mucronate, exserted; seeds 3.0–3.5 mm long, conspicuously appendaged. In our area ssp. **americanus** (Buchen.) Hult. Bog pools. Circumpolar: seAlas, nBC, swMack, nSask, Huds Bay to Nfld, NY s to Lk Superior. Map 335

J. tenuis Willd.
Stems densely or loosely tufted, 1–6 dm tall; leaves basal, flat or involute, usually a third to a half as long as the stem, the sheaths auricled; inflorescence commonly 2–6 cm long, compact to open and lax, exceeded by the involucral bract; perianth-segments lanceolate, acuminate, 3–5 mm long; capsule oblong-ovoid, shorter

than the perianth; seeds without appendages. $2n = 30, 32, 40, 80$. In our area var. **dudleyi** (Wieg.) Hermann (*J. dudleyi* Wieg.) with sheath auricles little-prolonged, cartilaginous, yellow brown, and var. **multicornis** E. Mey. (*J. macer* S.F. Gray) with sheath auricles prolonged 1–3 mm, thin, white-scarious, and the inflorescence usually more open and lax than in var. *dudleyi*. Wet ground. Europe, eAsia: seAlas to Calif, transcont s of 55. Map 336

J. torreyi Coville
Stems stout, 4–10 dm tall, the slender rootstocks with tuberous thickenings; leaves mainly cauline, the blades usually divaricate, thick, 1–3 dm long; sheaths with scarious margins, loose above, prolonged into conspicuous auricles; inflorescence of few–many spherical heads, usually congested, each head 10–15 mm thick, 25–100-flowered; perianth-segments 4–5 mm long, linear, acuminate with rigid tips, the outer longer than the inner; capsule about 5 mm long, slender, long-beaked, 3-angled. $2n = 40$. Wet ground. sBC to swQue, Me, NY s to Calif, Tex, Ala. Map 337

J. tracyi Rydb.
Resembling *J. ensifolius* but the stamens 6; the seeds larger and appendaged; styles long exserted, capsule slightly shorter than the perianth and sheaths with very conspicuous scarious margins. Wet ground. BC to Colo, Nev, Ariz. Map 338

J. vaseyi Engelm.
Stems 3–6 dm tall, tufted, from short rhizomes; leaves basal, very slender, up to 3 dm long; inflorescence 1–4 cm long, compact, the branches erect and 3–4-flowered; perianth-segments lanceolate, acute, 3.5–4.5 mm long; capsule narrowly oblong, 4–5 mm long, obtuse; seeds fusiform, with a slender appendage at each end. Moist shores and meadows. BC, swMack, James Bay, NS, Me, s to Colo, cSask, sMan, Ill. Map 339

Luzula DC. / Wood Rush

Low perennial herbs, tufted and grass-like; leaves mostly basal, with closed sheaths, the blades flat or channelled, often hairy on the margins; cymes few–many-flowered, sometimes condensed into a head; each flower with bractlets at the base, these often toothed or lobed; perianth as in *Juncus*; stamens 6; ovary with 3 ovules; seeds smooth and shiny, often with a conspicuous appendage (caruncle).

1 Flowers crowded in dense clusters — 2
  Flowers solitary (or 2–3 together) at the ends of branches — 4

2 Inflorescence stiffly erect; involucral bract often longer than the inflorescence; lowland species — **L. multiflora**
  Inflorescence arching; involucral bract much shorter than the inflorescence; alpine species — 3

3 Flowers in 2 or more small capitate spikes borne on filiform, often arching branches **L. arcuata**

Flowers crowded in a dense spike-like, generally interrupted head **L. spicata**

4 Inflorescence umbellate; flowers 2.5–4.5 mm long; seeds about 2.5 mm long with conspicuous appendages **L. acuminata**

Inflorescence cymose (usually much branched); flowers 2–3.0 mm long; seeds smaller, appendages lacking 5

5 Perianth-segments 2.5–3.0 mm long; anthers 1.2–1.5 mm long; capsule ovate, 2.8–3.8 mm long, with a conspicuous beak up to 1 mm long **L. hitchcockii**

Perianth-segments usually less than 2.5 mm long; anthers less than 1.0 mm long; capsule elliptic or rounded, less than 2.5 mm long, beak short, inconspicuous 6

6 Plants usually less than 30 cm tall; cauline leaves 2–3; bracts and bractlets strongly ciliate **L. piperi**

Plants usually more than 30 cm tall, cauline leaves (3) 4 or more; bracts entire or slightly ciliate-lacerate, the bractlets entire to lacerate but never ciliate **L. parviflora**

L. acuminata Raf.
Stems loosely caespitose, 1–4 dm tall; basal leaves hairy, up to 3 dm long, 3–12 mm wide, cauline leaves shorter and narrower; inflorescence umbellate, pedicels filiform, spreading, 3–6 cm long; flowers generally solitary, rarely in pairs or small clusters; perianth-segments 2.5–4.5 mm long, reddish brown with a scarious margin; capsule 3–4 mm long, mucronate, exserted; seeds, including caruncle, 1.5–2 mm long. $2n = 18$. *L. saltuensis* Fern. Disturbed moist woodland. Alta, cSask, sMan e to Nfld s to Ia, Gt Lakes, Apal. Map 340 ○

L. arcuata (Wahl.) Sw.
Stems tufted, from short rhizomes 0.5–2 (3) dm tall; leaves basal and cauline, mostly 5–15 cm long, 2–3 mm wide; inflorescence of 2–15 heads, each 3–6-flowered, borne on slender, often arcuate branches; involucral bract inconspicuous; perianth-segments 2–3 mm long, reddish brown; capsule about as long as perianth; seeds apiculate, 1.1–1.3 mm long. $2n = 36^*$, 42, 48. In our area ssp. **unalaschensis** (Buchan.) Hult. Gravelly alpine slopes and ledges. nEurope, eAsia: Alas, Yuk, wMack to Wash. Map 340 ●

L. hitchcockii L. Hämet-Ahti
Stems 2–3 dm tall; from well-developed scaly rhizomes; basal leaves 3–15 cm long, 5–6 mm wide, cauline leaves 3–4, 2–6 cm long, 6–10 mm wide; inflorescence nodding to erect, branches slender; flowers solitary, terminal; perianth-segments 2.5–3.5 mm long, blackish brown to deep chestnut, lanceolate, acute; capsule exserted, 2.8–3.8 mm long, beak about 1 mm long; seeds 1.5–1.6 mm long. $2n = 24$. *L. glabrata* (Hoppe) Desv. of Ed. 1. Montane coniferous woodland. sBC to Ore e to wMont. Map 340 ▲

L. multiflora (Retz.) Lej.
Densely tufted, the stems 2–5 dm tall, somewhat bulbous at base; leaves 2–6 mm wide, sparsely or arachnoid long-hairy; inflorescence somewhat um-

bellate, either open with the numerous ovate to oblong glomerules on elongate peduncles or congested with nearly sessile glomerules; flowers 2.5–3 mm long, chestnut-brown; seeds 1.5–2 mm long, with a white basal appendage about one-third the total length. $2n$ = 12, 24, 28, 30, 36, 48. *L. campestris* (L.) DC. ssp. *multiflora* (Retz.) Buchen. A polymorphic complex, variously treated. In our area ssp. **multiflora**. Low meadows and damp open woods. Circumpolar: Alas e to sBaffin Is s to Calif, Colo, sSask, sMan, Ky, Va. Map 341

L. parviflora (Ehrh.) Desv.
Stems 3–6 dm tall, single or a few in a tuft, from spreading rootstocks; cauline leaves generally 4 or more, 4–10 mm wide, glabrous except at the top of the sheath; inflorescence nodding, with very slender lax branches; perianth 1.8–2.5 mm long, pale to dark brown; seeds brown, 1.1–1.5 mm long. $2n$ = 22, 24, 36. In our area ssp. **fastigata** (E. Mey.) L. Hämet-Ahti (extreme sw) with lower internodes distinctly red, inflorescence few-flowered, anthers shorter than filaments, capsule exceeding perianth, and ssp. **parviflora** with only reddish tinge at stem-base, inflorescence many-flowered, anthers equal to filaments, capsule as long as perianth. Moist forests and marshy areas. Circumpolar: Alas, Yuk to Huds Bay, nQue, Nfld s to Calif, Ariz, NM, cSask, sMan, Gt Lakes. Map 342

L. piperi (Cov.) M.E. Jones
Caespitose, stem 2–3 (3.5) dm tall; basal leaves 5–10 cm long, 3–4 mm wide; cauline leaves 2–3, linear, up to 3–7 cm long, 3–5 mm wide; inflorescence nodding, bracts and bractlets brown and strongly ciliate with long, curly cilia; perianth-segments dark brown, 1.8–2.5 mm long, acute; capsule exserted, 1.8–2.5 mm long; seeds light yellowish brown, 1–1.3 mm long. $2n$ = 24. *L. wahlenbergii* Rupr. of Ed. 1. Alpine heaths and meadows. eAsia: s Alas, Yuk to Calif, Wyo. Map 343

L. spicata (L.) DC.
Small tufted alpine plant; stems very slender, 1–3 dm tall; basal leaves 1–4 mm wide, somewhat channelled, sparsely hairy; cauline leaves 1–3, rather distant; inflorescence arched or nodding, dense, spike-like, the sessile glomerules crowded or interrupted on the axis; perianth 2–3 mm long, brown, the segments bristle-pointed; seeds with a minute appendage at base. $2n$ = 12, 14, 18, 24, 36. Alpine slopes and among rocks. Circumpolar, arctic-alpine: Alas, Yuk, wMack to Calif, Colo; sBaffin Is to e Huds Bay, nQue, Nfld, NY; neMan. Map 344

LILIACEAE / Lily Family

Plants (mostly herbs) of diverse habit and structure; flowers solitary or clustered, regular or nearly so, usually perfect; perianth usually conspicuous (not chaffy), of 6 separate or united segments, the outer and inner whorls (calyx and corolla) commonly alike; stamens usually 6, free or adnate to the perianth; pistil 1; ovary commonly 3-loculed, with axile placentation, and superior (partly inferior in *Zigadenus elegans*); styles 3, separate or more or less united; fruit a capsule or a berry.

1 Leaves reduced to scales, the photosynthetic organs (cladodes) being axillary branches with needle-like segments; flowers axillary **Asparagus**
Leaves green, normal, not scale-like; cladodes absent; flowers usually terminal 2

2 Leaves linear or linear-lanceolate, usually many times longer than wide; fruit dry 3
Leaves lanceolate or broader, not more than 5 times as long as wide; fruit often juicy 12

3 Flowers 1–3 (rarely as many as 5) 4
Flowers more than 3, often very numerous 6

4 Petals yellowish white, bearded within, 2–2.5 cm long **Calochortus**
Petals usually reddish orange 5

5 Plants 3–6 dm tall; petals 5–7 cm long **Lilium**
Plants 1–3 dm tall; petals 1.5–2.5 cm long **Fritillaria**

6 Flowers in umbels; plants with an onion odour **Allium**
Flowers in dense or slender racemes 7

7 Petals purplish blue, about 2 cm long **Camassia**
Petals never blue 8

8 Leaves numerous, tough, fibrous 9
Leaves few, soft and easily broken 10

9 Flowers large, 3–5 cm long; leaves linear-attenuate, the margins whitish and fibrillose **Yucca**
Flowers smaller, 0.6–1 cm long; leaves linear, the margins rough **Xerophyllum**

10 Flowers drooping, bronze or greenish yellow **Stenanthium**
Flowers not drooping, white, yellowish white or greenish white 11

11 Leaves with their edges turned toward the stem; plants with rootstocks **Tofieldia**
Leaves with their surface to the stem; plants with bulbs **Zigadenus**

12 Stems not leafy 13
Stems leafy 14

13 Flowers yellow; fruit a dry 3-angled capsule **Erythronium**
Flowers white; fruit juicy, dark blue **Clintonia**

14 Flowers one on each stem **Trillium**
Flowers several on each stem 15

15 Flowers green; fruit dry; plants usually 1 m high or more **Veratrum**
Flowers white or yellowish; fruit juicy; plants usually much less than 1 m high 16

16 Flowers in racemes or panicles at the ends of the stems, white; stems not branched 17
Flowers 1–4 at the ends of the branches, or solitary or clustered in the leaf-axils; stems nearly always branched 18

17 Perianth parts in 3's **Smilacina**
Perianth parts in 2's **Maianthemum**

18 Flowers greenish white, or reddish, in the leaf-axils **Streptopus**
Flowers yellow or greenish yellow, 1–4 at the ends of the branches **Disporum**

Allium L. / Onion

Perennial herbs with a strong odour; stems scape-like, arising from tunicated bulbs, these sometimes borne on rhizomes; leaves slender, flat or terete, basal or on the lower part of the stem; flowers in an umbel or head-like cluster subtended by 1–3 thin bracts (spathe) and borne on the scape-like stem; in some species the flowers replaced by sessile bulblets; perianth-segments 6, white to pink or purple, in 2 whorls of 3, withering and persistent below the capsule; stamens 6, often adnate to the base of the perianth; anthers short, introrse; ovary superior, 3-loculed; style 1; fruit a small capsule, ovoid to obovoid, 3-lobed, loculicidal; seeds black, 1 or 2 in each locule, minutely alveolate.

1 Leaves terete or nearly so, hollow at least towards the base; inflorescence subcapitate; ovary crestless **A. schoenoprasum**
Leaves flat or channelled, not hollow; inflorescence umbellate; ovary crested with 6 flattened or knob-like processes 2

2 Outer bulb-coats without fibres, or with parallel fibres, not fibrous-reticulate; umbel nodding **A. cernuum**
Outer bulb-coats fibrous-reticulate; umbel erect or nearly so 3

3 Flowers usually sterile, some or all of them replaced by bulbils **A. geyeri**
Flowers fertile; bulbils not formed on the pedicels 4

4 Leaves 3 or more per scape; tips of inner perianth-segments erect; alveoli of seeds pustuliferous; flowers usually pink **A. geyeri**
Leaves usually 2 per scape; tips of inner perianth-segments spreading; alveoli of seeds not pustuliferous; flowers usually white **A. textile**

A. cernuum Roth / Nodding Onion
Bulb slender, crowning a barely persistent rhizome, the outer coat membranous, minutely striate; leaves several, shorter than the scape, commonly 2–4 mm wide; scape 1–5 dm tall, terete or ridged, arching at summit; umbel nodding or bent to one side; spathe usually caducous; perianth-segments 4–6 mm long, ovate or elliptic, obtuse or somewhat acute, pink or white; stamens exserted; ovary and capsule with 6 high crests near the summit. $2n = 14^*$. *A. recurvatum* Rydb. Parkland prairies, open slopes, thickets, rock slides. BC, Ore e to Man, sOnt, NY s to Ariz, Tex, and Apal. Map 345

A. geyeri S. Wats.
Bulb with outer coats fibrous-reticulate; leaves ordinarily 3 or more per scape, channelled, 1–5 mm broad; scape 2–5 dm tall, usually exceeding the leaves, terete or somewhat angled; bracts of spathe mostly 1-nerved; perianth-segments usually 6–8 mm long, erect, pink or rarely white; stamens usually shorter than the perianth; ovary inconspicuously crested with low knobs, these variously developed or obsolete in fruit; seeds with a minute pustule in the centre of each

alveolus. In our area var. **geyeri** with normal flowers, $2n = 14, 28$ and var. **tenerum** M.E. Jones (*A. rubrum* Osterh., *A. rydbergii* Macbr.), $2n = 28, 35, 42$, in which the flowers are largely replaced with ovoid, acuminate bulbils. Wet meadows and along streams. BC, Alta s to Ariz, NM, Tex. Map 346

A. schoenoprasum L. / Wild Chives
Bulbs 1 or 2, rarely more, oblong-ovoid, the outer coats membranous; leaves coarse, terete or nearly so, hollow; scape usually exceeding the leaves, 2–6 dm tall, coarse; umbel 2.5–5 cm wide, the pedicels shorter than the flowers; perianth-segments pink or purple with dark midveins, 8–14 mm long, ovate-lanceolate to linear-lanceolate, attenuate; stamens included. $2n = 16^*$. *A. sibiricum* L. In our area var. **sibiricum** (L.) Hartm. Wet meadows, banks, and shores. Circumpolar: Alas to Nfld s to Wash, Colo, nSask, nMan, Gt Lakes. Map 347

A. textile Nels. & Macbr. / Prairie Onion
Bulb with outer coats fibrous-reticulate; leaves ordinarily 2 per scape, channelled, 1–5 mm broad; scape 0.5–3 dm tall, terete or nearly so; bracts of spathe prominent, ovate, acuminate, mostly 1-nerved; perianth-segments 5–7 mm long, white or rarely pink, with reddish midribs; stamens shorter than the perianth; ovary crested with 6 rounded knobs, these variously developed or obsolete in fruit; alveoli of the seeds not pustuliferous. $2n = 14^*, 28^*$. Common on dry plains and hills. Alta to Man, Minn s to Utah, NM; Nev. Map 348

Asparagus L.

A. officinalis L. / Asparagus
Herb, rhizomatous; stem 1–3 m tall, erect, much branched; leaves whitish, scale-like; cladodes with narrow segments 8–15 (20) mm long; flowers axillary, solitary or paired, on jointed pedicels; corolla 3–5 mm long, campanulate, greenish white or yellowish; stamens 6; fruit a berry, red, about 8 mm wide. $2n = 20$. Dry banks, waste ground, an escape from cultivation. Introduced. Occasional.

Calochortus Pursh

C. apiculatus Baker / Mariposa Lily
Glabrous perennial herb, subscapose, 1–3 dm tall, from an ovoid bulb with membranous scales; leaf solitary, basal, linear, usually shorter than the stem, 5–15 mm broad; bracts lanceolate; flowers conspicuous, mostly 1–4, yellowish white, sometimes lined with purple, erect or spreading; sepals shorter than the petals, oblong-lanceolate; petals clawed, obovate to oblanceolate, fringed along the sides, sparsely hairy on the inner face with a gland near the base; gland small, nearly circular; stamens 6; anthers apiculate; ovary many-ovuled; stigmas 3; fruit a capsule, orbicular to oblong, 3-angled, nodding. $2n = 20$. Dry slopes. neWash, nIda, nwMont, seBC. Map 349

Camassia Lindl.

C. quamash (Pursh) Greene / Blue Camas
Glabrous perennial with bulbs and scape-like stems, 3–6 dm tall; bulbs 1–3 cm thick, with black or brown scales; leaves in a basal whorl, linear, 2–4 dm long; flowers in a terminal raceme, large, showy, deep purplish-blue, somewhat oblique; sepals and petals each 3, similar, narrow, about 2 cm long; stamens 6, adnate to the base of sepals and petals; filaments filiform; anthers versatile, pistil with filiform style and 3-lobed stigma; fruit an ellipsoid, 3-angled capsule, 15–18 mm long; seeds black, shining. *Quamasia quamash* (Pursh) Coville. $2n = 30$. In our area var. **quamash**. Wet meadows. BC to Calif e to Mont, Wyo, Utah. Map 350

Clintonia Raf.

C. uniflora (Schult.) Kunth
Stems 6–15 cm tall, from slender creeping rootstocks; leaves 2–5, large, oblanceolate, pilose beneath, 8–20 cm long, basal or nearly so; flowers bractless, usually solitary, white, showy, villose, about 2 cm broad; sepals and petals similar, distinct, oblanceolate, obtuse; stamens 6, inserted at base of perianth, shorter than the petals; ovary 2–3-loculed; style slender; fruit a berry, subglobose, deep blue, 8–10 mm thick, few-seeded, with little flavour. $2n = 28$. Moist woods in the Rocky Mts. seAlas to Calif, Mont. Map 351

Disporum Salisb. / Fairy-bells

Perennial herb with prominent horizontal rhizome; stems 3–8 dm tall, scaly below, leafy and with a few branches above; leaves sessile or clasping, ovate or lanceolate, somewhat asymmetrical; flowers 1–4 at tips of branches, drooping, whitish or greenish yellow, 1–1.5 cm long; perianth-segments 6, equal, narrow; stamens 6; anthers extrorse; ovary 3-loculed; fruit a berry, reddish at maturity, nearly flavourless.

Fruit ovoid, not papillose; stigma not 3-cleft; leaves hairy above **D. hookeri**
Fruit globose, the surface roughened with fine wart-like projections; stigma usually 3-cleft; leaves glabrous above **D. trachycarpum**

D. hookeri (Torr.) Britt.
Stems forked and pubescent above; leaves hairy when young; stamens mostly longer than the petals; berry with or without hairs, 10–15 mm long, turning lemon-yellow and finally orange-red. $2n = 18$. In our area var. **oreganum** (S. Wats.) Q. Jones (*D. oreganum* (S. Wats.) B. & H. of Ed. 1). Moist woods. cBC to nwOre, wMont. Map 352

D. trachycarpum (S. Wats.) B. & H.
Stems and leaves much as in *D. hookeri*, stamens equalling the petals or slightly longer; berry 8–10 mm broad, papillose, velvety in appearance, changing from

green through yellow and orange to deep red. $2n = 22^*$. Moist woods and thickets. cBC e to James Bay s to Ore, Ariz, Colo, SD, Neb. Map 353

Erythronium L.

E. grandiflorum Pursh / Glacier Lily; Dog-tooth Violet
Glabrous herb with simple, scape-like stem 1–4 dm high, from an elongated, deep-seated, bulb-like corm; leaves usually 2, borne near the ground level, the blades lanceolate to ovate-lanceolate, mostly acute, 1–2 dm long, narrowed to a short petiole; flowers large, usually bright yellow, nodding, solitary or a few together; sepals and petals similar, 2–5 cm long, lanceolate, with a nectariferous groove; stamens 6, free; anthers slender, reddish purple or yellowish, attached at the end of the filament; stigma 3-lobed; fruit an oblong or obovoid capsule, tapering at the base, bluntly 3-angled. $2n = 24$. Rich soil, grassy slopes, and open forests from foothills to timber-line. Comes into flower very promptly as the winter snow melts. sBC to Calif e to Mont, Colo, Utah. Map 354

Fritillaria L.

F. pudica (Pursh) Spreng. / Yellow-bell
Herb with a leafy stem and a deep-seated scaly bulb; stem 1–3 dm tall, simple; leaves few, narrowly oblong to lanceolate, blunt; flowers usually solitary, nodding, narrowly cup-shaped, 1.5–2 cm long; sepals and petals similar, yellow to orange, turning brick-red in age, each with an obscure basal gland; stamens included, the anthers basifixed; capsule obovoid-oblong, about 2 cm long. $2n = 24$, 26, 36, 39. *Ochrocodon pudicus* (Pursh) Rydb. Grassland and dry open forest areas. sBC to nCalif e to Mont, Wyo, Utah. Map 355

Lilium L.

L. philadelphicum L. / Western Wood Lily
Stem glabrous, 3–6 dm tall from a thick-scaled bulb; leaves linear to lanceolate, glabrous, alternate except the uppermost which are whorled; flowers large, 1–5 (commonly 1–3), umbellate; sepals and petals similar, long-clawed, 5–6 cm long, red or orange-red, marked at base of inner side with purplish black spots; anthers versatile; style 1; stigma capitate but 3-lobed; fruit a cylindric-obovoid capsule, 2–4 cm long, broad at summit and slightly tapering at base. $2n = 24^*$. In our area var. **andinum** (Nutt.) Ker. Open woodlands and prairie. seBC to wQue s to NM, ND, Mich, Ohio. Map 356

Maianthemum Web.

M. canadense Desf. / Wild Lily-of-the-valley
Low herb, creeping by slender branching rhizomes, producing erect flowering stems, each with 1–3 sessile or short-stalked leaves and an accompanying long-stalked basal leaf; stems 5–20 cm tall; leaf-blades 2–8 cm long, ovate to oblong-lanceolate, cordate at base, more or less pubescent; flowers small, 4–6 mm wide,

white, in a short terminal raceme; perianth-segments 4, spreading, about 2 mm long; stamens 4; ovary 2-loculed; style 2-lobed; fruit a pale red berry, globose, 3–4 mm thick, 1–2 seeded. $2n = 36, 54, 72^*$. In our area var. **interius** Fern. Moist woods. Common. wMack, eBC to Lab and Nfld s to cSask, ND, Gt Lakes, Apal. Map 357

Smilacina Desf. / False Solomon's-seal

Perennial herbs with creeping rhizomes, the annual stems bearing few–many leaves, these broad and sessile or short-petioled; flowers small, white or yellowish white, in a terminal raceme or panicle; perianth-segments 6, equal, spreading; stamens 6; anthers ovate, introrse; ovary globose, 3-loculed, each locule with 2 ovules; style very short; stigma obscurely 3-lobed; fruit a globose berry, mostly 1–2 seeded.

1	Flowers in a panicle	**S. racemosa**
	Flowers in a raceme	2
2	Leaves 6 or more, pubescent beneath; raceme nearly sessile	**S. stellata**
	Leaves commonly 3, glabrous; raceme long-peduncled	**S. trifolia**

S. racemosa (L.) Desf. / False Solomon's-seal
Rhizome thick, fleshy; stem 3–10 dm tall, erect or ascending, slightly zig-zag, more or less puberulent; leaves usually 5–12 on the stem, ovate to broadly lanceolate, acute or short-acuminate, 5–15 cm long, 4–8 cm wide, sessile or with very short petioles; panicle ovoid or pyramidal, 5–15 cm long, many-flowered; sepals and petals 1.5–3 mm long, linear to spatulate, usually white, exceeded by the stamens; berries red, often dotted with purple. $2n = 36, 72$. In our area var. **amplexicaulis** (Nutt.) S. Wats. Damp woods and thickets. seAlas to sSask, sMan to Que, NS s to Calif, Ariz, Colo, Mo, Miss, Ga. Map 358

S. stellata (L.) Desf. / Star-flowered Solomon's-seal
Rhizome extensively creeping, branching; stems 2–6 dm tall; leaves sessile and somewhat clasping, lanceolate or oblong-lanceolate, often folded along the midrib, minutely hairy beneath, 3–12 cm long, 1–5 cm wide; raceme 1.5–5 cm long, with few–several flowers; sepals and petals white, 3–5 mm long, exceeding the stamens; berries green with dark stripes, turning black. $2n = 36$. Moist shores, open woods, and prairies. Alas se to Nfld, transcont s. Map 359

S. trifolia (L.) Desf. / Three-leaved Solomon's-seal
Rhizome slender, branching; stems slender, 0.5–2 dm tall; leaves 2–4, sessile, oval to oblong-lanceolate, glabrous, 4–10 cm long; raceme slender, long-peduncled, the axis often zig-zag, usually with 3–8 flowers; sepals and petals about 3 mm long; stamens 6; berries dark red. $2n = 36$. Bogs and wet woods, usually in sphagnum moss. eAsia; wMack, nBC e to Nfld s to cSask, sMan, Gt Lakes. Map 360

Stenanthium A. Gray

S. occidentale A. Gray / Bronze-bells
Erect glabrous herb with an elongate bulb; stem slender, weak, 3–5 dm tall; leaves few, mainly basal, linear or linear-oblanceolate, narrowed at both ends; inflorescence a simple raceme or somewhat branched below; flowers green to brownish red, 10–15 mm long, narrowly cup-shaped, nodding on slender pedicels; bracts lanceolate, scarious; petals and sepals similar, narrowly lanceolate, with reflexed tips, withering-persistent; stamens 6, included; anthers reniform; capsule lanceolate, 3-beaked, 15–20 mm long. $2n = 16$. *Stenanthella occidentalis* (A. Gray) Rydb. Moist woods and banks. sBC to nCalif e to Ida, Mont. Map 361

Streptopus Michx.

Perennial, rhizomatous; leaves alternate, sessile, often clasping; flowers small, greenish white to purple, solitary or in pairs, peduncles slender, perianth campanulate to rotate; stamens 6; included; ovary sessile; 3-loculed, style 3-lobed or entire; fruit a berry, greenish to dark red; seeds 3–15.

1 Plants 5–10 dm tall; flowers greenish white; nodes glabrous **S. amplexifolius**
Plants less than 3 dm tall; flowers reddish, nodes fringed 2

2 Perianth campanulate; leaf-margins entire or with irregularly spaced, multicellular cilia **S. roseus**
Perianth rotate; leaf-margins with unicellular, hyaline, closely crowded teeth **S. streptopoides**

S. amplexifolius (L.) DC. / Twisted-stalk
Herb with rather stout stems, 3–10 dm high, coarsely pubescent below the first branches; rhizome usually short and thick; stem forking or sometimes unbranched; leaves ovate-lanceolate, 4–12 cm long, 1–6 cm wide, cordate-clasping, acuminate, glaucous beneath, entire or denticulate; flowers axillary, solitary or sometimes in pairs, their stalks jointed and abruptly bent; perianth whitish green, 8–12 mm long, recurved-spreading from a cup-shaped base, the perianth-segments lanceolate and acuminate; stamens with short filaments and relatively long anthers, these extrorse and pointed; ovary 3–4 mm long with several ovules; stigma entire or barely 3-lobed; fruit a red berry, ellipsoid to globose, 1–1.5 cm long, insipid and inedible, several-seeded. $2n = 16, 32$. In our area chiefly var. **americanus** Schultes (includes var. *denticulatus* Fassett), as described above. Also occurring, var. **chalazatus** Fassett with glabrous stems. Moist woods and thickets. Europe, eAsia: sAlas, Yuk, swMack to Calif, NM, Sask, Man e to Nfld s to Gt Lakes. Map 362

S. roseus Michx.
Herb, stems mostly simple, 1.5–3.0 (5.0) dm tall, nodes fringed; leaves sessile, ovate-elliptic, 3–10 cm long, 1–5 cm wide, margin entire or ciliate; flowers axillary, mostly solitary, campanulate, mottled rose with whitish tips; perianth-segments 6–10 mm long, oblong-lanceolate, curving outwards towards the apex; anthers

2 mm long, 2-beaked; ovary obovoid, style 4 mm long; fruit a red globose berry 5–8 mm across, several-seeded. $2n = 16, 48$. In our area ssp. **curvipes** (Vail) Hult., $2n = 48$. The eastern ssp. *roseus*, $2n = 16$, produces a sterile hybrid with *S. amplexifolius* (*S.* × *oreopolus* Fern., $2n = 24$), but no western equivalent has been recorded. Moist montane forests. seAlas to Ore; sMan to Nfld s to Apal. Map 363 ○

S. streptopoides (Ledeb.) Frye & Rigg
Perennial herbs 1–3 (4) dm tall; stems simple to twice branched, nodes fringed; leaves sessile, 3–5 cm long, 1–2 cm wide, ovate-elliptic to ovate-lanceolate, apex acuminate to acute; flowers solitary; perianth small, rotate, perianth-segments rose to reddish brown with yellowish green tips, oblong-lanceolate with acute to acuminate recurved tips, 2.8–4.2 mm long; stamens with filaments 1.4–2.2 mm long, anthers minutely apiculate, 1 mm long; ovary 3 mm long; style short, bulbous-conical; stigma not divided, but 3-faced; berry red, rarely maroon or black, ovoid-globose, 4.5–6.5 mm long, few-seeded. $2n = 16, 54$. In our area ssp. **brevipes** (Baker) Calder & Taylor. Coniferous woodlands. eAsia: cBC to Wash, nIda. Map 363 ●

Tofieldia Huds. / False Asphodel

Slender perennials with short rootstocks; leaves linear, somewhat 2-ranked, at or near the base of the scape-like stem; flowers small, white or greenish, in a dense raceme or spike; perianth-segments separate, without glands, oblong or obovate; stamens 6, with short basifixed anthers; ovary 3-loculed, many-ovuled; styles 3, short, awl-shaped; fruit a many-seeded capsule, subtended by the persistent perianth.

Plant hairy or glandular above; pedicels in clusters of 2 or 3, each with 3 small bractlets below the flower; seeds with tail-like appendages **T. glutinosa**
Plant glabrous; pedicels solitary, without bractlets below the flowers; seeds not appendaged **T. pusilla**

T. glutinosa (Michx.) Pers.
Stem 1–5 dm tall, viscid above, often with a bract-like leaf near the middle; basal leaves several, broadly linear, 5–20 cm long; raceme 2–5 cm long, becoming 1–2 cm thick; flowers whitish, fascicled, on glutinous stalks; perianth-segments about 4 mm long; capsule ovoid, 5–6 mm long, yellowish or red; seeds with a contorted appendage at each end. $2n = 30$. Calcareous marshes and shores. In our area var. **glutinosa**. Alas to Lab, Nfld s to Calif, Wyo, nSask, nMan, Gt Lakes, and Apal. Map 364

T. pusilla (Michx.) Pers.
Stem 0.5–2 dm tall, very slender, bractless or occasionally 1-bracted; basal leaves numerous, strongly equitant, in short tufts, mostly 2–5 cm long; raceme 0.5–3 cm long; flowers whitish or greenish; perianth-segments about 2 mm long; capsule 2–3 mm long; seeds unappendaged. $2n = 30$. Calcareous marshes, banks,

and wet ledges. Circumpolar, arctic-alpine: Alas, nBC e to Baffin Is s to Mont, James Bay, Nfld; Lk Superior. Map 365 ●

Trillium L. / Wake-robin

T. ovatum Pursh
Low glabrous herb, 2–4 dm tall, from a short, fleshy rootstock; stem stout, bearing 3 leaves and a flower at the summit; leaves broadly ovate to rhombic, 6–12 cm long, short-acuminate; flower single, long-stalked; sepals 3, green, linear-lanceolate; petals 3, white or pinkish, turning purplish; stamens with rather short filaments; stigmas 3, elongate; fruit a berry-like capsule, broadly ovoid, narrowly winged. $2n = 10$. Moist woods. For 60 years known in our area only from a photograph taken in 1915 by R.H. Riggall in Waterton Lakes NP, but recently rediscovered there. sBC to cCalif e to Mont, Wyo, Colo. Map 365 ○

Veratrum L. / False Hellebore

V. eschscholtzii (R. & S.) A. Gray
Tall coarse herb, 1–2 m tall, from a thick rootstock; stem stout, leafy to the top, somewhat hairy; leaves 1–2 dm long, oval to elliptic, acute, sessile or nearly so, often clasping, plaited and strongly veined; flowers numerous in a large diffuse panicle, the branches elongate, spreading or drooping; perianth-segments yellowish green, oblong or oblanceolate, acute, 8–10 mm long, ciliolate-serrulate, adnate to the lower part of ovary; stamens 6, opposite the perianth-segments; anthers obcordate, extrorse; ovary 3-lobed, each lobe with a short style; fruit an ovoid capsule, 2–2.5 cm long, many-seeded; seeds large, flat. $2n = 32$. *V. viride* Ait. ssp. *eschscholtzii* (R. & S.) Löve & Löve. Moist woods and open slopes. Alas to Calif e to Ida, Mont. Map 366

Xerophyllum Michx.

X. tenax (Pursh) Nutt. / Bear Grass
Stout perennial with a thick woody rootstock; stem 5–12 dm tall, from the centre of a large clump of leaves and bearing a conspicuous raceme; leaves mostly basal, narrowly linear, tough, 2–4 dm long, rough-margined; flowers creamish white, in a large dense raceme; perianth-segments 6–9 mm long, spreading; filaments filiform; ovary 3-grooved; style filiform, reflexed; fruit a small capsule, ovoid, 5–7 mm long; seeds oblong, 2 in each locule. $2n = 30$. Dry mountain slopes and open woods. seBC to Calif e to Ida, Mont. Map 367 ●

Yucca L. / Spanish Bayonet

Y. glauca Nutt. / Soapweed
Coarse plant, subacaulescent; the short woody base often decumbent and branching, forming 1 or more crowns; leaves numerous, linear-attenuate, rigidly pointed, spreading, 2–4 dm long, the margins inrolled, whitish and with a few thread-like fibers; flowers large, drooping, in a racemiform cluster, the flowering stalk

and axis stiff, 5–10 dm high; perianth greenish white, oval to lanceolate, 3–5 cm long, withering-persistent; stamens 6, shorter than the sepals and petals; style stout, swollen, green; stigmas erect or spreading; fruit a dehiscent capsule, oblong, many-seeded; seeds thin, flat. This species and its pollinating insect, the Pronuba moth, are interdependent, with closely correlated life-cycles. $2n = 52$. Dry slopes. Mont to Ariz, Tex e to Daks, Ia, Mo. Map 367 ○

Zigadenus Michx. / Death Camas

Glabrous herbs, the erect stems from bulbs, bearing elongate linear leaves near the base, bract-like leaves above and a racemose inflorescence; flowers white to yellowish or greenish white; perianth-segments 6, similar, each with a gland at the base, widely spreading, withering-persistent; stamens 6; anthers extrorse; ovary superior or partly inferior, deeply 3-lobed; styles 3; fruit a 3-lobed capsule, tipped with the persistent styles; seeds numerous, angled.

Ovary partly inferior; glands of perianth obcordate **Z. elegans**
Ovary superior; glands of perianth obovate or semicircular **Z. venenosus**

Z. elegans Pursh / White Camas
Plants 2–6 dm tall, somewhat glaucous; stems slender, from long-ovoid bulbs; leaves linear, keeled, usually acute or attenuate at apex; inflorescence usually a loose raceme, rarely branched below, the bracts scarious-margined; perianth-segments greenish white or yellowish white, usually 7–10 mm long, narrowly oblong-obovate, each with a dark gland just above the short claw; capsule 1.5–2 cm long, ovoid-oblong; seeds 5–6 mm long. $2n = 32$. *Anticlea elegans* (Pursh) Rydb. Moist meadows and open woods. Slightly poisonous to livestock. Alas, Yuk, wMack, Sask, sMan, to Gt Lakes, s to Ore, Nev, NM, Tex, and Daks. Map 368

Z. venenosus S. Wats / Death Camas
Plants 2–3.5 dm tall, rather slender, from an ovoid bulb; leaves linear, 2–3 mm wide, all sheathing at the base; inflorescence a raceme, sometimes with a few branches at the base, the bracts scarious; perianth-segments creamish white, narrowly ovate, obtuse or rounded at apex, 4–5 mm long, the gland thin and ill-defined; stamens more or less adnate to the base of the petals and sepals; capsule oblong-elliptic. $2n = 22$. *Toxicoscordion venenosus* (S. Wats.) Rydb., *Z. gramineus* Rydb. of Ed. 1. In our area var. **gramineus** (Rydb.) Walsh. Plains and hillsides. Very poisonous to livestock. sBC to Calif e to swSask, Daks, Neb, Colo. Map 369

IRIDACEAE / Iris Family

Perennial herbs with equitant 2-ranked leaves; flowers perfect, regular, subtended by spathe-like bracts; perianth of 6 segments, in 2 series of 3 each;

stamens 3, opposite the outer perianth-segments; anthers extrorse; ovary inferior, 3-loculed; style single, 3-cleft; fruit a many-seeded loculicidal capsule.

Leaves 5–10 mm wide; flowers about 6 cm long **Iris**
Leaves 1–4 mm wide; flowers about 1 cm long **Sisyrinchium**

Iris L. / Flag

I. missouriensis Nutt.
Rhizomatous; stem 2–5 dm tall, simple; leaves basal, 2–4 (5) dm long, 5–10 mm wide; flowers 2–4, blue, pedicels stout, up to 6 cm long; perianth-segments 5–6 (8) cm long, obovate to oblanceolate, the outer recurved, the inner erect; ovary, style-branches petaloid, 2–4 cm long; capsule 3–5 cm long. $2n = 38, 48$. Marshy ground. BC, Alta s to Calif, Ariz, NM, SD. Map 370 ○

Sisyrinchium L. / Blue-eyed Grass

Slender plants with short rootstocks and fibrous roots; stems more or less flattened, 2-edged; leaves narrowly linear; flowers delicate, blue-violet or nearly white, in a small cluster from a pair of erect greenish bracts; perianth mostly 8–10 mm long, the segments spreading, nearly alike, retuse or aristulate at apex; filaments united to near the apex; style-branches filiform; capsule globose, more or less dehiscent at summit; seeds globose to ovoid, black.

Leaves 1–4 mm wide; flowers blue; pedicels mostly longer than the inner bract; perianth-segments emarginate or abruptly narrowed to an aristulate apex **S. montanum**
Leaves 1–2 mm wide; flowers often white; pedicels mostly shorter than the inner bract; perianth-segments rounded at the apex, aristulate **S. septentrionale**

S. montanum Greene
Tufted, erect plant; stems 1–5 dm tall, simple, stiff, 1–4 mm wide, distinctly flattened and winged; leaves 1–3.5 mm wide; bracts of spathe very unequal, the outer 2–6 cm long and twice as long as the inner, pale green or purplish; flowers about 1 cm long; perianth blue-violet; capsule 3–6 mm high. $2n = 96^*$. Moist open areas. sYuk, wMack to Nfld s to Colo, sSask, sMan, Gt Lakes. Map 371

S. septentrionale Bicknell
Forming small tufts; stems 1–2 dm tall, simple, 1–2 mm wide; leaves mostly basal, 1–2 mm wide; bracts of spathe narrow, the outer one 2.5–4 cm long and usually considerably longer than the inner bract, both broadly hyaline-margined; flowers about 8 mm long; perianth pale violet or nearly white, the segments not emarginate, slenderly aristulate; capsules 3–4 mm high. $2n = 32^*$. *S. sarmentosum* Suksd. *ex* Green of Ed. 1. Moist grassy areas. neWash, BC e to Man. Map 370 ●

ORCHIDACEAE / Orchid Family

Perennial herbs with corms, bulbs or rootstocks and often with tuberous roots; leaves simple, sheathing and entire, sometimes reduced to scales or sheaths; flowers solitary, or few–many in a raceme or spike, perfect, zygomorphic, epigynous, often showy; sepals usually 3, green or coloured, often petaloid; petals 3, commonly white or coloured, the middle one (*lip*) usually very unlike the lateral ones, sometimes with a *spur* at the base; stamens variously united with the style into a central column; fertile anthers usually solitary, rarely 2; pollen often coherent in granular or waxy masses (*pollinia*); style often terminating in a beak (*rostellum*); stigma a viscid surface beneath the rostellum or in a cavity between the anther-sacs; ovary inferior, usually long and twisted, 1-loculed; ovules numerous, on 3 parietal placentae; fruit a 3-valved capsule, with many minute seeds.

1 Lip a large inflated sac or pouch 2
Lip otherwise 3

2 Plants with a single (basal) leaf **Calypso**
Plants with 2 or more leaves **Cypripedium**

3 Plants saprophytic, with reduced non-green leaves; rhizome coralloid **Corallorhiza**
Plants not saprophytic; leaves with well-developed blades 4

4 Spur of flower present and conspicuous 5
Spur of flower lacking 6

5 Petals and lip usually spotted with pink or purple; leaf solitary and basal **Orchis**
Petals and lip white, yellowish or green; leaves more than 1 (solitary and basal in *H. obtusata*) **Habenaria**

6 Leaves 2, opposite, or nearly so, at about the middle of the stem; lip lobed or bifid at the apex **Listera**
Leaves basal or cauline and alternate 7

7 Stem naked between the foliage-leaves and the raceme **Malaxis**
Stem with scale-leaves between the foliage-leaves and the raceme 8

8 Leaves green or often blotched with silvery white, forming a rosette at the base of the stem; lip saccate **Goodyera**
Leaves not blotched with white, not in a rosette; lip fiddle-shaped; flowers sweet-scented **Spiranthes**

Calypso Salisb.

C. bulbosa (L.) Oakes / Venus'-slipper
Low herb from a solid bulb with coralloid roots, producing a solitary basal leaf and a showy flower on a scape 5–15 cm high; leaf petiolate, the blade round-ovate, 2–4 cm long, acute at apex, rounded or subcordate at base; flowers variegated, purple, pink and yellow; sepals and lateral petals 10–15 mm long,

ascending over the lip, pale purple; lip large, 15–20 mm long, saccate, whitish with purple markings, yellow-hairy within; column convex, bearing a lid-like anther just below the apex; pollinia 2 in each sac, united, all sessile on a broad gland. *Cytherea bulbosa* (L.) House. $2n = 28$. Coniferous forests. Circumpolar: Alas, Yuk, wMack to Nfld s to Calif, Ariz, Colo, sSask, Man, Gt Lakes. Map 372

Corallorhiza Châtelain / Coral-root

Plants yellow, brown or purplish, the scape-like stems lacking green foliage; leaves reduced to sheathing scales; rhizome coralloid, the branches short and thick; roots none; flowers in a spike-like raceme; sepals nearly equal, the lateral ones united at the base and more or less adnate to the ovary, often forming a short spur; lateral petals similar to the sepals; lip adnate to the base of the flattened column, often with 2 lateral lobes; pollinia 4; capsule reflexed.

1 Sepals and petals 1-nerved; lip 4–5 mm long; stems yellowish, usually less than 3 mm wide **C. trifida**

Sepals and petals 3-nerved; lip over 5 mm long; stems usually purplish, often over 3 mm wide 2

2 Lip with 2 lateral lobes, usually white and purple-spotted; spur conspicuous **C. maculata**

Lip entire or undulate, usually purple-striate; spur none, but petals and sepals somewhat gibbous at base **C. striata**

C. maculata Raf. / Spotted Coral-root
Scape (with raceme) 2–5 dm tall, stout, purplish or yellow-brown; raceme 5–20 cm long, 10–40-flowered; sepals and petals 6–9 mm long, linear-lanceolate, whitish, usually spotted with purple or red; lip 6–8 mm long, white, blotched with purple, bearing 2 lateral lobes and 2 short ridges on the face; spur prominent, yellowish. *C. multiflora* Nutt. $2n = 42$. Woods. cBC, sSask, sMan e to Lab, Nfld s to Calif, NM, Ariz, Gt Lakes, Apal. Map 373

C. striata Lindl. / Striped-Coral-root
Scape (with raceme) 1.5–4 dm tall, stout, purplish or yellow-brown, rarely pale yellow; raceme 5–15 cm long, 10–25-flowered; sepals and petals 8–16 mm long, yellowish or purplish with 3 conspicuous purple stripes; lip 8–12 mm long, white, usually purple-striate, tongue-shaped, declined, entire or somewhat undulate; spur none. $2n = 42$. Woods. cBC, sSask, sMan to Nfld s to Calif, Utah, NM, Gt Lakes. Map 374

C. trifida Châtelain / Pale Coral-root
Scape (with raceme) 1–3 dm high, slender, yellowish; raceme 2–7 cm long, 3–12-flowered; sepals and petals about 5 mm long, linear, yellowish white to dull purple or greenish; lip slightly shorter than the petals, white, sometimes dotted with red or purple, oblong, usually with a small tooth on each side near the base and 2 short ridges on the face; spur adnate to the ovary and not

noticeable. $2n = 42$. *C. innata* R.Br. Thickets, woods, and bogs. Circumpolar: Alas to nLab, Nfld s to Ore, Utah, NM, sSask, Man, Gt Lakes. Map 375

Cypripedium L. / Lady's-slipper; Moccasin-flower

Herbs with short or elongate rhizomes and thickened coarse roots; stems pubescent; leaves large, plaited, sessile, sheathing at the base; flowers 1–3, large, showy; sepals and lateral petals somewhat unlike, widely spreading, the two lower sepals commonly united under the large pouch-like lip; margins of lip somewhat inrolled; column declined over the opening of the lip, bearing 2 fertile stamens and a thick petal-like staminode (sterile stamen) above; stigma broad, obscurely 3-lobed; pollinia granular.

1 Flowering stem a naked scape, with 2 leaves at base; lip pink with red veins, cleft down the front, drooping **C. acaule**
Flowering stem with alternate leaves nearly to the top; lip yellow or white, with a rounded opening 2

2 Sepals obovate or oval, not longer than the lip, the latter 1–2 cm long; staminode elliptic-cordate **C. passerinum**
Sepals lanceolate, attenuate, often longer than the lip, the latter 2–4 cm long 3

3 Flowers yellow, generally 1; staminode triangular **C. calceolus**
Flowers brownish, generally 2; lip white, staminode ovate or obovate **C. montanum**

C. acaule Ait. / Stemless Lady's-slipper
Leaves 2, basal, nearly opposite, narrowly elliptic or obovate, sparsely hairy; scape 1–3 dm tall, pubescent, bearing a solitary flower; sepals and petals lanceolate, yellow-green to greenish brown, 3–5 cm long; lip 3–6 cm long, usually pink with red veins, drooping, cleft down the front. $2n = 20$. Bogs, woods, sand-dunes. wMack, nSask e to Lab, Nfld s to Gt Lakes, Apal. Map 376 ○

C. calceolus L. / Yellow Lady's-slipper
Stem 1–4 dm tall; leaves mostly 3 or 4, ovate to lanceolate, 5–15 cm long; flowers 1 (2); sepals ovate-lanceolate, 3–5 cm long, yellowish or greenish, striped with purple; petals linear-lanceolate, 3–6 cm long, usually undulate and spirally twisted, greenish yellow to purplish brown; lip pouch-shaped, yellow, often marked with purple near the orifice, 1.5–4 cm long; staminode triangular. $2n = 20$. *C. parviflorum* Salisb. Moist woodlands and banks. Circumpolar: Yuk, wMack, e to Nfld s to Ore, NM, Tex, sSask, Man, and most of eUSA. Map 377 ●

C. montanum Dougl. *ex* Lindl. / Mountain Lady's-slipper
Plant 2–5 dm high, more or less glandular-pubescent throughout; leaves 4–6, ovate to broadly lanceolate, 5–16 cm long; flowers 1–3; sepals and petals lanceolate, acuminate, 4–6 cm long, brownish purple or greenish brown; lip globose-saccate, white, tinged and veined with purple, 2–3 cm long; staminode oblong-ovate or obovate, yellow with purple spots. Moist woods. seAlas to Calif, Mont, Wyo. Map 377 ○

C. passerinum Richards.
Stem 1–2.5 dm tall, often retrorsely villose; leaves oval to lanceolate, 5–15 cm long, more or less viscid-pubescent; flowers 1–3; sepals 1–1.5 cm long; the lower one slightly 2-cleft; petals oblong, obtuse, about 1.5 cm long; lip obovate, white or pink, with purple spots inside, about 1.5 cm long. $2n = 20$. Woods. Alas, Yuk, Mack to James Bay s to nwMont, cSask, cMan; Lk Superior; sQue. Map 376 ●

Goodyera R.Br. / Rattlesnake Plantain

Plants scapose, with a basal cluster of leaves, creeping rootstocks and thick roots; leaves deep green, often reticulated or blotched with white; scape erect, with several scale-like bracts and a terminal spike-like raceme of small, white or greenish, flowers; upper sepal connate with the lateral petals to form a hood over the lip; lateral sepals free, their tips somewhat spreading; lip concave or saccate, entire, prolonged into a reflexed beak; spur absent; pollinia 2, granular.

Raceme 6–10 cm long; lip scarcely saccate, with incurved margins, leaves 3–6 cm long **G. oblongifolia**

Raceme 3–6 cm long; lip deeply saccate, with recurved or flaring margins; leaves 1–3 cm long **G. repens**

G. oblongifolia Raf.
Scapes 1–4 dm tall, densely glandular-pubescent; leaves with wide stalks, the blades ovate-lanceolate, 3–6 cm long, dark green, often blotched with white or white-reticulate; raceme spiral or 1-sided; sepals and petals 6–9 mm long; anther acuminate. $2n = 30$. *Peramium menziesii* (Lindl.) Morong. Woods. sAlas to Calif, Ariz, NM, Mont; Gt Lakes, Que, NS. Map 378 ○

G. repens (L.) R.Br.
Scapes 1–2 dm tall, slender, glandular-puberulent; leaf-blades broadly ovate to oblong, 1–3 cm long, green throughout or white-reticulate; raceme one-sided; sepals and petals 3–5 mm long; anther blunt. $2n = 30$ (28, 32). *Peramium repens* (L.) Salisb. In our area var. **repens**. Mossy woods. Circumpolar: Alas, Yuk, swMack to Nfld s to sBC, sSask, Man, Gt Lakes, Apal; SD; Colo; NM; Ariz. Map 378 ●

Habenaria Willd. / Bog Orchid

Small glabrous plants with one or more leaves and usually fleshy roots; flowers small, greenish or white, in a terminal bracted spike or raceme; sepals and lateral petals similar in form and colour or the petals smaller, spreading or the petals erect and connivent with the upper sepal; lip entire or 3-lobed, linear to ovate or obovate, prolonged backward into a saccate or elongate spur; stigmatic discs exposed, often widely separate.

1 Foliage leaves basal, usually 1 or 2; stem naked above or with a few bracts only 2
Foliage leaves cauline, spread along the stem or clustered near the base 4

2 Leaves solitary, or rarely 2; plants 1–3 dm tall **H. obtusata**
Leaves 2, or rarely 3; plants 2–5 dm tall 3

3 Leaves broadly elliptic to orbicular, spreading on ground **H. orbiculata**
Leaves narrowly oblanceolate, erect or spreading but not prostrate **H. unalascensis**

4 Lip 2–3-toothed, oblong; spur pouch-like, 2–3 mm long, almost concealed **H. viridis**
Lip entire, variously shaped 5

5 Leaves clustered near the base of stem, withering at flowering time; sepals 1-nerved **H. unalascensis**
Leaves spread along the stem, not withering at flowering time; sepals 3-nerved 6

6 Flowers white, strongly spicy-fragrant; lip conspicuously dilated at base **H. dilatata**
Flowers greenish; lip linear or lanceolate 7

7 Spur shorter than the lip, saccate; inflorescence open **H. saccata**
Spur as long as the lip, club-shaped or narrowly cylindrical; inflorescence congested **H. hyperborea**

H. dilatata (Pursh) Hook. / Tall White Orchid
Stems stout or slender, 2–7 dm high, hollow; leaves several, mostly lanceolate, 4–20 cm long, reduced upwards; inflorescence 1–3 dm long, dense or lax; flowers strongly spicy-fragrant, white or yellowish white, occasionally greenish; sepals and petals ovate to narrowly lanceolate; lip 5–8 mm long, broadly lanceolate, conspicuously widened at the base; spur cylindrical, shorter than or equalling the lip. $2n = 42$. *Platanthera dilatata* (Pursh) Beck. Wet woods, thickets, bogs. Greenland: sAlas, Yuk, swMack se to Lab, Nfld s to Calif, Nev, NM, SD, sSask, Man, Gt Lakes. Map 379

H. hyperborea (L.) R.Br. / Northern Green Orchid
Stems slender or stout, 1–7 dm high; leaves several, linear-lanceolate to oblanceolate, 2–10 cm long, passing upwards into bracts; inflorescence usually compact, 2–10 cm long, the bracts and flowers about equal; flowers green or yellowish green; upper sepal ovate, shorter than the lanceolate lateral ones; petals lanceolate, erect; lip usually lanceolate, 4–7 mm long, gradually and uniformly widened toward the base; spur slenderly clavate or cylindrical, about as long as the lip. $2n = 42, 84$. *Platanthera hyperborea* (L.) Lindl. Bogs, wet meadows, and woods. Similar distribution to *H. dilatata* but extends somewhat farther n and to Iceland. Map 380

H. obtusata (Pursh) Richards. / Blunt-leaved Orchid
Plant scapose, 1–3 dm tall, the stem naked or rarely 1-bracted; leaf solitary (rarely 2) at the base of the stem, obovate to oblanceolate, 4–12 cm long, rounded or blunt at apex, tapering at the base; inflorescence 2–8 cm long, few-flowered, slender; flowers greenish white; upper sepal broad, rounded; lateral sepals and

petals lanceolate-oblong; lip linear-lanceolate, 5–8 mm long, deflexed; spur slender, tapering from the base, about as long as the lip. Moist woods. More or less circumpolar: Alas e to Lab, Nfld s to neOre, Mont, neUtah, Colo, sSask, Man, Gt Lakes. Map 381

H. orbiculata (Pursh) Torr. / Round-leaved Orchid
Plant scapose, 3–6 dm tall, the stem stout, rising from fleshy roots; leaves 2 (rarely 3) at the base of the stem, orbicular to broadly elliptic-oblong, fleshy, spreading flat on the ground; inflorescence 5–20 cm long, loosely flowered, the pedicels 5–10 mm long; flowers greenish white; sepals and petals 8–12 mm long, orbicular to lanceolate; lip linear, 10–15 mm long, directed downward; spur slenderly club-shaped, 15–30 mm long, reflexed, often arching at tip. $2n = 42$. *Platanthera orbiculata* (Pursh) Lindl. Woods. seAlas to nOre, e to nwMont, swMack e to Lab, Nfld s to cSask, sMan, Gt Lakes, Apal. Map 382 ●

H. saccata Greene / Slender Bog Orchid
Stems 2–5 dm high; leaves several, 4–12 cm long, mostly lanceolate or oblanceolate; inflorescence slender, remotely flowered, usually over 10 cm long, the bracts linear-lanceolate; flowers many, green, sometimes tinged with purple or brown, odourless; sepals and petals ovate to lanceolate, 3–6 mm long; lip linear to lanceolate or ovate-elliptic, 4–7 mm long, often purplish; spur shorter than the lip, pouch-shaped or broadly clavate, often purplish. $2n = 42$. *Platanthera saccata* (Greene) Hult. Wet meadows, bogs, and forests. Alas to Calif, Nev, NM. Map 388 ▲

H. unalascensis (Spreng.) S. Wats. / Alaska Bog Orchid
Stems slender, bracted above, 2–5 dm tall, from rounded tuberous bases; leaves 1–4, basal, erect or spreading, oblanceolate to lanceolate, 6–12 cm long, withering at flowering time; inflorescence 1–3 dm long, slender, remotely flowered; flowers greenish or yellowish green, often marked with purple, with an unpleasant odour; sepals and petals ovate to lanceolate, 1-nerved; lip ovate-lanceolate, widened at base, about 5 mm long; spur slender or clavate, about as long as the lip. $2n = 42$. *Platanthera unalaschensis* (Spreng.) Kurtz, *Piperia unalaschensis* (Spreng.) Rydb. Moist woods and meadows. sAlas to Baja Calif, Nev, Utah, Colo; SD; Gt Lakes; eQue. Map 382 ○

H. viridis (L.) R.Br. / Bracted Orchid
Stems 1.5–5 dm tall, very leafy up to the inflorescence; leaves 4–12 cm long, oblanceolate to lanceolate; inflorescence at first compact becoming more open, 2–20 cm long, the bracts foliaceous, lanceolate, the lowest up to 5 cm long; flowers greenish; sepals ovate, spreading; petals narrowly lanceolate; lip narrowly oblong or somewhat spatulate, 6–8 mm long, the truncate apex 2–3-toothed; spur shorter than the lip, saccate. $2n = 40, 42$. *Coeloglossum viride* (L.) Hartm. In our area var. **bracteata** (Muhl.) Gray. Moist meadows and woods. Circumpolar, boreal: sAlas to Lab, Nfld s to Wash, NM, sSask, sMan, Gt Lakes, Apal. Map 383

Listera R.Br. / Twayblade

Small plants with slender stems and fibrous roots; leaves 2, opposite, or nearly so, sessile, borne near the middle of the stem; inflorescence a slender lax raceme; flowers small, greenish or purplish; sepals and lateral petals similar, spreading or reflexed; lip longer than the sepals, drooping, 2-notched at apex, or deeply cleft; spur none; stigma with a rounded beak; anther erect at the summit of the column; pollinia 2, powdery.

1 Lip linear-oblong, cleft about half-way to the base into slender lobes; leaves cordate to deltoid, mucronate **L. cordata**

Lip broader, notched at apex, the short lobes rounded; leaves rarely cordate 2

2 Lip nearly oblong, somewhat narrowed in the middle, with 2 ear-shaped appendages at base **L. borealis**

Lip broadest at tip, cuneate to obovate, without basal appendages 3

3 Lip 8–10 mm long, with a short basal claw and a minute tooth on each side at the base, margin ciliolate **L. convallarioides**

Lip about 5 mm long, sessile, with a prominent tooth on each side at the base, margin glabrous **L. caurina**

L. borealis Morong / Northern Twayblade
Plants 5–25 cm tall; leaves ovate to elliptic, 1–5 cm long; flowers pale green or yellow-green, the veins of the strongly reflexed sepals and petals deep green; sepals and petals linear to lanceolate, 4–7 mm long; lip broadly oblong, somewhat narrowed in the middle, broadest and auriculate at base, 7–12 mm long. *Ophrys borealis* (Morong) Rydb. $2n = 56$. Mossy woods, moist meadows, and mountain slopes. Alas, Yuk, swMack to Huds Bay s to BC, neWash, Colo, cSask, sMan; Que (Gaspé). Map 384

L. caurina Piper
Plants 10–20 cm high, glandular-puberulent above; leaves broadly ovate to nearly orbicular, sometimes acutish, 3–6 cm long; flowers greenish or yellowish; sepals and petals spreading, linear-lanceolate, 3–4 mm long; lip cuneate to obovate, 4–6 mm long, rounded to retuse at apex, and a tooth on each side at the base. *Ophrys caurina* (Piper) Rydb. $2n = 34$. Moist woods. seAlas to nCalif, wMont, Wyo. Map 385

L. convallarioides (Sw.) Nutt.
Plants 8–20 cm high, glandular-hairy above; leaves broadly oval, obtuse or acutish, 3–5 cm long; flowers yellowish green; sepals and petals ovate-lanceolate to linear, reflexed, 4–5 mm long; lip cuneate, 8–10 mm long, notched at apex, usually with a small tooth on each side near the base. $2n = 36$. *Ophrys convallarioides* (Sw.) Wight. Boggy woods and meadows. Alas, (Aleutian Is), BC to Calif, wMont to Colo, Utah; Lk Superior to Nfld, s to NY. Map 386

L. cordata (L.) R.Br. / Heart-leaved Twayblade
Plants 6–20 cm high, glabrous or minutely hairy above, the leaves usually below the middle of the stem; leaves broadly ovate to reniform, cordate to deltoid at base, mucronate, 1–4 cm long; flowers purplish to yellowish green; sepals and petals ovate-oblong, 2–3 mm long; lip linear-oblong, purplish green, 4–5 mm long, cleft to or below the middle into linear lobes, with 2 subulate teeth near the base. $2n = 36, 38, 40, 42$. *Ophrys cordata* L. Mossy woods, thickets, and bogs. Circumpolar, boreal: Alas, Yuk, swMack e to Lab, Nfld s to Calif, Nev, Utah, NM, cSask, sMan, Gt Lakes, and Apal. Map 387

Malaxis Sw. / Adder's-mouth

Small plants from solid bulbs (corms); leaves solitary or 2–5, basal or near the middle of the low scape; flowers small, mostly greenish, in a very slender raceme; sepals lanceolate to ovate; petals much smaller, linear or lanceolate-ovate; lip variously shaped, often auricled at base; spur absent; column very small; pollinia 4, in pairs, waxy.

Leaf solitary, basal or near the middle of the scape; lip abaxial, drooping **M. monophylla**

Leaves 2–5, basal, enclosing a small daughter-bulb; lip adaxial, erect **M. paludosa**

M. monophylla (L.) Sw.
Stem 5–25 cm high, from a solid bulb (corm); leaf solitary, clasping, 3–5 cm long, elliptic-oval or ovate, basal or near the middle of the low scape; raceme very slender, tapering to the tip; flowers small, greenish to yellowish; sepals ovate to lanceolate, 2–2.5 mm long; petals linear, about 1 mm long; sepals and petals finally reflexed and appressed to the capsule; lip ovate, with a slender acuminate tip, 2–3 mm long, drooping. $2n = 28, 30$. *M. brachypoda* (A. Gray) Fern. of Ed. 1. *Microstylis monophyllos* (L.) Lindl. In our area var. **brachypoda** (A. Gray) Morr. & Eames. Damp woods, banks, and bogs. Circumpolar, boreal: sAlas, BC, cSask, sMan e to Nfld s to Gt Lakes; sCalif; Colo. Map 388 ●

M. paludosa (L.) Sw.
Stem very slender, 5–15 cm high, with 2–5 leaves at base, connected below with parent bulb by a short vertical stolon; leaves sheathing at base and broadly rounded at apex, enclosing a daughter bulb and often a few minute bulbils; raceme very slender, lax; flowers minute, yellowish green or pale green; sepals ovate-lanceolate, 2 of them erect, the third one drooping; petals half as long as the sepals; lip ovate, erect, conspicuously green-nerved. $2n = 28$. *Hammarbya paludosa* (L.) O. Kuntze. Wet bogs, in sphagnum moss. Circumpolar, boreal: scattered records, sAlas, BC, swMack, nSask, wOnt, Minn. Map 388 ○

Orchis L.

O. rotundifolia Banks *ex* Pursh / Round-leaved Orchid
Stem scape-like, 10–25 cm high, from a scaly rootstock with thickened roots; leaf solitary, elliptic to ovate or obovate, 3–7 cm long, near the base of the scape; raceme 2–8-flowered, bracted; flowers 12–15 mm long; sepals and lateral petals rose-coloured to white, the lateral sepals spreading, the upper sepal and the petals ascending; lip 6–9 mm long, white, purple-spotted, 3-lobed, the middle lobe large and notched at the apex; spur slender, curved, much shorter than the lip; pollinia granular. $2n = 42$. *Amerorchis rotundifolia* (Pursh) Hult. Moist woods. Greenland: Alas, Yuk, Mack, Huds Bay e to Nfld s to Wyo, sSask, sMan, Gt Lakes. Map 389

Spiranthes L.C. Rich.

S. romanzoffiana Cham. & Schlecht. / Ladies'-tresses
Plant rather stout, with leafy stem, 1–4 dm tall, from fleshy roots; leaves oblong-lanceolate to linear, 5–15 cm long, progressively reduced upward; flowers 6–10 mm long, white, sweet-scented, crowded, 3-ranked, in a twisted spike-like raceme; sepals and lateral petals similar, ovate-lanceolate, somewhat united forming an upward-arching hood; lip 9–12 mm long, strongly deflexed, constricted beyond the middle and then widened into a rounded terminal lobe, the margin wavy-crenulate; spur none; pollinia 2, each 2-cleft and split into delicate plates of granular pollen connected by elastic threads. $2n = 30$. Bogs and wet meadows. British Is: Alas se to Lab, Nfld s to Calif, Ariz, NM, sSask, Man, Gt Lakes. Map 390

SALICACEAE / Willow Family

Dioecious trees or shrubs with soft wood and bitter bark; leaves alternate, undivided, stipulate; flowers in catkins (aments), each flower solitary in the axil of a bract (scale); true perianth wanting; stamens 1–many; ovary solitary, 1-loculed, with 2–4 basal or parietal placentae; stigmas 2–4, often 2-lobed; fruit a dehiscent capsule; seeds numerous, furnished with tufts of silky hairs.

Winter-buds covered by several scales; catkins soon drooping; bracts of catkins mostly lacerate; flowers with a broad or cup-shaped disk; stamens usually more than 10; trees **Populus**

Winter-buds covered by a single scale; catkins erect or divergent, rarely drooping; bracts entire or merely toothed; disc none; stamens few, often 2; mostly shrubs **Salix**

Populus L. / Poplar; Cottonwood

Trees or sometimes shrubby, the bark smooth to furrowed; buds often resinous; leaves broad, often ovate, rounded or cordate at base, the margins commonly toothed; stipules small, caducous; flowers in drooping catkins, appearing in early spring and maturing while the leaves are expanding; staminate and pistillate catkins on separate trees; scales lacerate or lobed; stamens several to many, on a broad disc; ovary conic or ovoid, borne on a cup-shaped disc; stigmas 2–4; capsule opening by 2–4 valves.

Populus species hybridize readily and several hybrids occur in southern Alberta. Apart from *P.* × *acuminata* the following have received taxonomic recognition: *P.* × *andrewsii* Sarg. (*P. deltoides* × *P. tremuloides*), *P.* × *brayshawii* Boivin (*P. angustifolia* × *P. balsamifera*), *P.* × *dutillyi* Lepage (*P. balsamifera* × *P. tremuloides*), *P.* × *jackii* Sarg. (*P. balsamifera* × *P. deltoides*), *P.* × *sennii* Boivin (*P. angustifolia* × *P. tremuloides*). These hybrids are generally morphologically intermediate between the parental species, but may resemble one more closely than the other.

The following treatment deals only with native species.

1 Petioles strongly flattened laterally — 2
  Petioles circular in cross-section, not flattened — 3

2 Leaf-blades 2–6 cm long, suborbicular, crenulate or nearly entire; bark of trunk (well above base) smooth, light green or whitish — **P. tremuloides**
  Leaf-blades 5–10 cm long, broadly deltoid to cordate, coarsely toothed, often crenate; bark of trunk deeply furrowed, usually grey — **P. deltoides**

3 Petioles one-third the length of the blade or shorter; blades 3–7 times as long as broad, gradually tapered to apex — **P. angustifolia**
  Petioles longer and blades broader than above — 4

4 Blades green on both sides, abruptly acuminate, rhombic-lanceolate to ovate, margin crenulate to distinctly toothed — **P. x acuminata**
  Blades paler beneath, acute or somewhat acuminate, not at all rhombic, margin entire to crenulate — **P. balsamifera**

P. x acuminata Rydb. / Lance-leaf Cottonwood
Tree reaching 20 m tall, with a broad crown; twigs greenish brown; leaves often rhombic-lanceolate, abruptly acuminate, cuneate at base, the margin crenate; petioles 3–7 cm long; capsules ovoid, papillose, the pedicels 3–5 mm long. $2n = 38$. A fairly distinctive taxon but probably a hybrid between *P. angustifolia* and *P. deltoides*. River valleys. sAlta to NM, Utah, Ariz e to SD, Neb. Map 391

P. angustifolia James / Narrow-leaf Cottonwood
Tree reaching 12 m tall, usually slender, with ascending branches; twigs orange-brown; leaves short-petioled, 5–8 cm long, usually rounded at base and narrowly to broadly lanceolate, gradually narrowed to the apex, crenulate; capsules narrowly ovate, rugose. $2n = 38$. River valleys. sAlta to sSask s to Mont, Ida, Ore, Calif, Ariz, Nev, e to SD, Neb. Map 392

P. balsamifera L. / Balsam Poplar
Tree up to 25 m tall, with broad crown and dark furrowed bark; twigs brown; buds coated with a viscid gum; leaf-blades 6–12 cm long, ovate to ovate-lanceolate,

rounded to cuneate or somewhat cordate at base, deep green above, silvery-whitish or pale brownish beneath, the margins crenate-serrate; petioles 2–5 cm long; capsules oblong-ovoid, glabrous or pubescent, 2- or 3-valved. $2n = 38$. *P. tacamahacca* Mill. In our area, hardly distinguishable without flowers and fruits, ssp. **balsamifera** (*P. balsamifera* L. of Ed. 1) with 12–20 stamens, capsule generally 2-valved, glabrous, mostly c and nAlta, and ssp. **trichocarpa** (T. & G. *ex* Hook.) Brayshaw (*P. trichocarpa* Hook. of Ed. 1) with 40–60 stamens, capsule generally 3-valved, capsule often pubescent, mostly swAlta s of Bow R. Hybrids between the two subspecies occur throughout wAlta. Boreal forest, riverbanks, and alluvial flats. Alas, Yuk, sMack to Huds Bay, Nfld s to Calif, Utah, Wyo, sSask, sMan, Gt Lakes. Map 393

P. deltoides Marsh. / Western Cottonwood
Tree reaching 25 m high, with deeply furrowed bark on the trunk and yellowish twigs; leaves coriaceous, broadly deltoid or somewhat cordate, 5–10 cm long, short-acuminate, glabrous and shiny, the margin coarsely serrate-crenate; capsules ovoid, 5–10 mm long, on pedicels 4–6 mm long. $2n = 38$. *P. sargentii* Dode of Ed. 1. In our area var. **occidentalis** Rydb. Riverbanks and moist sand-dunes. Alta to Que s to Colo, Tex, generally distributed e. Map 394

P. tremuloides Michx. / Aspen
Slender tree, reaching a height of 30 m, the bark smooth, light green or greyish, becoming dark and furrowed in age; leaves with slender flattened petioles; blades small, orbicular to broadly ovate, finely serrate-crenate to nearly entire; capsules slenderly conic, 3–5 mm long, warty. $2n = 38^*$. Forest and parkland. Alas, Yuk, sMack, to James Bay, Nfld s to Calif, NM, Mo, Tenn, NJ. Map 395

Salix L. / Willow

Plants dioecious; erect shrubs to dwarf or trailing alpine shrubs or trees; buds covered by a single scale; leaves alternate, simple and usually stipulate; flowers unisexual, in spike-like aments which appear before, with or after the leaves, sessile on branches of the previous year or on short, leafy branchlets; each flower subtended by a bract and one to several nectaries; staminate flowers with 1–7 stamens and accompanying nectaries; pistillate flowers a single pistil and its nectaries; pistils sessile or borne on short stipes; styles single or bifurcate; stigmas 2; fruits a bivalved capsule containing several seeds surrounded by a coma of fine hairs.

Willows are pollinated by insects or wind. Hybridization is possible between some species, but hybrids are not conspicuous in nature.

The introduced species *S. alba, S. daphnoides, S. fragilis*, and *S. pentandra*, often cultivated as windbreaks or amenity trees in settlements, are not included in this treatment.

1	Leaves green beneath, sometimes paler but non-glaucous	2
	Leaves glaucous beneath or the leaf surface obscured by a dense indumentum	9

2 Leaves linear or nearly so, margins entire to distantly denticulate; aments often branched ... 3

Leaves narrowly elliptic to broadly ovate, margins serrulate to crenate; aments unbranched ... 4

3 Leaves linear, the sides parallel, lower surface green; styles absent; nectaries shorter than stipes ... **S. exigua**

Leaves very narrowly elliptic, the sides not parallel, lower surface pale or rarely glaucescent; styles evident; nectaries longer than stipes ... **S. melanopsis**

4 Leaves sparsely villous to villous-lanate ... **S. commutata**

Leaves glabrous beneath ... 5

5 Pistils sericeous-villous; leaves coriaceous ... **S. maccalliana**

Pistils glabrous; leaves thin ... 6

6 Leaf-apex long-acuminate, blade very narrowly elliptic, petiole glandular at base of blade ... **S. lucida**

Leaf-apex acute or short-acuminate, blade narrowly elliptic to broadly so, petiole non-glandular ... 7

7 Shrubs, low, decumbent, less than 1 m tall; stipules minute to 2 mm long; young leaves glabrous; styles 0.3–0.5 mm long ... **S. myrtillifolia**

Shrubs, erect, up to 6 m tall; stipules conspicuous, up to 12 mm long; young leaves pubescent; styles 0.4–0.9 mm long ... 8

8 Stipules 1–5 mm long, semi-ovate, apex not sharply acute; leaf-apex broadly acute to round ... **S. myrtillifolia**

Stipules 5–12 mm long, apex sharply acute; leaf-apex abruptly short-acuminate ... **S. boothii**

9 Dwarf shrubs, rarely exceeding 2 dm tall, trailing or forming compact mats ... 10

Erect shrubs from 2 dm tall to trees ... 13

10 Aments terminating prominent vegetative branchlets; pistils less than 3 mm long; styles 0.2–0.3 mm long ... **S. reticulata**

Aments terminating small lateral branchlets; pistils 4–9 mm long; styles 0.8–1.5 mm long ... 11

11 Pistils glabrous or sparsely pubescent on beak ... **S. stolonifera**

Pistils villous-lanate or sparsely so ... 12

12 Bracts with long straight hairs; styles red in life; leaves with stomata above; branchlets glabrous or very sparsely pubescent ... **S. arctica**

Bracts with shorter wavy hairs; styles green in life; leaves lacking stomata above; branchlets pubescent to villous-lanate ... **S. glauca**

13 Leaves essentially glabrous beneath ... 14

Leaves sericeous to sparsely villous or villous-lanate beneath ... 31

14 Pistils and capsules glabrous ... 15

Pistils and capsules sericeous to sparsely pubescent ... 27

15 Petioles glandular at base of blade ... 16

Petioles not glandular ... 17

16 Stipules minute glandular lobes or clusters of glands; capsules 7–9 mm long; flowering in late summer **S. serissima**

Stipules foliar, semi-ovate; capsules 5–7 mm long; flowering in early summer **S. lucida**

17 Stipules distinct, ovate to narrowly elliptic 18

Stipules absent or minute glandular lobes 24

18 Bracts deciduous after flowering, yellow or greenish; leaves with stomata above **S. melanopsis**

Bracts persistent, tawny to dark brown; leaves without stomata above 19

19 Leaves broadly elliptic to ovate; floral buds swollen; styles up to 2.5 mm long **S. lanata**

Leaves narrower; floral buds not swollen; styles shorter 20

20 Young leaves greenish; aments on leafy branchlets 21

Young leaves reddish; aments sessile or on very short branchlets 22

21 Young leaves sparsely villous with white hairs, the margins usually glandular-serrulate; capsules tawny or greenish **S. barclayi**

Young leaves glabrous or with fine white and ferruginous hairs only on midrib, the margins usually entire; capsules reddish or greenish **S. farriae**

22 Branches yellow to greyish-yellow; leaves shallowly serrulate to subentire **S. lutea**

Branches reddish brown to pale brown; leaves serrulate to coarsely glandular-crenate 23

23 Styles 0.7–1.2 mm long; leaves narrowly to broadly elliptic to obovate **S. pseudomonticola**

Styles 0.3–0.6 mm long; leaves narrowly oblong-obovate **S. prolixa**

24 Leaf-apex tapering to a long slender tip; trees **S. amygdaloides**

Leaf-apex short acuminate, acute to rounded; shrubs 25

25 Young leaves translucent, older leaves usually cordate or asymmetrical at base **S. pyrifolia**

Young leaves thicker, base round 26

26 Leaves, including midrib, glabrous, often glaucous on both sides, apex round **S. pedicellaris**

Leaves glabrous to sparsely villous, midrib finely pubescent with erect hairs, non-glaucous above, apex acute **S. farriae**

27 Aments sessile; stipes 0.5–2.5 mm long 28

Aments on leafy branchlets; stipes 0.8–5 mm long 29

28 Leaves dull or satiny above, veins widely spaced and irregular; stipes 2–2.5 mm long; stigmas equal to or longer than style **S. discolor**

Leaves highly glossy above, veins closely spaced and parallel; stipes 0.5–0.9 mm long; stigmas shorter than styles **S. planifolia**

29 Aments compact; stipes 0.8–1.2 mm long; ferruginous hairs on leaves **S. athabascensis**

Aments loose; stipes 2–5 mm long; lacking ferruginous hairs 30

30 Leaves long and slender, narrowly elliptic, serrulate; bracts dark brown **S. petiolaris**

Leaves shorter and broader, elliptic to obovate, entire to crenate; bracts tawny **S. bebbiana**

31 Pistils and capsules glabrous 32

Pistils sericeous to villous 33

32 Aments appearing before the leaves, sessile; stipes 0.2–0.5 mm long; leaves broadly elliptic to broadly ovate **S. lanata**

Aments appearing with the leaves, on leafy branchlets; stipes 0.5–1.5 mm long; leaves elliptic to obovate **S. barclayi**

33 Aments sessile, leafless or with a few bract-like leaves at base 34

Aments on leafy branchlets 40

34 Leaves densely white villous-lanate beneath 35

Leaves glabrate, sericeous or lanate 37

35 Branchlets puberulent to glabrous, often glaucous; styles 0.7–1.3 mm long; stigmas 0.3–0.4 mm long **S. drummondiana**

Branchlets densely white-lanate or velutinous, not glaucous; styles and stigmas various 36

36 Leaves white-villous beneath, veins evident beneath; styles 0.2–0.6 mm long; stigmas 0.6–1.2 mm long **S. scouleriana**

Leaves densely white- or grey-lanate beneath, veins obscured; styles 1.3–2.8 mm long; stigmas 0.5–1.8 mm long **S. alaxensis**

37 Buds and stipules oily, stipules broadly ovate and prominently glandular **S. barrattiana**

Buds and stipules not oily, stipules linear to ovate 38

38 Leaves obovate, ferruginous hairs prominent beneath; branchlets velutinous; styles 0.2–0.6 mm long **S. scouleriana**

Leaves narrowly elliptic to sometimes obovate, ferruginous hairs lacking or only on young leaves; styles longer, 0.6–1.5 mm long 39

39 Leaves densely sericeous or lanate beneath, margins revolute **S. drummondiana**

Leaves glabrous or sparsely pubescent beneath, margins flat **S. planifolia**

40 Stipes 2–5 mm long, about 10 times as long as nectaries 41

Stipes 0–1.6 mm long, shorter than to twice nectary length 42

41 Leaves long and slender, narrowly elliptic, serrulate; bracts dark brown **S. petiolaris**

Leaves shorter and broader, elliptic to obovate, entire to crenate; bracts tawny **S. bebbiana**

42 Leaves sericeous beneath with short, straight, appressed hairs 43

Leaves densely lanate to glabrescent beneath 44

43 Leaves 5–7 times as long as wide, narrowly ovate, serrulate, some sericeous hairs ferruginous, styles 0.2–0.5 mm long **S. arbusculoides**

Leaves 2.5–3 times as long as wide, narrowly elliptic to obovate, entire, satiny beneath with white hairs; styles 0.4–0.8 mm long **S. sitchensis**

44 Leaves densely white-lanate beneath, sparsely floccose above, midrib yellowish, prominent; styles red in life **S. candida**

Leaves grey-lanate to villous-lanate or glabrescent beneath, indumentum more sparse and not floccose; styles greenish-yellow in life 45

45 Aments terminating prominent, 3-leaved branchlets; petiole with dark glands; leaves reticulate above **S. vestita**

Aments on short leafy branchlets; petiole lacking glands; leaves plane above 46

46 Leaves with scattered ferruginous hairs, especially the young **S. athabascensis**

Leaves without ferruginous hairs 47

47 Petioles shorter than or up to 3 times bud length, often reddish; stipes usually absent; pistillate aments short-cylindrical to subspherical **S. brachycarpa**

Petioles longer than 3 times bud length; stipes 0.5–1.6 mm long; pistillate aments long-cylindrical **S. glauca**

S. alaxensis (Anderss.) Coville
Gnarled shrubs prostrate to 1 m tall or up to 3.7 m; branches reddish brown clothed with persistent grey indumentum or glabrescent; branchlets densely white to grey villous-lanate; leaves elliptic to obovate or narrowly so, 4–10 cm long, apex acute, base cuneate, margins entire and revolute, sometimes distantly glandular and undulating, the lower surface densely white-grey lanate, petioles 3–15 mm long, tawny, sometimes inflated around bud; stipules linear; aments appearing before the leaves and sessile; bracts dark brown to black, long villous on both sides; staminate aments 4 cm long, stamens 2; pistillate aments 6–10 cm long in fruit, pistils densely lanate, subsessile, capsules sparsely villous, sutures brownish, styles 1.3–2.8 mm long, stigmas 0.5–1.8 mm long. $2n = 38$. In our area ssp. **alaxensis**. Alpine slopes. eAsia: Alas, Yuk e to Lab s to BC, Alta, and nMan. Map 396 ○

S. amygdaloides Anderss.
Shrubs or trees up to 10 m tall; branches slender, yellow-grey; branchlets slender, reddish brown; leaves ovate to narrowly ovate, 5–12 cm long, apex abruptly tapering to a slender tip, margins serrulate to crenulate, the lower surface glabrous and glaucous, petioles slender, yellowish; stipules minute glandular lobes, not evident; aments 3–6 cm long borne on leafy branchlets 1.5–3 cm long, sometimes very short in staminate plants; bracts tawny, lanate inside or at the base on the outside, apex round and slightly erose, deciduous after flowering; stamens 4–7, filaments pubescent at base; capsules glabrous, 4–6 mm long, tawny, deciduous after fruiting, styles about 0.2 mm long, stipes 1.4–2 mm long, nectaries short reddish or yellowish lobes. $2n = 38$. River floodplains. eBC to Man; seOnt and swQue s to Wash, Utah, NM, Ark, NJ. Map 397 ○

S. arbusculoides Anderss.
Shrubs 1–4 m tall or trees to 6 m; branches slender, reddish brown, shiny, sparsely velutinous, consisting of short erect or sometimes curly hairs; leaves narrowly obovate to very narrowly so, 6–7 cm long, apex and base acute, margins glandular-serrulate, the upper surface glossy and glabrous, the lower surface glaucous and sericeous with spaced, short, white or ferruginous hairs; stipules 0.9–2.5 m long, glandular; aments appearing with the leaves and borne on short leafy branchlets; bracts oblong, brownish and pubescent; stamens 2, filaments glabrous, anthers about 0.5 mm long; pistillate aments 2–7.5 cm long, branchlets 0–5 mm long, capsules 4–6 mm long, sparsely sericeous with white and ferruginous hairs, styles 0.3–0.5 mm long, stipes 0.6–0.9 mm long and not much longer than the nectaries. $2n = 38$. Stream banks, floodplains, fens, bogs, and subalpine slopes. Alas, Yuk to Huds Bay s to BC; cQue. Map 397 ●

S. arctica Pallas
Dwarf, creeping shrubs; branches glabrous, stout or long, slender and appressed; branchlets glabrous or sparsely pubescent, sometimes glaucous; leaves narrowly to broadly elliptic or obovate, apex acute, obtuse or round, base acute to cuneate, margins entire, the larger leaves 2–6 cm long, young leaves loosely villous, the older leaves glossy and glabrate above or remaining sparsely pubescent, especially on the edges, usually stomatiferous on both sides, the lower surface glaucous and glabrescent or with sparse pubescence; petioles distinct; stipules absent, a minute lobe or narrow and up to 2 mm long; aments appearing with the leaves and borne on leafy branchlets; bracts light to dark brown or appearing almost black, sparsely clothed with long straight trichomes; pistillate aments 2–4 (7) cm long on branchlets 1.5–3 (4) cm long; pistils villous-lanate, becoming sparsely so; styles 0.8–2 mm long, often bifid; stipes 0.6 mm or less long; nectary 1, usually more than 2 times the stipes. $2n = 76, 114$. Includes var. **araioclada** (Schneid.) Raup. Hybridizes with *S. glauca*. Subalpine and alpine slopes and meadows. More or less circumpolar: Alas, Yuk to Ellesmere Is s to Calif, NM, sMack, Huds Bay, Nfld. Map 398

S. athabascensis Raup
Shrubs 0.3–1 or 2 m tall; branches grey-brown, exfoliating; branchlets chestnut-brown to greenish, shiny, densely to sparsely pubescent with wavy or curved hairs; leaves narrowly elliptic to obovate, 2–6 cm long, the upper surface of young leaves villous to sericeous with mixed white and ferruginous hairs, mature leaves glabrescent and shiny above, the lower surface glaucous and glabrescent with appressed hairs including some ferruginous hairs, margins entire or indistinctly glandular serrulate; stipules minute lobes to 0.5 mm long; aments appearing with the leaves and borne on leafy branchlets; bracts tawny, sparsely sericeous, apex round; staminate aments on short leafy shoots, stamens 2, nectaries often 2; pistillate aments 1–3.5 cm long, on branchlets 0.2–1.5 cm long; capsules tawny, sparsely pubescent with white or a mixture of white and ferruginous hairs, styles 0.5–1 mm long, stipes 0.8–1.2 mm long, 1 adaxial nectary. $2n = 76, 95, 114$. Marl bogs, fens, and muskegs. Alas, Yuk, sMack to Huds Bay s to BC. Map 399

S. barclayi Anderss.
Shrubs usually 1–3 m tall, but varying from decumbent to 0.7 m up to 5 m tall; branches dark reddish-brown, glabrate or with persistent indumentum; branchlets sparsely villous, occasionally glabrous; leaves elliptic to obovate, 3–7 cm long, apex acute to acuminate, base rounded to subcordate, margins glandular-serrate, rarely subentire, young leaves sparsely villous above, especially along midrib, older leaves glabrate or sparsely villous above, glabrous or sparsely villous and glaucous beneath; stipules ovate to narrowly elliptic, apex attenuate, sometimes round, margins and upper surface glandular-dotted; aments appearing with the leaves borne on leafy branchlets; bracts narrow, acute, light to dark brown, usually pubescent with long hairs; staminate aments 2–4 cm long on branchlets 0.5–1 cm long; pistillate aments 4–7 cm long on branchlets 1–3 cm long, capsules glabrous, 3–8 mm long, styles 0.5–2.5 mm long; stipes 0.5–1.5 mm long. $2n = 76$. Hybridizes with *S. barrattiana*, *S. commutata*, and *S. farriae* and possibly also with *S. brachycarpa* and *S. glauca*. Forms extensive alpine and subalpine thickets. Alas, Yuk, swMack s to Wash, Wyo. Map 400

S. barrattiana Hook.
Low, often depressed alpine shrubs 0.3–1.5 m tall; branches gnarled, dark reddish-brown, glossy and villous; branchlets straggly-villous with short internodes, buds sparsely villous, glossy and oily; leaves narrowly to broadly elliptic or obovate, 3.5–9 cm long, apex acute to obtuse, base round, acute or subcordate, margins entire, sometimes distantly glandular, upper surface grey-villous, lower surface villous to lanate becoming glabrescent; stipules broadly ovate, glandular on margin, oily, vegetative shoots with a balsamic fragrance; aments appearing with or just before the leaves, sessile on branches of the previous year; bracts dark brown to black, ovate to oblong and villous with long hairs; staminate aments stout, 3–5 cm long, stamens 2, filaments glabrous; pistillate aments stout, 4–11 cm long, pistils densely sericeous becoming villous in fruit, styles 0.6–1.6 mm long, stigmas short, stipes short. Includes var. *angustifolia* Anderss. Hybridizes with *S. barclayi*, *S. commutata*, and *S. myrtillifolia* var. *cordata*. Forms extensive alpine and subalpine thickets. Alas, Yuk, swMack s to Mont, Wyo. Map 401

S. bebbiana Sarg.
Shrubs or small trees 0.5–5 m tall; branches divaricate, reddish brown, pubescent to glabrescent; branchlets densely pubescent; leaves elliptic to obovate or narrowly so, 2–6 cm long, apex acute to obtuse, base round to acute, margins entire to crenate, the upper surface dull green, pubescent to glabrescent, the lower surface grey, sericeous-lanate to glabrescent, glaucous and varying from smooth to rugose-veiny, petiole 2–9 mm long; stipules small and deciduous except on vigorous shoots; aments appearing with or just before the leaves, borne on short, leafy branchlets; bracts narrowly oblong, tawny, rarely with a reddish tip, and sparsely pubescent; staminate aments subsessile on branchlets 1–3 mm long; pistillate aments 2–5 cm long, loosely-flowered and borne on branchlets 3–15 mm long, pistils long-beaked, sericeous, capsules 6–9 mm long, styles and stigmas short, stipes 2–5 mm long, pubescent, nectaries 1, about 0.1 times as long

as the stipe. $2n = 38$. Includes var. *perrostrata* (Rydb.) Schneid. Hybridizes with *S. petiolaris*. Occasional in aspen, coniferous, mixed woods and sand-dunes, commonly forming thickets on floodplains and slough margins. Eurasia: Alas, Yuk to Lab, s to Calif, NM, Ind, Md, Nfld. Map 402

S. boothii Dorn
Shrubs or trees up to 6 m tall; branches reddish brown, glossy; branchlets sparsely pubescent to glabrous; leaves usually narrowly elliptic to broadly so, apex usually abruptly short-acuminate, base round, margins glandular-serrulate, young leaves villous becoming glabrescent, the upper surface of older leaves glabrescent except on the midrib, stomata usually present on both sides, the lower surface glabrous and non-glaucous; stipules conspicuous, usually sharp-pointed, 5–12 mm long; aments appearing with or just before the leaves; bracts brown to dark brown or pale at base, pubescent with long wavy hairs, becoming glabrescent; staminate aments on short leafy branchlets or sessile; pistillate aments on leafy branchlets up to 1.5 cm long or subsessile, pistils glabrous, styles 0.4–0.8 (1.3) mm long, stipes 0.5–2 mm long. $2n = 76$. Includes *S. pseudocordata* (Anderss.) Rydb. of Ed. 1. Closely related to *S. myrtillifolia*. Possibly hybridizes with *S. myrtillifolia* var. *cordata*. Lake shores, riverbanks, bogs, and meadows. sBC, Alta s to nCalif, Colo. Map 403 ○

S. brachycarpa Nutt.
Low, erect shrubs, 0.3–1.5 m tall; branches slender, coarsely grey-villous to glabrescent; branchlets densely white or grey villous-lanate, internodes often short, resulting in a fan-like arrangement of leaves when pressed; leaves narrowly elliptic to elliptic-obovate, apex acute, base round, margins entire, villous-lanate on both sides with matted, greyish-white hairs, sometimes only sparsely pubescent, major veins impressed above, lower surface glaucous, petioles reddish to yellowish, often shorter than or equal to the bud, but varying to 3 times the bud; stipules minute lobes to 1 mm long; aments appearing with the leaves, borne on leafy branchlets, the leaves strap-shaped and blunt; bracts tawny with short straight hairs; anthers 0.4–0.5 mm long; pistillate aments short-cylindrical to subglobose, 1–2 cm long, branchlets 2–15 mm long, pistils lanate, styles 0.6–1.2 mm long, stipes usually absent, nectaries 1, 0.8–1.2 mm long. $2n = 38$. Includes var. *sansoni* Ball and var. *antimima* (Schneid.) Raup. In our area ssp. **brachycarpa** as described above and var. **glabellicarpa** Schneid., with glabrous to partly pubescent pistils and long slender beaks, which occurs mixed with the typical variety in Banff N.P. Hybridizes with *S. glauca* and possibly *S. barclayi*. Meadows and thickets on river floodplains, subalpine streams, and fens. sYuk, sMack to James Bay s to Ore, Utah, Colo, Sask; Ungava; Gaspé; Calif. Map 404

S. candida Fluegge *ex* Willd.
Low shrubs 0.4–1.5 m tall; branches light brown and glabrescent; branchlets densely lanate; leaves narrowly elliptic to narrowly ovate, 3–8 cm long, apex acute to obtuse, base acute, margins entire or rarely glandular-crenulate, revolute, upper surface dark green (often drying dark brown) and thinly lanate-

floccose, lower surface densely white-lanate, midrib prominent and yellowish, petioles 3–12 mm long; stipules narrowly elliptic, irregularly glandular-toothed; aments appearing with the leaves, borne on short leafy branchlets or the staminate aments subsessile; bracts brownish and wavy-pubescent; staminate aments 0.5–1.2 cm long, anthers reddish; pistillate aments 2–5 cm long, pistils densely lanate, capsules tawny, styles 0.8–1.6 mm long, red in life, stigmas short, spreading, stipes 0.5–1.2 mm long, nectaries 1, red in life. $2n = 38$. Includes var. *denudata* Anderss. Hybridizes with *S. pseudomonticola*. River floodplains, marl bogs, fens, and meadows. Alas, Yuk to Nfld s to Colo, Ia, and NJ. Map 405

S. commutata Bebb
Low shrubs 0.2–2 m tall; branches dark brown, sometimes remaining pubescent; branchlets densely white-lanate to sparsely villous; leaves elliptic to broadly so, apex acute to acuminate, base round, margins glandular-serrulate to entire; young leaves densely villous on both sides, older leaves usually villous-lanate with straggly trichomes, especially beneath, sometimes glabrescent, non-glaucous beneath; stipules small, half-ovate, glandular-serrulate and villous; aments appearing with the leaves, borne on definite leafy branchlets; bracts oblong, apex acute to round, tawny to dark brown and villous with short curly or long straight hairs; pistillate aments 3.5–7 cm long, leafy branchlets 1–3 cm long, pistils often reddish but capsules usually tawny, glabrous or rarely with patchy pubescence, styles 0.5–1 mm long, stipes 0.3–1.5 mm long. $2n = 38$. Includes var. *denudata* Bebb. Hybridizes with *S. barclayi* and *S. barrattiana*. Forms subalpine thickets. Alas, Yuk, swMack s to Ore and Mont. Map 406 ○

S. discolor Muhl.
Shrubs 2–11 m tall; branches dark brown to yellowish, glabrescent; branchlets divaricate, glabrous or very sparsely pubescent; leaves elliptic to obovate, 3–10 cm long, apex and base acute, margins entire and undulate to crenulate-serrate, young leaves densely sericeous with white and ferruginous hairs, becoming glabrescent, the upper surface glabrous or sparsely pubescent, dull to shiny, the lower surface glaucous and glabrescent, the upper side of petioles and midribs remaining finely pubescent, primary veins widely spaced and irregular; stipules usually minute, prominent on vigorous shoots; aments appearing before the leaves, sessile or rarely on short branchlets, bud-scale splits and remains at base of ament; bracts dark brown to black, villous; staminate aments 2–3 cm long; pistillate aments 4–7 cm long, loose in fruit, pistils long-beaked, finely sericeous with refringent hairs, capsules 8–10 mm long, styles 0.3–0.7 mm long, stigma-lobes as long as or longer than the styles, stipes 2–2.5 mm long, nectaries less than half as long as the stipes. $2n = 76, 95, 114$. Closely related to *S. scouleriana*. Hybridizes with *S. planifolia*. Stream banks, aspen and mixed woods, and thickets around prairie sloughs. BC to Nfld s to Ida, Tenn, NJ, and NS. Map 407 ●

S. drummondiana Barr. *ex* Hook.
Shrubs 1–3 m tall; branches dark reddish-brown to blackish, often prominently glaucous, brittle; branchlets yellow-green to brown, puberulent to glabrous, often glaucous; leaves narrowly elliptic or elliptic to narrowly obovate, 4.5–11 cm long,

apex and base acute, margin entire to distantly glandular or glandular-crenate, revolute, the upper surface dark green and sparsely villous becoming glabrescent, the lower surface densely sericeous or lanate, sometimes thinly sericeous, midrib tawny; stipules narrowly elliptic, 3–8 mm long or minute glandular-margined lobes; aments flowering before the leaves, sessile or with a few small bracts at the base; bracts black or dark brown, elongate, acute and clothed with long straight hairs; staminate aments 1–5 cm long, filaments glabrous, 5–6 mm long; pistillate aments 3.5–11 cm long, pistils sericeous, 3–6 mm long, capsules tawny and sparsely sericeous, styles 0.6–1.5 mm long, stigmas 0.3–0.8 mm long. $2n = 38$, 57, 76. Includes var. *bella* (Piper) Ball and var. *subcoerulea* (Piper) Ball. Subalpine thickets, river floodplains, and lake shores. Yuk, swMack s to Calif and NM. Map 408

S. exigua Nutt.
Colonial shrubs or small trees up to 4 m tall, forming thickets; branches greyish; branchlets yellow-brown to dark reddish-brown, sericeous becoming glabrescent; leaves linear with distinctly parallel sides, 5–13 cm long, 7–20 times as long as broad, apex sharply acute, base tapering, margins usually shallowly denticulate or glandular-dotted, but varying from distinctly denticulate to entire, young leaves sericeous to glabrescent, older leaves sericeous to glabrescent, non-glaucous beneath, stomata on both sides, veins distinct or not, new leaves following defoliation densely villous, petioles usually longer than the buds; stipules linear, up to 7 mm long; the first aments appearing with the leaves, others developing throughout the season, all borne on leafy branchlets, aments often branched; bracts tawny, apex pointed, rounded or erose, pubescent, caducous; staminate aments 3–4 cm long on short branchlets, filaments densely pubescent at base, anthers 0.5–1 mm long; pistillate aments 1–7 cm long on branchlets 3–6 cm long, pistils glabrous, rarely villous, 4–7 mm long, styles usually absent, stigmas short and stout, stipes 0.3–0.6 mm long, nectaries usually shorter than the stipes. $2n = 38$. Includes *S. interior* Rowlee and vars. *pedicillata* (Anderss.) Ball and *wheeleri* Rowlee of Ed. 1. Closely related to *S. melanopsis*. River floodplains and slough margins. Alas, Yuk, Mack to NB s to Calif, Tex, La, Ky, and NJ. Map 409 •

S. farriae Ball
Low shrubs 0.2–1.5 (2) m tall; branches dark reddish-brown; branchlets yellowish to brown, glabrous or sparsely pubescent; leaves elliptic to elliptic-obovate, 3–6 cm long, apex acute, base round, margins entire, sometimes glandular-dotted or serrulate, young leaves glabrous, rarely sparsely villous, but usually the midrib finely erect-hairy with white and ferruginous hairs, older leaves glabrous on both sides, except the midrib above, glaucous beneath, petioles pubescent above; stipules small glandular lobes up to 4 mm long; aments appearing with the leaves, borne on short, leafy branchlets; bracts brown to tawny, dark at tip, thinly lanate to glabrate; staminate aments 1.5–3 cm long; pistillate aments 1–4.5 cm long, branchlets 0.5–1.5 cm long, pistils glabrous, capsules tawny to reddish, styles 0.4–1.2 mm long, stipes 0.8–1.5 mm long, nectary 1, usually less than half as long as the stipes. Often confused with *S. barclayi* with which it

hybridizes. Subalpine river floodplains and thickets, open coniferous woods, alpine slopes, and marl bogs. BC, Alta s to eOre, Wyo. Map 410

S. glauca L.
Erect shrubs 0.5–1.5 m tall, rarely prostrate; branches reddish brown or greyish, epidermis often exfoliating, usually pubescent; branchlets densely to sparsely villous; leaves elliptic, obovate or narrowly obovate, 2–5 cm long, apex acute or obtuse, base tapered, margins entire, young leaves loosely to densely villous, older leaves glossy and dark green above, glaucous beneath, variously pubescent on both sides from sparsely to moderately villous or with straggly hairs to sometimes almost glabrate above, petioles yellowish to brown, usually more than 3 times the length of the buds; stipules small glandular lobes to ovate or sometimes narrow and up to 8 mm long; aments appearing with the leaves and borne on leafy branchlets; bracts tawny to light brown, pubescent with short hairs; staminate aments 1.5–2.5 cm long, branchlets 0.5–1.5 cm long; anthers 0.5–0.8 mm long; pistillate aments 2–5 (7.5) cm long, branchlets 0.5–2.5 cm long, pistils villous-lanate becoming sparsely lanate, capsules 3–8 mm long, styles 0.5–0.8 mm long, green or yellowish in life, stipes 0.5–1.6 mm long, nectaries 1, usually shorter than the stipes. $2n = 76, 95, 114$. In our area var. **villosa** (Hook.) Anderss., $2n = 114$*, as described above, which hybridizes with *S. arctica*, *S. brachycarpa*, and possibly *S. barclayi*. Common in alpine and subalpine thickets and slopes, moist coniferous woods, river floodplains, and marl bogs. Circumpolar: Alas, Yuk to sBaffin Is s to Utah, NM, cSask, nMan, James Bay, Gaspé, Nfld. Map 411

S. lanata L.
Low shrubs up to 5 dm tall; branches gnarled, reddish brown; branchlets villous becoming glabrescent; overwintering buds large and subglobose, villous; leaves broadly ovate to broadly elliptic; apex abruptly acute, base usually round, sparsely villous to glabrescent on both sides, glaucous beneath, margins glandular-serrulate to entire and glandular-dotted; stipules prominent, the apex round or sometimes acuminate, glandular-dotted on margins and on upper surface; bracts dark, long-villous; aments stout, 5–6 cm long and sessile, appearing before the leaves; filaments glabrous; capsules glabrous, reddish brown, styles up to 2.5 mm long, stipes 0.2–0.5 mm long, nectaries longer than the stipes. $2n = 38$. In our area ssp. **calcicola** (Fern. & Weig.) Hult. as described above. Floodplain of the N Sask R. Circumpolar: Alas, Yuk, to nBaffin Is s to nBC, nMan, Que, Nfld; Gaspé; Alta. Map 409 ○

S. lucida Muhl.
Large shrubs or small trees 2–9 m tall; branchlets reddish brown to yellowish, glabrous or sparsely lanate, brittle; well-developed leaves narrowly to broadly ovate, 6–15 cm long, apex narrowly acute to long-acuminate, base round, margins glandular-serrulate, the upper surface glabrous or glabrescent, glossy and non-stomatiferous, the lower surface glaucous, petioles pubescent, glandular at the distal end, tawny; stipules semi-ovate, glandular; aments appearing with the leaves and borne on leafy branchlets; bracts tawny, 2–3 mm long, apex erose or

denticulate, deciduous after flowering; stamens 4–5, filaments pubescent at base; pistils glabrous, capsules 5–7 mm long, yellowish, styles about 0.5 mm long, stigmas 2 horseshoe-shaped lobes, stipes about 1 mm long. $2n = 76$. In our area ssp. **lasiandra** (Benth.) Argus (*S. lasiandra* Benth. and var. *lancifolia* (Anderss.) Ball of Ed. 1) as described above and ssp. **caudata** (Nutt.) Argus (S. *caudata* (Nutt.) Heller and var. *parvifolia* Ball of Ed. 1) with the leaves narrowly ovate to very narrowly so, stomatiferous on both sides, pale green and non-glaucous below. Lake and slough margins, river floodplains, and sand-dune slacks. Alas, Yuk, swMack, Sask, Huds Bay, Nfld s to Calif, NM, SD, Ill, Ohio, Md. Map 412

S. lutea Nutt.
Shrubs 1.5–4 m tall; branches yellowish to greyish-yellow, glabrous; branchlets glabrous, reddish and sometimes thinly glaucous; leaves narrowly elliptic; apex abruptly acuminate to acute, base round to acute, margins shallowly serrulate to subentire, young leaves reddish, villous above and glabrous beneath, older leaves glabrescent above, glaucous and glabrous beneath, petioles reddish, velutinous on upper side; stipules ovate, auriculate; aments appearing well before or just before the leaves in staminate or with the leaves in pistillate, sessile or on short leafy branchlets; bracts narrow, tawny to brown, pubescent; staminate aments 2–4 cm long, stamens 2, sometimes united at base; pistillate aments 2–12 cm long on leafy shoots 0.5–1 cm long, pistils glabrous, styles 0.2–0.5 mm long, stipes 1.2–4 mm long, nectaries 1, 0.5–0.8 mm long, shorter than the stipes. $2n = 38$. Closely related to *S. prolixa* Anderss. with which it may hybridize. Forms thickets on riverbanks and slough margins; occasional in prairies, coulees, pine parklands, and active sand-dunes. Mack, Alta to Ont s to Calif, Colo, and Neb. Map 396 ●

S. maccalliana Rowlee
Shrubs 1–5 m tall; branches dark reddish-brown, glabrous and glossy; branchlets brown or yellowish, glabrous or puberulent; leaves coriaceous, narrowly elliptic to oblong, 5–7 cm long, apex acute, base acute to round, margins glandular-crenulate-serrulate, pubescent with matted reddish and white hairs that are soon lost, green on both sides, non-glaucous beneath; stipules minute glandular lobes or up to 0.5 mm long; aments 2–6 cm long, borne on short leafy branchlets; bracts narrowly oblong, tawny or lemon green, villous; stamens 2, anthers 0.8–1 mm long; capsules about 8 mm long, tawny, sericeous-villous, styles 0.8–1.2 mm long, stipes 1–2 mm long, nectaries half as long as the stipes. $2n = 224$. Riverbanks, wet thickets on slough margins and marshes, marl bogs, and rocky subalpine slopes. Yuk s to BC e to cQue. Map 413

S. melanopsis Nutt.
Colonial shrubs up to 2.5 m tall, forming thickets; branches greyish- to reddish-brown; branchlets sericeous, sparsely lanate or densely villous, becoming glabrescent; leaves narrowly elliptic, the sides not parallel, 4–7 cm long, 4–6 times as long as broad, apex and base acute, margins shallowly denticulate, distantly glandular-dotted or subentire; young leaves often reddish, sparsely pubescent to densely lanate-villous; older leaves glabrescent to sparsely lanate, pale be-

neath, rarely glaucescent, stomata on both sides, veins indistinct, petioles short, scarcely longer than the bud; stipules ovate to narrow; aments usually appearing with the leaves, apparently unbranched, borne on leafy branchlets; bracts yellow to greenish, broadly rounded at apex, deciduous after flowering; staminate aments slender, borne on short branchlets, filaments densely pubescent at base, anthers 0.6–0.7 mm long; pistillate aments 3–5 cm long on branchlets 1–2 cm long, pistils glabrous, stipes 0.2–0.6 mm long, styles short but evident, stigmas short and stout, nectaries usually exceeding the stipes. Includes var. *bolanderiana* (Rowlee) Schneid. and var. *tenerrima* (Hend.) Ball. Closely related to *S. exigua*. Mountain streams and river floodplains. BC, Alta s to Calif, Colo. Map 414 ○

S. myrtillifolia Anderss.
Low shrubs 1–10 dm tall, usually decumbent or trailing and rooting along the stem; branchlets greenish to reddish-brown, sparsely pubescent with short curved hairs; leaves narrowly elliptic to narrowly obovate, apex broad and acute but the tip blunt, base round, margins glandular-crenate, the upper surface glabrous and lacking stomata, the lower surface glabrous, pale green but not glaucous; stipules narrowly elliptic, blunt-tipped, 0.2–2 mm long; aments appearing with the leaves, borne on short leafy branchlets; bracts narrow, dark brown or pale at base, glabrous to sparsely pubescent; pistillate aments 1.3–4.2 cm long, pistils glabrous, styles 0.3–0.5 mm long, stipes 0.6–1.6 mm long, nectaries 1, 0.1–0.5 times as long as stipes. $2n$ = 38, 76. Closely related to *S. boothii*. Hybridizes with *S. barrattiana* and possibly *S. boothii*. In our area var. **myrtillifolia** as described above and var. **cordata** (Anderss.) Dorn (includes var. *pseudomyrsinites* (Anderss.) Ball *ex* Hult.), which is a taller shrub (1–4 m), and has juvenile and mature leaves pubescent, at least on the upper midrib, with stomata often present on both sides, stipules 1–5 mm long, and style 0.5–0.9 mm long. Lake and stream banks, floodplain thickets, fens, marl bogs, muskegs, and moist coniferous forests. Alas, Yuk, sMack to Huds Bay, Nfld s to BC, Ont. Map 415

S. pedicellaris Pursh
Low shrubs 2–15 dm tall; branchlets reddish brown, glabrous to minutely velutinous; leaves leathery, oblong to narrowly obovate or elliptic, 2–5 cm long, apex obtuse to round, margins entire, sometimes slightly revolute, young leaves glabrous, older leaves green and thinly glaucous above, glabrous and glaucous beneath, petioles 3–8 mm long; stipules glandular lobes, minute up to 1.5 mm long; aments appearing with the leaves, borne on leafy branchlets; bracts short, tawny and sparsely pubescent; pistillate aments 1.5–3.5 cm long, loosely-flowered, borne on prominent leafy branchlets; capsules slender, 5–7 mm long, glabrous, tawny, styles and stigmas very short, stipes 2–3 mm long, nectaries 1, less than half the length of the stipes. $2n$ = 38, 57, 76. Includes var. *hypoglauca* Fern. Sphagnum bogs, fens, and muskegs. Yuk to Nfld s to Ore, Ia, and NJ. Map 414 ●

S. petiolaris J.E. Smith
Tall shrubs or small trees 1.8–5 (6.5) m tall; branches dark reddish-brown, drying purplish; branchlets yellowish to reddish-brown, velutinous becoming sparsely

pubescent, rarely glaucous; leaves very narrowly elliptic, 5–10 cm long, apex narrowly acute, base attenuate, margins glandular-serrulate on later leaves, entire to subentire on the first leaves, young leaves densely sericeous with a mixture of white and ferruginous hairs, older leaves glabrescent on both sides, glaucous beneath; stipules linear or minute linear lobes, caducous; aments appearing with or just before the leaves, borne on very short leafy branchlets that fall with the aments; bracts light brown to blackish, sericeous; staminate aments 1–2.5 cm long, stamens 2, filaments sparsely pubescent at the base; pistillate aments short cylindrical or subspherical, loosely-flowered, 1–3 cm long, pistils pyriform, long-beaked, sericeous, styles very short, stipes 2–4 mm long, nectaries 0.3–0.6 mm long. $2n = 38$. Commonly forming thickets around sloughs, old shore lines, and sedge meadows. Mack, eBC to NB s to Colo, Ind, and NJ; Ky. Map 406 •

S. planifolia Pursh
Shrubs 0.5–4 m tall; branches dark reddish-brown, sometimes glaucous; branchlets divaricate and glabrous; leaves elliptic to narrowly elliptic, 3–5 cm long, base and apex usually acute, margins slightly revolute, subentire to remotely serrulate or crenate-serrulate, young leaves sparsely sericeous with white and ferruginous hairs, the lower surface sparsely sericeous and glaucous, primary veins numerous, closely spaced and parallel; stipules minute to 2.5 mm long; aments appearing before the leaves and sessile, buds usually dehiscing at the base and capping the elongating aments; bracts black or dark brown, villous; staminate aments 2–3.5 cm long; pistillate aments 2–6 cm long, compactly fruited, pistils sericeous, up to 8 mm long in fruit, styles 0.8–1.3 mm long, stigmas usually 2-lobed, less than half the length of the style, stipes 0.5–0.9 mm long, nectaries almost as long as the stipes. $2n = 57, 76$. Closely related to *S. scouleriana* with which it may hybridize. Also hybridizes with *S. discolor*. In our area ssp. **planifolia**. Forms extensive thickets on river and lake shores, meadows, muskegs, moist coniferous woods, marl bogs, and active sand-dunes. Asia: Alas, Yuk, wMack to Huds Bay, Nfld s to Nev, NM; Wash, Ore, Calif; Vt and NH. Map 416

S. prolixa Anderss.
Shrubs 1.5–5 m tall; branches reddish brown to pale brown; branchlets usually glabrous, sometimes sparsely pubescent to velutinous; leaves narrowly oblong-obovate, 3.5–6 times as long as wide, apex gradually to abruptly acuminate, base round, margins serrulate, young leaves reddish, villous to sparsely pubescent or glabrous above, older leaves glabrescent above, the midrib and upper petiole may remain velutinous, glaucous and glabrous beneath, petioles reddish; stipules usually ovate, auriculate; aments appearing with or just before the leaves, borne on a short leafy shoot or sessile; bracts brown, usually villous with curly or wavy hairs, sometimes glabrate; pistillate aments 3–9 cm long, pistils slender, reddish or greenish and glabrous, styles 0.3–0.6 mm long, stipes 1.2–3.2 mm long; nectaries 1, up to half the length of the stipes. Includes *S. mackenzieana* (Hook.) Barratt *ex* Anderss. of Ed. 1. Closely related to *S. lutea* with which it may hybridize. River and lake shores, aspen woods, fens, and lower mountain slopes. Yuk, sMack to Sask s to Calif, Utah, and Wyo. Map 417

S. pseudomonticola Ball
Shrubs 1–4 m tall; branches dark reddish-brown, the 3-year-old bark with light brown furrows; branchlets yellowish to reddish, glabrous or occasionally villous; leaves narrowly to broadly elliptic to obovate, 4–8 cm long, apex abruptly short-acuminate to acute, base cordate to subcordate or round, margins coarsely glandular-crenulate, young leaves reddish, translucent and sparsely villous above, the older leaves dull green, and glabrous or glabrescent above, midrib remaining pubescent, the lower surface glabrous and glaucous; stipules ovate, apex round or short-acuminate; aments appearing before or with the leaves, sessile or on leafy branchlets up to 5 mm long; bracts dark brown, pubescent with straight or curly hairs; pistillate aments 4–9 cm long, pistils about 4.5 mm long, glabrous, styles 0.7–1.2 mm long, stipes 0.9–2 mm long, nectaries 1, about half as long as the stipes. $2n = 38$. Includes var. *padophylla* Rydb. Hybridizes with *S. candida.* River and lake shores, slough margins, aspen woods, and muskegs. Alas, Yuk to wQue s to BC, Wyo. Map 418

S. pyrifolia Anderss.
Shrubs 1–3 m tall; branches divaricate, dark reddish-brown; branchlets and buds glabrous, buds and foliage with a balsamic fragrance; leaves elliptic or ovate to narrowly so, 3–10 cm long, apex acute to acuminate, base round to cordate, margins glandular-serrulate to subcrenate, young leaves thin and translucent becoming thicker and glaucous in age, first mature leaves villous beneath but later leaves glabrous and often rugose beneath, petioles 9–20 mm long, tawny or reddish with glands at base of blade; stipules small and caducous; aments appearing with the leaves on leafy branchlets; bracts tawny and villous; staminate aments 2.5–5 cm long on short leafy branchlets, stamens 2; pistillate aments 5–8 cm long on leafy branchlets 0.5–3 cm long, pistils glabrous, capsules 4.5–8 mm long, styles 0.3–0.5 mm long, stigmas short, stipes 1.6–28 mm long, nectaries 1, very short. $2n = 38$. Poplar and aspen woods, slough margins, muskegs, and sand-dune slacks. sNWT to Lab s to BC, Wis, Mich, nNEng, NS, and Nfld; nQue; Yuk. Map 403 •

S. reticulata L.
Diminutive, rhizomatous shrubs, 1–10 cm above ground; branchlets glabrous, sometimes sparsely glaucous, ascending branches consist of short-shoots bearing 3 leaves and sometimes terminated by an ament, bud-scale persistent; leaves obovate to elliptic or orbicular, (0.4) 0.8–2.0 cm long, apex round or rarely acute, upper surface usually light green, glabrous, glossy and impressed-veiny, lower surface glabrous or rarely with a few long silky hairs, glaucous and reticulate-veiny, leaves often persistent for more than one year, stipules absent; aments appearing with the leaves, borne on a naked peduncle terminating a vegetative shoot; bracts tawny to dark brown, glabrous on outer side, sparsely pubescent on inner side; staminate aments short, stamens 2, filaments pilose at base; pistillate aments 0.5–3 cm long, including the peduncle, 2–22-flowered, pistils white-sericeous, styles 0.2–0.3 mm long, stigmas minute, stipes minute to 0.5 mm long, shorter than the bifurcate nectaries. $2n = 38$. Includes *S. saxi-montana* Rydb. Open alpine and subalpine slopes and thickets. In our area, as

described above, ssp. **nivalis** Löve, Löve & Kapoor (common) (*S. nivalis* Hook. and var. *saximontana* (Rydb.) Schneid. of Ed. 1) and ssp. **reticulata** (rare), which is generally larger as seen in the longer shoots, larger leaves and aments, and more flowers. Leaves not persistent for more than one year, 2.0–3.2 cm long, usually dark green above, the lower surface with long silky hairs, at least on the first leaves, venation conspicuously reticulate; aments up to 4 cm long, including peduncle, flowers 20–30 or more. Alpine and subalpine slopes. Circumpolar: Alas, Yuk to Lab s to Calif, NM, nSask, nMan, and wNfld. Map 419

S. scouleriana Barr. *ex* Hook.
Shrubs or small trees 2–7 m tall; branches dark brown to yellowish brown, often remaining pubescent for 2–3 years, rarely glabrescent; branchlets velutinous to velutinous-lanate; leaves obovate to narrowly elliptic, 5–8 cm long, apex acute or round with acute tip, base cuneate, margins entire to glandular-serrulate, revolute, young leaves densely velutinous, older leaves dark green, dull to glossy and glabrescent above, the lower surface sericeous with appressed white and rusty-coloured hairs to densely white villous, petioles sparsely pubescent or velutinous above; stipules minute up to 3.5–10 mm long on vigorous shoots; aments appearing before or just with the leaves, sessile or on short branchlets with reduced leaves or bracts; bracts black to dark brown, long-villous; pistillate aments 2.5–6 cm long, sometimes drooping, pistils densely grey-sericeous, long-beaked, styles 0.2–0.6 mm long, stigmas as long as or longer than the styles, stipes 0.8–2 mm long, nectaries less than half the stipes. $2n = 76$. Includes var. *coetanea* Ball. Closely related to *S. discolor* and the eastern *S. humilis* Marsh. Hybridizes with *S. planifolia*. Dry coniferous woods, floodplains, sand beaches, and active sand-dunes. Alas, Yuk, swMack to wMan s to Calif, NM and SD. Map 420

S. serissima (Bailey) Fern.
Shrubs 1–4 m tall; branches grey- or yellowish-brown; branchlets usually reddish brown, highly glossy; leaves elliptic to narrowly so, 5–9 cm long, apex abruptly short-acuminate or acute, base acute to obtuse, margins glandular-serrulate, young leaves reddish and glabrous, older leaves glabrous and dark green above, pale or slightly glaucous beneath, coriaceous, petioles 6–10 mm long, glandular near base of leaf-blade; stipules minute single glands or a cluster of 2–3 glands; aments appearing with or after the leaves, stout-cylindrical to subspherical, 2–4 cm long, on prominent leafy branchlets; bracts pale tawny, densely sericeous on pistillate flowers, deciduous after flowering; stamens 3–6, filaments with spreading pubescence at the base; capsules 7–9 mm long, yellow-brown infused with red and firm, styles and stigmas short, stipes 0.8–1.6 mm long, nectaries cup-like, divided into 2–4 segments surrounding the stipes. Occasional in thickets on slough margins, meadows, and fens. Mack to Lab s to BC, Colo, SD, Ind, NJ, and Nfld. Map 421

S. sitchensis Sanson *ex* Bong.
Shrubs or small trees 1–8 m tall, sometimes prostrate; branches grey; branchlets velutinous becoming sparsely sericeous or puberulent, brittle; leaves narrowly obovate to narrowly elliptic or elliptic, 4–9 cm long, green above with greyish

appressed hairs, densely to sparsely sericeous beneath and appearing satiny, the veins beneath evident although covered with dense indumentum, margins entire or glandular-dotted, revolute; stipules half-ovate; aments appearing with the leaves and borne on leafy branchlets or rarely flowering before the leaves and sessile; bracts brown, darker toward the tip, villous, apex round; staminate aments 2.5–5 cm long on branchlets 2–10 cm long, stamens 1; pistillate aments 3–9 cm long on short branchlets, pistils sericeous, capsules tawny, styles 0.4–0.8 mm long, stigmas 0.1–0.3 mm long, stipes 0.5–1.4 mm long. $2n = 38$. Floodplain of the Athabasca R. sAlas, BC, Alta s to Calif and Mont. Map 422 ○

S. stolonifera Coville
Dwarf shrubs, often spreading by buried branches; branches reddish brown, glabrous and glossy, sometimes glaucous; branchlets glabrous; leaves elliptic, broadly so or obovate, apex broad, usually obtuse, base acute to round, margins usually entire, upper surface of leaves glabrous or pubescent near the margins, at least the lower leaves, the lower surface glabrous to sparsely pubescent, and glaucous, stomata on both sides; stipules minute; aments flowering with the leaves, borne on a leafy branchlet; bracts dark brown, sparsely pubescent; pistillate aments 1.5–4 cm long; branchlets 1–6 cm long, pistils usually reddish, glabrous or sparsely pubescent on the beak, styles (0.4) 0.8–1.4 mm long, stipes 0.4–1 mm long, nectaries 1, equalling or exceeding the stipes. Alpine, wet hummocks at 7,000–7,750 feet. sAlas, nwBC; Alta. Map 407 ○

S. vestita Pursh
Low shrubs up to 1.5 m tall; branches grey-brown, bark leathery and folded into ridges, often with numerous short internodes, glabrescent; leaves elliptic, obovate or narrowly elliptic, rarely suborbicular, firm, 1.8–7 cm long, apex round or obtuse, base round, truncate or subcordate, margins obscurely glandular-crenate or subentire and glandular, the upper surface dark green and plane or somewhat reticulate, the lower surface glaucous and densely to sparsely villous or often villous-lanate, petioles very short, 2–8 mm long with glandular dots on the margins; aments appearing after the flowers, terminating prominent 3-leaved branchlets; bracts ovate and densely short villous; staminate aments slender and tawny; pistillate aments 2.5–6 cm long, capsules ovate to obpyriform, densely pubescent with short silky trichomes, styles very short and branched, stipes 0.4–1.2 mm long, usually surrounded by several linear nectaries as long as the stipes. $2n = 38$. Rocky, open alpine and subalpine slopes, open coniferous woods, and stream margins. Central Asia: BC, Alta s to Ore and Mont; nMan to Nfld; Keewatin. Map 422 ●

MYRICACEAE

Myrica L.

M. gale L. / Sweet Gale
Low shrub, 0.5–1.5 m high, usually with ascending brown branches; leaves appearing after the flowers, oblanceolate, firm in texture, serrate towards the obtuse apex, cuneate at base to a very short petiole, resin-dotted and fragrant, pubescent to glabrous, 2.5–6 cm long, deciduous; flowers unisexual, aggregated into catkins, the staminate and pistillate on separate plants; staminate catkins thick-cylindric, 1–2 cm long, crowded, with large brown bracts; pistillate catkins ovoid, 8–10 mm long and cone-like in fruit; perianth none; staminate flowers with 4–16 stamens; pistillate flowers with a solitary 1-ovuled ovary, subtended by 2 short bractlets; styles 2; fruit an ovoid nutlet, 2.5–3 mm long, beaked, resin-dotted, 2-winged by the two bractlets adnate to its base. $2n = 48, 96$. Swamps and thickets. nwEurope, neAsia: Alas, Yuk, Mack to Huds Bay, Nfld s to BC, nAlta, nSask, sMan, Gt Lakes, NJ, Apal. Map 423

BETULACEAE / Birch Family

Monoecious trees or shrubs, without terminal buds, the branches extended by upper axillary buds; leaves deciduous, alternate, simple, pinnately veined, usually deeply serrate; flowers in catkins (aments), usually appearing before the leaves; staminate catkins drooping, with 1–3 flowers in the axil of each bract; calyx membranous, 2–4-cleft, or wanting; stamens 1–10; pistillate catkins erect or drooping, spike-like or capitate; calyx, when present adnate to the 1–2-loculed ovary; styles 2; fruit a small 1-seeded winged nutlet, or the nut larger and enclosed by an involucre formed by the enlarged bracts.

1 Fruit a nut enclosed in a hispid involucre with a long beak; pistillate flowers few in a short, bud-like catkin; staminate flowers solitary in the axils of the catkin bracts **Corylus**
Fruit a nutlet, often winged, in the axil of a catkin bract; pistillate catkin elongated; staminate flowers usually three in the axils of the catkin bracts 2

2 Bracts of pistillate catkins thin, 3-lobed, deciduous; stamens 2 **Betula**
Bracts of pistillate catkins becoming thick and woody, persistent; stamens 4 **Alnus**

Alnus Mill. / Alder

Shrubs or small trees with smooth, reddish or greyish-brown bark and 3-angled pith; leaves ovate or obovate, dentate or serrate; stiminate catkins long and pendulous, clustered at the ends of the branchlets; pistillate catkins erect, clustered near the base of the staminate group; short-stalked, peltate bracts of the staminate catkin each bearing 2–4 bractlets and 3 flowers, each flower consisting of a 4-parted calyx and 4 stamens; fleshy bracts of pistillate catkin each bearing 4 adnate bractlets and 2 flowers, these with pistils only; nutlets flat, ovate or oblong, with lateral wings or membranous border; fruiting catkin woody, ovoid or ellipsoidal, persistent.

Nutlet broadly wing-margined, the body comprising less than half the width; leaf-buds sessile, covered by imbricate scales or unequal length; flowers developed with the leaves on the twigs of the season; leaves shining; shrubs **A. crispa**

Nutlet wingless, merely margined; leaf-buds stipitate, covered by 2 or 3 scales of nearly equal length; flowers developed before the leaves in early spring; leaves dull; tall shrub or small tree **A. tenuifolia**

A. crispa (Ait.) Pursh / Green Alder
Shrub, usually 1–3 m high, ascending and bushy; branchlets and peduncles more or less glutinous and pubescent when young, becoming glabrous; leaves 2–8 cm long, ovate or broadly elliptic, mostly rounded or slightly cuneate at base, the margin irregularly serrulate or biserrulate with sharp teeth and often puckered; buds, flowers and nutlets as shown above; fruiting catkins 1–1.5 cm long. $2n = 28$. In our area ssp. **crispa** with leaves ovate, rounded to cuneate at base, serrulate or biserrulate, and ssp. **sinuata** (Regel) Hult., with leaves broadly ovate, truncate to subcordate at base, margin undulate and irregularly serrulate. Sand-hills, open woods, and edges of bogs. eAsia, Greenland: Alas, Yuk, Mack, Huds Bay, Nfld s to nCalif, Ida, Mont, cSask, Minn, Gt Lakes; NC. Map 424

A. tenuifolia Nutt. / River Alder
Large shrub or small tree, 2–8 m tall; leaves 4–10 cm long, oval to broadly ovate, thick, doubly serrate or serrate-dentate, sometimes undulate or shallowly lobed, somewhat pubescent beneath, especially along the veins; buds, flowers and nutlets as shown above; fruiting catkins 1–2 cm long. $2n = 28$. *A. incana* (L.) Moench ssp. *tenuifolia* (Nutt.) Breitung. Closely related to the eastern *A. rugosa* (Du Roi) Spreng. Borders of streams and lakes. Alas, Yuk, wMack s to Calif, Colo, wSask. Map 425

Betula L. / Birch

Trees and shrubs with thin bark (in ours) marked with conspicuous, horizontally elongated lenticels, slender twigs, and alternate, several-scaled buds, the terminal bud absent; leaves alternate, stipulate, simple, pinnately veined and variously toothed, the main lateral veins running into teeth; staminate and pistillate flowers in separate aments on the same plant; staminate flowers and their bractlets in triplets in the axil of a bract, each flower with a 2- or 3-lobed calyx and two stamens with bifurcated filaments bearing an anther chamber at the end of each branch; pistillate flowers in pairs or triplets in the axil of each 3-lobed bract, each flower consisting of a naked ovary with a pair of linear stigmas; fruit a samara consisting of a small single-seeded nutlet surrounded by a membranous wing; fruits and bracts shed together from the persistent ament axis.

All our species appear to be interfertile to some degree and hybrids commonly occur, particularly in disturbed areas, tending to blur the distinctions between the species. The following hybrids have received taxonomic recognition. *Betula* × *arbuscula* Dugle (*B.* × *sargentii* × *B. papyrifera*), *B.* × *eastwoodae* Sargent

(*B. glandulosa* × *B. occidentalis*), *B.* × *piperi* Britton (*B. occidentalis* × *B. papyrifera*), *B.* × *sandbergii* Britton (*B. papyrifera* × *B. pumila*), *B.* × *sargentii* Dugle (*B. glandulosa* × *B. pumila*), *B.* × *uliginosa* Dugle (*B. neoalaskana* × *B. pumila*), *B.* × *winteri* Dugle (*B. neoalaskana* × *B. papyrifera*). In addition, some of the varieties of species have been shown to be products of introgressive hybridization.

1 Trees or tall shrubs 3 m or more high; leaf membranous, usually over 3 cm long; petiole 1 cm or more long; samara wings at least as wide as nutlet body 2

Shrubs seldom over 2 m high; leaf relatively thick and stiff, usually under 3 cm long; petiole under 1 cm long; samara wings narrower than nutlet 4

2 Small tree or big shrub with tight, very dark bark; leaf oval to orbicular, 2.5–4 cm long; fruit bract lobes angular, acute, divergent; twig commonly glandular **B. occidentalis**

Trees with bark peeling easily, white or brown; lateral bract lobes rounded or obtuse 3

3 Twig pubescent with long hairs; leaf ovate, with tufts of hairs in vein axils beneath; lateral bract lobes ascending to divergent; samara with widely oval nutlet **B. papyrifera**

Twig glabrous and commonly densely glandular; leaf deltoid, glabrous or ciliate, without hair tufts in vein axils beneath; samara with narrowly oval to lanceolate nutlet **B. neoalaskana**

4 Twig densely glandular and indistinctly puberulent with short hairs; leaf orbicular, with 3 or fewer lateral veins and 10 or fewer teeth each side; samara wings less than half as wide as nutlet **B. glandulosa**

Twig pubescent, and somewhat glandular in ours; leaf obovate, with 4 or more lateral veins and 10 or more teeth each side; samara wing more than half as wide as nutlet **B. pumila**

B. glandulosa Michx. / Bog Birch; Dwarf Birch
Ascending or spreading shrub up to 2 m high, with dark grey-brown to blackish stems, twigs densely resinous-glandular and inconspicuously puberulent with very short hairs among the glands; leaves orbicular, sometimes wider than long, 1–2 cm long, rounded at base and apex, crenate-dentate with 6–10 teeth each side, and with 2 or 3 lateral veins each side of the midvein, relatively thick, stiff, and shiny, glandular-dotted beneath, on petioles 3–5 mm long; aments flowering with the appearance of leaves in spring; staminate aments up to 2 cm long, more or less drooping; pistillate aments erect, 10-15 mm long in fruit; fruit-bract with obliquely ascending lobes; samara with a nearly orbicular nutlet edged with wings less than half as wide. $2n = 28$*. Includes *B. nana* L. ssp. *exilis* (Sukatchev) Hult. Bogs and seepage areas, alpine slopes, and open subalpine forest at high altitudes. eAsia, Greenland: Alas, Yuk to sBaffin Is s to Calif, Nev, Colo, cSask, cMan, Gt Lakes, Nfld. Map 426

B. neoalaskana Sargent / Alaska Birch
Smallish to medium-sized tree up to 15 m tall, trunk to 30 cm in diameter, with white bark (brown in youth) with dark lenticels, twigs typically rough and

densely resinous-glandular, but sometimes with few glands on old trees, normally glabrous; leaf deltoid, acute, truncate to broadly cuneate at base, biserrate, 2.5–7 cm long, glabrous or sometimes ciliate, with scattered fine glands beneath; without hair tufts in the vein axils; aments pendulous; the staminate 2 or 3 together from the distal node of the previous year's growth; the pistillate single at the lower nodes, each catkin in the axil of a leaf on a short spur; fruit-bract with lateral lobes reflexed; samara with a lanceolate nutlet and broad wings. $2n = 28$*. *B. papyrifera* Marsh. var. *neoalaskana* (Sarg.) Raup of Ed. 1. Bogs, sandhills, and sunny slopes. Alas, Yuk, wMack to nwOnt s to neBC, cSask, cMan. Map 427

B. occidentalis Hook. / Water Birch; Black Birch
Small tree or tall shrub up to 10 cm high, with dark reddish-brown to black nonpeeling bark, branchlets slender and often drooping, sparsely to densely resinous-glandular, and puberulent, usually with short hairs only; leaves ovate to elliptic, rhombic or nearly orbicular, acute at apex, usually biserrate, thin but firm-textured, often shiny above, 2.5–6 cm long, glabrous or finely puberulent beneath; pistillate aments one or sometimes 2 together on a lateral spur, in various attitudes; fruit-bract with sharply angular, widely diverging lateral lobes; samara with a broadly obovate nutlet and broad wings. $2n = 28$*. In our area var. **occidentalis** with fruiting aments single, and twigs puberulent with short hairs, glandular; var. **inopina** (Jepson) C.L. Hitchc., with twigs pubescent with long and short hairs, densely resinous glandular; and var. **fecunda** (Britton) Fern., with fruiting aments often paired, twigs puberulent, somewhat glandular. Open areas along streams and lake shores. Alas, Yuk, wMack to Man s to Calif, Nev, NM, SD. Map 428

B. papyrifera Marsh. / White Birch; Paper Birch
Tree up to 30 m high and 60 cm in trunk diameter, with a columnar or oval crown of ascending branches, bark usually white, but varying to yellowish or reddish-brown, with conspicuous brown lenticels, readily peeling in sheets, juvenile bark brown, branchlets dark brown, twigs slender, pubescent with long hairs, and occasionally glandular; leaves ovate or rhombic, 4–9 cm long, serrate or biserrate, dull, the underside pubescent with long hairs, and with tufts of hairs in the vein axils; aments pendulous when mature; fruiting aments paired on spur shoots; fruit-bracts with long central axes and short rounded lateral lobes, ciliate, and pubescent; samaras 3 per bract, with broadly elliptic puberulent nutlets and wings at least as wide as nutlets. $2n = 56$*, 84*. In our area. var. **papyrifera** with leaves ovate, 5–7 cm long, rounded to cordate at base, apex acuminate; twigs pubescent, not glandular; lateral bract lobes ascending; mature bark white; var. **subcordata** (Rydb.) Sarg. with leaves 4–5 cm long, broadly ovate to suborbicular, truncate to cordate at base, apex acute; twigs glandular; sparsely pubescent; lateral bract lobes ascending; mature bark white or brown; and var. **commutata** (Regal) Fern., with leaves oval to rhombic, cuneate at base, biserrate; lateral bract lobes widely divergent; bark variously coloured. Boreal and montane forests, parkland. Alas, Yuk, swMack, nSask, cMan, Gt Lakes to Lab s to Wash, Mont, Colo, nNeb, Minn, NY, Pa. Map 429

B. pumila L. / Dwarf Birch
Ascending shrub up to 3 m tall, with dark brown stems, twigs pubescent, often densely so, and glandular in ours; leaves obovate to nearly orbicular, 1–3.5 cm long, cuneate to rounded at base, obtuse to rounded at apex, crenately toothed with 10–15 teeth each side, and with 3–5 lateral veins each side of the midvein, tough and shiny, on petioles 4–8 mm long; staminate aments single, pendulous from distal buds on previous year's growth; fruiting aments single, erect, from lateral spurs, 1–3 cm long; fruit-bract with upturned lateral lobes; samara with broadly obovate nutlet and wings narrower than but at least half as wide as the nutlet. $2n = 56^*$. In our area var. **glandulifera** Regel. Swamps and bogs. Yuk, wMack to Huds Bay, Nfld s to Ore, Wyo, ND, Ia, Ind, Ohio, NJ. Map 430

Corylus L.

C. cornuta Marsh. / Beaked Hazelnut
Shrub 1–3 m tall, the young branchlets villous, becoming glabrescent; leaves ovate-oblong to oblong-obovate, short-acuminate, rounded to somewhat cordate at base, coarsely biserrate, pale beneath, more or less pubescent, especially along the veins; flowering in early spring; staminate catkins elongate, the bracts each subtending a pair of small bractlets and a single naked flower; stamens 4, each filament 2-parted and bearing two 1-locular anthers; pistillate flowers few, in a short, bud-like catkin, concealed by the scale-like bracts, except the protruding reddish stigmas; ovary inferior, surmounted by a minute calyx; bracts united in fruit to form an involucre for the thin-shelled nut and greatly prolonged as a slender beak, this densely bristly near the base and narrowly lobed at the summit. $2n = 28$. *C. rostrata* Ait. Thickets and woods. BC transcont sCan to Nfld s to Calif, Kans, Ohio, Ga. Map 431

CANNABINACEAE / Hemp Family

Herbs; leaves lobed, stipulate; flowers dioecious, axillary; staminate pedicelled, the calyx of 5 distinct sepals, 5 stamens; pistillate sessile, the calyx entire, enfolding the sessile, unilocular ovary; ovule solitary; fruit an achene enveloped by the persistent calyx.

Stems twining, retrorse prickly	**Humulus**
Stems erect, not prickly	**Cannabis**

Cannabis L.

C. sativa L. / Hemp; Marijuana
Erect herb, 1–2 mm tall; leaves opposite below, alternate above, petioled, divided to base into 5–9 linear or lanceolate, serrate leaflets, 5–20 cm long; staminate flowers in paniculate clusters in the upper leaf-axils; pistillate flowers in clusters on short leafy lateral branches from the upper leaf-axils; achene 3–4 mm long. $2n = 20$. Waste ground. Introduced. Occasional.

Humulus L. / Hops

H. lupulus L. / Common Hop
Twining vine, 2–4 m long, herbaceous, perennial; stems and petioles retrorsely prickly; leaves opposite, the blades mostly palmately 3–5 lobed, longer than the petioles; stipules persistent, free; principal leaves ovate in outline, 3-lobed, broadly cordate at base, coarsely serrate, scabrous above, glandular-dotted beneath, the lateral lobes obliquely oblong-ovate, the terminal one ovate-lanceolate and constricted near the base; staminate flowers in branched axillary clusters; sepals 5 and stamens 5; pistillate flowers in a drooping bracted spike; bracts large, imbricated, each 2-flowered, in fruit forming a cone-like 'hop'; calyx cup-like, enveloping the superior ovary; stigmas 2, filiform, conspicuous; fruit an achene invested with the enlarged calyx, the fruiting calyx and achene sprinkled with yellow resinous granules. $2n = 20$. Thickets and waste ground. Introduced. Occasional.

URTICACEAE

Herbs, occasionally shrubs or small trees, mostly with stinging hairs; leaves opposite or alternate, generally stipulate; flowers small, hypogynous, imperfect, in cymose or spike-like clusters; calyx 3–5-lobed, petals none; stamens as many as calyx-lobes, inserted opposite them, inflexed in bud; ovary 1, superior, 1-loculed, ovule solitary, basal; fruit an achene or drupe.

Leaves alternate, extipulate, entire; stinging hairs absent	**Parietaria**
Leaves opposite, stipulate, toothed; stinging hairs present	**Urtica**

Parietaria L. / Pellitory

P. pensylvanica Muhl. *ex* Willd.
Annual, erect, ascending or decumbent, 5–50 cm tall; stem simple or branched, sparsely to densely pubescent, lacking stinging hairs; leaves entire, alternate, petiolate, lanceolate, ovate to elliptic, blades 1–8 cm long; inflorescence bracteate, with axillary clusters of intermixed staminate and pistillate flowers (occasionally some perfect); calyx-lobes 4, 1–2 mm long; stamens 4; achene reddish brown, shiny, c. 1.0 mm long. $2n = 16$. Shaded gravelly places and disturbed areas. BC to Que, generally distributed s. Map 432

Urtica L. / Nettle

Herbs with erect, simple or branching stems and stinging hairs; leaves opposite, petiolate, stipulate, the blades 3–7-nerved and toothed; monoecious or dioecious; flowers small, green, in axillary paniculate cymes; staminate flowers with a deeply 4-cleft calyx, 4 stamens and a rudimentary ovary; pistillate flowers with 4 sepals, the two inner ones larger and in fruit enclosing the ovary; pistil solitary, the ovary 1-loculed, 1-ovuled; stigma sessile, tufted; fruit a flattened achene, the seed often adherent to the pericarp.

Annual; staminate and pistillate flowers in mixed clusters shorter than petioles; achenes triangular, 1.5–2.5 mm long **U. urens**

Perennial; staminate and pistillate flowers mostly in separate clusters, the staminate as long as or longer than petioles, the pistillate shorter; achenes lenticular-ovate, 1.0–1.5 mm long **U. dioica**

U. dioica L. / Common Nettle
Stem 0.5–2 m tall, slender, sharp-angled, glabrous to bristly and sparingly pubescent; leaves narrowly lanceolate to ovate, slender-petioled, rounded to subcordate at base, acuminate, finely to coarsely serrate, glabrous on both surfaces or hairy beneath, the veins sometimes prominent; sepals about as long as the achenes. $2n = 26^*, 52^*$. Includes *U. lyallii* of Ed. 1. In our area ssp. **gracilis** (Ait.) Selander (*U. gracilis* Ait. of Ed. 1). Moist shady woodland, stream banks, roadsides, and mountain slopes. Circumpolar, boreal: Alas, Yuk, swMack to Huds Bay, Nfld, generally distributed s. Map 433

U. urens L. / Small Nettle
Erect annual, 2–4 dm high; stem with abundant stinging bristles; leaves slender-petioled, ovate to broadly elliptical, 3–5-nerved, glabrous or nearly so, deeply toothed; stipules about 4 mm long; flowers androgynous, mainly pistillate, both kinds in the same cluster; fruiting calyx bristly on the margins. $2n = 24, 26$. Gardens and waste places. Introduced. Rare.

SANTALACEAE / Sandalwood Family

Perennial herbs with horizontal rootstocks; stems slender, glabrous, leaves alternate, sessile or short-petioled, simple, entire, glabrous; flowers small, greenish white or bronze or rose-tinged, arranged in terminal or axillary cymules, these sometimes forming a paniculate cluster; flowers perfect or staminate, regular, epigynous; corolla none; calyx of 5 petaloid sepals, somewhat united above the ovary, the short tube lined with a fleshy disk; stamens 5, included, the anthers often connected with the sepals by tufts of hairs; ovary inferior, 1-loculed, few-ovuled; style 1; fruit a 1-seeded drupe (dry or fleshy), crowned by the persistent calyx. Somewhat parasitic on underground parts of other plants, by means of sucker-like organs formed by the rootlets.

Flowers in small cymules borne near the top of the plant, whitish; sepals much longer than wide; leaves mostly linear-lanceolate, acute; fruit dry, greenish **Comandra**

Flowers commonly 3, in cymules from the axils of the middle leaves, bronze or greenish; sepals triangular, about as long as wide; leaves thin, mostly oval, obtuse; fruit juicy, scarlet **Geocaulon**

Comandra Nutt. / Bastard Toad-flax

C. umbellata (L.) Nutt.
Stems 1.5–4 dm tall, usually branched; leaves numerous, linear to oblong-lanceolate, sessile or subsessile, usually firm, acutish, glabrous, pale and glaucous; flowers commonly in 3–5-flowered cymules, forming an ovoid or ellipsoidal cluster near the summit of the shoot; sepals whitish, lanceolate or narrowly oblong, 2–4 mm long; fruit firm, green or drab, ovoid to nearly globose, 4–8 mm long, narrowed above into a short neck, crowned by the persistent calyx. $2n = 26$, c.52. *C. pallida* DC. of Ed. 1. In our area var. **umbellata** with leaves thin, not glaucous or only slightly so; calyx-lobes to 2.5 mm long, ovate-lanceolate to ovate; fruit about 6 mm across; and var. **pallida** (DC.) M.E. Jones with leaves thick, strongly glaucous; calyx-lobes 3–4 mm long, narrowly lanceolate; fruit up to 10 mm across. Prairie grassland, gravelly slopes, and dry open pine woods. BC, nAlta to cMan, Lk Superior to Nfld s to Calif, Ariz, NM, Tex, Ga; Huds Bay. Map 434

Geocaulon Fern. / Bastard Toad-flax

G. lividum (Richards.) Fern.
Stems 1–2.5 dm tall, usually simple; leaves short-petioled, oval, obtuse or rounded at apex, narrowed at base, thin, reticulately veined, bright green or often yellow-variegated; flowers in slender-stalked, 2–4-flowered, axillary cymules; lateral flowers of a cymule staminate, usually only the middle one fertile; calyx spreading; sepals 1–1.5 mm long; style very short; fruit succulent, scarlet, nearly globose, 5–8 mm long. *Comandra livida* Richards. In damp humus and moss. Coniferous woods. Alas, Yuk, swMack, to Huds Bay, Nfld, s to Wash, nIda, cSask, nMinn, nMich, NY. Map 435

LORANTHACEAE / Mistletoe Family

Arceuthobium Bieb. / Dwarf Mistletoe

A. americanum Nutt.
Parasitic on the trunks and branches of *Pinus*, often causing witches' brooms and hypertrophy of branches; stems emerging from the host, 2–10 cm long, slender, fragile, greenish yellow, often much branched, the branches in pairs or whorls; leaves opposite, minute, scale-like, connate; dioecious; flowers solitary or clustered in the axils of the scales, regular, epigynous; corolla none; staminate calyx mostly 3-parted and with 3 stamens; anthers sessile on the calyx-lobes, circular, 1-loculed; pistillate calyx 2-cleft; ovary inferior, 1-loculed; fruit 1-seeded, berry-like, 2–3 mm long, blue-glaucous, on a short recurved pedicel, explosively dehiscent, the ejected seed often adherent by the mucilaginous pulp. $2n = 28$. In our area mostly on *P. banksiana* and *P. contorta*. BC, Alta to Sask, Man, wOnt, s to Calif, NM. Map 436

POLYGONACEAE / Buckwheat Family

Annual or perennial herbs, sometimes woody at base; leaves mostly alternate, sometimes opposite or whorled, simple, often all basal, generally entire; stipules (ocreae), when present, cylindric, sheathing the stem; flowers individually small, often aggregated in conspicuous clusters, regular, chiefly perfect, in some species unisexual; corolla none; calyx 4–6-lobed or of separate sepals, sometimes petal-like; stamens 4–9, distinct, inserted at the base of the calyx; anthers usually introrse; ovary superior, 1-loculed, 1-ovuled; styles 2 or 3, separate or united at base; fruit an achene, 3-angled or flattened, with abundant endosperm.

1 Dwarf annual; leaves subopposite; calyx 3-parted; stamens 3 **Koenigia**
Plants not as above 2

2 Ocreae lacking; flower clusters subtended by an involucre of bracts; stamens 9 **Eriogonum**
Ocreae present; flowers not in involucrate clusters; stamens 8 or fewer 3

3 Sepals 4 or 6; fruit winged 4
Sepals 5; fruit wingless 5

4 Sepals 4; leaves reniform **Oxyria**
Sepals 6; leaves not reniform **Rumex**

5 Leaves triangular-hastate to cordate; fruit exserted, twice as long as perianth **Fagopyrum**
Leaves various, if cordate, stems twining; fruit usually included **Polygonum**

Eriogonum Michx.

Perennials or annuals, mostly herbaceous, sometimes woody at base; leaves entire, without stipules, mostly basal, the flowering stem (scape) sometimes with a whorl of leaves at the summit; flowers small, surrounded by an involucre of united bracts, the involucres usually in heads or umbels; sepals 6, more or less coloured; stamens 9; filaments slender, often exsert; styles 3; achenes usually triangular, enclosed by the calyx.

1 Plants annual; stem diffusely branched, the branches slender; flowers in reflexed heads **E. cernuum**
Plants perennial; stems and flowers not as above 2

2 Inflorescence a compound umbel 3
Inflorescence a simple umbel 4

3 Perianth glabrous; leaves glabrous; bracts often reflexed **E. umbellatum**
Perianth pubescent at base; leaves hirsute above; bracts ascending **E. flavum**

4 Flowers attenuate, with a stipe-like base **E. androsaceum**
Flowers not attenuate, lacking a stipe-like base **E. ovalifolium**

E. androsaceum Benth.
Small plant, often forming dense mats; caudex tufted; scapes 2–10 cm high; leaves oblanceolate to spatulate, 1–2 cm long, densely villose or becoming glabrate and green on upper surface; involucres in a small umbel, sometimes reduced to as few as one; flowers sparingly pubescent, 4–5 mm long. Rock slides and exposed summits at high altitudes. sBC to Alta s to Mont. Map 437 ●

E. cernuum Nutt.
Annual plant, 1–4 dm high; leaves basal, petioled, orbicular to oval, 1–2 cm broad, white-tomentose beneath, less tomentose above; flowering scape with slender, spreading or recurved branches, glabrous; involucres glabrous, borne singly; calyx white or pinkish, about 2 mm long. Bad-lands and valley slopes. seAlta to swSask s to Mont, Ore, Calif, NM. Map 437 ○

E. flavum Nutt. / Yellow Umbrella-plant
Caudex stout and compact, usually multiple, the branches short; leaves basal, petioled, thick, 3–5 cm long, oblong to oblanceolate, densely white-pubescent, greenish in age above; stems scape-like, 1–3 dm tall, white-tomentose; inflorescence usually umbellate, subtended by large leaf-like bracts; involucres 5-toothed; flowers yellowish or rose-tinged, pubescent, 4–5 mm long; achenes villose, about 4 mm long. $2n = 38, 76^*–80$. Includes *E. piperi* Greene of Ed. 1. Dry plains and rocky outcrops. BC to sMan s to eWash, neOre, Colo, ND. Map 438

E. ovalifolium Nutt. / Silver-plant
Low plant with short, closely branched caudex, thickly set with leaves and densely white-tomentose; leaves oval, orbicular or obovate, about 1 cm long; scapes 5–15 cm tall; involucres small, cup-like, usually clustered in a single close head; bracts very small, united at base; flowers white, 4 mm long, the outer sepals oval to orbicular, the inner ones narrow, spatulate. $2n = 40$. In our area var. **ovalifolium**. Dry plains and rocky outcrops to alpine elevations. BC, Mont s to Calif, Colo. Map 439

E. umbellatum Torr. / Umbrella-plant
Forming short woody branches at the base and becoming loosely matted; leaves basal or on sterile shoots, elliptic to spatulate-oblong, densely white-tomentose beneath, glabrous above at maturity; scapes 1–3 dm tall, woolly, bearing a whorl of leafy bracts at the tip; inflorescence usually a large umbel, the rays 1–6 cm long; flowers pale yellow or greenish white, tinged with rose in drying, glabrous, 4–6 mm long; achenes hairy at tip. $2n = 76^*, 80$. In our area var. **subalpinum** (Greene) M.E. Jones (*E. subalpinum* Greene of Ed. 1). Dry plains and foothills to alpine ridges. BC, Mont s to Calif, Nev, Colo. Map 440 ●

Fagopyrum Mill.

Annual herbs; leaves alternate; ocreae present, obliquely truncate; flowers heterostylous, in axillary and terminal raceme-like clusters; calyx petaloid, 5-parted; stamens 8; styles 3; stigmas capitate; achene triquetrous, much exceeding the calyx.

Achene smooth **F. esculentum**
Achene rugose **F. tartaricum**

F. esculentum Moench / Buckwheat
Stems erect, mostly 15–60 cm tall, reddish at maturity, glabrous to puberulent; lower leaves petiolate, the upper sessile, cordate-sagittate, slightly longer, 3–10 cm, than wide, 2–8 cm; flowers white to pinkish; sepals 2–3 mm long; achene 5–7 mm long, surface smooth, angles entire. $2n = 16$. *F. sagittatum* Gilib. An occasional escape from cultivation. Introduced.

F. tartaricum (L.) Gaertn. / Tartary Buckwheat
Similar to *F. esculentum* but 10–80 cm tall, green at maturity; leaves generally slightly wider than long; flowers greenish, sepals about 2 mm long; achene rugose, angles irregular. $2n = 16$. Roadsides. Introduced, occasional.

Koenigia L.

K. islandica L.
Stems slender, flexuous, 1–4 (15) cm tall, often reddish or crimson; leaves 2–5 (9) mm long, broadly elliptic to oblanceolate, somewhat fleshy; ocreae hyaline; flowers few, greenish, in terminal or axillary clusters; calyx 3-parted, 1.0–1.4 mm long; stamens mostly 3, alternate with sepals; achenes 3-angled. $2n = 28$. Silty banks and mud-flats at high elevation. Circumpolar: Alas, Yuk to Devon Is s to nBC, Huds Bay, Lab; Alta, Mont, Colo. Map 440 ○

Oxyria Hill

O. digyna (L.) Hill / Mountain sorrel
Low perennial with a chaffy rootstock and stout root; stems often clustered, erect, glabrous, scape-like, 5–30 cm tall; leaves alternate but mostly basal, long-petioled, round-reniform or cordate, glabrous, the margin undulate; stipule-sheaths loose, oblique; flowers whorled, in a dense, panicled raceme, small, perfect, greenish or reddish; sepals 4, the inner erect in fruit and 4–6 mm long, the outer ones reflexed; stamens 6; stigmas 2, tufted; achene lens-shaped, broadly winged. $2n = 14^*$, 42. Moist rocky slopes at alpine elevations. Circumpolar: Alas, Yuk to nEllesmere Is s to Calif, Ariz, NM, swMack, Huds Bay, Que, Lab, Nfld, NS; Gaspé. Map 441

Polygonum L. / Knotweed; Smartweed

Annual or perennial herbs of diverse habit; stems commonly with swollen joints; ocreae usually prominent; leaves alternate, entire; flowers mostly perfect, axillary or spicate-racemose; calyx 4–6 (mostly 5)-parted, often petaloid, persistent and erect in fruit; stamens 3–9, usually 5; styles or stigmas 2 or 3; achenes 3-angled or lens-shaped; embryo in a groove around one side of the endosperm. Mature achenes are necessary for identification of many of the species.

1 Stems twining; leaves cordate 2
Stems and leaves not as above 3

2 Annual; outer sepals slightly winged in fruit, about 5 mm long **P. convolvulus**
Perennial; outer sepals strongly winged in fruit, up to 15 mm long **P. scandens**

3 Flowers in small axillary clusters (occasionally solitary in axils), or spike-like with conspicuous foliaceous bracts, ocreae generally bilobed, becoming lacerate; leaves mostly 0.5–3.0 cm long 4
Flowers mostly in terminal spikes or racemes; ocreae truncate, generally not lacerate; leaves mostly 2–20 cm long 12

4 Flowers in a dense terminal spike, about 1 cm long, the lower flowers occasionally remote; bracts foliaceous, linear **P. watsonii**
Flowers scattered along the stems in small axillary clusters 5

5 Fruit reflexed 6
Fruit erect or nearly so 8

6 Calyx and achene 3–4 mm long; plants 10–40 cm tall **P. douglasii**
Calyx and achene 2–3 mm long; plants 5–25 cm tall 7

7 Lower leaves linear or linear-lanceolate; upper bracts awl-shaped **P. englemannii**
Lower leaves ovate to obovate; upper bracts lanceolate or oblong **P. austiniae**

8 Plants with stems 2–10 cm tall, less than 1 mm thick, erect, often zig-zag; leaves mostly 5–10 mm long, elliptic to ovate **P. minimum**
Plants not as above 9

9 Achenes shining, often exserted, calyx yellowish green **P. ramosissimum**
Achenes dull, generally not exserted; calyx white, pink or yellowish green 10

10 Calyx bottle-shaped, constricted just below the apex; sepals yellowish green **P. erectum**
Calyx not constricted below the apex; sepals white or pinkish 11

11 Stem and branch leaves more or less equal; calyx-lobes about as long as tube; achene 1.5–2.8 mm long, with two convex faces and a shorter concave face **P. arenastrum**
Stem leaves larger than branch leaves (best seen in young vigorous plants); calyx-lobes much longer than tube; achenes 2.5–3.5 mm long, with three equal concave faces **P. monspeliense**

12 Leaves chiefly basal; stem simple, stiffly erect; raceme solitary, terminal, spike-like 13

Leaves chiefly cauline; stem often branched; commonly with several spike-like racemes 14

13 Raceme whitish, linear, 5–8 mm thick, bearing bulblets; achenes dull **P. viviparum**

Raceme pink, oblong, 1 cm or more thick, lacking bulblets; achenes shining **P. bistortoides**

14 Perennials, often aquatic; spike-like racemes terminal, solitary or in pairs; flowers red or pink 15

Annuals, terrestrial; spikes terminal and axillary, usually several; flowers red to white or greenish 16

15 Racemes usually more than 4 cm long, often long-tapering; peduncles pubescent **P. coccineum**

Racemes usually less than 4 cm long, ovoid to oblong; peduncles glabrous **P. amphibium**

16 Ocreae fringed with coarse bristles; peduncles non-glandular **P. persicaria**

Ocreae lacking bristles, or bristles minute; peduncles glandular **P. lapathifolium**

P. amphibium L. / Water Smartweed
Perennial with running rootstocks, aquatic or terrestrial, variable in habit and leaf characters; leaves large, petioled; ocreae cylindric, with or without a spreading collar; raceme spike-like, 1–3 cm long, about 1.5 cm wide, usually blunt, borne on a stout peduncle; calyx scarlet to pink, 4–5 mm long; stamens and styles dimorphic, some flowers with exserted stamens and short styles, other flowers with short stamens and exserted styles; achenes lens-shaped, 2.5–3 mm long, brown to black. $2n = 66, 88$. Includes *P. natans* (Michx.) Eat. and *P. hartwrightii* Gray. Shallow water, shores, swamps, and ditches. Circumpolar: Alas, Yuk, swMack to James Bay, Nfld, s to Calif, Utah, Colo, Neb, Pa, NJ. Map 442

P. arenastrum Jord. *ex* Bor. / Common or Yard Knotweed
Slender annual, or sometimes perennial, with pale or bluish green foliage; stems prostrate, ascending or sometimes erect when young or when growing among other plants, often much branched, the main branches 1–8 dm long; leaves thin or thick, mostly narrowly lanceolate to oblong-lanceolate, 5–15 mm long; ocreae silvery, becoming lacerate; axillary clusters 1–5-flowered; calyx 2–3 mm long, green with pink or white margins; achenes brown, dull, 2–2.5 mm long. $2n = 40$. *P. aviculare* L., in part, of Ed. 1. Our commonest member of the widely distributed *P. aviculare* complex. Waste places, yards, roadsides. Introduced. Common.

P. austiniae Greene
Annual, branched from the base, 4–12 cm high; lower leaves oval to obovate or ovate-lanceolate, acute at apex, the base narrowed into a slender petiole; upper leaves sessile, much reduced; flowers solitary or in pairs, in the leaf-axils, reflexed, calyx 2–2.5 mm long, green with white margins; achenes 2.5–3 mm long, black, smooth, shining. Dry ground. seBC to cOre, sIda, scMont, Wyo. Map 443

P. bistortoides Pursh / Bistort
Perennial with a thick, chaffy caudex or rhizome; stem erect, simple, 2–7 dm high; basal leaves long-petioled, the blades lanceolate-oblong or oblanceolate, 1–2.5 dm long, cuneate at base, glabrous on both sides or scabrous-puberulent below; cauline leaves sessile, mostly lanceolate, 2–12 cm long; ocreae oblique at apex; raceme spike-like, solitary, terminal, densely flowered, 1–6 cm long, 1–1.5 cm thick; calyx pink to white, about 5 mm long; stamens exserted; achenes 3-angled, pale brown, shining, 3.5–4 mm long. Mountain meadows and stream banks. BC s to Calif, NM. Map 444

P. coccineum Muhl. / Water Smartweed
Perennial with stout creeping rootstocks, aquatic or terrestrial; stems erect or decumbent, usually simple, 3–10 dm long; leaves 5–20 cm long, those of terrestrial plants short-petioled, broadly to narrowly lanceolate, acuminate, commonly pubescent, those of aquatic plants long-petioled, thinner, smoother; racemes 1 or 2, spike-like, slenderly cylindric, 3–12 cm long, often tapering at the tip; calyx scarlet to pink; flowers of two kinds as in *P. amphibium*; achenes lens-shaped, brown to black, 2.5–3 mm long. $2n$ = c.66. Includes *P. muhlenbergii* (Meisn.) S. Wats. and *P. rigidulum* Sheldon. Ponds, stream banks, marshy places. BC, swMack, transcont sCan to Nfld, generally distributed s. Map 445

P. convolvulus L. / Wild Buckwheat; Black Bindweed
Annual, glabrous but scurfy; stems branched, twining or trailing, 2–8 dm long, with long internodes; leaves long-petioled, the blades 2–5 cm long, ovate-sagittate, or the upper lanceolate-sagittate, acuminate; ocreae oblique, entire; flowers in axillary clusters or loose spike-like racemes; calyx greenish, about 4 mm long, the outer lobes keeled in fruit; achenes 3-angled, black, dull, granular. $2n$ = 40. *Bilderdykia convolvulus* (L.) Dum., *Tiniaria convolvulus* (L.) Webb. & Moq. Gardens and waste ground. Introduced. Common.

P. douglasii Greene
Annual, 1–4 dm high; stems slender, erect, usually with numerous ascending branches; leaves narrowly lanceolate or linear-oblong, 1–4 cm long, cuneate at base, nearly sessile, sometimes revolute-margined; upper bracts reduced and awl-shaped; axillary clusters 1–3-flowered; calyx about 4 mm long, white or pinkish on margins, soon strongly reflexed; achenes 3-angled, black, shining, 3–4 mm long, reflexed. $2n$ = 40. Plains and slopes. BC transcont sCan to Que, s to Calif, NM, NY, Vt. Map 446

P. engelmannii Greene
Annual, 5–15 cm high; stems slender, usually diffusely branched at base, the branches erect or spreading; leaves linear to lanceolate or oblanceolate, the upper reduced to slender bracts; axillary clusters 1–4-flowered; calyx 2–2.5 mm long, green or rose-coloured with whitish margins, reflexed in fruit; achenes 3-angled, black, shining, 1.5–2.5 mm long, reflexed. Moist banks and slopes at lower montane elevations. sBC, Alta, wIda s to Colo. Map 447 ○

P. erectum L. / Striate Knotweed
Annual; stem firm, ridged, freely branched, decumbent-ascending or erect; leaves 1–3 cm long, blue-green or bright green, elliptic to obovate, broadly rounded at apex; ocreae white-scarious; calyx about 3 mm long, the 3 outer lobes hooded and hiding the 2 short inner lobes; achenes somewhat 3-sided, dull, yellow-brown, about 2.5 mm long. $2n = 20, 40$. Includes *P. achoreum* Blake of Ed. 1. Roadsides, waste places, saline marshes. Alas, Yuk, swMack to James Bay, Nfld, s to eOre, Colo, Kans, Mo, Tenn, Ga. Map 448

P. lapathifolium L.
Annual, 2–8 dm tall; stems usually branched, erect or occasionally prostrate; leaves 5–20 cm long, lanceolate to oblong-lanceolate, often with a dark blotch, glabrous or densely tomentose, glandular-dotted; ocreae entire or short ciliate; spikes dense, 1–6 cm long; peduncles glandular; flowers greenish white to pink; calyx-segments about 2.5 mm long, strongly nerved; achenes 2–3 mm long, shining. $2n = 22$. Includes *P. scabrum* Moench of Ed. 1. Fields and waste places. Eurasia: BC to Nfld s to Calif, SD, Pa, NJ, introduced Alas, Mack. Map 447 ●

P. minimum S. Wats.
Annual, 3–15 cm high, erect or spreading, branched mostly at the base, leafy throughout; leaves ovate or obovate, sessile or short-petioled, 5–15 mm long, obtuse or acutish at apex; flowers 2 or 3 in the leaf axils, 1.5–2 mm long, erect, green with pinkish margins; achenes 2–2.5 mm long, smooth, shining. Dry ground. BC, Alta s to Calif, Colo. Map 449 ○

P. monspeliense Pers.
Similar to *P. arenastrum*; distinguished by the characters of the key. $2n = 60$. *P. aviculare* L., in part, of Ed. 1. A member of the widely distributed *P. aviculare* complex. Waste places, yards, roadsides. Introduced, Rare.

P. persicaria L. / Lady's-thumb
Annual, glabrous or somewhat puberulent; stems erect, 3–6 dm high, usually branched; leaves lanceolate, tapering to base and apex, without evident glandular dots, often with a dark blotch on the upper side; ocreae with conspicuous hairs on the upper margins; spikes 1–3 cm long, ellipsoid or thick-cylindric, mostly erect; calyx pink or purplish, rarely whitish, 2.5 mm long; achenes 3-angled or lens-shaped, 2.5–3 mm long, smooth and shining. $2n = 44$. Fields and roadsides. Introduced. Occasional.

P. ramosissimum Michx.
Annual, erect or ascending, 3–12 dm tall, often yellowish green; stems striate, mostly slender; leaves usually thin, lanceolate to linear, narrowed at each end, the lower sometimes broader, 1–6 cm long, the upper much reduced; ocreae silvery to brownish and much lacerate; flowers in small axillary clusters, greenish, yellowish, mostly 3 mm long, sometimes on exserted pedicels; achenes 3-angled, rounded at base, 2.5–4 mm long, usually dark reddish brown, smooth and shining or obscurely striate, often somewhat exserted from the calyx. $2n = 20$.

Includes *P. prolificum* (Small) Robins. of Ed. 1. Dry shores, mud-flats, and roadsides. BC transcont sCan to NS s to Calif, NM, Tex, Mo, Pa, Del. Map 449 •

P. scandens L. / Climbing False Buckwheat
Perennial twiner; stems glabrous or minutely scabrous; leaves broadly ovate, deeply cordate at base, acuminate, ocreae very oblique; racemes 4–10 cm long, interrupted; pedicels winged above the joint; calyx 1.5–2.5 mm long, whitish, the outer lobes broadly winged in fruit, the wings flat, 1.5–3 mm broad; mature fruit 7–10 mm long; achene 3–6 mm long, black, shining, closely covered by the calyx. $2n = 20$. *Bilderdykia scandens* (L.) Greene, *P. dumetorum* L. var. *scandens* (L.) Gray. Clearings and shores. Introduced. Occasional.

P. viviparum L. / Bistort
Perennial with a short thick caudex or rhizome; stem simple, erect, 1–3 dm tall; lower leaves lanceolate to oblong-ovate, cuneate to rounded or cordate at base; upper leaves nearly linear, short-petioled to sessile and much reduced, firm, usually glabrous; ocreae brown-scarious, open down one side; raceme solitary, spike-like, narrow-cylindric, 2–6 cm long, usually bearing ovoid bulblets below and sterile flowers above; calyx pink or white; stamens long-exserted; achenes 3-angled, brown, granular, dull. $2n = 66$, 80–c.132. Moist woods and meadows to high alpine elevations. Circumpolar: Alas, Yuk, to Ellesmere Is s to Wash, eOre, NM, cSask, cMan, Lk Superior, Nfld; SD, Vt. Map 450 •

P. watsonii Small
Annual, 3–15 (25) cm tall; stems erect, simple or branched, glabrous; leaves linear or linear-lanceolate, 1–3 cm long, glabrous; flowers crowded in a short terminal spike, the lower sometimes remote and axillary; floral-bracts similar to leaves, linear, mostly 5–15 mm long; calyx green, the lobes with white or pinkish margins; stamens 8, all fertile; achenes 3-angled, dark brown or dull black, striate. Moist meadows and flats. sBC to swSask s to Calif, NM. Map 450 ○

Rumex L. / Dock; Sorrel

Herbs, mostly coarse and perennial, usually with thick roots; leaves alternate, the lower often large; flowers small, perfect or unisexual, greenish or red-tinged, in small whorls aggregated into a compound inflorescence; calyx of 6 sepals, in two whorls of 3; outer sepals sometimes united at base, often spreading in fruit; inner sepals somewhat coloured, often becoming enlarged and veiny-winged in fruit (the valves), sometimes bearing a grain-like tubercle on the back; stamens 6; styles 3, tufted; achenes 3-angled, invested by the inner sepals.

1 Flowers unisexual; plants dioecious; leaves hastate or sagittate and/or plants alpine 2
Flowers perfect or mostly so; leaves cuneate, rounded, truncate or cordate at base; plants not alpine 4

2 Leaves rounded at base, not hastate or sagittate **R. paucifolius**
Leaves hastate or sagittate 3

3 Valves equalling the achene and closely investing it; stems slender, arising from slender rhizomes **R. acetosella**

Valves much exceeding the achene; stems stout, arising from a thick taproot **R. acetosa**

4 Plants rhizomatous; valves 1–3 cm across **R. venosus**

Plants not rhizomatous; valves smaller 5

5 Valves with long, marginal teeth 6

Valves more or less entire, lacking long, marginal teeth 8

6 Plants perennial; teeth short, less than the width of valve **R. stenophyllus**

Plants annual (biennial); teeth long, as wide as or wider than the valves 7

7 Inflorescence interrupted above; tubercle more than 0.5 mm across **R. dentatus**

Inflorescence continuous above; tubercle about 0.4 mm across **R. maritimus**

8 Tubercles 1–3, well-developed 9

Tubercles 0–1, inconspicuous 11

9 Base of tubercle not reaching edge of valve; pedicels obscurely jointed **R. britannica**

Base of tubercle abutting or exceeding edge of valve; pedicels conspicuously jointed 10

10 Stem mostly lacking axillary branches below the inflorescence; leaf-margins strongly wavy-curled **R. crispus**

Stem with axillary branches below the inflorescence; leaf-margins plane or undulate **R. triangulivalvis**

11 Pedicels obscurely jointed **R. occidentalis**

Pedicels conspicuously jointed 12

12 Lower cauline leaves only slightly longer than wide, base deeply cordate **R. confertus**

Lower cauline leaves much longer than wide, not cordate at base 13

13 Valves 3–5.0 mm wide; lower cauline leaves up to 8 times as long as broad **R. pseudonatronatrus**

Valves more than 5.0 mm wide; lower cauline leaves not more than 4 times as long as broad **R. longifolius**

R. acetosa L. / Green Sorrel

Glabrous perennial, with short rootstock and coarse roots; stems stout, ribbed, 2–8 dm high; leaves oblong or broadly lanceolate, cordate or sagittate at base, 2–10 cm long, the lower long-petioled, the upper sessile or nearly so; ocreae mostly entire; panicle 1–2 dm long; plants dioecious; staminate flowers with sepals 2–3 mm long; pistillate flower with outer sepals nearly 2 mm long, soon reflexed, and inner sepals (valves) 4–5 mm long in fruit and reticulate-veined, the midrib enlarged as a tubercle at base; achene about 2 mm long, dark brown. $2n = 14$ ♀, 15 ♂. A species complex variously treated. In our area, not easily distinguished morphologically, ssp. **acetosa** having leaves with a narrow sinus, ocreae fimbriate, and ssp. **alpestris** (Scop.) Löve, with a more open sinus and more or less entire ocreae. The former of gardens and waste places, introduced, the latter (Map 451) of moist banks and meadows to alpine elevations, native. More or less circumpolar: Alas, Yuk to Wyo.

R. acetosella L. / Sheep Sorrel
Annual, or perennial with creeping roots; stems slender, 1–5 dm high, simple or branched from the base; leaves mostly basal, variable, often hastate with 2 spreading basal lobes, the main portion of the blade narrowly lanceolate to oblong; racemes reddish to yellowish, slender, arranged in an open panicle; pedicels joined near summit; plants dioecious; outer sepals lanceolate; inner sepals (valves) of pistillate flowers about as long as the achene and covering it at first; achene becoming exserted, about 1.5 mm long, golden brown. A species complex variously treated. In our area ssp. **acetosella**, $2n = 42$, with the valves free from fruit, and ssp. **angiocarpus** (Murb.) Murb. (*R. angiocarpus* Murb.), $2n = 14$, with the valves adnate to fruit. Fields and waste places. Introduced. Common.

R. britannica L. / Water Dock
Glabrous perennial, 5–25 dm tall; stem stout, erect, simple; leaves 15–60 cm long, lanceolate, acute to rounded at base, flat or faintly undulate, entire or serrulate to erose-crenulate; panicle of several ascending branches, 2–5 dm long, dense, mostly leafless; pedicels obscurely jointed; valves 4–8 mm long, ovate to orbicular, truncate to shallowly cordate at base; tubercles 3. $2n =$ c.200. Marshes, lake shores, ditches. Alta transcont sCan to Nfld s to Neb, Ind, Pa, NJ. Map 452 ○

R. confertus Willd.
Perennial; stem 10–20 dm tall, stout, erect, simple; basal leaf-blades up to 5 dm long, generally somewhat longer than wide, triangular, deeply cordate at base, margin crisped, shorter than petioles; upper leaves similar but smaller, the petioles shorter; panicle axis flexuous, branches arcuate at base; flowers in interrupted clusters; pedicels 10–15 mm long, filiform, conspicuously jointed; valves 5–8 mm long, about as long as broad or slightly broader, scutate, base cordate to truncate, entire or obscurely toothed; tubercles 1 or none. $2n = 40$. Roadsides and clearings. Introduced. Rare.

R. crispus L. / Curled Dock
Perennial; taproot stout, yellow when cut across; stem simple, 3–8 dm tall; leaves 1–3 dm long, lanceolate to oblong, the margins strongly wavy-curled, often rounded to truncate and somewhat decurrent at base; inflorescence large, branched, leafless above, with dense whorls; valves 4–6 mm long in fruit, broadly ovate, obtuse; tubercles 3, often unequal, the larger plump, nearly globose, about half as long as the valve; achene 2.5 mm long. $2n = 60$. Moist ground, waste places. Introduced. Occasional.

R. dentatus L.
Annual; stem 2–4 (7) dm tall, erect, branched, leafy below, with flower clusters throughout; basal leaves petiolate, the blades 2–6 cm long, oblong, somewhat crisped, base cuneate to subcordate; flowers clusters, moniliform, seldom continuous; pedicels articulated and thickened below the middle; valves 4–6 mm long, ovate, with 3–4 marginal teeth 1–2 mm long; tubercles 1–3. $2n = 40$. Cultivated fields. Introduced. Rare.

R. longifolius DC.
Perennial, 6–12 dm tall; stem stout, erect, simple; leaves up to 80 cm long, broadly lanceolate to ovate, base subcordate to cuneate, margin crisped; panicle dense, lanceolate-elliptic, continuous or with lower axillary branches somewhat remote; pedicels articulated near the base; valves mostly 5–6 mm long and about as wide, orbicular-reniform, entire; tubercles generally absent throughout or occasionally single vestigial tubercles present. $2n = 60$. Moist open areas. Introduced. Rare.

R. maritimus L. / Golden Dock
Annual, often bushy-branched, with ascending stems, 2–6 dm long; lower leaves linear-lanceolate, cordate to truncate at base, the margins more or less crisped, usually puberulent beneath; panicle large and freely branched, golden brown at maturity, many of the whorls subtended by small leaves; valves with 1–3 bristle-like lobes on each margin; tubercles 3, narrowly lanceolate; achene about 1.4 mm long. $2n = 40$. In our area ssp. **fueginus** (Phil.) Hult. Saline shores and disturbed ground. Circumpolar: Alas, Yuk, swMack to Huds Bay, Nfld, generally distributed s. Map 453

R. occidentalis S. Wats. / Western Dock
Glabrous perennial with taproot; stem simple, stout, erect, 5–15 dm high, commonly tinged with red; lower leaves oblong-lanceolate, usually cordate at base, tapering to a narrow tip, the blades 5–20 cm long; panicle 2–5 dm long, narrow, dense; pedicels in fruit 5–10 mm long; valves 5–8 mm long, reddish, cordate, conspicuously reticulate; tubercles lacking; achene 2–4 mm long. $2n = 160$. *R. aquaticus* L. ssp. *occidentalis* (S. Wats.) Hult. Moist places. Alas, Yuk, wMack to Huds Bay, Nfld s to Calif, Nev, Utah, NM, SD, nLk Superior, sQue. Map 454

R. paucifolius Nutt.
Glabrous perennial with taproot and short rootstock, somewhat tufted; stems slender, 1–few from base, 2–6 dm tall; leaves mostly basal, lanceolate or elliptic, mostly obtuse, narrowed at base, 4–10 cm long, on long petioles or the upper reduced and sessile; panicle slender, naked; flowers usually tinged with red; pedicels jointed near the base; valves cordate-orbicular, 3–4 mm long, sometimes with a minute basal tubercle. $2n = 14^*, 28$. Moist meadows from montane to alpine elevations. BC to Calif, Colo. Map 452 •

R. pseudonatronatus Borbas
Perennial; stem 8–15 dm tall, slender; leaves narrowly lanceolate, 8–15 times longer than broad, base cuneate, margins strongly crisped; panicle long, narrow, dense, sometimes interrupted below, leafless; pedicels mostly 2–3 times longer than valves, articulated towards the base; valves heart-shaped, 3–5 mm long, generally somewhat narrower, margins entire; tubercles absent. $2n = 40$. *R. fennicus* (Murb.) Murb. of Ed. 1. Moist ground. Introduced. Rare.

R. stenophyllus Ledeb.
Perennial; stems mostly 3–10 dm tall, branching; leaves petiolate, mostly 5–20 cm long, oblong, lanceolate to linear, long-cuneate at base, margins wavy, finely serrulate; panicle with ascending branches, flower clusters more or less continuous or interrupted, those of the lower branches often moniliform; pedicels articulated towards the base; valves 3–5 mm long, deltoid, margin coarsely dentate; tubercles 3. $2n = 60$. Waste ground. Introduced. Rare.

R. triangulivalvis (Dans.) Rech. f. / Narrow-leaved Dock
Glabrous perennial, 3–6 dm high, often tufted; stem erect or decumbent, often with axillary branches; leaves pale green or glaucous, narrowly lanceolate, long-tapering at both ends, flat, often plicate, entire; panicle 1–3 dm long, with a few ascending branches; valves 3–6 mm long, triangular, acutish; tubercles 3, lanceolate, half to two-thirds the length of the valve; achene about 2 mm long. $2n = 20$. *R. mexicanus* Meisn. of Ed. 1. Moist open ground. Yuk, swMack, Huds Bay to cQue, s to Calif, Ariz, NM, Mo, Ind, NY. Map 455

R. venosus Pursh / Wild Begonia; Veined Dock
Perennial with running rootstocks; stems stout and leafy, 1–4 dm high, with ascending branches; ocreae dilated, conspicuous; leaves on short, slender petioles, the blades ovate to oblong-lanceolate, 5–10 cm long, entire; panicle short, very dense in fruit; pedicels of fruit 1–2 cm long; valves cordate-orbicular, with a deep sinus, 14–18 mm long, very veiny, usually rose-red; tubercles none; achene 4–7 mm long. $2n = 40$*. Dry sandy soil and dunes. sBC to sMan s to Calif, Colo, Kans. Map 456

CHENOPODIACEAE / Goosefoot Family

Annual or perennial herbs, or shrubs; leaves simple, without stipules or scarious bracts, usually alternate, often succulent; flowers small, greenish, perfect or unisexual, produced singly or in dense clusters; calyx of 5 sepals or fewer; corolla none; stamens 1–5, opposite the calyx-lobes; anthers introrse; pistil 1; styles 1–5; ovary superior; ovule 1; fruit small, 1-seeded, with a thin pericarp, usually enclosed by the persistent calyx, sometimes by a pair of bracteoles; seed horizontal or vertical, usually lenticular. Many of the species are weeds; a very few are ornamental; cultivated species include Swiss chard, beet, and spinach.

1 Stems fleshy, jointed; leaves scale-like, opposite; flowers axillary and sunk into the fleshy stem **Salicornia**

Stems scarcely fleshy, not jointed; leaves slender, elongate, mostly alternate; flowers otherwise 2

2 Spiny shrub, 1–2 m high; leaves narrowly linear, fleshy; flowers imperfect, the staminate in catkin-like spikes, without a perianth, but subtended by peltate bracts **Sarcobatus**

Herbs, or low spineless shrubs; flowers perfect or, if imperfect, the staminate flowers with a calyx 3

3 Plant prickly from the spine-tipped thread-like leaves and from bracts **Salsola**

Plants not prickly, leaves various 4

4 Flowers monoecious or dioecious, the pistillate with a pair of bracteoles, occasionally 3 5

Flowers hermaphrodite or polygamous, lacking bracteoles 8

5 Plants densely pubescent with yellowish brown stellate hairs 6

Plants with hyaline or white encrusting scales on stem and leaves 7

6 Annual herb; leaves lanceolate to elliptic, not revolute, 5–15 mm wide generally, 5–6 cm long; lateral branches largely naked in the lower half **Axyris**

Undershrub, woody at base; leaves linear, revolute, 2–3 mm wide, up to 4 cm long; branches naked only at the base **Eurotia**

7 Petioles as long as and generally much longer than the lamina **Suckleya**

Petioles shorter than the lamina **Atriplex**

8 Leaves broad, often hastate 9

Leaves narrow, linear 10

9 Stamen 1; fruit largely exposed, narrowly winged **Monolepis**

Stamens more than 1; fruit enclosed, not winged **Chenopodium**

10 Stems tomentose to densely glandular-villous 11

Stems not tomentose to densely villous 12

11 Calyx in fruit developing horizontal scarious wings **Kochia**

Calyx in fruit not developing horizontal scarious wings, but slender hooked spines **Bassia**

12 Plants heavily encrusted with white or hyaline scales **Chenopodium**

Plants lacking white or hyaline scales 13

13 Plants green, covered (more or less) with a fine stellate pubescence; bracts with a conspicuous hyaline margin **Corispermum**

Plants essentially glabrous, sometimes puberulent, greenish grey to purple, becoming almost black at maturity; bracts with a pale margin but not hyaline **Suaeda**

Atriplex L. / Orache; Saltbush

Herbs or low shrubs, usually mealy or scurfy, often silvery; leaves alternate or the lower opposite; flowers in dense axillary clusters or compound spikes; monoecious or dioecious; staminate flower without bracteoles but with a 3–5-lobed calyx and 3–5 stamens; pistillate flower usually lacking sepals but with a pair of appressed bracteoles enclosing the ovary; the bracteoles more or less united and enlarging in fruit, often hard and usually bearing tubercules on the surface.

1 Perennial shrubs with kranz-type leaf venation — 2
Annual herbs with kranz-type venation or normal dicotyledonous venation — 3

2 Bracteoles irregular in shape, with 4 projecting lateral wings — **A. canescens**
Bracteoles globose, with tubercles but without wings — **A. nuttallii**

3 Leaves with kranz-type venation — 4
Leaves with normal dicotyledonous venation — 7

4 Leaves commonly sinuate-dentate and lobed, greyish green; branches and leaves variously positioned — **A. rosea**
Leaves commonly entire, whitish-grey; branches and leaves ascending — 5

5 Bracteoles obovate, 4–8 mm long, deeply dentate — **A. argentea**
Bracteoles cuneate or oblong, 2.0–3.5 mm long, not dentate — 6

6 Bracteoles cuneate, with 3 minute teeth at the apex, dorsal surface smooth or with 1 or 2 minute tubercles; leaves not conspicuously nerved — **A. truncata**
Bracteoles oblong, crowned with a flattened lobe, dorsal surface with small tubercles below the lobes; leaves conspicuously 3-nerved — **A. powellii**

7 Staminate flowers with calyx-lobes cup-shaped, fleshy crested on the back, pink; bracteole margins united up to the apex — **A. suckleyi**
Staminate flowers with calyx-lobes ovate, rounded on the back, green in the central parts becoming membranous toward the margins; bracteole margins united at the base or up to the middle — 8

8 Bracteoles orbicular, entire, lacking lateral angles — 9
Bracteoles not orbicular, frequently toothed, with or without lateral angles — 10

9 Inflorescence with ebracteolate horizontal seeds present among the more abundant bracteolate vertical seeds; bracteoles subcordate or broadly rounded at the base, the veins merging above the base — **A. hortensis**
Inflorescence with all seeds bracteolate and vertical; bracteoles broadly cuneate at base, the veins merging at the base — **A. heterosperma**

10 Lateral angles of bracteoles rounded or absent; leaves green above, whitish below — **A. oblongifolia**
Lateral angles of bracteoles usually present; leaves the same colour above and below — 11

11 Leaves hastate; seeds circular — **A. prostrata**
Leaves ovate-lanceolate or linear; seeds elliptic — **A. subspicata**

A. argentea Nutt. / Silver Saltbush
Bushy herb, 15–80 cm high, branched from base, usually rounded in outline, stems angled, rather stout, whitish-scurfy when young; lower leaves usually opposite and stalked, the others alternate, sessile or nearly so; leaves 1–3 cm long, ovate, deltoid or lanceolate, hastate to cuneate at base, entire or wavy-dentate, grey-scurfy, sometimes glabrate; flowers monoecious, the pistillate commonly in axillary clusters, the staminate often in interrupted terminal spikes; fruiting bracteoles united to the middle or above, firmly enclosing the fruit, the margins green and more or less toothed. $2n = 18$. Dry plains and alkaline flats. sBC to swMan s to seOre, neCalif, Tex, SD. Map 457

A. canescens (Pursh) Nutt.
Shrub, 1–4 (20) dm high, generally woody throughout, branches spreading; leaves 1–5 cm long, sessile or short-petioled, elliptic, oblong, to spatulate or linear, mostly grey and fine-scurfy, margin entire; dioecious; staminate flowers tawny in dense continuous or interrupted spikes or panicles; bracteoles of pistillate flowers mostly 4–6 (20) mm long, prominently 4-winged, the wings undulate or irregularly dentate. $2n = 18, 36$*. In our area var. **aperta** (A. Nels.) C.L. Hitchc. Saline flats. Alta s to Wash, Calif, Colo, SD, Tex. Map 458

A. heterosperma Bunge
Annual; stem 10–15 dm tall, stiffly erect; leaves about as long as wide, triangular, base hastate to slightly cuneate, often with a pair of auricles, farinose below, becoming glabrate and green on both surfaces; monoecious; flowers in a loosely branched terminal or axillary inflorescence; staminate flowers with 5-parted calyx; stamens usually 5; bracteoles of pistillate flowers orbicular-ovate, lacking appendages, of two sizes, the larger 5–6 mm long, enclosing a yellow-brown fruit 2–3 mm across, the smaller 2–3 mm long, enclosing a black fruit to 1.5 mm across; pericarp easily detached from seed. $2n = 36$. Saline sloughs. Introduced. Occasional.

A. hortensis L. / Garden Orache
Tall erect herb, up to 2 m high, green to yellowish or reddish, glabrous or mealy; leaves petioled, 5–25 cm long, ovate-oblong to deltoid, cordate-hastate to truncate at base, the margins entire or sinuate-dentate; flowers in a large spicate panicle; pistillate flowers of 2 kinds, one with a small calyx and no bracteoles, producing small black shiny seeds, the other kind with no calyx but 2 large bracteoles producing large flat brownish seeds. $2n = 18$. Waste places and gardens. Introduced. Common.

A. nuttallii S. Wats. / Nuttall's Atriplex; Salt Sage
Perennial sub-shrub with a woody base, erect or spreading, often much branched, 2–6 dm high; leaves 1–5 cm long, sessile or short-stalked, linear-oblong, oblanceolate or obovate, often grey and fine-scurfy, margins usually entire; dioecious; staminate flowers in dense yellow spikes or panicles; bracteoles of the pistillate flowers 4–7 mm long, toothed and tubercled; fruit rounded and somewhat spiny.

$2n = 18$. Bad-lands, eroded slopes, and alluvial flats. Alta to swMan s to Calif, Ariz, NM, SD. Map 459

A. oblongifolia Waldst. & Kit.
Annual; stem erect, up to 12 dm tall; leaves up to 7 cm long, 2.5 cm wide, oblong-lanceolate, widest near the cuneate base, farinose below; monoecious; flowers in a loosely branched, finally elongate terminal or axillary inflorescence; bracteoles of pistillate flowers ovoid-rhombic or triangular-ovate, lacking appendages, of two sizes, the larger up to 13 mm long, enclosing a yellow-brown fruit about 3 mm across, the smaller enclosing a black fruit about 1.5 mm across; pericarp easily detached from seed. $2n = 18, 36$. Roadsides and waste ground. Introduced. Occasional.

A. powellii S. Wats.
Annual; stem 2–4 (5) dm tall, erect, branched from the base, the lower branches longer and curving upwards; leaves petiolate, the upper sessile, 0.5–2.5 cm long, 0.4–1.8 cm wide, usually 3-nerved, ovate to deltoid, base truncate to cuneate, margin entire or undulate, grey-farinose; monoecious, some plants almost entirely pistillate; flowers in glomerules in terminal and axillary spikes; bracteoles of pistillate flowers oblong, 2–3 (4) mm long and nearly as wide, margin lobed, surface tuberculate, rarely smooth; seeds light or yellowish brown. $2n = 18$*. Alkaline flats. Alta, Sask s to Ariz, NM, SD; eOre. Map 460 •

A. prostrata Bouch. *ex* DC.
Annual; stem 1–10 dm tall, green or striate; leaves scurfy to glabrous below, the upper 2–10 cm long, 2–9 cm wide, triangular-hastate, occasionally ovate, entire to serrate or irregularly toothed; monoecious; inflorescence 2–9 cm long, terminal or axillary, spike-like, glomerules continuous or interrupted; bracteoles of pistillate flowers green, becoming brown or black, 3–5 mm long, triangular-hastate to triangular-ovate, margins united at base, entire or toothed near the apex, dorsal surface smooth or with two tubercles; seeds of two kinds, brown, 1.5–2.5 mm across, and black, 1.0–1.5 mm across. $2n = 18$. *A. patula* L. var. *hastata* (L.) Gray of Ed. 1. Saline flats. Introduced or possibly native. Occasional.

A. rosea L.
Annual; stem 2–12 (20) dm tall, erect, simple or branched; leaves petiolate or the upper sessile, 2–5 (7) cm long, lanceolate to ovate or rhombic, base cuneate to truncate, sinuate dentate, grey-farinose or glabrate above; monoecious; flowers in glomerules in terminal or axillary spikes; bracteoles of pistillate flowers mostly rhombic, 4–8 (12) mm long, margin dentate, surface with several tubercles; seeds dark brown, 1.5–2 mm across. $2n = 18$. Roadsides and waste ground. Introduced. Occasional.

A. subspicata (Nutt.) Rydb.
Annual; stems 3–15 dm tall, erect, angular, sometimes reddish; leaves 2–12 cm long, 0.5–6.0 cm wide, narrowly to broadly lanceolate to lance-linear, margin irregularly toothed to entire, often with a pair of narrow, straight or forward-

curving lobes toward the base; inflorescence of terminal and axillary branches; monoecious; flowers in glomerules, continuous or interrupted; bracteoles of pistillate flowers 3–10 mm long, triangular to ovate-triangular, usually longer than wide, margin entire or short-toothed, dorsal surface with one or more tubercles, rarely smooth; seeds of two kinds, brown, dull, 1.5–3.0 mm across, and shiny, black, 1–2 mm across. $2n$ = 36*, 54*. Saline and alkaline soils, waste ground. BC, sMack to swOnt s to Calif, Utah, Okla, Kans; Nfld, St Lawrence s to NC; Huds Bay. Map 461

A. suckleyi (Torr.) Rydb.
A low branching annual, 1–3 dm high, often with purplish stems and foliage, almost glabrous; leaves 2–3 cm long, sessile, 1-nerved, lanceolate, entire; staminate flowers in axillary clusters and short terminal spikes; stamens 5; pistillate flowers solitary or clustered in the leaf-axils, each flower with 2 enveloping bracteoles; calyx of 2–3 hyaline sepals, shorter than the bracteoles; fruit ovate, compressed. $2n$ =18. *Endolepis suckleyi* Torr. of Ed. 1. Dry, eroded slopes. Alta to swSask s to Wyo, Neb. Map 462

A. truncata (Torr.) A. Gray
Annual; stem 1.5–4 (10) dm tall, erect, often freely branched; leaves 1–2.5 (4) cm long, sessile or short-petioled, broadly ovate to deltoid, base cuneate to subcordate or somewhat hastate, margin entire, sometimes undulate, grey-farinose or the upper surface glabrate; monoecious; flowers in glomerules in terminal and axillary spikes; bracteoles of pistillate flowers triangular, 2–2.5 (3.5) mm long, generally somewhat narrower, mostly with 3 teeth at the apex, surface smooth or with 1–2 tubercles; seeds about 1.5 mm across. $2n$ = 18. Alkaline flats and disturbed ground. BC to swSask s to Calif, Nev, Utah, Colo. Map 460 ○

Axyris L.

A. amaranthoides L. / Russian Pigweed
Annual herb with alternate leaves; stem erect, 4–8 dm high, with numerous short ascending branches; leaves short-stalked, lanceolate to elliptic-ovate, stellate-pubescent beneath, up to 6 cm long, gradually reduced to bracts in the flowering branches; leaf-margins entire or somewhat wavy; the whole plant becoming pale straw-coloured at maturity; flowers monoecious, the staminate in short terminal spikes, the pistillate axillary or intermixed with the staminate; bracteoles of the pistillate flowers commonly in 3's; calyx 3–5-parted, hyaline; stamens 2–5; stigmas 2, filiform; ovary compressed, at maturity enclosed in the whitish calyx; pericarp membranous; seeds of 2 kinds, the larger 2.5–3 mm long, flattened, oblong-oval, brown and silky, with a 2-lobed membranous wing at the apex; the smaller 1.5–2 mm across, somewhat flattened, circular, greyish brown and minutely notched at the apex. $2n$ = 18. A weed of fields, shelter belts, and waste ground. Introduced. Common.

Bassia All.

B. hyssopifolia (Pall.) Ktze.
Annual herb with a habit much like *Chenopodium album*; stem 5–12 dm high, erect, freely branched, often reddish or whitish, finely tomentose or villose; leaves alternate, linear-oblong to narrowly lanceolate, entire, fleshy, 1–2 cm long; flowers perfect, about 1 mm broad, solitary or in small axillary clusters arranged in leafy-bracted, interrupted spikes; bracts linear-oblong, blunt, grey-hairy; calyx white-villose or tomentose, 5-lobed; each of the sepals at maturity bearing a hooked spine. $2n = 18$. A weed of waste ground in the south. Introduced. Occasional.

Chenopodium L. / Goosefoot; Pigweed

Chiefly weed-like annuals, with alternate, linear to ovate leaves, often mealy-coated; flowers small, greenish or reddish, perfect, produced singly or in dense clusters (glomerules), these axillary or forming spikes or panicles: calyx commonly 5-lobed, persistent and enclosing the fruit; stamens 2–5; styles and stigmas 2–5; fruit thin-walled, containing one lens-shaped seed.

1 Seeds vertical or occasionally horizontal in the same inflorescence 2
Seeds horizontal, rarely vertical in the same inflorescence 5

2 Plants perennial; stigmas more than 0.8 mm long; seeds averging more than 1.5 mm wide by 1.5 mm long **C. bonus-henricus**
Plants annual; stigmas less than 0.8 mm long; seeds averaging less than 1.5 mm wide by 1.5 mm long 3

3 Leaves 2–3 cm long, densely white, farinose beneath, at least when young **C. salinum**
Leaves 3–10 cm long, sparsely or not at all farinose beneath 4

4 Flowers in small clusters; calyx not both fleshy and bright red in fruit; pericarp non-adherent **C. rubrum**
Flowers in large spicate glomerules; calyx fleshy and bright red in fruit; pericarp adherent **C. capitatum**

5 Primary leaves linear, narrowly lanceolate, lanceolate with 1 or 2 lobes above the base, deltoid-rhombic or oblong-ovate, less than 1.5 cm wide, 1–5 times longer than wide 6
Primary leaves rhombic, broadly lanceolate, truncate, generally more than 1.5 cm wide, 1–3 times longer than wide 10

6 Plants glabrous to very sparingly farinose, primary leaves 1-veined from the base; pericarp loosely attached (non-adherent); seeds averaging 1.5 mm long **C. subglabrum**
Plants moderately to densely farinose; primary leaves 3-veined from the base (except *C. leptophyllum*); pericarp attached or loosely attached to seed; seeds averaging 0.9–1.2 mm long 7

7 Calyx-lobes covering the seed at maturity 8
Calyx-lobes not covering the seed at maturity 9

8 Pericarp loosely attached (non-adherent); primary leaves 3-veined; plants branched from the base, semi-erect or spreading **C. desiccatum**

Pericarp attached (adherent); primary leaves 1-veined from base; plants erect or occasionally semi-erect **C. leptophyllum**

9 Primary leaves with 1 or 2 lobes or teeth above the base; pericarp non-adherent; plants strictly erect; flowers in long spikes **C. pratericola**

Primary leaves entire or with basal lobes; pericarp generally adherent; plants open and branched, not strictly erect; flowers in interrupted glomerules **C. atrovirens**

10 Seed more than 1.5 mm wide by 1.5 mm long; leaves truncate or ovate-cordate at base, bright green; pericarp non-adherent; calyx-lobes not covering the seed at maturity **C. gigantospermum**

Seed less than 1.5 mm wide by 1.5 mm long; leaves rhombic to broadly lanceolate, not truncate or cordate at base, light-green to mealy underneath; pericarp adherent; calyx-lobes covering or nearly covering the seed at maturity 11

11 Pericarp areolate reticulate or honeycombed **C. berlandieri**

Pericarp not areolate reticulate (except *C. watsonii*, the plants then only 2–15 cm tall) or honeycombed 12

12 Plants up to 5 dm high at full maturity 13

Plants usually more than 5 dm high at full maturity 14

13 Pericarp adherent, plants 2–15 cm tall; ill-smelling; stems distinctly angular **C. watsonii**

Pericarp non-adherent, plants up to 50 cm tall; not ill-smelling; stems not distinctly angular **C. incanum**

14 Leaves broadly triangular to rhombic; testa smooth with light streaks **C. fremontii**

Leaves lanceolate, ovate to triangular hastate; testa marked by faint reticulations **C. album**

C. album L. / Lamb's-quarters
Stout annual, erect, usually much branched, 4–12 dm high; often reddish late in the season; leaves usually green above and densely mealy beneath, ovate, rhombic or lanceolate, broadly cuneate at base, the margins of the larger leaves wavy or irregularly toothed, or somewhat lobed; flower clusters in a dense spicate panicle, often with a bluish tinge; calyx more or less white-mealy, its lobes keeled; pericarp thin and closely adherent to the seed, the latter black, shining, not pitted, 1.1–1.6 mm long, 1.1–1.5 mm wide. $2n = 54$. Gardens, roadsides, waste ground. Introduced. Common.

C. atrovirens Rydb.
Annual; stems 1–5 dm tall, erect, with many ascending branches; leaves 1.5–3.0 cm long, 0.5–1.5 cm wide, oblong, triangular-oblong to ovate, obtuse to rounded at apex, rounded to cuneate at base, entire or with 2 basal lobes, farinose becoming glabrate; inflorescence of dense glomerules, simple or in interrupted paniculate spikes; calyx-lobes sharply carinate-obovate, not covering the seed at maturity; pericarp adherent or occasionally free; seeds 1.0–1.3 mm long, 0.9–1.3 mm wide, dark reddish-brown, shining, obscurely rugulose. $2n = 18$. Open dis-

turbed areas, generally at higher elevations. BC, Alta s to eOre, Calif, Mont, Colo, Tex. Map 463 ○

C. berlandieri Moq.
Annual, often ill-smelling; stems 4–15 dm tall, erect, striate, generally simple at base, with many spreading or ascending branches above; leaves 1–6 cm long, 0.5–3.5 cm wide, rhombic-ovate, ovate, to oblong, apex acute to obtuse, base cuneate, margin sinuate-dentate to irregularly dentate, occasionally entire, farinose becoming glabrate; inflorescence leafy, of small glomerules in slender, continuous or interrupted paniculate spikes; calyx-lobes strongly keeled (in ours) covering the fruit; pericarp adherent; seeds 1.0–1.4 mm long, 1.1–1.3 mm wide, black, shining. $2n = 36$. In our area ssp. **zschackei** (Murr.) Zobel. Grassland and disturbed areas. Alas, Yuk, wMack to Gt Lakes, Pa, generally distributed s. Map 463 ●

C. bonus-henricus L.
Perennial; stems 2–6 (8) dm tall, erect, simple; leaves 5–12 cm long, 4–9 cm wide, long-petioled, triangular hastate, margin entire, undulate; inflorescence terminal, leafless except at base, the glomerules forming a dense, narrowly pyramidal spike-like panicle; seeds 1.5–1.8 mm long, 1.5–1.8 mm wide, $2n = 36$. Waste ground, yards. Introduced. Rare.

C. capitatum (L.) Aschers. / Strawberry Blite
Glabrous succulent herb, 2–5 dm high, erect or spreading; leaves 3–10 cm long, broadly triangular, acute, often hastate, entire to coarsely sinuate-dentate; flowers in globose clusters, at the leaf-axils or in terminal interrupted spikes, the clusters becoming bright red, pulpy and 10–15 mm thick, then resembling ripe strawberries; calyx 3–5-lobed, becoming fleshy and red at maturity, enclosing the fruit; seeds 0.7–1.2 mm long, 0.6–0.9 mm wide, vertical, black, separating from the pericarp. $2n = 18$. Clearings, waste ground, burns. Europe: Alas, Yuk, wMack to James Bay, NS s to Calif, NM, sSask, sMan, Minn, Pa, NJ, Mass. Map 464 ●

C. desiccatum A. Nels.
Annual, densely farinose; stem 1–5 dm tall, semi-erect or spreading, branched from the base; leaves mostly 1–4 cm long, 0.5–1 cm wide, fleshy, 3-veined, lanceolate to ovate-oblong, entire, farinose above and below or becoming glabrate above; inflorescence generally leafy, paniculate, glomerules densely crowded; calyx-lobes not strongly keeled, enclosing fruit; pericarp non-adherent; seeds 0.9–1.1 mm long, 0.8–1.0 mm wide. $2n = 18$. Undisturbed saline soils. BC to swMan s to Calif, NM, Tex, Mo. Map 464 ○

C. fremontii S. Wats.
Annual; stems 1–10 dm tall, erect, branches slender, ascending, often flexuous; leaves 1.5–6.5 cm long, about as wide, broadly triangular to rhombic, the apex rounded, mucronate, margin sinuate-dentate generally with paired lobes to-

wards the base, occasionally more or less entire, mostly green above, sparsely grey-farinose below; inflorescence of terminal and axillary slender paniculate spikes, glomerules often widely spaced; calyx-lobes keeled, nearly covering the fruit; pericarp non-adherent; seeds 1.1–1.3 mm long, 1.0–1.2 mm wide. $2n = 18$. Moist places, often shaded. BC to Sask s to Calif, Nev, Ariz, NM, Tex. Map 465

C. gigantospermum Aellen / Maple-leaved Goosefoot
Tall, erect, bright green annual, 3–15 dm high; stem grooved, freely branched, slender; leaves 3–15 cm long, thin, slender-petioled, ovate to triangular-ovate, acuminate, the base rounded, subcordate or truncate, sinuate-dentate with 1–4 large teeth on each side, prominently-veined; flowers in small glomerules, arranged in a loose spicate panicle; seeds 1.3–1.9 mm long, 1.2–1.9 mm wide, black, obscurely grooved. $2n = 36$. *C. hybridum* L. var. *gigantospermum* (Aellen) Rouleau of Ed. 1. Dry to moist woodlands and waste ground. BC, swMack, s of Huds Bay to Que, NB, s to nCalif, NM, nWyo, Tex, Mo, Ky, Va; Yuk. Map 466 ●

C. incanum (S. Wats.) Heller
Annual; stem 1–5 dm tall, erect, angular, striate with many spreading or ascending branches, densely farinose; leaves 1–3 cm long, triangular-rhombic to rhombic-orbicular, cuneate at base, distinctly lobed, densely farinose above and below or glabrous above; inflorescence with or without leaves; glomerules in dense, crowded, paniculate spikes; calyx farinose; pericarp free; seeds 1.0–1.3 mm long, 0.8–1.1 mm wide, black, smooth, shining. $2n = 18$. Dry plains, hillsides, waste sandy ground. Alta to Sask s to Calif, Nev, Utah, Tex, Neb. Map 466 ○

C. leptophyllum (Nutt. *ex* Moq.) S. Wats.
Annual; stem up to 4 dm tall, erect or semi-erect, branching from the base, greyish green to white-farinose; leaves linear, up to 3 cm long, less than 1.5 cm wide, 1-nerved, entire, fleshy; inflorescence open, glomerules widely spaced; calyx-lobes densely farinose, covering the fruit at maturity; pericarp adherent; seeds 0.9–1.1 mm long, 0.9–1.1 mm wide, black, testa rugulate. $2n = 18$. Open, slightly disturbed sandy areas. BC to Man, wOnt s to Calif, Ariz, NM, adventive elsewhere. Map 467 ○

C. pratericola Rydb.
Annual, moderately to densely farinose; stem up to 8 dm tall, erect or semi-erect, simple or branched above the base; leaves 2–6 cm long, 0.4–1.8 cm wide, thin, 3-veined, linear to lanceolate or oblong-elliptic, with 1 or 2 lobes near base, margin entire or toothed; inflorescence generally leafy, paniculate, glomerules densely crowded; calyx-lobes not strongly keeled, tending to spread and expose fruit; pericarp non-adherent; seeds 0.9–1.3 mm long, 0.9–1.2 mm wide. $2n = 18$. Slough margins and open disturbed moist alkaline areas. Yuk, BC to swOnt s to Calif, NM, Tex, Mo. Map 467 ●

C. rubrum L. / Red Goosefoot
Erect, rather fleshy annual, glabrous or glabrate throughout, often reddish; stems 20–80 cm tall, more or less branched at base or above; leaf-blades 3–10 cm long, rhombic-ovate to deltoid-ovate, usually coarse and irregularly sinuate-dentate, the leaves of the inflorescence smaller and often entire; flowers crowded in axillary spikes, forming a leafy panicle; calyx slightly fleshy, not crested, becoming reddish; seeds 0.7–0.9 mm long, 0.6–0.8 mm wide, easily separating from pericarp, brownish. $2n = 36$. Lake shores and dried slough beds. Circumpolar: BC, swMack, transcont sCan to Nfld s to Calif, Ariz, NM, Neb, Mo, NY, NJ; Alas, Yuk. Map 468 ●

C. salinum Standl. / Oak-leaved Goosefoot
Low, somewhat succulent annual; stems glabrous, often reddish, spreading or erect, 1–4 dm long; leaf-blades broadly lanceolate to oblong, 2–3 cm long, sharply sinuate-dentate, glabrous above, white-mealy beneath; glomerules usually in dense axillary spikes; calyx-lobes not keeled or mealy, obovate; pericarp free; seeds 0.9–1.1 mm long, 0.8–1.1 mm wide, vertical and horizontal, dark brown, shining. $2n = 18$. *C. glaucum* L. ssp. *salinum* (Standl.) Aellen of Ed. 1. Bare saline soil. swMack to wOnt s to seBC, Calif, Nev, NM, Kans; Alas, Yuk. Map 469 ●

C. subglabrum (S. Wats.) A. Nels.
Annual, glabrous to sparsely farinose in the inflorescence; stems 1–5 (6) dm tall, erect or semi-erect with many ascending branches, light green; leaves 1–5 cm long, 1–4 mm wide, linear, entire, 1-veined, fleshy; inflorescence open, leafy, glomerules small, widely spaced; calyx-lobes covering the fruit; pericarp non-adherent; seeds 1.3–1.6 mm long, 1.2–1.4 mm wide. $2n = 18$. Sand-dunes. Alta to swMan s to Ore, Wyo, SD, Mo. Map 469 ○

C. watsonii A. Nels.
Annual, ill-smelling; stems 2–15 cm tall, angular, much branched, erect or ascending, sparsely to densely farinose; leaves 1.5–3.5 cm long, broadly rhombic-ovate, rounded or obtuse at the apex, base cuneate, entire, thick, densely farinose above and below; inflorescence leafy, glomerules large, in short dense paniculate spikes; calyx-lobes ovate or oblong, densely farinose, enclosing fruit at maturity; pericarp adherent; seeds 1.0–1.3 mm long, 0.9–1.3 mm wide; testa coarsely rugulate. $2n = 18$. Open areas, mainly in the bad-lands. Alta, swSask s to Mont, NM, Ariz. Map 468 ○

Corispermum L. / Bugseed

Annual herb with spreading branches and narrow leaves, hairy or nearly glabrous; stems 1–6 dm tall, stiffish, green or reddish, much branched, often divaricate from near the base; leaves alternate, sessile, linear, entire, 1-nerved; flowers perfect, in the axils of more or less leaf-like bracts, arranged in a terminal spike, the bracts scarious-margined; sepals 1–3, usually 1, minute, scarious; stamens 1–3, rarely 5; ovary flattened, erect; styles 2, short, persistent; fruit oval, planoconvex, the pericarp firmly attached to the enclosed seed.

Spikes dense, 4–6 mm wide; lower bracts usually wider than fruit, which is 3–4 mm long **C. hyssopifolium**

Spikes open, less than 4 mm wide; lower bracts narrower than fruit, which is 2–3 mm long **C. nitidum**

C. hyssopifolium L.
Stems glabrous or somewhat stellate-villose; leaves 1–7 cm long, 1–3 mm wide, glabrous or hairy; bracts 5–9 mm long, ovate or lanceolate; fruit 3–4 mm long, 2.5 mm wide, narrowly to broadly winged. $2n = 18$. Includes *C. orientale* Lam. var. *emarginatum* (Rydb.) Macbr. of Ed. 1. Eurasia: Alas, swMack, BC to Que s to Ariz, Colo, Tex, Kan, Mo. NY. Possibly introduced. Map 470 ●

C. nitidum Kit.
Similar to *C. hyssopifolium*, differing in the characters of the key and the shorter bracts, rarely over 5 mm long. $2n = 18$. Sandy banks and dunes. Eurasia: BC to Ont s to Ariz, Tex, Mo, Ind. Possibly introduced. Map 470 ○

Eurotia Adans.

E. lanata (Pursh) Moq. / Winter Fat
Low perennial, 2–8 dm high, woody at base, with stiff, erect branches, white or reddish brown tomentose; leaves alternate, sessile or short-petioled, linear, with entire, revolute margins, densely stellate-pubescent mixed with long hairs, white to reddish brown; flowers mostly in axillary clusters, the staminate uppermost in the plant; staminate flowers ebracteolate; calyx 4-parted; stamens 4; pistillate flowers with 2 enclosing bracteoles, united into a villose 2-horned tube; fruiting bracteoles 4–6 mm and horns about 1 mm long; ovary hairy, stigmas 2, elongate; seeds obovate. $2n = 18$. *Krascheninnikovia lanata* (Pursh) A. Meeuse & Smit. A valuable range plant. Prairies and foothills. eWash to swMan s to Calif, Colo. Map 471

Kochia Roth

K. scoparia (L.) Schrad. / Summer Cypress; Burning Bush
A bushy annual herb, turning reddish purple in autumn; branches numerous, erect or ascending, pubescent or glabrous; leaves alternate; linear, or narrowly lanceolate, entire, narrowed to a short petiole, usually hairy; flowers solitary or in small clusters, in the leaf-axils, perfect or pistillate; bracts linear-lanceolate, attenuate, acute or acuminate, greenish; calyx 5-cleft, slightly hairy, persistent, each sepal with a short, wedge-shaped, scarious wing in fruit; stamens 3–5; ovary ovoid; stigmas 2; pericarp free from the seed. $2n = 18$. An ornamental that has escaped from cultivation and become a noxious weed. Introduced. Common.

Monolepis Schrad.

M. nuttalliana (Schultes) Greene / Spear-leaved Goosefoot
Low branching annual, mostly glabrous; stem 1–3 dm long, prostrate or ascending, with spreading branches, somewhat fleshy and often reddish; leaves long-stalked or nearly sessile, the blades thick, 1–5 cm long, lanceolate with a pair of divergent lobes near the cuneate base and sometimes a few teeth above; flowers in axillary clusters; sepals 1, green, bract-like; stamen 1; stigmas 2; pericarp pitted when dry, more or less adherent to the flattened seed. $2n = 18$. Saline soils, also a common weed of cultivated land and roadsides. sBC to sMan s to Calif, NM, Mo, introduced elsewhere. Map 472

Salicornia L.

S. europaea L. / Samphire
A very succulent annual, 5–25 cm high, with jointed stems, opposite branches and scale-like leaves, the whole plant becoming bright red at maturity; stem branches erect or ascending; scale-like leaves inconspicuous, broadly triangular, wider than long; flowering spikes 1–5 cm long; flowers perfect or pistillate, in groups of 3 and deeply sunk in the axis of the spike, the middle flower higher than the others; calyx fleshy, not lobed, opening by a small slit; stamens 1 or 2; styles 2; fruit minutely pubescent, enclosed in the spongy calyx; seed about 1 mm long. $2n = 18$. *S. rubra* A. Nels. of Ed. 1. In our area ssp. **rubra** (A. Nels.) Breitung. Saline lake shores and sloughs. Circumpolar: sAlas, swYuk, sMack to Huds Bay, Nfld s to Calif, Nev, NM, Kans, Ill, NY, Ga. Map 473 •

Salsola L.

S. kali L. / Russian Thistle
Bushy, spiny annual, 1–8 dm high; often breaking off at the ground level and blown about as a 'tumbleweed'; stems glabrous or pubescent; lower leaves thread-like, 2–6 cm long, about 1 mm wide; upper leaves shorter, awl-shaped, spine-tipped, becoming hard and stiff; flowers usually solitary in the leaf-axils, with a pair of awl-shaped, spinose bracts; calyx 5-parted, becoming winged or keeled on the back; stamens mostly 5; styles 2; fruit flattened and enclosed by the calyx. $2n = 18$. A weed of dry sandy ground. Introduced. Common.

Sarcobatus Nees

S. vermiculatus (Hook.) Torr. / Greasewood
Erect spiny shrub, 3–30 dm high; stems widely branched, the younger branches yellowish white, glabrous or white-hairy; leaves fleshy, narrowly linear, pale green, 1–4 cm long; staminate flowers in terminal catkin-like spikes, each flower consisting of 3 stamens and at the base of a peltate scarious bract; pistillate flowers solitary in leaf-axils, the calyx united with the ovary and forming a horizontal wing; stigmas 2, recurved; fruiting calyx about 6 mm long, with a

broad, membranous wing. $2n = 36$. Saline sloughs and flats. sBC to sSask s to Calif, Tex. Map 474

Suaeda Forsk.

Herbs or shrubs; glabrous to pubescent; leaves alternate, fleshy; flowers small, perfect (ours) or imperfect, solitary or in glomerules, axillary or in spikes, each flower usually with 2 small membranous bracts; calyx 5-lobed, fleshy; stamens 5; styles 2 (3–5), fruit mostly horizontal, enclosed by calyx.

Annual; calyx-lobes with one or more larger than the others, hooded and keeled; pericarp easy to remove **S. calceoliformis**

Perennial; calyx-lobes of equal size, rounded, not hooded or keeled; pericarp difficult to remove **S. moquinii**

S. calceoliformis (Hook.) Moq.
Erect or spreading, somewhat fleshy glabrous, annual; stems up to 5 dm tall, usually much branched at base; leaves 1–4 cm long, crowded, fleshy, narrowly linear, subterete, enlarged at base; flowers greenish, solitary or clustered in the axils of bracts on the upper branches; calyx-lobes unequal, one or more becoming keeled or crested; seeds black, smooth or slightly reticulate, 1.0–1.7 mm wide; plants usually greenish, occasionally purplish, turning almost black at maturity. $2n = 54$*. *S. depressa* (Pursh) S. Wats. of Ed. 1. Saline flats and slough borders. sYuk, Mack, sSask, swMan, wMinn s to Calif, Nev, Mont, Tex, Mo; James Bay, St Lawrence, NS, Nfld. Map 475

S. moquinii (Torr.) Greene
Erect or ascending perennial, branching from a woody base; stem up to 7 dm tall, glabrous or puberulent; leaves 1–2 cm long, crowded, linear, flat, occasionally subterete, narrow at base; flowers solitary or clustered in the axils of bracts on upper branches; calyx-lobes equal, rounded; seeds black, smooth or slightly reticulate, 1.5–2.0 mm wide. $2n = 18$*. *S. intermedia* S. Wats. Moist saline or alkaline soil. eWash to sAlta, sSask s to Calif, NM, Tex. Map 473 ○

Suckleya A. Gray

S. suckleyana (Torr.) Rydb.
Low spreading annual; stems 2–4 dm long, nearly prostrate, with ascending branches, often reddish, glabrous or somewhat scurfy; leaves petioled, blades circular or rhombic-ovate, 1–3 cm broad, sinuate-dentate, cuneate at base, glabrate; flowers in axillary clusters, the staminate with a lobed calyx, the pistillate with 2 bracteoles but no calyx; stamens 3 or 4; fruiting bracteoles 5–6 mm broad, flattened, with toothed ridges or wings. $2n = 18$. Saline shores. sAlta to sSask s to eMont, Colo, SD. Map 476

AMARANTHACEAE / Amaranth Family

Amaranthus L.

Annual herbs, erect, ascending or prostrate; leaves alternate, petiolate, entire or rarely undulate; plants monoecious or dioecious; flowers small, green or purplish, 3-bracted, in small axillary clusters or aggregated in axillary and terminal spiked clusters; corolla wanting; calyx segments 2–5, distinct; stamens 2–5, opposite the sepals; ovary superior, 1-loculed; stigmas 2 or 3; fruit an ovoid 1-seeded capsule, 2–3-beaked, opening transversely (circumscissile); seeds flattened, smooth, the embryo annular.

1 Leaf-blades 0.5–3 (4) cm long; plant much branched from base; flowers all in axillary clusters 2
Leaf-blades 4–15 cm long; plant usually not branched at base, flowers in terminal or axillary simple or paniculate spikes 4

2 Sepals 4–5; stamens 3–4; seeds about 1.5 mm wide **A. graecizans**
Sepals mostly 3; stamens 1–3; seeds usually less than 1 mm wide 3

3 Bracts 2–3 mm long; pistillate sepals all about same size **A. albus**
Bracts less than 2 mm long; pistillate sepals of unequal size **A. californicus**

4 Upper stem coarsely villous; leaves hairy below, at least along veins; sepals clearly longer than ovary, midvein green **A. retroflexus**
Upper stem glabrous to sparsely pubescent; leaves generally glabrous below; sepals shorter than ovary, midvein usually not green **A. powellii**

A. albus L. / Tumbleweed
Stems whitish, rigid, erect or ascending, diffusely branched, the plant bushy or the lower branches prostrate, 2–6 dm high, glabrous or sparingly hairy; leaves mostly 1–4 cm long, slender-petioled, oblong to obovate, veiny, obtuse to rounded at apex, often mucronate; flowers in short axillary clusters; bracts awl-shaped, 2–3-times as long as the flowers; sepals commonly 3; fruit wrinkled when dry, longer than the calyx; seeds lens-shaped, 0.7–1 mm wide. $2n$ = 32, 34. Fields and waste ground. Eurasia: BC to NS, generally distributed s. Map 477

A californicus (Moq.) S. Wats.
Annual, prostrate, mat-forming; stems often reddish; leaves rhombic-elliptic to obovate, 5–15 mm long, with margin undulate or crisped, the base abruptly narrowed to a slender petiole 5–15 mm long; monoecious; flowers in small axillary clusters; bracts lanceolate, 1.0–1.5 mm long; sepals 3, those of staminate flowers equal, above as long as bracts, acute, the pistillate sepals with 2 reduced or absent; stamens 1–2 (3); seeds about 0.7 mm across. Lake shores, waste ground, roadsides. eWash to swSask s to Calif, Nev, Mont. Perhaps introduced in our area. Map 478

A. graecizans L. / Prostrate Amaranth
Stems prostrate, much branched, 3–7 dm long, often purplish; leaves spatulate-oblong to obovate, 1–4 cm long, tapering into a slender petiole; flowers in dense axillary clusters; bracts ovate-oblong, short-acuminate, about as long as the flowers; sepals usually 5, unequal in length; fruit scarcely wrinkled, about equalling the longest sepal; seeds 1.3–1.5 mm wide. $2n = 32$. A weed of disturbed ground. Eurasia: BC to Man s to Calif, Tex, introduced elsewhere. Map 479

A. powellii S. Wats.
Similar to *A. retroflexus*, differing in the characters of the key; in addition, plants monoecious; the inflorescence narrower and not so dense; sepals acute to obtuse mostly with an excurrent midvein; stamens 3 (4). $2n = 34$. Waste ground. Introduced. Rare.

A. retroflexus L. / Red-root Pigweed
Stout annual with a reddish root; stems erect or strongly ascending, 3–10 dm tall, more or less branched, usually rough and pubescent; leaves ovate to rhombic-ovate, 4–10 cm long, with entire to undulate margins, narrowed at the base and long-petioled; dioecious; flowers in stout, dense, erect spikes in a terminal cluster and in the axils of the upper leaves; bracts lanceolate, rigid, awn-tipped, twice as long as the scarious sepals; sepals mostly spatulate, apex rounded or emarginate, often mucronate; stamens 5 (4); fruit 1.5–2 mm long; seeds about 1 mm wide. $2n = 34$. Gardens and waste places. Introduced. Common.

NYCTAGINACEAE

Herbs, shrubs, trees; leaves entire, mostly opposite, exstipulate; flowers regular, hypogynous, bracteate; commonly in cymes, sometimes racemes or umbellate; bracts often forming a calyx-like involucre, sometimes petaloid; sepals 5, petaloid, fused below; petals absent; stamens 1–many; ovary 1-loculed, solitary; fruit an achene, often closely enveloped by the base of the persistent calyx.

Bracts of involucre free; flowers sessile; fruit conspicuously winged **Abronia**
Bracts of involucre fused; flowers pedicellate; fruit not winged **Mirabilis**

Abronia Juss. / Sand Verbena

A. micrantha Torr.
Annual, decumbent, trailing; branches up to 6 dm long, the tips ascending; leaves opposite, blades 2–6 cm long, 1–3 cm wide, elliptic to ovate or ovate-rhombic; flowers greenish white, in umbellate clusters, peduncles shorter than leaves; bracts 5–10 mm long, lanceolate; calyx 10–12 mm long; stamens 5; fruit mostly about 2 cm long, the papery wings strongly veined. Sandy ground. Alta s to eMont, Utah, Nev, NM, wKan, nTex. Map 480 ○

Mirabilis L.

Perennial herbs, 2–10 dm tall, sometimes woody at base; stems usually swollen at the joints; leaves opposite; flowers perfect, regular, borne singly or in a cluster of 2–4, subtended by a calyx-like involucre of united bracts; calyx corolla-like, gamosepalous, cup-shaped or funnelform, 5-lobed, the short tube closely surrounding the ovary and constricted above it; stamens 3–5, unequal; ovary enclosed by the tube of the persistent calyx; stigma capitate; fruit and the enveloping calyx-tube 5-angled or 5-ribbed; involucre rotate, becoming much enlarged and membranous.

Leaves mostly linear-lanceolate, sessile or only the lower short-petioled; inflorescence more or less glandular-pubescent **M. hirsuta**

Leaves ovate and cordate or deltoid, distinctly petioled; inflorescence not glandular **M. nyctaginea**

M. hirsuta (Pursh) MacM. / Umbrellawort
Stems 2–8 dm tall, solitary or few, erect or decumbent, densely hirsute or pilose to nearly glabrous, but always hirsute near the nodes and glandular-hairy in the inflorescence; leaves sessile or the lower ones short-petioled, the blades linear-lanceolate to oblong or ovate-lanceolate, cuneate to subcordate at base, densely hirsute to viscid-puberulent or glabrate; inflorescence both axillary and cymose-paniculate; involucres commonly 3-flowered, viscid-hairy; calyx pinkish, 8–10 mm long; fruit 4–5 mm long, 5-angled, densely pubescent, the sides rather rugose. *Oxybaphus hirsutus* (Pursh) Sweet. *Allionia hirsuta* Pursh. Includes *A. pilosa* (Nutt.) Rydb. Dry open ground, especially sandy prairie and hills. sBC to sMan s to Ariz, Tex, Mo. Map 480 •

M. nyctaginea (Michx.) MacM. / Umbrellawort
Stem 3–10 dm tall, glabrous or nearly so below, slightly pubescent above, becoming repeatedly forked; leaves with petioles 1–3 cm long, the blades broadly ovate, usually cordate or truncate at base, acute or acuminate at apex; inflorescence axillary or cymose; involucres usually 3-flowered, saucer-shaped, about 1 cm broad in fruit, pilose on the margins; calyx pink or purplish, 8–10 mm long; fruit narrowly obovoid, 5-angled, 4–5 mm long, densely hairy, rough on the sides. $2n = 58$. *Oxybaphus nyctagineus* (Michx.) Sweet. Dry plains, roadsides, and railway embankments. sBC to sMan, Wisc s to Colo, Tex, La, Ala. Map 481

PORTULACACEAE / Purslane Family

Annual or perennial herbs, mostly small and succulent; leaves simple, entire, opposite or alternate, often basal; flowers perfect, actinomorphic; sepals 2, separate or united at base, often caducous; corolla of 4–6 petals (or 3 in *Montia*), distinct or united basally; stamens 2–5 or more, borne on the base of the petals; pistil composed of 2–3 carpels, the ovary superior (partly inferior in *Portulaca*), 1-loculed; styles and stigmas 2–8; fruit a capsule with 2–many seeds.

1 Plants annual, weedy, generally prostrate; flowers yellow; ovary partly inferior **Portulaca**
Plants not as above 2

2 Annual or perennial with fibrous roots or rhizome; cauline leaves several, alternate **Montia**
Perennial with taproot or tuber-like corm; cauline leaves generally 2, opposite 3

3 Stamens 4–12; stigmas 3–6; capsule generally dehiscing transversely (circumscissile) **Lewisia**
Stamens 5 (6); stigmas 3; capsule dehiscing vertically (valvate) **Claytonia**

Claytonia L. / Spring Beauty

Low perennial herbs, usually succulent and glabrous; cauline leaves 2, opposite or subopposite, often sessile; basal leaves none, one or many, petioled; flowers in loose terminal racemes, white or pinkish; sepals 2, persistent; petals 5; stamens 5; style 3-cleft; capsule 3-valved, 3–6-seeded.

Plant with few or no basal leaves and a deep-seated rounded corm; cauline leaves lanceolate **C. lanceolata**
Plant with a dense cluster of basal leaves, a thick caudex and fleshy root; basal leaves spatulate or obovate **C. megarhiza**

C. lanceolata Pursh / Western Spring Beauty
Stems 5–20 cm tall, from globose corms 1–2 cm thick; basal leaves 1–2, sometimes absent, petioled, oblanceolate to ovate; stem leaves 2, opposite, linear to oblong-lanceolate, sessile or narrowed into a short petiole; flowers 3–15 in a short raceme; sepals ovate; petals 6–12 mm long, rose-coloured with purple lines or whitish; pedicels recurved in fruit; capsule 4 mm high; seeds black, shining, 2 mm long. $2n$ = 16, 24, 32, 48, 64, 74, c.90. Moist woods, meadows, and banks. BC, Alta s to Calif, NM. Map 482

C. megarhiza (A. Gray) Parry
Plant with a thick fleshy taproot and stout caudex, not over 1 dm high; basal leaves numerous, fleshy, spatulate to obovate, 1–8 cm long including the winged petiole; flowering stalks numerous, almost as long as the leaves; sepals 4–7 mm long; petals 6–8 mm long, white or pink; capsule 4–6 mm long; leaves and stems usually tinged with red. $2n$ = 32, 34*, 36. Rock crevices and talus at alpine elevations. wMack; BC, swAlta s to Nev, Utah, NM; wMack. Map 483

Lewisia Pursh / Bitter-root

L. pygmaea (A. Gray) Robins.
Perennial herb with short caudex and fleshy taproot; leaves mostly basal, spatulate to narrowly linear, the bases scarious-dilated; flowering scapes several, 1–3-flowered, usually shorter than or equalling the leaves; sepals obtuse, glan-

dular-denticulate; petals 6–8, pink or rose-red; capsule circumscissile near the base, then splitting upwards; seeds 10–20. $2n$ = c.66. Gravelly and rocky slopes at montane to alpine elevations. sYuk, BC, swAlta s to Calif, NM. Map 484

Montia L.

Plants annual, or perennial by slender rootstocks; cauline leaves alternate (in the species below), more than 2, thin or fleshy; flowers 1–10, loosely racemose at the ends of the branches; sepals 2, persistent; petals 5, pink or white; style 3-parted above; capsule 3-valved, 1–3-seeded, the valves spreading in age.

Annual with fibrous roots; leaves thin, linear; flowers nodding **M. linearis**
Perennial with thick caudex and slender rootstocks; leaves thick, the basal ones obovate or spatulate; flowers nearly erect **M. parvifolia**

M. linearis (Dougl.) Greene
Annual, erect or decumbent, 5–25 cm high; stems branched at or near the base; leaves linear of filiform, 1–4 cm long, alternate, scarious at the base; racemes several, 1-sided, loosely flowered, the pedicels recurved; petals white, as long as or exceeding the sepals; stamens 3; seeds black, shining, 2 mm wide. *Claytonia linearis* Dougl. *Montiastrum lineare* (Dougl.) Rydb. Moist ground, plains, and hills. BC, Alta s to Calif, Utah, nMont; swSask. Map 485

M. parvifolia (Moc.) Greene
Perennial with a short caudex and several very slender stems 5–20 cm long, erect or decumbent; often producing axillary bulblets; basal leaves several, petioled, spatulate or rhombic-ovate, fleshy; cauline leaves much smaller, narrow, alternate, reduced upward into bracts; racemes few-flowered; calyx 1.5–3 mm long; corolla 6–10 mm long, rose-coloured or white; seeds black, minutely roughened. $2n = 22$*. *Claytonia parvifolia* Moc., *Naiocrene parvifolia* (Moc.) Rydb. Moist montane slopes. Alas, swYuk, BC, swAlta s to Calif, Utah. Map 486

Portulaca L.

P. oleracea L. / Purslane
Mat-forming annual weed; stems prostrate, fleshy, glabrous, usually purplish red, repeatedly branched; leaves wedge-shaped to obovate, 1–3 cm long, mostly alternate, occasionally opposite; flowers sessile, solitary or in small terminal clusters; calyx 2-cleft; petals 4–6, yellow, opening only in the sunshine; stamens 7–12; style 4–6-parted; ovary partly inferior; capsule conical, 4–9 mm long, circumscissile near the middle, many-seeded; seeds black, shining, about 0.8 mm broad, tuberculate, with a white scar at one end. $2n = 36, 54$. Persistent weed of gardens and waste ground. Introduced. Common.

CARYOPHYLLACEAE / Pink Family

Annual or perennial herbs; stems usually swollen at the nodes; leaves entire, mostly opposite, often united at the base; flowers regular, hypogynous or perigynous, commonly in cymes; sepals 4 or 5, separate or united, persistent; petals 4 or 5 (or rarely wanting), often with a spreading blade and a distinct claw; stamens distinct, commonly twice as many as the petals; ovary usually 1-loculed, with free-central placentation; styles 2–5, or rarely united into one; fruit a capsule with few to many seeds or rarely a one-seeded utricle; capsule dehiscent, at least at the apex, by valves or teeth of the same number or twice the number of the styles; seeds reniform.

1	Fruit 1-seeded, indehiscent; petals absent	2
	Fruit several-seeded, dehiscent; petals usually present	3
2	Stipules present, hyaline; calyx-lobes much longer than the tube	**Paronychia**
	Stipules absent; calyx-lobes about as long as tube	**Scleranthus**
3	Calyx of united sepals; petals clawed; ovary stipitate	4
	Calyx of separate sepals; petals lacking claws; ovary not stipitate	8
4	Styles 2	5
	Styles 3–5	**Silene**
5	Calyx with sheathing bracts at base	**Dianthus**
	Calyx lacking sheathing bracts at base	6
6	Flowers generally less than 1 cm long	**Gypsophila**
	Flowers more than 2 cm long	7
7	Petals appendaged; calyx tubular	**Saponaria**
	Petals not appendaged; calyx ovoid, strongly winged	**Vaccaria**
8	Leaves with scarious stipules	9
	Leaves without stipules	10
9	Leaves opposite, but often with secondary fascicles of leaves; styles and valves of capsule 3	**Spergularia**
	Leaves whorled; styles and valves of capsule 5	**Spergula**
10	Corolla present	11
	Corolla absent	16
11	Petals entire or emarginate	12
	Petals deeply notched or 2-cleft	15
12	Styles 3	13
	Styles 4 or 5	**Sagina**
13	Leaves elliptic to oblong-lanceolate, 1–4 cm long, generally more than 3 mm wide; seeds with a fleshy appendage (strophiole)	**Moehringia**
	Leaves narrowly linear-lanceolate or subulate, if ovate to elliptic less than 1 cm long, less than 3 mm wide; seeds not strophiolate	14

14 Capsule dehiscing by 6 valves or teeth **Arenaria**
Capsule dehiscing by 3 entire valves **Minuartia**

15 Styles usually 3; capsule ovoid or ellipsoid, dehiscing by valves **Stellaria**
Styles usually 5; capsule cylindric, dehiscing by teeth (10) **Cerastium**

16 Plants caespitose; leaves 3–10 mm long, subulate, generally with axillary fascicles; flowers solitary; alpine **Minuartia**
Plants not as above **Stellaria**

Arenaria L. / Sandwort

Low, annual or perennial herbs, usually tufted, sometimes mat-forming; leaves without stipules, mostly sessile and narrow; flowers small, solitary or in terminal cymes; sepals 5; petals mostly 5, entire or slightly emarginate, usually white; stamens normally 10; styles 3, rarely 2–5; ovary 1-loculed; capsule globose or short-oblong, few–many-seeded; valves of capsule twice as many as styles.

1 Leaves less than 1 cm long, about twice as long as broad 2
Leaves more than 1 cm long, needle-like 3

2 Stem short, 2–4 cm high, with 1–2 pairs of leaves, flowers solitary **A. longipedunculata**
Stem taller, most 3–20 cm high, stem-leaves numerous, flowers several in a branching inflorescence **A. serpyllifolia**

3 Flowering stems glandular-puberulent above; inflorescence open when mature, more or less forked; sepals 3–4 mm long **A. capillaris**
Flowering stems glabrous throughout; inflorescence more or less congested and umbellate; sepals 4–5 mm long **A. congesta**

A. capillaris Poir.

Plant densely tufted, 1–2 dm high, with a somewhat woody caudex; flowering stems glabrous below, glandular-puberulent above; leaves linear, erect and often curved, minutely ciliolate; inflorescence open (at maturity), more or less forked; sepals 3–4 mm long, broadly scarious-margined, 1-nerved, obtuse or acutish; petals 5–7 mm long; capsule about twice as long as the sepals, the valves 2-cleft; seed 2–2.5 mm long. $2n = 22$*. In our area var. **americana** (Maguire) Davis. Mountain meadows and open slopes. eAsia: Alas, Yuk, wMack; BC, Alta s to Ore, nNev, Mont. Map 487 ●

A. congesta Nutt.

Plant loosely matted, 1–3 dm high, with a somewhat woody caudex; flowering stems simple, with 2–4 pairs of leaves, glabrous; leaves 1–6 cm long, filiform-linear, sharp-pointed, glabrous, erect or ascending, inflorescence of one or more rather dense or somewhat loose and irregular cymes; sepals 4–5 mm long, scarious-margined, 1-nerved, obtuse or acutish, glabrous; petals and capsule longer than the sepals; capsule-valves 2-cleft. $2n = 22$. In our area var. **lithophila** (Rydb.) Maguire. Dry plains and slopes. Alta, swSask s to Calif, Ida, Utah, Colo. Map 487 ○

A. longipedunculata Hult.
Matted, with filiform, subterranean, light-coloured runners; leaves lanceolate to ovate-lanceolate, blunt or somewhat acute, slightly ciliate at base; flowering stems thick, erect, glandular-puberulent, 1-flowered, with pair of ovate, acute bracts below middle; sepals ovate, obtuse, or somewhat acute, glabrous or glandular-puberulent, indistinctly 3-nerved, about as long as petals; styles 3; capsule ovate, narrower at tip, slightly longer than calyx; seeds brown, suborbicular, shiny, rugulose, 0.7–0.8 mm long. $2n = 80$. Moist gravelly areas at higher elevations. Often treated as *A. humifusa* Wahl. Alas, Yuk, wMack s to BC, Alta. Map 488

A. serpyllifolia L. / Thyme-leaved Sandwort
Annual, with delicate to wiry slender stems, puberulent, 5–15 cm tall, often branched from the base; leaves usually 8–10 pairs, ovate, 3–6 mm long, acute, 3–5-nerved, sparsely puberulent, sessile or the lowest short-stalked; flowers in open leafy-bracted cymes; pedicels 3–8 mm long; sepals 2–3 mm long, puberulent; petals shorter than the sepals; capsule firm, the valves 2-cleft at apex; seed wrinkled, about 0.5 mm wide, grey-black. $2n = 10, 20, 40$. Dry, gravelly roadsides and waste ground. Introduced. Rare.

Cerastium L. / Mouse-ear Chickweed

Hairy herbs, annual or perennial; flowers in cymes; sepals 5, or rarely 4; petals of the same number as the sepals, 2-cleft at the apex, rarely lacking; stamens 10 or rarely fewer; styles mostly 5, rarely 4 or 3, opposite the sepals; capsule cylindrical, often curved, membranaceous, opening at the apex by twice as many teeth as the style number; seeds rough.

1 Annual, lacking basal offshoots; petals about as long as sepals or slightly longer; capsule 2–3 times longer than sepals **C. nutans**
  Perennial, with basal offshoots; petals much longer than sepals or somewhat shorter (sometimes about as long); capsule 1–2 times longer than sepals 2

2 Petals equalling or shorter than sepals **C. vulgatum**
  Petals much longer than sepals 3

3 Leaves linear to linear-lanceolate, the upper with axillary fascicles; stems erect or ascending; petals twice as long as sepals **C. arvense**
  Leaves oblong to lanceolate or spatulate, lacking axillary fascicles; stems often spreading or prostrate; petals usually less than twice as long as sepals **C. beeringianum**

C. arvense L.
Tufted or matted perennial, 1–3 dm tall, the stems branched above, densely villose to glabrous, glandular or non-glandular; leaves 1–4 cm long, linear-subulate to lanceolate or narrowly ovate, usually hairy, sometimes glandular; primary leaves usually with conspicuous axillary fascicles or short sterile shoots; inflorescence few–many-flowered, the pedicels slender, erect; bracts scarious-margined; sepals 5–7 mm long; petals white, twice or thrice length of sepals, the lobes broad and spreading; capsule cylindric, equalling to twice the length

of the calyx; seeds reddish brown, obovoid, 0.5–1 mm long, tuberculate. $2n = 36$, 72. An extremely variable species. Dry ground of plains, hills, and mountains. Circumpolar: Alas, Yuk, wMack, Huds Bay to nQue, Nfld s to Calif, NM, Neb, Minn, Ga. Map 489

C. beeringianum Cham. & Schlecht.
Similar to *C. arvense* but of lower stature and the flowers fewer; main stems 0.5–2.5 dm long, spreading or ascending, usually short-hairy, glandular and viscid; leaves of flowering stems commonly oblong to lanceolate or spatulate, 1–2 cm long, mostly obtuse; inflorescence with 1–few flowers; bracts not scarious; sepals 3.5–6 mm long, the inner scarious-margined and often tinged with purple; petals 6–8 mm long; capsule about twice as long as the calyx. $2n = 72$. Alpine slopes and ledges. Eurasia: Alas, Yuk to Ellesmere Is to Nfld s to Calif, Ariz, Utah, Colo, nSask, Huds Bay, nQue; Lk Superior. Map 490

C. nutans Raf.
Annual, with stems 5–25 (30) cm tall, tufted, erect, glandular-hairy; lower leaves commonly oblanceolate, 5–25 (40) mm long, short-petiolate; upper leaves linear or linear-oblong, sessile; inflorescence leafy, open; pedicels slender, 5–20 mm long, usually reflexed near tip; sepals about 4 mm long, broadly lanceolate; petals 4–6 mm long, rarely absent; capsule 2–3 times as long as sepals. $2n = 36$. Moist ground and woodlands. BC, swMack, cMan to swQue s to Ariz, Tex, Fla. Map 491

C. vulgatum L.
Short-lived perennial; stems 1–4 dm long, usually tufted, erect or ascending, hirsute or glandular and viscid; lower leaves oblong to ovate-lanceolate, hirsute; upper leaves oblong-lanceolate, 1–2 cm long; bracts similar but smaller and often scarious-margined; mature inflorescence rather open; sepals 4–6 mm long, scarious-margined; petals narrow, about as long as the sepals; capsule 7–12 mm long, sometimes curved; seeds reddish brown, angular-obovoid, about 0.7 mm long, wrinkled-papillate. $2n = 144$. A weed of fields, roadsides, and waste ground. Introduced. Common.

Dianthus L. / Pink

Herbs, mostly perennial; leaves generally linear; exstipulate; flowers solitary or in a cyme, perfect or imperfect; calyx gamosepalous, tubular, 20–40-nerved, invested at the base by 1–several pairs of imbricate bracts; petals present, clawed, lacking appendages; stamens 10; styles 2; capsule dehiscing by 4 valves; seeds numerous.

1 Flowers in clusters; subtending bracts about as long as calyx; inflorescence contracted **D. barbatus**
  Flowers not in clusters; subtending bracts shorter than calyx; inflorescence open 2

2 Stems pubescent **D. deltoides**
  Stems glabrous 3

3	Petals laciniate	**D. plumarius**
	Petals dentate	**D. seguieri**

D. barbatus L. / Sweet William
Perennial, glabrous; stems 2–6 dm tall, erect, simple or branched above; leaves 4–9 cm long, 1–2 cm wide, oblong-lanceolate to lanceolate, sessile; flowers in dense cymose clusters; calyx about 15 mm long, about 40-nerved; petals red, pink or white, variously spotted and barred in some forms, claw about as long as calyx, blade 6–10 mm long, broadly obovate, dentate; capsule shorter than calyx. $2n = 30$. An escape from cultivation. Introduced. Occasional.

D. deltoides L. / Maiden Pink
Perennial; stems several, 1–4 dm tall, from a slender branched caudex, decumbent at base ascending to erect, densely puberulent; leaves 10–25 mm long, linear to narrowly oblanceolate, entire, margins stiff hairy; flowers mostly solitary or paired, in any open cyme; calyx 12–17 (20) mm long, about 30-nerved; petals generally pink to rose, claw about as long as calyx, blade 6–10 mm long, often pilose above, obovate, irregularly toothed; capsule as long as or slightly longer than calyx. $2n = 30$. Roadsides, probably an escape from cultivation. Introduced. Rare.

D. plumarius L. / Garden Pink
Perennial, generally glaucous; stems simple, 1–4 dm tall; leaves 2–8 cm long, 1.5–3.0 mm wide, linear, 3-nerved; inflorescence open, 1–(5)-flowered; calyx 15–20 mm long, 40-nerved; petals white or pink, blade 1–2 cm long, obovate, divided to about the middle, hairy above; capsule slightly longer than calyx. $2n = 30, 60, 90$. An escape from cultivation. Introduced. Rare.

D. seguieri Vill.
Perennial; stems simple or branched, 1–6 dm tall; leaves 2–8 cm long, 1–4 mm wide, linear, flat, 3-nerved; inflorescence open, generally several-flowered; calyx 15–20 mm long, about 40-nerved; petals pink, blade 1–2 cm long, cuneate, apex dentate, hairy above. $2n = 90$. Roadsides, probably an escape from cultivation. Introduced. Rare.

Gypsophila L.

Annual or perennial herbs; inflorescence a dichasial cyme, open or contracted; leaves exstipulate; calyx gamosepalous, membranous between the lobes; petals exceeding calyx, not conspicuously clawed, lacking appendages; stamens 10; styles 2; capsule 4(6)-valved.

1	Plants annual; petals 8–10 mm long	**G. elegans**
	Plants perennial; petals shorter	2
2	Inflorescence glandular hairy; calyx-teeth 3–4 mm long	**G. scorzonerifolia**
	Inflorescence glabrous; calyx-teeth 1.5–2.0 mm long	**G. paniculata**

G. elegans Bieb.
Annual, glabrous; stem 2–5 dm tall, branched above; leaves 2–7 cm long, 0.3–1.0 cm wide, linear-lanceolate, sessile, entire, obscurely 1–3-veined; inflorescence open; calyx 3–5 mm long, petals (6) 8–10 mm long, white or pink; seeds with obtuse tubercles. $2n = 34$. Ditches and roadsides. Introduced. An escape from cultivation. Occasional.

G. paniculata L. / Baby's-breath
Perennial, glaucous; stem 4–10 (12) dm tall, diffusely branched, with scant foliage when in bloom; leaves 2–10 cm long, up to 1 cm wide, linear to lanceolate, entire, 1-veined; inflorescence open, flowers numerous; calyx about 2 mm long; petals slightly longer, white; seeds with obtuse tubercles. $2n = 34$. Waste ground and roadsides. Introduced. An escape from cultivation. Occasional.

G. scorzonerifolia Ser.
Perennial, glaucous; stem 3–10 dm tall, glabrous below, branched and glandular hairy above; leaves 2–8 cm long, 1.0–3.5 cm wide, ovate to oblong-lanceolate, sessile, entire, obscurely 1–3-veined, glabrous; inflorescence open, calyx 3–4 mm long, petals mostly 4–6 mm long, pink to purple; seeds with an obtuse tubercle. Waste ground. Introduced. Rare.

Minuartia L.

Annual or perennial; small tufted or mat-forming herbs with usually linear or subulate, rarely broader, opposite, exstipulate leaves; flowers small, solitary or in cymes; sepals 5; petals 5, sometimes absent, usually white, entire or slightly emarginate; stamens normally 10; styles 3, rarely more; capsule ovoid or oblong, opening by the same number of teeth as styles; seeds reniform, papillose or spiny.

1 Sepals oblong, rounded to obtuse at apex — 2
Sepals acuminate, acutish or acute — 3

2 Flowering stems 3–8 (10) cm tall; sepals 4–5 mm long; petals obovate or spatulate, usually much longer than sepals — **M. obtusiloba**
Flowering stems 1–2 (3) cm tall; sepals 2–3 (4) mm long; petals narrowly oblong, usually not much longer than sepals — **M. biflora**

3 Plants glabrous — 4
Plants glandular-pubescent — 6

4 Flowers several, in a loose cyme — **M. dawsonensis**
Flowers solitary — 5

5 Sepals purple, weakly 3-nerved; petals as long as sepals, rarely absent — **M. elegans**
Sepals green, strongly 3-nerved; petals shorter than sepals, usually absent — **M. austromontana**

6 Leaves mostly 8–10 mm long, about 1 mm wide, often curved; sepals 4–6 mm long, obscurely 3-nerved **M. nuttallii**

Leaves generally shorter, narrower, not curved; sepals 2–4 mm long, conspicuously 3-nerved **M. rubella**

M. austromontana Wolf & Packer

Plants caespitose; glabrous throughout; leaves 3–6 mm long, somewhat fleshy and keeled, subulate, obtuse, spreading to ascending; small leafy fascicles sometimes present in axils of primary leaves; flowers solitary at the ends of stems; pedicels 5–15 mm long; sepals green, 2–3 mm long, linear to lanceolate, acute, strongly 3-nerved, sometimes scarious-margined; petals usually lacking or, if present, rudimentary, linear to oblong, shorter than sepals; capsule as long as calyx. $2n = 30$*. *Arenaria rossii* (R.Br. *ex* Richards) Graebn. in part, of Ed. 1. Dry alpine slopes. Alta s to Wyo, Utah; neOre. Map 492

M. biflora (L.) Schinz & Thell.

Resembling *M. obtusiloba*, but generally a smaller plant, 2–5 cm high, and differing in the flower characters shown in the key; stems usually minutely puberulent; leaves thin and with an obscure midvein; capsule 3–6 mm long; seed 0.5–0.6 mm broad, smooth or nearly so. $2n = 26$*. *Arenaria sajanensis* Willd. of Ed. 1. Alpine slopes and ridges. More or less circumpolar: Alas to cBaffin Is s to BC, Alta, sMack, Huds Bay, nQue; Colo. Map 493

M. dawsonensis (Britt.) House

Glabrous, loosely tufted plant, often diffuse, with a weak taproot; stems 0.5–2 dm high, filiform, wiry, leafy only at base, much branched above, often more or less spreading; leaves linear or awl-shaped, firm, the midrib stronger than the 2 lateral veins; primary leaves 5–15 mm long, often with secondary leaves fascicled in their axils; inflorescence loosely cymose, the pedicels rigid and ascending or slightly divergent; sepals narrowly ovate, subulate-tipped, glabrous but granular-roughened, hyaline-margined, usually 3-nerved; petals shorter than the sepals; capsule longer than the calyx; seeds oblong, prominently granulate. $2n = 30, 60$. *Arenaria dawsonensis* Britt. of Ed. 1. Moist open woods, clearings, and stream banks. Alas, Yuk, wMack, Huds Bay, nQue, Nfld s to BC, Alta, Sask, Gt Lakes, Gaspé. Map 494 ●

M. elegans (Cham. & Schlecht.) Schischk.

Plants loosely tufted, often purplish; glabrous throughout; leaves 3–10 mm long, linear, obtuse, ascending, small leafy fascicles present in the axils of primary leaves; flowers solitary at the ends of stems; pedicels 1–4 cm long; sepals purple, 2–4 mm long, ovate-lanceolate, acute, somewhat keeled, weakly 3-nerved; petals oblong to obovate, obtuse to emarginate, equal to or shorter than calyx, rarely rudimentary or lacking; capsule as long as calyx. $2n = 30, 60$*. *Arenaria rossii* (R.Br. *ex* Richards) Graebn., in part, of Ed. 1. Moist alpine slopes. Siberia (Chukchi Peninsula): Alas, Yuk, sMack s to BC, Alta. Map 495 ○

M. nuttallii (Pax) Briq.
Coarse perennial, loosely matted, with a strong caudex and taproot; stems decumbent and ascending, 6–12 cm long, glandular-hairy; leaves awl-shaped, 5–12 mm long, 3-nerved, ascending or often recurved-spreading, viscid-pubescent; cymes few–many-flowered; pedicels 5–15 mm long; sepals lanceolate, 4–5 mm long, acuminate; petals shorter than the sepals; capsule shorter than the calyx; seed 1–1.3 mm long, brown-black. $2n = 36$. *Arenaria nuttallii* Pax of Ed. 1. In our area var. **nuttallii**. Dry open areas to alpine elevations. BC, Alta s to Calif, Utah, Wyo. Map 494 ○

M. obtusiloba (Rydb.) House
Densely caespitose and matted, the caudex branched and somewhat woody; stems 3–10 cm long, decumbent to erect, clothed near the base with old leaves, glandular-pubescent; leaves narrowly linear, 4–8 mm long, ascending, obtuse or acutish, with a prominent midvein; flowers solitary or 2–3 on a stem; sepals 4–5 mm long, blunt, 3-nerved, hairy; petals mostly 4–8 mm long, obovate or spatulate; capsule 4–7 mm long, rather firm-walled, the valves entire or merely emarginate at apex; seeds about 0.8 mm broad, indistinctly papillate-tuberculate or nearly smooth. $2n = 26^*$, c.52, 78. *Arenaria obtusiloba* (Rydb.) Fern. of Ed. 1. Dry alpine slopes. Alas, Yuk, nwMack s to Ore, Ida, NM. Map 495 ●

M. rubella (Wahl.) Graebn.
Tufted, mat-forming perennial with slender taproot, more or less glandular-hairy; leaves linear-subulate, flat, 3-ribbed, 3–10 mm long; flowering stems filiform, decumbent at base, 2–18 cm high, 1–7-flowered; pedicels 2–2.5 mm long; sepals 3 mm long, lanceolate, acuminate, with 3 uniform ribs; petals usually shorter than the sepals; capsule 3–5-valved, equalling or exceeding the calyx; seeds reddish brown, rugose, 0.4–0.7 mm broad. $2n = 24^*$. *Arenaria rubella* (Wahl.) J.E. Smith of Ed. 1. Dry, sandy, gravelly, open areas. Circumpolar: Alas to Ellesmere Is, Nfld s to Calif, Nev, NM, nwSask, James Bay, nQue; Gaspé, Vt. Map 496

Moehringia L.

M. lateriflora (L.) Fenzl.
Perennial with slender rhizome, widely creeping; stems slender, decumbent at base, ascending to erect, 5–20 cm tall, densely puberulent to glabrate; leaves in remote pairs, thin, commonly 3–10 mm wide, oval to narrowly oblong, obtuse or acutish; cymes 1–5-flowered, axillary or terminal; pedicels filiform, 5–15 mm long; sepals 2–2.5 mm long, obtuse or acutish, glabrous, faintly 3–5-nerved; petals white (rarely pinkish), 3–8 mm long, obtuse; styles 3; capsule 3–6 mm long, 3-valved, the valves deeply 2-cleft; seed 1–1.3 mm wide, black, shining, with a pale appendage (strophiole) at the hilum. $2n = 48$. *Arenaria lateriflora* L. of Ed. 1. Moist meadows, gravelly shores, thickets, and woods. Circumpolar: Alas, Yuk, wMack, Huds Bay, Nfld s to Calif, NM, Sask, Man, Gt Lakes, Pa. Map 497

Paronychia Mill.

P. sessiliflora Nutt. / Low Whitlow-wort
Low, tufted or matted, perennial with a woody base; branches very densely crowded, spreading, hairy; leaves opposite, imbricated, linear to awl-shaped, 4–6 mm long, erect or recurved-spreading, often acute or mucronate, with large silvery scarious stipules; flowers terminal, solitary (or in pairs), small, with conspicuous silvery scarious bracts; sepals 5, linear-oblong, 2–3 mm long, concave and awned at the apex; petals none; stamens 5; style 2-cleft; fruit an ovoid utricle, 1-seeded, included in the calyx. $2n = 64$. Dry hills and ridges. Alta, Sask s to Utah, Tex. Map 498

Sagina L. / Pearlwort

Diminutive, annual or perennial herbs, tufted or mat-forming; leaves linear to subulate, exstipulate; flowers small, 4–5-merous, solitary or in cymes; petals entire, white, often absent; stamens equal to or twice the number of sepals; styles 4–5, alternating with sepals; capsule 4–5-valved, valves opposite sepals; seeds many.

1 Petals 3–5 mm long, about twice as long as sepals; leaves of the upper main stem generally with leafy fascicles in their axils **S. nodosa**
Petals less than 2.5 mm long, shorter or only slightly longer than sepals; leaves of the upper main stem lacking axillary fascicles 2

2 Flowers 4-merous, sometimes with 5-merous flowers in association **S. nivalis**
Flowers 5-merous 3

3 Perennial, basal rosette generally present; sepal margins white; capsule dehiscing to base **S. saginoides**
Annual, basal rosette generally absent; sepal margins often purple; capsule dehiscing to middle or less **S. decumbens**

S. decumbens (Ell.) T. & G.
Annual; branches 5–20 cm long, slender, ascending or decumbent; basal rosette of leaves lacking or early deciduous; cauline leaves mostly 4–22 mm long, linear, leaf-bases connate, margins hyaline, apices apiculate, upper cauline leaves becoming subulate; pedicels filiform, glabrous or glandular-pubescent; flowers 5-merous, rarely 4-merous; calyx glabrous or glandular-pubescent at base; sepals 1.0–2.0 (3.0) mm long, ovate, hyaline margin conspicuous, margins or apex frequently purple; petals elliptic, slightly exceeding sepals at anthesis, equal to or shorter than sepals during capsule development; stamens 10 or fewer; capsule 2.0–3.0 (3.5) mm long, dehiscing less than half the length of the capsule, valves thin, sepals remaining appressed following dehiscence; seeds light tan, obliquely triangular, with dorsal groove, surface smooth or tuberculate, 0.3–0.4 mm long. $2n = 36$. In our area spp. **decumbens**. Dry to moist open sandy places. Perhaps introduced in our area. sBC to Calif, Ariz; seUS; Alta, swSask. Map 499 ●

S. nivalis (Lindbl.) Fries
Caespitose, perennial, forming low cushions; lowering stems numerous, from axils of basal rosette leaves, 1–5 cm tall, slender, sometimes purple-tinged; cauline leaves subulate, with connate leaf-bases forming shallow scarious cup, often purplish, becoming shorter toward stem apex; pedicels long, filiform, glabrous; flowers generally 4-merous, sometimes 4- and 5-merous; sepals 1.5–2.0 mm long, orbicular to elliptic, rounded at tip, glabrous, frequently purplish, hyaline margins nearly always purple, sometimes only at tip; petals equalling to slightly shorter than sepals; stamens 8 or 10; capsule 2.0–3.0 mm long, dehiscing to base, valves 4–5, thick, sepals remaining appressed following dehiscence; seeds brown, obliquely triangular, with dorsal groove, distinctly notched at hilum, lateral surfaces frequently with elongate ridges, dorsal surface appearing smooth to pebbled, 0.5 mm long. $2n = 84, 88^*$. *S. intermedia* Fenzl. Moist gravelly areas and alluvium at alpine elevations. Circumpolar: Alas to Ellesmere Is s to nBC, sYuk, nMack, Huds Bay, nQue, Lab; Alta. Map 499 ○

S. nodosa (L.) Fenzl.
Perennial; basal leaves tufted, 15–30 mm long, linear, rosettes lacking; main stems 5–25 cm long, ascending to loosely spreading to prostrate, simple or branched, glabrous or pubescent at nodes; cauline leaves linear-subulate, abruptly shorter above, the upper 1–2 mm long with axillary leaf clusters (fascicles), giving a knotted appearance; pedicels glabrous or pubescent; flowers showy, mostly 6–10 mm in diameter, 5-merous or sometimes 5- and 4-merous; calyx glabrous or glandular-pubescent at base; sepals elliptic, 2–3 mm long; tips frequently purplish; petals generally much exceeding sepals; stamens 10 or 8; capsule 3.0–4.0 mm long, valves thick, sepals remaining appressed following dehiscence; seeds dark brown, 0.5 mm long, ovoid to reniform, a distinct notch present at hilum, dorsal groove present or absent, smooth to distinctly pebbled. $2n = 44, 56$. In our area ssp. **borealis** Crow. Moist gravel, sand, rock crevices. Eurasia: wMack to sBaffin Is, Nfld s to nAlta, nSask, cMan, Lk Superior, Mass. Map 499 ▲

S. saginoides (L.) Karst. / Pearlwort
Low perennial herb, matted or loosely tufted, 2–8 cm high, glabrous or nearly so; stems slender, decumbent or ascending; leaves linear or filiform, 5–12 mm long, mucronate, the lower ones in basal rosettes; flowers on slender stalks in the leaf-axils; sepals, petals and styles, usually 5 each; petals white, entire, shorter than the sepals, the latter 2–3 mm long, obtuse; stamens 10; capsule 3–5 mm long, the valves normally 5, sometimes 4. $2n = 22^*$. Moist gravel, shores, and ledges. Circumpolar: Alas, Yuk, swMack s to Calif, NM; James Bay, nQue, Gaspé. Map 500

Saponaria L.

S. officinalis L. / Soapwort; Bouncing Bet
Stout perennial herb with a horizontal rhizome; stems 3–8 dm tall, erect, simple or branched, glabrous or puberulent; leaves elliptic or ovate to lanceolate, acute,

narrowed at the base, usually glabrous; flowers in a dense terminal head-like cyme; calyx-tube 15–20 mm long, cylindrical or nearly so, the lobes triangular; petals pale pink or white, well exserted and spreading, crowned with an appendage at the top of the claw, the blade obcordate; stamens 10; styles 2; capsule 4-toothed at apex. $2n = 28$. Roadsides and waste ground. Introduced. Occasional.

Scleranthus L.

S. annuus L. / Knawel
Annual, mat-forming; stems 2.5–25 cm long, prostrate or ascending; leaves 5–20 mm long, subulate, acute, often ciliate, scarious-margined above the shortly connate base; exstipulate; flowers perigynous, greenish, in terminal or axillary, contracted, bracteated cymes; bracts usually longer than flowers; calyx 3–4 mm long, lobes lanceolate, about as long as the strongly ribbed tube; petals absent; stamens 10 or fewer, shorter than sepals; styles 2; fruit an ovoid, 1-seeded utricle, enclosed by the hardened calyx-tube. $2n = 44$. Cultivated fields and waste ground. Introduced. Rare.

Silene L. / Catchfly; Campion

Herbaceous annuals or perennials with opposite entire leaves; flowers solitary or in cymes; calyx tubular, 5-toothed 10–many-nerved; petals 5, the claw narrow and sometimes with auricles, the blade usually bilobed or 4-cleft, mostly crowned by a scale at the junction of claw and blade; stamens usually 10; styles 3–5 (sometimes varying in the flowers of a single plant); ovary usually on a stipe (carpophore); capsule opening by as many teeth as styles or twice the number; seeds reniform to globose, variously roughened. Includes *Lychnis* L. of Ed. 1.

1 Styles 3 — 2
  Styles 5 (4) — 9

2 Calyx glabrous, lobes sometimes ciliate — 3
  Calyx pubescent — 6

3 Plants less than 8 cm tall, cushion-forming — **S. acaulis**
  Plants taller, not cushion-forming — 4

4 Upper stem usually with broad glutinous bands; calyx mostly 8 mm long or less, prominently 5–10-nerved — **S. antirrhina**
  Upper stem lacking glutinous bands; calyx mostly more than 8 mm long, 10–20-nerved — 5

5 Calyx 15–20 mm long, inflated, prominently net-veined above the middle — **S. cucubalus**
  Calyx mostly 8–12 mm long, not inflated, obscurely net-veined above the middle — **S. cserei**

6 Nerves of calyx 20–30 — **S. conoidea**
  Nerves of calyx 10 — 7

7 Plants annual, weedy; calyx 15–30 mm long; petal-claws more than 12 mm long **S. noctiflora**
Plants perennial; calyx and petal-claws generally shorter 8

8 Flowers mostly less than 10 mm long; inflorescence leafy **S. menziesii**
Flowers more than 10 mm long; inflorescence not leafy **S. parryi**

9 Plants dioecious; weedy **S. pratensis**
Plants not dioecious; not weedy 10

10 Petals purple; calyx somewhat inflated; flowers generally nodding **S. uralensis**
Not as above 11

11 Stem generally branched, somewhat lax and bushy; calyx less than 1 cm long **S. menziesii**
Stem little branched, generally stiffly erect; calyx more than 1 cm long 12

12 Petals barely exserted from calyx, lobes 1–3 mm long; calyx tubular, not inflated **S. drummondii**
Petals conspicuously exserted from calyx, lobes longer; calyx campanulate, somewhat inflated 13

13 Plants generally less than 15 cm tall; flowers generally solitary; calyx pilose, hairs with conspicuous purple septa; styles 5 (4) **S. furcata**
Plants generally more than 15 cm tall; flowers generally more than 1; calyx glutinous, hairs lacking conspicuous purple septa; styles 3 (4–5) **S. parryi**

S. acaulis L. / Moss Campion
Low perennial, growing as a dense mat or cushion; stems short and crowded, mostly 3–6 cm long, the old leaves persisting on the stem bases; leaves linear-lanceolate, 4–12 mm long, mostly glabrous, the margins somewhat ciliolate; flowers solitary at the ends of branches, commonly imperfect and dioecious; calyx cup-shaped, 4–6 mm long, often purplish; petals purple or lavender, exceeding the calyx, mostly notched at apex; capsule 3-loculed, barely exserted. $2n = 24$*. In our area var. **exscapa** (All.) DC. with calyx to 10 mm long and var. **subacaulescens** (Williams) Fern. & St. John with calyx about 6 mm long. Alpine slopes and meadows. Circumpolar: Alas, Ellesmere Is to Nfld s to Ore, Ariz, NM, swMack, James Bay, nQue, NH. Map 501 ●

S. antirrhina L. / Sleepy Catchfly
Annual; stems simple or sparingly branched, slender, 1–6 dm high, commonly puberulent below and glabrous above, often with dark glutinous bands below the upper nodes; leaves oblanceolate to linear, glabrous or puberulent; flowers on slender pedicels, in a forked inflorescence; calyx 4–8 mm long, 10-nerved, glabrous, somewhat contracted at apex, the teeth often purplish; petals white or pink, shorter or slightly longer than the calyx, frequently obsolete; stamens included; capsule 4–8 mm long, 3-loculed; seed 0.5–0.8 mm long, with 3 or 4 rows of dorsal papillae. $2n = 24$. Dry open areas. sBC; transcont sCan to NB, generally distributed s. Perhaps introduced in our area. Map 501 ○

S. conoidea L.
Annual; stem 2–8 dm tall; basal leaves 3–12 cm long, 5–12 mm wide, oblanceolate; inflorescence glandular, open, several-flowered; calyx 2–3 cm long, about 30-nerved, tubular, tapering, becoming globose below, long-attenuate above, lobes 8–12 mm long, lance-attenuate; corolla white to purple; petals entire or slightly emarginate; styles 3; seeds 1.2–1.5 mm long. $2n = 20, 24$. Waste ground. Introduced. Rare.

S. cserei Baumg.
Biennial or short-lived perennial, 2–8 dm tall, usually glabrous and more or less glaucous; leaves oblanceolate to ovate-lanceolate, thick, the cauline often somewhat clasping; inflorescence narrow, the flowers numerous, fascicled; calyx 8–11 mm long, little inflated at maturity, glabrous, the nerves obscure, the orifice constricted; petals white, the blade deeply cleft; ovary 3-loculed; capsule slightly exceeding the calyx; carpophore puberulent; seeds 0.6–1 mm long, conspicuously papillate. $2n = 24$. Railway yards and waste places. Introduced. Rare.

S. cucubalus Wibel / Bladder Campion
Perennial with persistent roots, resembling *S. cserei* but leaves usually thinner and narrower; calyx larger, much inflated at maturity, conspicuously net-veined, the orifice not at all constricted; capsule almost spherical and enclosed in the calyx; carpophore glabrous; seeds 1–1.5 mm long, warty-papillose. $2n = 24, 48$. *S. vulgaris* Garcke. Fields and waste ground. Introduced. Common.

S. drummondii Hook.
Perennial, 2–5 dm tall, with a stout vertical root; stems 1 or more from the crown, stiffly erect, little branched, puberulent, glandular above; basal leaves oblanceolate, the upper narrowly lanceolate to linear; flowers few, on stiff, often elongated, appressed stalks; calyx 10–12 mm long, 5-toothed, 10-nerved; petals white or purplish, included or barely exserted, the blade narrower than the auricled claw; capsule 10 mm long, narrowly oblong. $2n = 48$. *Lychnis drummondii* (Hook.) S. Wats. of Ed. 1. Dry hillsides and plains. BC to sMan s to Nev, Ariz, Colo, Neb. Map 502

S. furcata Raf.
Perennial, often purple-tinged; stem erect, 0.5–2.5 (3.5) dm tall, densely glandular-villous above, the hairs with purple septa; basal leaves 1.5–5.5 cm long, 2–8 mm wide, oblanceolate, apex acute to obtuse, margin ciliate; cauline leaves 1–2 (3) pairs, narrower; flowers 1 (2–3), erect; calyx generally 1–2 cm long, campanulate to elliptic, 10-nerved, pubescent like the upper stem; petals usually white, bilobed, exserted; styles 5, occasionally 4; seeds 1.0–1.5 mm long, with a thin, narrow wing. $2n = 48$. Gravelly alpine slopes. Circumpolar: Alas to Ellesmere Is s to BC, swAlta, sMack, Huds Bay, nQue. Map 503 ○

S. menziesii Hook.
Perennial with taproot and slender rootstocks; stems several, slender, 1–3 dm tall, forked above, usually glandular-pubescent; leaves 2–5 cm long, lanceolate

or elliptic to oblanceolate, often acuminate, tapering to the base, puberulent or grey-pubescent, usually ciliolate on the margins; flowers on slender pedicels, in the axils of the upper reduced leaves; plants more or less dioecious; calyx 5–8 mm long, glandular-hairy; petals white, 2-cleft, slightly exserted, appendages minute or wanting; ovary 1-loculed; capsule equalling the calyx. $2n = 24, 48$. Includes *S. williamsii* Britt. In our area ssp. **menziesii**. Woodlands and clearings. Alas, Yuk, wMack s to Calif, wMont, NM. Map 503 •

S. noctiflora L. / Night-flowering Catchfly
Annual or winter annual, 2–8 dm tall, simple or branched, coarsely hairy throughout and viscid in the inflorescence at least; leaves ovate-lanceolate to oblanceolate, 4–10 cm long, the basal somewhat narrowed to a broad petiole, the cauline sessile; flowers fragrant, opening at night, often unisexual, in an open cyme; calyx about 15 mm long at anthesis, in fruit becoming inflated-ovoid, 25–40 mm long, 10-nerved, long-hairy and glandular-viscid; corolla white to pinkish, the blades 5–9 mm long, deeply cleft, the appendages about 1 mm long; styles 3; carpophore 1–3 mm long; capsule 3-loculed; seeds 0.8–1 mm long, uniformly papillose-rugose. $2n = 24$. Cultivated ground and waste places. Introduced. Common.

S. parryi (S. Wats.) C.L. Hitchc. & Maguire
Perennial, with few–many stems from a simple or branched caudex, the stems simple, 2–5 dm tall, puberulent, glandular above; basal leaves several, linear to oblanceolate, 3–8 cm long, narrowed to long petioles, puberulent and usually glandular; cauline leaves mostly 2 or 3 pairs; inflorescence commonly 3–7-flowered; calyx 12–16 mm long, with 10 prominent greenish or purplish nerves, glandular-pubescent, the hairs sometimes purple-banded; corolla white or tinged with green or purple, the blade usually bilobed, each lobe usually with a prominent lateral tooth, the appendages commonly 2 and 1.5–2 mm long; stamens scarcely exserted; styles mostly 3, less commonly 4 or 5, usually exserted. $2n = 48, 96$. Grassy slopes and talus to alpine elevations. sBC s to Wash, cIda, nwWyo. Map 504

S. pratensis (Rafn) Godron & Gren. / White Cockle; Campion
Dioecious biennial or perennial, with a stout root; stems 4–10 dm tall, branched, coarsely pubescent, glandular and viscid above; cauline leaves in several pairs, oblong-lanceolate to broadly elliptic, 2–8 cm long, puberulent to hirsute, the lower petioled, the upper sessile; flowers characteristically unisexual, opening in the evening and closing in the forenoon; calyx 15–20 mm long, tubular, 10-nerved in the staminate flower, 20-nerved in the pistillate, the lobes 3–5 mm long; petals white, 2–4 cm long, auriculate, with erose appendages above, the blade deeply lobed, capsule ovoid, 10–15 mm long, enveloped by the distended calyx, dehiscent by 10 teeth. $2n = 24$. *Lychnis alba* Mill. of Ed. 1. Roadsides and waste ground. Introduced. Common.

S. uralensis (Rupr.) Bocq.
Dwarf perennial, 5–15 cm tall, the stems more or less hairy, often with reddish viscid pubescence above; leaves linear-oblanceolate, sparingly pubescent, the basal attenuate to a margined petiole; flowers solitary (rarely 2 or 3), nodding in anthesis; calyx 10–15 mm long, narrowed at the mouth, somewhat inflated, densely pubescent; petals purplish, included or somewhat exserted, the blade narrow and bifid. $2n = 24$. *Lychnis apetala* L. of Ed. 1. In our area ssp. **attenuata** (Farr) McNeill. Rocky alpine slopes. Circumpolar: Alas to Ellesmere Is s to BC, nMont, sMack, Huds Bay, nQue; Utah, Colo. Map 505

Spergula L.

S. arvensis L. / Corn Spurry
Annual; stems several from the base, 2–4 dm tall, slender, sparsely puberulent and often viscid; leaves 2–5 cm long, linear-filiform, somewhat fleshy, clustered at the nodes in two opposite sets of 6–8, appearing whorled; stipules small, connate at first; inflorescence cymose, forked and open, the pedicels often reflexed; sepals 5, separate except at the base, about 3 mm long, glandular-puberulent; petals white, shorter than to exceeding the sepals; stamens 10 or sometimes 5; styles 5; capsule usually 5-valved, nearly twice as long as the sepals; seeds blackish, rounded, 1–1.5 mm wide, with a very narrow wing-margin, usually with minute light-coloured papillae. $2n = 18$. Fields and waste ground. Introduced. Common.

Spergularia J. & C. Presl / Sand Spurry

Low annuals; stems branched, erect or prostrate, mostly 0.5–2 dm long, commonly glandular-puberulent; leaves opposite, somewhat fleshy linear or filiform, mucronate, glandular-puberulent or glabrous; stipules deltoid, scaly-membranaceous; flowers in spreading terminal cymes; sepals 5; petals 5, white or pinkish, shorter than the sepals; stamens 2–7; styles usually 3; capsule equalling or surpassing the calyx, 2.5–6 mm long, normally 3-valved, dehiscent to the base.

Seeds black, 0.4–0.5 mm long, minutely sculptured; capsules globose; leaves 0.5–1 mm wide **S. diandra**

Seeds brown, 0.5–0.9 mm long, smooth; capsules ovoid; leaves 0.6–1.5 mm wide **S. marina**

S. diandra (Guss.) Boiss.
Plant diffuse or prostrate, with stems mostly 0.5–1 dm long; leaves 1–2.5 cm long, glandular; sepals 2.5–3.5 mm long, glandular; stamens 4–7; capsule 2.5–4 mm long; seeds black, with a silvery tinge, pear-shaped, minutely sculptured or papillate, wingless. $2n = 18$. Saline depressions. Introduced. Rare.

S. marina L.
Similar to *S. diandra* but usually with longer stems and larger leaves; stems and leaves glandular-puberulent or glabrous; sepals 2.5–5 mm long; stamens 2–5;

capsule 3.5–6 mm long; seeds pale brown or reddish, smooth, wingless or with a thin friable wing. $2n$ = 18, 36. In our area var. **leiosperma** (Kindb.) Gürke. Saline areas. Eurasia: BC to Man s to Calif, Ariz, NM, ND; seQue, NS to Nfld s to Fla, Gulf Coast; James Bay. Map 506 •

Stellaria L. / Chickweed; Starwort

Slender herbs, mostly somewhat tufted or matted; leaves broad or narrow; flowers solitary in the forks of the stem or in terminal cymes; sepals usually 5, separate or nearly so; petals 5 (or wanting), white, deeply 2-lobed, without claw or crown; stamens 10 or fewer; styles commonly 3; capsule 1-loculed, several–many-seeded, usually opening by 6 valves; seeds mostly oblong-obovate, commonly tubercled, sometimes nearly smooth.

1 Plants more or less glandular-hairy and viscid; low perennial with ovate leaves **S. americana**
Plants not glandular-viscid 2

2 Basal leaves petioled, ovate or rhombic-ovate; plants annual **S. media**
Basal leaves sessile; plants perennial 3

3 Bracts of inflorescence, at least the uppermost, scarious, whitish 4
Bracts of inflorescence all green, not scarious 7

4 Petals as long as the sepals or slightly longer 5
Petals minute or none; leaves oblong-lanceolate; pedicels reflexed in age **S. umbellata**

5 Leaves linear, acute at both ends, spreading or loosely ascending; pedicels widely spreading **S. longifolia**
Leaves linear-lanceolate, mostly broadest below the middle, erect, rigid; pedicels erect or ascending 6

6 Capsule straw-coloured, teeth reflexed **S. arenicola**
Capsule generally purplish black, teeth erect **S. longipes**

7 Leaves linear or elliptic-lanceolate, usually much more than 4 times as long as broad 8
Leaves ovate-lanceolate, less than 4 times as long as broad 10

8 Petals much shorter than the sepals or none; stems angled **S. calycantha**
Petals equalling or longer than the sepals 9

9 Leaves soft, slightly fleshy; plant freely branching, often matted **S. crassifolia**
Leaves firm, keeled; plant with a few short, erect or ascending stems **S. longipes**

10 Sepals acute, with scarious margins **S. crispa**
Sepals obtuse, the margins not scarious **S. obtusa**

S. americana (Porter) Standl.
Perennial with rootstocks, forming loose prostrate mats; stems very leafy, 1–2 dm long, viscid-puberulent; leaves sessile, ovate, 1–3 cm long, usually obtuse, viscid; cymes very leafy; sepals obtuse, 3–4 mm long, glandular; petals longer than the sepals. Rocky slopes, mostly at high altitudes. sAlta to scMont. Map 506 ○

S. arenicola Raup
Similar to the several-flowered lowland populations of *S. longipes;* distinguished by the characters of the key. $2n = 52$. Perhaps just one of the more distinctive morphological and ecological forms of that variable complex. Shifting sand-dunes. neAlta, nwSask. Map 507 ○

S. calycantha (Ledeb.) Bong.
Perennial with slender rootstocks; stems weak, decumbent or erect, angled, glabrous or sometimes pubescent, mostly 1–4 dm tall; leaves narrowly lanceolate to elliptic or ovate-lanceolate, 1–4 cm long, 2–8 mm wide, usually narrowed at base, thin, often ciliolate at base; flowers in leafy-bracted cymes or solitary on slender pedicels, these usually reflexed at maturity; sepals 2–4 mm long, scarious-margined; petals shorter than the sepals or absent; capsule dark, longer than the sepals; seed obscurely sculptured. $2n = 26, 52$. *S. borealis* Bigel. Includes *S. alpestris* Fries, *S. sitchana* Steud. A morphologically plastic species, variously treated by taxonomists but requiring further critical study. Moist places, often shaded. More or less circumpolar: Alas, Yuk, wMack to Huds Bay, nQue, Nfld, s to Calif, Utah, Wyo, sSask, sMan, Gt Lakes, Apal. Map 507 ●

S. crassifolia Ehrh.
Stems slender, weak, ascending or diffuse, often matted, freely branching, glabrous; leaves numerous, soft, 6–16 mm long, lanceolate to elliptic-lanceolate, narrowed at base, acute or obtuse; cyme leafy, few-flowered, or the flowers solitary; sepals 2.5–4 mm long; petals usually longer than the sepals; capsule usually exceeding the sepals; seed rugose-tuberculate. $2n = 26$. Wet ground. Circumpolar: Alas to cBaffin Is to Nfld s to BC, Ida, Colo, SD, Minn, James Bay, nQue, NS. Map 508

S. crispa Cham. & Schlecht.
Stems 1–3 dm long, weak, decumbent, from creeping rootstocks; leaves ovate, sessile or very short-petioled, thin, acute or short-acuminate, the margins often crisped; flowers axillary, solitary; sepals 3–4 mm long, lanceolate, scarious-margined; petals minute or absent; capsule longer than the calyx; seeds reticulate. $2n = 26$. Moist montane woods. sAlas, Yuk s to Calif, Mont, Wyo. Map 509

S. longifolia Muhl. / Long-leaved Chickweed
Stems 1–5 dm tall, slender and weak but erect or ascending, sharply 4-angled; leaves linear or narrowly lanceolate, 2–5 cm long, acute at each end, glabrous or sometimes ciliate at base; cyme usually many-flowered, open, the long filiform pedicels spreading or reflexed; bracts scarious; sepals lanceolate, 3–5 mm long;

petals longer than the sepals; capsule straw-coloured, exceeding the sepals; seeds obscurely roughened, essentially smooth. $2n = 26^*$. Moist woods, meadows, shores. Circumpolar: Alas, Yuk, swMack to Huds Bay, Nfld s to Calif, NM, Neb, eKan, Mo, SC. Map 510

S. longipes Goldie / Long-stalked Chickweed
Stems 0.5–3 dm tall, glabrous, tufted, erect or ascending, from extensive, branched rootstocks; leaves linear-lanceolate, usually broadest near the base, 1–3 cm long, erect or ascending, rigid, usually glabrous and shining, often blue-glaucous; cymes forking, several-flowered or sometimes the flowers solitary; upper bracts scarious or green and leaf-like; pedicels unequal, usually erect; sepals ovate-lanceolate, 3–4.5 mm long, glabrous or pubescent, sometimes ciliate; petals longer than the sepals; capsule much longer than the sepals, mostly dark brown to purplish black. $2n = 52^*, 78^*–104$. A highly plastic species variously treated by taxonomists, but the morphological variation completely intergrading and largely clinal. In our area, alpine populations with leafy bracts and glabrous sepals, the plants often glaucous (*S. monantha* Hult.), are common and some with leafy bracts and pubescent sepals (*S. laeta* Richards.) less so. Dry to moist open areas, open woodland, and alpine slopes. Circumpolar: Alas to Ellesmere Is to Nfld s to Calif, Ariz, NM, wSD, ND, Minn, Gt Lakes, NY. Map 511

S. media (L.) Cyrill. / Common Chickweed
Annual; stems trailing, matted or ascending, 1–5 dm long, branched, leafy, puberulent in lines; leaves mostly ovate, 1–3 cm long, glabrous, often pustulate, the upper sessile, the lower ones petioled and commonly ciliate; flowers solitary or in few-flowered leafy cymes; sepals 4–6 mm long, more or less hairy and pustulate; petals shorter than the sepals or none; stamens 3–10; seeds 1–1.2 mm long, reddish brown, tuberculate. $2n = 40$. A weed of yards, gardens, waste ground. Introduced. Common.

S. obtusa Engelm.
Matted plant with numerous, branched, prostrate stems, 5–15 cm long, glabrous throughout; leaves ovate, 4–8 mm long, acute; flowers solitary in the leaf-axils; sepals obtuse, 2–3 mm long; petals none or minute; capsule obtuse, longer than the sepals; seeds minutely tuberculate. $2n =$ c.78. Damp meadows and stream sides. BC, Alta s to Calif, Colo. Map 512

S. umbellata Turcz.
Stems from creeping rootstocks, erect or ascending, 1–3 dm high, very slender; leaves 5–20 mm long, linear-lanceolate to elliptic-oblong, glabrous; flowers in the upper axils and in terminal subumbellate cymes; pedicels capillary, commonly recurved at the tip; bracts scarious, at least on the margins; sepals 2–3 mm long, scarious-margined; petals none or minute; capsule 4–5 mm long. $2n = 26$. *S. gonomischa* Boivin. Moist montane forests. Eurasia: Alas, BC, Alta to Calif, Mont, Colo. Map 513

Vaccaria Medic.

V. pyramidata Medic. / Cow-cockle
Annual with a slender taproot; stem 3–10 dm tall, glabrous and glaucous, dichotomously branched above; leaves lanceolate to ovate-lanceolate, 3–10 cm long, clasping or the lower ones connate; inflorescence an open paniculate cyme; calyx 10–15 mm long, flask-shaped, strongly 5-ribbed, enlarged and wing-angled in fruit; petals bright pink, about 2 cm long, without appendages or auricles, the blades obovate, retuse; stamens 10, exserted; styles 2; capsule 6–8 mm long, opening at apex by 4 teeth; seed globose, 2–2.5 mm long, dull black. $2n = 30$. *V. segetalis* (Neck.) Garcke of Ed. 1. Fields and waste ground. Introduced. Occasional.

NYMPHAEACEAE / Water-lily Family

Aquatic perennial herbs with horizontal rootstocks; leaves long-petioled, the blades usually floating, deeply cordate; flowers solitary, long-peduncled, perfect; sepals 4–6; petals numerous, often small; stamens numerous; pistil compound, syncarpous, 8–30-loculed, with numerous ovules; stigmas radiate, on a disc-like structure; fruit many-seeded, with a leathery rind.

Leaves 1 dm or more wide; flowers yellow; sepals petal-like; petals small **Nuphar**
Leaves less than 7 cm wide; flowers white; sepals green; petals large **Nymphaea**

Nuphar J.E. Smith

N. variegatum Engelm. / Yellow Pond-lily
Petioles and peduncles arising from a stout rhizome; leaf-blades broadly ovate, 1–3 dm long, deeply cordate, with rounded basal lobes; flowers 4–7 cm wide; sepals usually 6, yellow, more conspicuous than the small petals; stamens many, yellow; fruit about 4 cm long; seeds 2.5–4 mm thick. $2n = 34$. Lakes, ponds, slow streams. Alas, swMack to James Bay, Nfld s to nIda, Mont, Neb, Ohio, Del. Map 514

Nymphaea L. / White Water-lily

N. tetragona Georgi
Rhizome erect, the petioles and peduncles arising from its summit; leaf-blades 5–12 cm long, 3–7 cm wide, green or purplish beneath, the sinus open and the basal lobes acute; flowers white, 2.5–4 cm broad, inodorous; petals 8–17, white with purplish lines; stigmas 6–9; seeds 3 mm long. $2n = 84, 112, 120$. In our area ssp. **leibergii** (Morong) Porsild. Acidic lakes and ponds, generally in deep water. Circumpolar: Alas, Yuk, BC; nIda; swMack, nAlta; eSask to James Bay s to nMinn, nMich; Que to RI. Map 515 ○

CERATOPHYLLACEAE / Hornwort Family

Ceratophyllum L.

C. demersum L. / Hornwort
Herb, growing in water, usually without roots; stems 3–15 dm long, freely branched, often forming large masses; leaves in whorls of 5–12, fairly rigid, 2–3 times forked into slender segments, these often spiny-denticulate; flowers minute, sessile in the axils, monoecious; perianth lacking but the flower is enveloped in a calyx-like involucre of 8–12 bracts; stamens 12–16, the filaments very short; ovary 1, with a single ovule; style filiform; fruit an achene, 4–6 mm long, ellipsoid, with a persistent style and 2 basal spines. $2n = 24$. Ponds and slow streams. Circumpolar: Alas, Yuk, swMack, neMan, cOnt to NS, generally distributed s. Map 515 ●

RANUNCULACEAE / Crowfoot Family

Herbs or climbing shrubs, mostly with acrid juice; leaves simple or compound, often ternate, mostly without stipules; flowers hypogynous, regular or zygomorphic, with all parts free and distinct; sepals 3–15, sometimes petal-like; petals present or absent; stamens usually numerous; carpels numerous or few, rarely solitary, 1–many-ovuled; fruits achenes or follicles, rarely fleshy.

1	Plants woody climbers; leaves opposite	**Clematis**
	Plants not as above	2
2	Flowers irregular, showy, deep blue or purple	3
	Flowers regular	4
3	Upper sepal helmet-shaped, enclosing petals, not spurred	**Aconitum**
	Upper sepal spurred, not enclosing petals	**Delphinium**
4	Petals spurred; flowers showy	**Aquilegia**
	Petals not spurred	5
5	Carpel 1; fruit a berry	**Actaea**
	Carpels 2 or more; fruit an achene or follicle	6
6	Ovaries 1-ovuled; fruit an achene	7
	Ovaries several-ovuled; fruit a follicle	10
7	Petals present, mostly yellow	8
	Petals absent, the sepals often petal-like but seldom yellow	9
8	Leaves basal; sepals spurred; achenes in elongate, tapering heads	**Myosurus**
	Leaves basal and cauline; sepals not spurred, achenes in globose or cylindrical heads	**Ranunculus**

9 Cauline leaves alternate; sepals inconspicuous, greenish; flowers mostly imperfect, the plants dioecious **Thalictrum**
Cauline leaves whorled, forming an involucre; sepals conspicuous, not greenish; flowers perfect **Anemone**

10 Follicles stipitate; leaves compound, trifoliolate **Coptis**
Follicles not stipitate; leaves simple, round or palmate 11

11 Leaves palmately divided; sepals white **Trollius**
Leaves entire or merely toothed; sepals white, pink, or yellow **Caltha**

Aconitum L. / Aconite; Monkshood

A. delphinifolium DC.
Perennial herb with short tubers; stems erect, 3–7 dm tall, finely pubescent or glabrate, leaf-blades glabrous, palmately lobed and cleft, the segments of the basal leaves lanceolate; flowers conspicuous, blue, perfect, irregular, in a short raceme; calyx corolla-like, the 5 sepals unequal; upper sepal (the helmet) hood-shaped and prolonged into a very short beak; upper 2 petals covered by the helmet, clawed at base, their upper ends comprising a nectary; lower 3 petals small or absent; stamens numerous; carpels 3–5, each with a short style; fruit 3–5 short follicles, several-seeded. $2n = 16, 32$. Poisonous, especially the roots and seeds. Montane woods and alpine meadows. eAsia: Alas, Yuk, wMack s to BC, Alta. Map 516 ○

Actaea L. / Baneberry

A. rubra (Ait.) Willd. / Red and White Baneberry
Perennial herb, 3–8 dm high, with large, 2–3-ternately compound leaves; leaflets ovate, lobed and sharply toothed, commonly pubescent on the veins beneath; flowers white, on slender pedicels in a short terminal raceme; sepals 3–5, obovate, petal-like, soon falling away; petals clawed at base, deciduous; stamens numerous; carpel 1; stigma 2-lobed; fruit a several-seeded berry, 6–8 mm long, ellipsoidal to spherical, lustrous, red or white; berries appearing as a conspicuous cluster. $2n = 16$. Poisonous, especially the berries. Includes *A. arguta* Nutt. Moist woods. Alas, Yuk, wMack to Huds Bay, Nfld s to Calif, Ariz, NM, SD, Ohio, NJ. Map 516 ●

Anemone L.

Perennial herbs with long-stalked basal leaves from the caudex and 2–5 stem (cauline) leaves comprising an involucre below the flower; leaf-blades palmately lobed or parted; peduncles 1-flowered, solitary or others from secondary involucres; sepals 5 or more, usually white or purplish, petal-like; petals none; carpels numerous; achenes pointed or tailed.

1 Styles plumose, 1.5–3.5 cm long; sepals 2–4 cm long 2
Styles not plumose, much shorter; sepals generally less than 2 cm long 3

2 Flowers blue or purple **A. patens**
Flowers white, sometimes purple-tinged **A. occidentalis**

3 Involucral leaves petiolate 4
Involucral leaves sessile 6

4 Plants rhizomatous, 1–2 dm tall; achenes hirsute **A. quinquefolia**
Plants not rhizomatous, 2–8 dm tall; achenes densely woolly 5

5 Penduncles with secondary involucres; leaf-lobes serrate on one side for half their length or more **A. riparia**
Peduncles naked, lacking secondary involucres; leaf-lobes merely coarsely toothed at apex **A. cylindrica**

6 Flowers yellow; styles reflexed; plants with slender rhizomes **A. richardsonii**
Not as above 7

7 Leaves trifoliolate, mostly 1–3 cm wide, the segments wedge-shaped **A. parviflora**
Leaves much divided, the segments linear to lanceolate-ovate, if wedge-shaped the leaves much wider than 3 cm 8

8 Achenes glabrous or strigose; leaf-segments wedge-shaped to lanceolate-ovate, generally more than 5 mm wide; sepals white or cream **A. canadensis**
Achenes woolly; leaf-segments linear to lanceolate-ovate, generally less than 5 mm wide; sepals often blue or reddish on the outer side 9

9 Plants generally 1–2 dm tall; 1-flowered; sepals often bluish **A. lithophila**
Plants generally 2–5 dm tall; 1–several-flowered; sepals often reddish **A. multifida**

A. canadensis L. / Canada Anemone
Perennial herb, 2–7 dm tall; basal leaves several, long-petioled, their blades strongly veined, deeply 3–5-parted, the divisions mostly 3-cleft and sharply toothed; involucral leaves similar but sessile and more deeply cleft; primary involucre subtending a long-peduncled flower; secondary involucres, formed in vigorous plants, smaller and each bearing a single flower; sepals 5, white, 1–2 cm long; head of achenes globose; achenes flat, strigose to glabrate, abruptly beaked, the erect, slender style 2–5 mm long. $2n = 14$. Damp meadows, thickets, and sandy shores. swMack to Huds Bay, NS s to BC, NM, Mo, Ohio, NJ. Map 517 •

A. cylindrica A. Gray / Long-fruited Anemone
Perennial herb, the stems 2–8 dm high, grey-villose, and stiffly erect from a stout caudex; basal leaves few to several, petioled, deeply 5-parted into rhomboid divisions, cuneate to the base, incised and sharply toothed; involucral leaves similar and long-petioled, commonly twice as many as the peduncles; inflorescence usually of 2–6 erect peduncles, 1–3 dm long, some of them often with a secondary involucre; sepals greenish or cream-white, silky outside, 0.5–1.5 cm long; head of achenes cylindrical, dense, woolly, 2–4 cm long; achenes densely woolly, the style about 0.5 mm long, usually with recurved tip. $2n = 16$. Prairies and open dry woods. BC to swQue s to Ariz, NM, Kans, Mo, NJ. Map 518

A. lithophila Rydb.
Plants low, 1–2.5 dm tall, from a stout woody caudex, the stems and leaves silky-hirsute; basal leaves long-petioled, 3–4-ternate, the segments linear-oblong; involucral leaves sessile or short-stalked; inflorescence simple, usually 1-flowered; sepals white, or bluish on outer side, appressed pubescent, 8–10 mm long; filaments and styles yellowish; achenes woolly, in a globose head about 1 cm thick, the mature styles 4–6 mm long. $2n = 48$*. *A. drummondii* S. Wats. of Ed.1. Alpine slopes and ridges. BC, Alta s to nwWash, nIda, Mont, nwWyo. Map 517 ○

A. multifida Poir. / Cut-leaved Anemone
Plants 1.5–5 dm tall, from a stout upright caudex, the stems and leaves silky-villose; basal leaves several, 2–3-ternate, the segments cleft into linear-oblong acuminate lobes; involucral leaves 3, short-petioled or nearly sessile; inflorescence 1–4-flowered, the primary peduncle 1–flowered, the secondary peduncles with small involucres and 1-flowered; sepals white, yellowish or red, often bluish or reddish on outer side; heads of achenes ovoid or globose, about 1 cm wide; achenes silky-villose, the styles 1–2 mm long. $2n = 32$*. Prairie grassland and dry open woods. Alas, Yuk, swMack to Huds Bay, Nfld s to Calif, Nev, NM, Minn, Me; SAm. Map 519

A. occidentalis S. Wats. / Chalice-flower
Perennial herb of meadows and slopes at high altitudes; plants 1–6 dm tall, densely villose when young, sometimes glabrate in age, the flowering stalks at first short but greatly elongated in fruit; basal leaves few, 3-parted, the divisions pinnately divided into numerous linear-lanceolate segments; involucral leaves 3, short-stalked, often tinged with purple, dissected like the basal leaves; sepals white, sometimes tinged with purple, 1.5–3 cm long; achenes numerous, in age reflexed, the plumose styles 2–4 cm long; head of fruit often broad above and narrow below. $2n = 16$*. *Pulsatilla occidentalis* (S. Wats.) Freyn. Moist alpine meadows. BC, Alta s to Calif, nIda, Mont. Map 520

A. parviflora Michx.
Plants 1–3 dm tall, from slender elongate rootstocks, sparingly hairy; basal leaves few, long-petioled, their blades thick, deeply 3-parted into broadly cuneate-obovate segments, these crenately lobed or incised into broad and usually obtuse divisions; involucral leaves similar but nearly sessile; peduncle solitary and 1-flowered; sepals 5 or 6, very rarely more, white or tinged with blue or rose; head of achenes globose or ovoid, densely villose. $2n = 16$* Moist banks and slopes from montane to alpine elevations. neAsia: Alas, Yuk, Banks Is, Victoria Is to nHuds Bay, Lab, Nfld s to nWash, neOre, Mont, Colo, nAlta, nSask, sHuds Bay, sQue. Map 521

A. patens L. / Pasque Flower; Prairie Crocus
Plants 1–4 dm tall, silky-villose throughout, from a short erect or ascending caudex with brownish sheaths; basal leaves several, long-petioled, ternately divided, the lateral segments 2-cleft, the middle one stalked and 3-cleft, the

various divisions further incised into linear or lanceolate acute lobes; involucral leaves similar but smaller and sessile; blooming in early spring, the stems being then quite short and the leaves still not expanded; sepals 5–7, blue, purple or nearly white, 2–4 cm long, villose on the back; stamens often accompanied by very small staminodia; achenes with a persistent plumose style, 2–4 cm long. $2n = 16$*. *Pulsatilla patens* (L.) Mill. Praries and dry open woods. Eurasia: Alas, Yuk, wMack, nSask, swMan, wOnt s to Wash, Mont, NM, Tex, Mo, Ill. Map 522 ●

A. quinquefolia L. / Wood Anemone
Delicate herb, 1–2 dm high, from a slender horizontal rhizome; basal leaves long-petioled, with 3–5 leaflets, coarsely and unevenly toothed or incised; involucral leaves commonly 3, petioled, each with 3–5 leaflets, cuneate-obovate to lanceolate, variously incised and cleft; peduncle solitary, villose with spreading hairs; sepals commonly 5, white or pinkish, 1–2 cm long; achenes fusiform, densely short-hirsute, 3–4 mm long, the stiff hooked beak 1–2 mm long. $2n = 32$. *A. nemorosa* L. var. *bifolia* (Farw.) Boivin. In our area var. **bifolia** Farw. Moist woods. BC; Alta; eSask to NS s to Ia, Ky, NC. Map 522 ○

A. richardsonii Hook.
Delicate herb, 8–15 cm high when in flower, the solitary basal leaves and the flowering stems arising from slender, horizontal, golden yellow or light brown rhizomes; blades of basal leaves circular-reniform, deeply 5-cleft into cuneate-obovate incised segments; involucral leaves 3, sessile or nearly so, each with 3 cuneate, dentate lobes; sepals sulphur-yellow, elliptic, 8–14 mm long; achenes glabrous, rather few, forming a loose globose head, the styles elongating and becoming reflexed. $2n = 14$. Moist woods and slopes. Siberia, Greenland: Alas, Yuk, Mack, sVictoria Is, Huds Bay, nwQue s to BC, Alta, nSask. Map 523 ●

A. riparia Fern.
Stems coarse and stiffly erect, 3–8 dm tall, from a stout caudex; basal leaves several, involucral leaves 3, both types long-petioled, deeply 3–5-parted into rhombic-ovate segments with cuneate bases, deeply incised and sharply toothed; plants sparingly pubescent to glabrate; primary peduncles 1–3, one or 2 of them often bearing a secondary involucre with 1–3 flowers; sepals 4–6, white, greenish white or rarely red, 0.7–2 cm long; head of achenes slenderly ovoid or oblong, usually about 8 mm thick; achenes densely woolly, the styles 1–1.5 mm long, ascending and the stigmas often incurved. $2n = 16$. Thickets and woods. BC to Nfld s to Minn, Ill, NEng. Map 523 ○

Aquilegia L. / Columbine

Perennial herbs with a caudex-like rhizome, large compound leaves and a few conspicuous flowers; leaves petiolate, 2–3 times ternate; sepals 5, petal-like; petals 5, the blade expanded and prolonged backward as a hollow spur, which is nectar-producing; stamens numerous, the inner staminodial; carpels commonly 5, erect, each with a slender style; fruit of 5 follicles; seeds black.

1 Flowers blue or purple, the petals often whitish 2
Flowers yellow, red or both 3

2 Plants mostly less than 15 cm tall; leaves basal only; usually 1-flowered **A. jonesii**
Plants more than 15 cm tall; some leaves cauline; several-flowered **A. brevistyla**

3 Flowers yellow, sometimes tinged with pink; spurs somewhat incurved **A. flavescens**
Flowers red; spurs straight **A. formosa**

A. brevistyla Hook. / Blue Columbine
Plants 20–80 cm high, slender, sparingly branched, more or less pubescent and glandular above; basal leaves long-petioled, biternate, the leaflets lobed and crenate; stem leaves few, nearly sessile, lobed; flowers 1.5–2.5 cm long, nodding or ascending; sepals blue or purple; petals yellowish white, their blades nearly as long as the sepals; the spur bluish, hooked, 6–8 mm long; styles and stamens barely exceeding the petals; follicles pubescent, 2–2.5 cm long, their beaks 2–5 mm long. $2n = 16$. Meadows, open woods, rock crevices. Alas, Yuk, wMack, Huds Bay s to BC, cSask, SD, Minn. Map 524 ●

A. flavescens S. Wats. / Yellow Columbine
Plants 20–60 cm high, branching freely, glabrous to pilose or glandular-pubescent; basal leaves long-petioled, biternate to triternate, the leaflets cleft and coarsely toothed; flowers usually nodding, lemon-yellow, sometimes tinged with pink or scarlet; sepals usually yellow, spreading or reflexed, longer than the obovate blades of the petals; spurs about as long as the petal-blades, more or less curved, knobbed; stamens longer than the petals, much exserted; follicles pilose or glandular-pubescent, about 2 cm long, their beaks slender, usually over 5 mm long. *A. formosa* var. *flavescens* Hook. Open woods and rocky slopes. BC, Alta s to Ore, Utah, Colo. Map 525

A. formosa Fisch. / Sitka Columbine
Similar to *A. flavescens*; differing in the characters of the key and in being somewhat taller, to 10 dm. $2n = 14$. Hybridizes with *A. flavescens*. Open woods and rocky slopes at somewhat lower altitudes than *A. flavescens*. sAlas to Calif e to Alta, Mont, Utah. Map 526 ○

A. jonesii Parry / Blue Columbine
Plants 5–20 cm high; nearly all the leaves basal, crowded, from a short caudex; leaflets small, crowded, biternate, the divisions broadly cuneate, finely pubescent, often glaucous; flowering stalk short, usually 2–5 cm long and with a single erect flower; sepals about 1.5 cm long, blue; petals often whitish, their blades only half as long as the sepals, the spurs shorter than the sepals, somewhat incurved; follicles 1.5–2 cm long. $2n = 14^*$. Alpine screes. Alta to nwWyo. Map 524 ○

Caltha L. / Marsh Marigold

Low perennial herbs of wet places; stems soft and glabrous; leaves simple, mostly basal, the blades elliptic or oval, and cordate or reniform at the base; sepals 5–9, petal-like; petals none; stamens numerous; carpels 5–many; ovules numerous in 2 rows in the ovary; follicles dehiscent on the ventral suture.

1 Stems erect, single-leaved or scapose; sepals oblong, white or bluish **C. leptosepala**
Stems ascending, decumbent or floating, with several leaves; sepals obovate, white, pink or yellow 2

2 Stems ascending or decumbent; flowers yellow, 1.5–4 cm wide; carpels 6–12 **C. palustris**
Stems floating or creeping, slender; flowers white or pink, smaller; carpels about 30 **C. natans**

C. leptosepala DC. / Mountain Marigold
Plants 1–4 dm high, with a short erect rootstock; basal leaves usually several, long-petioled, the blades oval or oblong-oval, cordate at base, the margin coarsely and sinuately toothed or crenate; flowering stems erect, scapose or bearing a single leaf; 1- or 2-flowered; flowers 2.3–3.3 cm broad; sepals 6 or more, oblong, white or tinged with blue on the outside; follicles about 15 mm long, the beak nearly 1 mm long. $2n = 48$. Wet alpine meadows. sAlas, sYuk s to Ore, Nev, Utah, Colo. Map 527 ○

C. natans Pallas
Plants with slender floating stems, or creeping on mud or sand, rooting at nodes; leaves cordate-reniform, with a narrow sinus, entire or crenate; flowers small, about 1 cm broad; sepals white or pinkish; follicles about 30, in a dense head, about 5 mm long; beaks minute. $2n = 16$*, 32. Ditches, ponds, lakes. Eurasia: Alas, Yuk, sMack to wHuds Bay s to BC, Sask, cMan, Minn. Map 527 ●

C. palustris L. / Marsh Marigold
Plants with ascending or decumbent, hollow stems, 2–6 dm long, branched above; basal leaves long-petioled, the uppermost sessile or nearly so; leaf-blades circular to reniform, usually broadly cordate at base, the sinus closed, narrow or broad, the margin dentate, crenate or subentire; flowers 1.5–4 cm wide, on short or elongate peduncles; sepals obovate or elliptic, yellow; follicles 6–12, curved, 10–15 mm long; beaks often recurved-spreading. $2n = 32$*, 48, 56–c.80. A species complex variously treated. In our area ssp. **palustris**. Marshy ground in meadows, swamps, woods. Circumpolar: Alas, Yuk to Melville Is to Huds Bay, Nfld s to BC, Sask, ND, Neb, Ill, Tenn, Apal, Md. Map 526 ●

Clematis L. / Virgin's-bower

Climbing plants with somewhat woody stems, grasping supporting bushes by means of twisted leaf-stalks; leaves opposite and composed of 3–7 leaflets; flowers solitary or panicled, perfect or unisexual; sepals 4, petal-like; petals none or present as small staminodia; stamens numerous; carpels numerous; fruit a cluster of achenes, each terminated by the persistent style as a hairy tail.

1	Flowers white, borne in clusters	**C. ligusticifolia**
	Flowers yellow, blue or purple, solitary	2
2	Flowers blue or purple; leaves with 3 leaflets	**C. occidentalis**
	Flowers yellow; leaves with 5 leaflets	**C. tangutica**

C. ligusticifolia Nutt. / Western Clematis
Semi-woody vine, 3–6 m high, climbing over bushes and sometimes into trees, often forming a dense growth; leaves opposite, pinnate, the 5–7 leaflets stalked, commonly ovate to lanceolate, rounded or cuneate at base, variously cleft and toothed, with appressed hairs; flowers numerous, mostly dioecious, in large paniculate clusters; sepals white, about 1 cm long; petals none; stamens numerous; pistillate flowers with sterile stamens; achenes numerous, hairy, the persistent plumose style 3–5 cm long. $2n = 16$. Coulees and river valleys. sBC to sMan s to Calif, NM, SD. Map 528 ●

C. occidentalis (Hornem.) DC. / Purple Clematis
Low vine, trailing or climbing, up to 2 m long; leaflets 3, long-stalked, thin, ovate, cordate at base, acute or acuminate, usually entire; flowers solitary, usually on long peduncles; sepals 4, purple or blue, 3–5 cm long, sharp-pointed, softly villose; petals absent, but staminodia between the sepals and numerous fertile stamens; styles erect, filiform; achenes villose, in a dense head, the persistent styles flexuous, plumose, 3–5 cm long. $2n = 16$*. *C. verticellaris* DC. of Ed. 1. In our area var. **grosseserrata** (Rydb.) Pringle. Open woods. BC to Sask s to Wash, Ida, Utah, Colo; Ont to Que, NB s to Ia, Ohio, Apal, Del. Map 529

C. tangutica (Max.) Korsh.
Trailing or climbing plants; leaflets 5, variously lobed, margins toothed; flowers 4–6 cm across, yellow, occasionally drying purplish; styles 5–6 cm long in fruit. $2n = 16$. Waste ground, railway grades. An escape from cultivation. Introduced. Occasional.

Coptis Salisb. / Goldthread

C. trifolia (L.) Salisb.
Low perennial herb with slender, elongate, bright yellow rhizomes; leaves basal, evergreen, lustrous, slender-petioled, ternately divided into 3 leaflets, the latter with very short stalks, cuneate-obovate, obscurely 3-lobed, sharply toothed; peduncle slender, 5–15 cm high, 1-flowered; flowers 10–16 mm wide; sepals 5–7, narrowly elliptical to lanceolate or spatulate, whitish; petals none; staminodia (may be regarded as petals) 5–7, small, clavate, hollow and nectariferous at the apex; stamens numerous; carpels 3–7, stipitate; follicles 3–7, long-stalked and with persistent erect beaks, the mature follicle with its beak being 7–12 mm long. $2n = 18$. Includes *C. groenlandica* (Oeder) Fern. Damp, mossy woods.

eAsia, Greenland: Alas; Mack, Huds Bay, sBaffin Is to Nfld s to BC, Alta, Man, Gt Lakes, Apal, NJ. Map 528 ○

Delphinium L. / Larkspur

Erect herbs, usually with palmately lobed or cleft leaves and conspicuous blue or purple flowers arranged in a terminal racemose cluster; flowers zygomorphic and spurred; sepals 5, petal-like, the upper sepal prolonged backward into a long spur; petals 4, irregular, the upper pair extending backward into the sepal spur; stamens numerous, the filaments often dilated; carpels few, commonly 3, ripening into follicles; seeds black, angled or winged.

1 Stems hollow, 6–20 dm tall; flowers numerous; raceme generally more than 15 cm long **D. glaucum**
Stems solid or nearly so, usually less than 6 dm tall; flowers 1–15; raceme seldom more than 15 cm long 2

2 Lower petals notched more than one-fifth the length of the blade, deep blue; leaves mostly cauline; roots often tuber-like **D. nuttallianum**
Lower petals notched less than one-fifth the length of the blade, often white or pale blue; leaves mostly basal; roots often fleshy but not tuberous **D. bicolor**

D. bicolor Nutt. / Low Larkspur
Low perennial; stems arising from a cluster of thickened but fibrous roots; stems 2–5 dm tall, simple, thinly hairy to glabrate below, markedly pubescent upwards; leaves few, mostly basal, often long-petioled, villose; blades thick, reniform to circular in outline, palmately dissected into linear-oblong and linear divisions; racemes few-flowered, short-pyramidal, the flowers large, bluish violet; sepals unequal, the 2 lower being larger, widely variable in size and shape, often about 15 mm long; upper petals white (yellow or brownish on drying), marked with purple lines, the lower petals variable, with a few white hairs; follicles 15–18 mm long, more or less hairy. $2n = 16$. Grassland and open woods. sBC to sSask s to neWash, cIda, Mont, Wyo, SD. Map 530

D. glaucum S. Wats. / Tall Larkspur
Tall perennial herb, the stems 1–2 m tall, occasionally taller, glabrous, sometimes blue-glaucous, usually hollow, from a stout woody caudex; leaf-blades glabrous or puberulent, often palmately cleft into about 5 primary divisions, these further cleft into narrow lobes; flowers deep blue or purplish, numerous in long racemes; sepals and petals blue or purplish, the sepals puberulent on the back, the upper petals white-edged; follicles erect, 12–14 mm long, glabrous or hairy. $2n = 16$. Meadows, stream banks, moist woods to subalpine elevations. Alas, Yuk, swMack to wcSask s to Calif, Nev. Map 531

D. nuttallianum Pritz. *ex* Walp.
Stem 1–4 dm tall, quite slender, finely puberulent, at least above, weakly connected with the tuber-like roots; leaves few, thin, the basal withering early, the cauline more persistent, these dissected into a few narrow divisions; raceme lax,

few-flowered; flowers blue-purple to light blue, resembling *D. bicolor* but generally smaller; sepals 7–14 mm long; follicles 10–18 mm long, pubescent. $2n = 16$, 32. Dry open mountain slopes. BC, Alta s to Calif, Ariz, Colo, Neb. Map 532

Myosurus L. / Mousetail

Small annual herbs with a tuft of basal leaves and a few leafless 1-flowered scapes; roots fibrous; leaves linear or filiform, entire; scapes 3–10 cm long; flowers small, perfect; sepals usually 5, spurred at the base; petals 5 or fewer, inconspicuous, linear, 1–3 mm long, nectariferous at the tip; stamens 5–10; carpels numerous, borne on a slender receptacle, which becomes greatly elongated in fruit; spikes of achenes 1–6 cm long; achenes crowded on the cylindrical axis, tardily separating.

Beak of achene extended less than 0.5 mm beyond body, not divergent from spike; mature spike 1–6 cm long **M. minimus**

Beak of achene extended 0.5–1.5 mm beyond body, divergent from spike; mature spike 0.5–2.5 cm long **M. aristatus**

M. aristatus Benth. *ex* Hook.
Similar to *M. minimus*; differing by the characters of the key, and the usually 1-nerved sepals. $2n = 16$. Prairie mud-flats and moist depressions. BC to Sask s to Calif, Utah, Wyo. Map 533 ○

M. minimus L. / Mousetail
Leaves 2–8 cm long; sepal blades 2–3 mm long, 3-nerved, the sepal-spurs 1–3 mm long; achenes 1–3 mm long, more or less flat on the back, their beaks very short and closely appressed. $2n = 16$*. Prairie mud-flats and moist depressions. Eurasia: BC to sOnt s to Calif, Tex, Fla. Map 534

Ranunculus L. / Buttercup; Crowfoot

Mostly low perennial herbs; cauline leaves alternate, the blades entire to variously lobed or dissected; flowers solitary or somewhat corymbed, yellow, rarely white; sepals usually 5; petals commonly 5, often broad and conspicuous, each with a nectariferous spot or pit at the base; stamens usually numerous, occasionally few; carpels 5 to numerous, each with a single basal ovule; fruit a globose or oblong head of achenes, the latter often beaked.

1 Petals white (sometimes yellowish at base); aquatic plants with all leaves submersed and finely dissected 2

Petals yellow 3

2 Free petiole about as long as the stipular base of the leaf; blades usually collapsing when taken from the water, not circinate; achenes usually 15–25, the beaks minute, less than 0.3 mm long, or lacking **R. aquatilis**

Free petiole none or shorter than the stipular base; blades usually not collapsing when taken from the water, circinate; achenes 30–80, the beaks (in ours) 0.3–0.5 mm long **R. circinatus**

3 Sepals normally 3; leaves reniform, deeply 3-parted; beak of achene curved or hooked; usually in boggy woods **R. lapponicus**
Sepals usually 5 4

4 Leaves entire or merely toothed; plants with stolons or runners 5
Leaves all, or in part, deeply lobed to compound 6

5 Leaves cordate-ovate to reniform, coarsely crenate, achenes longitudinally ribbed **R. cymbalaria**
Leaves linear to oblanceolate, entire; achenes not ribbed **R. reptans**

6 Aquatic or creeping mud plants, the leaves mostly floating or submersed 7
Terrestrial plants, mostly with erect stems 8

7 Leaves circular in outline, palmately lobed or dissected, often into linear segments **R. gmelinii**
Leaves reniform, 3-lobed **R. hyperboreus**

8 Basal and cauline leaves distinctly unlike; some or all of the basal leaves merely toothed (or entire), most of the cauline deeply cleft and sessile or subsessile 9
Basal and cauline leaves all deeply incised or divided into distinct segments, the cauline smaller or with few segments, or with shorter petioles 13

9 Petals 1.5–3 mm long, shorter and narrower than the sepals; basal leaves reniform, sometimes 3-lobed **R. abortivus**
Petals 3–5 mm long, longer and wider than the sepals 10

10 Achenes in a globose head, glabrous; plants generally less than 1.5 dm tall; early-flowering species 11
Achenes in a cylindrical head, densely hairy; plants generally more than 1.5 dm tall 12

11 Plants hairy; sepals long-villose; petals 5–9 mm long **R. rhomboideus**
Plants glabrous or nearly so; sepals lavender-tinged on the back, petals 6–15 mm long **R. glaberrimus**

12 Flowering stalks slender, often shorter than the conspicuous upper leaves, petals 3–8 mm long; nectary scales glabrous **R. inamoenus**
Flowering stalks stout, elongated; petals 8–15 mm long; nectary scales ciliate **R. cardiophyllus**

13 Terminal segment of the larger leaves stalked 14
Terminal segment of the larger leaves sessile, usually joined to the lateral by thin tissue 16

14 Petals shorter than the sepals; heads of achenes cylindric-oblong **R. pensylvanicus**
Petals equalling or longer than the sepals; head of achenes ovoid to globose 15

15 Petals usually less than 7 mm long, as long as or slightly longer than the sepals **R. macounii**
Petals usually more than 7 mm long, much longer than the sepals **R. repens**

16 Petals mostly 5 mm long or less 17
Petals mostly more than 5 mm long 21

17 Plants mostly 0.2–1.0 dm tall; alpine 18
Plants taller 19

18 Stems often arcuate in fruit; achenes 2–2.5 mm long **R. grayi**
Stems not arcuate in fruit; achenes about 1 mm long **R. pygmaeus**

19 Achenes almost beakless; plants of marshy places **R. sceleratus**
Achenes with conspicuous curved beaks 20

20 Stems mostly 1–2 dm tall; achene beaks about 0.5 mm long; alpine species **R. verecundus**
Stems mostly 2–6 dm tall; achene beaks 1–2 mm long; woodland species **R. uncinatus**

21 Achenes plump, without wing-like margins; basal leaves ternately divided to digitately cleft 22
Achenes flattened, with distinct margins; basal leaves palmately 3–5-lobed 24

22 Plants more or less hairy; beaks of achenes short and recurved; basal leaves deeply cleft in digitate fashion; widespread **R. pedatifidus**
Plants glabrous or sparsely brown-hairy; basal leaves 3-lobed and cleft; alpine 23

23 Backs of sepals hirsute with coarse brown hair **R. nivalis**
Backs of sepals lacking coarse brown hair **R. eschscholtzii**

24 Sepals reflexed in fully expanded flower; beak of achenes long, slender, hooked at apex; mountain species **R. occidentalis**
Sepals spreading; beak of achenes very short; introduced weed **R. acris**

R. abortivus L. / Small-flowered Crowfoot
Stems erect, 2–8 dm tall, forked above, glabrous or pubescent; basal leaves long-stalked, simple or divided, the simple ones rounded-ovate to reniform, cordate at the base, their margins crenate or dentate; cauline leaves short-stalked or sessile, commonly divided into 3 narrow segments, mostly toothed; petals shorter than the spreading or reflexed sepals, the latter 2.5–5 mm long; achenes about 1.5 mm long, smooth, with minute beaks, the fruiting head globose or ovoid. $2n = 16^*$. Open woods, stream banks, and moist places. Alas, Yuk, sMack to James Bay, Nfld s to Wash, Ida, Colo, Tex, nFla. Map 535 ●

R. acris L. / Tall Buttercup
Tall perennial, up to 1 m high, branched above, more or less pubescent; basal leaves long-stalked, the blades 3–5-palmately parted, the divisions cleft into narrow, lanceolate, toothed segments; cauline leaves mostly short-stalked and 3-lobed; flowers large, on long stalks; petals broadly obovate, 8–16 mm long, deep yellow or sometimes cream-coloured; achenes smooth, flattish, margined, 2–3 mm long, the beaks minute, the fruiting head subglobose. $2n = 14^*, 28, 56$. Roadsides, damp meadows, clearings. Introduced. Common.

R. aquatilis L. / White Water Crowfoot
Stems usually in water and elongate, branched, sometimes on mud and shorter; leaves submersed, usually collapsing when removed from water, the primary divisions arising usually well above the non-dilated stipular base; blades repeatedly trichotomous into filiform or narrowly linear segments, usually not circinate; fruiting pedicels not recurved; petals white, 4–8 mm long; achenes usually 15–25, glabrous or somewhat hairy, the beaks very small, less than 0.3 mm long or absent; receptacle usually hirsute. $2n$ = 16, 32, 48. *R. trichophyllus* Chaix. In our area var. **capillaceus** (Thuill.) DC. Ponds, lakes, ditches. Circumpolar: Alas, Yuk, Mack to Huds Bay, Nfld, generally distributed s. Map 533 •

R. cardiophyllus Hook. / Heart-leaved Buttercup
Erect perennial, little branched, 1–4 dm high, the stems stout, more or less villose; basal leaves stalked, mostly ovate-cordate and coarsely crenate, occasionally lobed; cauline leaves short-stalked to sessile, 3–7-cleft or parted into linear divisions; flowers few to several, rather large, usually on long thick stalks; petals ovate-orbicular, 8–15 mm long; achenes pubescent, in a large oblong-cylindric head, the beaks short. $2n$ = 32*, 64. *R. pedatifidus* J.E. Smith var. *cardiophyllus* (Hook.) Britt. Moist meadows and open woods. BC to Sask s to Wash, Ariz, NM; SD. Map 536

R. circinatus Sibth. / White Water Crowfoot
Resembling *R. aquatilis*; the leaves less flaccid, often firm, tending to maintain their rounded form when taken from water, sessile or nearly so at the dilated stipular base; blades once or twice trichotomous, then dichotomous, circinate; fruiting pedicels markedly recurved, at the base; petals white, 5–9 mm long; achenes 30–80, glabrous or hairy, each achene 1–1.5 mm long (usually smaller than in *R. aquatilis*), the style persistent as a beak, 0.2–0.5 mm long; receptacle hairy. $2n$ = 16*. In our area var. **subrigidus** (Drew) L. Benson. Ponds, lakes, ditches, streams. Europe: swYuk, Mack, Victoria Is to Prince Charles Is, Huds Bay to Nfld s to Calif, Ariz, NM, Tex, Ia, Mich, Mass. Map 537

R. cymbalaria Pursh / Seaside Crowfoot; Creeping Buttercup
Low perennial, glabrous or hairy, spreading by runners, rooting and producing new plants at the nodes; leaves long-petioled, mostly basal, the blades small, ovate to reniform, usually cordate at the base, more or less crenate; flowers few, on scapes 5–20 cm high; petals 3–5 mm long, slightly longer than the sepals; fruiting heads oblong-cylindric; achenes about 2 mm long, longitudinally striate, their beaks erect. $2n$ = 16*. Borders of ponds, lakes, stream banks. Asia, Greenland: Alas, Yuk, Victoria Is, Mack to Huds Bay, Nfld s to Calif, Ariz, NM, Tex, Ark, Ill, NJ. Map 538

R. eschscholtzii Schlecht.
Glabrous perennial with a short stout caudex; stems erect, 1–3 dm high, 1–3-flowered; basal leaves petioled, deeply 3–5-parted and again cleft, the blades usually with rounded or truncate bases, the ultimate lobes and sinuses rounded; sepals lavender-tinged beneath; petals 5–10 mm long, cuneate-obovate, often

retuse; head of achenes ovoid to oblong, the achenes glabrous, the beaks slender, not recurved. $2n = 32^*$. Moist alpine slopes and meadows. eAsia: Alas, Yuk, wMack s to Calif, Ariz, NM. Map 539

R. glaberrimus Hook. / Early Buttercup
Low glabrous perennial, usually less than 1 dm high; roots thick and fleshy; stems prostrate or ascending, not rooting, 4–18 cm long, 1–6-flowered; basal leaves petioled, the blades elliptic, ovate, obovate, oblanceolate or 3-lobed, the margins entire, undulate or toothed; flowering very early in the spring; petals 6–15 mm long, yellow or in age turning white; achenes 75–150 in a large globose head, each achene obovoid, smooth, with a short beak. $2n = 64, 80, 128$. Prairie grassland and meadows. BC to sSask s to Calif, NM, Neb. Map 540 ○

R. gmelinii DC. / Yellow Water Crowfoot
Aquatic or creeping mud plant, the stems floating or reclining, freely rooting at nodes; submersed leaves delicate, circular or reniform in outline, 3–5-cleft into cuneate-linear divisions; floating and emersed leaves similar but smaller, thicker and with wider lobes; flowers 1–4, on long stalks; petals 4–7 mm long; fruiting heads 4–7 mm long; achenes about 1.5 mm long, not markedly thickened on the margin but may be enlarged at the base, the beaks thin. $2n = 16^*, 32^*, 64^*$. A species complex variously treated taxonomically. In our area var. **gmelinii**, sparsely to markedly hirsute, with sepals 2.5–3.0 mm long, petals 1.5–3.0 mm broad; var. **limosus** (Nutt.) Hara (*R. limosus* Nutt.), markedly pubescent, with sepals 2.5–3.5 mm long, petals 3.5–5.0 mm broad; and var. **hookeri** (D. Don) L. Benson (*R. purshii* Richards.), glabrous essentially, with sepals 4–6 mm long, petals 3–6 mm broad. Ponds, ditches, streams, and lake shores. Asia, neEurope: Alas, Yuk to Prince Patrick Is, Mack to Huds Bay, Nfld s to Ore, NM, SD, Ia, Mich, Me. Map 541

R. grayi Britt.
Perennial, essentially glabrous, 5–12 (15) cm tall; stems straight or curved, simple or bifurcating, with a leafy involucre at or somewhat below the middle; leaves basal; petioles 1–6 cm long, white, often curved; lamina up to 1.5 cm long, 2 cm wide, lobed or dissected, the segments ovate, oblong to oblanceolate, glabrous; flowers 1–3; sepals 5, 2.5–4.5 mm long, usually purple-tinged, deciduous; petals 5, yellow, 3.5–5.0 mm long; fruiting heads 5–9 mm long, globose to ovoid-cylindric; achenes 2–2.5 mm long, plump; beaks 0.5 mm, curved. $2n = 16^*$. *R. gelidus* Kar. & Kir. ssp. *grayi* (Britt.) Hult. Dry alpine screes and slopes. eAsia: Alas, Yuk, nwMack; BC, Alta to Mont; Colo. Map 535 ○

R. hyperboreus Rottb.
A delicate, glabrous aquatic plant with floating stems, rooting at the lower nodes, freely branching; leaves alternate, the blades simple, reniform, cordate, truncate to rounded, lobed or cleft into 3 main segments, the lateral of which are 2-lobed, all lobes rounded and entire or toothed; petals obovate, 3–4 mm long, about as long as the spreading sepals; achenes (10) 20–60 in a globose or ovoid head, the receptacle (1.5) 3–5 mm long; each achene about 1 mm long, glabrous, the beak

stout, not recurved. $2n = 32^*$. Includes *R. natans* C.A. Mey. of Ed. 1. Ponds and lakes. Circumpolar: Alas to Ellesmere Is to Nfld s to BC, Ida, Mont, Colo, nSask, Huds Bay, nQue. Map 542 ●

R. inamoenus Greene
Erect perennial, 1–3 dm high, the stems hairy and rather leafy; basal leaves petioled, the blades circular, obovate, fan-shaped or 3–5-lobed, often crenate, appressed pubescent; upper cauline leaves usually divided into 3 narrow segments; flowering stalks rather slender, at first short, elongating in fruit; sepals spreading or turned downward; petals 3–8 mm long, about as long or twice as long as the sepals; achenes densely short pubescent, in a large oblong-cylindric head, the beaks slender. $2n = 32^*$, 48. Moist banks. sBC, Alta, swSask s to Wash, Nev, Ariz, Colo. Map 543

R. lapponicus L.
Stems arising singly, 1–2 dm high, from slender rootstocks; each stem with one basal leaf (rarely 2) and one small cauline leaf (or none); leaves petioled, the blades reniform in outline, deeply 3-parted, the cuneate or obovate segments coarsely crenate to shallowly lobed; stems 1-flowered; sepals 3; petals 6–10, oblong, yellow or whitish; achenes about 3 mm long, plump below and flattened above, the beaks long, sharply curved and hooked. $2n = 16^*$. Mossy woods and bogs. Circumpolar: Alas to seBaffin Is to seLab s to cBC, Alta, Sask, Man, Minn, nOnt; St Lawrence R. Map 540 ●

R. macounii Britt. / Macoun's Buttercup
Coarse annual or perennial herb, branched and leafy, 3–7 dm high, erect or somewhat prostrate and often rooting at the nodes, usually hirsute or hispid with long hairs; basal leaves petioled and hispid, ternately compound, the leaflets stalked and 2–3-cleft into incised or toothed lobes; sepals reflexed; petals 4–6 mm long, slightly longer than the sepals; heads of achenes globose or ovoid; achenes about 3 mm long, sharply margined, glabrous, the beaks stout, triangular-subulate, straight or bent. $2n = 32^*, 48^*$. Alluvial thickets, moist woods, and meadows. seAlas, Yuk, swMack, Huds Bay, Gaspé, Nfld s to Calif, Ariz, NM, Kans, Minn, Mich. Map 544 ●

R. nivalis L.
Perennial, glabrous or with sparse brown hairs; stems 0.3–1.5 (3) dm tall, simple, erect; basal leaves with petioles 1–5 cm long, blades 1–2 cm long, generally wider, deeply 3-lobed, the lateral often shallowly divided, base subcordate to truncate; cauline leaves 2–3, generally sessile, similar to basal but often larger and more deeply lobed, base truncate to broadly cuneate, lobes more or less entire; flowers solitary; sepals 5, 5–9 mm long, densely brown-hairy on back; petals yellow, 9–12 mm long; fruiting head 5–15 mm long, conical to cylindric; receptacle naked; achenes 1.5–2.0 mm long; beak slender, 1.0–1.5 mm long. $2n = 48$. Moist alpine slopes. Circumpolar: Alas to Ellesmere Is to nLab s to nBC, swMack, arct coast to neHuds Bay; Alta. Map 542 ○

R. occidentalis Nutt.
Terrestrial perennial, hairy or sometimes glabrous, the stems erect or somewhat reclining, 2–7 dm long, freely branching above; basal leaves petiolate, 3-parted or sometimes divided, the lobes cuneate, again lobed or parted, appressed pubescent; cauline leaves small with 3 linear divisions; sepals reflexed, soon deciduous; petals 6–12 mm long, at least twice as long as the sepals; achenes 8–15 in a hemispherical head, glabrous or hispid, each achene about 3 mm long, the beak slender, curved or nearly straight and hooked only at the apex. $2n = 28$. In our area var. **brevistylis** Greene. Dry to moist subalpine slopes. Alas, Yuk, nwMack s to Calif, BC, Alta. Map 544 ○

R. pedatifidus J.E.Smith / Northern Buttercup
Terrestrial perennial, the caudex with 1–10 erect stems, 1–4 dm high, bearing 1–8 flowers; basal leaves long-petioled, cordate in outline, pedately divided or cleft into 5–7 linear divisions some of which are incised, the blades cordate at base, thinly hairy or glabrous; cauline leaves sessile or subsessile, divided into 3–7 linear segments; sepals greenish yellow, villose on the back, 4–6 mm long; petals bright yellow, 8–10 mm long, sometimes absent; fruiting head cylindric or ellipsoid, 8–10 mm long; achenes 2 mm long, finely hairy or glabrate, the beaks slender, recurved. $2n = 32^*, 48$. A species complex variously treated taxonomically. In our area ssp. **affinis** (R.Br.) Hult. (*R. affinis* (R.Br.). Moist meadows to alpine elevations. Asia, Greenland: Alas, Yuk to Ellesmere Is to Nfld s to BC, Alta, sSask, nMan, nOnt, nQue; Ariz, Utah, Wyo, Colo. Map 545

R. pensylvanicus L.f.
Erect annual or perennial, 3–7 dm high, hirsute to hispid, not rooting at the nodes; basal leaves ternately compound, at least the terminal leaflet stalked, the leaflets cuneate, 2–3-parted; cauline leaves similarly 3-ternate, lobed and incised; flowers few, short-peduncled; sepals reflexed; petals 2–3 mm long, half as long as the sepals; heads of achenes cylindric-oblong; achenes about 3 mm long, sharply margined, glabrous, the beaks flat-subulate, straight or somewhat curved. Marshy ground, wet thickets, ditches. BC to Nfld s to Wash, Ariz, NM, Neb, Ohio, Pa, Del, adventive Alas, Mack. Map 546 ●

R. pygmaeus Wahlenb. / Dwarf Buttercup
Dwarf perennial with scape-like stems, erect or ascending, 1–5 cm long, bearing 1 or 2 flowers; leaf-blades semicircular to reniform, 3-cleft, the lateral divisions of the basal leaves 2-lobed, rounded, the middle division entire; uppermost leaves sessile, undivided; flowering stalks up to 8 cm long in fruit; sepals 2–3.5 mm long; petals 1.5–3.5 mm long; head of achenes ovoid to cylindric, about 3.5 mm long; achenes plump, glabrous, about 1 mm long, the beak short. $2n = 16$. Moist alpine slopes and ledges. Circumpolar: Alas to Ellesmere Is to Lab s to BC, Alta, nMack, Huds Bay; Gaspé. Map 546 ○

R. repens L. / Creeping Buttercup
Perennial with stems commonly creeping and rooting at the nodes, or somewhat erect; leaves petioled, the blades 3–8 cm broad, commonly ternate, stigose or

glabrate, the divisions stalked or sessile, variously cleft and toothed; petals 8–15 mm long, at least twice as long as the sepals; achenes 20–25 in a subglobose head, sharply but narrowly margined, each achene about 3 mm long, the beak triangular, usually somewhat curved, about 1 mm long. $2n = 16, 32$. Wet meadows, fields, roadsides. Introduced. Rare.

R. reptans L. / Creeping Spearwort
Small creeping herb, the stems filiform, rooting at the nodes; leaves petioled, or rarely sessile, the blades 1–3 cm long, filiform, linear, lanceolate, oblanceolate, or rarely ovate; petals obovate, 3–5 mm long, about twice as long as the sepals; achenes 5–20 in a hemispherical head, the receptacle in fruit 1 mm long or less, the beaks very short and curved. $2n = 32$*. *R. flammula* L. of Ed. 1. Borders of ponds and lakes. Circumpolar: Alas, Yuk, Mack to nHuds Bay, Nfld s to Calif, Ariz, NM, sSask, Minn, Pa, NJ. Map 547

R. rhomboideus Goldie / Prairie Buttercup
Low perennial with erect stems, 1–2 dm high, 3–12-flowered, rather densely hirsute; basal leaves long-petioled, simple, ovate-oblong to broadly ovate, crenate mostly above the middle, more or less tapering at the base; cauline leaves sessile or subsessile, cleft into 3–5 linear divisions; flowers on villose stalks, appearing early in the spring; sepals yellowish, tinged with lavender beneath, pilose; petals narrow, 7–8 mm long, much longer than the sepals; achenes in a globose head, 5–6 mm wide, each achene flattened at the base, plump above the middle, 2 mm long, glabrous, the beak minute. $2n = 16$*. Dry to moist open woods and grassland. BC, swMack, Alta to Ont s to Neb, Ia, Ill, Mass. Map 548 •

R. sceleratus L. / Cursed Crowfoot
Annual or short-lived perennial, the stems erect, 2–6 dm high, stout, hollow, usually glabrous, freely branched above and bearing many flowers; basal and lower cauline leaves long-petioled, succulent, reniform in general outline, deeply 3-parted, the divisions again cleft or lobed; upper leaves much smaller, of 3 narrow segments or undivided; petals light yellow, 2–4 mm long, usually shorter than the sepals; achenes about 1 mm long, plump, very numerous in a short-cylindric head, the beaks minute; surface of achene glabrous, often with minute transverse ridges or a circle of 'pin-prick' depressions. $2n = 32$*, 64. In our area ssp. **multifidus** (Nutt.) Hult. Ditches, marshes, pond and lake shores. Circumpolar: Alas, Yuk, wMack to Huds Bay, Nfld s to Calif, NM, La, Fla. Map 549

R. uncinatus D. Don
Perennial herb; stems 3–8 dm high, hispid, the hairs commonly reddish brown; basal leaves with hispid petioles, the blades 3–10 cm wide, cordate at the base, 3-parted, the divisions lobed and acutely toothed; cauline leaves similar but smaller; sepals about as long as the petals, reflexed; petals small, 3 mm long or less; achenes 8–20 in a globose or subglobose head, about 4 mm wide; achenes obovate, 1.8 mm long, flattened, hispid, the beaks slender but broad at base,

1.7–2 mm long, recurved and hooked. $2n = 28$. Includes *R. bongardii* Greene. Moist, shaded woodland. sAlas, swMack s to Calif, Ida, Colo. Map 548 ○

R. verecundus B.L. Robins.
Stout perennial with a branching caudex; stems 1–3-flowered, 5–15 cm high, decumbent-ascending, not rooting, glabrous or villose above; basal leaves cordate or reniform, usually 1–4 cm broad, 3-cleft, the lobes again crenately divided, the middle one 3-cleft; the lobes and sinuses obtuse or rounded; cauline leaves divided into 3–4 oblong segments; sepals pubescent, tinged with lavender beneath, about as long as the petals, the latter 3–5 mm long; achenes numerous in a narrowly oblong head, 4–8 mm long; each achene 1–1.5 mm long, glabrous, the beak rather stout, recurved. Moist alpine slopes. BC, Alta s to Ore, Ida, Mont. Map 550 ●

Thalictrum L. / Meadow Rue

Erect perennial herbs with rootstocks and yellow roots; basal and cauline leaves 2–3-ternately compound, the divisions and the leaflets stalked; petioles dilated at the sheathing bases; flowers small, usually numerous and in large clusters, perfect, or imperfect and plants dioecious, monoecious or polygamous; sepals 4–5, petal-like or greenish, early deciduous; petals none; stamens numerous, the filaments elongating; carpels several, becoming grooved or ribbed achenes.

1 Flowers perfect; achenes strongly flattened, on stipes to 3 mm long; anthers 1 mm long or less — **T. sparsiflorum**
Flowers imperfect; achenes more or less inflated, on stipes up to 1.5 mm long; anthers generally more than 1 mm long — 2

2 Leaflets 3-lobed, longer than broad, mostly dark green above, hairy below; achenes usually hairy — **T. dasycarpum**
Leaflets several-lobed, often broader than long, pale green above, glabrous or glandular-puberulent below; achenes glabrous or glandular-puberulent — 3

3 Peduncles wide-spreading, all of about the same length; achenes 5–6 mm long, spreading or reflexed; filaments 5–10 mm long — **T. occidentale**
Peduncles ascending, of markedly varying length; achenes 3–5 mm long, ascending or erect; filaments 3–5 mm long — **T. venulosum**

T. dasycarpum Fisch. & Avé-Lall. / Tall Meadow Rue
Plants 1 m or more tall; stems stout, often purplish; leaves few but large; upper leaves sessile; leaflets firm, oblong or obovate, commonly 3-toothed or lobed at tip, dark green above, pale and usually puberulent beneath; achenes 4–6 mm long, obtuse at base. $2n = 154$. Damp meadows and thickets. sBC to wQue s to Wash, Ariz, NM, Tex, La, Ohio. Map 550 ○

T. occidentale A. Gray / Western Meadow Rue
Plants 5–10 dm high, glandular-puberulent or glabrate; leaves mostly 2–3-ternate, the leaflets commonly obovate-cuneate or orbicular, 3-lobed and coarsely crenate;

achenes 5–8 mm long, tapering at both ends, with 8–12 thin ribs, the grooves shallow and rounded. $2n = 56$. Moist woods. seAlas, Yuk, BC to cSask s to Calif, Utah, Colo. Map 551

T. sparsiflorum Turcz. / Flat-fruited Meadow Rue
Plants slender, 3–10 dm high, branched above, glabrous or minutely glandular and scented; leaves 2–3-ternate, the uppermost sessile; leaflets thin, orbicular with cordate base to cuneate-obovate, 3-cleft to coarsely crenate; flowers perfect; sepals whitish; achenes 6–15, short-stipitate, flat, 5–8 mm long, 3–4 mm wide, half-rhombic with a straight back, the ribs few and inconspicuous. $2n = 14, 42$. Moist woods and meadows. Asia: Alas, Yuk, swMack to wHuds Bay; James Bay s to BC, Utah, Colo; Ore, Calif. Map 552 ●

T. venulosum Trel. / Veiny Meadow Rue
Plants 2–9 dm tall, little branched, glabrous; all leaves petioled except those subtending the flower-cluster; leaflets firm, orbicular to cuneate, 3-lobed and toothed, often bluish green, veiny and pale beneath; panicle narrow and dense; achenes oblong-ovoid, 3–5 mm long, the ribs thick, the grooves narrow. Includes *T. turneri* Boivin. Woods, thickets, and moist prairie. Yuk, Mack to Huds Bay, Que s to Ore, Ida, neNM, Neb, SD, Minn, Wisc. Map 553

Trollius L.

T. albiflorus (A. Gray) Rydb. / Globe-flower
Perennial herb with a thick caudex, fibrous roots and erect stems, 2–4 dm high; stems glabrous, 2–4-leaved, 1–few-flowered; leaves all petioled except the uppermost, 4–8 cm broad, deeply 5–7-parted, the segments obovate or rhombic and variously cleft or toothed; flowers terminal, dirty-white, large, 2–4 cm broad; sepals 5–7, petal-like, white or tinged with pink; petals 5–8, inconspicuous, linear; stamens numerous; carpels 10–20, becoming follicles, each about 1 cm long, several-seeded and slender-beaked. $2n = 16^*, 32$. Moist alpine and subalpine banks and meadows. BC, Alta s to Wash, Colo. Map 552 ○

BERBERIDACEAE / Barberry Family

Berberis L.

B. repens Lindl. / Creeping Mahonia
Low glabrous shrub; stem trailing or erect, 1–3 dm long; leaves pinnate, with 3-7 leaflets, these leathery, ovate to oblong, spinulose-dentate; flowers yellow, in racemes; sepals 6, yellow, subtended by a few small bracts; petals 6, in 2 series, each with 2 basal glands; stamens 6, opposite the petals, bending inward when the base is touched; anthers opening by uplifted valves; pistil simple, with few ovules; stigma sessile, circular; fruit a blue berry, juicy, sour, with a few large seeds; leaves persistent, in autumn tinted with red or purple. $2n = 28$. *Mahonia repens* (Lindl.) G. Don. Mountain woods. BC, Alta s to Calif, Ariz, Utah, wTex. Map 554

PAPAVERACEAE / Poppy Family

Papaver L. / Poppy

Herbs with milky juice and pinnately dissected leaves; flower-buds nodding; flowers large, solitary on long peduncles, perfect, hypogynous; sepals 2, deciduous; petals normally 4, large, thin in texture; stamens numerous; pistil of 4-several carpels, the compound ovary partially divided by intruding parietal placentae bearing numerous ovules; stigmas united in a flat radiate crown surmounting the ovary; fruit a many-seeded capsule opening by small pores near the margin of the stigmatic disc.

1 Cauline leaves present; capsules glabrous **P. somniferum**
Cauline leaves absent; capsules hispid 2

2 Plants more than 15 cm tall; flowers more than 3 cm across; leaves glabrous **P. nudicaule**
Plants generally less than 15 cm tall; flowers less than 3 cm across; leaves mostly pubescent 3

3 Flowers brick-red **P. freedmanianum**
Flowers yellow to orange or pink 4

4 Leaves essentially glabrous below, sparsely setose-hispid above **P. pygmaeum**
Leaves densely hirsute above and below **P. kluanensis**

P. freedmanianum D. Löve
Caespitose perennial; stems 6–10 cm tall, erect, scapose, pubescent, basal hairs white, yellowish brown at the apex; leaves 3–4 cm long; sheaths light greyish-brown with ciliate margins; lamina 1–1.5 cm long, 3-lobed, ovate in outline, centre lobe divided into 3 obtuse lobes, petiole and lamina covered with a fine white woolly pubescence; flowers 2–2.5 cm across; petals 4–5, brick-red with yellow spot at base, about as long as the pistil; ovary barrel-shaped, covered with dull brown non-bulbous bristles; stigma flat, rays 5–6. Shaly alpine slopes. swYuk; Alta. Map 555 ○

P. kluanensis D. Löve / Alpine Poppy
Caespitose, scapose perennial, 6–9 cm tall; leaves basal, numerous, 2–3 cm long, divided into about 5 ovate-lanceolate lobes, both petiole and blade with long white hairs; scapes with spreading hairs, these yellowish brown above; oval bud covered with dark brown hairs; flower 2.5–3 cm broad; petals sulphur-yellow, drying to pale green; stamens about 25 and as long as the oval pistil; capsule oval, covered with stiff brown bristles. A segregate of the *P. radicatum* complex. $2n = 42, 56^*$. Shaly alpine slopes. Yuk, BC, Alta; Wyo to NM. Map 555 ●

P. nudicaule L. / Iceland Poppy
Caespitose perennial, the flowering scapes mostly 2–3 dm tall; leaves all basal and petioled, pinnately lobed or cleft, usually glabrous and glaucous; flowers fragrant, 3–5 cm broad, commonly orange or reddish; scape with dark brown

appressed hairs; capsule 1–2 cm long, densely hispid. $2n$ = 14, 28, 42. Waste ground, roadsides. Introduced. An escape from cultivation. Occasional.

P. pygmaeum Rydb.
Caespitose perennial; stems 5–10 cm tall, erect, scapose, pubescent with stiff yellowish, sometimes purple mottled hairs; leaves 2–5 cm long, sheaths grey-green to dark brown, ciliate when young, lamina 1–1.5 cm long, ovate in outline, deeply lobed, much shorter than petiole, grey-green, glabrous below, often with appressed strigose hairs above and on petioles; flowers 1–2 cm across; petals 4–j, yellow to orange, generally drying pinkish; ovary elliptic, with stiff yellow bristles that have a bulbous base; stigma-rays 4–5; capsule 1–1.3 cm long, obconic. $2n$ = 14*. Alpine screes. BC, Alta, nwMont. Map 555 ▲

P. somniferum L. / Opium Poppy
Annual; stem generally 3–10 dm tall, simple, erect, glaucous, mostly glabrous; leaves sessile, clasping, ovate-oblong, undulate, shallowly toothed; flowers up to 18 cm across, white to purple; capsule globose to ovoid. $2n$ = 22. Waste ground. An escape from cultivation. Introduced. Occasional.

FUMARIACEAE / Fumitory Family

Corydalis Medic.

Biennial herbs, leafy-stemmed, glabrous throughout, often glaucous; stems rather soft and with watery juice; leaves alternate, bipinnately dissected; flowers in racemes, zygomorphic; sepals 2, small; corolla irregular; petals 4, one of the outer pair spurred at the base, both of the outer pair somewhat dilated and keeled at the apex, the two inner petals similarly modified at the apex and coherent over the stigma; stamens 6, in 2 sets, opposite the outer petals, the middle anther of each set 2-celled, the lateral ones 1-celled; pistil of 2 carpels, 1-loculed, with 2 parietal placentae; stigma 2-lobed; fruit a long, slender, 2-valved capsule; seeds black, shining, crested or arillate.

Flowers yellow; capsules spreading or pendent — **C. aurea**
Flowers purplish pink, with yellow tips; capsules ascending — **C. sempervirens**

C. aurea Willd. / Golden Corydalis
Plants usually 1–3 dm high; stems decumbent or ascending, glabrous and more or less glaucous, diffusely branched and leafy at base; leaves bipinnate, the leaflets pinnatifid; flowers golden yellow; corolla 12–15 mm long, the spur 4–5 mm long; capsules loosely spreading or pendent, more or less constricted and torulose; seeds 2–2.5 mm wide, smooth. $2n$ = 16. Open woods, clearings, shores. Alas, Yuk to swMack to nMan, nOnt, Que s to Calif, Tex, Mo, La, Ohio, Vt. Map 556

C. sempervirens (L.) Pers. / Pink Corydalis
Plants commonly 3–6 dm high; stems slender, erect, branched above, pale green; leaves glaucous, bipinnately dissected, the lower petioled, the upper nearly sessile; racemes loosely panicled; flowers pink to purplish, with yellow tips; corolla 12–15 mm long, the spur 2.5–5 mm, rounded; capsules erect or ascending; seeds 1.3–1.5 mm wide, low tuberculate. $2n = 16$. Open woods and clearings. Alas, Yuk, wMack to Huds Bay, Nfld s to sBC, Mont, Sask, Man, Gt Lakes, Ga. Map 557

CAPPARIDACEAE / Caper Family

Annual herbs, often ill-scented; leaves mostly with 3 leaflets, sometimes simple above; leaflets usually entire; flowers in terminal racemes, whitish to pink or rose; sepals 4; petals 4, nearly equal or somewhat irregular; stamens 4–many, long-exserted; ovary sessile or stipitate, 2-carpellate, 1-loculed with 2 parietal placentae; style short; receptacle bearing a gland at the base of the ovary; fruit a 2-valved, pod-like, 1-loculed capsule.

Plants glabrous or nearly so; stamens usually 6; stipe of capsule elongated **Cleome**
Plants viscid-puberulent; stamens usually over 10; stipe none or very short. **Polanisia**

Cleome L.

C. serrulata Pursh / Bee Plant
Stems 3–8 dm tall, freely branched above, glabrous and pale; leaflets 3, oblanceolate or lanceolate, 2–7 cm long, lower leaves long-petioled, upper sessile or nearly so; racemes elongating in fruit, the pedicels spreading or recurved; sepals united near base; petals pink-rose or whitish, short-clawed, 8–12 mm long; stamens long-exserted; stipe of capsule 1–2 cm long; body of capsule 2.5–5 cm long. $2n = 34, 60$. Disturbed prairie and roadsides. sBC to sMan s to Calif, NM, Tex, Mo, Minn. Map 558

Polanisia Raf.

P. dodecandra (L.) DC. / Clammyweed
Plant viscid-pubescent, clammy and with a disagreeable odour; stems usually branched; leaves 3-foliolate, with petioles 1–4 dm long; leaflets oblanceolate to narrowly ovate, 1–4 cm long; sepals lanceolate, tinged with purple; petals 8–12 mm long, yellowish white, notched at apex; stamens 5–16, purple, well exserted, unequal in length; style 4–6 mm long; capsule 3–5 cm long, sessile or nearly so, turgid, veiny; seeds areolate and often rough. $2n = 20$. *P. trachysperma* T. & G. of Ed. 1. In our area ssp. **trachysperma** (T. & G.) Iltis. Gravelly or sandy soil. sBC to swQue s to Calif, Tex, Ark, Mo. Map 559

CRUCIFERAE / Mustard Family

Herbs, often with pungent juice; leaves alternate, without stipules; flowers regular, hypogynous, perfect, in terminal racemes or corymbs; sepals 4; petals 4, or rarely absent, usually narrowed below into a claw; stamens normally 6, tetradynamous (the outer 2 inserted lower down than the inner 4); pistil 1, consisting of 2 united carpels; ovary 2-loculed by a septum which extends across from the parietal placentae, rarely 1-loculed; stigma discoid or 2-lobed; fruit, in most genera, a pod-like capsule (silique or silicle) dehiscent by 2 valves and with a persistent septum, sometimes indehiscent and nut-like or 1-loculed or transversely jointed.

Both flowers and mature fruit are usually necessary in identifying species of this family.

1 Fruit 2–3 times longer than broad or shorter — 2
Fruit much longer than broad — 3

2 Petals white — **Group I**
Petals yellow — **Group II**

3 Petals white, pink-tinged, pink or purple — **Group III**
Petals yellow — **Group IV**

GROUP I

1 Plants of alpine habitats — **Draba**
Plants of lowland habitats — 2

2 Fruit triangular; basal leaves pinnate, cauline leaves auriculate, clasping — **Capsella**
Fruit not triangular — 3

3 Pod flattened at right angles to septum; cauline leaves often auriculate, clasping, if not, the pubescence not stellate — 4
Pod flattened parallel to septum, if at all; cauline leaves not auriculate and clasping — 6

4 Plants glabrous — 5
Plants not glabrous (puberulent to pubescent) — **Cardaria**

5 Fruit 8–18 mm long, orbicular, broadly winged all round — **Thlaspi**
Fruit shorter, ovate to orbicular, wingless or only narrowly winged towards the apex — **Lepidium**

6 Plants glabrous; basal leaves 1–4 dm long, irregularly crenate — **Armoracia**
Plants pubescent; basal leaves 1–4 cm long, entire to denticulate — 7

7 Stem unbranched, leafless — **Draba**
Stem generally branched, leafy — 8

8 Petals bilobed; fruit generally more than 5 mm long, wingless, valves convex; style 2–3 mm long — **Berteroa**
Petals entire; fruit less than 5 mm long, sometimes winged, valves flat; style 1 mm long — **Alyssum**

GROUP II

1 Plants alpine; fruits flattened **Draba**
Plants not as above 2

2 Upper stem-leaves auriculate, clasping; entire to denticulate 3
Upper stem-leaves, if any, not auriculate and clasping; entire to divided 5

3 Cauline leaves ovate, the basal lobes enveloping stem, appearing perfoliate **Lepidium**
Cauline leaves, if any, linear to lanceolate, not appearing perfoliate 4

4 Fruit globose, about 2 mm long, slightly broader; 1 seed per locule **Neslia**
Fruit obovate, about 5 mm long; several seeds per locule **Camelina**

5 Leaves more or less entire 6
Leaves deeply serrate to pinnate or bipinnate 9

6 Petals 9–12 mm long; fruit 8–15 mm long, conspicuously inflated; leaves mostly spatulate, up to 1 cm wide **Physaria**
Petals 4–8 mm long; fruit often less than 8 mm long, not inflated; leaves lanceolate to oblanceolate, less than 1 cm wide 7

7 Fruits flattened, sometimes twisted; plants mostly annual 8
Fruits not flattened; plants perennial **Lesquerella**

8 Ovules and seeds 1 or 2 per locule **Alyssum**
Ovules and seeds more than 2 per locule **Draba**

9 Plants of marshy habitats; glabrous or with simple hairs **Rorippa**
Plants of dry habitats; pubescent with stellate and sometimes glandular hairs **Descurainia**

GROUP III

1 Leaves more or less entire 2
Leaves deeply serrate to pinnate 12

2 Cauline leaves auriculate, clasping 3
Cauline leaves, if any, not auriculate and clasping 6

3 Plants glabrous; glaucous 4
Plants pubescent; generally not glaucous 5

4 Stem 0.5–1.5 dm tall, branchy **Thellungiella**
Stem 2–7 dm tall, seldom branched **Conringia**

5 Fruit evidently compressed **Arabis**
Fruit terete or nearly so **Halimolobos**

6 Flowers pink or purple 7
Flowers white or tinged with pink 9

7 Plants 5–15 dm tall; leaves 5–20 cm long, lanceolate, spreading; fruit 4–10 cm long, slightly torulose **Hesperis**
Plants not as above 8

8 Dwarf alpine plants, appearing almost acaulescent in flower; pods taller than plants **Erysimum**
Taller plants, stems evident in flower; pods much shorter than the plants **Arabis**

9 Fruit torulose **Braya**
Fruit not torulose 10

10 Stem 1–5 (10) cm tall, barely exceeding the leaves, generally scapose; fruit linear, more than 10 times longer than broad; plants glabrous **Cardamine**
Not as above 11

11 Fruits linear, straight **Arabis**
Fruits lanceolate to oblong, often twisted **Draba**

12 Petals pink or purple **Cardamine**
Petals white, sometimes purple-veined or tinged 13

13 Fruit moniliform, 4.5–6.0 cm long, conspicuously grooved, indehiscent **Raphanus**
Fruits not moniliform, dehiscent 14

14 Plants aquatic or of marshy habitats; mostly glabrous 15
Plants of drier habitats, generally pubescent 16

15 Fruit straight; seeds uniseriate **Cardamine**
Fruit curved; seeds biseriate **Nasturtium**

16 Petals 15–20 mm long, conspicuously purple-veined; fruit 15–25 mm long, 3–5 mm wide, with a flat sword-like beak, often approaching the length of the valves **Eruca**
Petals much shorter; fruits narrower 17

17 Plants pubescent above; cauline leaves pinnate; fruit 5–11 mm long, 1.5–2.5 mm wide, prominently 1-nerved **Smelowskia**
Plants more or less glabrous above; cauline leaves not pinnate; fruit 20–40 mm long, 1.0–1.5 mm wide, nerveless **Arabis**

GROUP IV

1 Leaves more or less entire 2
Leaves divided 5

2 Cauline leaves auriculate, clasping 3
Cauline leaves not auriculate and clasping 4

3 Plants glabrous; cauline leaves oblong-lanceolate; sepals 6–8 mm long; petals 7–12 mm long **Conringia**
Plants generally hirsute below; cauline leaves lanceolate to ovate-lanceolate, strongly imbricate; sepals 2–3 mm long; petals 5–6 mm long **Arabis**

4 Fruits not flattened; plants 2–10 dm tall; lowland areas **Erysimum**
Fruits flattened; plants generally less than 2 dm tall; mostly alpine areas **Draba**

5 Upper cauline leaves auriculate, clasping 6
Upper cauline leaves not auriculate and clasping 7

6 Petals 6–14 mm long; fruit 2.5–3.5 mm wide, beak one-third to half as long as body; upper leaves more or less entire **Brassica**

Petals 3–8 mm long; fruit narrower, beak shorter; upper leaves deeply lobed to pinnate **Barbarea**

7 Plants stellate-pubescent **Descurainia**

Plants not stellate-pubescent 8

8 Stem glabrous 9

Stem hirsute, at least below 10

9 Fruit more than 1.5 cm long; plants glaucous **Schoenocrambe**

Fruit less than 1.5 cm long; plants not glaucous **Rorippa**

10 Pods moniliform, 4.5–6.0 cm long, 3–6 mm wide, indehiscent **Raphanus**

Pods not moniliform, dehiscent 11

11 Leaves essentially basal, oblanceolate, coarsely toothed to pinnate; fruits spaced all along stem, mostly 1–3 cm apart **Diplotaxis**

Not as above 12

12 Beak of fruit more than 5 mm long, generally flattened **Brassica**

Beak absent or less than 5 mm long 13

13 Seeds biseriate; plants of wet habitats **Rorippa**

Seeds uniseriate; plants of dry habitats 14

14 Fruit-valves 3-nerved, and/or the fruits appressed, and tapered from the base **Sisymbrium**

Fruit-valves 1-nerved 15

15 Pods 2.0–4.5 cm long, about 1.5 mm wide, often somewhat torulose **Erucastrum**

Pods 5–10 cm long, about 1.0 mm wide, linear **Sisymbrium**

Alyssum L.

Annual or perennial, pubescent with simple and forked hairs; leaves simple, entire or serrate; sepals non-saccate; stamens 6, usually with appendaged filaments; fruit an elliptic or circular silicle, valves flat; seeds winged, 1–2 (6) in each locule.

1 Ovules and seeds 1 per locule **A. murale**

Ovules and seeds 2 per locule 2

2 Fruit glabrous; style about 1 mm long **A. desertorum**

Fruit stellate-pubescent; style about 0.5 mm long **A. alyssoides**

A. alyssoides L.
Annual, 5–25 cm high, densely stellate-hairy; stems simple or several from the base, arched-ascending; leaves simple, 8–15 mm long, linear-oblong to oblanceolate, entire, silvery, dull; flowers in racemes; sepals more or less spreading;

petals white to pale yellow, 3.5 mm long; filaments commonly dilated at base; pedicels spreading; fruit orbicular in outline, 3 mm wide, stellate-pubescent, notched at apex, flat at the margin, dehiscent, flattened parallel to the septum; seeds 2 in each locule. $2n = 32$. Waste ground. Introduced. Occasional.

A. desertorum Stapf
Similar to *A. alyssoides*; differing in the characters of the key, the somewhat smaller flowers, and the deciduous sepals that fall shortly after anthesis. $2n = 32$*. Waste sandy ground. Introduced. Rare.

A. murale Waldst. & Kit.
Perennial; stems 2–7 dm tall, stellate-pubescent; leaves 10–20 mm long, 3–6 mm wide, oblanceolate, entire, green above, greyish below, stellate-pubescent; inflorescence a somewhat corymbiform raceme; sepals 2–2.5 mm long, petals 2.0–3.5 mm long, yellow; fruit 2.0–5.5 mm long, 1.5–4.0 mm wide, stellate-pubescent; seeds 1 per locule. $2n = 16$. Pasture. Introduced. Rare.

Arabis L. / Rock Cress

Biennial or perennial herbs, sometimes woody at the base, mostly pubescent with branched hairs; basal leaves petiolate; cauline leaves smaller, usually sessile, often auricled or sagittate; inflorescence racemose; petals white, yellowish white, pink or purple; pistil with slender ovary, short style and truncate stigma; siliques long-linear, flat or sometimes terete, erect to reflexed; seeds flattened, often with a marginal wing.

1 Siliques erect, often tightly appressed, less than 1.5 mm wide; seeds wingless or the wing less than 0.3 mm wide; flowers usually white or cream, sometimes pink to lavender-tinged — 2
Siliques erect to reflexed, generally more than 1.5 mm wide; seeds winged, the wing mostly over 0.3 mm wide; flowers white, pink or purple — 5

2 Plants mostly well over 3 dm tall, hirsute at base; cauline leaves usually auriculate — 3
Plants usually less than 3 dm tall, glabrous or only sparsely hirsute at base; cauline leaves not auriculate — 4

3 Seeds uniseriate; siliques mostly 3–5 cm long, clearly compressed; upper stem often hairy, not glaucous — **A. hirsuta**
Seeds irregularly biseriate; siliques mostly 5–9 cm long, only slightly compressed; upper stem glabrous, glaucous — **A. glabra**

4 Basal leaves (at least some) lyrate-pinnatifid, not ciliate — **A. lyrata**
Basal leaves entire, generally strongly ciliate — **A. nuttallii**

5 Basal leaves usually felty with greyish, dendritic hairs; plants mostly 1–2 dm tall; siliques ascending, spreading or slightly reflexed — **A. lemmonii**
Basal leaves either not felty or, if so, the plants more than 2 dm tall — 6

6 Plants mostly under 2 dm tall; leaves somewhat fleshy, usually glabrous; siliques ascending or spreading but not reflexed **A. lyallii**
Plants mostly well over 3 dm tall 7

7 Seeds biseriate, the wing nearly 1 mm wide; siliques strictly erect, 2–3 mm broad; pubescence (if any) of the lower stem simple to bifurcate, often appressed; basal leaves with malpighiaceous hairs **A. drummondii**
Seeds mostly uniseriate, the wing not over 0.5 mm wide; siliques erect to reflexed, 1–2.5 mm broad; pubescence of the lower stem often stellate to dendritic; basal leaves mostly with trifurcate to stellate or dendritic hairs 8

8 Hairs of the lower stems and leaves predominantly trifurcate; siliques straight or arcuate, stiffly erect to spreading **A. divaricarpa**
Hairs more freely branched; pubescence felty; siliques straight, mostly conspicuously reflexed, or pendulous **A. holboellii**

A. divaricarpa A. Nels.
Biennial or, rarely perennial; stems 4–8 dm tall, simple or branched above, usually glabrous above and sparsely pubescent below; basal leaves oblanceolate or spatulate, dentate to nearly entire 1–6 cm long, pubescent with 3–several-rayed trichomes; cauline leaves narrowly oblong to lanceolate, mostly entire, auriculate and often sagittate, commonly glabrous; sepals 3–5 mm long, scarious-margined; petals pale pink to purple, 6–10 mm long; siliques loosely ascending at an angle of about 45 degrees, or spreading (rarely pendulous), 2–8 cm long; seeds broadly oblong to nearly orbicular, narrowly winged. $2n = 14, 21, 28$. Dry sandy slopes. Alas, Yuk, wMack, nMan, Ont, Que s to Calif, Colo, nwNeb, Ia, Mich, Vt. Map 560

A. drummondii A. Gray
Biennial or perennial; stems 1–few, 1–7 dm tall, from a simple caudex, glabrous or sparingly appressed-pubescent below; basal leaves oblanceolate, glabrous to pubescent, often with ciliate petioles, the trichomes appressed 2-forked; cauline leaves oblong-lanceolate, sessile, auriculate, glabrous, 1–6 cm long, entire to sparingly dentate; sepals 3–5 mm long; petals white to pinkish (or purple in alpine plants), 5–9 mm long; siliques erect, crowded, glabrous, 4–8 cm long, 1.5–3 mm wide; seeds in 2 rows, oblong, prominently winged on distal end and on one side. $2n = 14^*, 20$. Moist wood margins to dry open areas. Alas, Yuk, nwMack to cSask, Man, Ont, to Nfld s to Calif, Colo, nwNeb, Ia, Ohio, Del. Map 561

A. glabra (L.) Bernh. / Tower Mustard
Biennial; stem erect, stout and tall, 4–10 dm high, glabrous and glaucous above, hairy below; basal leaves oblanceolate to oblong, usually toothed or cleft, coarsely pubescent with forked hairs, rarely almost glabrous; cauline leaves lanceolate to ovate, sessile, auricled or sagittate, mostly glabrous and entire; sepals yellowish to rarely purplish; petals yellowish white or with purplish tips; siliques erect, numerous, nearly terete, slender, about 1 mm thick, the valves 1-nerved to the middle or beyond; seeds very narrowly winged or wingless, biseriate or nearly

so. $2n = 12, 16$. *Turritis glabra* L. Possibly introduced. Thickets and fields. Alas, Yuk, swMack, cSask to Que s to Calif, NM, Neb, Minn, Ga. Map 562

A. hirsuta (L.) Scop.
Biennial or perennial; stems erect, 2–7 dm high, one to several from a simple or branching caudex, slender, hirsute to summit with spreading, mostly simple hairs; basal leaves short-petioled, hirsute; cauline leaves oblong to lanceolate, sessile and more or less clasping, entire to dentate, at least the lower ones pubescent; racemes slender, lax; petals white to yellowish white, 3–5 mm long; siliques erect, flat, 3–5 cm long, the valves 1-nerved to or well beyond the middle; seeds in one row, winged. $2n = 16, 18, 32$. In our area ssp. **pycnocarpa** (Hopkins) Hult. Ledges, banks, open woods, moist grassland. Circumpolar: Alas, Yuk, wMack, to James Bay, sQue s to Calif, Ariz, NM, Kans, Ark, Minn, Ga. Map 563

A. holboellii Hornem.
Biennial or perennial, the caudex simple or branched; stems erect, pubescent throughout with appressed or spreading hairs to glabrous above, 2–7 dm tall; basal leaves linear-oblanceolate, entire to somewhat dentate, densely pubescent with dendritic hairs; cauline leaves 1–4 cm long, entire, lower densely hairy, upper hairy to glabrous; sepals 2–4 mm long, scarious-margined; petals pinkish to white, 5–10 mm long; fruiting pedicels abruptly deflexed at base or gently curved downward; siliques reflexed, somewhat spreading or pendulous, the valves prominently 1-nerved below the middle; seeds orbicular, narrowly winged all around. $2n = 14^*, 21, 28, 42$. In our area mostly var. **retrofracta** (Graham) Rydb. Dry open woods, rocky slopes, and prairies. Greenland: Alas, Yuk, Mack to nMan, Gt Lakes; Que s to Calif, Colo, Neb, Minn, Mich. Map 564

A. lemmonii S. Wats.
Perennial with a branching caudex; stems few to many, slender, 0.5–30 cm high, pubescent throughout or only below; basal leaves spatulate, densely pubescent with minute branched hairs; cauline leaves sessile, oblong-lanceolate, auriculate and slightly clasping, glabrous or the lower pubescent; sepals often purplish, 2–3 mm long; petals pink to purple, 4–6 mm long; siliques usually horizontal, 2–5 cm long, 2–3.5 mm wide. $2n = 14^*$. Alpine slopes. swYuk; BC, Alta s to Calif, Colo. Map 565

A. lyallii S. Wats.
Caespitose perennial, the caudex usually branched; stems few to numerous, glabrous, 5–25 cm high, dwarfed at high altitudes; basal leaves narrowly oblanceolate, entire, petioled, glabrous or pubescent with small branched trichomes; cauline leaves few, sessile, without auricles or slightly auricled, mostly glabrous; petals purplish or rose-coloured; siliques erect to divergent, 2–5 cm long, 2–3 mm wide, glabrous. $2n = 21^*$. Dry alpine slopes and ridges. BC, Alta s to Calif, Utah, Wyo. Map 566

A. lyrata L.
Tufted biennial or perennial, 1–4 dm high; stems slender, often flexuose, simple or branched; basal leaves spatulate or oblanceolate, usually lyrate-pinnatifid or dentate, 2–4 cm long, glabrous or sometimes with simple or forked hairs; cauline leaves narrow, tapering at base, glabrous, entire to coarsely toothed; petals white to pinkish, 3–8 mm long; siliques erect to spreading-ascending, 1.5–4 cm long, 1–1.5 mm wide; seeds oblong, wingless. $2n = 16^*, 32$. In our area ssp. **kamchatica** (Fisch.) Hult. Open sandy woods and dunes. eAsia: Alas, Yuk, wMack s to Wash, Mont, nSask; Ont, Que s to Minn, Mo, Tenn, Ga. Map 567

A. nuttallii Robinson
Perennial; stems slender, several or many from a branching caudex, hirsute below, glabrous above, 6–30 cm high; basal leaves spatulate-obovate, entire glabrous to ciliate and hirsute on both sides; cauline leaves oblong to somewhat elliptical, hirsute to glabrous, sessile, not auriculate; petals white, 5–8 mm long; siliques erect to spreading, 1–3 cm long; seeds uniseriate, wingless, less than 1 mm broad. Moist banks and ridges. BC, Alta s to Wash, Nev, Utah, Wyo. Map 568

Armoracia Gaertn., Mey. & Scherb.

A. rusticana Gaertn. / Horse-radish
Glabrous perennial herb, 2–10 dm high, often branched, with a thick taproot, propagating mainly by rootstocks; basal leaves 1–4 dm long, petioled, oblong, crenate; cauline leaves smaller, sessile, lanceolate; racemes panicled; petals white, 5–7 mm long, obovate; fruits obovoid, 4–6 mm long. $2n = 32$. The roots furnish the table relish. Roadsides and around dwellings. Introduced and spread from cultivation. Occasional.

Barbarea R.Br. / Winter Cress

Biennial herbs, with basal rosettes, smooth or nearly so; basal leaves pinnatifid with a large terminal lobe and 2–several small lateral lobes; cauline leaves smaller, with clasping bases, mostly lyrate-pinnatifid or toothed; racemes elongate in fruit; sepals oblong, ascending; petals yellow; ovary cylindric; style slender; silique terete or somewhat 4-angled, tipped by the persistent style, the valves with strong midnerve; seeds flattish in a single row in each locule.

Siliques with thick pedicels, the beaks 0.3–1 mm long; uppermost leaves lyrate-pinnatifid with few basal lobes; petals pale yellow, 2.5–5 mm long; stem and lower leaves often purplish **B. orthoceras**

Siliques slender-pedicelled; beaks 1.8–3 mm long; uppermost leaves often angulately toothed, rarely pinnatifid; petals bright yellow, 5.5–8 mm long; stem and lower leaves not purplish **B. vulgaris**

B. orthoceras Ledeb.
Stems simple or branched above, 2–5 dm tall; siliques 2–4 cm long, strongly ascending, the pedicels clavately thickened, as much as 1 mm thick; seeds brown-

ish, minutely rugulose. $2n = 16$. *B. americana* Rydb. Stream banks and moist woods. Asia: Alas, Yuk, wMack, nMan, Ont to Nfld s to Calif, Ariz, Colo, Sask, Man, Minn, NH. Map 569

B. vulgaris R.Br. / Yellow Rocket
Similar to *B. orthoceras*; differing by the characters of the key and the lustrous grey seeds. $2n = 16$. Moist disturbed ground. Introduced. Occasional.

Berteroa DC.

B. incana (L.) DC. / Hoary Alyssum
Annual, grey-green with fine branched pubescence; stems erect or ascending, 3–6 dm high, usually branched above; leaves lanceolate or oblong, 1–4 cm long, entire, the lowest narrowed into a petiole; flowers racemose, numerous, 2–3 mm across, white; petals 2-cleft; filaments 2-toothed at the base; fruit 5–8 mm long, elliptic, plump, canescent, with a persistent style; seeds winged, 3–6 in each locule. Differing from *Alyssum* in the lobed petals and winged seeds. $2n = 16$. Fields and waste ground. Introduced. Rare.

Brassica L. / Mustard

Introduced annual or biennial herbs; lower leaves pinnatifid; flowers racemose, yellow, usually conspicuous; sepals often saccate at base; petals yellow, clawed, mostly about 1 cm long; filaments slender; anthers oblong; ovary nearly cylindric, scarcely narrowed to a short style; stigma capitate; silique slender or thickish, nearly terete or angled, with a prominent beak; valves conspicuously nerved; seeds globose, one row in each locule.

1 Beak of silique flat or conspicuously angled, often containing a seed; each valve of silique with 3 parallel nerves of about equal prominence; cauline leaves not clasping — 2
  Beak of silique terete or conical, not 2-edged, seedless; each valve with a prominent midvein, the other nerves weaker and scarcely parallel — 3

2 Leaves mostly pinnatifid; silique bristly, about 4 mm thick; beak sword-shaped, longer than the body of the fruit — **B. hirta**
  Leaves (at least the middle and upper) merely toothed; silique glabrous or rarely bristly, about 2 mm thick; beak angular, 2-edged, about half as long as the body — **B. kaber**

3 Leaves not clasping — **B. juncea**
  Leaves (at least the middle and upper) clasping and sessile — 4

4 Buds overtopping opened flowers; leaves glabrous to sparsely hirsute — **B. napus**
  Buds not overtopping opened flowers, leaves bristly — **B. campestris**

B. campestris L. / Rape, Colza
Annual or biennial, commonly glaucous, glabrous or nearly so; stems 4–8 dm tall; basal leaves lyrate-pinnatifid; middle and upper leaves sessile, oblong-

lanceolate, entire or dentate, completely clasping the stem by the deeply cordate base; raceme not elongating near the tip so that opened flowers overtop the buds; petals 6–8 mm long, pale yellow; pedicels at maturity 1–2 cm long, spreading-ascending; siliques 3–7 cm long, with a long tapering beak about one-third as long as the rest of the fruit; seeds reddish brown or grey-black, minutely roughened. $2n = 18, 20$. Fields and waste ground. Widely cultivated in the province for its oil seed. Introduced. Common.

B. hirta Moench / White Mustard
Annual, 3–7 dm tall, more or less hirsute; lower leaves lyrate-pinnatifid; upper leaves reduced, less lobed or merely dentate, not clasping; petals about 1 cm long; fruiting pedicels spreading, about 1 cm long; siliques about 3 cm long, including the flattened, sword-shaped beak, the valves commonly bristly and much distended over the seeds; seeds light yellow to pale brown, smooth. $2n = 24$. Fields and waste ground. Introduced. Occasional.

B. juncea (L.) Cosson / Indian Mustard
Annual, 4–10 dm tall, glabrous, often glaucous, branched above; lower leaves lyrate-pinnatifid, long-petioled; upper leaves lanceolate or linear, sessile or nearly so, mostly entire; petals 7–9 mm long; fruiting pedicels 8–15 mm long; siliques 2–5 cm long, including the slender tapering beak; seeds yellow, conspicuously reticulate. $2n = 36$*. Fields and waste ground. Introduced. Common.

B. kaber (DC.) L.C. Wheeler / Charlock; Wild Mustard
Annual, 2–8 dm tall, more or less hirsute, the branches purplish at their junction with the main stem; lower leaves usually lyrate-pinnatifid with a large rounded terminal lobe; upper leaves sessile, usually undivided but coarsely toothed; petals 6–8 mm long; fruiting pedicels about 5 mm long; siliques ascending, smooth or somewhat hairy, the beak about half as long as the body, angular and 2-edged; seeds black or dark brown, reticulate at high magnification. $2n = 18$. A weed of fields and waste ground. Introduced. Occasional.

B. napus L. / Rutabaga; Swede
Similar to *B. campestris*; differing in the characters of the key and the larger petals, 10–14 mm long. $2n = 38$. Often cultivated in the province but apparently rarely escaping. Introduced. Occasional.

Braya Sternb. & Hoppe

Low, caespitose perennials with branched caudex and stout taproot; leaves mostly tufted at the base of the stems, more or less pubescent, often densely so; flowers white or purplish; sepals ovate, equal at the base; style short, stigmas capitate; siliques cylindric to lance-subulate, often torulose, the valves faintly 1-nerved, the septum appearing cross-striolate under a strong lens.

Flowering stems leafy; pods linear, about 1 mm wide	**B. humilis**
Flowering stems leafless; pods plump, 2–3 mm wide	**B. purpurascens**

B. humilis (C.A. Mey.) Robins.
Stems several from the caudex, 0.5–3 dm long, erect or spreading, purplish, simple or forking, more or less pubescent; basal leaves rather thick, linear-oblanceolate, 1–3 cm long, entire or distantly toothed, glabrous or somewhat pubescent; cauline leaves smaller, remote; flowers white or purplish; petals 3–4 mm long; pedicels erect or closely ascending; siliques 1–3 cm long, narrowly cylindric, torulose, minutely pubescent; style slender, 0.5–1 mm long. $2n = 28^*$, 42^*, 56, 70. Includes *B. richardsonii* (Rydb.) Fern. and *B. glabella* Richards. of Ed. 1. Moist to dry open woods, banks, gravel bars, to alpine elevations. Asia, Greenland: Alas, Yuk, wMack s to BC, Alta; Colo; Huds Bay, Gt Lakes; eQue, Nfld; NH. Map 570 ●

B. purpurascens (R.Br.) Bunge
Perennial, from a branching caudex; stems 0.5–1.5 dm tall, scapose, generally pubescent; leaves 1–3 cm long, somewhat fleshy, spatulate to oblanceolate, mostly glabrous, margin often ciliate; flowers clustered; sepals 1.5–2.5 mm long, purple; petals 3–4 mm long, white or pinkish; fruit about 10 mm long, 2–3 mm wide, straight; style stout, 1 mm long. $2n = 56^*$. Includes *B. americana* (Hook.) Fern. Alpine screes. Circumpolar: Alas, swYuk to Ellesmere Is to nQue, Lab; Alta. Map 570 ○

Camelina Mill. / False Flax

Annuals or winter annuals, with erect stems 3–9 dm tall, branched above, with both simple and branched hairs; basal leaves simple, tapering to petioles; cauline leaves sessile, lanceolate or linear, sagittate-clasping; flowers small, yellow, the raceme greatly elongating in fruit; sepals erect, obtuse; petals 3–4 mm long; anthers ovate; style persistent; fruits (silicles) short, obovoid, slightly flattened parallel to the septum, margined, short-stipitate.

1 Fruit pear-shaped, apex rounded; seeds oblong; widespread — 2
Fruit top-shaped, apex flat; seeds round; rare — **C. parodii**

2 Plant pubescent at least below, with long simple hairs as well as stellate; fruits 4–6 mm long; seeds mostly less than 1 mm long — **C. microcarpa**
Plant glabrous or, if pubescent, with simple hairs not exceeding the stellate; fruits 6–9 mm long; seeds 1–2 mm long — **C. sativa**

C. microcarpa Andrz. *ex* DC. / Small-seeded False Flax
Stems and leaves grey-green with dense pubescence, composed of long simple and short stellate hairs; leaves entire or denticulate; flowers pale yellow, small; fruits pear-shaped, strongly margined, 4–6 mm long, on slender, spreading pedicels; seeds reddish brown, oblong-angular, mostly 0.8–1 mm long. $2n = 40^*$. A weed of fields and waste ground. Introduced. Common.

C. parodii Ibarra & La Porte
Quite similar to *C. sativa;* differing in the characters of the key, the larger seeds, and incumbent cotyledons. $2n = 40$. Waste ground. Introduced. Rare.

C. sativa (L.) Crantz / Large-seeded False Flax
Similar to *C. microcarpa*; differing by the characters of the key and the pale yellow-brown seeds. $2n = 40$. A weed of fields and waste ground, less prevalent than *C. microcarpa*. Introduced. Occasional.

Capsella Medic.

C. bursa-pastoris (L.) Medic. / Shepherd's-purse
Annual or winter annual, with a rosette of basal leaves and a few reduced cauline ones on the branched stems, 1–5 dm high, smooth or somewhat hairy; basal leaves mostly 3–8 cm long, usually deeply cleft, sometimes merely toothed; cauline leaves sessile, lanceolate to linear, entire or denticulate, auriculate at base; flowers small, white, racemose; petals 2–4 mm long, white, exceeding the sepals; style very short; fruits (silicles) obcordate-triangular, flattened at right angles to the septum, 5–8 mm long. $2n = 32$. A weed of fields, gardens, lawns, and paths. Introduced. Common.

Cardamine L. / Bitter Cress

Herbs, commonly with leafy stems; leaves pinnately lobed or pinnately compound (simple in one species); pubescence none or of simple hairs; flowers usually racemose, white; siliques linear, tipped with the persistent style, usually flattened, opening elastically from the base, the valves nerveless or nearly so; seeds in one row in each locule, wingless.

1 Leaves simple, ovate to elliptical, entire; dwarf alpine, 3–10 cm tall; flowers 1–5 **C. bellidifolia**

Leaves pinnate or pinnatifid; taller plants and usually more flowers 2

2 Petals 8–13 mm long, generally pink, rarely white; plants of northern bogs and springy places **C. pratensis**

Petals 1.5–4 mm long, white 3

3 Petals 3–4 mm long; inflorescence subumbellate, contracted; stem few-leaved **C. umbellata**

Petals 1.5–3 mm long; inflorescence elongate; stem very leafy 4

4 Stem somewhat hairy at base (except submersed bases); leaflets with bases mostly decurrent; terminal leaflet elliptic to reniform; style of silique 0.5–1.5 mm long **C. pensylvanica**

Stem glabrous throughout; leaflets not decurrent at base; terminal leaflet linear to oblanceolate; style of silique 0.2–0.7 mm long **C. parviflora**

C. bellidifolia L.
Glabrous perennial, tufted from a branched caudex; stems 3–10 cm high; leaves long-petioled, ovate or elliptical, entire or with 1 or 2 lateral teeth; flowers 1–5, white; petals 3–4 mm long; siliques erect, 2–4 cm long; style stout, about 1 mm long. $2n = 16^*$. Moist alpine banks and ledges. Circumpolar: Alas, Yuk to Ellesmere Is to nQue, Lab s to nCalif, BC, Alta, nMack, Huds Bay; Gaspé. Map 571 ●

C. parviflora L.
Stems usually solitary, 1–3 dm tall, simple to much branched, glabrous; terminal leaflet of basal leaves linear-oblong to somewhat oblanceolate; cauline leaves with 3–6 pairs of narrow segments, the lateral ones linear to spatulate or oblong, usually entire, not decurrent on the rachis; flowers about 4 mm wide; siliques ascending, 1–3 cm long; seeds 0.7–0.9 mm long. $2n = 16$. Sandy ground and dry open woods. Eurasia: Ore; BC; Alta; swMack; Man to Nfld s to Minn, Tex, Fla. Map 571 ○

C. pensylvanica Muhl.
Commonly biennial; stems erect or spreading, branched, sometimes decumbent at base, 1–5 dm high, often hispid near the base, otherwise glabrous; rosette leaves 2–8 cm long, with 1–6 pairs of rounded glabrous leaflets, the terminal obovate or elliptical, the lateral leaflets more or less confluent with the rachis or decurrent; cauline leaves with linear-oblanceolate or oval leaflets, their bases oblique and decurrent; flowers and siliques like *C. parviflora* except as noted in the key; seeds 1–1.5 mm long. $2n = 32, 64$. Wet ground. Yuk, swMack, Huds Bay, Nfld s to Calif, Colo, Okla, Fla. Map 572 ●

C. pratensis L.
Erect perennial, 1–4 dm high; leaves pinnate with numerous segments, the basal leaves long-petioled; terminal leaflet of basal leaves entire or obscurely toothed; lateral leaflets of cauline leaves usually distinctly stalked; petals 8–13 mm long, obovate; siliques 2–3 cm long; beaks 1–2 mm long. $2n = 16–96$. A species complex. Moist meadows and swamps. Circumpolar: Alas, Yuk to Ellesmere Is to Nfld s to nBC, Alta; cSask, Man, Gt Lakes, Apal. Map 572 ○

C. umbellata Greene
Perennial with rootstocks; stems several, erect, 2–5 dm high, glabrous or nearly so; leaves pinnate, with 3–7 oval or oblong leaflets, or the terminal one reniform; flowers in a short corymb, appearing subumbellate; petals 3–4 mm long, spatulate; siliques crowded, erect, 2–2.5 cm long; beak minute. $2n = 48$. Moist places in mountains. eAsia: Alas, Yuk, wMack s to Ore, Alta. Map 573

Cardaria Desv. / Hoary Cress

Perennials, spreading by creeping roots and forming dense patches as noxious weeds; stems erect or decumbent, 1–5 dm high, branched above; basal leaves petioled, oblong, irregularly toothed to nearly entire; cauline leaves sessile, ovate to narrowly oblong, sagittate-clasping, puberulent; racemes in flat-topped clusters; sepals blunt; petals white, 3–4 mm long, exceeding the sepals; pods (silicles) ovoid, cordate or nearly globose, firm-walled, indehiscent or tardily dehiscent, tipped by a short persistent style; seeds 1 or 2 in each locule.

1 Fruit and sepals with short simple hairs — **C. pubescens**
  Fruit and sepals glabrous — 2

2 Silicles cordate **C. draba**
Silicles not cordate **C. chalepensis**

C. chalepensis (L.) Handel-Mazzetti
Similar to *C. draba* but differing mainly in the larger, ovate, obovate to somewhat reniform silicles, the bases truncate to cuneate, never cordate. $2n = 80^*$. Fields and roadsides. Introduced. Common.

C. draba (L.) Desv. / Heart-podded Hoary Cress
Stems rather stout, somewhat hoary, 1–5 dm tall, from large white rootstocks, much branched above; leaves puberulent with short unbranched hairs; racemes in dense terminal corymbs; silicles glabrous, reniform to cordate, but often scarcely notched at the base, 3–4 mm long, 3–5 mm wide; seeds oval, slightly flattened, reddish brown. $2n = 64^*$. Produces hybrids ($2n = 72^*$) with the preceding species in sAlta. Fields and roadsides. Introduced. Common.

C. pubescens (C.A. Mey.) Rollins / Globe-podded Hoary Cress
Perennial; stem 1–4 dm tall, stout, branched above, sparsely to densely pubescent; leaves 2–8 cm long, 1.5–2.0 mm wide; silicles sparsely to densely pubescent, 3.0–4.5 mm long, 2.5–4.5 mm wide, strongly inflated, obovoid to subglobose. $2n = 16^*$. Fields and roadsides. Introduced. Common.

Conringia Adans.

C. orientalis (L.) Dum. / Hare's-ear Mustard
Annual or winter annual, glabrous throughout and slightly succulent; stem erect, usually simple, 3–6 dm high; leaves oval to elliptical, broadly rounded at apex, 4–10 cm long, entire, bluish green, the cauline ones sessile and deeply cordate-clasping; flowers in elongate racemes; petals yellowish white, 8–12 mm long; pods (siliques) ascending, 8–12 cm long, 4-angled, short-beaked; seeds in one row in each locule, oblong, 2.5–3 mm long, the surface granular. $2n = 14$. A weed of fields, roadsides, and waste ground. Introduced. Occasional.

Descurainia Webb & Berthelot / Tansy Mustard

Annual or biennial herbs, more or less pubescent, often greyish green, the pubescence including forked or stellate hairs; leaves twice pinnatifid or finely dissected, the basal rosette usually withered at fruiting time; flowers 2–4 mm wide, yellow or whitish, in terminal racemes, these much elongated in fruit; petals small, often barely surpassing the sepals; siliques linear-cylindric to clavate, slender-pedicelled, the valves 1-nerved; seeds elliptic or oblong, marginless, in 1 or 2 rows in each locule.

1 Pedicels much shorter than the fruit; replum 2–3-nerved; seeds uniseriate, mostly more than 20; some leaves bi- or tri-pinnate **D. sophia**
Pedicels more than half the length of the fruit; replum 0–1-nerved; seeds partially biseriate or less than 20; leaves generally pinnate 2

2 Fruits somewhat clavate, not or only slightly torulose, about as long as or shorter than pedicels; replum nerveless **D. pinnata**

Fruits not clavate, generally distinctly torulose, mostly longer than pedicels; replum 1-nerved **D. richardsonii**

D. pinnata (Walt.) Britt. / Green Tansy Mustard
Stems simple or variously branched, 2–6 dm tall, glandular and moderately pubescent; lower leaves bipinnate or pinnate with deeply cleft segments, greenish, glabrous or somewhat pubescent; upper leaves reduced and less divided, the segments narrow and mostly obtuse; siliques 6–13 mm long, narrowly clavate, 1–2 mm wide; pedicels at maturity widely divergent, about as long as the siliques. $2n = 14, 28^*, 42$. Dry open areas and waste ground. BC, swMack, Alta to Que, generally distributed s. Map 574

D. richardsonii (Sweet) O.E. Schulz / Grey Tansy Mustard
Stems 3–8 dm high, erect, branched above, more or less pubescent throughout; leaves grey with close pubescence, pinnate, the segments often lobed or deeply pinnatifid; siliques linear, 5–12 mm long, about 1 mm wide; pedicels erect or ascending, distinctly shorter than the siliques; style 0.5 mm or more long. $2n = 14, 28, 42$. Dry open areas and waste ground. Alas, Yuk, wMack to Huds Bay, sLab s to BC, Ida, Mont, NM, Kans, Minn, Me. Map 575

D. sophia (L.) Webb / Flixweed; Tansy Mustard
Stems 3–10 dm tall, erect, usually much branched, greyish green with close pubescence; leaves greyish green, pubescent, 2–3-pinnate into linear or narrowly oblanceolate segments; siliques narrowly linear, 15–30 mm long, 0.5–1 mm wide, erect or ascending, slightly curved; pedicels spreading-ascending, 7–12 mm long. $2n = 14, 28^*, 56$. Fields and waste ground. Introduced. Common.

Diplotaxis DC.

D. muralis (L.) DC. / Sand Rocket
Annual or biennial, branched from the base, sparingly pubescent or glabrous; stems 2–5 dm high, erect or arched-ascending; leaves mostly basal, oblanceolate in outline, pinnatifid or coarsely toothed, long-attenuate to the base; sepals 3–5 mm long; petals yellow, 5–10 mm long; fruiting raceme elongate, open, the pedicels 8–20 mm long; siliques ascending, flattish, 2–3 cm long, about 2 mm wide, attenuate at both ends, short-beaked; seeds in 2 rows in each locule, small, smooth, marginless. $2n = 42$. Roadsides and waste ground. Introduced. Common.

Draba L.

Low tufted herbs with scapose or leafy stems, and usually a prominent leaf-rosette at the base; leaves simple, entire or toothed, mostly stellate-pubescent; flowers racemose; petals yellow or white, longer than the sepals; pods flat, elliptic to linear, sometimes twisted, rarely as much as 5 times as long as wide; stigmas almost capitate; seeds in 2 rows in each locule.

1	Plants completely glabrous	2
	Plants at least partly pubescent	3
2	Petals yellow; style less than 0.15 mm long	**D. crassifolia**
	Petals white; style more than 0.15 mm long	**D. longipes**
3	Style less than 0.15 mm long	4
	Style more than 0.15 mm long	10
4	Annual or biennial plants of lowland habitats	5
	Perennial plants of montane and alpine habitats	7
5	Petals deeply bilobed, white; cauline leaves absent	**D. verna**
	Petals emarginate, yellow; cauline leaves present	6
6	Silicles about 6 times as long as broad	**D. reptans**
	Silicles less than 3 times as long as broad	**D. nemorosa**
7	Lower surface of leaves glabrous or with simple to multi-forked hairs; cruciform to stellate hairs absent	**D. crassifolia**
	Lower surface of leaves predominantly with cruciform to stellate hairs	8
8	Lower surface of leaves predominantly with stellate hairs	**D. praealta**
	Lower surface of leaves predominantly with cruciform hairs	9
9	Upper surface of leaves with mainly trifid and cruciform hairs; stems mainly with cruciform hairs	**D. stenoloba**
	Upper surface of leaves with mainly simple or bifid hairs; stems mainly with simple hairs	**D. albertina**
10	Silicles mainly with short-stalked cruciform or stellate hairs.	**D. cana**
	Silicles glabrous or with simple to forked hairs	11
11	Lower surface of leaves predominantly with simple to 3-forked hairs	12
	Lower surface of leaves predominantly with cruciform to stellate hairs	13
12	Stem glabrous; pollen fertile	**D. fladnizensis**
	Stem pubescent; pollen not fertile	**D. densifolia**
13	Lower surface of leaves mainly with cruciform hairs	14
	Lower surface of leaves mainly with stellate hairs	18
14	Stem-leaves (2) 3–6 (10); fruiting stems usually more than 15 cm high	**D. borealis**
	Stem-leaves 0–2 (3); fruiting stems usually less than 15 cm high	15
15	Lowest pedicels shorter than silicles; basal leaves compacted	16
	Lowest pedicels usually about same to twice length of silicles; basal leaves loosely tufted	17
16	Petals white; stem leafless or with 1 leaf; upper leaf-surface with some stellate hairs	**D. porsildii**
	Petals yellow; stem leafless; upper leaf-surface lacking stellate hairs	**D. macounii**

17 Leaf hairs sessile; petals yellow **D. kananaskis**
Leaf hairs short-stalked; petals white **D. longipes**

18 Lower surface of leaves mainly with long-stalked stellate hairs 19
Lower surface of leaves predominantly with short-stalked or sessile stellate hairs 20

19 Long-stalked stellate hairs on both leaf-surfaces **D. ventosa**
Long-stalked stellate hairs only on lower surface; upper surface with long, simple or once- or twice-forked hairs **D. paysonii**

20 Stellate hairs on lower surface of leaves mainly with 9 or more rays 21
Stellate hairs on lower surface of leaves mainly with 8 or less rays 25

21 Stellate hairs on lower surface of leaves sessile; pollen not fertile **D. oligosperma**
Stellate hairs on lower surface of leaves stalked; pollen fertile 22

22 Stellate hairs less than 0.2 mm in diameter **D. nivalis**
Stellate hairs more than 0.2 mm in diameter 23

23 Cauline leaves 2–10, rarely fewer; stem over 10 cm high **D. glabella**
Cauline leaves 0–1, stem less than 10 cm high (occasionally taller in *D. incerta*) 24

24 Petals yellow; lowest pedicel usually twice length of silicle; stellate hairs mostly with one longer central axis **D. incerta**
Petals white; lowest pedicel less than twice length of silicle; stellate hairs without a longer central axis **D. lonchocarpa**

25 Stems with 0–2 leaves 26
Stems with 4 or more leaves 27

26 Flowers yellow, stem-leaves 0–1 **D. incerta**
Flowers white; stem-leaves 0–2 **D. porsildii**

27 Stems with more than 10 leaves; pedicels appressed to stem; petals bright yellow **D. aurea**
Stems with less than 10 leaves; pedicels spreading; petals creamy yellow to white 28

28 Lowest pedicel about same length to twice length of silicle; some of cruciform and stellate hairs long-stalked; petals creamy yellow **D. praealta**
Lowest pedicel shorter than silicle; all cruciform and stellate hairs short-stalked; petals white 29

29 Style more than 0.5 mm long; mostly cruciform hairs on leaf-surfaces **D. borealis**
Style less than 0.5 mm long; stellate hairs on leaf-surfaces mostly with 7 or more rays **D. glabella**

D. albertina Greene
Biennial or perennial; stem 0.4–3.0 dm tall, erect, mostly with simple or bifid hairs; basal leaves 0.3–3.5 cm long, 1.5–7.0 mm wide, obovate to oblanceolate, entire to slightly denticulate; cauline leaves (0) 1–3 (4), 0.4–3.0 cm long, 2–18 mm wide, sessile, lanceolate to broadly lanceolate, entire to slightly denticulate;

leaves with simple to bifid hairs above, rarely glabrous, short-stalked bifid, trifid or cruciform hairs below, rarely some simple hairs; raceme open; flowers yellow, 2–30; pedicels 2–15 mm long; sepals 1.5–2.0 mm long, glabrous or with sparse simple hairs; petals 2.5–3.0 mm long; fruit 5–15 mm long, 1–2 mm wide, narrowly elliptic to narrowly ovate, generally glabrous, dehiscent. $2n = 24^*$. Dry to moist open areas at alpine elevations. sYuk, wMack; sBC, Alta s to Calif, Colo. Map 576

D. aurea Vahl
Perennial; stems 1–several, 1–5 dm tall, erect or somewhat decumbent at base, simple or branched, with mixed simple, bifid and cruciform hairs; basal leaves rosulate, 1–5 cm long, 2–10 mm wide, oblanceolate, entire to slightly denticulate, densely pubescent above and below with mostly cruciform hairs; cauline leaves 3–30, sessile and sometimes clasping, lanceolate to ovate, entire to denticulate; flowers numerous, yellow; sepals 2–3 mm long; petals 4.5–6.0 mm long; fruit 7–20 mm long, 2–4 mm wide, lanceolate, often twisted, pubescent with simple and bifid hairs, rarely glabrous; style 0.3–1.5 mm long. $2n = 74^*$. Rocky slopes, open woods, clearings, and disturbed areas. Greenland: Alas, Yuk, wMack to Huds Bay, nQue s to Ida, Ariz, NM, nSask; Lk Superior. Map 577

D. borealis DC.
Perennial; stems 1–several, 0.5–4.0 (5.5) dm tall, erect, simple, pubescent with simple, forked and branched hairs; basal leaves 0.5–5.0 cm long, 3–18 mm wide, oblanceolate, entire to denticulate, pubescent with stalked cruciform hairs below, mostly simple or bifid hairs above; cauline leaves (2) 3–6 (10), 0.5–4.0 cm long, 2–18 mm wide, lanceolate, entire to dentate, with simple and branched hairs above and below; flowers mostly 5–20, white or cream; sepals 2–3 mm with simple and bifid appressed hairs; petals 4–6 mm long; fruit 7–14 mm long, 2–4 mm wide, lanceolate to oblong, often twisted, usually with simple and forked hairs; style 0.5–1.0 mm long. $2n = 80^*$. Moist banks and ridges mostly at alpine elevations. eAsia: Alas, Yuk, nwMack, Victoria Is s to BC, Alta. Map 578

D. cana Rydb.
Perennial; stems several, 0.5–2.5 dm tall, simple to freely branched, densely pubescent with intermixed short stellate hairs and long simple or bifid hairs; basal leaves 1–3 cm long, 2–8 mm wide, oblanceolate, entire to denticulate, densely canescent with branched stellate hairs above and below, the margins sometimes with simple or bifid hairs; cauline leaves (3) 5–8 (12) often wider and more evidently toothed than the basal; raceme in fruit about half the length of stem; flowers 10–50, white; sepals about 2 mm long with simple and forked hairs; petals 3–5 mm long; fruit 4–12 mm long, 1.5–3.0 mm wide, elliptic-lanceolate to oblong, often twisted, stellate-pubescent; style 0.4–1.0 mm long. $2n = 32^*$. *D. lanceolata* Royle of Ed. 1. Dry open areas from foothills to alpine elevations. Asia, Greenland: Alas, Yuk, swMack, nwSask s to Ida, Nev, Utah, Colo; Gt Lakes, Que, NB. Map 579

D. crassifolia R. Grah.
Perennial; stem 0.2–1.2 dm tall, erect, glabrous; basal leaves 0.2–3.0 cm long, 1–6 mm wide, narrowly elliptic to obovate or oblanceolate, entire to slightly denticulate, glabrous or with sparse simple or bifid hairs; cauline leaves 0–1, glabrous; raceme open; flowers yellow, 2–15; pedicels 2–6 mm long; sepals 1.0–1.5 mm long, glabrous; petals 1.5–2.0 mm long; fruit 4–10 mm long, 1.5–2.5 mm wide, narrowly ovate, glabrous, dehiscent. $2n = 40^*$. Dry to moist open areas at alpine elevations. Greenland: Alas, Yuk, Mack to sBaffin Is, nQue, Lab, s to Calif, Ariz, Mont, Colo. Map 580 ●

D. densifolia Nutt. *ex* T. & G.
Perennial, caespitose; stems leafless, 0.3–15.0 cm tall, erect, pubescent with simple and forked hairs; leaves 2–18 mm long, 0.5–3.0 mm wide, densely imbricate, linear to linear-oblanceolate, sparsely pubescent below with forked to stellate hairs, generally glabrous above, margins often crisped, ciliate with long, straight, simple hairs; flowers 3–15, yellow; petals 2–6 mm long; fruit 2–7 mm long, 2.0–3.5 mm wide, ovate to ovate-elliptic, with simple to stellate hairs, occasionally glabrous; style 0.5–1.0 mm long. $2n = 36^*$. An agamospermous species. Alpine ridges. sBC, swAlta s to Calif, Ida, Utah, wWyo; cAlas. Map 580 ○

D. fladnizensis Wulfen
Perennial, caespitose; stems 1–9 cm tall, erect, simple, essentially glabrous; basal leaves 3–10 mm long, 1–2 mm wide, oblanceolate, entire, with simple to trifid hairs above and below, margins ciliate, midrib prominent; cauline leaves 0–2; flowers 3–12, white; sepals 1.0–1.5 mm long, with simple hairs; petals 2–3 mm long; fruit 3–6 mm long, 1.5–2.0 mm wide, oblong ovate, mostly glabrous; style 0.1–0.2 mm long. $2n = 16$. Alpine slopes. Circumpolar: Alas, Yuk to Southampton Is; neBaffin Is s to nBC; Alta. Map 581 ●

D. glabella Pursh
Similar to *D. borealis*; differing in the characters of the key. $2n = 64, 80$. Moist alpine banks and ledges. Circumpolar: Alas, Yuk to nBaffin Is to Nfld s to nBC, cMack, nMan, James Bay, NS; Alta; Gt Lakes. Map 581 ○

D. incerta Payson
Perennial, rather loosely tufted; stems mostly 0.2–1.5 (2.0) dm tall, simple, pubescent with simple, stellate and irregularly branched hairs; basal leaves 7–15 (25) mm long, 1.0–3.5 mm wide, imbricate, linear-oblanceolate, entire, pubescent especially below with stalked, irregularly branched multi-rayed hairs, margins ciliate with simple and forked hairs; cauline leaves usually absent, occasionally 1; flowers generally 2–8, yellow; sepals 2–3 mm long, glabrous or with a few simple hairs; petals 4–5 mm long; anthers dehiscent, pollen well-formed; fruit 6–10 mm long, 2.5–3.8 mm wide, elliptic, ovate to suborbicular, pubescent with simple or branched hairs, occasionally glabrous. $2n = 112^*$. Generally dry, open areas from montane to alpine elevations. sAlas, swYuk s to Wash, Ida, Wyo. Map 582

D. kananaskis Mulligan
Perennial, loosely tufted; stem 0.3–2.0 dm tall, with sparse simple, forked or stellate hairs; basal leaves 0.5–2.0 cm long, 2–9 mm wide, with slender petioles, obovate, entire to denticulate, with nearly sessile, mostly cruciform hairs above and below, the margins with simple, forked or stellate hairs; cauline leaves 1–2, ovate to narrowly ovate; sepals about 2.0 mm long; petals 3.5–4.0 mm long, yellow, spatulate-oblanceolate; fruit 6–10 mm long, 2–3 mm wide, longer than pedicels with a few simple hairs; style 0.3–0.5 mm long. $2n = 64$*. Alpine slopes. Alta (endemic). Map 583 ○

D. lonchocarpa Rydb.
Similar to *D. nivalis* but generally somewhat larger plants; differing in the character of the key and the following: stems glabrous, occasionally with a few scattered simple or stellate hairs towards the base; upper leaf-surface with appressed simple and bifid hairs as well as the stellate; fruits longer, mostly about 10 mm. $2n = 16$*. *D. nivalis* var. *elongata* S. Wats. Dry alpine slopes and screes. seAlas, sYuk, swMack s to Wash, neOre, Ida, Colo. Map 583 ●

D. longipes Raup
Perennial; stem 1.0–1.5 dm tall, with sparse simple to branched hairs; basal leaves 1.5–3.0 cm long, 1–5 mm wide, oblanceolate, entire to denticulate with stalked bifid, trifid and cruciform hairs above and below, sometimes glabrous above; cauline leaves 1–3 (5), elliptic to lanceolate; raceme somewhat dense, subcorymbose in flower, open in fruit; flowers 3–15, white; sepals 2.5–3.0 mm long, glabrous or with sparse simple hairs; petals 4–5 mm long; fruit 4–13 mm long, 2–3 mm wide, narrowly ovate to oblong-lanceolate, glabrous; style 0.6–0.9 mm long. $2n = 64$. Moist alpine banks and ledges. Alas, Yuk, swMack s to BC, Alta. Map 584

D. macounii O.E. Schulz
Perennial; stem 0.1–0.8 dm tall, erect, leafless, with simple and forked hairs; leaves basal, 3–15 mm long, 2–5 mm wide, entire, upper surface glabrous or with simple hairs, lower surface predominantly with trifid to cruciform hairs; raceme 1–15-fruited; fruiting pedicels 1–4 mm long; sepals about 1.5 mm long; petals yellow, about 4 mm long; fruit 3–6 mm long, 2–3 mm wide, usually glabrous, elliptic to ovate, up to 8 ovules in each locule; seeds oval, light brown. $2n = 64$. Alpine slopes. eAlas, sYuk, swMack s to BC, Alta. Map 585

D. nemorosa L.
Annual; stem 0.3–3.5 dm tall, erect with simple and forked hairs; leaves 0.2–4.5 cm long, 0.1–2.0 cm wide, generally confined to lower third of stem, ovate-lanceolate to obovate, denticulate, mostly with simple and fewer long-stalked bifid, trifid and cruciform hairs above and mostly long-stalked cruciform hairs below; raceme open, flowers yellow, 10–50; sepals 2.0–2.5 mm long, glabrous or with simple hairs; petals 2.5–4.0 mm long; fruit 3–10 mm long, 1–2 mm wide, elliptic, glabrous or pubescent, dehiscent. $2n = 16$*. Moist to dry open

ground, sometimes weedy. More or less circumpolar: sAlas, sYuk, sMack to Huds Bay, swQue s to Calif, Colo, Neb, Minn. Map 586

D. nivalis Liljebl.
Perennial, caespitose, often cinereous; stems 1–10 cm tall, erect, simple, sparsely to densely stellate-pubescent; leaves 2–10 mm long, 1–2 mm wide, oblanceolate to obovate, entire, marcescent, densely stellate-pubescent above and below, midrib prominent, cauline leaves absent, rarely 1; flowers 2–10, white; sepals about 2 mm long, with simple and forked hairs; petals 2–4 mm long; fruit mostly 6–8 mm long, 1–2 mm wide, narrowly lanceolate to elliptic, glabrous, often twisted; style 0.2–0.4 mm long. $2n = 16$*. Dry alpine slopes and screes. Circumpolar: Alas, Yuk to Ellesmere Is to Nfld s to BC, Alta, sMack, Huds Bay, nQue; Gaspé. Map 587

D. oligosperma Hook.
Similar to *D. incerta* but generally more compact, less loosely tufted, and the leaves more tightly imbricate. In addition to these and the differences in the key are the following: plants shorter, 2–6 (8) cm in ours; stems leafless; leaves generally less than 1 mm wide, the margins often lacking hairs or with only a few simple hairs at the base; anthers indehiscent, pollen misshapen; fruit 2.5–8.0 mm long. $2n = 64$*. An agamospermous species. Alpine slopes and screes. seAlas, sYuk, nwMack s to Calif, Nev, Utah, Colo. Map 588

D. paysonii Macbr.
Perennial, caespitose; stems 0.5–4 (6) cm tall, scapose, with simple and forked hairs; leaves 4–14 mm long, 1.5–2.0 mm wide, marcescent, imbricate, linear to linear-oblanceolate, entire, with mostly simple and bifid hairs above and long-stalked stellate hairs below, these lacking above, margins ciliate, midrib prominent; raceme somewhat congested, flowers 3–10, yellow; sepals 2–3 mm long, hairy; petals 2–4 mm long; fruit 3–8 mm long, 2.5–4.5 mm wide, ovate-lanceolate, ovate to nearly orbicular, with dense simple to forked hairs, somewhat inflated; style 0.5–1.0 mm long. $2n = 42$*. An agamospermous species. Alpine screes. BC, Alta s to Calif, Nev, Utah, Wyo; eAlas. Map 589

D. porsildii Mulligan
Perennial; stem 2.0–6.5 cm tall, erect, glabrous to sparsely pubescent; leaves 3–12 mm long, 1.5–2.0 mm wide, oblanceolate, entire, with simple bifid, cruciform and 4–8-rayed stellate hairs above and below; cauline leaves 0–1; raceme compact, mostly 2–5-flowered; sepals 1.5–2.0 mm long; petals white, 2–3 mm long; fruit 4–8 mm long, 2–3 mm wide, obovate, glabrous; style about 0.25 mm long. $2n = 32$*. Moist banks and turfy slopes at alpine elevations. sAlas, sYuk, swMack, nBC; Alta. Map 590

D. praealta Greene
Biennial or perennial; stems 1–several, 1–3 dm tall, simple or branched, pubescent with simple to cruciform and stellate hairs; basal leaves 10–30 mm long, 2–6 mm wide, oblanceolate, entire to denticulate, pubescent above and more densely

below with simple, forked and stellate hairs; cauline leaves 1–6, similar to basal, the margins often more distinctly toothed; raceme open, flowers 5–30, white to yellow; fruit 8–14 mm long, 1.5–2.5 mm wide, lanceolate to elliptic, with simple and forked hairs; style 0.1–0.2 mm long. $2n = 56$*. Moist montane and subalpine banks. Alas, Yuk, Mack s to Ore, Wyo; Nev. Map 591 ●

D. reptans (Lam.) Fern.
Annual; stem 0.5–2.0 dm tall, erect, simple or branching basally, hispid below, glabrous above; basal leaves 1–3 cm long, 2–10 mm wide, ovate, obovate to spatulate, entire, bristly with simple, and mostly long-stalked bifid and trifid hairs above and below; cauline leaves 1–3, near base; racemes subumbellate, flowers white, 3–12, dimorphic, those of branches smaller and often apetalous; sepals 1.5–2.5 mm long with simple hairs; petals 3–4 mm long; fruit 5–20 mm long, 1–2 mm wide, linear-lanceolate, slightly curved, glabrous or hispid; style mostly wanting. $2n = 30$. Dry sandy or gravelly ground, especially disturbed areas. sAlta to sOnt, Pa s to Calif, Ariz, NM, Tex, Mo, Mich, Ga. Possibly introduced in our area. Map 591 ○

D. stenoloba Ledeb.
Biennial or perennial; stem 0.3–3.2 dm tall, erect, mostly with short-stalked cruciform hairs; basal leaves 0.4–3.0 cm long, 1.5–9.0 mm wide, obovate to oblanceolate, entire to slightly denticulate; cauline leaves (0) 1–2 (4), 0.3–1.6 cm long, 2–10 mm wide, elliptic to broadly lanceolate, generally with shallow teeth; leaves mostly with short-stalked trifid and cruciform hairs above and below, rarely glabrous above; raceme open, flowers yellow, 2–15; pedicels 3–15 mm long; sepals 1.5–2.5 mm long, glabrous or with simple to bifid hairs; petals 3.0–3.5 mm long; fruit 10–20 mm long, 1.5–2.0 mm wide, narrowly elliptic to narrowly ovate, glabrous or rarely pubescent, dehiscent. $2n = 40$. Moist banks and gravel at alpine elevations. Alas, Yuk s to BC, Alta. Map 592

D. ventosa A. Gray
Similar to *D. paysonii*; differing in the character of the key. $2n = 36$*. An agamospermous species. Alpine screes. Alta s to Wyo, Utah, Colo; Yuk. Map 593

D. verna L.
Annual; stems 0.5–2.0 (2.5) dm tall, scapose, simple, glabrous or pubescent near base; basal leaves rosulate, 10–25 mm long, 1–5 mm wide, spatulate to oblanceolate, entire to denticulate, with short and long-stalked branched hairs above and below, sometimes more or less glabrous above; racemes open, flowers 3–20, white; sepals 1–2 mm long, sparsely hirsute; petals 2–3 mm long; bifid nearly to middle; fruit 3–10 mm long, 2–3 mm wide, elliptic, glabrous; style 0.1 mm long. $2n = 14$–94. *Erophila verna* (L.) Chevall. Open, dry areas. Introduced. Rare.

Eruca Mill.

E. sativa Mill. / Garden Rocket
Annual herb; stem 1–6 dm tall, branched above, stiff hairy; leaves 5–15 cm long, lyrate-pinnatifid to pinnate, coarsely toothed, the terminal lobe oblong or obovate, the lower petiolate, upper sessile; inflorescence a raceme, ebracteate; sepals 7–10 mm long, erect, inner pair saccate; petals 12–20 mm long, white to yellow with conspicuous veins; silique 12–25 mm long, including flat seedless beak, 3–5 mm wide, erect, valves 1–nerved; seeds biseriate. $2n = 22$. Fields and waste ground. Introduced. Rare.

Erucastrum Presl

E. gallicum (Willd.) Schulz / Dog Mustard
Annual, or winter annual, erect or ascending, branching, 2–6 dm high; stem retrorsely pubescent, the hairs stiff and simple; leaves glabrous or hairy, oblong or oblanceolate in general outline, deeply pinnatifid, the lobes rounded and variously toothed, the sinuses broad; flowers in elongated racemes, the pedicels slender and in fruit 6–10 mm long, the lower flowers subtended by reduced leaves or leaf-like bracts; petals pale yellow, 5–7 mm long; pods (siliques) 4-angled, 2–4 cm long, usually curved upward, the valves sharply 1-nerved, the beak about 3 mm long; seeds 1 row in each locule. $2n = 30$. Fields, roadsides, and waste ground. Introduced. Occasional.

Erysimum L.

Annual, biennial or perennial herbs, pubescent with appressed 2–3-forked hairs; leaves simple, not clasping; flowers yellow or purple, in terminal racemes; siliques linear, 4-angled, the valves keeled by a strong nerve; style very short; stigma broadly lobed; seeds 1 row in each locule, oblong, marginless.

1 Petals 3–5 mm long; siliques 1–3 cm long; pubescence of leaves mostly of trifid hairs **E. cheiranthoides**
Petals 6–20 mm long; siliques 2–10 cm long; pubescence of leaves mostly of bifid hairs 2

2 Flowers pink-purple; plants alpine **E. pallasii**
Flowers yellow; plants not alpine 3

3 Petals 6–12 mm long; style 1.0–1.5 mm long **E. inconspicuum**
Petals 15–25 mm long; style generally more than 1.5 mm long **E. asperum**

E. asperum (Nutt.) DC. / Prairie Rocket
Stiff perennial, ash-grey with close pubescence; stems erect, stout, 1–4.5 dm tall; leaves thick and firm, the basal ones numerous, undulate-dentate, the cauline ones narrowly lanceolate to linear, entire or somewhat toothed; limb of petal 5–10 mm long; pedicels loosely ascending or divergent; siliques stout, stiffly

erect or spreading, 6–10 cm long, 2 mm thick. $2n$ = 18, 36. Dry sandy prairies. sBC to sMan s to Calif, NM, Okla, Minn. Map 594 ●

E. cheiranthoides L. / Wormseed Mustard
Annual or winter annual, 2–5 dm tall; stem simple or branched, 2–6 mm thick at base, green to purplish, sparsely pubescent with 2-branched hairs; leaves somewhat spreading, narrowly lanceolate, entire to denticulate, broadest in the middle, finely pubescent with trifid hairs; flowers small, on divergent to spreading-ascending filiform pedicels; petals yellow, 2–4 mm long; siliques 1–3 cm long, erect or nearly so, glabrous or sparsely pubescent. $2n$ = 16. In our area the introduced ssp. **cheiranthoides** as described above and the native ssp. **altum** Ahti, a taller, more robust, and densely pubescent plant, 4–10 dm tall, stem-base purplish, up to 15 mm thick, leaves broadest towards the base and appressed, petals 3–5 mm long. Fields, roadsides, disturbed ground. More or less circumpolar: Alas, Yuk, swMack to Huds Bay, Nfld s to Calif, Utah, NM, Kans, Mo, NC. Map 595

E. inconspicuum (S. Wats.) MacM. / Small-flowered Rocket
Perennial, 3–6 dm tall; stem stiff, simple or branched, grey-pubescent and scabrous with 2–3-branched hairs; leaves oblanceolate to linear, entire to denticulate, pubescent with bifid hairs; pedicels thick, with enlarged tip; petals pale yellow, the limb 2.5–3 mm long; siliques erect, cinereous, 2–5 cm long, 1 mm thick. $2n$ = 54, 81*. Dry, open areas. Alas, Yuk, wMack to cMan, Gt Lakes, eQue s to Ore, Nev, Colo, SD, Minn. Map 596

E. pallasii (Pursh) Fern.
Biennial or short-lived perennial, with prominent basal rosette; leaves linear-lanceolate, 5–7 cm long, entire or sinuate-dentate, the blade gradually tapering into a narrow petiole; flowering axis at first very short, elongating greatly in fruit; sepals scarious-margined; petals bright purple, 10–16 mm long; pods stiffly erect, dark purple, 4–10 cm long, 2 mm thick; seeds 2 mm long. $2n$ = 24, 28, 36, c.42. Alpine shale. nAsia, Greenland: Alas, Yuk to Ellesmere Is s to Gt Bear Lk, nBaffin Is; Alta. Map 594 ○

Halimolobos Tausch

H. virgata (Nutt.) Schulz
Biennial (annual); stems 1 or more, 1–4 dm tall, simple or branched, pubescent with a mixture of longer, straight, simple or forked hairs and shorter, branched hairs; leaves of the basal rosette 3–6 cm long, 5–15 mm wide, lanceolate to oblanceolate, denticulate to dentate; cauline leaves several, reduced above, at least the upper sessile and auriculate; inflorescence an ebracteate raceme; flowers white with pinkish veins, on slender pedicels 7–11 mm long; sepals 2–3 mm long, often purplish; petals 3–4 mm long; stamens 6; fruit a glabrous, erect or nearly so, linear, terete, silique 1.5–4.0 cm long, 1 mm wide; seeds irregularly biseriate. Dry prairies. seAlta to sSask s to eIda, Utah, Colo. Map 597 ○

Hesperis L.

H. matronalis L. / Dame's Rocket
Tall perennial, pubescent with both simple and branched hairs; stem erect, 3–10 dm high, simple, branched above; leaves simple, oblong-oblanceolate to ovate-lanceolate, dentate, pubescent on both sides; flowers large, fragrant, racemose; sepals erect, the outer narrow, crested near the apex; petals purple, varying to pink or white, 15–25 mm long, the claws exceeding the calyx; stigma 2-lobed; silque linear, 5–10 cm long, ascending or spreading, nearly terete, somewhat constricted between the seeds, scarcely dehiscent; seeds in a single row in each locule. $2n = 12, 14, 16, 24, 28, 32$. An escape from cultivation. Introduced. Occasional.

Lepidium L. / Peppergrass

Annual to perennial herbs, glabrous to hirsute with simple hairs; leaves entire, toothed or pinnatifid; flowers small, usually whitish, sometimes apetalous, in dense bractless racemes; petals sometimes shorter than the sepals or none; stamens 2, 4 or 6; ovary with 2 ovules; style short or none; fruit a silicle, strongly flattened at right angles to the septum, orbicular, ovate or obovate, the valves keeled or winged, often notched at the apex; seeds solitary in each locule, hanging from the apex.

1 Upper leaves sagittate or deeply cordate-clasping at the base — 2
Upper leaves tapering at base, not sagittate or cordate-clasping — 3

2 Leaf-auricles acute, much shorter than the rest of the blade; lower leaves essentially like the upper but larger — **L. campestre**
Leaf-auricles rounded, appearing perfoliate; lower leaves strikingly different (pinnately dissected) — **L. perfoliatum**

3 Perennial; 5–12 dm tall; silicle not notched at apex — **L. latifolium**
Annual or biennial; generally less than 5 dm tall; silicle notched at apex — 4

4 Fruit 2.5–3.5 mm long, orbicular to obovate, widest above the middle, glabrous, apical teeth rounded — **L. densiflorum**
Fruit 2.5–3.0 mm long, elliptic to ovate, widest at or below the middle, glabrous or pubescent, apical teeth acute — 5

5 Silicle glabrous — **L. bourgeauanum**
Silicle puberulent or ciliate, at least along margin — **L. ramosissimum**

L. bourgeauanum Thell.
Similar to *L. ramosissimum*; differing in the characters of the key and the fewer racemes per branch, mostly less than 5. $2n = 32$. Open areas and waste places. Alas, Yuk, swMack to Man s at least to sBC, sAlta, sSask, sMan, introduced eCan. Map 597 ●

L. campestre (L.) R.Br. / Pepperwort
Annual or biennial; stem 2–5 dm high, simple to profusely branched above the middle, grey-green with dense spreading hairs; basal leaves entire or lyrate, soon falling; lower cauline leaves petiolate; middle and upper leaves sessile, triangular, denticulate, pubescent, clasping the stem with narrow pointed basal lobes; flowers small; sepals about 1.5 mm long; petals white, little longer than the sepals; stamens 6; silicles oblong-ovate, about 5 mm long, densely covered with small white vesicles, the upper surface concave; style 0.2–0.6 mm long, in the notched apex. $2n = 16$. Fields, roadsides, waste ground. Introduced. Rare.

L. densiflorum Schrad. / Common Peppergrass
Annual or winter annual, 2–5 dm tall, with single erect stem, branched above, grey-green with dense short hairs; basal leaves petiolate, deeply toothed to pinnately lobed, grey-green with pubescence; upper leaves sessile, linear-lanceolate, entire or slightly toothed; racemes numerous, 5–15 cm long, densely flowered; sepals about 1 mm long; petals filiform, shorter than the sepals, or none; silicles 2–3.5 mm long, obovate to oblong-obovate, glabrous or puberulent, broadly rounded at the winged summit; style obsolete, the stigma practically sessile in the apical notch. $2n = 32$. In our area var. **densiflorum** as described above and var. **macrocarpum** Mulligan (var. *bourgeauanum* (Thell.) C.L. Hitchc. of Ed. 1), with larger fruits, 3.0–3.5 mm long, and conspicuously flattened pedicels. A weedy species of dry open areas and disturbed habitats. Europe: Alas, Yuk, wMack to Huds Bay, Nfld s to Calif, Ariz, Colo, Okla, NC. Map 598

L. latifolium L.
Tall perennial, 5–10 dm high, with thick branched rootstock and spreading roots; stems and leaves glabrous or nearly so; leaves entire to dentate, the basal petiolate, as much as 3 dm long, the upper nearly sessile, ovate to lanceolate, much reduced; racemes dense, numerous in a large panicled cluster; pedicels slender, much longer than the fruits; sepals broadly white-margined, pilose; petals white, about 1.5 mm long; silicles ovate-orbicular, 2 mm long, with no or very slight apical notch, tipped by stigma, the style very short. $2n = 24$*. Waste ground, Lethbridge. Introduced. Rare.

L. perfoliatum L.
Annual, with single erect stem, 1–5 dm tall, sparsely hairy, usually branched above; basal leaves long-stalked, bipinnate, the ultimate segments less than 1 mm wide, pubescent to glabrous; middle and upper leaves broadly ovate, 1–1.5 cm long, acute, glabrous, clasping the stem with large basal lobes, the uppermost perfoliate; pedicels longer than the fruits; sepals about 1 mm long, pilose; petals pale yellow, a little longer than the sepals; silicles variable in shape, about 4 mm long, rhombic-ovate, very narrowly winged above; style usually projecting beyond the apical notch. $2n = 16$*. Roadsides and waste ground. Introduced. Rare.

L. ramosissimum A. Nels. / Branched Pepperwort
A biennial similar to *L. densiflorum;* differing in the characters of the key. $2n = 64^*$. Dry slopes and plains. seBC to Man s to Ida, Mont, NM, ND, introduced wMack, wOnt. Map 599

L. sativum L. / Garden Cress
Annual with stiffly erect stem, 2–4 dm tall; leaves mostly dissected, pubescent or glabrous, the basal ones long-stalked, soon falling; sepals slightly over 1 mm long; petals white or reddish, twice as long as the sepals; silicles 5–7 mm long, oblong-ovate, narrowly winged above, glabrous and more or less glaucous; style not projecting beyond the deep apical notch. $2n = 16, 24$. Roadsides and waste ground. Introduced. Occasional.

Lesquerella S. Wats. / Bladder-pod

Low herbs, tufted, with prominent rosettes of leaves at the base, stellate-pubescent; stems usually unbranched; radical leaves linear to oblanceolate, entire or remotely toothed; cauline leaves similar, entire; flowers racemose; fruiting pedicels often curved; sepals oblong; petals yellow, spatulate; anthers linear, sagittate; pods (silicles) globose, ellipsoid or ovoid, usually inflated, stellate-pubescent; styles 2–5 mm long; seeds 2–7 in each locule; funiculi attached to the nerved septum for about half their lengths.

1 Fruit ovate; pedicels generally sigmoid **L. alpina**
Fruit more or less globose; pedicels not sigmoid 2

2 Inflorescence often secund; pedicels distinctly recurved; styles generally 3.0–5.5 mm long **L. arenosa**
Inflorescence not secund; pedicels erect, straight or slightly curved; styles generally 1.0–2.5 mm long **L. arctica**

L. alpina (Nutt.) S. Wats.
Caespitose perennial; stems erect, 3–12 cm long; basal leaves oblanceolate to ovate, 1–4 cm long, entire; cauline leaves few, linear; petals yellow, narrowly spatulate, 5–7 mm long; fruiting pedicels 5–10 mm long; pods 4–5 mm long, acute; septum frequently perforated. $2n = 10$. In our area ssp. **alpina**. Dry plains. sAlta to sSask s to Mont, Utah, Colo, Kans, wND. Map 600 •

L. arctica (Wormskj. *ex* Hornem.) S. Wats.
Perennial; stems 0.5–2.5 dm long, erect, spreading or prostrate; basal leaves obovate to oblanceolate, mostly 2–6 cm long, entire; petals yellow, 4–7 mm long, obovate; fruiting pedicels generally 5–20 mm long; pods 4–6 (9) mm long. $2n = 60$. Sandy slopes and ridges to alpine elevations. Siberia, Greenland: Alas, Yuk to Ellesmere Is s to nBC, swMack, Huds Bay, sBaffin Is, nLab; Alta; eQue, Nfld. Map 600 ○

L. arenosa (Richards.) Rydb.
Annual to short-lived perennial; tufted; stems slender, decumbent or ascending, 0.1–2.5 dm long; basal leaves narrowly oblanceolate to oval, 1–6 cm long, entire or remotely dentate, usually narrowed into a slender petiole; cauline leaves linear to oblanceolate; petals yellow or variously tinged with red or purple, about 7 mm long; fruiting pedicels 7–10 mm long; pods mostly 3.5–5.5 mm long. $2n = 10, 26, 30$. In our area var. **arenosa**. Dry sandy plains. sAlta to sMan s to Mont, Wyo, nwNeb, SD. Map 601

Nasturtium R.Br.

N. officinale R.Br.
Perennial herbs, essentially glabrous; stem 10–60 cm long, erect to procumbent and floating, simple to sparingly branched; leaves 4–12 cm long, pinnate, pinnae ovate to orbicular, the terminal lobe ovate to round-cordate, margins more or less entire; inflorescence of corymbose racemes elongating in fruit; flowers white, generally on spreading pedicels 8–20 mm long; sepals pale, outer pair saccate; petals 3–4 mm long, about twice as long as sepals; stamens 6; fruit a silique 10–18 mm long, erect or ascending; style about 1 mm long; seeds biseriate. $2n = 32$. *Rorippa nasturtium-aquaticum* (L.) Schinz & Thell. Ditches and streams in shallow water. Introduced. Rare. A sterile hybrid ($2n = 48$) between this species and the similar and likewise introduced species (but not in our area) *N. microphyllum* (Boenn.) Reichenb. ($2n = 64$) occurs at Banff.

Neslia Desv.

N. paniculata (L.) Desv. / Ball Mustard
Annual or winter annual; stem erect, slender, 2–8 dm tall, branched above, stellate-pubescent; leaves entire or nearly so, sparsely pubescent with small, branched hairs, the cauline leaves sessile, lanceolate, sagittate-clasping; raceme elongate in fruit; flowers bright yellow, small, about 1.5 mm wide; sepals short, equal at base; petals 2 mm long; style elongate; ovules 2 in each locule; pod (silicle) subglobose, firm, thick-walled, indehiscent, reticulate and pitted, about 2 mm thick, usually 1-chambered by obliteration of the septum; seed 1 (rarely 2) yellowish orange. $2n = 14$. A weed of fields and waste ground. Introduced. Common.

Physaria A. Gray

P. didymocarpa (Hook.) A. Gray / Double Bladder-pod
Caespitose perennial, silvery-stellate; stems several from a thick crown, simple, decumbent, about 1 dm long, rather leafy; basal leaves numerous, petiolate, broadly obovate, 1–4 cm wide, often repand or dentate, densely silvery-pubescent with fine branched hairs; cauline leaves few and reduced; inflorescence racemose; sepals pubescent; petals yellow, 8–12 mm long, spatulate; pods didymous (2-lobed), inflated, coarsely pubescent, erect, the valves 8–12 mm long; styles persistent, 7–9 mm long; seeds 4 in each locule. $2n = 8, 16^*, 24$. Dry slopes and plains. seBC to swAlta s to Wash, Ida, Wyo. Map 602 ●

Raphanus L.

R. raphanistrum L. / Jointed Charlock; Wild Radish
Annual or winter annual; stem stout, 3–8 dm tall, freely branched, sparsely hispid especially at the base; lower leaves obovate-oblong in general outline, deeply pinnatifid into several oblong segments and with a large terminal segment; upper leaves progressively smaller, few-lobed to dentate or entire; sepals somewhat saccate at base; petals 10–15 mm long, yellow at anthesis, becoming white, sometimes purplish, conspicuously veined; mature pedicels ascending, 8–15 mm long; pods long-beaked, 2–6 cm long, nearly cylindrical when fresh, when ripe constricted between the seeds, longitudinally ribbed, separating into barrel-shaped segments; seeds oval, finely reticulate, reddish brown, one in each segment. $2n = 18$. Fields and waste places. Introduced. Occasional.

Rorippa Scop. / Yellow Cress

Annual to perennial herbs; stems ascending or erect, usually branched; leaves simple to pinnate; flowers in terminal or axillary racemes; sepals spreading in anthesis; petals yellow or whitish; stamens sometimes fewer than 6; pods short, terete or nearly so, globose to narrowly cylindrical, tipped with a short style and persistent stigma; seeds in 2 irregular rows in each locule (except in *R. sylvestris*).

1 Annual or biennial; petals shorter than sepals — 2
 Perennial; petals much longer than sepals — 4

2 Basal and lower cauline leaves entire, crenate or serrate, never pinnately divided or regularly toothed — **R. curvipes**
 Basal and lower cauline leaves lobed, often pinnately divided or lyrate — 3

3 Petals more than 1 mm long; pedicels more than 3 mm long — **R. palustris**
 Petals less than 1 mm long; pedicels less than 3 mm long. — **R. tenerrima**

4 Middle and upper leaves serrate to nearly entire; pedicels at maturity 3–4 times as long as the fruit — **R. austriaca**
 Middle and upper leaves pinnatifid; pedicels at maturity little, if any, longer than the fruit — 5

5 Leaf-divisions usually obtuse, entire to denticulate; beak of pod 2–3 mm long — **R. sinuata**
 Leaf-divisions usually acute, toothed to incised; beak of pod 0.5–1 mm long — **R. sylvestris**

R. austriaca (Crantz) Besser / Austrian Cress
Perennial, with rhizomes; stems glabrous or nearly so, up to 6 dm high; leaves oblanceolate, 3–10 cm long, variously toothed to nearly entire, tapered to an auriculate base; flowers 3–5 mm long; pedicels ascending, 7–17 mm long at maturity; pods ellipsoid to nearly globose, 1.5–3 mm long, tipped by a persistent style. $2n = 16$. Wet meadows and ditches. Introduced. Rare.

R. curvipes Greene
Annual; stem 1–5 dm long, prostrate, decumbent or erect, simple or branched from base; leaves 2–13 cm long, 0.5–3.0 cm wide, sessile or short-petiolate, entire, crenate or serrate, glabrous, occasionally sparsely pubescent above; pods cylindrical, straight or curved, 2–5 (8) mm long, about as long to twice as long as the pedicels. $2n = 16$. Moist ground. BC to Sask s to Calif, Ariz, NM, Kans, SD. Map 603 ●

R. palustris (L.) Besser / Yellow Cress
Stems mostly 2–6 dm high, commonly much branched above, glabrous or somewhat hispid; leaves oblong-lanceolate in general outline, at least the lower lyrate-pinnatifid, the oblong or lanceolate divisions variously toothed; upper leaves auricled at base; petals about 2 mm long, yellow, spatulate; pedicels at maturity spreading, 4–8 mm long; pods ellipsoid-oblong to nearly globose, 3–7 mm long, often curved. $2n = 32$*. *R. islandica* (Oeder) Borbes of Ed. 1. In our area ssp. **palustris** (includes ssp. *glabra* (O.E. Schulz) Stuckey and ssp. *fernaldiana* (Butters & Abbe) B. Jonsell) with the leaves glabrous below and stems glabrous or sparsely pubescent basally, and ssp. **hispida** (Desv.) B. Jonsell with the leaves hirsute below, strongly hirsute stems and usually nearly globose pods, 3–5 mm long, distinctly shorter than the pedicels. Marshy ground. Circumpolar: Alas, Yuk, Mack to Huds Bay, nQue, Nfld, generally distributed s. Map 604

R. sinuata (Nutt.) A.S. Hitchc.
Plant usually branched from the base, the stems ascending, 1–4 dm long, glabrous or nearly so; leaves glabrous, 2–7 cm long, little reduced upward, pinnately lobed or incised, the divisions sometimes denticulate; cauline leaves auriculate-clasping; petals about 4 mm long; pods straight or curved upward, 7–15 mm long, 2 mm thick, the beak slender, 1–3 mm long. Shores, stream flats, and roadsides. Alta, Sask, ND, Ill s to Calif, Ariz, NM, Tex, Ark. Map 602 ○

R. sylvestris (L.) Besser / Creeping Yellow Cress
Stems ascending from rootstocks, slender, 2–5 dm tall, simple or sparingly branched upwards; leaves glabrous, deeply pinnatifid, the divisions often incised or sharply toothed; lower leaves up to 12 cm long, the upper ones progressively reduced; petals about 4 mm long; pods straight or slightly curved, 8–15 mm long, about 1 mm thick, the beak 0.5–1 mm long. $2n = 16, 32, 40, 48$. Moist ground, Introduced. Rare.

R. tenerrima Greene
Glabrous annual, 1–2 dm high, with numerous slender branches, decumbent or prostrate from the crown; leaves deeply pinnatifid, with rounded, sinuately toothed divisions; sepals 1.5 mm long; petals 0.6–0.8 mm long; pods mostly 4–8 mm long, minutely papillose, 2–4 times the length of pedicels. *R. obtusa* (Nutt.) Britt. of Ed. 1. Moist ground. BC, Alta s to Calif, Utah, NM, Neb, Mo. Map 603 ○

Schoenocrambe Greene

S. linifolia (Nutt.) Greene
Perennial, rhizomatous; stems 2–7 dm tall, erect to somewhat decumbent, usually branched, glabrous, mostly glaucous; leaves 2–9 cm long, 1–6 mm wide, linear, entire to pinnate, the pinnae linear-lanceolate; inflorescence an ebracteate raceme; flowers yellow, on slender pedicels 5–10 mm long; sepals 4–5 mm long, slightly saccate; petals 6–10 mm long; stamens 6; fruit a linear, terete, non-stipitate, scarcely torulose silique 2.5–5.0 cm long; seeds uniseriate, about 1.5 mm long. $2n = 14$. *Sisymbrium linifolium* Nutt. Dry fields and plains. Introduced in our area at Fort Saskatchewan. Rare.

Sisymbrium L.

Annuals, winter annuals or biennials, more or less pubescent with simple hairs; stems erect, tall, simple or branched above; leaves, at least the lower ones, deeply pinnatifid; flowers in elongate racemes; sepals obtuse, ascending; petals small, yellow, with a narrow claw; filaments slender, unappendaged; pod elongate, linear or awl-shaped, terete or slightly 4-sided, tipped with a minute beak; valves 3-nerved, with a prominent midrib and weaker lateral nerves; seeds numerous, 1 row in each locule.

1 Pods awl-shaped, closely appressed to the rachis; pedicels erect, 2–3 mm long **S. officinale**
Pods linear, spreading or ascending; pedicels divergent, 5–10 mm long 2

2 Pods 5–10 cm long; pedicel at maturity nearly as thick as the pod; upper leaves with narrowly linear divisions **S. altissimum**
Pods 1–4 cm long; pedicel at maturity slender; upper leaves with lanceolate divisions **S. loeselii**

S. altissimum L. / Tumbling Mustard
Stem 6–12 dm tall, usually branched above, glabrous or sparsely pubescent with long spreading hairs; leaves pinnatifid, the lower petiolate and with 5–8 pairs of oblong or lanceolate segments, the upper with a few narrowly linear segments; petals pale yellow or whitish, 6–8 mm long; fruits as shown in the key. Seeds angled, appearing greasy. $2n = 14$. A weed of fields and waste ground. Introduced. Common.

S. loeselii L.
Stem 4–10 dm tall, pubescent toward the base with stiff reflexed hairs; lower leaves lyrate-pinnatifid, the segments triangular to ovate and toothed, hirsute beneath or glabrous; middle and upper leaves with lanceolate divisions; petals bright yellow, 5–6 mm long; fruits as described in the key; seeds yellow-brown shiny, the surface obscurely wavy. $2n = 14$. Fields and waste ground. Introduced. Occasional.

S. officinale (L.) Scop. / Hedge Mustard
Stem 3–8 dm tall, stiffly erect, branched above; lower leaves petiolate, deeply pinnatifid with a rounded terminal lobe and 4–5 pairs of smaller lateral lobes, all toothed; upper leaves sessile or nearly so, with a long terminal lobe and few oblong lateral lobes; racemes stiffly erect; petals yellow, 3 mm long; fruits 1–1.5 cm long, acuminate, glabrous or hairy, stiffly erect and appressed to the axis; seeds ovoid, orange-brown. $2n = 14$. Fields and waste ground. Introduced. Rare.

Smelowskia C.A. Meyer

S. calycina (Stephan.) C.A. Mey.
Perennial, forming dense tufts, the caudex branched and clothed with persistent leaf-bases; stems 0.5–2 dm long, more or less hairy; leaves pinnately divided into many narrow lobes, grey-pubescent with chiefly branched hairs; basal leaf-bases strongly ciliate; flowers small, in terminal racemes; sepals oblong, 2–3 mm long, soon falling; petals white to cream-coloured, rarely pinkish, 5–6 mm long; pods (siliques) narrowly oblong to linear, 5–9 mm long, tapering at each end, the valves sharply keeled; stigma expanded in fruit; seeds 6–10. $2n = 12$, 22*, 24. In our area var. **americana** (Rydb.) Drury & Rollins. Alpine screes. Asia: Alas, Yuk, wMack s to Wash, Nev, Utah, Colo. Map 605 ●

Thellungiella O.E. Schulz

T. salsuginea (Pallas) Schulz
Annual, glabrous, glaucous; stem 0.5–1.5 dm tall, erect, branched above; basal leaves mostly absent; cauline leaves 3–20 mm long, 1–4 mm wide, lanceolate to oblong-lanceolate, entire, sessile, auriculate-clasping at base; inflorescence an ebracteate raceme; flowers white, on spreading pedicels 1–9 mm long; sepals about 1 mm long, pinkish, deciduous; petals 2–3 mm long; stamens 6; fruit a linear, subterete or slightly compressed silique 7–16 mm long, 1 mm wide; seeds uniseriate. $2n = 14$. *Arabidopsis salsuginea* (Pallas) Busch. Saline shores and flats. Asia: sYuk, wMack, sMan s to BC, nAlta. Map 605 ○

Thlaspi L.

T. arvense L. / Pennycress; Stinkweed
Glabrous annual, 2–5 dm high; plants turning yellow soon after flowering; lowest leaves petiolate, narrowly obovate, soon dropping; middle and upper leaves sessile, oblong, sagittate-clasping, toothed to entire; flowers small, white, in elongating racemes; petals 3–4 mm long; pods (silicles) flat, rounded, 1–2 cm long, broadly winged, deeply notched at apex; seeds 2–8 in each locule, compressed-ovoid, blackish with concentric ridges. $2n = 14$*. Fields and waste ground. Introduced. Common.

SARRACENIACEAE / Pitcher-plant Family

Sarracenia L.

S. purpurea L. / Pitcher-plant
Insectivorous perennial herb with a basal cluster of hollow leaves ('pitchers') and a solitary flower on an erect scape; leaves 1–2 dm long, often marked with red or purple, curved, broadly winged on the adaxial side, with a broad hood-like blade at the top; scape 2–4 dm tall; flower bracteolate, nodding, 5–7 cm wide, commonly deep purple; sepals 5, spreading; petals 5, incurved, usually dark purple-red; stamens numerous; pistil of 5 united carpels; ovary superior, 5-loculed; style slender at base, greatly dilated and umbrella-like above and bearing minute stigmas at the angles; ovules numerous, on axile placentae; fruit a loculicidal capsule. $2n = 26$. Peat bogs and muskeg. wMack, neBC to nSask, James Bay, Nfld s to cAlta, cMan, Minn, Ill, Ohio, Del. Map 606

DROSERACEAE / Sundew Family

Drosera L.

Small insectivorous perennial herbs with glandular-viscid leaves borne in a basal rosette; leaves petioled, the blades densely glandular and fringed with long reddish glandular hairs (tentacles) which exude drops of viscid fluid; flowers 3–10, on a naked scape in a circinate cyme, nodding at apex; flower regular, perfect, hypogynous, the sepals, petals and stamens each of 4–8 members; pistil of 3–5 united carpels; ovary 1-loculed, with parietal placentae; styles 3–5, deeply 2-parted; fruit a many-seeded, loculicidal capsule; seeds fusiform, with a loose testa.

1 Leaf-blades nearly round or broader than long — **D. rotundifolia**
Leaf-blades much longer than broad — 2

2 Leaves linear; flowers 1–4; seeds 0.5–0.8 mm long — **D. linearis**
Leaves narrowly obovate to elongate-spatulate; flowers 1–9; seeds 1.0–1.5 mm long — **D. anglica**

D. anglica Huds. / Sundew
Leaves often reddish, more or less erect, the petioles 2–6 cm long, the blades 1–3 cm long and 3–4 mm wide; stipules 5 mm long, adnate, fimbriate along the upper half; scape erect, much longer than the leaves, arising from the centre of the rosette; sepals about 5 mm long; petals white, about 6 mm long. $2n = 40$. Swamps and bogs. Circumpolar: Alas, Yuk, Mack to James Bay, Nfld s to Calif, Nev, Ida, Mont, Sask, Man, Gt Lakes. Map 607

D. linearis Goldie
Similar to *D. anglica* except for leaf shape; distinguished by the characters of the key; leaves 2–5 cm long, about 2 mm wide, linear, erect. $2n = 20$. Bogs. BC to Nfld s to Minn, Mich, Me. Map 608

D. rotundifolia L. / Sundew
Similar to *D. anglica* except for leaf shape; distinguished by the characters of the key; leaves 3–10 cm long, blades orbicular to broadly obovate, about 1 cm across. $2n = 20$. Swamps and bogs. Circumpolar; Alas, Yuk, Mack to Huds Bay, Nfld s to Calif, Nev, Ida, Mont, ND, Gt Lakes, Apal, Ala, Fla. Map 609

CRASSULACEAE

Herbs or shrubs, usually succulent; leaves simple, exstipulate; flowers actinomorphic, hermaphrodite, rarely dioecious, hypogynous or perigynous, 4–5-merous, occasionally more; petals free or fused; stamens as many or twice as many as petals; carpels as many as petals, free or fused at base; fruit a cluster of follicles or a capsule.

Petals bright yellow; flowers perfect, the parts in whorls of 5 **Sedum**
Petals purple; flowers often imperfect, the parts in whorls of 4 or 5 **Tolmachevia**

Sedum L. / Stonecrop

Perennial herbs with leafy stems; leaves opposite or alternate, simple, succulent, glabrous; flowers regular, hypogynous, in cymes; sepals and petals 4 or 5, the sepals united at base; stamens 8 or 10, free or adnate to petals below; carpels as many as petals, free or united basally, each with a nectary at the base; fruit a group of follicles.

1 Cauline leaves mostly opposite, broadly oval to orbicular **S. divergens**
Cauline leaves alternate, linear to linear-lanceolate 2

2 Inflorescence with bulblets partly replacing flowers; leaves flattened, becoming scarious or brown; follicles widely divergent **S. stenopetalum**
Inflorescence lacking bulblets; leaves terete, green or reddish; follicles erect **S. lanceolatum**

S. divergens S. Wats.
Mat-forming with prostrate stems and ascending to erect flowering branches, generally 0.5–1.5 dm tall; leaves opposite or subopposite, 4–8 (10) mm long, oval, broadly obovate to orbicular on the sterile branches, more oblong to spatulate on the flowering branches; flowers 3–15 in a loose to dense cyme; petals oblong-lanceolate, shortly apiculate, longer than the triangular sepals; follicles strongly divergent. $2n = 16$. Moist alpine ridges. BC, Alta s to Ore. Map 610 •

S. lanceolatum Torr. / Common Stonecrop
Stems 5–15 cm high, tufted; leaves 3–12 mm long, mostly near the base and crowded on sterile shoots; cyme with several branches, dense; petals lanceolate, nearly twice as long as the ovate sepals, $2n$ = 16, 32, 48. *S. stenopetalum* Pursh of Ed. 1. Dry slopes and ridges. seAlas, sYuk, Alta to sSask s to Calif, NM, SD, Neb. Map 611

S. stenopetalum Pursh / Stonecrop
Stems 1–2 dm high, simple or branching from near the base, the branches often densely leafy; leaves glaucous, 1–2 cm long, flattish, acuminate; cyme broad or sometimes much reduced; petals lanceolate-acuminate, longer than the ovate-acuminate sepals. *S. douglasii* Hook. of Ed. 1. Dry rocky slopes. BC, Alta s to Calif, Mont. Map 610 ○

Tolmachevia Löve & Löve

T. integrifolia (Raf.) Löve & Löve
Perennial; rootstock branching, each branch terminating in a stout scaly caudex and producing annual stems, 5–15 cm tall; leaves obovate to oblong, entire or toothed; flowers polygamous, dark purple, rarely yellow, in a dense cyme; petals nearly twice as long as the sepals; follicles more or less divergent, red or purplish. $2n$ = 18, 36. *Sedum rosea* (L.) Scop. ssp *integrifolium* (Raf.) Hult. of Ed. 1. Moist alpine screes and ridges. neAsia: Alas, Yuk, wMack s to Calif, Nev, Colo. Map 612

SAXIFRAGACEAE / Saxifrage Family

Perennial herbs with crimson root tips; leaves simple (sometimes lobed), usually alternate, often all basal; flowers usually perfect, regular or nearly so, solitary, racemose or cymose; hypanthium (floral tube) usually well developed, often with a disk; sepals 4–5, separate or partly united; petals 4–5, separate, sometimes lacking; stamens twice as many as the petals, or just as many; carpels commonly 2, rarely more, separate or more or less united; ovary superior to inferior; styles and stigmas usually 2, often partly united; ovules numerous, on axile or parietal placentae; fruit of follicles or a capsule.

1	Sepals 4; petals absent; stamens 2–8, mostly 4–7	**Chrysosplenium**
	Sepals 5; petals generally present; stamens 5 or 10	2
2	Stamens 5	3
	Stamens 10	7
3	Petals absent	**Heuchera**
	Petals present	4
4	Petals pinnatifid or 3-cleft	**Mitella**
	Petals entire	5

5 Rootstock elongate, scaly; ovary 1-loculed with 2 parietal placentae 6
Rootstock corm-like bulblet-bearing; ovary 2-loculed with axile placentation **Suksdorfia**

6 Flowers turbinate; inflorescence a simple raceme with the flowers widely spaced **Conimitella**
Flowers not turbinate; inflorescence spike-like, the flowers densely crowded, or branching, the flowers with conspicuously exserted stamens **Heuchera**

7 Styles 3; fruit 3-beaked; plants often bearing bulblets; petals laciniate **Lithophragma**
Styles 2; fruit 2-beaked; plants generally lacking bulblets; petals not laciniate 8

8 Petals pinnatifid **Mitella**
Petals entire 9

9 Basal leaves leathery, 3–15 cm long, more or less elliptic; carpels separate almost to base **Leptarrhena**
Basal leaves not leathery; carpels united below 10

10 Capsule valves of markedly unequal size; petals linear **Tiarella**
Capsule valves of about equal size; petals spatulate or wider 11

11 Styles united at base; petals pink to purple; leaves petiolate, the blades 1–4 cm wide **Telesonix**
Styles free at base; petals white or yellow (if purple, the leaves sessile, small and tightly imbricate) **Saxifraga**

Chrysosplenium L.

Perennial herbs, often with stolons or bulbils; leaves opposite or alternate, simple, lobed, petioled, exstipulate; flowers terminal, solitary or generally in a cyme surrounded with bracteal leaves, small, green, yellow, white or purplish; hypanthium generally more or less adnate to ovary; sepals 4 (5) persistent; petals absent; disc around style, more or less developed, nectariferous; stamens mostly 8 (10) or 4, inserted on margin of hypanthium; ovary inferior or superior, 1-loculed, of 2 carpels, with 2 parietal placentae; styles 2, free; stigma punctate; capsule subinferior or subsuperior, more or less 2-lobed, flattened, dehiscent along inner suture; seeds numerous, oblong-globose with a carina on one side, thick-walled, smooth, sometimes hairy or longitudinally ridged or verrucose.

Sepals green, erect, of about equal size; stamens 4; central flowers 2–3 mm broad **C. tetrandrum**
Sepals golden-yellow, spreading, the outer pair wider; stamens 2–8; central flowers 3–5 mm broad **C. iowense**

C. iowense Rydb.
Stem 3–15 cm tall, erect, generally branching only in the uppermost quarter of its length; stolons leafy; leaves orbicular to reniform, conspicuously veined above in living material, 5–20 mm across, crenate, often pubescent above and below with coarse, erect, white hairs; petioles 1.0–3.5 cm long, mostly glabrous, some-

times villous with reddish haris; cauline leaves (1) 2 (3) reniform to flabellate, lobed, glabrous; bracts yellow, flabellate to ovate or obovate, distinctly lobed; flowers 5–12, short-campanulate, 2.5–5.0 mm across, the centre flowers 3.5–5.0 mm, the lateral flowers 2.5–3.5 mm; sepals 4, golden-yellow, 3-nerved, the outer pair in the centre flowers 1.5–2.0 mm across, usually somewhat broader than long, rounded or obtuse at the apex, the tips markedly recurved, the inner pair of sepals similar but narrower and usually rather longer than broad, sepals of the lateral flowers proportionately smaller, the two pairs not so obviously unequal and frequently not recurved; stamens 2–8, varying in flowers of the same inflorescence, 5–8 in the centre flowers, 2–4 in the lateral; disc narrow, entire; styles 0.3–0.4 mm long; dehisced capsule equalling the sepals; seeds smooth, 0.7–0.85 mm long, light chestnut-brown. $2n$ = c. 120*. Stream banks and marshy ground, in shade. sMack, Alta to cSask, swMan; neIa. Map 613

C. tetrandrum (Lund) T. Fries
Stem 2–20 cm tall, erect, often branching from below the middle; stolons leafy; leaves orbicular to reniform, the venation not conspicuous, 3–15 mm across, crenate, mostly glabrous, occasionally sparsely pubescent above around the sinus; petioles 1–3 cm, glabrous or sparsely villous below the lamina; cauline leaves 1–4, subreniform to obovate, lobed, glabrous; bracts green, often purple-dotted, ovate to obovate, weakly lobed to more or less entire; flowers 3–10, turbinate, 2–3 mm across; sepals 4, mostly green, often purple-dotted, obscurely 1-nerved, triangular to ovate, 1 mm long and about as broad, erect, the two pairs generally alike; stamens 4 (3), opposite the sepals; disc obsolete; styles 0.2–0.3 mm long; capsule turbinate to campanulate, the dehisced edges longer than the sepals; seeds 0.6–0.7 mm long, light reddish-brown. $2n$ = 24*. Moist, shady banks and ledges. Circumpolar: Alas to sEllesmere Is to nQue, Lab s to Wash, Alta, sMack, nMan, James Bay; Colo. Map 614

Conimitella Rydb.

C. williamsii (D.C. Eat.) Rydb.
Perennial from short ascending rootstocks; flowering stems leafless, 2–4 (6) dm tall, finely glandular-puberulent; leaves with slender petioles (1) 1.5–6 (10) cm long, the blades reniform, rather leathery, usually purplish on the lower surface, 1–4 cm broad, shallowly and broadly bicrenate, stiffly ciliate, otherwise usually glabrous; inflorescence a loose raceme, the bracts more or less laciniate; flowers regular, perfect; calyx 4–6 mm long, turbinate, adnate to the ovary for about half to one-third its length, the hypanthium about equalling the 5 erect to slightly spreading ovate sepals; petals 5, white, 4–5 mm long, with an entire blade and slender claw, inserted with the stamens near the base of the hypanthium; stamens 5, opposite the sepals, the filaments about equalling the anthers; pistil 2(3–5)-carpellate, 1-loculed with 2 parietal placentae, almost completely inferior; stigmas nearly sessile, capitate; seeds numerous, about 1 mm long, black. $2n$ = 14. Open montane slopes. sBC, Alta s to Ida, eMont, Wyo. Map 615

Heuchera L. / Alum-root

Perennial herbs with stout scaly rootstocks; leaves basal, stipulate, long-petioled, round-cordate, palmately veined and lobed or coarsely toothed; inflorescence a very narrow panicle, borne on a long naked stem; flowers greenish, yellow or purplish; hypanthium cup-shaped or saucer-shaped, adnate to the lower part of the pistil, often more or less oblique; sepals 5, often unequal; petals 5, small, sometimes absent; stamens 5; ovary approximately half inferior, 1-loculed, with 2 parietal placentae; styles 2, elongate; capsule 2-beaked, opening between the divergent beaks.

1 Calyx 2.0–3.5 mm long, from tip of pedicel — 2
Calyx more than 4.5 mm long, from tip of pedicel — 3

2 Stamens strongly exserted; inflorescence open — **H. glabra**
Stamens not exserted, shorter than sepals; inflorescence spike-like and somewhat congested — **H. parvifolia**

3 Flowers pink; petals about as long as sepals; stamens more or less exserted — **H. richardsonii**
Flowers yellowish; petals much shorter than sepals, often absent; stamens not exserted, shorter than sepals — **H. cylindrica**

H. cylindrica Dougl. *ex* Hook.
Flowering stems 2–5 dm tall from a branching caudex; leaf-blades mostly glabrous, 2–5 cm long, 1.5–4 cm wide, deeply cordate at base, 5–7-lobed, the lobes with aristulate teeth; petioles glandular-puberulent to glabrous; floral bracts ovate-lanceolate, fimbriate or aristulate; panicle dense, narrow, glandular-pubescent; base of hypanthium densely glandular-puberulent; petals usually absent; beaks of carpels exserted, the styles recurved. $2n = 14, 28^*$. Rocky slopes and ledges. BC, Alta s to Calif, Nev, Mont, Wyo. Map 616

H. glabra Willd. *ex* R. & S.
Flowering stems 1–6 dm tall, glabrous below, glandular-pubescent above; leaves 2–9 cm long, slightly narrower, palmate, margins irregularly dentate, glabrous or glandular-puberulent below; inflorescence bracteate, open, branches filiform; calyx mostly 2–3 mm long, glandular-pubescent, lobes 1 mm long, ovate-oblong, base truncate; petals 2 mm long, white; stamens strongly exserted; beaks of capsule exserted. $2n = 14^*$. Moist alpine screes. Alas s to Ore e to Alta. Map 617 ○

H. parvifolia Nutt. *ex* T. & G.
Flowering stems usually short-hairy and glandular, 2–4 dm high from a branched caudex and matted growth of basal leaves; leaf-blades glabrous or puberulent beneath, 1–3 cm wide, deeply cleft, the lobes more or less crenate; panicle at first narrow and dense, later elongating and expanding, becoming 1–3 dm long and interrupted; hypanthium saucer-shaped with a turbinate base, about 2 mm long, glandular-puberulent; petals white or yellowish, minute; stamens very

short; ovary covered by a glandular disc; beaks of capsule slightly exserted. $2n = 14$. *H. flabellifolia* Rydb. of Ed. 1. Gravelly montane slopes and talus. Alta s to NM, Nev, Ariz. Map 617 •

H. richardsonii R.Br.
Flowering stems hispid, glandular-hairy above, 3–4 dm high from a stout caudex; leaf-blades glabrous to sparsely hairy above, hispid beneath, with rounded lobes and broadly ovate teeth; petioles more or less hispid and glandular; floral bracts somewhat scarious; flowers zygomorphic and oblique, about 1 cm long, the hypanthium and calyx glandular-puberulent; petals spatulate, purplish, glandular-ciliate; capsule not exserted; seeds ovoid, with 1 or 2 flat faces, about 0.7 mm long. $2n = 14, 28$. Open sandy ground and rocky shores and slopes. sMack, BC to wOnt s to Colo, Neb, Ia, Gt Lakes. Map 618

Leptarrhena R.Br.

L. pyrolifolia (D. Don) R.Br. / Leather-leaved Saxifrage
Rootstock horizontal or ascending, rather slender but woody; leaves alternate, approximate on the caudex, distant and very few on the flower-stalk; leaf-blades 3–8 cm long, leathery, glabrous, deep green and shining above, pale or brownish beneath, oblong-obovate, crenate-serrate except towards the base, narrowed into stout winged petioles or sessile; flowering stalks 1–4 dm tall, more or less glandular-puberulent above, bearing 1 or 2 clasping leaves; flowers in dense cymules arranged in a short panicle; hypanthium flattened, sepals ovate; petals white or pinkish, persistent, 2–2.5 mm long; stamens 10; carpels almost separate, the ovary slightly inferior, 8–10 mm long, purple or pinkish, erect with slightly divergent tips. $2n = 14$*. Stream banks and moist slopes at subalpine and alpine elevations. Alas s to Wash, nIda, wMont, Alta. Map 619

Lithophragma (Nutt.) T. & G.

Slender perennial herbs, the rootstocks often bulblet-bearing, the flowering stems 1–3 dm tall, sparingly leafy; leaves mostly basal, petiolate, palmately or digitately lobed to divided, the petioles with stipule-like dilated bases; flowers in short racemes; hypanthium cup-shaped or obconic; sepals 5, short; petals 5, white or pink, much longer than the sepals, 3–5-cleft, stamens 10, included; pistil 1-loculed, with 3 parietal placentae and 3 styles; capsule many-seeded, 3-valved at apex.

Hypanthium cup-shaped, mostly rounded at the base, no longer than 5 mm in fruit, adnate only to the base of the ovary; leaf-axils usually bulblet-bearing, the stipules broad and fimbriate, often pinkish **L. glabrum**

Hypanthium obconic, narrowed at the base, 5–8 mm long in fruit, adnate to the lower half of the ovary; leaf-axils without bulblets **L. parviflorum**

L. glabrum Nutt. *ex* T. & G.
Stems 1–2 dm high, glandular-puberulent; basal leaves mostly glabrous, ternately divided to the base, the divisions 5–15 mm long, deeply 3-lobed; cauline leaves

few similar to the basal ones, the axils usually bulblet-bearing, the stipules short, broad, pinkish, fimbriate; flowers 3–6, some often replaced by bulblets; hypanthium short-hairy and glandular; petals 4–7 mm long, white or pinkish. $2n = 14$, 28. *L. bulbifera* Rydb. of Ed. 1. Rocky slopes and dry montane meadows. BC, Alta s to Calif, Nev, Utah, Colo, SD; swSask (Cypress Hills). Map 620

L. parviflorum (Hook.) T. & G.
Stems 1–3 dm high, more or less densely glandular-puberulent, often scabrous; basal leaves sparsely white-pubescent, 3–5-divided to the base, the divisions ternately cleft into entire or toothed lobes; cauline leaves few, similar to the basal ones; raceme 3–9-flowered; hypanthium densely pubescent and glandular, narrowed at base; petals 4–10 mm long, white or pinkish, deeply 3–5-cleft, narrowed to slender claw. $2n = 14$, 21, 28, 35. Gravelly to grassy montane slopes. BC, Alta s to Calif, Nev, Utah, Wyo, SD. Map 621

Mitella L. / Bishop's-cap; Mitrewort

Low and slender herbs with rootstocks; leaves mostly basal, long-petioled, heart-shaped or kidney-shaped, shallowly lobed and toothed; flowering stem usually naked, with a terminal spike-like raceme of small flowers; hypanthium shallow, adherent to a portion of the ovary; sepals 5, short; petals 5, pinnatifid or trifid; stamens 5 or 10, included; pistil 1-loculed, with 2 placentae (parietal or nearly basal); capsule 2-valved at the summit, opening widely, exposing the shining black or brown seeds.

1 Stamens 10 **M. nuda**
Stamens 5 2

2 Petals 3-cleft near the apex; petioles with rigid deflexed white hairs and numerous sessile or subsessile glands **M. trifida**
Petals pinnatifid; petioles lacking deflexed white hairs and glands except near leaf-blade 3

3 Stamens opposite sepals; petioles with abundant, long, flexous, more or less decurved brown hairs **M. breweri**
Stamens opposite petals; long, flexuous, brown hairs few, except near leaf-blade **M. pentandra**

M. breweri A. Gray
Rootstock brown-bracted, leaves orbicular to reniform, 4–8 cm wide, with a deep basal sinus, glabrous to sparsely pubescent on both sides, obscurely lobed, irregularly crenate-dentate with short apiculate teeth; petioles with long, tawny, weak hairs; scapes 1 or more, slender, 8–25 cm high, minutely glandular-puberulent; hypanthium almost entirely adnate to the depressed ovary; sepals broad, soon reflexed; petals usually with 3 pairs of filiform divisions; stamens opposite the sepals, stigmas 2, nearly sessile, 2-lobed. $2n = 14$. Moist montane woods. BC, Alta s to Calif, nIda, nwMont. Map 622

M. nuda L. / Bishop's-cap; Mitrewort
Rootstocks long, slender, creeping, often stolon-like; basal leaves rounded reniform, cordate at base, obscurely lobed, crenate, sparsely hairy; cauline leaf solitary, or none; raceme 2–10 cm long, glandular-puberulent; flowers yellowish green; hypanthium adnate to the base of the ovary; petals pectinately pinnatifid, about twice as long as the sepals; stamens 10; seeds few, black, smooth, shining, about 1 mm long. $2n = 14, 28^*$. Moist woods. eAsia: seAlas, sYuk, swMack to Huds Bay, nQue, Nfld s to Wash, Mont, Sask, ND, Gt Lakes, Pa. Map 623

M. pentandra Hook.
Similar to *M. breweri*; differing by the characters of the key. $2n = 14^*$. Moist montane woods. Alas, Yuk s to Calif, Colo. Map 624

M. trifida Graham
Rootstock scaly; leaves basal, long-petioled; blades round-reniform or cordate, 2–4 cm broad, 7–9-lobed, with broad shallow rounded lobes, crenate, obscurely glandular-toothed, sparsely strigose with conspicuous white hairs; petioles retrorsely hirsute to glabrous, obscurely glandular; scapes 1–several, 1.5–4 dm high, somewhat hirsute and minutely glandular, often scabrous; racemes often 1-sided, the flowers subtended by small fimbriate bracts, the pedicels 1 mm or less long; hypanthium cup-shaped, adnate to the lower half of the ovary; sepals erect, whitish; petals wedge-shaped, 3-cleft, about twice as long as the sepals, whitish, often tinged with violet or pink; styles short; stigmas capitate. Moist montane woods. BC, Alta s to Calif, sMont. Map 625

Saxifraga L. / Saxifrage

Perennial herbs with naked or leafy stems; leaves alternate, basal or rarely opposite, entire, toothed or lobed; flowers rather small, perfect, regular or slightly irregular, borne in cymes or panicled cymes or rarely solitary; hypanthium well developed, adnate to the lower part of the pistil; sepals 5; petals 5, usually narrow (rarely absent); stamens 10; carpels normally 2, united completely or only at base, the ovary superior to partly inferior; styles 2; ovules numerous, on axile placentae; fruit a capsule or pair of follicles, 2-beaked, the beaks often divergent.

1	Leaves opposite; petals purple	**S. oppositifolia**
	Leaves alternate or basal; petals white or yellow	2
2	Cauline leaves present	3
	Cauline leaves absent	10
3	Petals yellow; leaves entire	4
	Petals white or cream, sometimes spotted; leaves various	5
4	Leaves with stiff marginal bristles; naked stolons generally present	**S. flagellaris**
	Leaves lacking stiff marginal bristles; naked stolons absent	**S. aizoides**
5	Basal leaf-blades broad, orbicular to reniform	6
	Basal leaf-blades narrower, lanceolate, oblong-cuneate or flabelliform	7

6 Inflorescence of a single terminal flower and lateral bulblets **S. cernua**
Inflorescence of several flowers; bulblets absent **S. hyperborea**

7 Basal leaves all entire, margins spinulose **S. bronchialis**
Basal leaves, some or all, toothed or lobed; margins not spinulose 8

8 Basal leaves stiff, usually with 3 apical teeth produced into spiny tips **S. tricuspidata**
Basal leaves not stiff, the lobes not spiny 9

9 Plants strongly caespitose; basal leaves ligulate to narrowly flabelliform, trifid, mostly 1.0–1.5 mm wide at base **S. caespitosa**
Plants not strongly caespitose, the stems arising from a basal rosette; leaves broadly cuneate to ovate, the apex shallowly toothed or entire, the base mostly more than 1.5 mm wide **S. adscendens**

10 Leaves subcordate to cordate at base 11
Leaves narrowly to broadly cuneate at base 13

11 Inflorescence with bulblets; lobes of leaves 3-toothed **S. mertensiana**
Inflorescence lacking bulblets; leaves with simple teeth 12

12 Petals broadly elliptic to suborbicular, with conspicuous claws, white with 2 yellow or green spots near the base **S. odontoloma**
Petals narrow to broadly elliptic, the base cuneate or only slightly clawed, often tinged with pink, lacking spots at base **S. nelsoniana**

13 Inflorescence with bulblets; petals dimorphic, 3 broader, 2 narrower; sepals reflexed **S. ferruginea**
Inflorescence lacking bulblets; petals not dimorphic 14

14 Leaves cuneate to flabellate, toothed at the apex; inflorescence open; ovary superior **S. lyallii**
Leaves not as above; inflorescence congested; ovary partially inferior 15

15 Plants generally less than 10 cm tall; leaves less than 2 cm long; inflorescence reddish **S. nivalis**
Plants generally more than 10 cm tall; leaves more than 2 cm long; inflorescence greenish-white 16

16 Leaves crenate-serrate; stamens inserted on a narrow disc, surrounding but not covering the top of the ovary **S. occidentalis**
Leaves entire to sinuate or remotely denticulate; stamens inserted on a broad, lobed disc that more or less covers the ovary **S. oregana**

S. adscendens L.
Small tufted plant, 2–8 cm high, glandular-pubescent, often densely leafy; leaf-blades, 3–8 mm long, cuneate to oblong-spatulate, 3-toothed at apex or entire; flowers in a compact cyme; petals 3–4 mm long, whitish with green nerves, about twice as long as the brown or purple sepals. $2n = 22$. In our area ssp. **oregonensis** (Raf.) Bacigalupi. Moist alpine banks and ledges. Europe: Alas, Yuk, swMack s to Ore, Utah, Colo. Map 626

S. aizoides L.
Matted caulescent herb with leafy stems; leaves 4–8 mm long, linear-oblong, thick, succulent, spine-tipped; flowering stems ascending 4–10 cm long, glandular-puberulent or glabrous; petals yellow, about 4 mm long, more or less erose at apex. $2n = 26$*. Alpine streams and talus. Europe, Greenland: seYuk, Bank Is to cBaffin Is, sEllesmere Is to Nfld s to BC, Alta, sMack, James Bay, Gaspé, NS. Map 627

S. bronchialis L.
Tufted or matted, the stems more or less horizontal and densely leafy; leaves narrowly lanceolate, 5–14 mm long, acuminate and spine-tipped, the margin spinulose; flowering stalks erect, with scattered small leaves, glandular-puberulent or glabrate; inflorescence open and short or elongated; sepals obtuse, 1.5–3 mm long; petals 4–6 mm long, whitish often spotted with maroon or orange. $2n = 26$*. In our area ssp. **austromontana** (Wieg.) Piper. Rocky outcrops from montane to subalpine elevations. Asia: cBC, Alta s to Wash, neOre, Ida, Utah, NM. Map 628

S. caespitosa L.
Dwarf tufted or matted plant; stems 4–10 cm high, glandular-pubescent; leaves crowded on the caudex, the blades 6–14 mm long, usually cuneate and 3-lobed at apex, the lobes lanceolate or linear; cualine leaves 2 or 3 and reduced; flowers 1–few on a stem; hypanthium glandular-puberulent; sepals 2–3 mm long; petals longer than the sepals, white, clawless. $2n = 80$. Most alpine banks and ledges. Circumpolar: Alas, Yuk to Ellesmere Is, Nfld s to Ore, Nev, Utah, NM, nMack, Huds Bay, nQue, Gaspé. Map 629

S. cernua L.
Rootstock short, often with bulblets; stems erect, 6–14 cm tall, more or less glandular-pubescent, solitary or somewhat gregarious; leaves long-petioled, basal and scattered on the stem, reduced above; blades round-reniform in outline, 1–2 cm wide, palmately 3–7-lobed; flowers, except the terminal one, commonly represented by axillary bulblets; sepals 2–4 mm long; petals 6–10 mm long, white, cuneate-obovate. $2n = 36–72$. Moist alpine banks, screes, ledges: Circumpolar: Alas to Ellesmere Is to Lab s to Wash, Utah, NM, sMack, Huds Bay, nQue; Nev; Lk Superior; Gaspé. Map 630

S. ferruginea Graham
Plant 1–3 dm high, more or less glandular-pubescent; leaves crowded on the short caudex, the blades spatulate to oblanceolate, 2–6 cm long, copiously pubescent, especially beneath, prominently toothed above the middle; flowering stems rather widely paniculate, the branches ascending; bracts usually much smaller than the leaves; lateral pedicels sometimes bearing bulblets; sepals 1.5–2 mm long, reflexed; petals unequal, white, the 3 upper ones with 2 yellow spots at the base. $2n = 20$*, 38*. Moist alpine banks and ledges. sAlas s to Calif e to Alta, wMont. Map 631 ●

S. flagellaris Willd.
Rootstock slender; stems solitary, 0.3–2.0 dm tall, glandular-pubescent, simple, arising from a compact basal rosette that produces naked, arching stolons; basal leaves 5–10 mm long, obovate to oblanceolate, glabrous, margins ciliate with white, bristle-like setae; cauline leaves several, 5–10 mm long, elliptic to oblanceolate or oblong, glabrous above, glandular-pubescent below; flowers 1–3 (5); calyx campanulate, lobes 2–5 mm long, narrowly ovate to oblong, glandular-pubescent below; petals 6–11 mm long, obovate, yellow, the inner base often pink-spotted; stamens inserted at base of calyx-lobes, filaments linear-lanceolate; ovary barely one-fourth inferior; capsules 5–10 mm long, beaks divergent. $2n = 16$*. In our area ssp. **setigera** (Pursh) Tolm. Turfy alpine slopes and ridges, on limestone. Asia: Alas, Yuk, wMack s to BC, Alta. Map 631 ○

S. hyperborea R.Br.
Low tufted plant, glabrous to glandular-puberulent; leaves mostly basal, slender-petioled, reniform, palmately 3–5-lobed; flowering stems 3–8 cm high, 1–3-leaved, soft-pubescent, bearing 1–5 small flowers; sepals about 2 mm long, obtuse; petals white or pale pink, 3–5 mm long, oblong-ovate. $2n = 26$*. *S. rivularis* L. of Ed. 1. Includes *S. debilis* Engelm. Alpine rocks and ledges. More or less circumpolar: Alas, Yuk, wMack (Arct Is uncertain) s to Calif, Ariz, Utah, Colo. Map 632 ●

S. lyallii Engler
Acaulescent from a prominent caudex; leaves fan-shaped or cuneate-obovate, tapering into the petiole, glabrous, sharply and coarsely toothed usually only above the middle, 1–4 cm long; scapes 1–3 dm tall, glabrous or sparingly glandular-puberulent above, often purplish; inflorescence open, few–many-flowered; sepals soon reflexed; petals 2–4 mm long, white with 2 greenish yellow blotches that soon fade; filaments narrowly petaloid; capsule purplish, the beaks 2 or often 3–5. $2n =$ c.56. Hybridizes with *S. odontoloma*, the plants generally having a nearly orbicular blade but with the upper part of the petiole flabellate. Stream banks and moist mountain slopes. Alas, Yuk, swMack s to Wash, wMont. Map 633

S. mertensiana Bong.
Rootstock erect, stout, scaly, often bulbiferous; leaves basal, petioled, glandular-pubescent to glabrous, the blades orbicular to ovate-reniform, cordate at base, with many shallow and commonly 3-toothed lobes; flowering stems 1–3 dm tall, mostly hairy and minutely glandular, widely branched, bearing 1 or 2 leaves toward the base and bractlets above; cymules few-flowered, often all but the terminal flower modified as bulblets, the pedicels becoming abruptly recurved. $2n = 48$–50. Moist banks. sAlas to Calif e to Alta, wMont. Map 634 ○

S. nelsoniana D. Don
Perennial, rhizomatous; stem 0.5–3.0 dm tall, more or less glabrous below, villous above with white, often purple-septate hairs; leaves basal, long-petiolate, blades reniform to orbicular, mostly 1–4 cm wide, cordate, dentate, sparsely punctate, more or less glabrous above and below, margins hairy; inflorescence paniculate,

open, branches generally ascending; sepals 1.3–3.0 mm long, glabrous, margins ciliate, reddish; petals 2.4–4.6 mm long, elliptic, base cuneate, white, often strongly rose-tinged; filaments filiform to clavate or elliptic; disc evident; ovary more or less superior; capsule 3.2–8.8 mm long, reddish, the beaks moderately to strongly divergent. $2n = 70$. *S. arguta* D. Don (in part) of Ed. 1. In our area ssp. **porsildiana** (Calder & Savile) Hult. Moist rocks, ledges, and stream banks at alpine elevations. Asia: Alas, nMack (to Bathurst Inlet) s to Ore, BC, Alta, sMack. Map 636 ●

S. nivalis L.
Scape 4–15 cm tall, from a short stout caudex; leaves in a basal rosette; blades 1–3 cm long, rather thick, ovate to orbicular, obtuse, crenate-serrate, purple beneath, glandular-pubescent, near the margins, about as long as the broad petiole; scape glandular-pubescent, often purplish; flowers 3–12 in a dense cluster, the pedicels very short; sepals ovate to triangular, often purplish; petals greenish white, reddish or purplish on the back, about 3 mm long, obovate, fruits purplish, with divergent tips. $2n = 60$. Alpine slopes and ridges. Circumpolar: Alas to Ellesmere Is to Lab s to BC, Alta, Mack, Huds Bay, nQue. Map 634 ●

S. occidentalis S. Wats.
Plants with a short caudex and often a single scape; leaves basal, spreading, the blade ovate to oblong, dentate to crenate, often rusty-hairy beneath, narrowed to a broad petiole; scapes 5–20 cm tall, glandular-pubescent; inflorescence at first compact, then lengthening into a corymbose cyme, glandular-pubescent; sepals glabrate, obtuse; petals 2.5–3 mm long; filaments narrowly petaloid. $2n = 38$. Moist to dry open montane slopes. sBC, Alta s to Ore, Nev, nwWyo. Map 635

S. odontoloma Piper
Perennial, rhizomatous; stems 1.0–7.5 dm tall, glabrous below, glandular above; leaves basal, long-petiolate, blades orbicular, 2.5–10.0 cm wide, shallowly to deeply cordate, dentate, glabrous or sparsely hairy between the teeth or at junction of petiole and lamina; inflorescence paniculate, open, branches erect or spreading, glandular; sepals mostly 1.8–2.8 mm long, somewhat longer in fruit, glabrous to glandular-ciliate; petals 2–5 mm long, broadly elliptic to orbicular, conspicuously clawed, white with 2 yellow or green basal spots; filaments clavate; disc evident; ovary more or less superior; capsule 5–9 mm long, the beaks slightly to moderately divergent. $2n = 48$. *S. arguta* D. Don (in part) of Ed. 1. Hybridizes with *S. lyallii*. Shady mountain stream banks to subalpine elevations. sBC, Ida, Mont s to Calif, Nev, Ariz, NM; Alta. Map 636 ○

S. oppositifolia L. / Purple Saxifrage
Densely matted, with numerous prostrate crowded branches; leaves 2–5 mm long, in 4 rows, overlapping, oval to obovate, concave, fleshy, hairy on the margins; flowers solitary, on peduncles 1–2 cm long; sepals oblong to ovate, 2.5–3 mm long, ciliate; petals 6–8 mm long, purple, sometimes white; filaments and styles purplish. $2n = 26^*, 52$. Alpine slopes and ridges. Circumpolar: Alas

to Ellesmere Is to Nfld s to Ore, Ida, nwWyo, Mack, Huds Bay, nQue, Gaspé; Vt. Map 637

S. oregana Howell
Stems 1.5–12 dm tall, glabrous to pilose below, glandular-pubescent above; leaves basal, mostly 6–20 cm long, generally oblanceolate, entire to sinuate-denticulate or serrate, glabrous to sparsely hirsute, the basal margins pilose; inflorescence paniculate, open or contracted; calyx 2–4 mm long, adnate to ovary at base, lobes ovate to oblong-lanceolate, usually reflexed; petals 2–4 mm long, greenish white, sometimes vestigial or absent; stamens inserted on the edge of a lobed disc that more or less covers the ovary, the styles exserted; ovary mostly one-third to half inferior; fruit 3–5 mm long, the beaks divergent. $2n = 38$. In our area var. **montanensis** (Small) C.L. Hitchc. Moist alpine slopes. Wash to Mont s to Calif, Wyo, Colo; Alta. Map 632 ○

S. tricuspidata Rottb.
Matted or tufted, with densely leafy branches; leaves mostly cuneate-oblong, 3-toothed at apex, 7–14 mm long, stiff, ciliate, the teeth spine-tipped; flowering stalks 6–16 cm tall, with few small leaves, sparingly glandular-hairy or glabrous; sepals 1.5–3 mm long; petals 6–7 mm long, whitish, often spotted with maroon or orange. $2n = 26$*. Dry, open slopes and ridges. Greenland: Alas to Ellesmere Is to Lab, s to BC, Alta, nSask, cMan, James Bay, nQue; seMan, wcOnt. Map 638 ●

Suksdorfia A. Gray

Herbaceous perennials, with short, bulblet-bearing rootstocks; leaves alternate, stipulate; flowers perfect, more or less actinomorphic; sepals 5, petals 5, stamens 5, inserted opposite sepals; pistil 2-carpellate; ovary 2-loculed, placentation axile, about half to almost wholly inferior.

Petals generally pink or violet, erect; calyx turbinate; disc obsolete **S. violacea**
Petals white, spreading; calyx campanulate; disc present **S. ranunculifolia**

S. ranunculifolia (Hook.) Engl.
Perennial herb, glandular-pubescent, with a thick caudex and fibrous roots, forming bulblets at the base; stem 1–3 dm high; basal leaves petioled, ternately divided to the base, the middle division more or less wedge-shaped and 3-lobed, the lateral divisions oblique and lobed; uppermost leaves sessile, often with broad stipules, 3-lobed at the apex; inflorescence corymbiform, short and congested at first, the lower branches elongating and the flowers tending to be secund; hypanthium cup-shaped, adnate to the lower half of the ovary; sepals 5, triangular-lanceolate, often purplish; petals 5, white, often purplish at base, becoming yellowish, 4–6 mm long; stamens 5, on a thick disc; ovary 2-loculed, with axile placentae; styles distinct; stigmas capitate. *Hemieva ranunculifolia* (Hook.) Raf. of Ed. 1. Wet rocks in the mountains. BC, Alta s to Calif, Ida, Mont. Map 638 ○

S. violacea A. Gray
Slender perennial, 1–2 dm high, with leafy flowering shoots from a small bulblet-bearing rootstock; stem slender, more or less glandular-pubescent; lower leaves long-petioled, the blades reniform, with 5–7 shallow, slightly toothed lobes, more or less puberulent; cauline leaves stipulate, the upper ones subsessile, wedge-shaped and 2–4-toothed at the tip; flowers few, in an elongate loose panicle, the branches erect; hypanthium deeply cup-shaped, adnate to most of the ovary, densely glandular; sepals 2–3 mm long; petals 5–8 mm long, violet, almost erect, the long claws whitish or yellow; ovary almost wholly inferior; style short, nearly erect. $2n = 14$*. Wet rocks in mountains. BC, Alta, s to Wash, nIda, nwMont. Map 639 ○

Telesonix Raf.

T. heucheriformis Rydb.
Low glandular-pubescent perennial with a thick scaly rootstock and flowering shoots 5–15 cm high; basal leaves reniform-orbicular, 2–4 cm broad, deeply and doubly crenate-dentate, bearing sessile glands; upper cauline leaves fan-shaped or cuneate, tapering into winged petioles or sessile; flowers in a leafy, short, somewhat secund panicle; hypanthium often purplish, deeply cup-shaped, adnate to the lower half of the ovary; sepals ovate-lanceolate; petals deep pink, equalling or slightly exceeding the sepals, short-clawed; stamens 10; beaks of fruit united about half-way and slightly divergent. $2n = 14$*. Alpine rocks and ledges, limestone. Alta to Wyo; Nev; SD. Map 639 ●

Tiarella L / False Mitrewort

Perennial herbs with mostly basal leaves and leafy flowering shoots from a scaly rootstock; leaves simple or trifoliate, petioled, stipulate, flowers delicate, whitish, borne in an elongate panicle; hypanthium small, cup-shaped, nearly free from the ovary; sepals 5, ovate or lanceolate; petals 5, linear-subulate, much longer than the sepals; stamens 10, conspicuously exserted, with long filiform filaments; ovary 1-loculed, with 2 parietal or nearly basal placentae; styles 2, elongated; capsule membranous, with 2 very unequal valves; seeds few.

Leaves ternately compound **T. trifoliata**
Leaves simple, merely lobed **T. unifoliata**

T. trifoliata L.
Flowering stems slender, 2–4 dm tall, hairy or glabrous at base, glandular-hairy above, 2–4-leaved; leaves usually trifoliate, the middle leaflet more or less 3-lobed and dentate, the lateral leaftlets mostly 2-cleft, hirsute to glabrous; sepals white or pinkish, about 1.5 mm long, glandular-dotted on back; petals white or pinkish, 3–4 times as long as the sepals; valves of fruit 6–10 mm and 3–5 mm long. $2n = 14$*. Moist woods. seAlas s to nOre e to Alta, wMont. Map 640 ○

T. unifoliata Hook.
Flowering stems 1–4 dm tall, glabrous or hairy below, glandular above, 1–4-leaved; leaves broadly cordate in outline, 3–5-lobed, double crenate to crenate-dentate, the teeth mucronate, 4–10 cm broad, hirsute or glabrate; sepals white or pinkish, about 2 mm long; petals white, fully 3 times as long as the sepals; valves of fruit 9–12 mm and 4–6 mm long. $2n = 14$. Moist woods. seAlas s to Calif e to Alta, wMont. Map 640 ●

PARNASSIACEAE / Grass-of-Parnassus Family

Parnassia L. / Grass-of-Parnassus

Glabrous perennial herbs; leaves entire, mostly clustered at the base of the plant; flowers solitary on long scape-like stems which often bear a single sessile leaf or bract; hypanthium short; sepals 5, slightly united at the base; petals 5, white, with greenish or yellowish veins; staminodia gland-tipped, small, united in 5 scale-like clusters, opposite the petals; stamens 5, alternating with the staminodial scales; ovary superior or nearly so, 1-loculed, with 4 projecting parietal placentae; stigmas usually 4, nearly sessile; capsule many-seeded, usually 4-valved, loculicidal; seeds winged.

1 Petals fringed on the sides, at least below; leaves reniform to reniform-cordate **P. fimbriata**
Petals not fringed; leaves ovate or cordate 2

2 Petals 3-veined, usually shorter than the sepals; stem naked or with a leaf close to its base **P. kotzebuei**
Petals 5–9-veined, usually longer than the sepals; stem usually with a leaf at or below the middle 3

3 Leaf-blades acute at the base; staminodia usually 5–7 on each scale; cauline leaf small, narrowed to base **P. parviflora**
Leaf-blades cordate at base; staminodia 7–15 on each scale; cauline leaf larger, often cordate-clasping **P. palustris**

P. fimbriata Konig
Leaf-blades reniform or broadly cordate, 2–4 cm wide, with a broad basal sinus; scapes 1–3 dm high, with a single ovate-cordate leaf at the middle or somewhat above; sepals 5–6 mm long, often erose at the apex; petals short-clawed, about 10 mm long, 5-nerved, fimbriate below the middle; staminodial scales rather fleshy, with 5–9 short blunt segments. $2n = 36$*. Stream banks and springy places. Alas, Yuk, swMack s to Calif, Mont, NM. Map 641

P. kotzebuei Cham. & Schlecht.
Leaf-blades ovate or oval, subcordate or rounded at base but narrowed to the dilated petiole; scapes about 1 dm high, slender, furrowed, leafless or rarely with an oval leaf near the base; petals 3-nerved, shorter than to equalling the

sepals, 3–6 mm long; staminodia 3–5 on each scale; capsule depressed at apex. $2n = 18$. Moist alpine slopes and ledges. eAsia, Greenland: Alas, Yuk, nMack to sBaffin Is, nQue, Nlfd, s to Wash, Wyo, nSask, cMan, James Bay, Gaspé. Map 642

P. palustris L.
Leaf-blades broadly ovate, more or less cordate at base; scapes 1–many, 1–3 dm high, with an ovate, cordate-clasping leaf at or near the middle; petals oval, 8–12 mm long, sessile, faintly 5–9-nerved, up to twice as long as the sepals; staminodia slender, 7–15 on each scale; capsule ovoid, about 1 cm long. $2n = 18$, 27, 36, 54. Includes *P. montanensis* Fern. & Rydb. of Ed. 1. A confused species complex. Wet shady places. Circumpolar: Alas, Yuk, nMack to Huds Bay, nQue, Nfld s to Calif, Nev, Utah, Colo, ND, Minn, Gaspé; nMich. Map 643

P. parviflora DC.
Leaf-blades oval to ovate, tapering into the petiole; scapes slender, 0.5–3 dm high, with a small narrow leaf at or below the middle; petals 5–7-nerved, elliptical to oblong, 5–9 mm long, usually slightly longer than the sepals; staminodia usually 5–7 on each scale; capsule ovoid, about 1 cm long, rounded at apex. Bogs and stream banks. BC to Nfld s to Ida, Mont, SD, Minn, NS. Map 644

HYDRANGEACEAE / Hydrangea Family

Philadelphus L.

P. lewisii Pursh / Mock Orange; Syringa
Shrub, 1–3 m high, freely branched, the branches spreading and ascending, the young twigs reddish, glabrous; leaves opposite, without stipules, short-petiolate, the blades ovate-lanceolate, 2–6 cm long, acute at apex, 3–5-ribbed, entire or denticulate, glabrous or more or less pubescent, especially on the veins beneath; flowers showy, white, fragrant, in racemosely disposed cymes; hypanthium top-shaped, adnate to the ovary for nearly its total length; sepals 4, rarely 5, pubescent at the tips within; petals white, 4 or 5, 1–2 cm long; stamens 20–40; anthers bright yellow; ovary inferior, mostly 4-loculed; styles united half their length or more; capsule firm, 4-valved, with numerous minute seeds. $2n = 26$. Montane woods. BC, Alta s to Calif, Ida, Mont. Map 645

GROSSULARIACEAE / Currant or Gooseberry Family

Ribes L.

Shrubs with erect or procumbent branches, sometimes prickly; leaves alternate, lobed, palmately veined, often fascicled on the branches; flowers racemose or solitary, the pedicels often with bractlets at about the middle; hypanthium adnate to the globose ovary and more or less produced above it as a saucer-shaped or ubular structure; sepals usually 5, often coloured; petals 5, shorter than the sepals; stamens 5, inserted in the mouth of the hypanthium; pistil of 2 united carpels, 1-loculed, with 2 parietal placentae; ovary inferior; styles 2, separate to united; fruit a berry.

1 Stems more or less prickly and usually bristly; inflorescence commonly 1–5-flowered; pedicels not jointed below the ovary (except *R. lacustre*) 2
Stems not prickly or bristly; inflorescence a raceme of few–many flowers; pedicels jointed below the ovary. Currants 5

2 Calyx saucer-shaped; pedicels jointed beneath the ovary; berries black, bristly glandular **R. lacustre**
Calyx tubular to cup-shaped; pedicels not jointed; fruit red to black, smooth. Gooseberries 3

3 Stamens longer than the petals and about equalling the sepals; fruiting stems rarely with bristles on the middle and upper internodes 4
Stamens about as long as the petals and shorter than the sepals; fruiting stems usually bristly **R. oxyacanthoides**

4 Older stems dark-coloured, often with nodal spines; leaves commonly cuneate at base, mostly glabrous **R. hirtellum**
Older stems whitish, often without nodal spines; leaves truncate to cordate at base, sometimes grey-pubescent **R. inerme**

5 Flowers golden-yellow or reddish, slightly irregular; leaves convolute in the bud **R. aureum**
Flowers not golden-yellow, regular; leaves plicate in the bud 6

6 Leaves dotted beneath with resinous glands or bearing stalked glands; fruits black when mature 7
Leaves without resinous glands; fruit reddish or blue-black when mature 9

7 Plants with leaves, young stems, inflorescence and fruits all glandular-puberulent **R. viscosissimum**
Plants not as above 8

8 Hypanthium-tube broadly cup-shaped; sepals white; ovary usually with a few sessile glands **R. hudsonianum**
Hypanthium-tube tubular to narrowly cup-shaped, sepals greenish white or yellowish; ovary lacking glands **R. americanum**

9 Ovary and fruit not bristly; pedicels often minutely glandular **R. triste**
Ovary and fruit bristly with glandular hairs 10

10 Fruit red; sepals 2–2.5 mm long, glabrous on outside **R. glandulosum**
Fruit blue-black; sepals 3–4 mm long, pubescent on outside **R. laxiflorum**

R. americanum Mill. / Wild Black Currant
Shrub, about 1 m high; branches erect; leaves nearly orbicular in outline, sharply 3–5-lobed, broadly truncate to shallowly cordate at base, somewhat hairy and resinous-dotted beneath, the lobes serrate-dentate; racemes drooping, many-

flowered; pedicels exceeded by the slender bracts; flowers greenish white o yellowish; hypanthium-tube cylindrical, about 4 mm long; sepals 4.5–5 mm long petals 2.5–3 mm long; fruit black, glabrous, 6–8 mm wide. $2n = 16$. Moist woods Alta to NB, NS s to Mont, NM, Neb, Ia, Minn, Del. Map 646 ●

R. aureum Pursh / Golden Currant
Erect shrub; leaves deeply 3–5-lobed and dentate, cuneate to subcordate at base glabrous or nearly so; flowers slightly irregular; bracts 5–10 mm long; hypan thium-tube yellow, cylindrical, 6–10 mm long; sepals more than half as long a the tube; petals yellow, often tipped with red; fruit black, red or yellow. $2n = 16$ Riverbanks and rocky slopes. BC to swSask s to Calif, NM, SD. Map 646 ○

R. glandulosum Grauer / Skunk Currant
Stems reclining or ascending; leaves orbicular in outline, 5–7-lobed, the ovat lobes acute and serrate to doubly serrate, cordate at base, glabrous or sparsel hairy on the veins beneath; racemes ascending; glandular-puberulent; ovary an fruit glandular-bristly; hypanthium-tube deeply saucer-shaped, about 1 mm long sepals 2–2.5 mm long, glabrous; petals 1–1.5 mm long, white or reddish; styl bifid nearly to the base; fruit dark red; bruised plants with odour of skunk $2n = 16$. Moist woods and clearings. Alas, Yuk, swMack to James Bay, nQue Nfld s to BC, Alta, Sask, Man, Gt Lakes, NY; NC. Map 647

R. hirtellum Michx. / Wild Gooseberry
Low shrub; nodal spines few, 3–8 mm long, often absent; new canes commonl prickly; fruiting canes mostly without prickles on the middle and upper inter nodes; leaves 3–5-cleft, the lobes crenate or dentate, commonly cuneate at th base, glabrous or thinly pubescent, without glands on the surfaces; bracts villose ciliate, lacking marginal glands; hypanthium-tube obconic, 2.5–3 mm long; sepal 3–4 mm long; petals 2–2.5 mm long; stamens and style just equalling the sepals $2n = 16$. Moist woods. Alta to Nfld s to eMont, neNM, SD, Minn, Pa. Ma 648 ●

R. hudsonianum Richards. / Wild Black Currant
Stems erect or ascending; leaves reniform-ovate, often broader than long 3–5-lobed, the terminal lobe usually rather short, cordate at base, dentate o serrate, glandular-dotted beneath and usually pubescent; racemes erect or as cending, 6–12-flowered, the bracts 2–5 mm long, hairy; hypanthium-tube ope cup-shaped, about 1 mm long; sepals white, about 4 mm long, hairy; petal white, 1.5–2 mm long; ovary glabrous, usually with a few sessile resinous glands fruit black. Moist woods. Alas, Yuk, sMack to James Bay s to Wash, Ida, wMont Alta, Sask, Man, Minn, sOnt. Map 649 ●

R. inerme Rydb.
Coarse shrub, resembling *R. hirtellum* in flower and stem characters, but differing as shown in the key. $2n = 16$. Montane woods. BC, Alta s to Calif, NM. Ma 649 ○

R. lacustre (Pers.) Poir. / Bristly Black Currant
Stems erect to ascending, 6–10 dm high, very spiny and prickly on the internodes, with clusters of longer nodal spines; leaves deeply 3–5-lobed, the lobes incised-toothed, cordate or truncate at base, glabrous or sparsely hairy; ovary glandular-bristly; hypanthium-tube broadly cup-shaped, about 1 mm long, glabrous; sepals about 2 mm long; petals about 1.3 mm long; stamens equalling the petals, fruit dark purple or black, bristly. $2n = 16$. Moist woods. Alas, Yuk, swMack to James Bay, Nfld, s to Calif, Utah, Colo, SD, Mich, Pa. Map 650

R. laxiflorum Pursh
Similar to *R. glandulosum*; differing by the characters of the key and the red to purple flowers. $2n = 16$. Moist subalpine woods. sAlas, sYuk s to Calif, nIda, Alta. Map 648 ○

R. oxyacanthoides L. / Wild Gooseberry
Stems 3–6 dm high, erect or prostrate, nodal spines usually present; internodes commonly bristly; leaves 3–5-lobed, subcordate or cordate at base, more or less pubescent, often somewhat glandular beneath and on the petioles; floral bracts glandular hairy; ovary, fruit and hypanthium glabrous; hypanthium-tube about 3 mm long; sepals oblong, 4.5–5 mm long; stamens equalling or somewhat exceeding the petals but shorter than the sepals. $2n = 16$. Includes *R. setosum* Lindl. of Ed. 1. Moist woods. sAlas, Yuk, Mack to James Bay s to BC, Alta, eWyo, Neb, Minn, Pa. Map 651

R. triste Pall. / Wild Red Currant
Stems reclining or ascending, the branches often rooting; leaves reniform-orbicular in outline, 3-lobed or rarely 5-lobed, shallowly cordate or rarely truncate at base, the lobes mostly broad-deltoid, coarsely dentate-serrate, glabrous above, pubescent to glabrate beneath; racemes drooping, the axis and pedicels mostly with short-stipitate glands; hypanthium-tube saucer-shaped; sepals greenish purple, about 2 mm long; petals about 1 mm long, truncate or notched; berries red, smooth. $2n = 16$. Moist woods. eAsia: Alas, Yuk, nMack to Huds Bay, nQue, Nfld s to Ore, Alta, Sask, Man, ND, Minn, NJ. Map 652 ●

R. viscosissimum Pursh / Sticky Currant
Stout shrub, mostly 5–10 dm high, with spreading branches; leaves sticky, orbicular or reniform in outline, 3-lobed or rarely 5-lobed, 4–7 cm broad, cordate at base, the lobes crenate-dentate, hairy and glandular on both surfaces; petioles densely glandular-hairy; racemes 3–7-flowered, glandular-puberulent; bracts large, oblanceolate; flowers greenish white, tinged with pink; hypanthium-tube deeply cup-shaped, 6–7 mm long; sepals about as long as the tube; petals about half as long as the sepals; fruit black, glandular-hairy, about 1 cm long. $2n = 16$. Montane woods. BC, Alta s to Calif, Ariz, wWyo, nwColo. Map 652 ○

ROSACEAE / Rose Family

Herbs, shrubs or trees; leaves alternate, simple or compound, usually stipulate, the stipules often deciduous; flowers usually in racemes or cymes, perfect, regular, perigynous or epigynous, the hypanthium saucer-shaped to urn-shaped; sepals usually 5, often with 5 alternate bractlets; petals commonly 5; stamens many, seldom fewer than 10; carpels 1 to many, the ovaries superior or inferior, separate or united; fruits of various kinds–achenes, follicles, drupes, pomes–the modified hypanthium often an integral part.

1 Herbs, or with a woody base and otherwise herbaceous ... 2
Shrubs or trees ... 12

2 Leaves simple, usually toothed ... 3
Leaves compound ... 5

3 Petals usually 8; carpels many, becoming achenes with long plumose styles; low spreading plants ... **Dryas**
Petals 5 ... 4

4 Carpels numerous, becoming drupelets (raspberry-like); leaves lobed ... **Rubus**
Carpels 3–5, becoming follicles; leaves entire to toothed ... **Spiraea**

5 Hypanthium with hooked prickles; carpels 2; inflorescence spicate-racemose ... **Agrimonia**
Hypanthium lacking hooked prickles; carpels 5–many ... 6

6 Calyx-bractlets lacking; carpels becoming drupelets (raspberry-like) ... **Rubus**
Calyx-bractlets present (appearing like 10 sepals); carpels becoming achenes (sometimes on a fleshy receptacle) ... 7

7 Stamens 5; carpels 5–20; petals small, little if any longer than the sepals ... 8
Stamens 10–many; carpels 10–many; petals usually longer than the sepals ... 9

8 Leaves 2–4, ternately divided into linear lobes; petals white ... **Chamaerhodos**
Leaves 3-foliolate, the leaflets wedge-shaped, 3–5-toothed at apex; petals yellow ... **Sibbaldia**

9 Basal leaves with 3 leaflets ... 10
Basal leaves with more than 3 leaflets ... 11

10 Petals white; plants with stolons; receptacle fleshy in fruit (a strawberry) ... **Fragaria**
Petals white or yellow; stolons generally absent; fruit dry ... **Potentilla**

11 Styles articulated at base and deciduous from the achenes, not jointed near the middle, not hooked or plumose ... **Potentilla**
Styles not articulated at base or deciduous but persisting in the fruit, often with an abrupt bend or joint near the middle, hooked or plumose in fruit ... **Geum**

12 Ovary or ovaries superior; hypanthium shaped like a saucer, cup or urn and free from the ovary ... 13
Ovary inferior, enclosed in and adnate to the hypanthium; fruit a pome but often berry-like ... 21

13	Carpel 1; fruit a drupe (cherry-like); leaves simple	**Prunus**
	Carpels 2 or more; leaves simple or compound	14
14	Leaves compound	15
	Leaves simple	18
15	Carpels mostly 5; leaves dissected into linear divisions; plants 5–15 cm tall	**Luetkea**
	Carpels numerous; leaves not linear-dissected; plants generally taller	16
16	Fruit dry (achenes); bractlets alternating with the sepals	**Potentilla**
	Fruit fleshy; sepal bractlets absent	17
17	Carpels becoming drupelets (raspberry-like)	**Rubus**
	Carpels becoming achenes, enclosed in a fleshy receptacle	**Rosa**
18	Flowers solitary; petals 8; fruit with plumose styles	**Dryas**
	Flowers in umbels, corymbs or short racemes	19
19	Carpels numerous; fruit fleshy (raspberry-like)	**Rubus**
	Carpels 2–5; fruit dry, dehiscent	20
20	Leaves more or less entire, serrate; calyx not stellate-pubescent	**Spiraea**
	Leaves palmately 3–5-lobed; calyx stellate-pubescent	**Physocarpus**
21	Leaves pinnately compound; flowers in broad compound corymbs	**Sorbus**
	Leaves simple; flowers not as above	22
22	Leaves mostly incised-toothed or lobed; stems with stout thorns	**Crataegus**
	Leaves entire to toothed; stems not thorny	**Amelanchier**

Agrimonia L.

A. striata Michx. / Agrimony
Erect perennial herb with stout rhizome and fibrous roots; stems coarse and stiff, sometimes forked above, 5–12 dm tall, hirsute to pubescent; leaves interruptedly pinnate, the 5–9 main leaflets interspersed with very small ones; leaflets lanceolate to narrowly obovate, long-acuminate, coarsely toothed, veiny, soft-pubescent and glandular-dotted beneath; stipules large, usually deeply toothed; flowers small, crowded, in a slender spike-like raceme, the bracts 3-cleft; hypanthium obconic,grooved, the margin beset with hooked prickles; calyx with 5 lobes, closing after flowering; petals 5, yellow, about 3 mm long; stamens 5–15; carpels 2, becoming achenes; fruiting hypanthium glandular-dotted, reflexed and deeply furrowed on the surface, with a crown of hooked bristles, and enclosing 2 seed-like achenes. $2n = 56$. Borders of woods. BC to Nfld s to Ariz, NM, Neb, Ia, Ohio, NJ. Map 653

Amelanchier Medic. / Service-berry; June-berry

A. alnifolia Nutt. / Saskatoon
A shrub or small tree, 1–6 m high, often spreading by stolons and forming colonies, the branches smooth, greyish or brownish; leaves alternate, simple, at

first tomentose beneath, somewhat pubescent or glabrous at maturity, elliptic to obovate, broadly truncate to rounded at the apex, the margin coarsely serrate-dentate or, towards the base, entire; flowers white in an erect, dense raceme, appearing early in the season; calyx 5-cleft; petals white, oblanceolate, 6–10 mm long, occasionally longer; stamens about 20; carpels 5, styles usually 5, united below; ovary inferior, tomentose on top, 5-chambered, each cavity with 2 ovules and an intruding false partition; fruit a pome, but berry-like, globose to obovoid, blue-purple with a bloom, 6–10 mm in diam., juicy and sweet. $2n = 34, 68$. A constituent of pemmican; makes excellent pies. Open woodlands. Alas, Yuk, Mack to James Bay s to Calif, Ariz, NM, Neb, Minn. Map 654

Chamaerhodos Bunge

C. erecta (L.) Bunge
Biennial herb, with a woody taproot, forming a basal tuft of leaves the first year and a flowering shoot, 1–3 dm tall, during the second year; stem erect, usually simple below and branching above, sometimes branched from the base, hirsute and glandular; basal leaves petioled, comprising a rosette in the first year, followed by more upright leaves in the second year, 2–4-ternately dissected into narrow divisions, hirsute and glandular-pubescent; cauline leaves similar to the basal ones but somewhat smaller, the uppermost reduced to sessile but deeply dissected bracts; flowers white, numerous in bracted cymes; hypanthium small, cup-shaped; sepals 5, lanceolate, with shining hairs; petals 5, white, slightly longer than the sepals; stamens 5, opposite the petals; ovaries 5–10, superior; style nearly basal, slender; achenes ovoid-pyriform, brownish. $2n = 14$. In our area ssp. **nuttallii** (Pickering) Hult. Dry slopes and plains, often on sandy or gravelly soils. Asia: sAlas, sYuk, sMack, BC to swMan s to Colo, ND, Minn, Mich. Map 655

Crataegus L. / Hawthorn

Shrubs or small trees, usually with stout thorns on the branches; leaves alternate, petioled, simple, serrate and often shallowly lobed; flowers perfect, regular, usually in small corymbs; hypanthium cup-shaped, adnate to the ovary; sepals 5; petals 5, white; stamens 5–25; carpels 2–5, syncarpous, the compound ovary inferior; fruit a berry-like pome, globose or ovoid, red to black at maturity, with 1–5 bony nutlets.

Fruit black or deep purple when ripe; sepals ovate, hairy above, often without glands; leaves nearly glabrous beneath, more or less pubescent above; thorns stout and short; nutlets pitted **C. douglasii**

Fruit red or red-purple; sepals lanceolate, acuminate, glandular-margined; leaves pubescent beneath, at least on the veins; thorns long; nutlets not pitted **C. rotundifolia**

C. douglasii Lindl.
Shrub or small tree, 1–5 m high, armed with stout thorns, usually less than 2 cm long; leaf-blades ovate to oval, glabrous or nearly so, sometimes with

appressed hairs on the upper surface, irregularly toothed, more or less lobed above the middle, the lobes in 2–4 pairs, small, rounded or acute, irregular; fruit purplish black, juicy, said to be edible but of rather poor quality; nutlets irregularly pitted on the inner surfaces. $2n = 51$. Open woods and rocky slopes. BC, Alta s to Calif, eNev, Utah, Colo. SD; Alas; swSask; Ont, Mich. Map 656

C. rotundifolia Moench
Shrub or small tree with rounded crown, intricately branched, 1–5 m high, usually with numerous thorns, 2–7 cm long; leaf-blades broad, orbicular, oval or rounded-obovate, doubly serrate except near the base, more or less lobed, the lobes in 3 or 4 pairs and triangular, usually shallow; petioles commonly with dark red raised glands and the teeth of the blades red-tipped; fruit red or yellow-orange; nutlets not pitted. $2n = 34, 51, 68$. An agamospermous complex. Our material of the complex is generally treated as *C. chrysocarpa* Ashe (Ed. 1.). Coulees, river valleys, and open woods. BC to NB, NS s to Ore, Ida, Mont, NM, SD, Minn, Wisc, NEng. Map 657

Dryas L. / Dryad; Mountain Avens

Low prostrate shrubs, the horizontal stems much branched and forming dense mats, freely rooting; leaves alternate, petioled, leathery, simple, crenate or entire, white-tomentose beneath; stipules linear-lanceolate; flowers usually solitary, on tomentose peduncles, these naked or with 1–4 vestigial bracts; hypanthium saucer-shaped, tomentose and glandular; sepals 8–10, linear-lanceolate, persistent; petals 8–10, white or pale yellow; stamens numerous; carpels numerous, becoming achenes, the styles terminal, persistent, in fruit much elongated and long-plumose; the styles are spirally twisted together when immature and in damp weather.

1 Flowers yellow; filaments hairy; leaves cuneate at base — **D. drummondii**
Flowers white; filaments glabrous; leaves truncate to cordate at base — 2

2 Leaves crenate to incised-lobed, strongly rugose above; midvein prominent below, glandular — **D. octopetala**
Leaves more or less entire or with a few teeth towards base, smooth above; midvein not prominent below, lacking glands — **D. integrifolia**

D. drummondii Richards. / Yellow Dryad
Loosely caespitose, mat-forming undershrub; leaf-blades elliptic to oblong-obovate, 1.5–3 cm long, coarsely crenate, cuneate at base, rounded at apex, glabrous or slightly tomentose and strongly rugose above, white-tomentose beneath; peduncles 5–25 cm high, with 1–4 vestigial bracts, the uppermost sometimes supporting reduced flowers; hypanthium and calyx covered with black, gland-tipped hairs and somewhat tomentose; sepals ovate; petals pale yellow, strongly ascending. $2n = 18$*. Gravelly slopes, river bars, and roadsides. Alas, Yuk, swMack, nwSask s to Wash, neOre, Mont; n shore Lk Superior; Que, Nfld. Map 658

D. integrifolia M. Vahl / White Dryad
Densely caespitose, matted plant with numerous leaves; leaf-blades lanceolate-oblong, 0.8–1.5 cm long, broadest below the middle, the margins entire or with a few teeth in the lower half and usually somewhat revolute, the base cordate or truncate, the upper surface usually glabrous and not rugose, the lower surface thinly white-tomentose, the non-prominent midvein lacking glandular-stipitate dark hairs; peduncles up to 10 cm long in fruit, bractless; hypanthium sparingly tomentose and with a few black glandular hairs; sepals linear-lanceolate; petals white, 1–1.3 cm long; fruiting styles 2–4 cm long. $2n = 18$*. Hybridizes with *D. octopetala*, the plants with leaves smooth and shiny above but the margins crenate-serrate. Alpine slopes. NeSiberia (Chukchi Pen), Greenland: Alas to Ellesmere Is to Nfld, NB s to BC, Alta, sMack, Huds Bay, nQue. Map 659

D. octopetala L. / White Dryad
Dwarf undershrub, forming small dense mats; leaf-blades oblong-ovate, 0.8–2.5 cm long, the margins coarsely incised-crenate and somewhat revolute, the base subcordate or rarely truncate, obtuse at apex, the upper surface strongly rugose, with wart-like excrescences or merely glandular-viscid, the lower surface white-tomentose between the prominent veins; the veins, also the petioles and stipules, covered with yellowish brown glands; peduncles 5–20 cm high, white-tomentose, bearing black glands, bractless or with one reduced bract; hypanthium and calyx white-tomentose and black-hairy; sepals linear-lanceolate; petals white, 8–15 mm long; styles in fruit 2–3 cm long; achenes about 3 mm long. $2n = 18$*. In our area ssp. **hookeriana** (Juz.) Hult. (*D. hookeriana* Juz. of Ed. 1). Alpine slopes. More or less circumpolar: Alas to Melville Is s to Wash, neOre, Ida, Mont, Colo. Map 660

Fragaria L. / Strawberry

Perennial, acaulescent herbs, with short scaly rootstocks and producing long stolons which root and form new plants; leaves basal, long-petioled, stipulate, trifoliolate, the 3 leaflets short-stalked to sessile, obovate-cuneate, coarsely serrate, usually silky-hairy beneath; flowers white, usually several on a scape-like peduncle; hypanthium almost flat; bractlets, sepals and petals, 5 each; petals white, longer than the sepals; stamens about 20; carpels numerous; style attached near the middle of the ovary; receptacle hemispheric or conic, becoming much enlarged, juicy and red in fruit, with the numerous minute seed-like achenes scattered over it.

Leaf-teeth sharply divergent, the terminal one normally projecting beyond the adjacent lateral ones; upper surface somewhat silky-pilose; inflorescence at maturity usually longer than the leaves — **F. vesca**

Leaf-teeth ascending, the terminal one normally shorter than the adjacent lateral ones; upper surface glabrous; inflorescence at maturity usually shorter than the leaves — **F. virginiana**

F. vesca L. / Woodland Strawberry
Flowering scapes 0.5–2.5 dm tall, generally somewhat longer than the leaves, pubescent; leaves with petioles 3–20 cm long, often reddish; leaflets light-green, mostly 10–30 mm long, 10–20 mm wide, obovate, base generally cuneate, margin serrate, at least above the middle, teeth divergent; flowers 3–15; sepals 4–5 mm long; petals 8–11 mm long; fruit about 1 cm across; achenes often superficial. $2n = 14^*$. Dry to moist open woods, often sandy. Eurasia: BC to Nfld s to Calif, NM, Neb, Mo, Ill, Va. Map 661

F. virginiana Duchesne / Wild Strawberry
Similar to *F. vesca;* distinguished by the characters of the key and the somewhat thicker leaves that are generally bluish green above with the leaflets often rounded at the base; the often smaller flowers; the achenes that are generally sunken. $2n = 56^*$. In our area ssp. **glauca** (S. Wats.) Staudt. Woods and plains to subalpine elevations. Alas, Yuk, wMack to Huds Bay, Nfld s to Calif, Colo, Okla, Tenn, Ga, Mex. Map 662

Geum L. / Avens

Perennial hairy herbs with stout crowns or rootstocks; leaves alternate, chiefly basal and petioled; lower leaves pinnate, often lyrate, usually with smaller segments interposed between the larger ones; upper leaves much smaller, trifid or bract-like; hypanthium cup-shaped or saucer-shaped; calyx-lobes 5, persistent, with 5 alternate bractlets; petals 5; stamens many; carpels many, superior, on a hemispheric or club-shaped receptacle; style terminal; fruit a cluster of achenes, each tipped by a long persistent style, this plumose or sharply curved and jointed, its upper part often deciduous.

1 Style jointed, hooked in fruit; cauline leaves several, alternate — 2
Style not obviously jointed, plumose in fruit; leaves mostly basal, cauline leaves few (2), opposite — **G. triflorum**

2 Petals purplish or purple-veined; sepals erect, 7–10 mm long; flowers nodding — **G. rivale**
Petals yellow; sepals reflexed, 3–8 mm long; flowers erect — 3

3 Style glandular-pubescent, at least below; terminal segment of basal leaves 3-lobed, cordate or rounded at base — **G. macrophyllum**
Style not glandular; terminal segment of basal leaves deeply incised, cuneate at base — **G. aleppicum**

G. aleppicum Jacq. / Yellow Avens
Coarse, erect herb, 4–10 dm high, usually hirsute throughout, the leaves with large stipules; basal leaves petioled, lyrately pinnate, with 5–7 toothed or cleft leaflets and often a few very small interspersed leaflets; cauline leaves short-petioled or sessile, smaller and with fewer divisions, the principal leaflets rhombic-ovate to oblong and cuneate at base; inflorescence divaricately branched; sepals lanceolate, nearly as long as the yellow petals; receptacle of fruit densely

hairy; head of achenes globose-ovoid, about 1 cm broad, each achene with a hooked beak about twice as long as the body, the latter glabrous along the sides, otherwise hairy. $2n = 42$*. Moist woods and meadows. More or less circumpolar: swMack, aC to Nfld s to Calif, NM, Neb, Minn, Pa, NS; Alas; Yuk. Map 663

G. macrophyllum Willd. / Yellow Avens
Differing from *G. aleppicum* as shown in the key and as follows: terminal leaflet cordate to truncate rather than cuneate at its base, also large in relation to the lower leaflets; leaflets crenate or dentate rather than sharply incised; sepals broadly triangular; fruiting receptacle sparsely hairy; achenes pubescent, not glabrous on the sides. $2n = 42$. In our area ssp. **perincisum** (Rydb.) Hult. Moist woods and meadows. Japan, Kamchatka: Alas, Yuk, Mack to James Bay, Nfld s to Calif, NM, SD, Minn, Gt Lakes, NY. Map 664

G. rivale L. / Purple or Water Avens
Slender herb, 4–10 dm tall, with simple or little-branched stem; basal leaves up to 3 dm long, lyrately pinnate, the principal leaflets usually 3, the terminal one broadly obovate and more or less 3-lobed, serrate; additional minute leaflets remote and variable in shape; cauline leaves much smaller, variously divided and toothed; flowers few to several, nodding; sepals erect, 7–10 mm long, purple or yellowish; petals erect, flesh-coloured to yellowish and purple-veined, abruptly contracted to claw; style jointed near the middle, the upper half plumose, the lower half hirsute; head of fruit stalked, erect, the achenes spreading in the head. $2n = 42$. Possible hybrids with *G. aleppicum* (*G.* × *pervale* Boivin) and *G. macrophyllum* (*G.* × *aurantiacum* Fries) have been reported, the latter from our area. Stream banks, marshes, wet meadows. Eurasia: BC to Nfld s to Wash, NM, Mo, Ind, NJ. Map 665

G. triflorum Pursh / Old Man's Whiskers; Prairie Smoke
Low herb, 2–4 dm tall, with a stout rootstock bearing old leaf-bases and a tuft of new leaves, soft-hairy to hirsute throughout and somewhat glandular; leaves mostly basal, interruptedly pinnate, the 9–19 leaflets rather crowded, cuneate-obovate, cleft into narrow, toothed divisions; cauline leaves 1 or 2, small, usually pinnatifid; cymes 1–5-flowered, the flowers often nodding; hypanthium deep cup-shaped, red-purple; bractlets longer than the sepals and as long as the erect, oblong, petals, these yellowish white to pinkish; styles not jointed, slender, elongated in fruit, 2–4 cm long, plumose except at the extreme tip, bronze to purplish; achenes about 3 mm long. $2n = 42$. Foothills and prairie. sMack, BC to Nfld s to Calif, Nev, Utah, NM, SD, Ia, Ill, NY. Map 666

Luetkea Bong.

L. pectinata (Pursh) Kuntze / Partridge-foot
Semi-shrubby caespitose plant, with prostrate, stoloniferous branches; leaves crowded, glabrous, twice 3-cleft into linear, acute divisions; flowers in a short raceme on a leafy shoot, 5–15 cm high; hypanthium hemispheric; sepals 5; petals 5, white, longer than the sepals; stamens about 20, the filaments united below;

carpels usually 5, distinct, ripening into hard dehiscent follicles, about 4 mm long. $2n = 18^*$. Moist, shady areas around timberline. sAlas, sYuk, wMack s to Calif, Ida, Mont. Map 667 ●

Physocarpus Maxim. / Ninebark

P. malvaceus (Greene) Kuntze
Erect shrub, 1–2 m high, with bark exfoliating in strips; leaves rounded-ovate or orbicular in outline, usually more or less 3-lobed and doubly crenate, glabrous or with branched hairs; inflorescence a terminal corymb, few–many-flowered; stellate-tomentose; sepals 5, 4–5 mm long, persistent; petals 5, white; stamens many; carpels 2, united more than half their length, becoming pod-like fruits, flattened, keeled, tomentose, about 5 mm long, dehiscent on both sutures, each with usually 2 seeds. Montane woods. BC, Alta s to Ore, Utah, Wyo. Map 667 ○

Potentilla L. / Cinquefoil

Mostly perennial herbs, a few annual or biennial, one species a shrub; leaves alternate, compound; flowers cymose or solitary; hypanthium saucer-shaped to hemispheric; sepals 5, alternating with as many bractlets; petals 5, yellow, white or purplish; stamens few to many; carpels usually numerous, the styles terminal or lateral, deciduous; fruit a head of achenes, usually enveloped by the persistent, enlarged calyx.

1 Shrub; flowers yellow — **P. fruticosa**
Herb, or woody at base — 2

2 Flowers reddish purple; a decumbent plant of aquatic and marshy habitats — **P. palustris**
Flowers white or yellow; plants of drier habitats — 3

3 Flowers white; leaves trifoliolate, dark green above; leaflets usually 3-toothed at apex; low plants with woody base — **P. tridentata**
Flowers mostly yellow, if white the leaves pinnate — 4

4 Flowers solitary on naked peduncles, mostly from the nodes of stolons; leaves pinnate, silver-white below — **P. anserina**
Flowers few–many, in cymes; if solitary plants dwarf and caespitose — 5

5 Basal leaves pinnate — 6
Basal leaves trifoliolate or 5–9-digitate — 20

6 Petals white to yellowish white; styles slender-fusiform, nearly basal in origin; plants glandular-hairy — 7
Petals yellow; styles tapered from base, arising near top of ovary; plants rarely glandular-hairy — 8

7 Cymes contracted; petals white, little longer than sepals; plants robust — **P. arguta**
Cymes open; petals white to yellow, longer than sepals; plants slender — **P. glandulosa**

8 Lower surface of leaves densely white tomentose, frequently overlain with sericeous hairs 9

Leaves glabrous to variously pubescent but not densely white tomentose below 12

9 Plants caespitose; mostly 4–10 cm tall; leaflets 1–2 cm long **P. concinna**

Plants not caespitose; 15–50 cm tall; leaflets longer 10

10 Leaves distinctly pinnate; leaflets mostly 7–11, divided about half-way to midrib **P. hippiana**

Leaves often indistinctly pinnate; leaflets approximate (subdigitate), mostly 5–7, divided almost to the midrib 11

11 Plants generally less than 20 cm tall; stipules of upper cauline leaves entire **P. multifida**

Plants generally more than 20 cm tall; stipules of upper cauline leaves toothed, at least basally **P. bipinnatifida**

12 Leaflets serrate or lobed, generally divided only half-way to midrib or less 13

Leaflets divided almost to the midrib, or pedately divided or pinnatisect with narrow segments 16

13 Leaves densely woolly-hairy below, conspicuously veined above 14

Leaflets sparsely hairy, mostly on veins 15

14 Bractlets much longer than the sepals, as wide as or wider than sepals; petals 3–5 mm long, as long as or slightly shorter than the sepals **P. finitima**

Bractlets about as long as sepals, mostly narrower than sepals; petals 4–6 mm long, generally longer than sepals **P. pensylvanica**

15 Leaflets crenately lobed; stem-leaves several, pinnate; petals about equalling sepals **P. paradoxa**

Leaflets serrate-dentate; stem-leaves 0–2, subdigitate; petals much longer than sepals **P. drummondii**

16 Leaflets often approximate, the leaves appearing subdigitate, leaves sparsely sericeous hairy above and below (more densely on veins) or essentially glabrous 17

Leaflets separate, the leaves distinctly pinnate; generally densely sericeous below or with plump appressed strigose hairs throughout 18

17 Plants 20–50 cm tall **P. bipinnatifida**

Plants 5–20 cm tall **P. multisecta**

18 Plants with appressed strigose hairs throughout; stems generally decumbent or spreading; leaves green above and below **P. plattensis**

Plants not appressed strigose hairy throughout; leaves with abundant appressed to spreading sericeous hairs on lower surface 19

19 Stems mostly 10–50 cm high; basal leaves generally with 5–9 leaflets; woolly velutinous to densely sericeous below, but lacking underlying crinkly hairs; petals somewhat longer than sepals **P. pensylvanica**

Stems mostly 5–15 cm high; basal leaves generally with 9–21 leaflets, divided nearly to base into 3–5 linear segments 0.5–1.5 mm wide; sericeous below with underlying or marginal crinkly hairs; petals much longer than sepals **P. ovina**

20 Basal leaves trifoliolate 21
Basal leaves digitate with 5–9 divisions 27

21 Lower leaf-surface glabrous to variously pubescent, generally sparsely, but not tomentose 22
Lower leaf-surface densely white tomentose 24

22 Plants generally 5–10 cm tall; petals evident, longer than sepals **P. hyparctica**
Plants taller, generally 20–60 cm; petals obscure, shorter than sepals 23

23 Petals about half as long as sepals; stamens 5–15; achenes smooth, 0.5–0.8 mm long; leaves glabrous above **P. rivalis**
Petals almost equalling sepals; stamens 15–20; achenes ribbed, 0.8–1.3 mm long; leaves with scattered appressed strigose hairs above **P. norvegica**

24 Stem and petiole tomentose, often floccose, lacking any other kind of pubescence; terminal leaflets often elliptic **P. nivea**
Stem and petiole with or without tomentum, but never floccose, and some additional pubescence present (in the form of curly or wavy hairs, ascending or spreading straight hairs); terminal leaflets obovate, base cuneate 25

25 Flowers generally 3 or more; stem and petiole lacking tomentum; pubescent, often sparsely, with short, curly or wavy hairs and much longer straight or nearly straight ascending or spreading hairs **P. hookeriana**
Flowers generally 1 or 2; stem and petiole tomentose, with overlying short curly or wavy hairs and much longer straight or nearly straight ascending or spreading hairs 26

26 Plants generally 4–8 cm tall; stem usually not branched; flowers mostly 1, 15–20 mm across; veins on lower surface of leaves obscured by the dense tomentum **P. uniflora**
Plants generally 8–12 cm tall; stem often branched about middle; flowers mostly 2 (or more), 10–15 mm across; veins on lower surface of leaves prominent above the dense tomentum **P. villosa**

27 Plants 15–70 cm tall, leaflets divided almost to midrib, densely white tomentose below, overlain with abundant sericeous hairs 28
Plants mostly shorter, 5–30 cm tall, but sometimes up to 60 cm; leaflets mostly dentate, incised only half-way to midrib or less 30

28 Leaves mostly cauline, 1–5 cm long **P. argentea**
Leaves mostly basal, longer 29

29 Stipules toothed, at least towards their base; petals shorter than sepals; inflorescence compact **P. bipinnatifida**
Stipules entire; petals much longer than sepals, inflorescence open **P. gracilis**

30 Leaves densely white tomentose below; upper and lower surfaces of strongly contrasting colour 31
Leaves not densely white tomentose below; the upper and lower leaf-surfaces often of similar green colour or the lower greyish-brown 33

31 Plants generally less than 10 cm tall; leaflets 1–2 cm long, cuneate at base, mostly entire below the middle **P. concinna**
Plants generally at least 15 cm tall 32

32 Larger leaflets about 2 cm long, obovate; styles warty-papillose at base **P. hookeriana**
Larger leaflets at least 3 cm long, oblanceolate to elliptic; styles not warty-papillose at base **P. gracilis**

33 Petals shorter than sepals; stamens 5–15 **P. rivalis**
Petals longer than sepals; stamens more than 15 34

34 Achenes striate or strongly reticulate-veiny; petals 9–12 mm long; basal leaves not conspicuous **P. recta**
Achenes more or less smooth; petals 5–9 mm long; basal leaves clearly evident 35

35 Anthers mostly 0.5 mm long; leaflets sharply toothed apically, base entire, cuneate, mostly fresh green on both surfaces, secondary veins not conspicuously raised **P. diversifolia**
Anthers mostly 1.0 mm long; leaflets elliptic-lanceolate, toothed to the base, often a greyish-brown colour below, somewhat greener above, secondary veins conspicuously raised **P. gracilis**

P. anserina L. / Silverweed
Perennial herb, at first acaulescent, soon producing one or more long runners which root and form leaf-clusters at the nodes; leaves basal, tufted, 1–2 dm long, pinnate, with 7–21 main leaflets and interspersed smaller ones; leaflets sharply serrate, silky-tomentose beneath and usually green and glabrous above; flowers borne singly on long stalks from the leaf-clusters; bractlets often toothed; petals yellow, 6–10 mm long; stamens 20–25; style lateral; achenes grooved on the back and top. $2n$ = 28, 42. Lake shores and river flats. Circumpolar: Alas, Yuk, swMack to Huds Bay, Nfld s to Calif, NM, Neb, Ia, Ind, NY. Map 668

P. argentea L. / Silvery Cinquefoil
Perennial herb, freely branched, the stems 1–5 dm long, ascending or erect, white-woolly; lower leaves mostly 5-digitate; leaflets white-tomentose beneath, linear-oblanceolate in outline, long-cuneate and entire at base, incise-toothed to pinnatifid above the middle, the margins revolute; cyme leafy, many-flowered; petals yellow, about equalling the sepals. $2n$ = 14, 28, 35, 42, 56. Roadsides. Introduced. Rare.

P. arguta Pursh / White Cinquefoil
Erect glandular-hairy perennial herb, the sticky pubescence often brownish; stem stiff, 4–10 dm tall, often reddish; basal leaves pinnate, long-petioled; leaflets 7–11, fewer in the upper leaves, mostly ovate, incised-serrate, densely to sparsely hairy; cyme crowded; bractlets of flower lanceolate; sepals ovate, elongating in fruit; petals white or cream-coloured, little longer than the sepals; stamens 25–30, in 5 groups on a thick disc; style nearly basal; achenes finely striate, light

brown. $2n = 14$. *Drymocallis arguta* (Pursh) Rydb. Prairies. Alas, Yuk, swMack to cMan, sQue, NB s to Ore, Ariz, NM, Okla, Mo, Minn, Pa, NJ. Map 669

P. bipinnatifida Dougl. *ex.* Hook.
Perennial, from a stout, branching caudex; stems 2–5 dm tall, erect, often decumbent at base, pubescent with short and long, spreading or ascending hairs, occasionally more or less glabrous; basal leaves mostly 10–25 cm long, pinnate, often appearing digitate; petiole as long as or usually much longer than blade; leaflets 5–7 (9), crowded, the larger 3–6 cm long, 1–3 cm wide, pectinately divided almost to midrib, segments 1.5–3.0 mm wide, upper surface grey-green, sparsely to densely pubescent with short and long hairs, occasionally glabrous, lower surface white, densely tomentose overlain with long sericeous hairs, occasionally green and glabrous; stem-leaves similar, short-petiolate; inflorescence open to congested, generally many-flowered; flowers 6–10 mm across; sepals 3–4 mm long, ovate; bractlets of epicalyx oblong-lanceolate, shorter and narrower than the sepals; petals yellow, obcordate to more or less quadrate, shorter than sepals; achenes about 1 mm long. $2n = 56$*. Dry prairie and open slopes. Alas, Yuk, swMack to Huds Bay, Nfld s to BC, Nev, Ariz, NM, Neb, Minn, Ind, NH. Map 670

P. concinna Richards. / Early Cinquefoil
Low, spreading perennial herb, generally less than 1 dm high; caudex bearing a cluster of basal leaves and several prostrate or ascending stems; leaves 5–9-foliolate, digitate to pinnate; digitate leaflets (rarely the pinnate) divided about half-way to midrib, at least distally, pinnate leaflets mostly divided to the midrib into linear-lanceolate segments about 1 mm wide, sericeous strigose above, densely white tomentose below, with overlying sericeous hairs; inflorescence open, mostly 2–5-flowered; flowers generally 10–15 mm across; sepals villose, longer than the bractlets but surpassed by the yellow petals. The pinnate material, var. **macounii** (Rydb.) C.L. Hitchc. (var. *divisa* Rydb. of Ed. 1) closely approaches *P. ovina* but that species lacks tomentum. Dry, sandy prairies. Alta to Man s to Nev, Ariz, NM, SD. Map 671 •

P. diversifolia Lehm.
Perennial herb; stem 1–4 dm high, from a stout caudex, silky-hairy to glabrous; leaves mostly basal, digitate leaflets 5–7, 1–3 (5) cm long, obovate to oblanceolate, often narrowly cuneate at the base, commonly toothed above the middle, glabrous to sparsely strigose above, generally strigose below, especially on veins; flowers few–several, in an open cyme; sepals and bractlets exceeded by the yellow, obovate petals. $2n = 42–101$, c.140. Moist montane meadows to alpine slopes. Map 672

P. drummondii Lehm.
Perennial, from a stout caudex; stems mostly 1.5–5.0 dm tall, generally glabrous below, hirsute above; leaves mostly basal, 5–15 cm long; petiole as long as or longer than blade; leaflets mostly 5–9, approximate, 2–5 cm long, cuneate at

base, cleft above half-way to midrib, glabrous above, sparsely strigose hairy below, mainly on the veins and margin; cauline leaves 1–3, similar, short-petiolate; inflorescence open to somewhat congested below; flowers 1.0–2.0 cm across; petals 6–11 mm long, yellow, obovate, the tips at times emarginate, longer than sepals; achenes about 1.5 mm long. $2n$ = c.63–c.105. In our area ssp. **drummondii**. Moist subalpine to alpine meadows. BC, Alta s to Calif. Map 671 ○

P. finitima Kohli & Packer
Perennial, from a short, often branched caudex; stem 2–4 dm tall, stout, tomentose with overlying coarse spreading hairs; leaves pinnate, feather-like; petioles 5–15 cm long, about half to as long as blade; leaflets 9–19, the larger 3–6 cm long, 1.0–1.5 cm wide, oblanceolate to elliptic, divided half-way to midrib, reticulately veined above and often densely pubescent with flexuous to sericeous hairs, densely woolly velutinous below; cauline leaves similar, with (5) 7–9 (11) leaflets, short-petiolate; inflorescence congested, many-flowered; flowers 10–16 mm wide; hypanthium 6–10 mm wide, generally lanate; sepals 4–5 mm long, ovate; bracklets of the epicalyx 6–10 mm long, elongate-rhomboid, much longer than sepals, as wide or usually wider; petals 3–5 mm long, yellow, as long as or slightly shorter than sepals; achenes about 1 mm long. $2n$ = 14*. Sandy prairies, hills, and dunes. sAlta to sMan s to ND, wMinn. Map 673 ▲

P. fruticosa L. / Shrubby Cinquefoil
A freely branched shrub, 3–10 dm high, the outer bark shredding; leaves numerous, short-petioled, mostly 5-pinnate; leaflets linear-oblong, tapering at each end, entire, often revolute, more or less hairy on both sides, the upper 3 often united at base; flowers solitary or cymose; petals yellow, longer than the sepals and bractlets; style lateral, clavate; achenes and receptacle densely hairy. $2n$ = 14, 28. *Pentaphylloides floribunda* (Pursh) A. Löve. Plains and open woods to subalpine elevations. Circumpolar: Alas, Yuk, Mack to Huds Bay, Nfld s to Calif, Nev, Utah, NM, SD, Minn, Ind, Pa. NJ. Map 674

P. glandulosa Lindl.
Perennial herb with glandular-pubescent shoots 15–30 cm high, from a scaly rootstock; leaves mainly basal, the 5–9 leaflets mostly obovate, cuneate at base, coarsely serrate, glandular and somewhat hairy; flowers 1.5–2 cm broad on slender stalks; petals whitish or yellow, much longer than the sepals. $2n$ = 14. *Drymocallis glandulosa* (Lindl.) Rydb. In our area ssp. **pseudorupestris** (Rydb.) Keck. Montane woods to shaly alpine slopes. BC, Alta s to Calif, nNev, Ariz, Wyo, nwColo. Map 673 ○

P. gracilis Dougl. *ex* Hook. / Graceful Cinquefoil
Perennial herb with a thick scaly caudex, often multiple; stems erect, often decumbent at base, 3–7 dm high, more or less branched; basal leaves long-petioled, digitate, the 5–7 leaflets narrowly to broadly oblanceolate, coarsely serrate to deeply incised or lobed, strigose to tomentose, sometimes densely white-tomentose beneath; upper leaves few and reduced in size; cyme often many-flowered with erect branches; flowers 1.5–2 cm broad; sepals broadly

lanceolate, bractlets slightly shorter; petals yellow, longer than the sepals. $2n$ = 51–109. A polymorphic species. In our area var. **gracilis** has narrowly oblanceolate leaflets, canescent-tomentose beneath and with elongate triangular-lanceolate teeth; var. **rigida** (Nutt.) S. Wats. (*P. nuttallii* Lehm.) differing in having green leaves, sparingly hirsute, barely if at all tomentose beneath; var. **pulcherrima** (Lehm.) Fern. (*P. pulcherrima* Lehm.) with somewhat broader leaflets, which are densely white-pubescent beneath and the margins crenate or crenate-serrate; and var. **flabelliformis** (Lehm.) Nutt. (*P. flabelliformis* Lehm.) differing from var. *pulcherrima* in that the leaflets are pectinately cleft into narrow divisions. Prairie grassland and open woods. Alas, Yuk to Alta, Sask, Man s to Calif, NM, Neb, SD, introduced Ont, Que. Map 675

P. hippiana Lehm.
Perennial with several ascending stems, 2–4 dm tall, from a thick brown-scaly crown; stems and leaves white-tomentose and silvery-hairy; leaves pinnate, the basal ones with 5–11 leaflets, 2–5 cm long, cuneate-oblong, coarsely toothed, white-tomentose on both sides, often also strigose below; flowers many in an open cyme; sepals tomentose, 4–6 mm long, as long or nearly twice as long as the slender bractlets; petals yellow, barely longer than the sepals. Includes *P. effusa* Dougl. of Ed. 1. Dry prairie grassland and foothills. cBC to sMan, wOnt s to Ariz, NM, Neb. Map 676

P. hookeriana Lehm.
Perennial, from a stout caudex; stems 1–3 dm tall, pubescent with short curly hairs overlain by much longer, straight, ascending or spreading hairs; leaves mostly basal, tufted, 3–15 cm long, 3- or 5-digitate or 5-pinnate; petioles as long as to several times longer than the blade, pubescence like stem; leaflets 10–20 mm long, cuneate, divided one-third to two-thirds to the midrib, upper surface olive-green, sparsely appressed hirsute with short and long hairs; lower surface white, densely tomentose, the veins sericeous-hirsute; inflorescence few- to several-flowered; flowers mostly 10–12 mm across; petals about 5 mm long, obcordate, as long as sepals or slightly longer; achenes 1 mm long. $2n$ = 42, 49. Includes *P. quinquefolia* Rydb. and *P. furcata* Porsild. *P. nivea* L. ssp. *hookeriana* (Lehm.) Hiitonen of Ed. 1. Dry rocky slopes to alpine elevations. nAsia: Alas to Victoria Is s to BC, Alta, nMont, nSask. Map 677

P. hyparctica Malte
Perennial, tufted or caespitose, from a stout caudex; stems 0.2–1.0 (2.0) dm tall, pubescent with short and long spreading hairs; leaves basal, 2–6 (10) cm long, trifoliolate; petioles as long as or longer than the blade; leaflets 5–10 mm long, obovate, base cuneate, coarsely toothed, upper surface green, glabrous to hirsute, lower surface paler, the veins sparse to densely hirsute, sometimes glandular; inflorescence 1–2-flowered; flowers about 10 mm across; calyx pilose, often glandular, lobes 3–4 mm long, ovate-triangular, bractlets of epicalyx ovate, elliptic to obovate, as long as or longer than sepals, generally as wide or wider; petals yellow, longer than sepals; achenes about 1.5 mm long. $2n$ = 42. Alpine slopes and ridges. Circumpolar: Alas to Ellesmere Is to Lab s to BC, Alta, swMack, Huds Bay; James Bay; sQue. Map 673 •

P. multifida L.
Perennial, densely tufted from a stout caudex; stems 1–2 (3) dm tall, erect or spreading, sparsely to densely pubescent with short and long spreading hairs; leaves mostly basal, 5–10 cm long, pinnate; petioles shorter to much longer than the blade; leaflets mostly 5–7, pectinately divided, segments linear, mostly 1.0–1.5 mm wide, the margins revolute, often strongly so, sparsely strigose above, densely white tomentose below, midribs sericeous; inflorescence open, 3–15-flowered, flowers 6–10 mm across; calyx campanulate, lobes 3–5 mm long, ovate, sericeous; petals yellow, about as long as sepals to shorter; achenes about 1.2 mm long. $2n = 28$. Gravel bars and open slopes. Eurasia: swMack to James Bay s to nAlta, nSask, cMan, nLk Superior; Alas, Yuk; nQue. Map 678 ○

P. multisecta (S. Wats.) Rydb.
Perennial, from a stout branching caudex, stems 0.5–1.5 (2.0) dm tall, pubescent with short, spreading, and long appressed, hairs; basal leaves 3–4 cm long, pinnate, rarely digitate; petioles about as long as blade; leaflets 5–9, 5–15 mm long, divided to midrib into linear to oblong segments (0.5) 1.0 (2.0) mm wide, glabrous to sparsely strigose above, similar below except for densely sericeous-strigose veins; inflorescence open, mostly 2–10-flowered; flowers mostly 10–15 mm across; sepals 3–5 mm long, lanceolate; bractlets of epicalyx narrower and shorter; petals longer than sepals; achenes about 1 mm long. Often treated as a variety of *P. diversifolia,* as in Ed. 1, and it might well be an alpine ecotype of that species. However, it is distinct morphologically and appears in several ways to be transitional to *P. ovina* with which it often occurs. Dry alpine slopes. BC, Alta s to Nev, Utah, Colo. Map 678 ●

P. nivea L.
Perennial, loosely to densely tufted from a stout branching caudex; stems 0.5–1.5 (2.0) dm tall, tomentose, usually densely so, often floccose, other hair types lacking; basal leaves mostly 2–10 cm long, trifoliolate, rarely 5-foliolate; petioles generally much longer than blade, pubescent like stem; leaflets 1–2 cm long, the terminal one larger, obovate to frequently elliptic, lobed or divided one-third to midrib, upper surface green, olive-green to brownish, sparsely (larger specimens) to densely sericeous, lower surface densely white tomentose, other hair types lacking; stem-leaves 1–2, often distinctly petiolate; inflorescence mostly 1–5-flowered; flowers 10–15 mm across; sepals 3–4 mm long, elongate-triangular; bractlets of the epicalyx narrowly lanceolate, half as wide as sepals, about as long; petals yellow, obcordate, longer than sepals; achenes about 1.8 mm long. $2n = 14, 28, 49, 56$. Alpine slopes. More or less circumpolar: Alas to sBaffin Is to Nfld s to BC, Utah, Colo, sMack, James Bay, nQue; Gaspé. Map 679

P. norvegica L. / Rough Cinquefoil
Coarse annual or biennial, 2–8 dm high, erect, or somewhat spreading, the stems leafy and hirsute with stiff spreading hairs, often reddish; lower leaves long-petioled, 3-foliolate, more or less hirsute; leaflets obovate, elliptic or oblanceolate,

coarsely serrate; cyme dense and leafy; petals light yellow, broadly obovate, or rounded, shorter than the sepals; stamens 15–20; achenes numerous, often with curved ridges, enveloped by the much enlarged calyx. $2n = 56$, c.63, 70*. Wet meadows, roadsides, and waste ground. Circumpolar: Alas, Yuk, swMack to Huds Bay, nQue, Nfld s to Calif, Ariz, NM, Tex, NC. Map 680

P. ovina Macoun
Perennial, from a stout branching caudex; stem 0.5–1.5 (2.5) dm tall, erect or spreading; leaves mostly basal, tufted, dark olive-green to grey, 2–5 (10) cm long, mostly 1.0–1.5 cm wide, pinnate; petioles 1–2 cm long, occasionally longer; leaflets crowded, ascending, divided nearly to base into mostly 3–5 linear segments, densely sericeous-hirsute above and below, underlain by sparse to dense crinkly hairs; inflorescence open, mostly 1–3(5)-flowered; sepals 4–5 mm long; petals yellow, somewhat longer than sepals, obcordate; achenes 1.5–2.0 mm long. Included in *P. plattensis* in Ed. 1., but really quite distinct. In the mountains plants occur that approach *P. multisecta* morphologically; they may be hybrids. In the foothills *P. ovina* is, at least superficially, similar to *P. concinna* var. *macounii*, but lacks the tomentum of that species. Dry, rocky slopes from foothills to alpine elevations. BC, Alta s to Calif, Utah, NM. Map 681 •

P. palustris (L.) Scop. / Marsh Cinquefoil
Stems from a long woody, rooting rhizome, ascending or erect, reddish brown, more or less glandular or fine-hairy above, 2–6 dm high; leaves pinnately 5–7-foliolate; leaflets oblong-lanceolate to oblanceolate, usually green and glabrous above, paler and fine-hairy beneath, sharply toothed; cyme leafy, few-flowered; flowers reddish purple, about 2 cm broad; sepals purplish, much longer than the linear bractlets; petals reddish, much shorter than the sepals; achenes smooth, brown, borne on the enlarged, spongy, hairy receptacle. $2n = 28, 35, 42$. *Comarum palustre* L. Marshes, bogs, lake margins. Circumpolar: Alas, Yuk, arct coast to nQue, Nfld s to Calif, Wyo, Ia, Ohio, Pa, NJ. Map 682 •

P. paradoxa Nutt.
Annual, biennial or short-lived perennial herb; stems spreading or ascending, 2–5 dm long, leafy, sparingly hairy to glabrous; larger leaves short-petioled, pinnately 7–11-foliolate; leaflets oblong to obovate and cuneate at base, crenate or sharply toothed; cyme leafy, many-flowered; flowers yellow, 5–7 mm broad; bractlets, sepals and petals about equal in length; achenes numerous, smooth, with a large corky swelling on the inner side. Moist flats and shores. sBC to sOnt s to Calif, NM, Tex, Mo, Ohio, NY. Map 682 ○

P. pensylvanica L.
Perennial, from a stout, branching caudex; stems 2–5 (7) dm tall, erect, rarely decumbent at base, tomentose with longer spreading hairs intermixed; basal leaves 5–15 (20) cm long, pinnate; petiole slightly shorter to much longer than the blade; leaflets 5–15, the larger 2–5 cm long, 1 cm wide, divided half-way to midrib, grey-green to green above, sparsely tomentose to sparsely to densely sericeous, veins often obscure and seldom reticulate, grey-green to white below,

densely woolly-velutinous to densely sericeous; stem-leaves similar, short-petiolate; inflorescence congested, several-flowered; flowers mostly 0.8–1.2 mm across; sepals 3–6 mm long, ovate; bractlets of epicalyx elliptic-lanceolate, about as long as sepals or slightly longer; petals 4–6 mm long, yellow, as long as sepals or longer; achenes about 1 mm long. $2n = 28^*$. Dry prairie and open slopes. Alas, Yuk, swMack to cMan, Ont, sQue s to BC, wMont, Utah, Neb, SD, Minn; Calif. Map 683

P. plattensis Nutt.
Perennial herb with a thick woody caudex, strigose throughout with fat, tapering, appressed white hairs; stems slender, 1–2 (3) dm long, erect, ascending or spreading; leaves mostly basal, green, pinnate, the 7–15 leaflets incised to near the midrib into linear, oblong or narrowly elliptic lobes; flowers 2–10 on slender stalks; sepals 4–5 mm long, strigose; bractlets narrower and shorter; petals yellow, longer than the sepals; style slender. Prairie grassland, coulees, dry flats. Alta to swMan s to Ariz, NM, SD. Map 681 ○

P. recta L.
Perennial herb with erect leafy stems, 3–7 dm tall; lower leaves long-petioled, digitately 5–7-foliolate; leaflets narrowly oblanceolate, 3–14 cm long, deeply toothed, more or less hirsute on both surfaces; cyme many-flowered; flowers about 2 cm across; sepals and bractlets about equal; petals yellow, about 1 cm long, deeply emarginate; stamens mostly 30; achenes marked with low curved ridges. $2n = 28$, 42. Roadsides and waste ground. Introduced. Rare.

P. rivalis Nutt.
Slender annual or biennial with a taproot and rather leafy stems, freely branched; leaflets 3–5-foliolate, digitate or somewhat pinnate, narrowly obovate-oblong, coarsely toothed, softly hairy to hirsute; cyme leafy; petals pale yellow, cuneate or narrowly obovate, shorter than the sepals; stamens 5–15; achenes smooth, about 0.7 mm long. $2n = 14$, 70. Riverbanks, lake shores, waste ground. BC to Ont s to Calif, NM, Tex, Okla, Ark, Ill. Map 684

P. tridentata Ait. / Three-toothed Cinquefoil
Low perennial with long creeping rootstocks and tufted leafy shoots, somewhat woody at the base; flowering stems 1–3 dm high; leaves mostly near the base, trifoliolate, the 3 leaflets of firm texture, oblong-oblanceolate, entire except for 3 teeth near the truncate apex, glabrous or nearly so; cyme stiff, few–several-flowered; bractlets lanceolate; sepals somewhat longer and ovate-triangular; petals white, or rarely pinkish; style slender, lateral; achenes and receptacle densely hairy. $2n = 14$, 28. Open sandy woods. Greenland: swMack to Huds Bay, nQue, Nfld s to Alta, Sask, ND, Ia, Minn, Mich, NY, Apal. Map 685 ●

P. uniflora Ledeb.
Perennial, caespitose, from a stout branching caudex; stem 0.2–1.0 dm tall, sparse to densely pubescent, generally tomentose with overlying, short and long, spreading hairs; leaves 1–2 cm long, crowded, trifoliolate; petioles about as long

as blade, pubescent like stem; leaflets 5–10 mm long, obovate, base cuneate, incised to one-third to two-thirds to midrib, the segments acute, often bearded at apex, upper surface brownish to olive-green, sparse to densely sericeous, lower surface white, densely tomentose, with overlying sericeous hairs; inflorescence 1-, rarely 2–3-flowered; flowers 15–20 mm across; calyx sericeous; sepals 3–5 mm long, triangular; bractlets of epicalyx lanceolate to ovate-lanceolate, narrower than sepals, as long as or somewhat shorter; petals 6–9 mm long, 8–10 mm wide, obcordate to reniform, yellow; achenes 1.2–1.3 mm long. $2n = 28$*. *P. ledebouriana* Porsild of Ed. 1. Rocky alpine slopes. Siberia: Alas, Yuk, Mack s to BC, wMont, Colo; nwHuds Bay. Map 685 ○

P. villosa Pallas *ex* Pursh
Perennial, tufted, from a stout caudex; stems 0.2–1.5 (2.0) dm tall, often branching about the middle, the upper stem frequently curved or flexuous, lanate to tomentose, overlain with short and long spreading hairs; leaves mostly 2–6 dm long, trifoliolate; petioles about as long as or longer than blade; leaflets 5–20 mm long, obovate, base narrowly to broadly cuneate, somewhat irregularly lobed or divided one-third to two-thirds to midrib, upper surface brownish to olive-green, sparsely sericeous, lower surface white tomentose with overlying sericeous hairs, veins generally evident; stem-leaves 1–3, short-petiolate with conspicuous stipules about 10 mm long, ovate-lanceolate, entire; inflorescence (1)2(5)-flowered; flowers 10–15 mm across; calyx densely sericeous, the hairs ascending, spreading and at base reflexed; lobes 3–5 mm long, triangular; bractlets of epicalyx ovate to elliptic, narrower than to about as wide as sepals, about as long; petals yellow, obcordate, longer than sepals; achenes 1.7–2.0 mm long. $2n = 14$. Alpine slopes and ridges. neAsia: Alas, sYuk s to Wash, BC, Alta. Map 686

Prunus L.

Shrubs or small trees, with alternate, simple, serrate leaves; flowers white, in racemes or umbel-like clusters; hypanthium cup-shaped; sepals 5, soon deciduous; petals 5, white, spreading; stamens about 20, exserted; pistil solitary, 2-ovuled; stigma small, capitate; fruit a 1-seeded drupe.

Flowers in umbel-like clusters; leaves lanceolate, long-acuminate, irregularly serrulate with rounded teeth; petioles usually glandular at the upper end **P. pensylvanica**

Flowers in terminal racemes; leaves elliptic to obovate, abruptly short-acuminate, sharply serrate; petioles not glandular **P. virginiana**

P. pensylvanica L.f. / Pin Cherry
Slender shrub or small tree, up to 8 m high, with reddish brown bark; leaves lanceolate, oblanceolate or narrowly ovate, 5–10 cm long, usually long-acuminate, finely and irregularly serrate with small rounded teeth, lustrous above, glabrous; flowers in umbel-like corymbs, expanding with the leaves, from axillary buds or short lateral branches; sepals glabrous; petals white, about 6 mm long; fruit globose, light red, 5–7 mm in diameter. $2n = 16, 32$. Woods and clearings,

often dry and sandy. swMack, eBC to Nfld s to Mont, Colo, SD, La, Ind, Va. Map 687

P. virginiana L. / Choke Cherry
Slender shrub or small tree, up to 10 m high, with dark bark, the branches reddish brown; leaves broadly elliptic to obovate or ovate, 5–10 cm long, abruptly short-acuminate, acute or obtuse at apex, sharply serrate, glabrous or somewhat pubescent beneath; flowers in racemes, 5–15 cm long, terminating leafy twigs of the season; sepals blunt, glandular-erose; petals white, about 4 mm long; fruit deep red, becoming red-purple or black, astringent, about 8 mm in diameter. $2n = 16, 32$. Woods and clearings, frequently with the previous species. swMack, BC to Nfld s to Calif, NM, nTex, Okla, Ark, Tenn, NC. Map 688

Rosa L. / Rose

Shrubs, usually prickly, with odd-pinnate leaves and stipules adnate to the petiole; leaves alternate, 5–11-pinnate, the leaflets serrate; flowers large, perfect, solitary or corymbose, perigynous; hypanthium globose or urn-shaped, contracted at the mouth; sepals normally 5, usually long-attenuate, often persistent in fruit; petals usually 5, large, usually pink; stamens numerous; carpels numerous, inserted on the bottom or also on the sides of the hypanthium: fruit red or orange, composed of the enlarged pulpy or fleshy hypanthium and enclosed achenes, the latter hairy. Hybridization between our species sometimes makes identification problematical.

1 Stems mostly dying back each year to near ground level; flowers (2) 3–many, borne on the generally unbranched stem(s) of the current year which are usually 1–4 dm tall; stem-prickles to 3 mm long; larger infrastipular prickles absent; leaflets (7) 9–11, the lower surface and teeth not glandular; in flower late June to August **R. arkansana**

Stems persistent, branched, generally more than 5 dm tall; flowers 1–3, rarely more; stem-prickles to 5 mm long; larger infrastipular prickles present or absent; leaflets (3) 5–7 (9), the lower surface and teeth generally glandular; in flower mid-June to mid-July 2

2 Infrastipular prickles present, conspicuously larger and stouter than the stem-prickles; stipules sparsely glandular; leaflets mostly 5–9, generally less than 25 mm long, base acute; fruit globose, lacking a neck below **R. woodsii**

Infrastipular prickles absent; stem-prickles uniformly dense throughout; stipules densely glandular; leaflets 3–7, generally more than 25 mm long, base obtuse or rounded; fruit globose (lacking a neck) to pear-shaped, with a distinct neck below **R. acicularis**

R. acicularis Lindl. / Prickly Rose
A bushy shrub, commonly 0.5–1.5 m high, or up to 2.5 m tall in the shade; stems stout, densely covered with straight slender thorns; stipules pubescent, glandular on the margins; leaf-rachis usually pubescent and glandular; leaflets 3–7, commonly 5, rarely 9, coarsely dentate-serrate, commonly 3–4 cm long,

occasionally up to 5 cm; flowers usually solitary; petals pink, 2–3 cm long; fruit pear-shaped to ellipsoid or globose. $2n$ = 42, 56. In our area ssp. **sayi** (Schw.) Lewis. Includes *R. bourgeauiana* Crépin. Woods, banks, roadsides, flowering before *R. woodsii*. Eurasia: Alas, Yuk, Mack to Huds Bay, Que s to BC, Ida, Mont, Colo, Neb, Ia, Minn, Vt. The floral emblem of Alberta. Map 689

R. arkansana Porter / Prairie Rose
A low shrub, commonly 2–4 dm high, not much branched; stems more or less bristly, often glaucous, tending to die back annually to near the ground; stipules glabrous to pubescent, entire or glandular-dentate near the top; leaflets commonly 9–11, sharply serrate, glabrous or puberulent, mostly 1–3 cm long; flowers mostly corymbose; petals rose to white, 2–3 cm long; fruit globose. $2n$ = 28. Includes *R. alcea* Greene and *R. suffulta* Greene. Plains and prairies. seBC to sMan s to Mont, NM, Tex, Okla, Mo. Map 690

R. woodsii Lindl. / Common Wild Rose
A bushy shrub, commonly 0.5–1.5 m high; stems with scattered bristles and thorns or prickly at the base only; infrastipular thorns often prominent; leaf-rachis often puberulent, sometimes glandular or bristly; leaflets 5–7 in some plants, 7–9 in other plants, oval to obovate or elliptic, often cuneate at base, glabrous to puberulent, sometimes glaucous, serrate, the teeth acuminate or acute; flowers usually corymbose; petals pink or rose, 1.5–2 cm long; fruit globose. $2n$ = 14. Includes *R. fendleri* Crépin, *R. macounii* Greene, *R. ultramontana* Heller. Open woods and plains, often in association with one or other of the above species. sYuk, swMack to Man s to Calif, nNM, nTex, Neb, Ia; James Bay, Gt Lakes. Map 691

Rubus L. / Raspberry; Bramble

Perennial shrubs or herbs, often with prickly stems; leaves simple or compound, lobed or toothed; flowers regular, perfect or unisexual, white, pink or red; hypanthium saucer-shaped with a convex or conical central receptacle; sepals usually 5; petals commonly 5, erect or spreading; stamens numerous; carpels usually numerous, becoming drupelets and forming an aggregate fruit on the enlarged receptacle.

1 Shrubs, 0.5–2 m high, armed or unarmed, leaves compound or merely lobed — 2
Plants less than 3 dm high, herbaceous or somewhat woody at base, unarmed — 3

2 Principal leaves compound, 3–5-foliolate; stems (canes) armed with prickles or bristles, and often glanduliferous; petals 4–12 mm long — **R. idaeus**
Principal leaves simple and lobed, 1–2 dm wide; stems unarmed; petals 15–20 mm long — **R. parviflorus**

3 Leaves rounded, shallowly several-lobed; flowers unisexual — **R. chamaemorus**
Leaves 3-foliolate or pedate; flowers mostly perfect — 4

4 Leaves pedate; delicate plant with long, slender, branched, rooting runners; flowers white — **R. pedatus**
Leaves 3-foliolate; plants with or without runners; flowers white or reddish — 5

5 Petals white, 4–8 mm long, erect; main stems prostrate, elongated, often semi-woody **R. pubescens**

Petals rose-coloured, 10–15 mm long, spreading at maturity; leafy stems tufted, herbaceous, from a perennial underground base **R. arcticus**

R. arcticus L. / Dwarf Raspberry
Low perennial herb, usually less than 15 cm high, the annual shoots tufted from a slender rootstock; stem slender, 2–4-leaved; leaflets 3, ovate to obovate, the lateral ones somewhat rhomboid, mostly rounded at apex, margins unevenly dentate-serrate; flowers usually solitary, some or all perfect; petals rose-coloured, 10–15 mm long, with a distinct claw; fruit red, globose, the 20–30 drupelets coherent, richly flavoured. $2n = 14^*$. In our area ssp. **acaulis** (Michx.) Focke. (*R. acaulis* Michx. of Ed. 1.). A possible hybrid between *R. arcticus* and *R. pubescens* **(R. × paracaulis** Bailey) occurs in our area. It is generally larger than *R. arcticus*, often with stolons and with the pink to rose-coloured flowers usually extending beyond the leaves. Boggy woods and marshes. More or less circumpolar: Alas, Yuk, wMack to Huds Bay, nQue, Nfld s to BC, Mont, Colo, Sask, Man, Minn, sQue, Gaspé. Map 692

R. chamaemorus L. / Cloudberry; Baked-apple Berry
Perennial herb with a creeping rootstock, growing in sphagnum bogs; upright stems 1–3 dm tall, unbranched, the lowest nodes with stipules only; leaves 1–3, long-petioled, the blades circular to reniform, shallowly 5–7-lobed, coriaceous, serrate; flowers solitary, long-peduncled, unisexual; petals white, about 1 cm long; fruit at first red-tinged, becoming yellowish and soft, edible; drupelets large with stones 4–5 mm long. $2n = 56$. Bogs, often in sphagnum moss. Circumpolar: Alas, Yuk to sBaffin Is to Nfld s to BC, Alta, cSask, sMan, Lk Superior, Que, NS. Map 693

R. idaeus L. / Wild Red Raspberry
Erect shrub, perennial with biennial stems (canes) up to 2 m tall, spiny and bristly, often glandular-hairy; first-year stems with strong broadly based prickles, often accompanied by slender bristles and glandular hairs, sometimes glabrous and glaucous below the prickles; leaves of first-year stems pinnately 3–5-foliolate, those of flowering branches 3-foliolate; leaflets ovate to lanceolate, short-acuminate or acute, the margins rather evenly double-serrate, green above, densely white-tomentose beneath; flowers in small terminal or axillary racemes, the stalks and hypanthium bristly and glandular-hispid; petals white, erect, 5–6 mm long; fruit red, ovoid, falling intact from the dry receptacle. $2n = 14^*, 21, 28, 35, 42$. Includes *R. strigosus* Michx. and *R. melanolasius* Focke of Ed. 1. In our area ssp. **melanolasius** Focke. Open woods. Circumpolar: Alas, Yuk, swMack to Huds Bay, nQue, Nfld s to Calif, Ariz, NM, SD, Mo. Map 694

R. parviflorus Nutt. / Thimbleberry; Salmon Berry
Erect shrub, 0.5–2 m tall, unarmed, with shredding bark, the young growth more or less glandular-hairy; leaves petioled, the blades 1–2 dm broad, simple,

circular to reniform in outline, with a few shallow, triangular lobes, the margin coarsely serrate, glabrous or pubescent beneath; flowers conspicuous, few, in a long-peduncled cluster; sepals caudate-acuminate; petals white, 1.5–3 cm long; styles club-shaped; fruit red, juicy, hemispheric, 1.5–2 cm wide, with pubescent coherent druplets. $2n = 14$. Moist open woods. seAlas to Calif, NM; SD; Gt Lakes. Map 695

R. pedatus J.E. Smith
Low perennial with slender creeping branches up to 1 m long and very short, erect flowering shoots, the latter 2–4-leaved; leaves mostly pedately 5-foliolate, the two lateral leaflets often cleft almost to the base; leaflets nearly sessile, obovate to rhombic, thin, irregularly toothed and incised; flowers solitary; sepals toothed, reflexed in fruit; petals white, oblong; carpels 1–6; fruit reddish, composed of 1–6 druplets, short-stipitate and adherent to the receptacle. $2n = 14$. Moist woods and mossy banks. sAlas to Ore, nIda, wMont, Alta. Map 696

R. pubescens Raf. / Dewberry; Running Raspberry
Perennial herb with slender trailing stems and erect flowering branches, the trailing stems 1–10 dm long, rooting at the tips and becoming semi-woody; leaves usually ternate, occasionally 5-foliolate, glabrous or nearly so; leaflets oval to obliquely ovate, tapered at either end, often markedly acuminate, sharply and doubly serrate; flowers 1–4, from the crown of the plant or from nodes of runners; sepals 2–4 mm long, reflexed in fruit; petals white or pale pink, 6–10 mm long; fruit red, globose, the juicy drupelets rather few, loosely coherent, tardily separating from the spongy receptacle, the stones shallowly wrinkled or pitted. $2n = 14, 28$. May hybridize with *R. arcticus*. Moist woods. nYuk, Mack to James Bay, nQue, Nfld s to Wash, wMont, Sask, SD, Minn, Pa, NJ; Colo, Neb. Map 697

Sibbaldia L.

S. procumbens L.
Dwarf, densely tufted perennial, less than 1 dm high, with short creeping rootstocks; often forming dense carpets in alpine meadows; leaves slender-stalked, ternate, the leaflets 1–3 cm long, obovate to oblanceolate, cuneate at base, truncate and 3–5-toothed at apex, sparsely hairy; flowers few, in dense cymes; hypanthium saucer-shaped; sepals 5, with 5 alternating bractlets; petals 5, yellow, shorter than the sepals; stamens 5; carpels 5–15, becoming small achenes on a dry receptacle. $2n = 14^*$. Moist alpine slopes and ridges. Circumpolar: Alas, Yuk, swMack s to Calif, Utah, Colo; Huds Bay, nQue to Nfld, Gaspé. Map 698

Sorbus L. / Mountain Ash

Shrubs or small trees; leaves alternate, pinnately compound, with numerous serrate leaflets; flowers perfect, white, numerous in compound cymes; hypanthium urn-shaped; sepals 5; petals 5, white, spreading; stamens many; carpels

3–5, their ovaries adnate to the hypanthium, the styles free; fruit a small reddish berry-like pome, with 1 or 2 seeds in each locule.

1 Trees; leaflets 13 or more; winter-buds densely white-villose; styles 2–3 **S. aucuparia**

Shrubs; leaflets generally less than 13; winter-buds glabrous, glutinous or rufous-pilose; styles 3–5 2

2 Leaflets usually 11–13, acute or acuminate at apex, serrate to near the base; young twigs and inflorescence white-pubescent **S. scopulina**

Leaflets usually 9–11, rounded at the apex, entire toward the base; young twigs and inflorescence with rust-coloured hairs **S. sitchensis**

S. aucuparia L. / Rowan Tree; European Mountain Ash
Small tree, up to 10 m tall, round-topped, the young branches more or less hairy; winter buds white-villose, not sticky; leaflets 11–17, oblong, acute or obtuse, 3–5 cm long, serrate, glabrous above, paler and usually long-hairy beneath; inflorescence flat-topped or rounded, 8–18 cm broad; hypanthium densely white-hairy; fruit bright red, about 1 cm in diam. $2n = 34$. A cultivated ornamental, occasionally escaping in moist woods. Introduced.

S. scopulina Greene
Shrub, 1–4 m tall, with reddish-brown branched stems; young twigs white-pubescent; leaflets 9–17, commonly 11–13, elliptic-lanceolate or narrower, acute or acuminate, serrate almost to the rounded base; stipules soon deciduous; inflorescence flat-topped or slightly rounded, 9–15 cm broad, the branches white-hairy; stamens equalling or longer than the petals; fruit orange to scarlet, about 8 mm in diam. Woods and moist open slopes. Alas, Yuk, swMack s to Calif, Utah, NM; SD. Map 699

S. sitchensis Roemer
Shrub, 1–3 m tall; stems dull brown in colour, sparsely branched; young twigs with rust-coloured hairs; leaflets commonly 9–11, oblong, mostly rounded at apex, serrate usually only above the middle; stipules persistent; inflorescence small, convex, the branches with minute rust-coloured hairs; stamens shorter than the petals; fruit orange to red or purplish, about 8 mm long. $2n = 34$. In our area ssp. **sitchensis**. Montane woods and open slopes. sAlas to Calif e to Alta, nwMont. Map 700

Spiraea L. / Meadowsweet

Shrubs with simple leaves, usually toothed and without stipules; inflorescence of numerous small flowers, flat-topped or elongate; flowers perfect; hypanthium cup-shaped; sepals 5; petals 5, white or reddish; stamens 15–many; carpels commonly 5, distinct, inserted on the bottom of the hypanthium; ovules 2–several; follicles firm in texture, glabrous, shining, opening along the inner face; seeds usually 4, narrow, tapering at both ends.

1 Inflorescence elongate, conical or ovoid in outline; leaves mostly oblanceolate, tapering at both ends, finely serrate **S. alba**
Inflorescence flat-topped; leaves broader 2

2 Petals pink or rose-coloured **S. densiflora**
Petals white **S. betulifolia**

S. alba Du Roi / Narrow-leaved Meadowsweet
Slender shrub, up to 2 m high, with brownish twigs; leaves narrowly oblanceolate, rather firm, 4–6 cm long, acute at both ends, sharply and finely serrate; inflorescence finely pubescent, paniculate, 5–20 cm long, the numerous white flowers crowded on the branches; sepals triangular; petals white, about 2 mm long. $2n = 36$. Wet meadows and shores. Alta to nQue, Nfld s to SD, Mo, Ill, Ohio, Del. Map 701 ●

S. betulifolia Pallas / White Meadowsweet
Low shrub, 2–8 dm high, with creeping rootstocks; stems and branches erect, often dying down annually to near the base; leaves short-petioled, the blades oval to obovate, 2–6 cm long, thin, glabrous, shining green above, paler beneath, coarsely serrate or incised, especially above the middle; inflorescence broad, flat-topped; flowers white, but turning brownish, nearly odourless; sepals triangular; petals white, about 2 mm long. $2n = 36$. *S. lucida* Dougl. of Ed. 1. Thin woods and open slopes. Asia: BC, Alta s to Ore, Wyo; swSask; SD. Map 702

S. densiflora Nutt. / Pink Meadowsweet
Slender shrub 0.5–1.5 m high, with ascending reddish-brown branches; leaves oval to elliptic, 1.5–3 cm long, rounded at both ends, crenate or serrate above the middle, bright green above, paler beneath, glabrous or nearly so; inflorescence 2–5 cm broad, dense; flowers showy, very sweet-scented; sepals ovate, obtuse; petals 1.5 mm long, pink or rose. Moist montane woods and meadows. BC, Alta s to Calif, Ida, Mont. Map 701 ○

LEGUMINOSAE / Pea Family

Herbs or woody plants; leaves alternate, stipulate, usually pinnately or palmately compound; flowers somewhat perigynous, typically zygomorphic and papilionaceous; hypanthium small; calyx cup-shaped or tubular, its lobes or teeth 5, or fewer; corolla of 5 petals, the upper median one (standard) large, two lateral ones (wings), two lowest ones (comprising the keel) often coherent and enclosing the stamens and pistil; stamens 10 (rarely 5), usually 9 united part-way by their filaments, rarely all united or all separate; pistil 1, simple, its ovary superior and usually 1-loculed, sometimes 2-loculed by intrusion of sutures; fruit commonly a pod (legume) dehiscent along both sutures, in some species breaking into 1-seeded segments, in some species 1–few-seeded and indehiscent. The roots often bear nodules with bacteria that fix free nitrogen.

1	Shrub, spiny	**Caragana**
	Herbs	2
2	Plants climbing; leaves with tendrils	3
	Plants not climbing; lacking tendrils	4
3	Style filiform; bearded at the apex	**Vicia**
	Style flattened; bearded on the upper side	**Lathyrus**
4	Fruit a loment; keel mostly truncate	5
	Fruit a legume; keel not truncate	6
5	Inflorescence a raceme; keel truncate	**Hedysarum**
	Inflorescence a loose, umbel-like head; keel acute	**Coronilla**
6	Leaves glandular-dotted or palmately 5-foliolate, the root tuber-like	7
	Leaves not glandular-dotted	9
7	Leaves palmately divided	**Psoralea**
	Leaves pinnately divided	8
8	Flowers in racemes; fruit prickly; stamens 10	**Glycyrrhiza**
	Flowers in dense, cylindrical spikes; fruit not prickly; stamens 5	**Petalostemon**
9	Leaflets more than 3	10
	Leaflets 3 (rarely 1)	14
10	Inflorescence umbel-like; flowers yellow; leaflets 5, pinnately arranged (the lower pair basal and resembling stipules)	**Lotus**
	Inflorescence, flowers and leaves not as above	11
11	Leaves palmate	**Lupinus**
	Leaves pinnate	12
12	Flowers pinkish to red or purple; calyx-lobes subulate, much longer than the tube; fruit only slightly longer than wide, with short, stout spines, 1-seeded	**Onobrychis**
	Flowers, calyx and fruits not as above	13
13	Leafy stems often present; keel rounded or obtuse at the tip	**Astragalus**
	Leafy stems usually absent; keel with a sharp point at the tip	**Oxytropis**
14	Leaflets finely toothed	15
	Leaflets entire	19
15	Flowers in globose heads; petals persistent in fruit; pods straight, beakless, enclosed by calyx	**Trifolium**
	Flowers and pods not as above	16
16	Pods conspicuously beaked	**Trigonella**
	Pods beakless	17
17	Inflorescence a slender raceme, 5–10 cm long, about 1 cm wide	**Melilotus**
	Inflorescence not as above	18

18 Stamens monadelphous; pods straight **Ononis**
Stamens diadelphous; pods falcate or coiled **Medicago**

19 Plants with elongate stems; flowers yellow **Thermopsis**
Plants tufted or cushion-like; flowers blue or whitish **Astragalus**

Astragalus L. / Milk Vetch

Perennial herbs with taproots or creeping rhizomes; leaves mostly pinnate, with numerous leaflets; flowers in axillary racemes, sometimes spicate; calyx 5-toothed; corolla usually long and narrow, the standard equalling or exceeding the wings and keel; stamens 10, diadelphous (9 united, 1 free); pods are diverse, few–many-seeded, 1-loculed or by a false partition partly or completely 2-loculed, the wall papery, leathery, woody or fleshy.

1 Leaves needle-like, about 10 mm long, 1 mm wide, distinctly spine-tipped **A. kentrophyta**
Leaves not needle-like 2

2 Plants low, 10 cm or less; densely caespitose or pulvinate; leaves many, crowded; flowers often included among leaves or barely exceeding them 3
Plants taller; stems evident, erect or decumbent, often branched and leafy 6

3 Flowers blue; leaves 1–4 cm long, linear to oblanceolate, simple or with 1 or 2 linear leaflets **A. spatulatus**
Flowers yellow, occasionally pale lavender; leaves not as above 4

4 Leaves trifid **A. gilviflorus**
Leaves pinnate with several pairs of leaflets 5

5 Leaflets about twice as long as broad, ends rounded; flowers 20–30 mm long **A. purshii**
Leaflets 4–5 times longer than broad, ends acute; flowers 7–10 mm long **A. lotiflorus**

6 Stem generally thick, 5–10 mm at base; leaflets 2–4 (6) cm long 7
Stems generally less than 5 mm thick at base; leaflets shorter 11

7 Flowers blue or purple 8
Flowers yellow or white 9

8 Raceme dense, spike-like; stem lacking black hairs; pod glabrous with two deep grooves on upper side **A. bisulcatus**
Raceme open; stem and pod generally black hairy, pod lacking grooves **A. robbinsii**

9 Leaflets narrow, linear, 2–6 cm long; flowers mostly 20–25 mm long **A. pectinatus**
Leaflets broad, ovate-oblong to elliptic; flowers shorter, mostly 10–15 mm long 10

10 Inflorescence a dense spike-like raceme, 30–150-flowered; pods erect; plants with malpighian hairs **A. canadensis**

Inflorescence a loose raceme, 10–40-flowered; pods reflexed, malpighian hairs lacking **A. americanus**

11 Petals blue, purple or pink or predominantly so (sometimes very pale with only flashes of deeper colour) 12

Petals mostly yellow, cream or orange, the keel sometimes tinged with blue or blue-tipped 25

12 Flowers 5–12 mm long 17

Flowers 12–30 mm long 13

13 Plants with malpighian hairs 14

Plants lacking malpighian hairs 15

14 Leaves silvery-grey; flowers mostly 3–9 in a loose raceme; stipules free **A. missouriensis**

Leaves green; flowers mostly in a dense, spike-like head (occasionally a loose raceme); stipules connate **A. striatus**

15 Flowers spreading or reflexed in a spike-like raceme 4–15 cm long, much exceeding the leaves; pods reflexed, with 2 deep grooves on upper side **A. bisulcatus**

Flowers and pods not as above 16

16 Flowers ascending to erect, in dense globose heads, 1–4 cm long; pods erect, about 1 cm long **A. dasyglottis**

Flowers in a loose raceme; pods 2–3 cm long, plum-like, indehiscent **A. crassicarpus**

17 Calyx-teeth linear, about as long as or longer than the calyx-tube **A. vexilliflexus**

Calyx-teeth deltoid to lanceolate, shorter than the calyx-tube 18

18 Stems with black hairs 19

Stems lacking black hairs 21

19 Plants 2.5–6.5 dm high; larger stems 1.5–3.0 mm thick at base; larger leaflets 1–2 cm long **A. robbinsii**

Plants 0.5–3.0 dm high; larger stems 1.0–1.5 mm thick at base; larger leaflets about 1 cm long 20

20 Stems prostrate or weakly ascending; leaflets rounded at the ends; stipe of pod about as long as calyx-tube **A. alpinus**

Stems erect; leaflets mostly acute; stipe of pod much shorter than calyx-tube **A. bourgovii**

21 Keel-petals less than 6 mm long **A. eucosmus**

Keel-petals longer than 6 mm 22

22 Pods glabrous or with only white hairs; plants of dry plains and foothills 23

Pods with at least some black hairs; plants of moist habitats in the north and the mountains 24

23 Leaflets narrowly oblong to oblanceolate, often truncate or retuse at the tip; apex of keel bluntly rounded **A. flexuosus**

Leaflets elliptic-ovate, acute at the tip, sometimes arcuate, the leaves appearing flexuous; keel-tip acuminate **A. miser**

24 Calyx almost half as long as the flower; pod short-stipitate, the stipe much shorter than the calyx-tube **A. bodinii**

Calyx only about a third the length of the flower; pod stipitate, the stipe about as long as the calyx-tube **A. alpinus**

25 Flowers about 20 mm long 26

Flowers 10–15 mm long 27

26 Keel-petals yellow, tips blue; pod dry, dehiscent **A. drummondii**

Keel-petals lavender-tinged; pod plum-like, indehiscent **A. crassicarpus**

27 Wing-petals deeply bidentate at the apex **A. aboriginum**

Wing-petals entire 28

28 Keel-petals less than 6 mm long, much shorter than the wings **A. tenellus**

Keel-petals longer than 6 mm, about as long as the wings 29

29 Pod globose, coarsely pilose with black and white hairs; leaves strigose above and below with basifixed hairs; flowers 10–12 (15) mm long, in a dense raceme **A. cicer**

Pod, leaves and flowers not as above 30

30 Flowers about 15 mm long, in dense globose or elongate heads 31

Flowers about 10 mm long, in loose racemes 32

31 Plants with malpighian hairs **A. striatus**

Plants lacking malpighian hairs **A. dasyglottis**

32 Pod glabrous, or with only white hairs **A. miser**

Pod with at least some black hairs 33

33 Calyx almost half as long as the flower; pod short-stipitate, the stipe much shorter than the calyx-tube **A. bodinii**

Calyx only about a third the length of the flower; pod stipitate, the stipe about as long as the calyx-tube **A. alpinus**

A. aboriginum Richards.
Herb with woody caudex and taproot; stems tufted, 1–4 dm high, villose, strigose or nearly glabrous; leaflets 7–15, linear-oblong to lanceolate or elliptic, villose or strigose on both sides, or glabrous above; raceme short; calyx black-hairy; corolla yellowish white, tipped with purple; pod flat, glabrous in age, the valves papery, long-stipitate, the stipe 4–8 mm long, the dorsal suture intruding slightly as a rudimentary septum. $2n = 16, 32$. Open areas to alpine elevations. eAsia: Alas, Yuk to Banks Is, Victoria Is s to BC, Nev, Colo, Sask, ND; neOre; Gaspé. Map 703

A. alpinus L.
Low, mat-forming herb, from creeping rhizomes, the stems slender, spreading or ascending, 1–3 dm high, glabrous or strigose; leaflets 13–25, oval to elliptic, retuse or obtuse, hairy on both sides or glabrous above; raceme short, 1–4 cm long; calyx black-hairy, 3–4 mm long; corolla purplish, the wings often whitish, 7–12 mm long, the keel as long as the standard; pods reflexed, short-stipitate, ellipsoid-oblong, about 1 cm long, black-hairy. $2n = 16, 32$. Moist banks and

rocky slopes. Circumpolar: Alas, Pr Patrick Is to nBaffin Is to Nfld s to Wash, neOre, neNev, NM, Alta, Sask, Man, wOnt, James Bay, Que. Map 704

A. americanus (Hook.) M.E. Jones
Tall herb, usually with a single erect stem, 5–10 dm high, from the caudex, sparsely hairy to glabrous; stipules conspicuous, reflexed; the lower nodes marked by stipules only; leaflets 7–15, elliptic-oblong, 2–4 cm long, glabrous above, sparsely hairy beneath; racemes several, long-peduncled; flowers white, turning yellowish or sordid, 13–15 mm long; pods drooping, inflated, ellipsoid or ovoid, membranous, 1.5–2 cm long, shiny, on stipes 6–10 mm long. $2n = 16$. *A. frigidus* (L.) Gray var. *americanus* (Hook) S. Wats. of Ed. 1. Stream banks and moist woods, seAlas, Yuk, nwMack to swHuds Bay s to BC, Mont, Colo, Sask, Man, SD. Map 705 ●

A. bisulcatus (Hook.) A. Gray / Two-grooved Milk Vetch
Large handsome herb, often in dense clumps, the stems stout, 4–10 dm high; leaflets 17–27, oblong-elliptic, 1–3 cm long, hairy or glabrous; flowers reddish purple (rarely white), 12–15 mm long, somewhat reflexed, borne in a long dense raceme; calyx gibbous at base, its teeth filiform; pods reflexed, about 2 cm long, linear-oblong, with 2 deep grooves on upper side. $2n = 24$. Prairies, coulees, and alluvial flats. Alta to sMan s to Mont, eIda, neNM, wOkla, Neb. Map 706 ●

A. bodinii Sheldon
Stems very slender 1–3 dm long, decumbent or ascending; leaflets 7–15, small, 4–12 mm long, retuse or rounded at apex, glabrous above, strigose beneath; flowers in a lax raceme or few-flowered head; corolla 7–10 mm long, light purple at tip; calyx black-hairy, the tube about 2.5 mm long and the teeth subulate; pod erect, black-hairy, 5–7 mm long, 1-locular. $2n = 24$. *A. yukonis* M.E. Jones of Ed. 1. Gravelly banks and moist sandy meadows. Alas, swYuk, wMack s to Alta, ecSask; Man; Nfld; Utah, Wyo, Colo, Neb. Map 720 ○

A. bourgovii A. Gray
Caespitose herb, with a stout caudex and taproot; stems densely tufted, slender, erect or ascending, 1–3 dm high; leaflets 13–23, oblong-elliptic to ovate-lanceolate, strigose on both surfaces; flowers purple or violet, 8–11 mm long, in a lax raceme; calyx black-hairy; pod oblong-elliptic, 1–1.5 cm long, flattened, membranous, black-hairy. Gravel bars and stream banks to subalpine elevations. seBC, Alta s to Ida, Mont. Map 706 ○

A. canadensis L.
Tall erect herb, from creeping rootstock, with stout stems 4–12 dm high, glabrous or thinly strigose; leaflets 13–29, elliptic to oblong, 2–4 cm long, glabrous to somewhat strigose; flowers greenish yellow or white, 12–15 mm long, in a long-peduncled dense raceme; pods oblong, rather woody, glabrous or nearly so, beaked, 2-loculed, crowded and persistent on the stiff erect stalk. $2n = 16$. Moist open woodland, banks, and shores. swMack, BC to Que s to Calif, Colo, Tex, Ark, Va. Map 708 ●

A. cicer L.
Perennial, with creeping roots; stem 4–6 (10) dm high; leaves mostly 10–15 cm long; leaflets 1–2 cm long, elliptic to oblong; raceme 4–6 cm long; flowers yellow, 10–15 mm long; pod 10–15 mm long, globose, inflated. $2n = 64$. Fields. Introduced. Rare.

A. crassicarpus Nutt. / Buffalo Bean; Ground Plum
Stems several, prostrate, spreading from a stout caudex and forming a wide mat, up to 7 dm across; leaves numerous; leaflets 17–25, narrowly elliptic to oblong, often blunt and emarginate at apex, strigose beneath, glabrous above; flowers in a loose raceme, yellowish white with a purple-tinged keel, about 2 cm long; pods ellipsoidal to globose, about 2 cm in diam., thick-walled, glabrous, indehiscent, fleshy and reddish at first, becoming hard and brownish when dry. $2n = 22$. Prairie grassland. seBC to sMan, Minn s to Ariz, NM, Tex, Ark, Tenn. Map 707

A. dasyglottis Fisch. *ex* DC.
Low herb, often forming large patches from slender rhizomes, the stems slender, decumbent or ascending, 5–20 cm long, glabrous or strigose, the hairs basifixed; leaflets 15–19, linear-oblong to elliptic, often retuse, strigose or glabrate; raceme dense and spike-like, 2–3 cm long; calyx-tube about 5 mm long, often with black hairs; corolla purple (bluish on drying), or occasionally white, about 18 mm long; pod erect, ovoid-oblong, about 1 cm long, densely hairy, 2-loculed. $2n = 16$. *A. agrestis* Dougl. *ex* Hook. of Ed. 1. Moist plains and meadows. eAsia: Yuk, wMack to sHuds Bay s to Calif, NM, Kans, Ia, Minn. Map 709

A. drummondii Dougl. *ex* Hook.
Tall erect herb, often with several stout stems from the caudex, 4–8 dm high, densely and woolly hairy; leaflets 15–33, oblong-elliptic, obtuse, pubescent beneath and nearly glabrous above; raceme many-flowered, 2–10 cm long; flowers reflexed, 1.5–2 cm long, yellowish white, the keel sometimes tinged with purple; calyx-tube gibbous at base; pods slender-cylindric, smooth, drooping, 2–2.5 cm long, nearly 2-locular. $2n = 22$. Dry hillsides and prairies. Alta, Sask s to Ida, Utah, NM, Kans. Map 710

A. eucosmus Robins.
Stems solitary or several from the caudex, slender and stiffly erect, 3–6 dm tall, glabrous or strigose; leaflets 9–15, oblanceolate to elliptic, sparsely strigose beneath, glabrous above or nearly so; racemes at first compact and 2–3 cm long, elongating to about 1 dm in fruit; flowers purple or nearly white, 6–8 mm long; calyx black-hairy; pods reflexed, sessile in calyx, obliquely ovoid to ellipsoid, about 1 cm long, densely hairy, with a narrow partial partition. $2n = 16, 32$. Moist meadows, gravel bars, and rocky montane slopes. Alas, Yuk, wMack, Banks Is s to BC, Mont, Colo; Huds Bay, sBaffin Is, Lab, Nfld, Gaspé. Map 711 ●

A. flexuosus Dougl. *ex* G. Don
Low herb, with slender, spreading or ascending stems, 3–10 dm long, ashy-puberulent; leaflets 15–21, narrowly oblong, glabrous above, strigose beneath; racemes 2–3 cm long at anthesis, much longer at maturity; flowers pink to purplish, sometimes whitish, 8–10 mm long; calyx-tube white-hairy; pods narrowly cylindric, 1–2 cm long, sparsely strigose, straight or curved, 1-locular. $2n = 22$. Prairies and dry gravelly areas. seBC to Man, Minn s to NM, nwKans, SD. Map 712

A. gilviflorus Sheldon / Cushion Milk Vetch
Low, cushion-like, early flowering; stems not over 5 cm long, densely covered with leaves and stipules; leaves trifoliolate; leaflets elliptic to narrowly oblanceolate, about 1 cm long, appressed silvery-silky hairy on both sides; flowers yellowish white, borne in the crown of the plant and amongst the leaves; calyx-tube 10–15 mm long, silky; corolla 2–3 cm long; pods ovoid, silky-hairy, few-seeded, partly included in the calyx. $2n = 24$*. *A. triphyllus* Pursh of Ed. 1. Dry prairies and rocky outcrops. Alta to swMan s to Colo, wKans, wOkla. Map 713

A. kentrophyta A. Gray / Prickly Milk Vetch
Low herb with a strong taproot and prostrate, much-branched stems, 1–4 dm long, comprising a loose or dense mat; leaves with spinose stipules; leaflets 5–7, linear to lanceolate, silky-strigose, stiff and spine-tipped; flowers small, in clusters of 1–4 in the leaf-axils; corolla yellowish white with a purplish tinge, 4–6 mm long, the standard arched; pods elliptic, 4–7 mm long, 1-loculed. $2n = 24$. Dry prairies. sAlta to sSask s to Nev, Ariz, NM. Map 705 ○

A. lotiflorus Hook.
Low herb, caespitose at base, with stems 2–10 cm long, spreading and erect; leaflets 5–15, oblong-elliptic, 5–15 mm long, grey-pubescent; flowers yellowish white, 7–9 mm long, borne in a short raceme, either on an elongated peduncle or close to the caudex; calyx white-pubescent, the teeth long and slender; pods ovoid-oblong, 1–2 cm long, densely pubescent, long-beaked. $2n = 22$. Dry slopes and prairie. seBC to swMan, Minn, s to Mont, NM, Tex, Ia. Map 711 ○

A. miser Dougl. *ex* Hook. / Timber Milk Vetch
Herb, caespitose at the base, the tufted stems slender, erect or ascending, 1–4 dm long, strigose; leaflets 9–21, linear to lanceolate-oblong, acute, often glabrous above, strigose beneath; racemes lax, 5–12-flowered; corolla yellowish white, often purplish, 6–14 mm long; pod drooping, sessile, linear-oblong, 2–2.5 cm long, 1-locular, strigose or glabrous. $2n = 22$*. *A. decumbens* (Nutt.) A. Gray of Ed. 1. In our area var. **serotinus** (Gray) Barneby. Moist meadows and montane slopes. BC, Alta s to Nev, Utah, Colo, SD. Map 714

A. missouriensis Nutt.
Herb with stout taproot and multiple caudex, the stems tufted, spreading, 2–10 cm high, densely silvery-pubescent, overtopped by the erect leaves and peduncles; leaflets 9–21, oval to obovate, 8–12 mm long, silky-hairy; racemes

short, on long peduncles; flowers bluish purple, or bright purple, 15–20 mm long; calyx-tube oblique, the lobes subulate; pods oblong, spreading, 1.5–2 cm long, sessile, firm and leathery, the surface strigose and cross-ribbed. $2n = 22$. Prairies and gravel flats. Alta to swMan, Minn s to NM, Tex. Map 715

A. pectinatus Dougl. *ex* Hook. / Narrow-leaved Milk Vetch
Herb with several stems, erect or spreading, forming a clump 2–6 dm high; leaflets 9–17, narrowly linear, 2–5 cm long, sparingly strigose; flowers yellowish white, about 2 cm long, in conspicuous racemes; pods oblong to elliptical, 8–15 mm long, circular in cross-section, fleshy when immature, becoming dry and hard, reflexed. $2n = 22$. Prairies, often alkaline. Alta to sMan s to Mont, Utah, Colo, Kans. Map 716

A. purshii Dougl. *ex* Hook.
Low tufted plant, usually less than 5 cm high, cushion-like or with a short prostrate stem, densely villose and tomentose throughout; leaflets 7–13, elliptic to oval, silky-villose; raceme short, 2–10-flowered; calyx-tube cylindric, the teeth short; corolla about 2 cm long, yellowish white, the keel purple-tipped; pods spreading, sessile, obliquely ovoid, villose-tomentose. $2n = 22, 44$. In our area var. **purshii**. Dry prairies. sBC to sSask s to Calif, NM, SD. Map 717 ○

A. robbinsii (Oakes) A. Gray
Stems from the caudex erect or ascending, 2–6 dm long; leaflets 9–15, ovate to narrowly elliptic, obtuse, glabrous above and sparingly strigose beneath; racemes 7–11-flowered; calyx 3–4 mm long; corolla 8–10 mm long, pinkish white to purple; pods 16–18 mm long, somewhat drooping, stipitate, the surface black-hirsute, the septum vestigial. $2n = 32$. *A. occidentalis* (S. Wats.) M.E. Jones of Ed. 1. In our area var. **occidentalis** S. Wats. Stream banks and montane woodland. Alas, Yuk, wMack s to Wash, Ida, Mont, Colo, eNev; Lab, Nfld, eQue, NS, Vt. Map 717 ●

A. spatulatus Sheldon
Acaulescent, caespitose, cushion-like plant, 2–8 cm high, densely silvery-grey; leaves occasionally 3-foliolate, usually reduced to single, linear or linear-lanceolate blades, 1–4 cm long, silky-canescent; racemes 3–7-flowered; calyx about 4 mm long; corolla bluish-purple, 6–9 mm long; pods erect, sessile, elliptic to oblong, 6–12 mm long, 1-locular. $2n = 24$. *A. caespitosus* (Nutt.) A. Gray of Ed. 1. Dry prairies. Alta to swSask s to Utah, Colo, wNeb. Map 708 ○

A. striatus Nutt.
Stems erect or spreading from a branched caudex, 2–4 dm high, strigose; the hairs malpighian; leaflets 13–25, elliptic to narrowly oblong, 1–2 cm long, silky-strigose beneath with malpighian hairs, strigose to glabrous above; flowers purplish or whitish, 15–18 mm long, ascending in a dense, short spike at the end of a long peduncle; pods oblong 7–10 mm long, densely strigose, deeply grooved on one side. $2n = 32$. Dry grassland and open slopes. Alas, Yuk, sMack to sMan s to Wash, NM, Neb, Ia. Map 718

A. tenellus Pursh
Stems few–several from a stout caudex, tufted, erect, 2–5 dm high, thinly strigose to glabrous; leaflets 9–19, linear or narrowly oblong, 8–12 mm long, nearly or quite glabrous; racemes numerous, loose, short, 3–6 cm long; flowers yellowish white, 7–9 mm long; pods oblong, pointed at both ends, flattened, 8–12 mm long, glabrous, stipitate, drooping. $2n = 24$. Prairies and gravel flats. sYuk, wMack to Man s to Nev, NM, Neb, SD, Minn. Map 719

A. vexilliflexus Sheldon
Mat-like, dwarf plant, with a woody taproot and multiple caudex; stems slender, strigose, ascending or prostrate, 1–3 dm long, branched, forming a loose or dense mat; leaflets 7–11, narrowly oval to oblanceolate, 5–12 mm long, often acute, densely strigose, at least beneath; racemes loosely 3–6-flowered; corolla purplish, rarely whitish, 6–8 mm long; pods sessile, loosely reflexed, elliptic-oblong, 6–10 mm long, strigose, flattened, 1-locular. $2n = 22$. Dry slopes, gravel banks, and roadsides. seBC, Alta, swSask s to Ida, Mont, Wyo, wSD. Map 720 •

Caragana Lam.

C. arborescens Lam. / Common Caragana
Shrub or small tree; bark of the young stems slightly winged; leaves even-pinnate, usually of 4–6 pairs of leaflets, the leaf-axis tipped with a bristle-like spine; leaflets entire, oval or obovate; stipules linear, spine-tipped, commonly developing into stiff spines and persisting; flowers bright yellow, 15–23 mm long, borne singly on slender pedicels, several from each of the stout scaly buds on the previous year's wood; calyx 5-toothed, with hairy margins; standard upright but curled backward at the sides; keel straight and obtuse; stamens diadelphous; pod 4–5 cm long, linear, terete, brownish, with several seeds. $2n = 16$. Open woods, clearings, roadsides. Cultivated as a hedge or windbreak, occasionally escaping. Introduced.

Coronilla L.

C. varia L.
Perennial, glabrous; stem 2–12 dm high, branched above; leaflets mostly 11–15, 1–2 cm long, elliptic to oblong, margin scarious; inflorescence of axillary, pedunculate, umbel-like heads; flowers mostly 10–20, white to pink or purple; calyx about 2 mm long; petals 10–15 mm long, auriculate; stamens diadelphous; fruit a linear loment 2–6 (8) cm long, 4-angled. $2n = 24$. Waste ground, railway tracks. Introduced. Rare.

Glycyrrhiza L.

G. lepidota (Nutt.) Pursh / Wild Licorice
Coarse perennial herb with deep-seated rootstocks and erect stems, 3–10 dm high; leaflets 11–19, lanceolate, 2–4 cm long, mucronate at apex, entire, glandular-dotted; racemes erect, dense, 3–4 cm long, the peduncles shorter than the sub-

tending leaves; flowers yellowish white, about 12 mm long; pods bur-like, oblong, 1–2 cm long, reddish brown, densely beset with hooked prickles; seeds 3–5, large, 4 mm long. $2n = 16$. Moist prairies and shores. seBC to Ont s to Calif, Ariz, Tex, Mo, Minn. Map 721

Hedysarum L.

Perennial herbs with erect or ascending stems and odd-pinnate leaves; leaflets numerous, nearly sessile, mostly obscurely punctate above; flowers pink, purple, yellow or whitish, borne in axillary, peduncled racemes, erect or reflexed; calyx cup-shaped, 5-lobed, pubescent; corolla standard and wings shorter than the keel, the wings narrow and auriculate at base, the keel more or less truncate at the end; stamens diadelphous; pod (loment) flat and jointed, usually pendent or spreading, indehiscent but breaking into segments at the narrow constrictions.

1 Leaflets with obscure veins; calyx-lobes about equal, longer than the tube, linear-subulate **H. boreale**
Leaflets conspicuously veined; calyx-lobes markedly unequal, the upper triangular 2

2 Flowers pinkish or reddish purple; segments of loment 3.5–5 mm broad, narrowly wing-margined **H. alpinum**
Flowers yellowish; segments of loment 6–8 mm broad, conspicuously wing-margined **H. sulphurescens**

H. alpinum L.
Stems erect or ascending, 2–7 dm high, somewhat strigose; stipules brown; leaflets 15–21, broadly lanceolate to oblong, apiculate, glabrous above, sparsely hairy beneath, the veins prominent; flowers usually deflexed, 12–18 mm long, pinkish to reddish purple, glabrous to appressed pubescent; loments glabrous to appressed hairy, the segments wing-margined and the areolae mostly polygonal. $2n = 14^*$, 28. In our area ssp. **americanum** (Michx.) Fedtsch. Moist open woods and slopes, gravelly banks to alpine elevations. Asia: Alas, Yuk, Banks Is, Victoria Is, nMack to Huds Bay, Nfld s to BC, Mont, Wyo, Sask, Man, cOnt, Vt, Me; SD. Map 722

H. boreale Nutt.
Stems 2–6 dm tall, pubescent, branched above; stipules brown; leaflets 9–13, linear-oblong to elliptical or obovate, densely hairy on both surfaces to glabrous above, the veins obscure; flowers numerous, erect, pale or dull reddish purple, 12–19 mm long; loments appressed pubescent, not wing-margined, the reticulations transversely elongated and short tubercles sometimes present. $2n = 16$. In our area var. **boreale**, usually greyish with mostly carmine to magenta flowers, 11–17 mm long, in a raceme up to 15 cm long, and var. **mackenzii** (Rich.) C.L. Hitchc. (*H. mackenzii* Rich. of Ed. 1), usually green, with mostly deep purple flowers, 17–21 mm long, in a compact raceme 2–4 (6) cm long. Plains and open slopes in mountains. Alas, Yuk, Bank Is, Victoria Is, to Huds Bay, Nfld s to neOre, Ariz, NM, nTex, wOkla, ND, Man, nOnt. Map 723

H. sulphurescens Rydb.
Stems from a woody base, 2–6 dm tall, appressed pubescent: stipules brown; leaflets 9–17, elliptical to ovate-oblong, usually apiculate, prominently veined, glabrous above, sparsely pubescent below; flowers sulphur yellow or whitish yellow, 15–18 mm long; calyx-teeth unequal, shorter than the tube; loments pendent, the segments 2–4, conspicuously wing-margined, glabrous, 6–8 mm broad, the areolae polygonal. Prairie grassland and mountain woods. BC, Alta, s to Wash, Mont, Wyo. Map 724

Lathyrus L. / Vetchling; Pea Vine

Perennial herbs with horizontal rootstocks and slender climbing stems, up to about 1 m tall; leaves alternate, pinnate, usually tendril-bearing at the tip; flowers in axillary racemes; calyx obliquely cup-shaped, gibbous at base; keel of corolla upwardly curved and shorter than the wings; stamens diadelphous, the tube usually truncate at apex; style somewhat enlarged and flattened upward, hairy along its inner side; pod linear, glabrous, more or less flattened, dehiscent, with several seeds.

Flowers yellowish white; leaves glabrous; leaflets 3–4 pairs; stipules large, often half as long as the leaflets **L. ochroleucus**
Flowers purple; leaves finely pubescent beneath; leaflets 4–6 pairs; stipules small **L. venosus**

L. ochroleucus Hook.
Stems slender, glabrous, up to 1 m long; stipules ovate or semi-cordate, about half as long as the leaflets, the larger ones toothed; leaflets usually 3 or 4 pairs, ovate to elliptic, 2.5–5 cm long, glabrous; racemes shorter than the subtending leaf, bearing 5–10 flowers; corolla yellowish white, about 15 mm long; pod about 4 cm long. $2n = 14$. Moist woods and clearings. seAlas, BC, wMack to Que s to Wash, Wyo, SD, nwNeb, Ohio, Pa, Vt. Map 725

L. venosus Muhl.
Stems stout, ridged and striate, more or less pubescent, often 1 m high, occasionally higher; stipules small, narrowly lanceolate or semi-sagittate; leaflets usually 4–6 pairs, narrowly to broadly elliptic, 3–6 cm long, very veiny beneath, rounded to the mucronate apex, finely pubescent beneath, glabrous above; racemes dense, with 10–20 flowers; corolla purple, about 15 mm long; pod 4–5 cm long. $2n = 28$. Moist woods. Alta to wOnt s to SD, Tex, Ark, La, Ga; Gaspé. Map 726 •

Lotus L.

L. corniculatus L.
Perennial herb; stems several, decumbent or ascending, 1–4 (6) dm long, glabrous to pilose; leaves 5-foliolate, the basal pair sessile and remote; leaflets 5–17 mm

long, elliptic to obovate, often apiculate, margins generally finely serrulate; stipules minute; inflorescence umbel-like, mostly 2–10-flowered; flowers 8–15 mm long, yellow, often red-tinged; pod 2–4 cm long. $2n = 12, 24, 36$. Ditches and roadsides. Introduced. Rare.

Lupinus L. / Lupine

Annual or perennial herbs, with alternate, palmate leaves, each leaf with 5–17 entire leaflets; flowers perfect, in terminal racemes; calyx deeply 2-lipped, the upper lip 2-toothed, the lower 3-lobed or entire; corolla blue, purple or whitish, sometimes bicoloured; standard broad, with strongly reflexed margins; wings curved and united near the tips; keel sickle-shaped and with a beak-like apex; stamens 10, monadelphous; anthers alternately larger and smaller; pod flattened, dehiscent, 2-valved; seeds 2–several.

1	Plants annual; 5–25 cm tall, decumbent at the base; pod 2-seeded	**L. pusillus**
	Plants perennial; generally taller, erect; pod 4–6-seeded	2
2	Standard conspicuously pubescent on back	**L. sericeus**
	Standard glabrous on back	3
3	Standard not strongly reflexed; basal leaves generally absent at flowering; flowers about 10 mm long	**L. argenteus**
	Standard strongly reflexed; leaves and flowers not as above	4
4	Plants generally less than 30 cm tall; upper calyx-lobe cleft to half its length or more	**L. lepidus**
	Plants more than 30 cm tall; upper calyx-lobe bifid or cleft less than one-third its length	5
5	Leaflets 5–9, oblanceolate to spatulate, ends rounded, mostly widest just below tip	**L. nootkatensis**
	Leaflets (5) 9–13 or more, elliptic-lanceolate, acute to acuminate, mostly widest at middle	**L. polyphyllus**

L. argenteus Pursh / Perennial Lupine
Perennial herb, diffusely branched, the stems erect, appressed pubescent, 5–10 dm tall; leaflets 7–8, oblanceolate, often narrowly so, 2–5 cm long, densely silvery-hairy beneath, sparsely hairy to glabrous above; raceme usually rather open; flowers about 1 cm long, light violet, occasionally whitish; standard glabrous or sparsely hairy on the back; pod about 2 cm long, 4–6-seeded. $2n = 48$. Prairie slopes and ridges. sBC to Sask s to neCalif, Ariz, NM, SD. Map 727

L. lepidus Dougl. *ex* Lindl.
Perennial; stems erect, 1–4 dm tall, exserted well beyond leaves, densely sericeous; leaves mostly basal, leaflets 5–10, 1–4 cm long, oblanceolate, apiculate, often conduplicate, densely sericeous above and below; raceme 2–15 cm long, dense; flowers about 10 mm long, blue, occasionally silvery-brown; keel ciliate; pod 2–3 cm long. $2n = 48$. Includes *L. minimus* Dougl. *ex* Hook. River flats and

open gravelly areas to alpine elevations. seAlas, BC, Alta s to Calif, Mont, Wyo, Colo. Map 728 ○

L. nootkatensis Donn *ex* Sims
Perennial; stems mostly 3–12 dm tall, erect or ascending, sparsely to densely pubescent with white spreading hairs; leaves cauline, leaflets 5–9, 2–5 (7) cm long, oblanceolate to spatulate, apex rounded, densely hirsute below, glabrous to hirsute above; raceme 5–20 (35) cm long, often open; flowers 15–20 mm long, blue, rarely white; calyx lower lip boat-shaped, sometimes reflexed; keel densely ciliate above; pod 3–6 cm long. $2n = 48$. Open forest areas at subalpine elevations. sAlas, swYuk s to BC, Alta. Map 729 ○

L. polyphyllus Lindl.
Perennial; stems mostly 5–10 dm tall, often fistulose, glabrous to strigillose or hirsute; basal and lower cauline leaves with petioles 3–6 times longer than blade; leaflets (5) 9–13 (18), mostly 3–12 cm long, elliptic to oblanceolate, apex acute to acuminate, glabrous above, strigillose to densely pilose below; raceme generally 10–40 cm long, dense; flowers 10–16 mm long, blue, rarely pink, glabrous, or the keel ciliate; pod 3–5 cm long. $2n = 48, 96$. Moist woods. sAlas, BC, Alta s to Calif, Colo. Map 726 ○

L. pusillus Pursh / Annual Lupine
Low annual, the stem stout, hirsute, 0.5–2.5 dm tall, with branches spreading from near the base and decumbent; cotyledons clasping and persistent; leaflets commonly 5, oblong-lanceolate, 2–3 cm long, sparsely hairy beneath, glabrous above, rounded at apex; calyx hirsute-ciliate; corolla tinged with rose or purple, or almost white, 8–10 mm long; pod nearly 2 cm long, 2-seeded. $2n = 48$. Dry sandy soil and dunes. Alta to sSask s to Calif, Ariz, Colo, Neb. Map 728 ●

L. sericeus Pursh / Perennial Lupine
Perennial herb with erect stems, 4–8 dm tall, densely silky pubescent, sometimes also villose, the longer hairs appressed or spreading; leaflets 5–10, oblanceolate, often conduplicate, 3–6 cm long, densely grey-pubescent on both sides with appressed silky hairs; pubescence, at least in the inflorescence, sometimes brownish in age; raceme commonly dense; flowers about 1 cm long, blue or bluish purple, rarely white; standard densely hairy on the back; pod 2–3 cm long, 4–6-seeded. $2n = 48$. Plains and open montane woods. BC, Alta s to Calif, Ariz, NM. Map 729 ●

Medicago L. / Medick

Annual or perennial herbs; stems slender, elongated and erect or prostrate; leaves 3-foliolate and usually toothed, the terminal leaflet stalked; flowers small, in axillary heads or short head-like racemes; calyx cup-shaped, the slender lobes nearly equal; corolla yellow or bluish purple; standard oblong; keel obtuse, shorter than the wings; stamens 10, diadelphous; pods curved or spirally coiled.

1 Annual plant; prostrate; corolla yellow, about 3 mm long; pod slightly curved, kidney-shaped, one-seeded, becoming black **M. lupulina**
Perennials; stems erect; corolla 6–10 mm long; pod several-seeded 2

2 Corolla yellow; pod curved or straight **M. falcata**
Corolla violet or blue; pod spirally coiled **M. sativa**

M. falcata L. / Yellow Lucerne
Deep-rooted perennial with a thick crown; stems erect or decumbent, up to 1 m long; leaflets linear to cuneate-oblong, serrate at the truncate apiculate apex; flowers about 7 mm long, yellow; pod 1–1.5 cm long, sickle-shaped or straight, slightly twisted. $2n = 32$. Roadsides and waste ground. Introduced. Occasional.

M. lupulina L. / Black Medick
Prostrate or decumbent annual, branched from the base, the stems wide-spreading, 2–8 dm long; leaflets narrowly to broadly obovate, cuneate at base, toothed; flowers small, yellow, crowded in a head up to 1 cm long; pod somewhat kidney-shaped, ridged, pubescent or glabrate, becoming black, 1-seeded. $2n = 16$. Waste ground and fields. Introduced. Common.

M. sativa L. / Alfalfa; Lucerne
Similar to *M. falcata* in habit; leaflets oblanceolate or oblong, slightly hairy, 1–3 cm long, sharply serrate and apiculate at the apex; flowers 7–10 mm long, in oblong racemes 2–3 cm long, usually blue-violet to purple, occasionally whitish; pod coiled 1–3 times, more or less pubescent. $2n = 16, 32, 48$. A forage plant, common in fields, waste ground, roadsides. Introduced.

Melilotus Mill. / Sweet Clover

Annual or biennial herbs, sweet-scented in drying; stems smooth, leafy, branched, 8–25 dm tall; leaves 3-foliolate, the leaflets denticulate; flowers in elongate, lax racemes; calyx with 5 teeth, nearly equal; corolla white or yellow; standard abruptly contracted at base; wings narrow, cohering with the short obtuse keel; stamens 10, diadelphous; pod ovoid or nearly globose, reflexed, scarcely dehiscent, 1–2-seeded.

Corolla white; standard somewhat longer than the wings; pod not wrinkled, but inconspicuously reticulate; seeds yellow (green when immature) **M. alba**
Corolla yellow; standard about equalling the wings; pod wrinkled; seeds often olive green or purple spotted **M. officinalis**

M. alba Desr. / White Sweet Clover
Plants 5–25 dm tall, freely branched above; racemes lax; flowers 4–5 mm long, standard somewhat longer than wings; pod 3–5 mm long, glabrous, reticulate. $2n = 16, 24, 32$. Roadsides and waste ground. Introduced. Common.

M. officinalis (L.) Lam. / Yellow Sweet Clover
Similar to *M alba*; differing in the characters of the key and the larger flowers, 5–6 mm long. $2n = 16$. Roadsides and waste ground. Introduced. Common.

Onobrychis Mill.

O. viciifolia Scop. / Sainfoin
Perennial, erect, 2–6 dm tall; leaflets 11–25, oblanceolate to oblong, 1–2 cm long; stipules reddish brown, scarious; racemes dense, 10–50-flowered; flowers 10–12 mm long; calyx campanulate; petals pink to purplish; stamens monadelphous; fruit ovate, 6–8 mm long, indehiscent. $2n = 14, 28$. Roadsides. Introduced. Rare.

Ononis L.

O. arvensis L.
Perennial, base somewhat woody; stem 5–10 dm tall, erect, pubescent with long spreading hairs, becoming villous above; leaves 3-foliolate; leaflets 1–3 cm long, elliptic to ovate, finely toothed, glabrous to sparsely glandular-hairy above, glandular-hairy below; flowers pedicellate, in pairs at each node, in a dense terminal raceme; petals 10–20 mm long, pink to blue; pod 6–9 mm long, about as long as the calyx. $2n = 30, 32$. Roadside ditches. Introduced. Rare.

Oxytropis DC. / Loco-weed; Oxytrope

Perennial herbs, acaulescent (short-caulescent in *O. deflexa*); leaves and scapes on a multiple caudex from a taproot; leaves ascending odd-pinnate with numerous entire leaflets, usually grey or silvery pubescent; stipules mostly adnate to the petiole (free in *O. deflexa*); flowers blue, purple, whitish or yellowish, in spike-like racemes on a long scape; calyx campanulate to cylindric, with 5 narrow teeth; wing-petals with long claws, often 2-lobed or emarginate; keel with a prominent ascending beak, generally shorter than the wings; stamens diadelphous, anthers uniform; pod sessile or stipitate, ovoid to cylindric, deeply grooved on the lower side and partially or completely 2-loculed by intrusion of the ventral suture, the wall membranous to woody, sometimes inflated, several-seeded; seeds reniform.

1 Plants caulescent (stems usually with one or more evident nodes); stipules only slightly adnate to the petioles; pods pendulous **O. deflexa**
Plants acaulescent; stipules adnate to the petioles; pods erect or spreading 2

2 Leaflets mostly fascicled, appearing whorled **O. splendens**
Leaflets not fascicled, opposite or subopposite 3

3 Flowers purplish or blue 4
Flowers yellowish, ochroleucous, or whitish (occasionally blue-tinted in *O. monticola*) 7

4 Bracts, scapes, and pods densely silky-pilose; calyx somewhat turgid or inflated at anthesis **O. lagopus**

Bracts, scapes, and pods thinly pilose to glabrate; calyx neither turgid nor inflated at anthesis 5

5 Racemes 1–3-flowered, commonly 2-flowered; pods ovoid, inflated, stipitate **O. podocarpa**

Racemes 6–many-flowered; pod ellipsoid to oblong-cylindroid, not inflated, sessile or subsessile 6

6 Herbage, calyx, and pod more or less viscid-glandular **O. viscida**

Herbage, calyx, and pod eglandular **O. jordalii**

7 Keel 15–20 mm long; pods coriaceous, rigid at maturity **O. sericea**

Keel 11–14.5 mm long; pods chartaceous, not rigid at maturity 8

8 Leaflets 17–33; racemes 10–30-flowered; robust plants of prairies, open woodlands, and low elevations in the mountains **O. monticola**

Leaflets 11–17; racemes 5–15-flowered; low plants of alpine and subalpine ridges and meadows **O. cusickii**

O. cusickii Greenm. / Alpine Loco-weed
Low plants with a thick and multiple caudex; stipules glabrous or sparingly pilose; leaves 4–12 cm long; leaflets 7–15, oblong-elliptic, variably pubescent; scapes erect, arcuate-ascending or prostrate, 2.5–15 cm long; racemes 3–15-flowered; calyx 7–9 mm long, with dark and pale hairs, corolla 12–18 mm long, ochroleucous; keel 11–15 mm long, concolorous or occasionally maculate; pods oblong-ovate, pubescent, often with black and white hairs, chartaceous, not rigid at maturity. $2n = 48^*$. *O. campetris* (L.) DC. var. *cusickii* (Greenm.) Barneby of Ed. 1. (*O. alpicola* (Rydb.) M.E. Jones). Rock slides, stony ridges, and alpine and subalpine meadows. BC, Alta s to Wash, Mont, Wyo, Utah, nColo; neOre. Map 730

O. deflexa (Pall.) DC. / Reflexed Loco-weed
Plants caulescent or subacaulescent, usually with at least one prominent internode; stems ascending, 10–35 cm long, loosely hairy; stipules pale green or purplish, adnate 1.5–3 mm to the petiole, more or less pilose dorsally; leaves 5–20 cm long; leaflets 15–41, arranged to give the leaf a flattened appearance, lanceolate, 5–15 mm long, loosely villose to appressed hairy; racemes 5–25-flowered, elongating in fruit; flowers suberect in bud, at length pendulous and commonly secund; corolla 6–9 mm long, bluish or white and tipped with purple; keel 4.5–8.5 mm long; pods oblong, reflexed or pendent, hairy. $2n = 16$. In our area var. **foliolosa** (Hook.) Barneby has the lateral sinuses of the calyx similar to the ventral, broad and obtuse, and the raceme of 7–10 purple flowers remains compact in fruit, while var. **sericea** T. & G. is characterized by narrow petals commonly only tinged with blue or purple and with the lateral sinuses of the calyx similar to the dorsal pair, narrow and acute. Open woods, moist thickets, and banks. nNorway, Asia: Alas, Yuk, Mack to wBaffin Is, to Nfld s to Wash, Ore, NM, ND, James Bay; Gaspé. Map 731

O. jordalii Porsild
Low plants from a branching caudex; stipules glabrate or loosely pilose, the margins with clavate processes; leaves 6–14 cm long; leaflets 9–25, narrowly lanceolate to oblong-elliptic, subinvolute, variable in vestiture; scape green to purplish green, 10–14 cm long; inflorescence capitate to subcapitate, 6–12-flowered; calyx campanulate, 5–7 mm long, with black and pale hairs; corolla pink or purple, 11–14 mm long; keel 10–12 mm long; pods chartaceous, 8–15 mm long, oblong-ovate, with black and white hairs. $2n$ = 32*. In our area ssp. **jordalii**. Alpine and subalpine meadows and dry ridges. Alas, Yuk, swMack, s to nBC; Alta. Map 732 ○

O. lagopus Nutt.
Caespitose, with a branching caudex; the herbage everywhere silky-pilose with long silvery hairs; stipules membranous, densely silky-pilose dorsally, sometimes glabrate in age; leaves 3–11 cm long, leaflets 5–17, ovate to oblong-elliptic, involute or with elevated margins 3–15 mm long; scapes erect, arcuate ascending, or subprostrate, 2–13 cm long; racemes 5–18-flowered, subcapitate to oblong; bracts shaggy-pilose dorsally, often with dark hairs; calyx 8–11 mm long, heavily silky-hirsute concealing the surface of the tube, turgid to inflated at anthesis; corolla pink-purple or blue-purple, 13–16 mm long; keel 11–14 mm long; pod erect, silky-pilose, shortly stipitate, exserted from the calyx, 6–15 mm long, chartaceous or submembranous. Sandy or gravelly knolls, grassy hillsides. Alta s to Ida, Mont, Wyo. Map 732 ●

O. monticola A. Gray / Late Yellow Loco-weed
Plants usually with a stout multiple caudex; stipules glabrate to loosely pilose dorsally; leaves 6–23 cm long; leaflets 17–33, linear-oblong, lanceolate or obovate, 6–25 mm long, appressed silky-hairy on both sides; scapes 10–40 cm long; racemes 10–30-flowered; bracts pilose dorsally; calyx appressed-villose, often with black hairs; corolla 12–17 mm long, whitish or ochroleucous, occasionally bluish, pinkish or purplish in the foothills, the wings rarely dilated upward; keel 10–14.5 mm long; pods ovate-oblong, pubescent, often with black and white hairs, membrano-chartaceous. Blooms later than *O. sericea*, in June or July. $2n$ = 32*, 48. *O. campestris* (L). DC. var. *gracilis* (A. Nels.) Barneby of Ed. 1. In our area ssp. **monticola**. Prairies, open woodland, and moist banks. seBC to swMan s to Wash, Mont, nColo, SD. Map 733

O. podocarpa A. Gray / Inflated Oxytrope
Low caespitose plant, with multiple caudex; stipules persistent, sparingly hirsute dorsally to glabrate, bristly-ciliate along the margins; leaves 1–5 cm long; leaflets 9–25, linear to oblong, silky-hairy, 2–8 mm long; scapes divergent or erect, 1–5 cm long; racemes 1–3-flowered, commonly 2-flowered; calyx 7–10 mm long, villose; corolla purple or blue, 11–14 mm long; keel 10–11 mm long; pods ovoid-ellipsoid, stipitate, bladdery-inflated, 1.5–3 cm long, thinly chartaceous, sparsely hairy. $2n$ = 16. Exposed rocky ridges and turfy alpine hillsides. nBC s to Mont, Wyo, Colo; sBaffin Is s to nQue, nLab. Map 734

O. sericea Nutt. / Early Yellow Loco-weed
Robust plants with a stout multiple caudex, stipules generally pilose dorsally, the free blades hispid-ciliate; leaves 4–30 cm long; leaflets 11–17, opposite or subopposite, oval to oblong-lanceolate, grey or silvery pilose; scapes erect or ascending, 7–30 cm long; racemes 6–27-flowered; calyx pubescent with dark and pale hairs, 8–12.5 mm long; corolla pale yellow to ochroleucous, 18–22 mm long, the wings conspicuously dilated upward and at least 5 mm wide at the apex; keel 15–20 mm long; pods oblong, beaked, short-hairy, the wall coriaceous and rigid at maturity. Blooms early, often in May. $2n = 48$. In our area var. **spicata** (Hook.) Barneby. Prairies and dry hillsides. sYuk, nBC, Alta s to Nev, Utah, NM, Tex, Okla, Sask. Map 735

O. splendens Dougl. *ex* Hook. / Showy Loco-weed
Caespitose herb, densely long-silky and villose throughout; leaves 7–26 cm long; leaflets numerous, mostly in whorls of 3–4 in 7–15 fascicles, a few sometimes opposite or scattered, lanceolate to oblong-elliptic; scapes erect, 10–36 cm long; racemes 12–35-flowered, dense, spike-like, 3–8 cm long; calyx densely villose; corolla 10–15 mm long, blue to reddish purple, drying violet; keel 10–11.5 mm long; pods ovoid, 10–17 mm long, densely villose, short-beaked, thinly chartaceous. $2n = 16$. Grassy slopes, open woods, gravelly banks. Alas, Yuk, Mack, to Lk Superior s to seBC, NM, ND, nMinn. Map 736

O. viscida Nutt. / Viscid Loco-weed
Acaulescent; low plant with a thick and often multiple spreading caudex, bracts, calyx-teeth, and pods more or less sticky with minute yellowish glands; stipules membranous, pale, densely pilose to glabrate dorsally, the margins ciliate; leaves 3–23 cm long; leaflets 19–35, subopposite, rarely geminate, lanceolate or narrowly elliptic-lanceolate, 8–15 mm long, glandular and sparsely hairy; scapes ascending, 10–25 cm long; racemes 3–20-flowered, spike-like, 2–3 cm long; bracts glabrous dorsally, glandular; calyx villose and glandular; corolla 11–15 mm long, blue-purple, yellowish at base; keel 8–11 mm long; pods oblong-ellipsoid, hairy, beaked, 8–12 mm long, membrano-chartaceous. $2n = 32$. Dry banks and hillsides. Alas, Yuk, wMack s to Calif, Colo. Map 737

Petalostemon Michx. / Prairie Clover

Perennial herbs with slender leafy stems, 2–8 dm long, sometimes branched near the base, commonly erect or spreading, occasionally prostrate; leaves alternate, odd-pinnate, the 3–9 leaflets glandular-dotted; flowers small, rose-purple or white, in a dense terminal spike; calyx 5-toothed; corolla is unusual for the family; standard broad and attached by a slender claw in the usual way, but the other 4 petals (sometimes interpreted as sterile stamens) are similar to each other, narrow, and inserted between the fertile stamens on the staminal tube; anther-bearing stamens 5, exserted, monadelphous near the base; ovary 2-ovuled, the style elongated; pods small, indehiscent, 1–2-seeded, enclosed by the persisting calyx.

Flowers white; calyx glabrous except the lobes; leaflets 5–9	**P. candidum**
Flowers rose or purple; calyx villose-tomentose; leaflets 3–5	**P. purpureum**

P. candidum (Willd.) Michx.
Stems usually several, 2–8 dm long, commonly erect, occasionally decumbent or prostrate, from the crown of the plant, glabrous; leaflets 5–9, linear, oblong or oblanceolate, glabrous; spikes compact, cylindric, 2–5 times as long as thick, the bracts with slender tips and conspicuous in the young spike, usually extending beyond the flower-buds; calyx-tube sharply ribbed, the lobes pubescent; corolla white. $2n = 14$. Prairie grassland, dry slopes, and gravel flats. Alta to wOnt s to NM, Tex, La, Miss. Map 738

P. purpureum (Vent.) Rydb. / Purple Prairie Clover
Stems usually several, 3–8 dm high, erect or ascending, glabrous or sparsely hairy, rarely densely villose; leaflets 3–5, narrowly linear, involute; spikes compact, ovoid to short-cylindric, 1–5 cm long; bracts abruptly acuminate; calyx densely villose-tomentose; corolla rose or purple. $2n = 14$*. Prairie grassland and eroded slopes. Alta to wOnt s to NM, Tex, Ark, Ala. Map 739

Psoralea L.

Perennial herbs with rhizomes or thickened roots, the stems and foliage more or less glandular-punctate; leaves alternate, 3–5-digitate, the leaflets entire; flowers spicate or racemose, white or blue-purple; calyx 5-cleft, persistent, the lowest lobe often longest; standard broad and auricled at base; wings scythe-shaped; keel shorter than the other petals, incurved; stamens usually diadelphous; pod short, thick, 1-seeded, indehiscent or irregularly dehiscent.

1	Flowers about 1.5 cm long	**P. esculenta**
	Flowers smaller, generally less than 1.0 cm long	2
2	Plants greyish-white to silver; silky-haired throughout	**P. argophylla**
	Plants green; glabrous or strigose	**P. lanceolata**

P. argophylla Pursh / Silver-leaf Psoralea
Plant with creeping rhizomes and erect, widely branched stems, 3–6 dm high; leaflets commonly 3, obovate to elliptic, 2–4 cm long, densely silvery-hairy; lower leaves often with 5 leaflets; flowers in a small interrupted spike of 1–3 whorls; calyx silvery, its tube about 2 mm long and the lobes unequal; corolla blue or purple, scarcely exceeding the longest calyx-lobe; pod ovoid, silky. $2n = 22$. Prairie grassland. Alta to sMan s to NM, Okla, Mo. Map 740

P. esculenta Pursh / Indian Bread-root
Low, bushy herb with a strong taproot, often greatly enlarged as a tuber-like body a few inches below the ground level; stem erect, 1–3 dm high, hirsute; leaflets 5, digitate, obovate to elliptic, 2–6 cm long, glabrous above, hairy beneath, not glandular-dotted; flowers in a dense thick spike, 2–8 cm long; calyx hirsute,

non-glandular; corolla bluish purple, 12–15 mm long; pod ovoid, long-beaked, membranous, rupturing about the middle; seed brown, flattened, about 5mm wide. $2n = 22$. Prairies and dry banks. Alta to sMan s to NM, Okla, Ark. Map 741

P. lanceolata Pursh / Scurf Pea
Plants with long, branching rootstocks; leafy shoots erect or ascending, 2–6 dm high, with many slender branches, glabrous or nearly so; leaves palmately 3–5-foliolate, yellowish green, conspicuously glandular-dotted; leaflets narrowly linear to oblanceolate, 1.5–3 cm long; flowers white or bluish white, 5–6 mm long, in short dense spikes; calyx sparingly hairy, glandular, 2–2.5 mm long; corolla white or purple-tinged at the tip, especially on the keel; pod globose, short-beaked, densely glandular, villose or glabrate. $2n = 22$. Sandy ground and dunes. sAlta, sSask s to Calif, NM, Kans. Map 742

Thermopsis R.Br. / Golden Bean

T. rhombifolia (Nutt.) Richards.
Perennial herb with creeping rootstocks; stems erect or ascending, 1–4 dm high; leaves alternate, with large ovate stipules, digitately 3-foliolate; leaflets obovate or elliptic, 2–3 cm long, entire, strigose or nearly glabrous; flowers golden yellow, 15–20 mm long, borne in a short raceme; stamens 10, distinct; pods flat, often curved in a semicircle, 4–7 cm long at maturity, pubescent to glabrous. $2n = 18$. Dry sandy grassland. seBC to sMan s to Colo, Neb. Map 743

Trifolium L. / Clover

Annual, biennial or perennial herbs; leaves 3-foliolate, usually serrulate; stipules adnate to the petiole; flowers in globose spikes or compact racemes; calyx-tube cup-shaped or tubular, the lobes often unequal; petals all separate or more or less united into a tube, often withering and persistent; stamens 10, diadelphous; fruit small, often included in the calyx, indehiscent or tardily dehiscent, 1–6-seeded.

1 Flowers yellow; plants annual **T. aureum**
Flowers white, pink, rose or purple; plants biennial or perennial 2

2 Flowers 12–20 mm long, sessile in the head, usually reddish; the head subtended by 1 or 2 short-petioled leaves **T. pratense**
Flowers 7–11 mm long, distinctly pedicelled, white or pink; the head without subtending leaves 3

3 Stems erect or ascending, not rooting; flowers pink or rose-tinged, sometimes white; leaflets oval to elliptic, 2–4 cm long **T. hybridum**
Stems creeping, often rooting at the nodes; flowers white or pink-tinged; leaflets obovate to broadly elliptic, 1–2 cm long **T. repens**

T. aureum Poll. / Yellow or Hop Clover
Stems ascending or erect, 2–5 dm tall, branched, appressed-hairy; petioles slender, 6–10 mm long, usually slightly longer than the oblong-lanceolate stipules; leaflets oblong-obovate, sessile or equally short-stalked; heads rounded or oblong, 1–2 cm long, on peduncles about 2 cm long; corolla yellow, diagonally striate, brownish in age. $2n = 14$. *T. agrarium* L. of Ed. 1. Roadsides and waste ground. Introduced. Rare.

T. hybridum L. / Alsike Clover
Glabrous perennial, the stems usually much branched, 2–8 dm tall; leaflets 2–4 cm long, oval to elliptic, or sometimes obovate, rounded or retuse at the apex; heads on peduncles 2–8 cm long; flowers 8–9 mm long, commonly pinkish; calyx with linear-subulate teeth, slightly longer than the tube; pod stipitate, 2–4-seeded; seeds lenticular, greenish to black. $2n = 16$. Roadsides, waste ground, fields. Introduced. Common.

T. pratense L. / Red Clover
Biennial or short-lived perennial, erect or spreading, widely branched, 4–8 dm tall; lower leaves long-petioled, the upper short-petioled to sessile; leaflets ovate or elliptical, 2–5 cm long, sparingly pubescent; heads sessile or on very short peduncles; flowers 12–20 mm long, red, pink or nearly white; pod 1-seeded, splitting around the middle; seeds rounded-triangular, yellow or purple. $2n = 14$. Roadsides, waste ground, fields. Cultivated for forage and freely escaping. Introduced. Common.

T. repens L. / White Clover; Dutch Clover
Low, glabrous perennial, the stems creeping, often rooting at the nodes; leaf-petioles and head-peduncles 5–20 cm long; leaflets 1–2 cm long, obovate to broadly elliptic, often retuse at the apex; flowers usually about 8 mm long, white or pinkish; pods 2–5-seeded; seeds squarish, yellow. $2n = 32$. Roadsides, gardens, fields. Introduced. Common.

Trigonella L.

T. caerulea (L.) Ser.
Plants annual; stems 2–6 (10) dm tall, glabrous or sparsely hairy; leaves trifoliolate; leaflets ovate to oblong, emarginate, denticulate; flowers in dense axillary racemes; calyx about 3 mm long; corolla about 6 mm long, white to blue or violet; pod about 7 mm long, abruptly contracted into a beak. $2n = 16, 32$. Gardens and fields. Introduced. Rare.

Vicia L. / Vetch

Climbing herbs with slender stems; leaves pinnate, stipulate, with tendrils at the tips; flowers in axillary, peduncled racemes; calyx often gibbous at base and the teeth subequal; standard with a broad claw overlapping the wings, the latter adherent to the curved keel; stamens 10, more or less diadelphous, the tube ending obliquely; style slender, with a tuft of hairs at or near the apex; pods commonly flat.

Racemes 1-sided, dense, 15–40-flowered **V. cracca**
Racemes not 1-sided, 2–9 flowered **V. americana**

V. americana Muhl. / Wild Vetch
Perennial herb with stems 3–10 dm high, climbing and spreading, often in tangled masses; leaflets 8–14, highly variable in size and shape, commonly elliptic to oblong, glabrous or pubescent, entire or sharply toothed towards the apex, the latter acute or truncate and commonly mucronate; tendrils well-developed, forking; stipules commonly semi-sagittate, sharply toothed; racemes shorter than the subtending leaves, lax, with 3–9 reddish purple flowers, drying bluish; pods about 3 cm long, flat, glabrous. $2n = 14$. Includes *V. sparsifolia* Nutt. of Ed. 1. Open woods, meadows, shores. seAlas, wMack to Sask, Man, Gt Lakes s to Calif, NM, Tex, Okla, Mo, Ohio, Gaspé. Map 744

V. cracca L.
Perennial, the stems weak, climbing or trailing, 4–10 dm long, glabrous or pubescent; leaflets 10–20, linear to narrowly oblong, 1–3 cm long, mucronate; racemes long-peduncled, mostly surpassing the leaves; flowers bluish purple, 9–13 mm long, numerous, crowded, secund, usually reflexed. $2n = 14, 28$. Roadsides, waste ground, fields. Introduced. Occasional.

GERANIACEAE / Geranium Family

Herbs with leaves opposite, stipulate, commonly lobed or incised; flowers perfect, regular, hypogynous, cymose or umbellate; sepals 5, persistent; petals 5, purple, pink or white; stamens 10 or 5; pistil of 5 carpels, 5-lobed, the thin styles united to a central column; ovules 2, mature seed 1, in each carpel; capsule long-beaked, at maturity splitting from below upward to form 5 parts, each 1-seeded and long-tailed.

Leaves pinnately veined and divided; anthers 5; stylar parts spirally twisted at dehiscence and bearded within **Erodium**
Leaves palmately veined and divided; anthers 10; stylar parts merely recoiled at dehiscence and glabrous within **Geranium**

Erodium L'Hér.

E. cicutarium (L.) L'Hér. / Stork's-bill
Winter-annual or biennial; stems low and spreading, up to 4 dm long; leaves mostly basal, hirsute, pinnately compound, the segments sessile, deeply and irregularly cleft; flowers in long-peduncled umbel-like clusters; sepals 3–7 mm long, bristle-tipped; petals pink to purple, somewhat longer than the sepals; stamens 5, alternating with 5 staminodia; style-column 3–4 cm long, its segments

spirally twisted when dry. $2n = 20, 40$. Fields and waste places. Introduced. Occasional.

Geranium L. / Crane's-bill

Annual, biennial or perennial herbs; leaves palmately lobed, cleft or divided; flowers cymose, often on axillary 1–2-flowered peduncles; stamens 10, in 2 sets of unequal length, rarely 5, the filaments usually broadened near the base and more or less adnate; capsule splitting from below upward into 5 segments, which are recurved or recoiled when dry; seeds often reticulate.

1 Plant annual or biennial; petals 2–8 mm long 2
Plant perennial, with rootstocks or caudices; petals 1–2 cm long 3

2 Fruiting pedicels much longer than the calyx; sepals broadly lanceolate to oblong-ovate; beak of mature style-column 3–5 mm long **G. bicknellii**
Fruiting pedicels shorter than to slightly longer than the calyx; sepals broadly ovate; beak of mature style-column 1–1.5 mm long **G. carolinianum**

3 Petals white or purple-tinged, pilose on the inner surface one-third to one-half their length **G. richardsonii**
Petals purple to pink, rarely white; pilose inside not one-fourth their length 4

4 Leaves uniformly pubescent below; flowers generally not overtopping leaves **G. erianthum**
Leaves pubescent below on veins only; flowers generally clearly overtopping leaves **G. viscosissimum**

G. bicknellii Britt.
Annual or biennial, usually much branched, 2–6 dm high, the stems with spreading hairs and sometimes glandular above; leaves 2–7 cm broad, deeply 5-parted, the cuneate divisions cleft into several narrowly oblong lobes; peduncles 2-flowered, the pedicels distinctly longer than the calyx and glandular-villose; sepals awn-tipped, hairy on margins and veins; petals rose-purple, 5–7 mm long, slightly longer than the sepals; mature fruit about 2 cm long, the filiform beak 3–5 mm long; seeds thick-cylindric, marked with a network of elongate areolae. Clearings, open woods, disturbed soil. Alas, Yuk, swMack to James Bay, NS, Nfld s to Calif, Utah, Colo, SD, Ia, Ind, Mass. Map 745 ●

G. carolinianum L.
Annual, freely branched, 1–5 dm high, the stems villose with spreading or retrorse hairs; leaves 2–7 cm broad, deeply cleft into 5–7 cuneate divisions, these lobed or toothed; flowers in crowded umbel-like corymbs; peduncles and paired pedicels minutely retrorse-hairy, shorter than the fruiting calyx; sepals broadly ovate, awned, with 5 pilose-hirsute nerves; petals rose-purple, about as long as the sepals; mature fruit 1–1.5 cm long, the subulate beak 1–1.5 mm long; seeds subspherical, faintly reticulate with about 50 rows of small square or rounded areolae. $2n = 52$. *G. sphaerospermum* Fern. of Ed. 1. Clearings and disturbed ground. BC to Ont s to Calif, Mont, SD, Kans, Mo, Tenn, NC. Map 746 ○

G. erianthum DC.
Perennial; stems 2–8 dm tall, erect or ascending, with retrorse hairs; leaf-blades 5–7-lobed, the 2 basal often contiguous, generally 3–10 cm long from sinus to apex, broader than long; flowers 2–several on axillary peduncles; sepals 8–12 mm long, aristate, pubescent with capitate hairs; petals 16–20 mm long, blue to purple; style-column 2.5–3.2 cm long. $2n = 28$. Moist open woods and meadows. eAsia: sAlas, sYuk s to BC, Alta. Map 745 ○

G. richardsonii Fisch. & Trautv.
Perennial herb with one or more erect stems from a short caudex; stems 4–8 dm tall, glabrous or sparsely pubescent; leaves deeply 3–7-parted, the divisions sharply lobed, sparsely strigose; peduncles commonly 2-flowered, 2–12 cm long, usually beset with stalked purple glands; sepals lanceolate to narrowly oval, 6–12 mm long, awned; petals 10–18 mm long, commonly white with pink or purple veins, occasionally pinkish; style-column 2–2.5 cm long; seeds coarsely reticulate. $2n = 26, 52$. Moist thickets and open woods. Yuk, swMack s to Calif, NM, SD. Map 747

G. viscosissimum Fisch. & Mey. / Sticky Purple Geranium
Perennial herb with stems 2–9 dm high from a woody caudex; leaves 5–11 cm broad, 5–7-parted and the divisions sharply and irregularly incised, densely pubescent and often glandular; inflorescence compactly cymose, the peduncles 2–6 cm long, densely glandular-villose; sepals 10–12 mm long, oblong-lanceolate, densely glandular-villose; petals 14–20 mm long, rose-purple or rarely white with dark veins; carpels glandular-pilose; style-column 2.5–3 cm long. $2n = 52$. Includes *G. nervosum* Rydb. of Ed. 1. Moist grassland slopes and coulees. sBC to sSask s to Calif, Colo, SD. Map 746 ○

LINACEAE / Flax Family

Linum L. / Flax

Annual or perennial herbs with slender but stiff stems; leaves simple, narrow, sessile, entire; flowers regular, blue or yellow, terminating the upper branches; sepals 5, usually persistent; petals 5, blue or yellow, very soon falling; stamens 5 (with or without alternating rudiments), the filaments united at their bases; pistil of 5 united carpels, the 5 slender styles distinct or partly united; ovary superior, each of its 5 locules divided by a false partition; fruit a rounded 10-seeded capsule, splitting from the top, the seeds flattened.

1 Flowers yellow; sepals, at least the inner ones, with marginal glands; stigmas terminal and capitate; fruiting stalks 0.5–1 cm long **L. rigidum**
Flowers blue; sepals glandless; stigmas introrse; fruiting stalks 1–4 cm long 2

2 Perennial; leaves linear; pedicels arching and spreading; sepals blunt to short-acuminate, not ciliate **L. lewisii**
Annual; leaves lanceolate or linear-lanceolate; pedicels erect; sepals long-acuminate, the inner ciliate **L. usitatissimum**

L. lewisii Pursh / Wild Blue Flax
Perennial; stems erect, 2–7 dm tall, glabrous, often striate, branching above; leaves numerous, linear, 1–2 cm long, ascending; flowers in small corymbs or scattered on the upper branches; sepals ovate, obtuse to short-acuminate or mucronate, entire; petals blue, 1–2 cm long; styles distinct or nearly so; capsule subglobose, longer than the sepals, the false septa incomplete, long-ciliate. $2n = 18^*$. *L. perenne* L. ssp. *lewisii* (Pursh) Hult. Grassland and dry open woods. Alas, Yuk, Victoria Is, wMack to James Bay s to Calif, NM, Tex, Okla, Wisc, Gt Lakes. Map 748

L. rigidum Pursh / Yellow Flax
Annual; stems 1–5 dm tall, glabrous or nearly so, simple or branched, the branches stiff and more or less angular; leaves soon deciduous, linear to linear-lanceolate 0.5–2.5 cm long; sepals lanceolate, acute or awn-pointed, glandular-margined; petals yellow, commonly 10–15 mm long, about twice as long as the sepals; styles united nearly to the top; capsule ovoid, a little shorter than the sepals, the false septa somewhat thickened at base. $2n = 30$. Open slopes and grassland. sAlta to sMan s to NM, Tex, wMo. Map 749

L. usitatissimum L. / Common Flax
Annual; stems 2–8 dm high, erect, usually solitary from a slender root, branching above; leaves linear-lanceolate, 3-nerved, 1–3 cm long; sepals 7–9 mm long, acuminate, the inner ciliate on the scarious margins; petals blue, 1–1.5 cm long; styles distinct or nearly so, the stigmas elongated; capsule globose, equalling or somewhat longer than the sepals, commonly indehiscent. $2n = 30$. Cultivated for its oil (linseed) and fibre (linen). An occasional escape in waste places. Introduced.

POLYGALACEAE / Milkwort family

Polygala L.

Small perennial herbs with simple, entire, alternate leaves and no stipules; flowers white or purplish, mostly in dense spike-like racemes at stem-tips; flowers perfect, hypogynous, zygomorphic, suggestive of legumes but actually quite different; sepals 5, free, the two inner ones (wings) much larger and often petaloid; petals 3, the middle one (keel) boat-shaped and with a fringed crest, all petals connected with each other and with the stamen-tube; stamens usually 8, filaments united below into a split sheath, anthers opening by apical pores; the ovary superior, 2-loculed, the style long, curved; fruit a small 2-seeded capsule, splitting lengthwise; seed obovoid with a small whitish 2-lobed appendage at the hilum end.

Flowers 1–4, rose-purple **P. paucifolia**
Flowers many, white **P. senega**

P. paucifolia Willd.
Stems 1–14 cm high, the shorter often bearing minute cleistogamous flowers; leaves short-petiolate or sessile; blades ovate to elliptic, the larger 1–5 cm long, 1–2 cm wide; chasmogamous flowers 1–4, terminal, pink to rose-purple; sepals dimorphic, 3 short, 3–6 mm long, 2 long, oblanceolate, 10–17 mm long; petals 10–18 mm long, crest about 6 mm long; stamens 6; ovary about 3.5 mm long; capsule 6–7 mm long, obovate, the valves distinct; seeds about 3.5 mm long. Moist coniferous woods. Alta to sQue s to Minn, Ill, Tenn, Ga, Del. Map 750 ○

P. senega L. / Seneca-root
Stems 1–5 dm high, usually unbranched, minutely puberulent; leaves 1–3 cm long, alternate, lanceolate or ovate, the margins scabrous or very inconspicuously serrulate; lowest leaves smaller, often scale-like; flowers greenish white, 4–5 mm wide, in a dense, narrowly conic or oblong raceme, 1–4 cm long; sepal wings orbicular-ovate; capsule suborbicular, broader than long; seeds pubescent. Open woods and moist prairies. Alta to Que, NB s to SD, Neb, Kans, Ark, Ga. Map 750 ●

EUPHORBIACEAE / Spurge Family

Euphorbia L. / Spurge

Annual or perennial herbs, with milky juice; leaves simple and alternate, opposite or whorled; flowers monoecious, small, usually in cyathia simulating perfect flowers, the cyathia borne in cymes, the latter often on whorled branches; the cyathium a cup-like involucre enveloping a single pistillate flower and few or many staminate flowers, the margin of the involucre often with nectariferous glands and in some species coloured appendages; pistillate flower stipitate, without calyx or corolla, ovary 3-loculed, styles 3, usually 2-cleft; staminate flowers in 5 fascicles, 1–several in a fascicle, each flower a single stamen and with a minute bract at its base; capsule 3-lobed, 3-seeded, usually nodding, splitting lengthwise and ejecting the seeds.

1 Perennial; leaves linear **E. esula**
Annual; leaves broader 2

Plants erect; leaves alternate below, opposite or whorled above 3
Plants prostrate; leaves opposite 4

3 Leaves entire **E. peplus**
Leaves serrulate **E. helioscopia**

4 Leaves linear-oblong, commonly entire; seeds with a few sharp transverse ridges **E. glyptosperma**
Leaves obovate or ovate to broadly oblong, usually serrate; seeds pitted and faintly ridged **E. serpyllifolia**

E. esula L. / Leafy Spurge
Perennial herb, deep-rooted, spreading by creeping roots; stems mostly erect, 4–9 dm high, hard, smooth, often branched above; leaves sessile, alternate except the whorled ones at the base of the umbel, entire, linear to lanceolate-oblong, 2–7 cm long, glabrous and somewhat glaucous; stipules absent; flowering stems often terminated by a primary umbel with 7–12 rays; larger floral leaves ovate to reniform, about 1 cm long, yellow-green; cyathia 2–3 mm high, the lobes with 4 crescent-shaped or horned glands, yellow or green, no appendages present; capsule about 4 mm wide, finely granular; seeds globose-ovoid, smooth, yellow-brown, about 2 mm long. $2n = 60, 64$. Fields, roadsides, and river flats. Introduced. Common.

E. glyptosperma Engelm.
Annual plant, glabrous throughout, with stems mostly prostrate, freely branched, often forming mats 2–6 dm across; leaves opposite, narrowly oblong to ovate, 4–12 mm long, entire or inconspicuously serrulate at apex, oblique at base; stipules subulate; cyathia 0.6–0.9 mm wide, with elliptic-oblong glands mostly depressed in middle, and white appendages; staminate flowers 1–5 in the cyathium; capsule about 1.5 mm long; seeds less than 1 mm long, white to tan, 4-angled, the faces with conspicuous transverse ridges. $2n = 22$. Dry sandy soil, waste ground. BC to NB s to Calif, Tex, Mo, Ohio, NY. Map 751

E. helioscopia L. / Wartweed; Sun Spurge
Annual, with erect stems 1–6 dm high, often branched at the base, glabrous; leaves alternate (except the terminal whorl), simple, obovate to spatulate, rounded or retuse at apex, finely serrate, smooth or sparsely hairy; primary umbel subtended by a whorl of leaves, composed of usually 5 stout rays, these often repeatedly branched or forked; cyathia about 2 mm long, the glands yellow, stalked; capsule about 3 mm long, smooth; seeds ovoid, 2–2.5 mm long, dark orange-brown, with coarse honeycomb-like markings. $2n = 42$. Fields and waste ground. Introduced. Occasional.

E. peplus L.
Annual; glabrous; stems 1–3 (5) dm high, freely branched above; lower leaves alternate, slender-petiolate; blades 1–3 cm long, rhombic-ovate to rhombic-obovate, abruptly narrowed at base; upper leaves opposite, ovate; cyathia 1.0–1.5 mm long; capsules about 2.5 mm long; seeds oblong, grey, 1.0–1.5 mm long, conspicuously pitted. $2n = 16$. Gardens and waste ground. Introduced. Rare.

E. serpyllifolia Pers.
Similar to *E. glyptosperma* in habit and general appearance; but with mostly broader leaves, their margins more conspicuously serrulate; pitted and faintly ridged seeds, cyathia having flat glands and 5–18 staminate flowers. $2n = 22$. Sandy and alluvial soil, along railroads, waste ground. BC to Man s to Calif, NM, Tex, Minn. Map 752

CALLITRICHACEAE / Water-starwort Family

Callitriche L. / Water-starwort

Small aquatic plants, usually tufted, submersed or partly floating, sometimes on mud; leaves small, opposite, obovate to linear, entire, without stipules; flowers minute, greenish, unisexual, 1–3 in the leaf-axils; calyx and corolla none; staminate flower of one stamen only; pistillate flower a single pistil of 2 carpels, with 2 filiform styles and a 4-loculed, superior ovary; fruit small, 4-lobed, separating at maturity into 4 one-seeded portions.

Leaves all linear, 1-nerved, submersed; flowers without bracts at base **C. hermaphroditica**

Leaves, at least the floating ones, obovate to spatulate, 3-nerved; flowers with a pair of bracts **C. verna**

C. hermaphroditica L.
Annual or perennial, mostly submersed, with delicate stems, 1–4 dm long; leaves dark green, linear, 5–15 mm long, broadly attached at the sessile base and not connected by a wing, with a single nerve; flower bracts, if present, very small; fruit 1–2 mm wide, circular, its broadly winged halves separated by a deep furrow. $2n = 6$. Ponds and streams. Circumpolar: Alas, Yuk, Mack to Huds Bay, Nfld s to Calif, NM, Neb, Minn, Gt Lakes. Map 753

C. verna L.
Annual or perennial, submersed or partly floating, sometimes on muddy shores; leaves pale green, their bases connected at the node by a narrow membranous wing; floating leaves petioled, narrowly obovate or broadly spatulate, 3-nerved, often crowded in a rosette at the stem-tip; submersed leaves sessile, linear, 1-nerved, 10–15 mm long; flowers with a pair of bracts 1 mm long; fruits about 1 mm wide, higher than broad, the narrowly winged halves separated by a wide groove. $2n = 20$. *C. palustris* L. of Ed. 1. Ponds and streams. Circumpolar: Alas, Yuk, Mack to Huds Bay, Nfld s to Calif, NM, Neb, Ill, Va. Map 754

EMPETRACEAE / Crowberry Family

Empetrum L.

E. nigrum L. / Crowberry
Low, heath-like evergreen shrub with procumbent, spreading branches up to 4 dm long; branchlets slender, minutely glandular and hairy or almost glabrous; leaves alternate, divergent or reflexed, linear-oblong, 4–8 mm long, glabrous, deeply grooved beneath, the margins revolute; flowers small, axillary, scaly-bracted, unisexual or perfect, sometimes polygamous; sepals 3; petals 3 or none; stamens 3, the pollen grains compound; pistil 1, ovary 6–9-loculed, style very short, stigmas 6–9; fruit a berry-like drupe, black or rarely purplish, with 6–9

nutlets each containing a seed. $2n = 26, 52$. In our area ssp. **hermaphroditicum** (Lange) Böcher. Acidic rocks, muskegs, coniferous forest. Circumpolar: Alas, Yuk, Victoria Is, Ellesmere Is to Nfld s to Calif, BC, Alta, Sask, Man, Gt Lakes, Me. Map 755

ANACARDIACEAE / Sumach Family

Rhus L.

Shrubs, or low herbs with woody base; leaves alternate, compound, trifoliolate or 3-lobed; flowers very small, greenish white or yellowish, in axillary or terminal panicles, perfect or unisexual and polygamous or dioecious; calyx 5-lobed; petals 5; stamens 5, inserted beneath a disc that surrounds the ovary; pistil 1, superior, composed of 3 carpels but unilocular and 1-ovuled; styles or stigmas 3; fruit a drupe, with a resinous mesocarp, 1-seeded.

Low plant with a single stem, woody at base; leaflets over 3 cm long; fruit white, smooth **R. radicans**

Shrub, 1–2 m high, much branched; leaflets usually less than 3 cm long; fruit red, hairy **R. trilobata**

R. radicans L. / Poison Ivy
Single-stemmed, erect plants, 1–4 dm high, slightly branched at summit, somewhat woody at base, growing in patches from creeping underground stolons; leaves few, long-petioled, with 3 leaflets, the terminal one on a long petiolule; blades of leaflets 3–15 cm long, ovate or rhombic-ovate, acute or abruptly acuminate, usually glabrous, irregularly toothed or entire; flowers in dense axillary panicles; sepals about 1 mm long; petals about 3 mm long, yellowish white; fruit globose, 5–6 mm wide, white or yellowish white, usually glabrous. $2n = 30$. In our area var. **rydbergii** (Small) Rehder (*Toxicodendron rydbergii* (Small *ex* Rydb.) Greene). Open woods, coulees, river flats. BC to NS s to Ore, Ariz, NM, Tex, Ark, Tenn, Va. Map 756

R. trilobata Nutt. / Skunk-bush
Shrub, 1–2 m tall, much branched, ill-scented, generally flowering before the leaves appear; twigs slender, usually puberulent at least when young; leaves with 3 leaflets or less commonly 3-lobed; leaflets sessile or nearly so, 1–4 cm long, obovate to elliptic, crenate, the terminal one often 3-lobed; flowers minute, in dense terminal panicles, yellowish green; sepals 1–1.5 mm long; petals about 2 mm long; fruit globose, 5–6 mm thick, red, glandular-pubescent. Closely related to the eastern and southern species *R. aromatica* Ait. Coulees and river valleys. Alta, Sask s to Calif, Ida, Mont. Map 757.

CELASTRACEAE / Bittersweet Family

Pachistima Raf.

P. myrsinites (Pursh) Raf. / Mountain Lover
Evergreen shrub, 2–10 dm high, erect or prostrate, the branches brown, 4-ridged, glabrous; leaves opposite or nearly so, short-petioled, oval or elliptic, 1–3 cm long, leathery, margins somewhat revolute and thickened, serrate; flowers small, numerous, in axillary clusters; sepals 4, about 0.7 mm long; petals 4, about twice as long as the sepals, reddish; stamens 4, inserted on the margin of a fleshy disc; pistil 1, the ovary superior, 2-loculed, the style short; fruit a small ovoid capsule, opening by 2 valves; seeds 1 or 2, with a white aril. Coniferous woods, mountain slopes. BC, Alta s to Calif. Map 758

ACERACEAE / Maple Family

Acer L. / Maple

Trees or shrubs with opposite, simple or compound leaves and small flowers in racemes, corymbs or umbels; flowers regular, perigynous, unisexual (in our species), the plants dioecious or monoecious; sepals 5, separate or connate; petals 5, small, separate, or lacking; stamens 4–10, often 8; pistil 1, the ovary superior, 2-carpelled and 2-loculed, flattened; styles 2, more or less united; fruit of 2 samaras, united at base, each 1-seeded and with a large wing.

Shrub or small tree; leaves simple, palmately 3–5-lobed	**A. glabrum**
Tree, up to 12 m tall; leaves pinnately 3–5-foliolate	**A. negundo**

A. glabrum Torr. / Mountain Maple
Shrub or small tree, up to 10 m tall, the young branches reddish and older stems grey; leaves opposite, slender petioled, glabrous, thin, 3–10 cm across, palmately 3–5-lobed, the lobes acute or acuminate, the margins irregularly and doubly serrate; flowers unisexual, monoecious or dioecious, 5–15 in a cluster, appearing with the leaves; samaras 2–4 cm long. Includes *A. douglasii* Hook. Montane slopes. seAlas, BC, Alta s to Calif, NM, Neb. Map 759

A. negundo L. / Manitoba Maple; Box Elder
Tree, up to 12 m tall, with rough greyish bark, the trunk soon divided into widely spreading branches, the twigs light green and puberulent; leaves pinnate, the leaflets usually 3 or 5, ovate to lanceolate, acuminate, coarsely and irregularly serrate, or with a few shallow lobes, sometimes pubescent, especially beneath; flowers unisexual, dioecious, apetalous, appearing before the leaves from lateral buds; staminate flowers in dense clusters, drooping; pistillate flowers in drooping racemes; fruit pendent, each samara about 3 cm long, falling in winter or spring. $2n = 26$. Widely planted, possibly native along streams in seAlta.

BALSAMINACEAE / Touch-me-not Family

Impatiens L. / Touch-me-not; Jewelweed

Herbs, 5–15 dm high, with succulent stems and watery juice; leaves alternate, simple, without stipules, petioled; blades thin, oval or ovate, coarsely toothed; flowers yellowish, perfect, irregular, zygomorphic, with a prominent pouch (sac) terminating in a spur; perianth of 6 segments, the large spurred sac and 2 small greenish parts being interpreted as sepals, and 3 lobed or notched parts being petals; stamens 5, more or less united around the stigma; pistil 5-locular, style obsolete; capsule pod-like, bursting elastically in 5 valves when ripe or at a touch and ejecting the seeds. Most of the seed capsules arise not from the large, conspicuous flowers but from very small cleistogamous flowers whose perianth parts do not expand.

Flowers usually bright orange or reddish yellow; spur bent back parallel with the sac **I. capensis**

Flowers pale yellow; spur recurved but not abruptly so **I. noli-tangere**

I. capensis Meerb. / Touch-me-not
Flowers bright orange to reddish or pale yellow, copiously spotted with red or purple, or occasionally unspotted; the saccate sepal 15–18 mm long, the pouch (sac) about two-thirds as broad as long and abruptly narrowed into the spur; the spur bent back parallel with the sac. $2n = 14$. Moist woods and stream banks. sMack, BC to Nfld, s to Mont, Kans, Okla, Ark, Fla. Map 760

I. noli-tangere L. / Touch-me-not
Flowers pale yellow, dotted with red or purple, or unspotted; the saccate sepal about 20 mm long, the sac about half as broad as long and gradually tapered into the spur; the spur recurved, but not abruptly so, as in the last species. $2n = 20^*, 40$. Moist woods and stream banks. Eurasia: sAlas to cMan s to Ore, Ida, Alta, cSask. Map 761

RHAMNACEAE / Buckthorn Family

Shrubs or trees with alternate, rarely opposite, simple leaves; flowers small, greenish or white, perfect or unisexual, regular, perigynous with an annular disc; sepals, petals and stamens 5 (or 4) each, or petals lacking; stamens inserted on the disc, opposite the petals; pistil 1, the ovary inserted on or embedded in the disc, 2–4-loculed, each locule with a basal ovule; fruit a capsule or berry-like drupe.

Leaves 3-nerved from the base; flowers in a dense terminal panicle; fruit dry **Ceanothus**

Leaves pinnately veined; flowers axillary, in small clusters or solitary; fruit juicy **Rhamnus**

Ceanothus L. / Tea Bush

C. velutinus Dougl. *ex* Hook. / Snow Brush; Deer Brush
Shrub, 1–3 m high, spreading, unarmed, usually growing in dense patches, the twigs olive-green; leaves odoriferous, evergreen, alternate, ovate to elliptic, with 3 prominent veins, 4–8 cm long, obtuse or rounded at apex, glabrous and shining-varnished above, velvety-puberulent to glabrescent and paler beneath, the margin closely glandular-serrulate; petioles rather short; flowers white, in a dense panicle; sepals, petals and stamens 5 each, the petals hooded and long-clawed; ovary immersed in the disc (partly inferior), 3-loculed, the style 3-cleft; capsule obovoid, 3-lobed, viscid-glandular, separating into 3 nutlets, these dehiscent, each with a flattened seed. $2n = 26$. Montane slopes. BC, Alta s to Calif, Colo, SD. Map 762

Rhamnus L. / Buckthorn

Trees or shrubs; leaves opposite or alternate; flowers solitary or umbellate in leaf-axils, greenish or greenish white, 4- or 5-merous, perfect or unisexual; ovary with 2–4 locules; fruit a drupe.

Plants spiny; flowers 4-merous	**R. catharticus**
Plants not spiny, flowers 5-merous	**R. alnifolia**

R. alnifolia L'Hér.
Shrub, 0.5–2 m high, with puberulent twigs; leaves alternate, short-petioled, elliptic or ovate, sometimes acuminate, 3–10 cm long, crenate-serrate, nearly glabrous, pinnately veined; flowers greenish, very small, commonly unisexual and dioecious, apetalous, borne singly or umbellate in leaf-axils; ovary free from the disc; fruit black, poisonous, 7–9 mm wide, containing 3 flat nutlets. Moist, shady woods. BC to Nfld s to Calif, Wyo, Neb, Ohio, Pa, NJ. Map 763

R. catharticus L.
Small tree or shrub up to 6 m tall; leaves opposite, 3–6 cm long, broadly ovate to elliptic-obovate; petals 0.6–1.3 mm long; flowers unisexual, dioecious; sepals 2 mm long; petals 0.6–1.3 mm long; fruit black, 5–6 mm wide, usually with 4 nutlets. $2n = 24$. Coulees, probably an escape from cultivation. Introduced. Rare.

MALVACEAE / Mallow Family

Herbs with alternate, palmately veined, usually lobed, leaves; stipules deciduous; flowers perfect, regular, produced singly or clustered in the leaf-axils; calyx of 5 united sepals, often with associated bractlets; petals 5, more or less joined at the base and to the staminal column; stamens numerous, monadelphous, the tube surrounding the style; pistil 1, of 5-many carpels, (in our species) united in a ring around a central axis and at the base of the single style, and then separating at maturity; style branched above; fruit splitting in several parts, each 1-seeded or with a few seeds.

1	Calyx-bractlets absent; flowers yellow	**Abutilon**
	Calyx-bractlets present; flowers not yellow	2
2	Stigmas extending along the inner surface of the filiform style branches	**Malva**
	Stigmas terminal, capitate	3
3	Stems erect, 5–20 dm tall	**Iliamna**
	Stems prostrate or ascending, 1–3 dm tall	**Sphaeralcea**

Abutilon Mill.

A. theophrasti Medic. / Velvet-leaf
Annual, softly pubescent throughout; stems 10–15 dm tall; leaves cordate, 10–15 cm long, acuminate, entire or obscurely toothed; flowers yellow, 15–25 mm across; fruiting heads 2–3 cm across, with 10–15 horned, several-seeded carpels. $2n = 42$. Waste ground. Introduced. Rare.

Iliamna Greene

I. rivularis (Dougl. *ex* Hook.) Greene / Wild or Mountain Hollyhock
Stout perennial herb, 0.5–2 m high, sparsely stellate-pubescent; leaves 5–15 cm broad, cordate to reniform in outline, 4–7-lobed, the lobes irregularly and coarsely toothed; flowers conspicuous, in interrupted spicate or corymbose racemes; calyx subtended by 3 bractlets; petals pink, rose-purple or nearly white, about 2 cm long; staminal column coarsely stellate-hirsute near the base; fruit-segments oblong, rounded at apex, covered with long stiff hairs and stellate-pubescent on the back, 2–3-seeded, dehiscent, but each remaining attached to the central axis by a stout fiber. $2n = 66$. Moist montane slopes. BC, Alta s to Ore, Mont, Colo. Map 764 ○

Malva L. / Mallow

Herbs with procumbent or erect stems; leaves alternate, orbicular or reniform, more or less lobed, toothed, pubescent; flowers solitary or clustered in leaf-axils; bractlets below the calyx 3, linear to narrowly lanceolate; calyx 5-lobed; petals 5, emarginate, white or tinged with mauve; fruit a schizocarp of 10–15 segments, around a central axis, each segment 1-seeded and indehiscent.

1	Petals mostly 2 cm long or more, deep bluish-purple	**M. sylvestris**
	Petals mostly 1 cm long or less, white or pale purple	2
2	Stems erect, 1–2 m tall	**M. verticillata**
	Stems decumbent or ascending, generally much shorter	**M. rotundifolia**

M. rotundifolia L.
Annual herb with spreading, ascending or erect stems, 2–8 dm long; leaves long-petioled, orbicular to reniform, usually 5-lobed, crenate, glabrous or sparingly hairy; flowers clustered in the axils; calyx becoming somewhat enlarged under the fruit and sometimes veiny-reticulate, the lobes ovate and usually spreading; petals pale blue or whitish, little longer than the sepals, commonly with bearded claws; carpels 8–11, pubescent or glabrate, with irregular low ridges on the back and the sutures more or less crested at maturity. $2n = 42$. Roadsides, yards, and waste ground. Introduced. Common.

M. sylvestris L. / High Mallow
Annual or biennial; stems 2–10 dm tall, hirtellous to glabrate; leaves mostly 3–10 cm long, cordate at base, 3–7-lobed, margin crenate-serrate; flowers in axillary clusters; bractlets ovate to oblong; petals 2.0–2.5 cm long, deep bluish-purple, obcordate; carpels rugose, sparsely hairy. $2n = 42$. Waste ground and yards. Introduced. Rare.

M. verticillata L. / Whorled Mallow
Annual; stems 1–2 m tall; leaves long-petiolate; blades 5–10 (15) cm long, orbicular to reniform in outline, somewhat irregularly 5–9-lobed, margin crisped, serrate; flowers more or less sessile, in axillary clusters; bractlets linear-lanceolate; petals about 6 mm long, white to pale purple; fruiting heads circular, with 8–11 carpels. $2n = 84$. Waste ground and yards. Introduced. Rare.

Sphaeralcea St. Hil.

S. coccinea (Pursh) Rydb. / Scarlet Mallow
Perennial herb with a woody base, spreading by rootstocks and often forming patches, the stems decumbent to ascending, 1–3 dm long, the entire plant densely stellate-pubescent; leaves deeply 3–5-parted, the divisions narrowly cuneate at the base and more or less lobed or forked, grey-pubescent; flowers in short racemes or spikes; calyx-lobes 3–5 mm long; petals yellowish-red or brick-red, 1–1.5 cm long, forming a saucer-shaped corolla; carpels 10 or more, densely stellate-pubescent on the back, reticulate on the sides, becoming 1-seeded segments in fruit. $2n = 10$. Prairie grassland and roadsides. BC to Man s to Ore, Utah, Tex, Okla, Ia. Map 764 ●

HYPERICACEAE / St John's-wort Family

Hypericum L. / St John's-wort

Herbs, shrubs or small trees; leaves opposite, exstipulate; flowers 5-merous, perfect, mostly yellow; sepals and petals often imbricate in bud; stamens usually many, often connate basally into fascicles; ovary superior, 1-loculed with 3–5 parietal placentae or 3–5-loculed with axile placentation; ovules numerous; fruit a septicidal capsule, rarely a berry.

Leaves ovate-oblong to suborbicular, black-dotted along the margin; petals about twice as long as sepals **H. formosum**

Leaves lanceolate, not black-dotted; petals not much longer than sepals **H. majus**

H. formosum HBK
Perennial herb with horizontal rootstocks and erect glabrous stems, 2–4 dm high; the buds tinged with red; leaves simple, opposite, sessile, 1–3 cm long, oblong-elliptic to ovate, entire, obtuse, punctate with black glandular dots at least along the margins; flowers yellow, perfect, regular, hypogynous, in loose cymes; sepals commonly 5, unequal, black-dotted along the margins; petals 5, yellow, distinct, 7–14 mm long, with black dots on the margins; stamens numerous, separate or nearly so; ovary 1-loculed with 3 parietal placentae; fruit a septicidal capsule, about 8 mm long, somewhat 3-lobed. $2n = 16$. In our area ssp. **scouleri** (Hook.) C.L. Hitchc. Moist slopes and ledges at alpine elevations. BC, Alta s to Calif, Mont, Wyo. Map 765 ○

H. majus (Gray) Britt.
Perennial, with short leafy rhizomes; stems 1–5 dm tall, erect, simple or branched above; leaves 1.0–3.5 cm long, lanceolate to narrowly oblong, 5–7-nerved; flowers 4–7 mm long, in cymes; stamens usually 15–35, separate; ovary 1-loculed; styles 3; fruit a capsule. $2n = 16$. Shores and marshes. BC to NS s to Wash, Colo, Kans, Ia, Ill, Pa, NJ. Map 765 ●

ELATINACEAE / Waterwort Family

Elatine L.

E. triandra Schk. / Waterwort
Small annual semi-aquatic plant, with slender stems rooting at the nodes, often forming tufts or mats on mud; leaves opposite, entire or nearly so, linear to obovate, glabrous, 3–8 mm long; flowers minute, regular, perfect, axillary; sepals, petals, stamens usually 3 each; pistil 1, the ovary superior, 2–3-loculed, with numerous ovules on axile placentae; fruit a septicidal capsule, many-seeded; seeds cylindric, straight or slightly curved, areolate with 9–25 pits in longitudinal rows. $2n = c.40$. Shallow water and mud-flats. Eurasia: swMack, BC to NB s to Calif, Tex, Okla, Mo, Va. Map 766 ○

CISTACEAE / Rockrose Family

Hudsonia L. / False Heather

H. tomentosa Nutt. / Sand Heather
Low shrub, tufted and heath-like, 6–12 cm high, forming mats or bushes, grey-tomentose and villose throughout; leaves alternate, small, ovate-lanceolate to subulate, closely appressed and imbricate, more or less concealed by pubescence;

flowers small but showy, yellow, crowded near the ends of the branches, perfect, hypogynous; sepals 5, the outer 2 much smaller than the inner ones; petals 5, yellow, conspicuous, fugacious; stamens 10–30; pistil with a slender style, the ovary with 3 parietal placentae; fruit a 1-loculed capsule, few-seeded, enclosed by the persistent calyx. $2n = 32$. Pine sand-hills and dunes. swMack, Alta to sLab, NB s to ND, Minn, Gt Lakes, WVa, NC. Map 766 •

VIOLACEAE / Violet Family

Viola L. / Violet

Low herbs, mostly perennial, some acaulescent; leaves alternate or basal, stipulate, simple; flowers borne singly on peduncles and usually nodding; two kinds of flowers, the spring ones showy, the later ones (cleistogamous) with reduced corolla or none, and self-fertilized within the closed calyx; the showy flowers perfect, hypogynous, irregular, sepals 5, auricled at the base, persistent; petals 5, the 2 lateral ones frequently bearded at base, the lowest one with a basal spur or sac and sometimes bearded at throat; stamens 5, their short filaments extending beyond the anthers, the 2 lowermost stamens with basal, nectar-bearing appendages that extend into the spur of the petal; pistil 1, the style projecting beyond the stamens, the ovary 1-loculed with 3 parietal placentae; fruit a 3-valved capsule, the seeds numerous, caruncled, ejected with some force.

1	Plants lacking stems, the leaves and peduncles arising from a caudex or stolons	2
	Plants with leafy stems; flowers axillary	8
2	Flowers yellow; leaves broadly rounded at apex, dark green, lying flat on the ground	**V. orbiculata**
	Flowers not yellow	3
3	Leaves deeply divided	**V. pedatifida**
	Leaves not deeply divided	4
4	Plants stoloniferous	5
	Plants not stoloniferous	6
5	Petals violet or blue; leaves 2.5–3.5 cm wide	**V. palustris**
	Petals white, the lower 3 with purple lines; leaves generally narrower	**V. macloskeyi**
6	Flowers white; leaves often more than 3 cm wide, often pilose below	**V. renifolia**
	Flowers violet-purple	7
7	Rhizome slender; leaves strigose above; spur more than 4 mm long	**V. selkirkii**
	Rhizome thick; leaves glabrous above; spur 4 mm long or less	**V. nephrophylla**
8	Stipules bract-like, more or less entire or fimbriate	9
	Stipules much larger, leaf-like, laciniate at base	12

9 Flowers yellow 10
Flowers not yellow 11

10 Basal leaves reniform, crenate-serrate; first peduncles from upper half of stem **V. glabella**
Basal leaves lanceolate to narrowly ovate, more or less entire; first peduncles from base of stem **V. nuttallii**

11 Stipules toothed; flowers deep violet to purple **V. adunca**
Stipules entire; flowers white to pale violet **V. canadensis**

12 Petals yellow, about as long as sepals **V. arvensis**
Petals blue, white, yellow, much longer than sepals **V. tricolor**

V. adunca J.E. Smith / Early Blue Violet
Stems tufted, from a slender rootstock, erect to spreading, 4–15 cm long, often quite short at early flowering; leaves long-petioled, ovate to round-ovate, usually cordate at base, crenulate, densely puberulent to nearly glabrous; stipules linear-lanceolate, spinulose-toothed or entire; corolla blue to violet, 8–15 mm long, the spur oblong, straight or hooked, 4–6 mm long; capsule 4–5 mm long; seeds dark brown. $2n = 20^*, 30^*, 40^*$. Includes *V. aduncoides* Löve & Löve. Grassland and open woods. Greenland: sAlas, Yuk, Mack to Huds Bay, nQue, Nfld s to Calif, Ida, Utah, Colo, SD, Minn, Gt Lakes, NEng. Map 767, $2n = 40$ ●, $2n = 30$ ★, $2n = 20$ ▲, Unknown ○

V. arvensis Murr. / European Field Pansy
Annual or short-lived perennial; stems angled, soft and brittle, and often branched from the base, 1–3 dm long, with reflexed hairs on the angles; lower leaves rounded-ovate, the upper oblong to narrowly elliptic and cuneate at base, the margins crenate; stipules pinnatisect at base; corolla about equalling the sepals, pale yellow throughout or with purplish tips; style globose-capitate at summit; capsule globose. $2n = 34$. Fields and gardens. Introduced. Occasional.

V. canadensis L. / Western Canada Violet
Plant 1–4 dm tall from an ascending rootstock, which also produces underground stolons, these ending in slender crowns; leaves mostly long-petioled, ovate to reniform, cordate, abruptly acuminate or acute, coarsely crenate, puberulent to nearly glabrous, somewhat rugulose; stipules scarious; flowers from axils of upper leaves; petals 8–12 mm long, white to violet, yellow at base; capsule ovoid; seeds brown, 1.5–2 mm long. $2n = 24^*$. *V. rugulosa* Greene of Ed. 1. In our area var. **rugulosa** (Greene) C.L. Hitchc. Moist woods and meadows. swMack, BC to NS s to Ariz, NM, Neb, Tenn, Ala, SC. Map 768

V. glabella Nutt.
Rootstock stout, usually horizontal, often branched; stems 5–30 cm tall, leafy near the tip; leaves reniform to ovate, short-pointed, cordate at base, crenate-serrate, 4–6 cm wide, nearly glabrous; stipules deltoid, serrulate; petals bright yellow, the lower purple-lined at base, the spur very short; capsule 1 cm long; seeds black. $2n = 24$. Moist woods. sAlas, BC, Alta s to Calif, Mont. Map 769

V. macloskeyi Lloyd
Similar to *V. renifolia*; differing in the character of the key, the more slender, horizontal rhizome, the generally smaller leaves, 1–2 (3) cm long, which are glabrous, and the usually bearded lateral petals. $2n = 24$*. In our area ssp. **pallens** (Banks *ex* DC.) Baker. Moist woods. swMack, BC, Alta s to Calif, Colo; Man to Nfld s to ND, Ohio, Ala, SC. Map 770 ○

V. nephrophylla Greene / Bog Violet
Plant acaulescent, with stout, ascending rootstock and no stolons; leaves ovate to reniform, cordate at base, obtuse to bluntly pointed, crenate, 3–5 cm wide, glabrous or nearly so; petals violet-purple, large, the lower 3 bearded at base, the spur 2–3 mm long; cleistogamous flowers on erect peduncles; capsule 8–10 mm long, green; seeds olive-brown. $2n = 54$. Moist meadows, stream banks, and moist open woods. wMack, BC to Nfld s to Calif, Ariz, NM, Okla, Ia, Minn, NY. Map 771

V. nuttallii Pursh / Yellow Prairie Violet
Plants caulescent, but sometimes appearing acaulescent, the leafy stems arising from a stout ascending rootstock; leaves 2–7 cm long, usually narrowly lanceolate and acute at apex, the base cuneate and tapering into a long margined petiole, the leaf-margin entire to repand-denticulate, the surface more or less pubescent, especially on the margins and veins; petals bright yellow, purple-veined at base and often reddish purple on outside, 8–12 mm long, the spur short; stigma bearded; cleistogamous flowers from the upper leaf-axils. $2n = 12$*, 24*, 36, 48*. In our area var. **nuttallii**, as described above; var. **linguaefolia** (Nutt.) Jepson, with broader leaf-blades, rounded at the apex and long hairy; and var. **vallicola** (A. Nels.) St. John, *V. vallicola* A. Nels. of Ed. 1, with truncate to subcordate leaf-bases and obtuse tips, the plants usually grey-puberulent. Prairie grassland and mountain slopes. sBC to swMan s to Calif, Ariz, NM, Kans. Map 772

V. orbiculata Geyer *ex* Hook. / Evergreen Violet
Plant acaulescent, the rootstock erect, stout, covered with old petiole bases, producing stolons; leaves orbicular or broadly ovate, deeply cordate, crenate-serrate, glabrous or nearly so, dark green, usually lying flat on the ground, persisting through the winter; petals yellow, purple-veined at base; stolons often flower-bearing; capsules ovoid to globose. $2n = 24$. Moist montane woods. BC, Alta, s to Ore, Ida, Mont. Map 773 ●

V. palustris L. / Marsh Violet
Plant acaulescent, with horizontal rootstock and slender stolons; leaves delicate, glabrous on both surfaces, ovate to reniform, cordate, crenulate, 2–5 cm long; petals pale lilac to nearly white and veined with purple, the lateral petals slightly bearded, the spur short; capsule ellipsoid, green; seeds dark brown. $2n = 48$*. Moist woods and bogs. swMack, BC to Lab, Nfld s to Calif, Utah, Colo, Man, NH, Me. Map 774

V. pedatifida G. Don / Crowfoot Violet
Acaulescent plant with a short vertical rootstock; leaves 3-parted, the divisions variously cleft and incised into linear lobes, pubescent to glabrate; petals bright violet, large, 1–2 cm long, the 3 lower petals bearded; capsule yellowish grey; seeds light brown. $2n = 54$. Dry gravelly hills and prairie grassland. sAlta to sOnt s to Ariz, NM, Okla, Mo, Ohio. Map 770 ●

V. renifolia A. Gray / Kidney-leaved Violet
Acaulescent plant, the ascending rootstock slender when young, becoming thick and scaly; leaves reniform to orbicular, cordate, the apex rounded or blunt-pointed, crenate-serrate, upper surface pubescent or glabrous; petals white, all beardless, the lower often with purplish veins, the spur short; peduncles of cleistogamous flowers arching or prostrate; capsule subglobose, usually purple; seeds brown. $2n = 24$. Moist woods and bogs. Alas, Yuk, wMack to James Bay, Nfld s to BC, Colo, Man, Minn, Wisc, Mich, Conn. Map 775

V. selkirkii Pursh / Great-spurred Violet
Delicate plant with a slender elongate rootstock; leaves ovate, cordate with a narrow sinus, glabrous below, strigose above, the margin crenate; petals pale violet, all beardless, the spur as long as the blade of the petal; capsule subglobose; seeds 1.5 mm long, buff. $2n = 24$. Moist woods. Circumpolar: Alas; Yuk; BC, Alta; Colo, NM; Sask, Man; Minn, Gt Lakes to Nfld. Map 773 ○

V. tricolor L. / Pansy; Johnny-jump-up
Resembling *V. arvensis* but stouter, the flowers variously marked with purple, yellow or white, the petals much longer than the sepals, the capsule ellipsoid. $2n = 26$. Waste ground. An escape from cultivation. Introduced. Occasional.

LOASACEAE / Loasa Family

Mentzelia L.

M. decapetala (Sims) Urban & Gilg / Sand Lily; Evening Star
Biennial herb; stem stout, erect, sparingly branched, 4–10 dm high, pale grey and very scabrous; whole plant armed with spine-like hairs; leaves alternate, mostly sessile, the lower oblong to oblanceolate, the upper lanceolate, sharply pinnatifid; flowers large, yellowish white, regular, perfect, epigynous, subtended by several pinnatifid bracts; sepals 5; petals 10, white or pale yellow, 4–8 cm long, about twice as long as the sepals; 5 of the petals often interpreted as petaloid stamens; anther-bearing stamens very numerous, the filaments filiform; pistil syncarpous, with an inferior, 1-loculed ovary and a 3-lobed style; ovules in 2 rows on each of 3 parietal placentae; fruit an oblong capsule, about 4 cm long; seeds numerous, very narrowly winged. $2n = 22$. Dry eroded hillsides and banks. sAlta to swMan s to Nev, Tex, Okla. Map 776

CACTACEAE / Cactus Family

Succulent spiny perennials, cushion-like or with jointed stems; leaves none or small and soon deciduous; spines arising from small hairy or bristly cushions (areolae); flowers large, perfect, regular, solitary, epigynous; sepals numerous, imbricate; petals many, in 2 or more series; stamens numerous, inserted on the hypanthium; pistil of several united carpels; ovary inferior, 1-loculed with numerous ovules on parietal placentae; style 1, elongate; stigmas as many as the placentae; fruit a fleshy or rather dry berry, many-seeded.

Plants cushion-like or globular; flowers red or purple **Coryphantha**
Plants branched and jointed, the stems more or less flattened; flowers yellow **Opuntia**

Coryphantha (Engelm.) Lemaire

C. vivipara (Nutt.) Britt. & Rose / Cushion or Ball Cactus
Plants globose, single or tufted, seldom over 5 cm high, the body covered with spine-tipped tubercles; leaves absent; flowers arising between the tubercles, 3–4 cm long, bright purplish-red; hypanthium tubular above the ovary; fruit pale green, becoming brown when ripe, edible, sweet. $2n = 22$. *Mamillaria vivipara* (Nutt.) Haw. of Ed. 1. Prairies, especially on stony ground. sAlta to swMan s to Mont, Colo, Kans, wMinn. Map 777

Opuntia Mill. / Cactus; Prickly Pear

Plants branched and jointed, the segments 3–10 cm long, more or less flattened; spines minutely barbed; leaves very small and deciduous; flowers lateral, yellow, fading into bronze tints; hypanthium bearing areolae, not extending above the ovary; stigma 5–7-lobed; fruit ovoid, about 3 cm long, spiny, edible and sweet; seeds flattened, with a bony aril.

Stem segments little flattened, 2–5 cm long, breaking apart readily **O. fragilis**
Stem segments much flattened, 5–12 cm long, not easily detached **O. polyacantha**

O. fragilis (Nutt.) Haw.
Stems prostrate, spreading, forming dense mats, the stem segments very fleshy, orbicular to obovate, somewhat flattened or nearly circular in cross-section, easily separating; areoles crowded, mostly armed with 3–7 spines; flowers pale yellow, about 5 cm broad. $2n = 66$. Prairie grassland, hillsides, and clay flats. BC to Ont s to Calif, Tex, Okla, Kans, Ia, Ill. Map 778

O. polyacantha Haw.
Stems prostrate, forming mats or clumps, the segments orbicular, 5–12 cm long, strongly flattened, about 1 cm thick; areoles of mature segments about 1 cm apart, almost all armed with 5–9 straight spines; flowers bright yellow, fading to bronze tints, 4–7 cm broad. $2n = 22, 44, 66$. Prairie grassland and denuded areas. sBC to sSask s to Ore, Ariz, Tex, Neb. Map 779

ELAEAGNACEAE / Oleaster Family

Shrubs or small trees with silvery or brown scurfy branches; leaves alternate or opposite, entire, often with silvery or rusty-brown scales; flowers perfect or unisexual, axillary, regular, apetalous, perigynous; hypanthium in the staminate flowers more or less cup-shaped, in pistillate or perfect flowers tubular and enclosing the ovary; sepals usually 4; stamens 4 or 8; ovary superior, 1-loculed with a single ovule; fruit drupe-like, consisting of the dry mature ovary and the fleshy or mealy hypanthium.

1 Leaves alternate; stamens 4 — 2
Leaves opposite; stamens 8 — **Shepherdia**

2 Flowers perfect or polygamous; sepals 4; tube much longer than the ovary — **Elaeagnus**
Flowers unisexual; sepals 2; tube shorter — **Hippophae**

Elaeagnus L. / Oleaster

Shrubs or small trees; winter-buds ovoid, small, with few outer scales; leaves alternate, short-petioled, dotted with silvery or brown scales; flowers in small lateral clusters on twigs of the current year; hypanthium tubular, in the pistillate and perfect flowers constricted above the ovary; stamens 4, on very short filaments, included; fruit mealy, with an ellipsoid stone.

Young twigs silvery; style enclosed at base by a tubular disc; fruit yellow — **E. angustifolia**
Young twigs brown; style not enclosed by a tubular disc; fruit silvery, dry — **E. commutata**

E. angustifolia L. / Russian Olive; Oleaster
Shrub or small tree to 7 m high, sometimes thorny; leaves oblong-lanceolate to linear-lanceolate, 4–8 cm long, an eighth to a third as wide, silvery; flowers short-stalked, in small clusters, about 1 cm long, yellow within, silvery outside, fragrant; fruit about 1 cm long, ellipsoid, yellow and coated with silvery scales. $2n = 28$. An escape from cultivation. Introduced.

E. commutata Bernh. *ex* Rydb. / Silver-berry; Wolf Willow
Upright shrub to 4 m, stoloniferous, unarmed, with rusty-brown twigs; leaves ovate to oblong or ovate-lanceolate, undulate, 2–10 cm long, a third to half as wide, silvery on both sides, sometimes with scattered brown scales beneath; flowers short-stalked, in clusters of 2 or 3, deflexed, 1.2–1.5 cm long, yellow within, silvery outside, sweet-scented; fruit about 1 cm long, broad-ellipsoid, silvery, dry and mealy, with an 8-striate stone. $2n = 28$. Grassland, shores, and valleys. Alas, Yuk, swMack to James Bay s to BC, Ida, Utah, Mont, ND, wMinn; Gaspé. Map 780

Hippophae L.

H. rhamnoides L. / Sea Buckthorn
Shrub or small tree, 2–5 m high, the branches grey and spiny, the young growth covered with stellate hairs or silvery scales; winter-buds golden brown; leaves alternate, short-petioled, linear to linear-lanceolate, 2–6 cm long, with silver-white scales on both surfaces; flowers appearing before the leaves, in short racemes, small, yellowish, dioecious; staminate flowers sessile, with reduced tube, 2 sepals and 4 stamens; pistillate flowers short-stalked, the hypanthium-tube with 2 small lobes, style filiform and stigma cylindric; fruit drupe-like, yellowish orange, attractive, persisting through winter. $2n = 12, 24$. An escape from cultivation. Introduced. Rare.

Shepherdia Nutt. / Buffalo-berry

Shrubs or small trees; winter-buds with one or two pairs of scales; leaves opposite, petioled, entire; flowers dioecious, small, in short clusters on twigs of the previous season, the pistillate ones usually few; hypanthium with an 8-lobed disc at its summit; sepals 4; stamens 8, alternating with the lobes of the disc; pistillate flower with the hypanthium orifice closed by the lobes of the disc; fruit drupe-like, ovoid.

Leaves silvery on both sides; plants thorny **S. argentea**
Leaves green and nearly glabrous above; plants unarmed **S. canadensis**

S. argentea Nutt. / Thorny Buffalo-berry
Shrub or sometimes tree-like, to 6 m, with spreading branches, often spiny, silvery when young; leaves oblong-lanceolate to oblong or oblanceolate, 2–5 cm long, obtuse, cuneate, silvery on both sides; flowers yellowish; sepals erect; fruit 4–6 mm long, scarlet or amber yellow, sour, edible. $2n = 22$. Open woods, shores, and valleys. sAlta to sMan s to seOre, eCalif, Mont, Wyo, Neb, wIa, Minn. Map 781

S. canadensis (L.) Nutt. / Canadian Buffalo-berry
Spreading shrub, 1–3 m tall, unarmed, with brown scurfy branches; leaves elliptic to narrowly ovate, 2–5 cm long, obtuse, green and nearly glabrous above, silvery to brown-scurfy beneath; flowers yellowish, about 4 mm wide; sepals spreading; fruit 4–6 mm long, red or yellow, scarcely palatable. $2n = 22$. Open woods and shores. Alas, Yuk, Mack to Huds Bay, Nfld s to Ore, NM, SD, ND, Minn, Ohio, NY. Map 782

LYTHRACEAE / Loosestrife Family

Lythrum L. / Loosestrife

L. salicaria L. / Purple Loosestrife
Perennial herb, 5–15 dm tall, more or less downy-pubescent; leaves opposite or whorled in 3's, lanceolate, sessile, rounded to cordate at base, 3–10 cm long; flowers showy, clustered in the axils of the upper leaves and thus forming interrupted spikes; flowers perfect, regular, perigynous; hypanthium tubular, nerved, bearing petals and stamens on its inner surface; sepals 4–7, alternating with subulate appendages in the sinuses; petals magenta-purple, 4–7; stamens twice as many as the petals; pistil 1, with a single style and a 2-loculed ovary; placentae axile; the flowers are trimorphous, those on different plants differing in the relative lengths of the stamens and style; fruit a septicidal, many-seeded, 2-loculed capsule, enclosed by the persistent hypanthium. $2n = 30, 50, 60$. Wet meadows and along streams. Introduced. Rare.

ONAGRACEAE / Evening Primrose Family

Perennial or annual herbs with simple, alternate or opposite, leaves; flowers perfect, epigynous, the parts mostly in 4's or 2's; stamens usually 8; hypanthium adnate to the ovary and often prolonged beyond; sepals, petals and stamens inserted at the summit of the hypanthium; pollen grains often connected by cobweb-like threads; ovary inferior, 4- or 2-loculed, with axile placentae; style 1; stigma capitate, discoid or 4-lobed; fruit a many-seeded capsule or 1-seeded and nut-like; flowers of most species open in the late afternoon or early evening.

1 Flower parts in 2's; fruit obovoid, bristly, indehiscent; leaves opposite **Circaea**
Flower parts in 4's 2

2 Fruit 1-seeded, nut-like, indehiscent; flowers somewhat irregular, white to scarlet **Gaura**
Fruit a many-seeded capsule, opening by valves 3

3 Seeds with a tuft of hairs (coma) at one end **Epilobium**
Seeds lacking a coma 4

4 Ovary 2-loculed; hypanthium not prolonged beyond the ovary; small annual with slender stems, linear leaves and white petals **Gayophytum**
Ovary 4-loculed; hypanthium prolonged beyond the ovary 5

5 Anthers attached near the base, erect; sepals erect; petals purplish; low, branching annual with lanceolate, denticulate leaves **Boisduvalia**
Anthers attached near the middle, usually versatile; sepals commonly reflexed; petals yellow, white or pink **Oenothera**

Boisduvalia Spach

B. glabella (Nutt.) Walp.
Low annual, the stems 1–3 dm long, often branching near the base, decumbent and spreading, glabrous to pubescent; leaves sessile, mostly lanceolate, denticulate or entire, 1–1.5 cm long, pubescent or glabrate; flowers solitary in leaf-axils, inconspicuous; petals pinkish or purplish red, 2–4 mm long; pollen in tetrads; capsule 6–8 mm long, slender, 4-ribbed, pointed at apex, puberulent; seeds about 1 mm long. $2n = 30^*$. Prairie mud-flats. BC; Alta, Sask s to seWash, Ore, Calif, Nev, Utah, Mont, SD. Map 783 ○

Circaea L. / Enchanter's Nightshade

C. alpina L.
Delicate, perennial, herb with tuberously thickened rootstock; stems 1–5 dm tall, simple or branched; leaves opposite, 2–5 cm long, ovate, base rounded or cordate, entire to coarsely dentate; flowers 8–12 in terminal racemes; sepals 2, reflexed, 1–2 mm long; petals 2, white, 1–2 mm long; stamens 2; fruit about 2 mm long, obovoid, covered with soft, hooked hairs; seeds 1 (2). $2n = 22^*$. In our area ssp. **alpina** (widespread) with the leaves cordate and dentate, the upper stem lacking strigose pubescence, and, in the Waterton Lakes area, ssp. **pacifica** (Asch. & Magnus) Raven (*C. pacifica* Asch. & Magnus of Ed. 1) with the leaves rounded at base, entire, and the upper stem strigose pubescent. Moist woods. Circumpolar: Alas, swMack to James Bay, Nfld s to Calif, Colo, Ga. Map 783 ●

Epilobium L. / Willow-herb

Annual or perennial herbs; leaves alternate or opposite, sessile or short-petioled, ovate to linear, entire or toothed; flowers in racemes or solitary in the leaf-axils; petals pink, purple or white, rarely yellow, usually notched; stamens 8, the alternate ones shorter; style short; capsule elongate, slender, 4-valved; seeds with a tuft of silky hairs (coma) at upper end.

1 Leaves alternate; flowers zygomorphic; style deflexed, at least before anthers dehisce; hypanthium absent — 2
  Leaves opposite or verticillate, at least below; flowers actinomorphic; style not deflexed; hypanthium present — 3

2 Inflorescence short, leafy; flowers few, 1–7 (15); style shorter than stamens; seeds 1.3–2.4 mm long — **E. latifolium**
  Inflorescence elongate, not leafy; flowers many; style longer than stamens at maturity; seeds 1.0–1.3 mm long — **E. angustifolium**

3 Petals yellow or cream; stigma distinctly 4-lobed — **E. luteum**
  Petals white, pink or rose-purple; stigma entire, cylindric, clavate or capitate — 4

4 Epidermis exfoliating at base of stem; seeds broadly obovoid, 0.8–1.2 mm wide — **E. paniculatum**
  Epidermis not exfoliating; seeds narrowly obovoid, 0.3–0.8 mm wide — 5

5 Plants with basal thread-like, remotely leafed stolons, terminating in compact, fleshy turions; inflorescence white-canescent, occasionally with a few glandular hairs; cauline leaves linear, entire, often revolute 6

Plants lacking stolons; inflorescence predominantly glandular-pubescent or subglabrous, rarely strigillose; cauline leaves either linear or entire, not both, never revolute 7

6 Leaves linear to lanceolate or oblong, glabrous to sparsely strigillose above; inflorescence often nodding in bud, eglandular **E. palustre**

Leaves linear, rarely wider, densely strigillose above; inflorescence erect or nearly so, often with a mixture of strigillose and glandular hairs **E. leptophyllum**

7 Stem lacking hairy, elevated lines, decurrent from the leaf-bases 8

Stem with hairy, elevated lines, decurrent from the leaf-bases 9

8 Flowers pink to rose-purple; stem glabrous, often glaucous; seeds 0.8–1.0 mm long, distinctly papillose **E. glaberrimum**

Flowers white; stem densely strigillose; seeds 1.7–2.2 mm long, reticulate **E. mirabile**

9 Plants clumped or caespitose, forming short, leafy, epigaeous stolons; stem-base decumbent to ascending 10

Plants not clumped, or if so loosely, forming sessile, leafy rosettes or fleshy turions; stem-base erect 13

10 Stem 3–22 cm high; cauline leaves entire; capsules 1.7–4.0 cm long 11

Stem 10–50 cm high; cauline leaves denticulate, rarely entire; capsules 4–10 cm long 12

11 Inflorescence nodding in bud, generally glabrous; leaves oblong-elliptic to narrow-ovate; seeds 0.7–1.4 mm long, usually reticulate, the coma persistent; roots gracile, loose, short **E. anagallidifolium**

Inflorescence erect, strigillose and usually glandular-pubescent; leaves ovate to narrow-elliptic; seeds 1.3–2.0 mm long, usually papillose, the coma readily detached; roots wiry, tangled, extended **E. clavatum**

12 Petals white; inflorescence mostly nodding in bud; pedicels in fruit 2.0–4.5 cm long; seeds 1.1–1.6 mm long, reticulate **E. lactiflorum**

Petals pink to rose-purple; inflorescence mostly erect in bud; pedicels in fruit 0.5–2.5 cm long; seeds 0.9–1.2 mm long, distinctly papillose **E. hornemannii**

13 Turions present; gemmae often present in the upper leaf-axils; inflorescence strigillose, eglandular; coma reddish, persistent **E. leptocarpum**

Turions present or absent; gemmae absent; inflorescence predominantly glandular mixed with strigillose hairs; coma white or dingy, readily detached 14

14 Plants 0.3–19.0 dm high, forming rosettes or large, hypogaeous turions; seeds 0.8–1.6 (1.9) mm long, longitudinally striate with hyaline crests or ridges but lacking distinct papillae **E. ciliatum**

Plants 0.2–6.0 dm high, lacking rosettes, only compact hypogaeous turions present; seeds 1.1–1.6 (1.8) mm long, distinctly papillose, the papillae often in longitudinal rows 15

15 Leaves sessile, clasping, mostly narrow-ovate, denticulate; turions fleshy, elongate; capsules subsessile, appressed; seed-collar conspicuous **E. saximontanum**

Leaves petiolate or subsessile, not clasping, lanceolate or narrower, entire or denticulate; turions compact, round; capsules on pedicels 0.8–3.8 cm long; seed-collar inconspicuous **E. halleanum**

E. anagallidifolium Lam.
Perennial, mat-forming, soboliferous; stem 0.3–2.0 dm high, ascending to erect, sparsely strigillose in decurrent lines from the petiole, or the upper stem strigillose with an admixture of glandular hairs; leaves mostly with petioles 1–6 mm long, blades 0.8–2.5 cm long, 0.2–0.9 cm wide, spatulate to oblong basally, elliptic to lanceolate above, entire to denticulate, glabrous or the midrib and margin of the upper leaves sparsely strigillose; inflorescence nodding in bud; sepals 1.5–5.0 mm long, glabrous or sparsely glandular; petals pink to rose-purple, rarely white, 2.5–6.5 (9) mm long; anthers 0.3–0.6 mm long; ovary 0.6–2.0 cm long, glabrous; style 1.2–2.5 mm long; stigma broadly clavate, rarely subcapitate; capsule 1.7–3.6 cm long, about 1 mm wide; seeds narrowly obovoid, 0.7–1.4 mm long, mostly reticulate, coma 1.5–4.0 mm long, dull white, persistent. $2n = 36$*. *E. alpinum* L. of Ed. 1. Moist or boggy alpine slopes. Circumpolar: Alas, Yuk to Baffin Is to Nfld s to Calif, Utah, Colo, sMack, Huds Bay, cQue; Me, NH. Map 784

E. angustifolium L. / Fireweed; Great Willow-herb
Perennial, often forming extensive colonies; stem 1.3–30.0 dm high, glabrous or densely strigillose above; leaves 1.5–20.0 cm long, mostly 0.5–3.5 cm wide, lanceolate, entire or denticulate, generally glabrous, short-petiolate; inflorescence an elongate, lax raceme, flowers many, nodding in bud; sepals 7–16 mm long, narrowly lanceolate; petals deep pink to magenta, rarely white, 10–20 mm long; anthers 1.5–2.2 mm long; ovary 0.8–1.8 cm long, densely white-canescent; style 10–20 mm long; stigma deeply 4-lobed; capsule 4–10 cm long, canescent; seeds fusiform, 1.0–1.3 mm long, irregularly reticulate, coma 9–14 mm long, white, not markedly deciduous. $2n = 36, 72, 108$. In our area ssp. **angustifolium** with leaves 2.5–15.0 (17.0) cm long, (3) 10–30 mm wide, the midrib of the lower leaf-surface glabrous, and ssp. **circumvagum** Mosquin with leaves 5–18 (22) cm long, (0.5) 2–5 cm wide, the midrib of the lower leaf-surface glabrous or densely strigillose. Open woods, clearings, gravel bars, waste ground, roadsides. Circumpolar: Alas, Yuk, Mack to sBaffin Is to Nfld s to Calif, Ariz, NM, SD, Ohio, NC. Map 785

E. ciliatum Raf.
Perennial; stem mostly 0.5–12.0 dm high, simple or branched, arising from turions, rosettes or the base of last year's plant, the lower stem with decurrent lines of strigillose hairs, becoming densely pubescent above with short non-glandular and long glandular hairs, occasionally villous or sericeous throughout; leaves mostly 3–12 cm long, 0.5–4.5 cm wide, the 4–8 lateral veins conspicuous, the upper narrowly lanceolate to ovate, denticulate, petioles 1–4 mm long, the lower often obovate to broadly elliptic, with longer petioles; inflorescence erect, rarely nodding; sepals 1.5–5.0 mm long, often strigillose and glandular; petals

white, pink or rose-purple, 1.5–12.0 mm long; anthers 0.4–1.8 mm long; ovary 1.0–4.0 cm long, densely glandular; style 1.1–8.2 mm long; stigma clavate to subcapitate; capsule generally 4–10 cm long, pubescent; seeds mostly 0.8–1.6 mm long, distinctly ridged, rarely reticulate or papillose, coma 2–8 mm long, white, readily detaching. $2n = 36$. *E. glandulosum* Lehm. of Ed. 1. In our area ssp. **ciliatum**, usually with epigaeous, leafy rosettes, the cauline leaves narrowly lanceolate to narrowly ovate, a non-leafy branched inflorescence, white to pink petals 1.5–5.0 (8.0) mm long, the seeds (0.6) 0.8–1.2 (1.5) mm long, and ssp. **glandulosum** (Lehm.) Hoch & Raven, usually with hypogaeous, fleshy, imbricate turions, broader cauline leaves, a mostly leafy, unbranched inflorescence, petals pink to rose-purple, rarely white, 3.5–10.0 (12.0) mm long, and seeds 1.1–1.6 (1.9) mm long. Moist, often disturbed, habitats. Japan, Korea, Kamchatka: Alas, Yuk, Mack, Huds Bay, Nfld s to Calif, Ariz, NM, Tex, Okla, Ia, Ill, Ind, Ohio, Tenn, NC. Map 786

E. clavatum Trel.
Perennial, forming large clumps and spreading by wiry, scaly soboles; stem 0.5–2.2 dm high, ascending, basally strigillose in decurrent lines from the petioles, strigillose above with an admixture of glandular hairs; leaves sessile or with petioles 2–3 mm long, blades 1.2–2.8 cm long, 0.6–1.6 cm wide, ovate to elliptic, entire to denticulate, glabrous or the midrib and margin sparsely hairy; inflorescence erect; sepals 2.5–4.2 mm long, glabrous or sparsely glandular; petals pink to rose-purple, 3.6–6.0 mm long; anthers 0.4–0.9 mm long; ovary 0.8–2.0 cm long; style 1.4–3.2 mm long; stigma narrowly clavate to subcapitate; capsule 2–4 cm long, 2–3 mm wide, glabrous or sparsely pubescent; seeds narrowly obovoid, 1.3–2.0 mm long, reticulate or papillose, coma 5–15 mm long, white, readily detaching. $2n = 36$. Moist alpine slopes. Alas, Yuk, swMack s to Calif, nNev, Utah, Colo. Map 787

E. glaberrimum Barbey
Perennial, densely clumped, spreading by wiry, scaly, branching soboles; stem 0.8–3.5 dm high, glaucous, glabrous, rarely with strigillose hairs in decurrent lines from the upper petioles and/or sparsely strigillose and glandular in the inflorescence; leaves mostly subsessile and clasping, 1.0–3.4 cm long, 0.7–1.8 cm wide, lanceolate to elliptic, entire to denticulate, glabrous, glaucous; inflorescence erect; sepals 1.6–4.0 mm long, glabrous to strigillose with some glandular hairs; petals pink to rose-purple, 2.5–7.0 mm long; anthers 0.5–1.0 mm long; ovary 1.0–3.4 cm long, sparsely strigillose or with an admixture of glandular hairs; style 2.2–3.6 mm long; stigma narrowly to broadly clavate; capsule 2.0–5.5 cm long, glabrous; seeds 0.8–1.3 mm long, distinctly papillose in longitudinal rows, coma 4–8 mm long, white, readily detaching. $2n = 36$. In our area ssp. **fastigatum** (Nutt.) Hoch & Raven. *E. platyphyllum* Rydb. of Ed. 1. Rocky alpine slopes. BC, Alta s to Calif, Ida, Utah, Wyo. Map 788 ○

E. halleanum Hausskn.
Perennial producing fleshy subterranean turions; stem 0.2–6.0 dm high, mostly simple, with lines of strigillose hairs decurrent from the petioles, generally with

strigillose and glandular hairs above; leaves sessile, 0.5–4.7 cm long, 0.3–1.4 cm wide, lanceolate to narrowly elliptic or ovate, entire or denticulate, mostly glabrous with strigillose margins; inflorescence nodding in bud; sepals 1.2–2.8 mm long, glabrous or sparsely glandular-hairy, green; petals white, often fading pink, 1.6–5.5 mm long; anthers 0.3–0.9 mm long; ovary 1.0–1.4 cm long, mostly densely glandular-pubescent and strigillose, rarely glabrous; style 0.8–5.0 mm long; stigma clavate; capsule 2.4–6.0 cm long, slender, glabrous to pubescent; seeds narrowly obovoid, 1.1–1.6 mm long, papillose, coma 3–6 mm long, white, readily detaching. $2n = 36$. Moist ground. BC, Alta, Sask s to sCalif, Ariz, Colo, SD. Map 789 ○

E. hornemannii Reichenb.
Perennial, often densely clumped, with short, leafy, epigaeous soboles; stem 1.0–4.5 dm high, with lines of dense strigillose hairs decurrent from the petioles, glandular-strigillose in the inflorescence; leaves with petioles 5–8 mm long below, sessile above, blades 1.5–5.5 cm long, 0.7–2.9 cm wide, broadly elliptic to narrowly ovate below, lanceolate to narrowly ovate above, denticulate, essentially glabrous, the margins occasionally strigillose; inflorescence erect; sepals 2.0–4.5 mm long, sparsely glandular, occasionally with strigillose hairs; petals rose to pink, rarely white, 3–9 mm long; anthers 0.4–1.2 mm long; ovary 1.5–2.5 cm long, glandular-pubescent; style 1.7–4.2 (8.3) mm long; stigma clavate or cylindrical; capsule 4.0–6.5 cm long, slender, green with scattered strigillose and glandular hairs; seeds narrowly obovoid, 0.9–1.2 mm long, distinctly papillose, coma 6–11 mm long, white, readily detaching. $2n = 36$. In our area ssp. **hornemannii**. Moist slopes, stream banks to alpine elevations. Circumpolar: Alas, Yuk, swMack s to Calif, Ariz, NM; eHuds Bay, nQue, Nfld s to Me, NH, Gaspé. Map 789 ●

E. lactiflorum Hausskn.
Perennial, often clumped, producing leafy epigaeous soboles; stem 1.5–5.0 dm tall, with lines of dense strigillose hairs decurrent from the petioles, strigillose and densely glandular-pubescent in the inflorescence; lower leaves with a winged petiole 3–12 mm long, the upper sessile, blades 2.0–5.5 cm long, 0.8–2.4 cm wide, elliptic or narrowly ovate to narrowly lanceolate, denticulate, glabrous, the margins strigillose; inflorescence nodding in bud; sepals 2.0–3.6 mm long, sparsely glandular-pubescent; petals white, rarely pink-tinged or with red veins; anthers 0.4–0.9 mm long; ovary 2–4 cm long, densely glandular-pubescent; style 1.4–4.6 cm long; stigma clavate, rarely subcapitate; capsule 5–10 cm long, sparsely glandular-hairy; seeds 1.1–1.6 mm long, attenuate to narrowly obovoid, coma 7–14 mm long, white, readily detaching. $2n = 36$. Montane stream banks and moist slopes to alpine elevations. More or less circumpolar: Alas s to Calif, Utah, Colo; eHuds Bay to Nfld s to NS, Me, NH, Gaspé. Map 790

E. latifolium L.
Perennial, often forming extensive colonies; stem 0.4–7.0 dm tall, often glaucous; leaves 1.0–5.5 (10.0) cm long, 0.5–2.5 cm wide, broadly lanceolate to elliptic, entire to sparsely denticulate, glabrous or the upper leaves strigillose, sessile or

short-petiolate; inflorescence a short raceme, 1–7 (15)-flowered, leafy; sepals 10–24 mm long, lanceolate, petals pink to rose-purple, rarely white, 1.3–3.2 cm long; anthers 1.4–3.2 mm long; ovary 1.2–2.8 cm long, densely canescent; style 3.5–10.0 mm long; stigma deeply 4-lobed; capsule 3.0–10.5 cm long, glabrous or strigillose; seeds attenuate to narrowly obovoid, 1.3–2.4 mm long, irregularly reticulate, coma 9–17 mm long, white, not markedly deciduous. $2n = 36, 72$. In our area ssp. **latifolium**. Gravel-bars, stream banks, scree slopes to alpine elevations. Asia, Greenland, Iceland: Alas, Yuk to Ellesmere Is to Nfld s to Calif, Colo, sMack, Huds Bay, nQue, Gaspé. Map 791

E. leptocarpum Hausskn.
Delicate perennial, often clumped, with small basal or subterranean turions and sometimes with gemmae in the upper leaf-axils; stem 0.8–3.0 dm high, ascending to erect, glabrous except for lines of strigillose hairs decurrent from the petioles, occasionally strigillose above; lower leaves with petioles 3–5 mm long, the upper sessile, blades 0.8–4.0 cm long, 0.4–1.3 cm wide, narrowly lanceolate to elliptic, entire below, denticulate above, essentially glabrous, the margins sparsely hairy; inflorescence nodding to suberect; sepals 2.9–4.0 mm long, narrowly lanceolate, strigillose; petals white, fading pink, 4.0–6.5 mm long; anthers 0.4–0.8 mm long; ovary 1.2–1.8 cm long, densely strigillose; style 3.2–3.8 mm long; stigma broadly clavate; capsule 2.5–5.5 cm long, 1–2 mm wide, sparsely strigillose; seeds 0.8–1.2 mm long, papillose, coma 3.5–6.5 mm long, tawny, persistent. $2n = 36$. Moist, open stony slopes. Alas s to nOre, cIda, Mont. Map 792

E. leptophyllum Raf.
Perennial, often robust, producing thread-like nearly leafless epigaeous stolons terminating in compact turions; stem erect, 1.5–9.5 dm high, simple or branched, terete, densely strigillose, sometimes glandular above, lacking lines of hairs decurrent from the petioles; leaves 2.0–7.5 cm long, mostly 0.2–0.7 cm wide, linear to narrowly elliptic, densely strigillose above and below, entire, sometimes revolute; inflorescence erect, leafy; sepals 2.5–4.5 mm long, strigillose; petals white to pink, 3.5–7.0 mm long; anthers 0.6–0.9 mm long; ovary 1.2–1.8 cm long, densely white canescent; style 2.0–3.8 mm long; stigma narrowly clavate; capsule 3.5–8.0 cm long, slender, canescent; seeds attenuate, narrowly obovoid, 1.5–2.2 mm long, densely papillose, coma 6–8 mm long, dingy white, persistent. $2n = 36$. Marshy ground, bogs. BC, swMack to James Bay, Nfld s to Calif, Colo, Kans, Mo, Ill, Ind, Ohio, WVa, NC. Map 788 •

E. luteum Pursh
Robust perennial with scaly, subterranean soboles, more rarely forming condensed basal turions; stems several, 1.5–7.5 dm high, often branched above; leaves 2.5–7.8 cm long, 1.2–3.5 cm wide, ovate, occasionally elliptic, finely denticulate with pellucid teeth, essentially glabrous, the margin and midrib sparsely strigillose; inflorescence nodding in bud, leafy, densely glandular-strigillose; sepals 10–12 mm long, lanceolate, densely glandular; petals yellow or cream, 12–22 mm long; anthers 2.2–3.0 mm long; ovary 2.0–3.5 cm long, densely glandular; style 15–22 mm long; stigma broadly 4-lobed, exserted; capsule 3.5–7.5

cm long, erect; seeds 1.0–1.2 mm long, attenuate, reticulate, coma 6.5–8.0 mm long, reddish, persistent. $2n = 36$. Forms a very distinctive hybrid, known as *E.* × *treleasianum* H. Lév., with several species, especially *E. ciliatum* and *E. hornemannii*. However, none of these plants, which have large purple flowers, has so far been recorded from our area. Moist montane woods and stream banks. Alas s to nCalif. Map 793 ○

E. mirabile Trel.
Perennial, often loosely clumped with each stem arising from a separate, compact, fleshy turion; stem 0.7–3.0 dm high, densely strigillose, lacking decurrent lines of hairs from the petioles; leaves sessile, coriaceous, 1.5–3.0 cm long, 0.7–1.2 cm wide, ovate and obtuse below, narrowly ovate and acute above, denticulate, sparsely strigillose above and below, densely pubescent on midrib and margins; inflorescence erect; sepals 2.0–3.2 mm long, sparsely glandular-pubescent, often purplish red; petals white, often red-tinged at tip, 3.8–5.0 mm long; anthers 0.4–0.6 mm long; ovary 1.0–1.8 cm long, strigillose and glandular-pubescent; style 2.0–2.3 mm long; stigma subglobose; capsule 3.0–4.5 cm long, 2–3 mm wide, glabrous or sparsely glandular; seeds narrowly obovoid, 1.7–2.2 mm long, coma 10–15 mm long, white, very full, readily detaching. $2n = 36$. Subalpine scree slopes. BC; Alta, Mont; Wash. Map 794 ○

E. palustre L.
Perennial, producing filiform epigaeous stolons terminating in small dark turions; stem 0.5–8.0 dm high, simple or branched, with lines of strigillose hairs decurrent from the petioles below, densely strigillose above; leaves sessile, 1.5–7.0 cm long, 0.2–1.9 cm wide, linear to lanceolate, entire or obscurely denticulate, glabrous, the margins and veins strigillose; inflorescence nodding in bud, densely strigillose; sepals 1.4–4.5 mm long, strigillose; petals white, rarely pink, 2–9 mm long; anthers 0.4–1.0 mm long; ovary 1.4–3.0 cm long, densely strigillose; style 1.0–4.5 mm long; stigma cylindrical to clavate; capsule 3–9 cm long, slender, strigillose; seeds elliptic-attenuate to narrow fusiform, 1.4–2.2 mm long, finely papillose, coma 5–7 mm long, white, persistent. $2n = 36$. Marshy ground, bogs. Circumpolar: Alas, Yuk, Mack to nHuds Bay, nQue, Nfld s to Calif, Nev, Colo, SD, Alta, Sask, Minn, Mich, Pa. Map 795

E. paniculatum Nutt. *ex* T. & G.
Annual, with a simple taproot; stem 2–20 dm high, generally branched above, glabrous below, with an exfoliating, shreddy epidermis, often glandular-pubescent above; leaves 1.5–5.0 cm long, 0.2–0.8 cm wide, linear to narrowly elliptic, often folded along the midrib, serrulate, mostly glabrous, sometimes glandular-pubescent in the inflorescence, short-petiolate, often with leafy fascicles in the axils; inflorescence paniculate or the flowers in lax racemes on filiform branches; sepals 2–8 mm long; petals white to rose-purple, 1–15 (20) mm long; anthers 0.5–4.0 mm long; ovary 0.4–1.6 cm long, glabrous or glandular-pubescent; style 1.3–2.5 mm long; stigma clavate to 4-lobed; capsule 1.5–3.2 cm long, 2–3 mm wide, fusiform, beaked, glabrous or glandular-pubescent; seeds broadly obovoid, 1.5–2.7 mm long, brown or grey, often flecked with darker spots, papillose,

coma 7–8 mm long, white readily detaching. $2n = 24$. Dry woods, prairies, shores, roadsides. cBC to Man s to Calif, NM, Neb, SD, Minn; sOnt, Que. Map 793 ●

E. saximontanum Hausskn.
Perennial, often with fleshy underground turions; stem 0.4–5.5 dm high, glandular-strigillose above, occasionally glabrous with lines of strigillose pubescence decurrent from the petioles; leaves generally sessile, clasping, occasionally with a petiole 1–3 mm long, blade 1.0–5.5 (6.6) cm long, 0.4–2.4 cm wide, lanceolate or narrowly elliptic to ovate, often obovate basally, entire to denticulate, glabrous with strigillose margins, the veins conspicuous; inflorescence erect, occasionally nodding in bud; sepals 1.2–3.5 mm long, strigillose and sometimes glandular; petals white, rarely pink to rose-purple, 2.2–5.0 (7.0) mm long; anthers 0.30–0.78 mm long; ovary 0.9–3.0 cm long, densely pubescent with strigillose and glandular hairs; style 1.6–2.8 mm long; stigma clavate, rarely subcapitate; capsule 2.0–5.5 (7.0) cm long, 1.5–3.0 mm wide usually sessile, densely strigillose and glandular; seeds narrowly obovoid, 0.9–1.8 mm long, rugose to papillose, coma 3–7 mm long, white usually readily detaching. $2n = 36$. Moist montane and subalpine meadows and stream banks. Alta s to eOre, Calif, Ariz, NM; Huds Bay, James Bay, nQue, Nfld; Gaspé. Map 794 ●

Gaura L.

G. coccinea Pursh / Scarlet Butterfly-weed
Perennial herb, 1–4 dm tall, the stems several to many, simple or branched above, decumbent to erect, usually strigose and puberulent; leaves numerous, alternate, sessile, 1–3 cm long, oblong-lanceolate to linear, entire or sinuate-dentate, puberulent; inflorescence a spike with persistent bracts; flowers somewhat irregular, whitish at emergence, becoming scarlet or brick-red; hypanthium extended beyond the ovary as a short slender tube; sepals reflexed at anthesis; petals 4–7 mm long, unequal and ephemeral; stamens commonly 8, declined, often with a scale at the base of each filament; fruit nut-like, obovoid, 4-angled, canescent, 1–4-seeded. $2n = 14, 28, 42$. Prairie grassland and roadsides. seBC to sMan s to Calif, Tex, Okla, Mo. Map 796 ●

Gayophytum Juss.

G. racemosum T. & G.
Low slender annual, 1–3 dm high, little branched or repeatedly branched from the base, glabrous or sparsely hairy; leaves linear or linear-oblanceolate, entire, 1–3 cm long, usually glabrous; flowers very small in the upper axils; sepals 4, reflexed, deciduous; petals 4, white to reddish, about 1 mm long; stamens 8, those opposite the petals shorter and sterile; capsule 8–14 mm long, slender, glabrous, erect; seeds without coma. $2n = 28$. Disturbed ground on open montane slopes. Alta s to Wash, Calif, Nev, Utah, Colo. Map 796 ○

Oenothera L. / Evening Primrose

Annual, biennial or perennial herbs, caulescent or acaulescent, rarely woody at base; leaves alternate or basal, mostly narrow; sepals 4, reflexed in anthesis; petals 4, yellow, pink or white, often aging to orange or red, usually large and conspicuous but soon withering; stamens 8, equal or those opposite the petals shorter, the anthers mostly versatile; ovary 4-loculed; capsule 4-valved, sometimes tardily dehiscent; seeds numerous, not comose.

1	Plants with leafy stems	2
	Plants lacking leafy stems, leaves forming a rosette	5
2	Plants suffrutescent, usually with several slender stems; flowers yellow; hypanthium funnelform	**O. serrulata**
	Plants herbaceous	3
3	Stems 3–10 cm tall, slender, branched; flowers minute	**O. andina**
	Stems taller, stout; flowers larger	4
4	Flowers yellow, the buds erect; stems green or reddish	**O. biennis**
	Flowers white or pinkish, buds nodding; stems white, shreddy	**O. nuttallii**
5	Stigma with 4 linear lobes; petals more than 1 cm long	6
	Stigma capitate; petals usually less than 1 cm long	**O. breviflora**
6	Petals more than 2 cm long, white, aging to pink; capsules not winged	**O. caespitosa**
	Petals less than 2 cm long; yellow, aging to pink; capsules distinctly wing-margined	**O. flava**

O. andina Nutt.
Low annual, 3–10 cm high; stems very slender, with spreading branches from near the base, finely canescent; leaves alternate, linear to narrowly oblanceolate, entire, 1–3 cm long, the petioles short; flowers axillary, in a congested cluster; petals yellow, 1–1.5 mm long; stigma capitate; capsule fusiform, 4-angled, canescent, about 6 mm long. $2n$ = 28, 42. *Camissonia andina* (Nutt.) Raven, *Sphaerostigma andinum* (Nutt.) Walp. Dry prairie slopes. BC to swSask s to Calif, Utah, Wyo. Map 797 ○

O. biennis L. / Yellow Evening Primrose
Biennial with a stout taproot; forming a leafy rosette in the first year and in the second year, an erect stiff stem, 5–15 dm tall, often branching, greyish or reddish, pubescent; leaves ascending or spreading, narrowly lanceolate to oblong, entire to denticulate, densely or sparsely hairy; flowers large, bright yellow, subtended by large bracts or reduced leaves, in terminal spikes, the buds erect and opening in the evening; hypanthium-tube about 3 cm long; sepals at first connate except for very short free tips, later separating and becoming reflexed; petals 12–25 mm long, yellow; capsule erect, 2–3 cm long, pubescent. $2n$ = 14*, 28. A polymorphic species variously treated taxonomically. Dry open areas. BC to Nfld s to Wash, Mont, Colo, Tex, Okla, La, Fla. Map 798

O. breviflora T. & G.
Acaulescent perennial with a taproot and a rosette of leaves close to the ground; leaves 5–10 cm long, deeply pinnatifid, puberulent; petals yellow, 5–8 mm long; capsule about 1.5 cm long, narrowly ovoid, tapering upwards, more or less 4-angled, roughened with many rounded projections, pubescent. $2n = 14$. *Camissonia breviflora* (T. & G.) Raven, *Taraxia breviflora* (T. & G.) Nutt. Clay flats. sBC to swSask s to Calif, Mont, Wyo. Map 799 ●

O. caespitosa Nutt. / Butte Primrose; Rock Rose
Caespitose perennial with a thick woody root, acaulescent or nearly so; leaves clustered, 1–2 dm long, oblanceolate, glabrous or pubescent on the margins, usually sinuate-dentate; flowers sessile in the crown of the plant, strongly sweet-scented; hypanthium-tube 3–10 cm long; petals 2–4 cm long, white on opening, soon turning pinkish; capsule about 2.5 cm long, woody, with 4 wavy-crested angles, sometimes tubercled. $2n = 14$. *Pachylophis caespitosa* (Nutt.) Raim. Clay flats, roadsides. sAlta, sSask s to eWash, Calif, Nev. Colo, wNeb, SD. Map 800

O. flava (A. Nels.) Garrett
Acaulescent perennial with a fleshy taproot and a rosette of leaves near the ground; leaves 1–2 dm long, oblanceolate, deeply pinnatifid and coarsely toothed, at least in the lower half, glabrous or pubescent on the margins, the midrib prominent on the lower side; hypanthium-tube 4–7 cm long; petals 12–18 mm long, yellow, turning pink; capsule 2–3 cm long, winged on the 4 angles. *Lavauxia flava* A. Nels. Clay flats. sAlta, sSask s to Wash, Ida, Calif, Colo, ND. Map 799 ○

O. nuttallii Sweet / White Evening Primrose
Robust perennial with fleshy roots; stems 4–10 dm tall, branched, glandular-puberulent in the inflorescence, whitish and smooth as if enamelled below and becoming shredded; leaves 5–10 cm long, linear to linear-oblong, entire or nearly so, tapering to the base, glabrous above, hairy beneath; flowers strongly scented, sessile in the upper axils, the buds nodding; hypanthium-tube 2–2.5 cm long; petals about 2 cm long, white, turning pink; capsule slender, 2–3 cm long. $2n = 14, 28$. *Anogra nuttallii* (Sweet) A. Nels. Sandy prairies and roadsides. BC to wOnt s to Colo, Kans, Minn, Wisc. Map 801

O. serrulata Nutt. / Shrubby Evening Primrose
Perennial with a somewhat woody base; stems slender, branched and leafy, 1–5 dm high, more or less hairy; leaves alternate, linear to oblong or spatulate, 2–5 cm long, serrulate to entire, commonly strigose; flowers sessile in the axils of the upper leaves; hypanthium-tube funnelform, 4-angled; sepals keeled on the back; petals yellow, crenulate; stigma discoid, shallowly 4-lobed; capsule slender, 15–25 mm long, canescent. $2n = 14$. *Calylophus serrulatus* (Nutt.) Raven, *Meriolix serrulata* (Nutt.) Walp. Sandy prairies and dunes. Alta to wOnt s to Ariz, NM, Tex, Mo, Wisc. Map 797 ●

HALORAGACEAE / Water-milfoil Family

Myriophyllum L. / Water-milfoil

Aquatic herbs with slender stems, largely submersed; leaves (in our species) whorled, mostly pinnatifid into filiform divisions; flowers in an emersed terminal spike, perfect or unisexual, the upper ones usually staminate; sepals 4; petals 4, soon deciduous; stamens 8; pistil with inferior ovary, 2–4-loculed; stigmas 2–4, plumose; fruit sub-globose, nut-like, splitting into 4 pieces, each 1-seeded.

Floral bracts ovate, entire or minutely serrulate, usually shorter than the flowers; stems white when dry **M. exalbescens**

Floral bracts dissected, usually longer than the flowers; stems scarcely white when dry **M. verticillatum**

M. exalbescens Fern.
Stems submersed, commonly about 1 m long, often forking, purple, when dry whitish; leaves whorled, commonly in 4's, 1–3 cm long, pinnatifid into 6–10 pairs of filiform divisions; flowers whorled in an interrupted spike, the bracts rarely as long as the fruit; sepals of the staminate flowers usually purplish. $2n = 42^*$. *M. spicatum* L. ssp. *squamosum* Laest. *ex* Hartm. f., *M. spicatum* L. ssp. *exalbescens* (Fern.) Hult. Sloughs, lakes, and streams. wGreenland, nEurope: Alas, Yuk, Victoria Is, Baffin Is, Mack to Huds Bay, Nfld s to Calif, Ariz, Tex, Okla, Ill, Ind, WVa. Map 802

M. verticillatum L.
Differs from *M. exalbescens* as shown in the key. $2n = 28$. Ponds and quiet streams. Circumpolar: Alas, Yuk, swMack to James Bay, Nfld s to Wash, Utah, Colo, Neb, Minn, Ky, Md. Map 803

HIPPURIDACEAE / Mare's-tail Family

Hippuris L.

Glabrous perennial, aquatic or amphibious, with several unbranched stems arising from a creeping rhizome; leaves in close whorls, sessile, linear, entire, acute, mostly 1–3 cm long; flowers green, sessile in the leaf-axils; calyx entire; petals absent; stamen 1, with short filament and large 2-loculed anther; ovary inferior, 1-loculed and 1-ovuled, the style filiform and stigmatic along one side, placed in the groove between the anther lobes; fruit nut-like, 1-seeded, indehiscent.

Leaves generally less than 1 mm wide and 1 cm long; flowers mostly imperfect **H. montana**

Leaves generally more than 1 mm wide and 1 cm long; flowers mostly perfect **H. vulgaris**

H. montana Ledeb.
Stems 1–10 cm tall; leaves 5–8 per whorl; flowers mostly imperfect, the staminate below the pistillate, sometimes together, occasionally perfect; filaments more or less absent to much longer than the anther; mature ovary about 1 mm long. $2n = 16$. Mountain streams and mossy banks. Alas, Yuk, wMack s to nwWash, BC, Alta. Map 804 ○

H. vulgaris L.
Stems 5–30 cm tall; leaves most 6–12 per whorl, 10–50 mm long, 1–2 mm wide; flowers mostly perfect; filaments about equalling anthers; mature ovary about 2 mm long. $2n = 32$. Streams, ponds, and lakes. Circumpolar: Alas, Yuk, Banks Is to nBaffin Is, Ellesmere Is to Nfld s to Calif, NM, Neb, Minn, Gt Lakes, NY. Map 804 ●

ARALIACEAE / Ginseng Family

Woody or herbaceous perennials; leaves simple or compound; flowers regular, perfect or unisexual, epigynous, variously disposed, the ultimate divisions of the inflorescence being umbellate, sepals 5, often minute or obsolete; petals 5, inserted on the margin of an epigynous disc; stamens 5, alternate with the petals; pistil with inferior ovary, of 2–5 united carpels; fruits berry-like drupes.

Herb with leaf and peduncle arising from the rhizome, the leaf ternate and 3–5-foliolate; carpels 5; fruit blackish at maturity **Aralia**
Shrub with thorny stems and simple, lobed leaves; carpels 2; fruit red at maturity **Oplopanax**

Aralia L.

A. nudicaulis L. / Wild Sarsaparilla
Acaulescent perennial herb, with creeping rhizome and short thick caudex bearing a single compound leaf and a shorter flowering stalk; leaf long-stalked, 3–6 dm high, twice compound, being ternate and each division pinnately 3–5-foliolate; leaflets ovate-oblong to elliptic-lanceolate, acuminate, serrate, 3–6 cm long; flowering peduncle with 2–7, commonly 3, umbels; flowers greenish or whitish; styles 5, distinct, spreading; fruit a fleshy, 5-loculed, berry-like drupe, green, turning purplish black at maturity, not edible; seeds flattened. $2n = 24$, 48. Moist woods. swMack, BC to Nfld s to Wash, Mont, Colo, Neb, Mo, Ga. Map 805

Oplopanax (T. & G.) Miq.

O. horridum (Sm.) Miq. / Devils's-club
Coarse shrub, with a strong odour, 1–3 m high, densely beset with spines and prickles; leaves alternate, long-petioled, the blades 1–3 dm broad, palmately 5–7-lobed, cordate at base, the lobes acuminate or cuspidate, the margin unevenly

serrate, the ribs prickly beneath; umbels of flowers in a large terminal cluster, 1–3 dm long; petals greenish; styles 2, filiform, distinct; fruit scarlet, 4–5 mm long, 2-seeded, not edible. $2n = 48^*$. In our area ssp. **horridum**. Moist woods. Japan: Alas, Yuk s to Ore, Ida, Mont; nLk Superior. Map 806

UMBELLIFERAE / Carrot Family

Herbaceous plants of diverse habit, often with a strong odour; stems usually hollow; leaves alternate or basal, compound or occasionally simple, the petioles usually expanded or sheathing at base; flowers small epigynous, usually in simple or compound umbels, rarely in head-like clusters; primary umbel with or without an involucre of bracts; secondary umbels commonly with involucels of bractlets; calyx-teeth small or lacking; petals 5, usually with inflexed tips; stamens 5, inserted on the disc which crowns the ovary; styles 2, sometimes enlarged at base, forming a stylopodium; ovary inferior, 2-loculed, with a single ovule in each locule; fruit a schizocarp, of 2 one-seeded mericarps, separating at maturity and usually suspended from the axis-tip (carpophore); each mericarp commonly marked lengthwise by 5 prominent ribs, sometimes with alternating secondary ribs, and oil-tubes between or under the ribs.

Mature fruit is usually required for species identification.

1 Plants with axillary bulblets largely replacing flowers **Cicuta**
Plants lacking axillary bulblets 2

2 Plants with a bluish tinge; flowers in dense globose heads; bracts spiny **Eryngium**
Plants not as above 3

3 Leaves palmate with 5–7 large leaflets; fruit covered with short, hooked spines **Sanicula**
Leaves pinnate, trifoliate, or simple; fruit not spiny 4

4 Lowest leaves simple 5
Lowest leaves (and the upper) compound 6

5 Leaves entire, linear-lanceolate to oblong **Bupleurum**
Leaves serrate, the lowest cordate at base, the upper commonly trifoliolate **Zizia**

6 Leaves usually trifoliolate; leaflets palmate, 10–30 cm wide **Heracleum**
Leaves not as above 7

7 Fruit several times longer than broad, linear to clavate, usually bristly-hairy (glabrous in *O. occidentalis*) **Osmorhiza**
Fruit generally 3 times as long as broad or less, oblong to elliptical 8

8 Leaves once pinnate 9
Leaves 2 or more times pinnate or ternate 11

9 Leaflets entire, mostly 1–2 mm wide; fruit about 2 mm long **Perideridia**
Leaflets serrate, more than 2 mm wide; fruit generally more than 2 mm long 10

10 Flowers white; fruit 2–3 mm long; leaflets linear to linear-lanceolate, much longer than broad **Sium**

Flowers yellow; fruit 5–6 mm long; leaflets elliptic to ovate in outline, divided, twice as long as broad or less **Pastinaca**

11 Plants less than 3 dm tall, usually 1–2 dm 12

Plants more than 3 dm tall 14

12 Fruits more or less rounded in cross-section, 3–4 mm long, tuberculate; flowers yellow **Musineon**

Fruits flattened dorsally (at right angles to the styles), more than 5 mm long; flowers white, purplish or yellow 13

13 Fruits conspicuously winged dorsally; flowers white **Cymopterus**

Fruits not winged dorsally; flowers yellow, white or purplish **Lomatium**

14 Fruits flattened laterally (parallel to the styles) 15

Fruits flattened dorsally (at right angles to the styles) 17

15 Stem purple-blotched **Conium**

Stem not purple-blotched 16

16 Plants generally 3–5 dm tall; with a taproot; fruit about twice as long as broad **Carum**

Plants generally more than 5 dm tall; with tuberous roots; fruit about as long as broad or slightly longer **Cicuta**

17 Stylopodium present 18

Stylopodium absent **Lomatium**

18 Leaflets ovate, serrate **Angelica**

Leaflets filiform, entire **Anethum**

Anethum L.

A. graveolens L. / Dill
Aromatic annual; stems 3–10 (15) dm tall, branched above, glabrous, glaucous; leaves petiolate; blades pinnately dissected into numerous filiform segments 5–20 mm long; umbels compound, rays mostly 15–30; bracts and bractlets absent; flowers yellow; sepals absent; fruit (3) 4–6 mm long, elliptical, strongly flattened dorsally, winged. $2n = 22$. Waste ground. An escape from cultivation. Introduced. Occasional.

Angelica L.

Tall perennials; stems glabrous, at least below; leaves large, glabrous, mostly twice-pinnate or ternate-pinnate, toothed or lobed, the upper progressively reduced in size; umbels very large, compound; primary rays numerous; secondary umbels densely many-flowered, the bractlets linear or filiform; sepals minute or none; petals white or yellow; fruit flattened dorsally, glabrous, the dorsal ribs narrowly winged, the lateral ones broader.

1 Flowers pale greenish yellow; umbel with several large toothed leaf-like bracts nearly as long as the rays **A. dawsonii**

Flowers white; primary umbel without involucral bracts 2

2 Main divisions of cauline leaves frequently reflexed and the rachis geniculate; inflorescence densely short-hairy; ovaries hispidulous; fruit glabrous **A. genuflexa**

Main division of leaves and rachis not as above; inflorescence glabrous or finely scabrous; ovaries and fruit glabrous **A. arguta**

A. arguta Nutt. / White Angelica
Stems very stout, 5–20 dm tall; petioles with very broad sheathing bases; leaflets ovate to lanceolate, 4–12 cm long, spinulose-serrate; rays of umbel and the pedicels conspicuously webbed; secondary umbels without bractlets or with a few filiform ones; flowers white or pinkish; fruit oval to obovoid; 4–6 mm long, smooth, often reddish or purplish. $2n = 22$. Moist woods and stream banks. BC, Alta s to Calif, Utah, Wyo. Map 807

A. dawsonii S. Wats. / Yellow Angelica
Stems rather slender, 3–10 dm tall; leaves 1–3-ternate, the leaflets lanceolate, 3–6 cm long, spinulose-serrate; bracts of umbel oblanceolate, laciniate, the bractlets much smaller; fruit oblong-oval, about 5 mm long. $2n = 22$. Moist montane woods. seBC, Alta s to Ida, wMont. Map 808 ○

A. genuflexa Nutt. / White Angelica
Resembling *A. arguta*; differing as shown in the key and as follows: leaflets coarsely serrate or incised; secondary umbels with numerous slender bractlets and having the pedicels inconspicuously webbed; fruit nearly orbicular, 3–4 mm long. In our area ssp. **genuflexa**. Moist woods and stream banks. eAsia: sAlas, BC, Alta s to nCalif. Map 808 ●

Bupleurum L.

B. americanum Coult. & Rose
Perennial herb, 1–3 dm tall, glabrous, often glaucous; leaves simple, clustered mainly near the base, oblong-lanceolate to linear, entire, 2–6 cm long; umbels compound, the bracts ovate to lanceolate, the bractlets smaller and united at the base; flowers yellow or purplish; fruit oblong, 3–5 mm long, smooth, somewhat flattened laterally, the ribs slender. $2n = 14$, c.28*. *B. triradiatum* Adams ssp. *arcticum* (Regel) Hult. Dry hillsides. eAsia: Alas, Yuk, nwMack; seBC, Alta s to Wyo. Map 809 ○

Carum L.

C. carvi L. / Caraway
Glabrous biennial, the stems 3–8 dm high, branched; leaves 3 or 4 times pinnately dissected, the leaflets cleft into linear lobes; umbels compound, borne on long peduncles; bracts and bractlets filiform or wanting; sepals none; petals white or rarely pink; fruit oblong to oval, 3–4 mm long, prominently ribbed, glabrous,

flattened laterally. $2n = 20^*$, 22. Roadsides and waste ground. An escape from cultivation. Introduced. Common.

Cicuta L. / Water Hemlock

Tall marsh plants with smooth stems, from stout rootstocks; leaves 1–3-pinnate or ternate-pinnate, the leaflets toothed or somewhat incised; umbels compound; bracts few or none; bractlets narrow; flowers white or greenish; sepals well developed, triangular; fruit oval to elliptical, flattened laterally, glabrous, with 5 thick ribs on each mericarp and oil-tubes in the intervals.

Roots of most species extremely poisonous to man and domestic animals.

1 Leaflets linear; axils of leaves usually with small bulblets **C. bulbifera**
Leaflets lanceolate; bulblets absent 2

2 Fruit usually about as long as broad or broader, gradually and evenly constricted along the commissure; vein enclosures of lower leaf-surface elongate-rectangular **C. virosa**
Fruit usually longer than broad, not constricted or abruptly and unevenly constricted at the commissure; vein enclosures of the lower leaf-surface squarish **C. maculata**

C. bulbifera L.
Plant slender, 3–8 dm tall; lower leaves 2–3-pinnate, the leaflets 2–6 cm long, sparsely toothed; bulblets commonly produced in the axils of the upper (small, simple) leaves; fruits orbicular, 1.5–2 mm long, rarely matured. $2n = 22$. Marshes and swamps. wMack, BC to Nfld s to Mont, Alta, Neb, La, Tenn, SC; Alas; Ore; Fla. Map 809 •

C. maculata L.
Plant stout, 6–18 dm tall, from a bulbous rootstock; leaves 1–3-pinnate or pinnate-ternate, the leaflets coarsely serrate to incised; umbels compound, 1–several; primary umbel with 18–28 rays; umbellets with 12–25 rays; fruit 2–4 mm long, generally somewhat narrower. $2n = 22^*$. *C. douglasii* (DC.) Coult. & Rose of Ed. 1. In our area var. **angustifolia** Hook. Marshes, streams, ditches. Alas, Yuk, swMack to James Bay, Que, NS s to Calif, Ariz, NM, Tex, La, Fla. Map 810

C. virosa L.
Similar to *C. maculata;* differing in the characters of the key, the narrower, longer leaflets; the fewer rays (9–21) in the primary umbel and the more numerous rays of the umbellets (50 or more); fruit 1.5–2.2 mm long, 2–3 mm broad. $2n = 22$. *C. mackenzieana* Raup of Ed. 1. Marshes, streams, ponds. Eurasia: Alas, Yuk, swMack to James Bay s to neBC, cAlta, cSask, neMan. Map 811

Conium L.

C. maculatum L. / Hemlock
Herb, biennial, glabrous, ill-scented; stem up to 20 dm tall, branched, furrowed, purple-blotched; leaves 2–4 dm long, pinnately compound, the ultimate segments ovate-oblong, toothed; umbels compound, numerous; bracts 2–5 cm long, few, ovate to lanceolate, reflexed; bractlets similar, shorter; flowers white; sepals absent; fruit about 3 mm long, laterally flattened, broadly ovoid. $2n = 22$. Very poisonous. Marshy ground. Introduced. Rare.

Cymopterus Raf.

C. acaulis (Pursh) Raf. / Plains Cymopterus
Low, acaulescent or short-stemmed, perennial herb, from a deep taproot; leaves pinnate or twice pinnate, 5–20 cm long; peduncles shorter than or equalling the leaves; umbels compact, with 3–5 short primary rays, the bractlets narrow, more or less united; flowers white; fruit 5–10 mm long, ovoid to oblong, flattened dorsally, with 1–3 dorsal wings and 1–3 thickened lateral wings. $2n = 22$. Dry prairie and hillsides. Alta to swMan s to Ore, Mont, Colo, Okla, wMinn. Map 812

Eryngium L.

E. planum L.
Perennial; stems 2–10 dm tall, bluish; basal leaves petiolate, blades 5–10 cm long, 3–6 cm wide, ovate, rounded to cordate at base, margin serrate; cauline leaves mostly sessile, clasping; inflorescence of several, pedunculate, ovoid to globose heads (of many, contracted single-flowered umbels), 1–2 cm long; bracts 15–25 mm long, linear-lanceolate, generally bluish above, margin spinescent; sepals rigid; petals notched; fruit scaly. $2n = 16$. Waste ground. Probably an escape from cultivation. Introduced. Occasional.

Heracleum L.

H. lanatum Michx. / Cow Parsnip
Coarse perennial herb, 1–2 m tall, pubescent or somewhat tomentose; leaves large, ternate, rarely pinnate, the few leaflets 1–3 dm broad, cordate at base, variously lobed and toothed; petioles with dilated sheaths; umbel large, compound, the primary rays unequal, the bracts lanceolate, deciduous; corolla white, usually irregular, larger in the outer flowers of the umbel; stylopodium present; fruit flat, obovate to obcordate, about 1 cm long, winged on the edges, puberulent, marked on the faces by narrow ribs and dark lines (oil-tubes) extending from the summit about half-way to the base. $2n = 22$. Moist woods, clearings, ditches. eAsia: sAlas, Yuk, swMack to Huds Bay, Nfld s to Calif, Ariz, NM, Kans, Ohio, Ga. Map 813

Lomatium Raf. / Prairie Parsley

Perennial herbs, mostly acaulescent or with very short stems, a few species with tall stems; leaves ternately or 2–5-pinnately compound, the leaflets commonly small, generally puberulent; umbel compound, usually spreading; flowers yellow, purple or white; involucral bracts mostly absent; bractlets of the secondary umbels usually present; petals with inflexed tips; stylopodium lacking; fruit strongly flattened dorsally, with well-developed lateral wings, these membranous or corky.

1	Leaves divided into a few linear segments	**L. triternatum**
	Leaves much divided	2
2	Ovary and fruit densely puberulent	3
	Ovary and fruit glabrous	4
3	Bractlets fused	**L. foeniculaceum**
	Bractlets free	**L. sandbergii**
4	Stems densely pubescent	**L. macrocarpum**
	Stems glabrous	5
5	Plants 0.5–3.5 dm tall; bractlets ovate-elliptic to oblanceolate	**L. cous**
	Plants 3–12 dm tall; bractlets linear	**L. dissectum**

L. cous (S. Wats.) Coult. & Rose
Plants generally acaulescent, 0.5–3.5 dm high, from a globose to elongate taproot; glabrous; leaves much dissected, the ultimate segments mostly narrowly elliptic to oblanceolate; umbels with 5–20 unequal rays; bractlets 2–5 mm long, ovate-elliptic to oblanceolate, usually free, occasionally some fused; flowers yellow; fruits 5–12 mm long, 3–5 mm wide, broadly elliptic, the wings about equalling or somewhat narrower than the body. $2n = 22$. Dry hillsides. seAlta, swSask s to eWash, eOre, Ida, Mont, nWyo. Map 814 ○

L. dissectum (Nutt.) Mathias & Constance
Stout perennial, caulescent, rarely acaulescent, 3–12 dm tall, with a thick caudex and spindle-shaped root; leaves 1–3 dm broad, ternate and 2–4-pinnate, the ultimate divisions linear, puberulent beneath; peduncles usually exceeding the leaves; flowers yellow or purplish, the pedicels 4–20 mm long; fruit oblong-oval, 10–12 mm long, flattened, the lateral wings narrow, thick and corky. $2n = 22$. In our area var. **multifidum** (Nutt.) M. & C. Open montane woods and rocky slopes. BC, Alta s to Calif, Ariz, Colo; swSask. Map 814 ●

L. foeniculaceum (Nutt.) Coult. & Rose
Plants acaulescent, 1–5 dm high, villose to glabrate, from a thickened taproot; leaf-blades 5–12 cm long, 3–4-pinnate, the ultimate segments linear; petioles sheathing about to the middle, the sheath purplish; peduncles longer than the leaves; bractlets united below, about equalling the yellow flowers; fruit ovate-oblong, 7–10 mm long, more or less pubescent, the wings narrower than the

body; oil-tubes 3 in the intervals. $2n = 22$. Prairie grassland and eroded slopes. BC to sMan s to Ore, Nev, Ariz, Tex, Okla, Mo. Map 815 ●

L. macrocarpum (Hook. & Arn.) Coult. & Rose
Plants short-caulescent, 1–5 dm high, densely hairy, becoming glabrate, purplish below, from a taproot; leaf-blades 3–12 cm long, ternate, then 2–3-pinnate, the ultimate divisions confluent, 1–7 mm long; petioles sheathing about to the middle; peduncles longer than the leaves; bractlets linear-lanceolate, equalling or greatly exceeding the white or yellow, sometimes purplish, flowers; fruit narrowly oblong, 9–20 mm long, the wings narrower than the body, the young fruit glabrous to villose, the mature fruit glabrous or glabrate; oil-tubes 1, rarely 2–3, in the dorsal intervals. $2n = 22$. Dry open hillsides. BC to Man s to Calif, Utah, Colo, nwSD, ND. Map 816

L. sandbergii Coult. & Rose
Plants caulescent, 7–30 cm tall, rough-puberulent to glabrate, from a slender taproot; leaves ternate, then tripinnate, the ultimate divisions 2–4 mm long and narrow; petioles 1–2.5 cm long, entirely sheathing, the sheath with a conspicuous white margin; umbels 5–14-rayed; bractlets filiform to linear; flowers yellow; fruits oblong, 7–10 mm long, more or less granulate-roughened. Alpine shales. seBC, swAlta s to Ida, Mont. Map 815 ○

L. triternatum (Pursh) Coult. & Rose
Plants caulescent or acaulescent, 2–8 dm tall, puberulent or glabrate, from a slender taproot; leaves commonly ternate, then pinnate to bipinnate; leaflets linear to lanceolate, entire; petioles sheathing about to the middle, purplish below; peduncles exceeding the leaves; bractlets filiform; flowers yellow; fruits oblong, 3–6 mm wide, the wings narrower than the body. $2n = 22, 44$. Includes *L. simplex* (Nutt.) Macbr. of Ed. 1. Prairie grassland and gravelly slopes. BC, Alta s to Calif, Utah, Colo. Map 817

Musineon Raf.

M. divaricatum (Pursh) Nutt.
Caulescent herb with taproot; stems erect or spreading, usually dichotomously branched, more or less scabrous, especially at the base of the umbel; leaves 1–2-pinnate or ternate-pinnate; leaflets oblong, pinnately lobed, the rachis more or less winged; bractlets linear; flowers yellow; fruits ovoid to narrowly oblong, 3–5 mm long, laterally flattened, prominently ribbed, scabrous-tuberculate. Dry prairies. sAlta to sMan s to Nev, Colo, Neb. Map 818

Osmorhiza / Sweet Cicely

Perennial herbs, with thick aromatic roots and more or less leafy stems; leaves ternately compound or ternate-pinnate, the lower petioled, the upper subsessile; leaflets ovate to lanceolate, lobed or toothed; flowers white; greenish white, yellowish, or purplish, in small umbels; pedicels divaricate, spreading or erect;

fruits linear-oblong or narrowly club-shaped, slightly flattened laterally, bristly-hispid (glabrous in one species), the ribs narrow.

1 Fruit glabrous, obtuse at base, not caudate; rays ascending or spreading-ascending; flowers pale yellow **O. occidentalis**

Fruit bristly-hispid, the base long-tapering, becoming caudate with conspicuous tails; rays spreading-ascending to divaricate; flowers whitish or purplish 2

2 Bractlets of inflorescence present and persistent; styles 2–3 mm long; fruit 18–22 mm long **O. longistylis**

Bractlets absent or very small and caducous; styles less than 0.5 mm long; fruit 10–20 mm long 3

3 Rays and pedicels wide-spreading or reflexed; fruit clavate, the apex convexly narrowed **O. depauperata**

Rays and pedicels usually ascending; fruit cylindrical, the apex concavely narrowed (constricted) into a beak 0.5–2.0 mm long 4

4 Flowers white or greenish; fruit 12–22 mm long; stylopodium longer than broad **O. chilensis**

Flowers purple, occasionally greenish white; fruit 8–13 mm long; stylopodium broader than long **O. purpurea**

O. chilensis Hook. & Arn.
Stems 3–10 dm tall, more or less hispid; leaves biternate; leaflets coarsely serrate, incised or lobed, 2–6 cm long, somewhat hairy, especially on the veins; rays 3–8; flowers and fruits as noted in key. $2n = 22$. Moist woods. sAlas, BC s to Calif, Ariz, Colo, SD; Gt Lakes; NH to Nfld. Map 819

O. depauperata Philippi
Stems 2–6 dm tall, hispid to glabrate; leaves biternate or ternate-pinnate; leaflets coarsely serrate, incised or lobed, hairy; rays 2–5; pedicels, flowers and fruits as noted in key. $2n = 22$. Moist woods. sAlas, swMack to sMan s to Calif, Nev, NM, SD; Gt Lakes; Vt to Nfld. Map 820

O. longistylis (Torr.) DC.
Stems 3–10 dm tall, glabrate; leaves 2–3-ternate; leaflets coarsely toothed, incised or lobed, more or less short-hairy, especially on the veins; rays 3–6; bracts 1 or more; bractlets several, reflexed; flowers and fruits as noted in key. $2n = 22$. Moist woods. Alta to NS s to NM, Tex, Ala, Ga. Map 821 ○

O. occidentalis (Nutt.) Torr.
Stems 3–10 dm tall, rather stout, usually hairy at the nodes at least; leaves 1–3-ternate or ternate-pinnate; leaflets oblong-lanceolate to ovate, 2–10 cm long, serrate and often incised or lobed, glabrous or somewhat hairy; pedicels, flowers and fruits as noted in key. $2n = 22$. Montane woods. seBC, swAlta s to Calif, Colo. Map 821 ●

O. purpurea (Coult. & Rose) Suksd.
Stems 2–6 dm tall, slender, hairy or glabrous; leaves 1–3-ternate; leaflets usually lanceolate, 2–6 cm long, serrate to incised or lobed; flowers and fruit as noted in key. $2n = 44$. Moist woods and mountain slopes. sAlas, BC s to Calif, Ida, nwMont. Map 822

Pastinaca L.

P. sativa L. / Parsnip
Biennial with a thickened taproot; stem 3–10 dm tall, stout and grooved, glabrous or nearly so; leaves large, pinnate; leaflets ovate to oblong, 4–10 cm long, coarsely serrate, incised or lobed; inflorescence compound, the umbels flat-topped, usually without bracts or bractlets; petals yellow; fruits broadly oval, 5–7 mm long, flattened dorsally, winged along the edges; oil-tubes large, solitary in the intervals, 2–4 on the commissure. $2n = 22$. Waste ground, fields, and roadsides. An escape from cultivation. Introduced. Occasional.

Perideridia Reichenb. / Squaw-root; Yampa

P. gairdneri (Hook. & Arn.) Mathias
Glabrous perennial with one or more tuberous or fusiform roots; stems slender, 3–6 dm tall, usually unbranched; leaves few, pinnate or occasionally bipinnate, with 3–7 linear or linear-lanceolate leaflets, 4–12 cm long; umbels compound, the bractlets linear; rays 8–20; flowers white or pinkish, very small; calyx-teeth conspicuous; stylopodium low-conic; fruits about 2 mm long, brown, orbicular, flattened laterally, glabrous. $2n = 28, 30, 80, 120$. In our area ssp. **borealis** Chuang & Constance. Woods and meadows. BC to swSask s to Calif, NM, SD. Map 823

Sanicula L.

S. marilandica L. / Snake-root
Glabrous perennial with a rootstock and fibrous roots; stem 3–6 dm high; basal leaves long-stalked, palmately 5–7-parted, the segments oblanceolate, 5–8 cm long, serrate to doubly serrate, often incised at apex; upper leaves short-petioled or nearly sessile; flowers greenish white, in head-like clusters (umbellets), each about 1 cm wide; umbellet with a few sessile or short-pedicelled perfect flowers, their hypanthia bristly, and several staminate flowers with smooth hypanthia, the latter flowers all or chiefly on longer pedicels; sepals narrow, 1–1.5 mm long, persistent; fruits about 5 mm long, narrowed toward the base, densely covered with hooked prickles; styles recurved, longer than the prickles. $2n = 16$. Moist woods. BC to Nfld s to Ida, Mont, NM, Neb, Mo, Fla. Map 824

Sium L.

S. suave Walt. / Water Parsnip
Glabrous perennial 0.5–1 m tall, with a fascicle of fusiform roots; stem stout, hollow, strongly ridged, usually branched; leaves mostly once-pinnate, with 5–many-linear or lanceolate leaflets, 5–10 cm long, sharply serrate, submersed leaves, if present, bipinnately dissected; umbels compound, the primary rays 6–many, the involucral bracts lanceolate and reflexed, the bractlets similar but smaller; sepals minute or obsolete; petals white; fruits oval, 2–3 mm long, slightly flattened laterally, with conspicuous corky ribs. $2n = 12$. Wet meadows and sloughs. BC, swMack to Huds Bay, Nfld s to Calif, Tex, Mo, SC; Alas. Map 825

Zizia Koch

Z. aptera (A. Gray) Fern. / Meadow Parsnip
Glabrous perennial, 2–6 dm high; basal leaves long-petioled, commonly oblong-ovate, cordate at base, usually simple; cauline leaves short-petioled or sessile, once or twice ternate, the leaflets ovate or lanceolate, toothed; umbels compound, with a few short, narrow bractlets; flowers bright yellow; fruit 2–4 mm long, oblong-ovate, somewhat flattened laterally, glabrous, the ribs narrow. $2n = 22$. Moist meadows. swYuk, BC to Que s to Ore, Nev, Colo, Okla, Mo, Ala, Ga. Map 826

CORNACEAE / Dogwood Family

Cornus L.

Shrubs or herbs; leaves opposite or, in one species, appearing whorled at the summit, without stipules, elliptic to ovate or obovate, entire, turning red or purplish in autumn; flowers small, epigynous, in open naked cymes, or in a dense head surrounded by a corolla-like involucre; sepals 4, minute; petals 4, white, valvate, spreading; stamens 4, with slender filaments; ovary inferior, usually 2-loculed; style solitary, slender; fruit a white or red drupe, with a 1–2-seeded stone (endocarp).

Herb; flowers in a dense head surrounded by an involucre of 4 whitish bracts; fruit red; leaves mostly whorled at summit of stem **C. canadensis**
Shrub; flowers in cymes, not surrounded by an involucre; fruit whitish; leaves opposite **C. stolonifera**

C. canadensis L. / Bunchberry
Low perennial herb, with a creeping rootstock, forming colonies; stems simple, erect, 8–18 cm high; principal leaves in an apparent whorl of 4–6 at the summit, subsessile, elliptic-ovate to obovate or rhombic, tapering to the base, pinnately veined with the secondary veins curved to the apex; lower leaves reduced and in remote pairs; flower-cluster on a short peduncle; bracts of involucre usually

4, white to cream-coloured, greenish when immature, ovate, 1–2 cm long; drupes bright red, globose, 6–8 mm broad. $2n = 22^*$, 44. *Chamaepericlymenum canadense* (L.) Aschers. & Graebn. Moist woods. eAsia, Greenland: Alas, Yuk, wMack to Huds Bay, Nfld s to Calif, NM, SD, Ohio, Pa, NJ. Map 827

C. stolonifera Michx. / Red Osier; Dogwood; Kinnikinnick
Shrub, 1–3 m high, the lower branches prostrate and often rooting, the twigs deep red, sometimes greenish or purplish, pubescent or nearly glabrous; leaves oval to ovate-lanceolate, acute to acuminate, entire, glaucous and appressed-hairy beneath; inflorescence a rather flat-topped cyme, minutely appressed-pubescent; petals white, 2–3 mm long; fruit white or tinged with blue, nearly globose, about 5 mm wide, juicy, not edible. $2n = 22^*$. Moist woods, riverbanks. Alas, Yuk, wMack to James Bay, Nfld s to Calif, Ariz, NM, Neb, Ohio, WVa, Pa. Map 828

PYROLACEAE / Wintergreen Family

Mostly low perennials with slender scaly rootstocks and simple, petioled, dark green leaves; flowers perfect, nearly regular, mostly white or purplish, racemose, corymbose or solitary; calyx 4–5-lobed; corolla of 5, or rarely 4, distinct or slightly united petals; stamens 8 or 10; filaments more or less dilated at the base; anthers inverted in anthesis, opening by basal pores or slits at the ends of small tubes; pollen grains commonly in tetrads; ovary superior, usually with a disc, 4–5-loculed; placentae axile; style often declined; stigma more or less 5-lobed; fruit a loculicidal capsule, nearly globose, 5-angled; seeds minute, numerous, with a large aril and a loose testa.

1	Flowers solitary	**Moneses**
	Flowers several	2
2	Inflorescence corymbose or umbellate; stems leafy	**Chimaphila**
	Inflorescence racemose; leaves basal	3
3	Raceme secund; corolla campanulate	**Orthilia**
	Raceme not secund; corolla not campanulate	**Pyrola**

Chimaphila Pursh / Prince's Pine; Pipsissewa

C. umbellata (L.) Bart.
Plants with long running rootstocks; stems 1–3 dm high, stout, slightly woody, leafy, often branched; leaves thick, evergreen, shining, 2–7 cm long, oblanceolate, short-petioled, serrulate except at the cuneate base, scattered along the stem or in whorls of 3–8; flowers white or purplish, fragrant, 4–8 in a corymb or umbel on a terminal peduncle; sepals 1–2 mm long, erose; petals 4–6 mm long; filaments enlarged near base; anther-sacs each opening with a wide terminal pore; style very short; capsule 6–7 mm wide, splitting from the apex downward. $2n = 26$. In our area ssp. **occidentalis** (Rydb.) Hult. Dry woods. Circumpolar: seAlas,

swMack to cMan, Nfld s to Calif, NM, SD, Sask, Minn, Mich, Ohio, Ga. Map 829

Moneses Salisb. / One-flowered Wintergreen

M. uniflora (L.) A. Gray
Small herb with slender creeping rootstock; stems simple, 3–10 cm tall; leaves mostly crowded at base of stem, opposite or in whorls of 3, short-petioled, ovate to obovate, rounded at apex, finely toothed or nearly entire, veiny, thin; flower solitary, terminal, nodding, waxy-white, fragrant, 1–2 cm wide; petals widely spreading; anther-tubes curved; style elongate; stigma large, peltate, with 5 lobes; capsule 6–8 mm wide, opening from apex downward. $2n = 26$. *Pyrola uniflora* L. Moist woods. Circumpolar: Alas, Yuk, Mack to Huds Bay, nQue, Lab, Nfld s to Calif, NM, Sask, Man, Gt Lakes; SD. Map 830

Orthilia Raf.

O. secunda (L.) House / One-sided Wintergreen
Plants with long creeping rootstocks, often forming dense colonies; leaves crowded or somewhat scattered on the forked caudex; leaf-blades longer than the petioles, 1.5–4 cm long, oval to broadly ovate or nearly orbicular, acutish and often mucronate at apex, rounded to cuneate at base, serrulate to denticulate or nearly entire; flowers crowded in a one-sided raceme; sepals 0.5–1 mm long, ciliolate; petals 4–5 mm long, greenish white; anthers without tubes; pollen-grains free; style long, exserted, straight; disc at base of ovary with 10 small glands. $2n = 38$. *Pyrola secunda* L. of Ed. 1. Woods. Circumpolar: Alas, Yuk, Banks Is, Victoria Is, Mack to nHuds Bay, nQue, Nfld s to Calif, Ariz, NM, SD, ND, Ohio, Va. Map 831

Pyrola L. / Wintergreen

Low, glabrous perennial herbs with creeping rootstocks; leaves basal, petioled, evergreen; flowers in a terminal raceme, short-pedicelled and mostly nodding, borne on a more or less scaly-bracted scape; calyx 5-parted; petals 5, distinct or very slightly united; stamens 10; filaments commonly incurved; anthers opening by a pair of pores at the somewhat 2-horned base (appearing as apex); style sometimes declined and curved upward; stigma 5-lobed or 5-rayed; capsule dehiscent from the base upward, the valves with cobwebby margins when opening.

1 Style straight, not thickened upward — **P. minor**
Style curved, gradually thickened to a truncate collar, deflexed at base — 2

2 Petals pink or purplish — 3
Petals white or greenish — 4

3 Leaves subentire, the apex rounded, dull above, thin — **P. asarifolia**
Leaves denticulate-margined, the apex acute, shiny above, firm — **P. bracteata**

4 Leaves ovate-lanceolate, acute, veins mottled-white above **P. picta**
Leaves rotund to elliptic, veins not mottled-white above 5

5 Sepals oblong to elliptic, much longer than wide; leaves subrotund **P. grandiflora**
Sepals more or less triangular, as wide as long; leaves elliptic to rotund 6

6 Leaves broadly elliptic to oblong, 3–7 cm long, mostly longer than the petioles **P. elliptica**
Leaves obovate to subrotund, 1–3 cm long, mostly shorter than the petioles **P. chlorantha**

P. asarifolia Michx. / Common Pink Wintergreen
Leaf-blades usually shorter than the petioles, 3–6 cm long, broadly elliptic or round-reniform, commonly cordate at base, obscurely crenulate, leathery, lustrous above; scapes 15–25 cm tall, usually with 1–3 scale-leaves below the raceme; sepals 2–3 mm long; petals pale to deep pink, 5–7 mm long; anther cuspidate at lower end, abruptly narrowed above into very short tubes. $2n = 46$. Moist woods. Asia: Alas, Yuk, swMack to James Bay, Nfld s to Ore, NM, SD, ND, Minn, Gt Lakes. Map 832

P. bracteata Hook. / Large Wintergreen
Leaf-blades 3–8 cm long, oval to broadly ovate, acute or rarely rounded at apex, rounded to subcordate at base, dark green and shining above, distinctly mucrodenticulate; scapes 15–30 cm tall; floral bracts conspicuous, lanceolate, acuminate, reddish; sepals 3–5 mm long; petals pink or purplish, 6–8 mm long. Wet woods. BC, Alta s to Calif. Map 833 ○

P. chlorantha Sw. / Greenish-flowered Wintergreen
Leaf-blades shorter than the slender petioles, 1–3 cm long, oval to obovate, often nearly orbicular, commonly rounded at both ends, crenulate or nearly entire, thick, rather dull green above; scapes 8–20 cm tall; raceme often few-flowered; sepals 1–1.5 mm long; petals 5–7 mm long, greenish white, converging; anther-tubes conspicuous; style curved upward at apex, 4–7 mm long at maturity. $2n = 46$. *P. virens* Schweig. of Ed. 1. Moist woods. Circumpolar: Alas, Yuk, swMack to Huds Bay, Nfld s to Calif, Ariz, NM, SD, Man, Gt Lakes, Pa. Map 834

P. elliptica Nutt. / White Wintergreen
Leaf-blades longer than the petioles, 2–8 cm long, elliptic to obovate, commonly rounded at apex, narrowed at base, thin, denticulate; scape 12–25 cm tall; flowers fragrant; sepals 1.5–2 mm long; petals whitish, 5–7 mm long. Rich woods. Japan: BC to Que s to NM, SD, ND, Ohio, Pa. Map 833 ●

P. grandiflora Radius / Arctic Wintergreen
Leaf-blades shorter than or equalling the petioles, 2–4 cm long, rounded at both ends or truncate at base, leathery, shining above, entire or obscurely crenulate; scapes 5–15 cm tall; floral bracts scarious, often brown; sepals 3–4 mm long; petals white, greenish white or tinged with pink, 5–10 mm long. $2n = 46$. Alpine

slopes and tundra. Circumpolar: Alas, Yuk, Banks Is, Victoria Is to Ellesmere Is to Nfld s to nBC, Alta, nSask, Huds Bay. Map 835 ●

P. minor L. / Lesser Wintergreen
Leaf-blades usually about as long as the petioles, 1–4 cm long, oval to nearly orbicular, rounded at both ends or slightly cordate at base, obscurely crenulate, dull green above; scape 5–18 cm tall; flowers nodding; sepals 1.5 mm long; petals white or pinkish, connivent, 4–5 mm long; anthers short, not produced into tubes; style straight, without a ring below the stigma, included; stigma deeply 5-lobed. $2n = 46$. Moist woods. Circumpolar: Alas, Yuk, swMack to Huds Bay, nQue, Nfld s to Calif, Utah, NM, Sask, Man, Minn, Gt Lakes. Map 836

P. picta J.E. Smith / White-veined Wintergreen
Leaf-blades 1–6 cm long, ovate-lanceolate to broadly ovate, commonly acute at apex and narrowed to the petiole at base, firm, mottled with white along the larger veins on the upper surface, denticulate to nearly entire; scape 10–20 cm tall; sepals about 1.5 mm long; petals 6–8 mm long, yellowish or greenish white. $2n = 46$. Coniferous woods. sBC to Alta s to Calif, Colo. Map 835 ○

MONOTROPACEAE / Indian-pipe Family

Saprophytic herbs, without chlorophyll; the whole plant white, pink, yellow or brownish red; stem arising from a dense rounded mass of matted roots, these mycorrhizal; leaves reduced to mere scales; flowers solitary or racemose, regular, perfect, usually drooping, sepals 2–5 or sometimes absent; petals mostly 4 or 5, distinct or partially united; stamens twice as many as the petals; anthers opening by transverse slits; pollen simple (not in tetrads); pistil syncarpous; ovary superior, usually 5-loculed; stigma disc-like, funnelform or capitate; fruit a loculicidal capsule; seeds minute, very numerous.

Petals distinct; anthers awnless; flowers solitary or in a short raceme **Monotropa**
Petals united; anthers with deflexed awns; raceme 1–3 dm long **Pterospora**

Monotropa L.

Plants white, yellow, pink or reddish when fresh, often turning blackish on drying; stems fleshy and bearing scale-like leaves; flowers nodding; sepals separate and bractlike; petals erect, scale-like, distinct, often saccate at base; anthers more or less reniform, 2-valved; disc at base of ovary; style short and thick; capsules erect, ovoid to globose.

Flowers less than 14 mm long, few–many **M. hypopitys**
Flowers 14–17 mm long, solitary **M. uniflora**

M. hypopitys L. / Pine-sap
Stems 1–3 dm tall, often clustered, yellow, pink or reddish, usually more or less pubescent; upper scale-leaves 1–1.5 cm long, the lower ones smaller and thicker; raceme dense, at first nodding, soon erect; terminal flowers largest and with 5 petals, the lower ones mostly with 4 petals; sepals lanceolate, erect; petals 10–12 mm long, pubescent, saccate at base; anthers opening into 2 unequal valves. $2n = 16, 48$. *Hypopitys monotropa* Crantz. Coniferous woods. Circumpolar: seAlas to swSask s to Calif, NM; Minn to Nfld s to Mo, Fla. Map 837

M. uniflora L. / Indian Pipe
Stems 1–2 dm tall, solitary or clustered; smooth, commonly waxy white, rarely pinkish, turning blackish in drying; flower solitary, nodding, white; calyx of 2–4 irregular bract-like sepals or lacking; petals usually 5, erect, broadly oblong, not saccate at base; anthers opening by 2 transverse clefts; stigma funnelform; capsule erect. $2n = 32$. Rich woods. Asia: seAlas, swMack to sJames Bay, Nfld s to Calif, Ariz, Colo, SD, ND, La, Fla. Map 838

Pterospora Nutt.

P. andromedea Nutt. / Pine-drops
Stems 2–8 dm tall, simple, stout, purplish or reddish brown, glandular-hairy and very sticky; scale-leaves lanceolate to linear-lanceolate, 1–3.5 cm long, thick, crowded near base of stem; raceme 1–3 dm long, erect, many-flowered, the bracts narrow, often longer than the pedicels; flowers nodding; calyx 5-parted, densely glandular-hairy, corolla urn-shaped, 5-lobed, whitish, 6–8 mm long; stamens 10, each anther with 2 retrorse awns; stigma capitate, 5-lobed; seeds with a broad apical wing. $2n = 16$. Coniferous woods. BC, Alta s to Calif, Colo; Ont, Que s to Mich, NY, Vt; swSask. Map 839

ERICACEAE / Heath Family

Shrubs; leaves simple, without stipules, often leathery and persistent; flowers regular or nearly so, perfect; sepals 4–5, often small, distinct or partly united; corolla 4–5-lobed, often urn-shaped, sometimes of separate petals; stamens commonly twice as many as the corolla-lobes; anthers introrse, usually opening by terminal pores, sometimes with awns; pollen in tetrads; pistil of 2–5 united carpels; ovary superior, or in one subfamily inferior; placentation axile; style 1; stigma capitate or peltate; fruit a capsule, drupe or berry.

1 Ovary inferior; fruit fleshy — 2
Ovary superior; fruit fleshy or dry — 3

2 Corolla divided nearly to the base, lobes reflexed; stems filiform and creeping — **Oxycoccus**
Corolla cup-shaped or urn-shaped; stems erect, or if creeping not filiform — **Vaccinium**

3 Petals free; leaves rusty-tomentose or resinous-dotted beneath; flowers white, in terminal clusters **Ledum**
Petals more or less united; leaves and flowers not as above 4

4 Leaves opposite 5
Leaves alternate 7

5 Stamens 5 **Loiseleuria**
Stamens 10 6

6 Flowers white; anthers awned **Cassiope**
Flowers pink; anthers awnless **Kalmia**

7 Fruit fleshy; stems mostly prostrate or decumbent 8
Fruit a dry capsule; stems mostly erect 9

8 Flowers axillary, solitary; calyx accrescent and fleshy in fruit **Gaultheria**
Flowers terminal, in clusters; calyx not as above **Arctostaphylos**

9 Leaves linear, less than 2 mm wide **Phyllodoce**
Leaves wider 10

10 Flowers 4-merous **Menziesia**
Flowers 5-merous 11

11 Anthers awned **Andromeda**
Anthers awnless 12

12 Inflorescence a secund raceme; corolla white, cylindrical **Chamaedaphne**
Inflorescence not a secund raceme; corolla purple (if white 10–20 mm wide), broadly cup-shaped **Rhododendron**

Andromeda L.

A. polifolia L. / Bog Rosemary
Stems erect, 1–3 dm high, from creeping rootstocks, little-branched; leaves persistent, linear to narrowly elliptic, 1–3 cm long, revolute, entire, acute, white-glaucous beneath; flowers few, nodding, at the ends of the branches; calyx-lobes about 1 mm long; corolla pinkish, urn-shaped, 5–7 mm long, 5-lobed, the lobes short; stamens 10; anthers opening by pores, 2-awned; capsule nearly globose, loculicidal. $2n = 48$. In our area ssp. **polifolia**. Peat bogs. Circumpolar: Alas, Yuk to nHuds Bay, nQue, Lab, Nfld s to BC, Ida, sSask, sMan, Gt Lakes, NJ. Map 840

Arctostaphylos Adans. / Bearberry

Creeping or tufted shrubs with alternate leaves; flowers in scaly-bracted terminal racemes or clusters; calyx deeply lobed; corolla 4–6 mm long, white or pinkish, globose, 4–5-toothed; stamens 10, included; anthers with 2 reflexed awns, opening by pores; fruit a berry-like drupe, red globose, 6–10 mm wide, with 5 nutlets.

Leaves entire, leathery, not conspicuously veiny, evergreen; drupe dry and mealy; branchlets puberulent or pubescent **A. uva-ursi**

Leaves crenate-serrate, thin, veiny, deciduous or marcescent; drupe juicy; branchlets glabrous **A. rubra**

A. rubra (Rehder & Wils.) Fern. / Alpine Bearberry
Prostrate or tufted, intricately branched, the branches ascending, 0.5–3 dm tall; twigs with persistent scales of winter-buds; leaves slender-petioled, oblanceolate to spatulate-obovate, 2–4 cm long, crenate-serrate, conspicuously reticulate-veiny, sometimes rugose, often turning purplish red in autumn and then withering-persistent; flowers 2–4 in a terminal cluster; corolla white, 4–5 mm long; drupe scarlet, very juicy, 6–10 mm in diam., the nutlets separate. $2n = 26^*$. *Arctous rubra* (Rehd. & Wilson) Nakai. Moist montane slopes and heaths. Asia: Alas, Yuk to nHuds Bay, wBaffin Is, sQue, Nfld s to BC, Alta, nSask, sMan, nOnt. Map 841

A. uva-ursi (L.) Spreng. / Common Bearberry; Kinnikinnick
Trailing evergreen shrub, often forming mats, the branches prostrate and rooting, commonly 5–10 dm long; branchlets glabrate or variously pubescent; leaves alternate, short-petioled, coriaceous, obovate to spatulate, entire, 1–2 cm long; flowers 3–10, nodding in a dense raceme; bracts thick and firm; corolla white to pink, 4–6 mm long; drupe dull red, mealy and flavourless, 6–8 mm in diam., the nutlets more or less united. $2n = 26^*, 39^*, 52^*, 78$. In our area ssp. **arcto-staphylos,** with the young branches and petioles lacking glandular hairs, though they may appear viscid; ssp. **adenotricha** (Fern & Macbr.) Calder & Taylor, with short and long hairs both glandular and non-glandular; ssp. **stipitata** Packer & Denford, with short-stipitate glandular hairs, occasionally with short non-glandular hairs; and ssp. **longipilosa** Packer & Denford, with short-stipitate glandular hairs associated with short and long non-glandular hairs. Moist to dry woods, open sandy areas, roadsides. Circumpolar: Alas, Yuk, swMack to Huds Bay, Nfld s to Calif, NM, SD, ND, Minn, Gt Lakes, Va. Map 842

Cassiope D. Don / White Mountain Heather

Low alpine shrubs, 1–3 dm high, heath-like, with numerous stems tufted on a thick base, often decumbent; leaves evergreen, 2–4 mm long, sessile, numerous, 4-ranked, closely imbricated, ovate-oblong, thick, blunt; flowers white or rose-coloured, bell-shaped, solitary, nodding on slender lateral peduncles; sepals 4 or 5; corolla cup-shaped, white or pinkish, the lobes reflexed; stamens 8 or 10, included; anthers awned, opening by large terminal pores; capsule loculicidal, 4–5-valved.

Leaves without groove on the back **C. mertensiana**

Leaves with a deep groove on the back **C. tetragona**

C. mertensiana (Bong.) D. Don
Stems rather slender; leaves round-keeled on the back, glabrous or nearly so; sepals 2–3 mm long, usually pinkish; corolla 5–8 mm long, white or tinged with rose. Moist subalpine and alpine slopes. seAlas, BC s to Calif, Nev, Mont. Map 843

C. tetragona (L.) D. Don
Somewhat coarser than the last species; leaves very thick, with a conspicuous furrow on the back, more or less pubescent; sepals 2–2.5 mm long; corolla 5–6 mm long, white or rose-coloured. $2n = 26$. In our area ssp. **saximontana** (Small) Porsild. Moist subalpine and alpine slopes. Circumpolar: Alas, Yuk to Ellesmere Is to Lab s to Wash, Mont, sMack, Huds Bay, nQue. Map 844 ●

Chamaedaphne Moench

C. calyculata (L.) Moench / Leather-leaf
Branching shrub, mostly 2–6 dm high, twigs puberulent or scurfy; leaves alternate, short-petioled, coriaceous, oblong to elliptic, 1–4 cm long, obtuse or acute, minutely crenulate or entire, scurfy (brown or whitish) especially beneath; flowers in a 1-sided leafy raceme; calyx deeply 5-parted, persistent in fruit, closely subtended by 2 bractlets; corolla white, nearly cylindric, slightly narrowed at throat, 6–7 mm long; stamens 10, included; anther with 2 erect tubes at apex, awnless; ovary subtended by a 10-lobed disc; capsule 3–4 mm wide, loculicidal. $2n = 22$. Peat bogs and ponds. More or less circumpolar: Alas, Yuk, swMack to Huds Bay, nQue, Nfld s to BC, Alta, cSask, sMan, Minn, Gt Lakes, Ga. Map 844 ○

Gaultheria L. / Creeping Wintergreen

Small shrubs; stems prostrate and rooting or ascending, often forming mats; leaves alternate, evergreen, entire or obscurely toothed; flowers axillary, perfect, the parts in 4's or 5's; corolla cup-shaped; stamens 8 or 10, included; anthers commonly awned; fruit berry-like, white or scarlet, essentially a capsule but enclosed in the fleshy calyx.

Stems with reddish pubescence; flower parts in 4's; fruit white	**G. hispidula**
Stems glabrous or nearly so; flower parts in 5's; fruit reddish	**G. humifusa**

G. hispidula (L.) Bigel. / Creeping Snowberry
Stems very slender, creeping, 1–4 dm long, very leafy, with appressed pubescence when young, the reddish hairs sharply bent towards the stem-apex; leaves short-petioled, 3–6 mm long, ovate to broadly elliptic, smooth above, with appressed brown hairs beneath; flowers few, solitary, axillary, subtended by 2 bractlets; calyx deeply 4-parted; corolla about 2 mm long, 4-lobed; pollen-sacs each with 2 erect appendages; ovary partly inferior; fruit white, mealy, hispid, 5–7 mm in diam., with a mild flavour of wintergreen. $2n = 22$. *Chiogenes hispidula* (L.) T. & G. Bogs and wet woods. BC to Nfld s to Ida, Minn, Mich, Pa, NC. Map 845 ●

G. humifusa (Graham) Rydb.
Stems prostrate or ascending, glabrous or nearly so, often forming mats, the flowering branches 5–10 cm long; leaves oval or orbicular, mostly 10–15 mm long, obtuse or rounded at apex, glabrous, entire or obscurely toothed; flowers solitary, the pedicels with 3–5 bractlets; corolla 5-toothed, about 3 mm long, little longer than the calyx; ovary superior; fruit scarlet, 5–6 mm in diam. Moist subalpine slopes. sBC, Alta s to Calif, Colo. Map 845 ○

Kalmia L.

Shrubs, ascending or matted; leaves simple, evergreen or deciduous, alternate or opposite; corolla rotate to campanulate; stamens 10, unappendaged; ovary superior, placentation axile; fruit a 5-loculed septicidal capsule.

Leaves 2–4 cm long, usually more than twice as long as wide; plants mostly more than 30 cm tall **K. polifolia**
Leaves 1–2 cm long, usually less than twice as long as wide; plants 10–20 cm tall **K. microphylla**

K. microphylla (Hook.) Keller
Plants 5–20 cm tall, branched, erect or matted, glabrous or puberulent; branchlets 2-edged; leaves opposite, nearly sessile, evergreen, coriaceous, 10–15 mm long, oval to elliptic-oblong, entire, revolute, glaucous-white beneath; flowers showy, rose-coloured, in terminal clusters, with reddish pedicels; calyx deeply 5-lobed; corolla rotate, 6–12 mm broad, pink to rose-purple, 10-saccate, the anthers released from the corolla-pouches at anthesis; style elongate; capsule about 6 mm long. $2n = 24$. Moist mountain meadows and bogs. Yuk, swMack s to Calif, Colo. Map 846 ●

K. polifolia Wang.
Similar to *K. microphylla;* distinguished by the characters of the key. $2n = 48$. Peat bogs. Yuk, swMack, BC to Nfld s to Calif, Alta, Sask, Man, Gt Lakes, Pa, NJ. Map 846 ○

Ledum L. / Labrador Tea

Erect or diffuse shrubs, freely branched, mostly 2–8 dm high; leaves alternate, sessile or short-petioled, entire, firm, persistent, fragrant when bruised; flowers white, in terminal umbel-like clusters from large scaly buds; new shoots of the season originating at the base of the flower cluster; pedicels erect in anthesis, nodding in fruit; calyx very small, 5-toothed; petals 5, distinct, spreading; stamens 5–10, with slender filaments; anthers short, opening by terminal pores; ovary 5-loculed; style elongate and persistent; capsule opening by 5 valves from the base; seeds minute, elongate, winged.

1 Leaves scaly-resinous and pale beneath, not revolute or only slightly so **L. glandulosum**
Leaves densely rusty-tomentose beneath, their margins strongly revolute 2

2 Leaves elliptic to oblong; stamens mostly 5–7 **L. groenlandicum**
Leaves linear; stamens mostly 8–11 **L. palustre**

L. glandulosum Nutt. / Glandular Labrador Tea
Stout shrub, mostly 4–8 dm high; twigs puberulent and minutely glandular; leaves petiolate, elliptic to oval, 1.5–4 cm long, pale and minutely resinous-granuliferous beneath; petals 5–7 mm long; stamens mostly 8; capsule about 4 mm long. $2n = 26$. Hybrids of intermediate morphology occur between this species and *L. groenlandicum* in the Banff area. Moist mountain meadows and open woods. BC, Alta s to Calif, Wyo. Map 847 ○

L. groenlandicum Oeder / Common Labrador Tea
Erect shrub, mostly 3–8 dm high; twigs densely rusty villose-tomentose; leaves oblong to elliptic, 1–5 cm long, obtuse, often mucronate, green above, densely rusty-tomentose beneath, strongly revolute, the petioles very short; pedicels finely pubescent and glandular; petals 5–8 mm long, oblong, narrowed at base; stamens mostly 5–7; capsule narrowly ovoid, puberulent, 5–7 mm long, tipped by the style. $2n = 26$. Bogs and wet coniferous woods. Greenland: Alas, Yuk, Mack to Huds Bay, nQue, Nfld s to Ore, BC, Alta, Sask, Man, Gt Lakes, Pa, NJ. Map 848

L. palustre L. / Northern Labrador Tea
Similar to *L. groenlandicum*; differing in the characters of the key, the basally ciliate filaments, and the smaller capsules, 3–5 mm long, which are abruptly deflexed. $2n = 26$, 52. In our area ssp. **decumbens** (Ait.) Hult. Bogs. Circumpolar: Alas, Yuk, Victoria Is to nBaffin Is to Lab s to nBC, Alta, Sask, Huds Bay, nQue. Map 847 ●

Loiseleuria Desv.

L. procumbens (L.) Desv.
Shrub, evergreen, essentially prostrate, mat-forming; leaves opposite, 2–8 mm long, elliptic to oblong, coriaceous, margin revolute; flowers 5-merous, in small terminal clusters; corolla 3–4 mm long, white to pink; ovary 2–3-loculed; seeds many. $2n = 24$. Moist alpine slopes. Circumpolar: Alas, Yuk, Mack to sBaffin Is to Nfld s to Wash, Alta, sMack, James Bay, sQue, NH. Map 849 ○

Menziesia J.E. Smith

M. ferruginea J.E. Smith
Branching shrub, often forming thickets, malodorous when bruised, 1–2.5 m tall, with loosely shredded bark; young twigs glandular-puberulent; leaves deciduous, oblong-elliptic to obovate, thin, 3–6 cm long, crenulate-serrulate, sparingly short-hairy or glabrous, pale beneath, tapering to short petioles; flowers on slender, glandular-puberulent pedicels, in terminal clusters, appearing with the leaves; calyx small, 4-lobed, ciliate; corolla 7–9 mm long, urn-shaped, 4-lobed, greenish yellow tinged with bronze; stamens 8, included; capsule ovoid,

4-loculed, septicidal; seeds linear. $2n = 26$. *M. glabella* A. Gray of Ed. 1. Montane woods. sAlas, sYuk s to Calif, Wyo. Map 850

Oxycoccus Adans. / Bog Cranberry

Trailing evergreen plants of wet bogs; stems slender, creeping and rooting, 1–5 dm long; leaves alternate, persistent, 4–10 mm long, ovate or lanceolate, entire, leathery, with more or less revolute margins; flowers 1–4, borne terminally, nodding; pedicels 1–4 cm long, usually with 2 red scale-like bracts at or near the middle; sepals 4; corolla 5–8 mm long, deeply 4-parted, pinkish, the petals slender, recurved; stamens 8; anthers exserted, awnless, with very long terminal tubes; ovary inferior, 4-loculed; fruit a globose or ellipsoidal berry, reddish, often spotted, 5–14 mm in diameter.

Leaves narrowly ovate to lanceolate, 6–11 mm long, 2–4 mm wide, acute to obtuse at apex, slightly revolute; pedicels puberulent; fruit 8–14 mm thick **O. quadripetalus**

Leaves elliptic-ovate, 2–6 mm long, 1.5–2 mm wide, pointed at apex, strongly revolute; pedicels usually glabrous; fruit 5–10 mm thick **O. microcarpus**

O. microcarpus Turcz. / Small Bog Cranberry
Differing from *O. quadripetalus* as shown in the key, and as follows: generally a more delicate species with filiform stems; flowers 1–3; pedicels with a pair of bracts below the middle; corolla 5–7 mm long. $2n = 24$. On wet moss in bogs. More or less circumpolar: Alas, Yuk, wMack to Huds Bay, sLab s to BC, Alta, cSask, sMan, James Bay. Map 849 •

O. quadripetalus Gilib. / Bog Cranberry
Stems and leaves as described above; flowers mostly 2–4 in a cluster; pedicels with reddish bracts above or below or at the middle; corolla 6–8 mm long. $2n = 48$. Mossy bogs. Circumpolar: seAlas, wMack s to Ore, Ida, Alta, sSask, sMan, Gt Lakes, Apal. Map 851

Phyllodoce Salisb.

Small heath-like shrubs, erect or spreading, 1–3 dm high, with densely leafy branches; leaves crowded, alternate, evergreen, linear, obtuse, grooved beneath, the margin revolute-thickened; flowers slender-stalked, nodding or erect, in terminal umbel-like clusters; calyx usually 5-parted, persistent; corolla urn-shaped to open cup-shaped, 5-lobed; stamens 10, included; anthers awnless, opening by oblique pores; style slender; capsule globose, septicidal, usually 5-valved.

Corolla rose-coloured to scarlet, bell-shaped; sepals ovate, blunt, without glands **P. empetriformis**

Corolla yellowish, urn-shaped, constricted at the throat; sepals lanceolate, acute, glandular-hairy **P. glanduliflora**

P. empetriformis (Smith) D. Don / Red or Purple Heather
Leaves subsessile, 5–12 mm long, glabrous or scabrous, sometimes denticulate; flowers fragrant; pedicels glandular-puberulent; sepals about 2 mm long; corolla 5–8 mm long, glabrous. $2n = 24$. Hybrids between this species and the next, **P. × intermedia** (Hook.) Camp, occur in our area (Map 855 •); they have dull white to pale rose corollas and other characters more or less intermediate between those of the species. Subalpine and alpine meadows. Yuk, swMack s to Calif, Ida, Mont. Map 852

P. glanduliflora (Hook.) Coville / Yellow Heather
Differing from *P. empetriformis* as shown in the key and as follows; pedicels glandular-pubescent; sepals 4–5 mm long; corolla about 8 mm long, sulphur-yellow, more or less glandular-puberulent. $2n = 24$. Subalpine and alpine meadows. sAlas, Yuk, swMack s to Ore, Wyo. Map 853

Rhododendron L.

Evergreen or deciduous shrubs; buds scaly conspicuous; leaves alternate, short-petioled, entire; flowers showy, in terminal umbel-like clusters or 1–2 from lateral buds; calyx saucer-shaped, 5-parted, small or foliaceous; corolla open cup-shaped, 5-lobed, slightly zygomorphic; stamens 5–10, slender, sometimes declined; ovary 5-loculed; style elongated; capsule septicidal, 5-valved; seeds minute, wing-margined.

Leaves deciduous, thin, pale green; flowers white, lateral on the stem; calyx-lobes 8–12 mm long **R. albiflorum**
Leaves persistent, leathery, densely brown-scurfy beneath; flowers bright purple, in terminal clusters; calyx-lobes minute. **R. lapponicum**

R. albiflorum Hook. / White-flowered Rhododendron
Stems mostly 5–12 dm tall; young branches covered with long, appressed, reddish brown hairs; leaves narrowly elliptic to oblanceolate, 2–6 cm long, acute, tapered to the strigose petiole, pale green, appressed rusty-hairy especially above and on the midrib beneath, entire or wavy-crenate; flowers white, nodding, 1–3 in the axils of previous year's leaves and lateral on the stem, the pedicels glandular-puberulent; sepals large, glandular and strigose; corolla open cup-shaped, 2 cm broad, with large rounded lobes; stamens 10, hairy at base; capsule ovoid, 6–8 mm high, glandular-hairy, woody. $2n = 26$. Moist montane and subalpine forests. BC, Alta s to Ore, Mont. Map 854

R. lapponicum (L.) Wahlenb. / Lapland Rose-bay
Low matted shrub, 5–25 cm high; young branches densely brown-scurfy with minute scales; leaves short-petioled, elliptic-oblong, obtuse, thick, 6–18 mm long, greenish above, brownish beneath, densely covered with rusty-brown scales; flowers 3–6 in a terminal cluster; pedicels 6–10 mm long; sepals triangular, ciliate; corolla open cup-shaped, 1–2 cm broad, royal purple; stamens 5–10, the filaments glabrous, exserted; capsules 4–8 mm long, the pedicels scurfy. $2n = 26$. Moist

alpine slopes. More or less circumpolar: Alas, Yuk, Banks Is, Victoria Is to Devon Is to Nfld s to BC, Alta, sMack, James Bay, sQue, NH; Wisc. Map 855 ○

Vaccinium L.

Shrubs of diverse habit, mostly deciduous; leaves entire or serrulate; flowers solitary or in short clusters; calyx spreading or appressed, the lobes often obsolete; corolla open cup-shaped to urn-shaped or globose, shallowly to deeply 4–5-lobed; stamens 8 or 10, commonly included; anthers prolonged into tubes, sometimes awned; style slender; ovary inferior; fruit a many-seeded berry, 4–5-loculed, or 8–10-loculed by false partitions.

1 Leaves thick, coriaceous, evergreen, revolute, shining above; corolla open cup-shaped, 4-lobed, rose-pink; fruit red; mat-forming, with creeping or trailing branches **V. vitis-idaea**

Leaves thin, deciduous; corolla urn-shaped or globose; fruit blue, black or red; plants mostly erect 2

2 Leaves very hairy, especially beneath; flowers in dense clusters **V. myrtilloides**

Leaves glabrous or sparsely hairy; flowers solitary or 2–3 together 3

3 Young branches reddish or brownish, not distinctly angled; leaves usually obovate or oblanceolate, markedly cuneate at base 4

Young branches often green, sharply angled; leaves lanceolate to ovate or elliptic, rarely obovate but then acute 5

4 Leaves firm, dull green above, entire; corolla-lobes commonly 4; stamens 8 **V. uliginosum**

Leaves thin, shining above, serrulate; corolla-lobes 5; stamens 10 **V. caespitosum**

5 Fruit bright red; leaves commonly about 8 mm long; branches numerous, slender, 'broomy' **V. scoparium**

Fruit black or blue 6

6 Leaves entire or nearly so, obtuse; fruit blue, with a bloom **V. ovalifolium**

Leaves serrulate, acute; fruit black, without a bloom 7

7 Plant low, 1–3 dm high; leaves 1–2 cm long; fruit 5–8 mm wide **V. myrtillus**

Plant 4–15 dm high; leaves 2–5 cm long; fruit 8–10 mm wide **V. membranaceum**

V. caespitosum Michx. / Dwarf Bilberry
Densely branched shrub, 0.5–2.5 dm high, the twigs terete, glabrous or obscurely puberulent; leaves nearly sessile, obovate or oblanceolate, tapered to the base, 1–3 cm long, serrulate in upper half, prominently reticulate-veined; sepals very short and broad; corolla 4–5 mm long, usually pink; fruit light blue (rarely white) with a pale bloom, about 6 mm wide, sweet. $2n = 24$. Woods and open slopes. Alas, Yuk to sMan s to Calif, Ida, Colo, Sask; Gt Lakes, James Bay to Nfld, NEng. Map 856

V. membranaceum Dougl. *ex* Hook. / Tall Bilberry
Large shrub, usually 5–10 dm high, widely branched; leaves 2–5 cm long, ovate to elliptic-oblong, pointed, serrulate, glabrous; petioles 2–3 mm long; flowers solitary, on pedicels 6–12 mm long; calyx-teeth nearly obsolete; corolla 5–6 mm long, subglobose, yellow-green or purplish; fruit 8–10 mm wide, black, sometimes purplish, of excellent flavour. Moist montane woods. swMack, BC s to Calif, Ida, Mont. Map 857

V. myrtilloides Michx. / Blueberry
Low shrub, 1–4 dm tall; twigs pubescent; leaves thin and soft, elliptic-lanceolate, 1–4 cm long, entire, softly pubescent; flowers in short clusters; calyx-lobes 5, triangular; corolla broadly cylindrical, about 4 mm long, greenish white or tinged with purple; fruit blue (rarely white), glaucous, 4–8 mm wide, edible. $2n = 24$. Dry woods, sandy ground. sMack, sBC to sHuds Bay, Nfld s to wMont, Alta, sSask, sMan, Ia, Gt Lakes, Va. Map 858 ●

V. myrtillus L. / Low Bilberry
Low shrub, mostly 1–3 dm tall, erect but branches wide-spreading; young branches green, sharply ridged, sometimes puberulent in the grooves; leaves mostly 1–2 cm long, ovate or oval, serrulate, acute or obtuse; scarcely paler beneath; calyx-teeth nearly obsolete; corolla about 4 mm long; fruit 5–8 mm wide, black or bluish. $2n = 24$. Montane to subalpine woods. Eurasia: BC, Alta s to Wash, NM. Map 859 ●

V. ovalifolium J.E. Smith
Shrub, 2–10 dm tall, diffusely branched; young branches ridged, glabrous, often reddish; leaves elliptic to oblong-oval, rounded or obtuse at apex, entire or obscurely undulate or serrulate below the middle, 2–4 cm long; petioles 1–2 mm long; flowers solitary, on pedicels 1–5 mm long; corolla urn-shaped, pinkish, 7–9 mm long; fruit about 9 mm wide, blue, with a bloom. Subalpine woods. eAsia: sAlas, BC s to Ore, Mont; Lk Superior; sQue to Nfld, NS. Map 859 ○

V. scoparium Leiberg / Grouse-berry
Shrub, 1–2 dm high, broom-like with many fine erect branches; leaves thin, 6–12 mm long, ovate or oval to ovate-lanceolate, acute, serrulate, shining, conspicuously veiny; corolla 3 mm long, pink; fruit 3–5 mm wide, bright red. $2n = 24$. Open mountain slopes near timber-line. BC, Alta s to Calif, Colo; SD. Map 860

V. uliginosum L. / Bog Bilberry
Stout shrub, 0.2–6 dm high, much branched and often spreading; leaves firm, deciduous, broadly obovate to oval or elliptic, blunt at apex, narrowed at base, entire; flowers solitary or 2–4 in a cluster, on pedicels 2–6 mm long; calyx-lobes about 1 mm long; corolla pink, ovoid, 4–5-lobed; fruit blue with a bloom. $2n = 24$, 48. Alpine slopes and bogs. Circumpolar: Alas, Yuk, Banks Is, Victoria Is to Ellesmere Is to Nfld s to Calif, BC, Alta, nSask, Man, Lk Superior, NY, Vt. Map 858 ○

V. vitis-idaea L. / Bog Cranberry; Cow-berry
Mat-forming plant, more or less shrubby, 1–2 dm high, many of the branches creeping or trailing; leaves evergreen, coriaceous, subsessile, narrowly elliptic to obovate, 6–15 mm long, rounded at apex, revolute, shining above, pale beneath and dotted with dark hairs; flowers few, in short terminal clusters; sepals 4; corolla cup-shaped, about 5 mm long, 4-cleft; stamens 8; fruit red, edible, 5–10 mm wide. $2n = 24$. In our area ssp. **minus** (Lodd.) Hult. Coniferous woods, dry bogs, alpine slopes. Circumpolar: Alas, Yuk, Victoria Is to sBaffin Is to Nfld s to BC, Alta, cSask, sMan, Minn, sJames Bay, sQue. Map 861

PRIMULACEAE / Primrose Family

Annual or perennial herbs; leaves mostly opposite or whorled, sometimes all basal, simple, entire or occasionally toothed, without stipules; flower perfect, regular, hypogynous; sepals commonly 4 or 5, usually more or less united; corolla (absent in *Glaux*) gamopetalous, shallowly lobed to deeply cleft, the divisions 4–9, commonly 5; stamens as many as the corolla-lobes and opposite them, usually inserted on the corolla-tube; anthers introrse; pistil 1; ovary superior; placentation free-central; style 1; stigma usually capitate; fruit a capsule, opening by 2–6 valves or circumscissile.

1 Corolla absent; calyx pinkish or white; flowers small, axillary; succulent perennial with small, opposite, sessile leaves **Glaux**
Corolla present 2

2 Leaves all or chiefly cauline; flowers rarely in definite umbels 3
Leaves in a basal rosette or clustered at the ends of short branches of the caudex; flowers commonly in terminal umbels 5

3 Plants annual; capsule circumscissile; depressed plants with alternate leaves and minute flowers **Centunculus**
Plants perennial; capsule opening lengthwise; erect plants with mostly opposite or whorled leaves 4

4 Leaves clustered near the top of the stem; corolla white **Trientalis**
Leaves chiefly in pairs along the stem; corolla yellow **Lysimachia**

5 Corolla-lobes sharply reflexed; stamens more or less united, the anthers connivent around the style **Dodecatheon**
Corolla-lobes erect or spreading; stamens separate 6

6 Corolla white, tube mostly shorter than the calyx; style minute **Androsace**
Corolla lilac or pinkish, tube equalling or longer than the calyx; style slender 7

7 Plant matted, with a branching, slightly woody, caudex; leaves small, crowded at the branch-tips **Douglasia**
Plant scapose, with a rosette of leaves **Primula**

Androsace L.

Small scapose plants; leaves in rosettes, either on the basal crown or at the tips of short branches of the caudex; flowering stems 1–several, each with a terminal umbel subtended by an involucre; flowers small; calyx 5-parted, persistent; corolla salver-shaped or funnelform, 5-lobed, the tube shorter than the calyx, its throat constricted; stamens 5, included; style very short; capsule with few or many seeds.

1 Perennial, caespitose; corolla much longer than the calyx and with a yellow eye; umbel subcapitate, the pedicels very short **A. chamaejasme**
Annuals; corolla of about the same length as the calyx; umbel with pedicels over 5 mm long 2

2 Bracts of involucre elliptic to narrowly ovate **A. occidentalis**
Bracts of involucre lanceolate or subulate **A. septentrionalis**

A. chamaejasme Host / Sweet-flowered Androsace
Small caespitose-tufted perennial, the leaves in dense rosettes at the ends of slender caudex-branches; leaves narrowly oblanceolate, 4–10 mm long, long-hairy especially on the margin, those of last year persistent below the new ones; scape 2–10 cm (commonly about 5 cm) long, white-villose; umbel subcapitate, mostly 4–5-flowered; involucral bracts lanceolate to elliptic-oblong, 2–4 mm long, hairy; flowers fragrant; calyx-lobes oval, obtuse; corolla white or cream-coloured with a yellow, orange or pinkish eye, short salver-shaped, the lobes oval, about 3 mm long; capsule few-seeded. $2n = 20$*. Mountain banks and slopes to alpine elevations. Eurasia: Alas, Yuk, Mack, Banks Is, Victoria Is s to BC, Mont. Map 862 ●

A. occidentalis Pursh / Fairy Candelabra
Small scapose annual (usually winter annual); leaves mostly oblanceolate, 6–12 mm long, entire or denticulate, stiff-puberulent; scapes 1–several, 2–5 cm tall, puberulent; involucral bracts elliptic to narrowly ovate, 3–4 mm long; pedicels 3–8, ascending, unequal, 3–15 mm long; calyx-tube cup-shaped, pale, about as long as the lanceolate lobes; corolla white, shorter than the calyx; capsule many-seeded. $2n = 20$. Dry sandy ground. sBC to wOnt s to Calif, NM, Tex, Ark, Ind. Map 862 ○

A. septentrionalis L. / Fairy Candelabra
Small scapose annual (usually winter annual) or possibly short-lived perennial; leaves linear-oblanceolate, 5–25 mm long, entire or toothed, glabrous or puberulent; scapes 1–numerous, 2–25 cm tall, glabrous to puberulent; involucral bracts lanceolate to subulate, 1–3 mm long; pedicels 3–20, ascending or spreading; calyx-tube cup-shaped, 5-ridged, pale, longer than the sharply triangular or subulate lobes; corolla white or pinkish, about as long as the calyx; capsule many-seeded. $2n = 20$*. Dry open ground to alpine elevations. Circumpolar: Alas,

Yuk, Mack, Bank Is, Victoria Is to James Bay, Que, Nfld s to Calif, Ariz, NM, SD, ND; Baffin Is, Ellesmere Is. Map 863

Centunculus L.

C. minimus L. / Chaffweed
Small annual herb, the branches erect or prostrate-ascending, 2–10 cm high; leaves alternate, subsessile, obovate or oblong, 4–8 mm long; flowers minute, nearly sessile, solitary in the leaf-axils, the parts mostly in 4's; sepals about 2 mm long, narrowly lanceolate, corolla pink, shorter than the calyx, the tube broad, withering at the top of the capsule; stamens inserted near the throat of the corolla; capsule nearly globose, transversely dehiscent, the top falling away as a lid. $2n = 22$. Moist soil or mud on pond margins. sBC to Sask, ND, Ind, Ohio, Del s to Calif, Ariz, Tex, Fla. Map 864 •

Dodecatheon L. / Shooting Star

Scapose perennial herbs, glabrous or glandular-pubescent; caudex short, with fibrous-fleshy roots; leaves simple, borne on the caudex in a basal rosette; scape terminating in an involucrate umbel; flowers showy, pink to purple, the parts in 5's; pedicels reflexed in flower, erect in fruit; calyx deeply parted, persistent in fruit; corolla cleft almost to the base, the lobes reflexed and exposing the stamens inserted on the tube; filaments short, usually yellow, free or united into a tube, with basally attached anthers, connivent around the style; connective prominent, smooth or transversely wrinkled; pollen-sacs yellow or dark maroon; capsule ovoid or cylindric, opening by 5 short terminal valves (valvate) or by an apical lid (operculate); seeds numerous, 2–3 mm long.

Leaves glandular-pubescent; connective rugose with transverse folds; capsule-teeth truncate **D. conjugens**
Leaves glabrous; connective smooth, usually longitudinally wrinkled when dry; capsule-teeth acute **D. pulchellum**

D. conjugens Greene / Shooting Star
Entire plant glabrous or the leaves sparingly glandular-pubescent; leaves including petiole 4–10 cm long, obovate to narrowly oblanceolate, entire, rounded to acute, tapered into petiole; scape mostly 1–2 dm tall, the umbel 2–6-flowered; bracts lanceolate to subulate; pedicels 1–4 cm long in flower, longer in fruit; corolla-tube maroon, yellow above, the lobes magenta to white; filaments 1–2 mm long, yellow to dark maroon, smooth; anthers 5–7 mm long, the pollen-sacs dark maroon to black, the connective rugose; capsule cylindric, operculate. $2n = 44$. Plains. seBC to swSask s to Calif, Wyo. Map 865

D. pulchellum (Raf.) Merr. / Shooting Star
Plant glabrous; leaves including petiole 4–20 cm long, oblanceolate, ovate or spatulate, acute to rounded, tapered into petiole, usually entire, occasionally sinuate; scape 5–50 cm tall, the umbel 3–20-flowered; bracts deltoid to lanceolate

or spatulate; pedicels 1–5 cm long in flower, longer in fruit; corolla-tube maroon, yellow above, the lobes magenta to lavender; filaments 0.5–3.5 mm long, yellow, smooth or rugose; anthers 3–8 mm long, the pollen-sacs usually yellow, the connective dark maroon to black, rarely yellow, smooth; capsule cylindric to ovoid, valvate. $2n$ = 44, 88, 132. *D. radicatum* Greene of Ed. 1. Wet meadows and saline flats. sAlas, Yuk, wMack to Sask, sMan s to Calif, Ariz, Colo, Neb. ND. Map 866

Douglasia Lindl.

D. montana A. Gray
Low herb with persistent, slightly woody stems, caespitose, the caudex slender-branched and terminating in short leafy shoots; leaves awl-shaped, about 4 mm long, sparsely and minutely ciliate, somewhat interruptedly imbricate-clustered; flowers small, solitary, on short stalks with 1–2 minute bractlets; calyx cup-shaped, 5-lobed; corolla purple or lilac, the tube as long as the calyx, the throat crested beneath the 5 sinuses; corolla-lobes about 4 mm long, obovate-cuneate; stamens included; style filiform; capsule 1–2-seeded. Dry alpine slopes. Alta s to Ida, nWyo. Map 864 ○

Glaux L.

G. maritima L. / Sea Milkwort
Small leafy-stemmed herb with slender running rootstocks; stems 5–10 cm high, diffusely branched, erect or spreading, succulent, glabrous, glaucous or pale; leaves opposite, sessile, entire, linear to oblong or narrowly elliptic, 6–12 mm long; flowers small, axillary, white to deep pink or purplish, apetalous; calyx cup-shaped, 5-parted, petaloid; stamens 5, inserted at the base of the calyx and alternate with its lobes; capsule ovoid, 2–3 mm broad, 5-valved at the top, few-seeded; seeds about 1.5 mm long. $2n$ = 30. Saline flats and marshes. Circumpolar: sAlas, Yuk, swMack to sMan s to Calif, NM, ND; James Bay; Nfld s to Va. Map 867

Lysimachia L / Loosestrife

Erect leafy-stemmed perennial herbs; leaves opposite, either sessile or petioled; flowers yellow, long-pedicelled and in pairs at nodes or in peduncled racemes from the leaf-axils; calyx 5–6-parted; corolla deeply 5–6-cleft; stamens 5 or 6, the filaments distinct, sometimes accompanied by staminodia; ovary sub-globose; style slender; capsule globose or ovoid, dehiscent by longitudinal valves.

1 Flowers in peduncled racemes from leaf-axils; corolla-lobes linear; leaves attenuate at base, scarcely petioled, glandular-dotted — **L. thyrsiflora**
Flowers long-stalked in leaf-axils, sometimes appearing in whorls of 4; corolla-lobes broad; leaves petiolate — 2

2 Petioles ciliate; plants rhizomatous — **L. ciliata**
Petioles not ciliate; plants not rhizomatous — **L. lanceolata**

L. ciliata L. / Fringed Loosestrife
Stem almost glabrous, 4–10 dm high, from a slender rhizome; leaves ovate-lanceolate, acuminate, rounded to subcordate at base, rough-margined, the blades 3–10 cm long, the petioles 5–20 mm long, ciliate-fringed; flowers about 15 mm broad, on filiform axillary peduncles; calyx-segments firm, pointed; corolla rotate, deeply 5-cleft, the broad lobes bright yellow, reddish and glandular at base, usually short-cuspidate and erose-dentate; stamens 5, separate, alternating with minute staminodia. $2n$ = c.72, 96. *Steironema ciliatum* (L.) Raf. Moist woods and shores. BC to NS s to Ore, Colo, Tex, Fla. Map 868

L. lanceolata Walt.
Similar to *L. ciliata*; differing in the characters of the key, the narrower leaves that gradually taper at the base, and the longer lower petioles, 10–30 mm long. In our area ssp. **hybrida** (Michx.) Ray (*L. hybrida* Michx. of Ed. 1.). *Steironema lanceolatum* (Walt.) Gray. Moist meadows and shores. Alta to Que s to Wash, Ariz, NM, Tex, La, Miss, Fla. Map 869 ○

L. thyrsiflora L. / Tufted Loosestrife
Stems glabrous or pubescent, erect, 2–6 dm high, from creeping rhizomes; leaves opposite, rarely whorled, the lower small and scale-like, the upper ones 5–10 cm long, lanceolate to narrowly elliptic-oblong, attenuate to base, densely dotted with dark glands; flowers in spike-like racemes on peduncles from leaf-axiles near the middle of the stem; calyx-teeth 2–3 mm long; corolla deeply cleft into linear segments, often purple-dotted; stamens longer than the corolla; small teeth (staminodia), alternating with the corolla-lobes, sometimes present; capsule gland-dotted. $2n$ = 42*, 54. Marshes, shores, and ditches. Circumpolar: Alas, wMack to Huds Bay, sQue, NS s to Calif, Colo, Mo, WVa; Lab. Map 869 ●

Primula L. / Primrose

Perennial herbs, scapose, the leaves in a basal rosette; leaves petioled or merely narrowed to base, sometimes covered with a meal-like powder (farinose); flowers showy, lilac or pinkish, borne in a terminal umbel, subtended by several bracts; calyx tubular, angled, 5-cleft; corolla salver-shaped, the tube longer than the calyx, the 5 lobes spreading and notched or deeply cleft; stamens 5, included, inserted on the corolla; style filiform; stigma capitate; capsule 5-valved at the apex.

1 Leaves entire or undulate, not farinose, with long slender petioles; limb of corolla 5–9 mm across, the lobes 1.6–4 mm broad, deeply cleft; mature capsule slender-cylindric, tapering at apex, much exceeding the shallow-lobed calyx **P. egaliksensis**

Leaves mostly dentate or crenate, often farinose beneath, petioled or merely narrowed to base; limb of corolla 5–20 mm across, the lobes 3–7 mm broad; mature capsule thick-cylindric to ellipsoid-ovoid 2

2 Low, very slender plants, rarely more than 12 cm tall; involucral bracts 2–6 mm long; mature calyx 3–6 mm long, cleft to about the middle; capsule 2–3 mm in diam.; seeds rounded, nearly smooth **P. mistassinica**

Larger plants, normally more than 12 cm tall; involucral bracts 3–14 mm long; mature calyx 4–11 mm long, not cleft to the middle; capsule 2.5–5 mm in diam.; seeds angular, roughened 3

3 Leaves green beneath or often somewhat farinose; involucral bracts subulate or tightly involute above the dilated base **P. stricta**

Leaves copiously farinose beneath; involucral bracts lanceolate to linear-oblong, flat **P. incana**

P. egaliksensis Wormskj.
Leaves thin, ovate to obovate or spatulate, the petiole often as long as the blade or longer; scapes commonly 6–18 cm high; involucral bracts 3–6 mm long, dilated and saccate at base; calyx cylindric, 4–6 mm long, cleft about one-quarter to base into glandular-ciliate lobes; corolla violet or deep lilac, lobes short-cuneate and deeply cleft; capsule becoming twice or thrice the length of the calyx; seeds pale, smooth. $2n = 36$. Marshy ground. Greenland: Alas, Yuk, Mack to James Bay, Nfld s to BC, Alta, sMack, cQue; Colo. Map 870 ○

P. incana M.E. Jones / Mealy Primrose
Leaves elliptic to oblong-obovate, strongly farinose beneath, shallowly denticulate, sessile or with winged petioles, commonly 3–6 cm long, scape mostly 1–4 dm high, stiff and stout, 1–2 mm thick near the top, more or less farinose; involucral bracts flat, 0.5–1 cm long, usually about as long as the short stout pedicels; umbel subcapitate, but with some fruiting pedicels elongating to 1–2.5 cm; calyx usually copiously farinose; corolla lilac; capsule ellipsoid, only slightly exceeding the calyx; seeds strongly angled, 0.5–0.7 mm long, reticulated. $2n = 72$. Marshy ground and shores. Alas, Yuk, swMack to neHuds Bay s to BC, Ida, Utah, Colo, sSask, sMan, James Bay. Map 871

P. mistassinica Michx. / Dwarf Canadian Primrose
Delicate plant, not farinose or only slightly so; leaves oblanceolate or cuneate-obovate, often dentate, sessile or short-petioled; scape very slender, mostly 5–12 cm high; involucral bracts usually non-saccate at base; calyx slenderly cup-shaped, 3–6 mm long; corolla-tube yellow, exserted, the limb bluish purple, lilac or pink, with a conspicuous yellow eye and broad, cuneate, cleft lobes. $2n = 18$. Marshy ground and shores, often calcareous. Yuk, Mack to James Bay, sLab, Nfld s to BC, Alta, cSask, sMan, Gt Lakes, NS. Map 870 ●

P. stricta Hornem.
Resembling *P. incana*, differing as shown in the key and as follows: generally a more slender plant; leaves oblanceolate to narrowly obovate, commonly 1–3 cm long, entire or obscurely dentate, green or sparingly farinose beneath; scape mostly 1–3 dm high, rather stiff, not farinose; calyx not farinose. $2n = 126$. Moist alpine slopes. nEurope: Alas, Yuk, Banks Is, Victoria Is to Huds Bay, nQue, Lab s to BC, Alta, sMack, James Bay. Map 870 ▲

Trientalis L. / Star-flower

Erect, unbranched herbs with slender roostocks; stems 0.5–2 dm tall, bearing the 5–9 principal leaves in a whorl at the top, with a few small alternate leaves or minute scales below; larger leaves lanceolate to obovate, 2–10 cm long; flowers white, 1 or more, on filiform pedicels from the leaf-axils; calyx 5–7-parted; corolla white, rotate with very short tube, 5–9-cleft; stamens mostly 7; filaments united into a short ring at the base; style filiform; capsule globose, 5-valved, few-seeded.

Principal leaves lanceolate, acuminate; reduced leaves minute, scale-like **T. borealis**
Principal leaves oblanceolate or obovate, acute or obtuse; reduced leaves larger, green **T. europaea**

T. borealis Raf.
Stems mostly 6–18 cm high, glabrous or minutely glandular, rarely with axillary leafy branches; leaves lanceolate, acuminate, thin, 3–10 cm long, entire or obscurely crenulate, narrowed to the base, sessile or short-petiolate; corolla 8–14 mm wide; petals ovate or lanceolate, long-acuminate. $2n = 96$. In our area ssp. **latifolia** (Hook.) Hult. Moist woods. BC, Alta to nQue, Nfld s to Calif, Ida, Sask, sMan, Gt Lakes, Ga, Va; Yuk. Map 872

T. europaea L.
Stems mostly 4–15 cm high, often minutely glandular above; leaves oblanceolate or obovate, acute or obtuse, 2–6 cm long, entire or nearly so, cuneate at base, sessile or short-petiolate; corolla 8–14 mm wide; petals ovate, abruptly acuminate or acute. $2n = 90, 100, 160$, c.170. In our area ssp. **arctica** (Fisch.) Hult. Moist woods. Eurasia: Alas, Yuk, swMack, nSask s to Ore, BC, Alta. Map 873

GENTIANACEAE / Gentian Family

Annual, biennial or perennial herbs, glabrous or nearly so; leaves mostly opposite, simple, sessile, entire, without stipules; flowers perfect, regular, hypogynous; calyx 4–5-parted, corolla tubular to rotate, 4–5-lobed; stamens as many as the corolla-lobes and alternate with them, partly adnate to the tube; pistil of 2 united carpels; ovary 1-loculed, superior, with 2 parietal placentae, or nearly the whole inner surface of the ovary bearing ovules; style 1 or obsolete; fruit a capsule, usually 2-valved.

1 Corolla rotate, the small tube much shorter than the lobes; stamens inserted at base of corolla **Lomatogonium**
Corolla tubular to funnelform or nearly cup-shaped, the tube as long as or longer than the lobes; stamens on the corolla-tube 2

2 Corolla, at least in the larger flowers, 4-spurred at base **Halenia**
Corolla not spurred at base 3

3 Corolla with plaits or folds between the lobes, the lobes 3-veined; nectaries at the base of the ovary **Gentiana**

Corolla without plaits between the lobes, the lobes 5–9-veined; nectaries upon the base of the corolla-tube alternate with the stamens **Gentianella**

Gentiana L.

Annual, biennial or perennial; flowers usually 5-merous; corolla-lobes with plaits or folds between them, throat naked; nectaries at base of ovary; anthers basifixed; capsule stipitate; seeds often winged.

1 Plants prostrate; leaves 3–5 mm long, crowded on the stem, the margins scarious or whitish; biennials **G. prostrata**

Plants erect or ascending; leaves not as above; perennials 2

2 Basal rosettes of leaves present at flowering time; corolla 12–18 mm long, greenish blue **G. glauca**

Basal rosettes none at flowering time; corolla 20–40 mm long, deep blue or purple 3

3 Leaves oblong or lanceolate, acute at apex; calyx-lobes acute; corolla 2–3 cm long **G. affinis**

Leaves broadly ovate, obtuse at apex; calyx-lobes obtuse; corolla 3–4 cm long **G. calycosa**

G. affinis Griseb. / Prairie Gentian
Perennial; stems usually tufted, 1–3 dm high, glabrous; leaves 6–12 pairs below the inflorescence, ovate to oblong-lanceolate, 1–4 cm long; flowers thyrsoid-racemose, numerous or very few; calyx-lobes unequal and variable; corolla blue or bluish, 2–3 cm long, narrowly funnelform, the lobes ovate, spreading, the plaits between the lobes cleft and laciniate-toothed. Moist prairie, meadows, and open woods. sBC to sMan s to Calif, Ariz, Colo, SD. Map 874

G. calycosa Griseb.
Caespitose perennial; stems 1–3 dm high, often tufted, glabrous; leaves crowded on the stem, mostly 1–3 cm long, ovate, 3–5-ribbed, the lower pairs smaller and connate-sheathing at base; flowers usually solitary, sessile, subtended by the upper broad leaves; lobes of calyx ovate, obtuse; corolla blue, 3–4 cm long, oblong-funnelform, the plaits between the lobes with a lanceolate, acuminate, and often laciniate prolongation. $2n = 26$. Moist subalpine and alpine meadows. BC, Alta s to Calif, Ida, Mont. Map 875 ○

G. glauca Pallas
Perennial with stems 3–10 cm high; basal rosette-leaves oval or obovate, about 1 cm long, rather succulent; cauline leaves 2–3 pairs, ovate or oval, 0.5–1.5 cm long; calyx-lobes lanceolate; corolla 12–18 mm long, greenish blue, the lobes ovate, the plaits between the lobes entire. $2n = 24$*. Moist subalpine and alpine banks and ledges. eAsia: Alas, Yuk, wMack s to BC, Alta. Map 875 ●

G. prostrata Haenke / Moss Gentian
Low biennial, 1–10 cm high, branched at base; stems few–many, prostrate to erect ascending, somewhat succulent; leaves numerous, small, imbricate, 3–5 mm long, glaucous with broad whitish margins, the basal ones orbicular or ovate, the upper oblong to linear-lanceolate, appressed; flowers solitary, terminal; calyx slender, 4–5-lobed; corolla 10–15 mm long, light greenish purple or whitish. $2n = 36$. Includes G. *fremontii* Torr. of Ed. 1. Turfy alpine and subalpine slopes. Eurasia: Alas, Yuk, wMack s to Calif, Nev, Utah, Colo. Map 876

Gentianella Moench

Annual or biennial; flowers 4- or 5-merous; corolla lacking folds between the lobes, throat usually ciliate; nectaries at base of corolla-tube, alternate with stamens; stamens versatile; capsule sessile or stipitate; seeds wingless.

1 Flowers usually more than 2 cm long; corolla-lobes fringed or toothed; calyx-lobes with thin hyaline margins 2
Flowers usually less than 1.5 cm long; corolla-lobes entire or sparingly toothed; calyx-lobes with green margins. 3

2 Base of the calyx near junction with pedicel and usually at least one pair of calyx-keels covered by papillae (under magnification), the keels green or purple; upper leaves acute **G. crinita**
Base of the calyx and the calyx-keels smooth or wrinkled, the keels glossy purple; upper leaves usually obtuse **G. detonsa**

3 Corolla with a fringed crown in the throat; corolla-lobes acute or obtuse **G. amarella**
Corolla without a fringed crown; corolla-lobes apiculate or setose at tips **G. propinqua**

G. amarella (L.) Börner / Felwort
Annual or biennial; stem simple or branched from the base, 0.5–5 dm high, the branches strongly ascending; basal leaves oval-spatulate, obtuse; upper leaves oblong-lanceolate to linear; flowers few–numerous, 1–3 in each axil or in dense clusters; calyx deeply cleft into lanceolate lobes; corolla 10–15 mm long, the slender tube fringed at the throat, the lobes light violet, pale blue, greenish yellow or whitish. $2n = 36$. *Gentiana amarella* L. In our area ssp. **acuta** (Michx.) Gillett. Moist woods, meadows, and roadsides. Circumpolar: sAlas, Yuk, swMack to Huds Bay, nQue, Nfld s to Calif, Ariz, NM, SD, ND, Minn, Vt. Map 877

G. crinita (Froel.) G. Don / Fringed Gentian
Annual, 0.5–4 dm high; stem stiffly erect, simple or with a few upright branches; basal leaves mostly oblanceolate, obtuse; cauline leaves linear-lanceolate, acute, 2–3 cm long; flowers blue, rarely whitish, 2–4 cm long, solitary or few, on long erect stalks; calyx 1.5–2.5 cm long, 4-angulate, 4-lobed, the lobes acuminate, unequal, the keels more or less granulose-roughened; corolla 2–4 cm long, the 4 lobes broadly obovate, with short teeth at the summit and a few slender marginal teeth; filaments usually hairy near the middle. $2n = 78$. *Gentiana crinita*

Froel. In our area ssp. **macounii** (Holm) Gillett. Moist shores, calcareous bogs. sYuk, swMack, to James Bay, Que s to BC, Mont, sSask, ND, Ia, Ohio, NC. Map 878 •

G. detonsa (Rottb.) G. Don / Fringed Gentian
Similar to *G. crinita;* distinguished as shown in the key. $2n = 44$. *Gentiana detonsa* Rottb. In our area ssp. **raupii** (Porsild) Gillett. Moist banks and meadows. Circumpolar: Alas, Yuk, Mack to Huds Bay s to Alta. Map 878 ○

G. propinqua (Richards.) J.M. Gillett
Differs from *G. amarella* as shown in the key and as follows: stems usually lower, 0.3–2.5 dm high, and with many basal branches, often purplish; calyx-lobes unequal; corolla blue, 10–20 mm long. *Gentiana propinqua* Richards., includes *G. arctophila* Griseb. Alpine slopes. eAsia: Alas, Yuk, Mack, Banks Is, Victoria Is to Huds Bay s to Mont; Lab, Nfld. Map 879

Halenia Borkh.

H. deflexa (Sm.) Griseb. / Spurred Gentian
Annual or biennial; stem erect, leafy, 1–5 dm high, simple or branched above; basal leaves oblanceolate or spatulate, petioled; cauline leaves lanceolate to ovate, 2–4 cm long, sessile or subsessile; flowers in terminal or axillary cymes; calyx deeply 4-cleft, the lobes lanceolate, acuminate; corolla purplish green or bronze, with a short tube, 4-lobed, each lobe usually prolonged at the base into a slender spur; stamens 4; stigmas 2. $2n = 22$. Moist woods. BC to Nfld s to Mont, SD, Mich, NY. Map 880

Lomatogonium A.Br.

L. rotatum (L.) Fries / Marsh Felwort
Slender annual, 1–3.5 dm high; stem simple or branched above, the branches strongly ascending; lower leaves oblanceolate or spatulate; middle and upper leaves mostly linear; acuminate, sessile; inflorescence racemiform, the flowers terminating the stem and on pedicels from the upper axils; calyx deeply cleft into narrow lobes about equalling the corolla; corolla white or bluish, 10–12 mm long, rotate, deeply 4–5-cleft, each division with a pair of scale-like appendages at the base; stamens inserted at base of corolla; style none; stigmas decurrent on the sides of the ovary. $2n = 10$. Wet meadows and saline flats. Circumpolar: Alas, Yuk, Victoria Is to Huds Bay, nQue, Nfld s to BC, Alta, Sask, Man, Me; Ida; Wyo to NM. Map 881

MENYANTHACEAE / Buck-bean Family

Menyanthes L.

M. trifoliata L. / Buck-bean
Perennial bog plant with stout creeping rootstocks; leaves alternate, crowded near the base of the flowering stem, 3-foliolate, long-petioled, the petioles sheathing at the base; leaflets oval or elliptic, 3–8 cm long, thick, entire, glabrous; flowers in a crowded raceme on a scape-like stem 1–3 dm high; calyx deeply 5-parted; corolla 10–15 mm long, short-funnelform, 5-cleft, its lobes white or tinged with pink, conspicuously bearded within; stamens 5, inserted on the corolla-tube; style persistent; stigma 2-lobed; capsule ovoid, 6–10 mm long, rupturing irregularly; seeds smooth, shining, 2–2.5 mm long. $2n = 54$. Bogs and pond margins. Circumpolar: Alas, Yuk, Mack to Huds Bay, nQue, Nfld s to Calif, Colo, Mo, Ind, Pa, Del. Map 882

APOCYNACEAE / Dogbane Family

Apocynum L. / Dogbane; Indian Hemp

Perennial herbs with milky sap (latex), a tough fibrous bark, forking stems and opposite, entire, leaves; flowers in cymes; calyx 5-lobed; corolla bell-shaped to tubular, 5-lobed; stamens 5, adnate to the very base of the corolla-tube; the latter with 5 small appendages, alternating with the stamens; anthers sagittate, coherent to the stigma; ovaries 2, joined above by a large, ovoid, obscurely 2-lobed stigma; pods (follicles) 2, long, slender, terete; seeds numerous, slender, each with a tuft of long silky hairs.

1 Leaves drooping or spreading; corolla pinkish, fully twice as long as the calyx, its lobes recurved or somewhat spreading; seeds 2.5–4 mm long — 2
  Leaves ascending or spreading; corolla greenish to whitish, less than twice as long as the calyx, its lobes ascending; seeds 4–6 mm long — **A. cannabinum**

2 Flower clusters both terminal and axillary, the clusters flowering simultaneously; flowers nodding; corolla-lobes recurving — **A. androsaemifolium**
  Flower clusters at the ends of leafy branches, the central clusters opening much earlier than the others; flowers upright or spreading; corolla-lobes not recurving — **A. × medium**

A. androsaemifolium L. / Spreading Dogbane
Stem erect, 2–10 dm high, well branched above; leaves ovate to oblong, mucronate, glabrous above, glabrous to pubescent beneath; flowers fragrant, cup-shaped, 6–9 mm long; corolla pink or pink-striped, the lobes becoming recurved; bracts lanceolate to oblong, somewhat persistent; pods 8–12 cm long; seeds 2.5–3 mm long. $2n = 16$. Sandy areas and dry woods. Alas, Yuk, swMack to James Bay, sQue, Nfld s to Calif, Ariz, Tex, Ga. Map 883

A. cannabinum L. / Indian Hemp
Stem erect, 3–8 dm high, with ascending branches; leaves ovate to lanceolate, glabrous to somewhat pubescent beneath, those of the main axis with short petioles, those of the branches often nearly sessile; calyx-lobes about equalling the corolla-tube; corolla tubular to urn-shaped, 2–4 mm long, with erect lobes, greenish white; flower-bracts green or scarious, linear, caducous; pods 6–12 cm long; seeds 4–6 mm long. $2n = 16$. Includes *A. sibiricum* Jacq. of Ed. 1. Moist woods and shores. swMack, BC to Nfld s to Calif, Tex, Fla. Map 884

A. × medium Greene / Dogbane
Similar to *A. androsaemifolium* but differing as shown in the key and as follows: flowers narrowly cup-shaped to short-tubular, 4–7 mm long; corolla greenish white to pinkish tinged, the spreading lobes not becoming recurved; bracts linear, caducous; seeds 3–4 mm long. A hybrid between the two previous species. Woods and shores. BC, Alta: Ont to Nfld s to NM, Tex, Mo, Tenn, NC. Map 885

ASCLEPIADACEAE / Milkweed Family

Asclepias L. / Milkweed

Perennial herbs with milky sap (latex) and mostly opposite leaves; flowers numerous in umbels; calyx persistent, the lobes becoming reflexed; corolla deeply 5-parted; stamens 5, attached at the base of the corolla and united to form a tube; seated on the stamen-tube is a 5-hooded corona, each part often with an incurved horn; anthers adherent to the stigma; pollen-grains in masses (pollinia), these adhering in pairs; ovaries 2, with short styles; pods (follicles) 2, one of them often abortive, large, soft, ovoid to spindle-shaped, on recurved stalks; seeds numerous, broad, flat, margined, each with a tuft of long silky white hairs.

1 Corona hoods without an incurved horn within **A. viridiflora**
Corona hoods each with an incurved horn within 2

2 Leaves ovate or lanceolate, tapered at base; pods lacking tubercles **A. ovalifolia**
Leaves oblong or oval, rounded or cordate at base; pods with soft tubercles **A. speciosa**

A. ovalifolia Dcne. / Low Milkweed
Stem slender, 2–6 dm high; leaves ovate to lanceolate, obtuse or somewhat acute at apex, short-petioled; umbels loosely 10–18-flowered; flowers less than 1 cm long, greenish white; pods without tubercles, downy. $2n = 22$. Open woods and slopes. Alta to wOnt s to Sask, Neb, Ia, Ill. Map 886 •

A. speciosa Torr. / Showy Milkweed
Stem coarse, up to 2 m high; leaves large, broadly lanceolate to oval-oblong, rounded or subcordate at base, the blunt apex often mucronate, the petioles very short; umbels large, rounded, many-flowered, the pedicels densely white-

tomentose; flowers over 1 cm long, purplish or pinkish; pods with soft tubercles, white-tomentose. $2n = 22$. Moist grassland, roadsides. sBC to sMan s to Calif, Tex, Mo. Map 886 ○

A. viridiflora Raf. / Green Milkweed
Stem 2–6 dm long, often decumbent, pubescent to glabrate; leaves opposite or alternate, oval, ovate or oblong-lanceolate, becoming leathery, with undulate margins; umbels many-flowered; pedicels pubescent; corolla greenish white; hoods yellowish green to purplish; pods spindle-shaped, puberulent, without tubercles. Dry hillsides. sAlta to sOnt s to wMont, Ariz, NM, Tex, Ga; BC. Map 887 ●

CONVOLVULACEAE / Morning-glory Family

Twining or trailing herbs, with alternate leaves (or scales) and tubular or funnel-shaped flowers; sepals 5, rarely 4, more or less united; corolla 5-lobed, rarely 4-lobed; stamens 5, attached to the base of the corolla; pistil of 2 united carpels; styles usually united; fruit a capsule, mostly 4-seeded.

Green leafy plants; flowers 2–6 cm wide	**Convolvulus**
Yellow or reddish parasitic plants, with a few very small scale-like leaves; flowers 1–5 mm wide, borne in clusters	**Cuscuta**

Convolvulus L. / Bindweed

Annual or perennial; leaves simple, large, green; flowers 5-merous, solitary or in axillary or terminal corymbs; ovary 1- or 2-loculed.

Calyx not enclosed by green bracts; corolla about 2 cm long	**C. arvensis**
Calyx enclosed by 2 large green bracts; corolla 3–8 cm long	**C. sepium**

C. arvensis L. / Field Bindweed
Slender, trailing or low twining, perennial herb with very deep roots; leaves narrowly oblong to ovate with hastate to sagittate bases; flowers usually solitary in the leaf-axils; flower-bracts small, remote; corolla white or pink, about 2 cm broad and 1.5–2 cm long; seeds grey, rough, 3.5 mm long. $2n = 48$. Fields, gardens, waste ground. Introduced. Occasional.

C. sepium L. / Wild Morning-glory; Hedge Bindweed
Coarse, twining, climbing or creeping, perennial vine; extensively branched; spreading underground by long rhizomes; leaves commonly ovate-triangular, the base hastate to sagittate with truncate or rounded lobes, the apex acute to acuminate, long-petioled; flowers from the leaf-axils; flower with 2 large green bracts enclosing the sepals; corolla white to pinkish, 3–6 cm broad; seeds black or brown, smooth, angular, 5 mm long. $2n = 22$. *Calystegia sepium* (L.) R.Br. Thickets, shores, waste ground. BC to Nfld s to Calif, NM, Tex, Fla. Map 888

Cuscuta L. / Dodder

C. gronovii Willd.
Parasitic annual, with orange-yellow twining stems attached by suckers to stems of various other plants; flowers 2.5–3 mm long, in small, compact clusters, whitish; corolla deeply cup-shaped, its lobes short; scales in the throat of the corolla much shorter than the corolla-tube, truncate or bifurcate, lacerate-fringed; capsule ovoid, 3–6 mm long; styles about one-fourth as long as capsule; stigmas capitate; seeds 2–3 mm long. Dodder plants grow from seed but do not take root in the soil, their young twining stems soon attaching to host plants. $2n = 60$. Includes C. *umbrosa* Hook. of Ed. 1. Alta to Nfld s to Utah, NM, Tex, Fla. Map 887 ○

POLEMONIACEAE / Phlox Family

Annual or perennial herbs; leaves alternate or opposite, simple or variously divided; flowers usually showy; calyx 5-lobed; persistent corolla convolute in the bud, 5-lobed, the tube usually well developed; stamens 5, epipetalous; pistil with a 3-lobed style and a 3-loculed superior ovary, the placentation axile; fruit a loculicidal capsule.

1	Leaves simple, entire	2
	Leaves pinnate or deeply dissected	4
2	Plants perennial, forming dense tufts or mats; leaves mostly opposite	**Phlox**
	Plants annual, not tufted; the upper leaves, at least, alternate	3
3	Calyx enlarging and covering capsule; leaves mostly alternate	**Collomia**
	Calyx not enlarging, not covering capsule; lower leaves opposite	**Microsteris**
4	Flowers blue, generally longer or wider than 1 cm; plants perennial	**Polemonium**
	Flowers white, pale blue or lavender, not as long or wide; plants annual	5
5	Leaves opposite, sessile, palmately divided into soft linear segments; calyx-lobes equal	**Linanthus**
	Leaves and calyx not as above	**Navarretia**

Collomia Nutt.

C. linearis Nutt. / Collomia
Annual; stem erect, simple or branched, 1–4 dm high; leaves linear-lanceolate, 2–5 cm long, puberulent, the upper leaves broader and often pale at the base; flowers inconspicuous and unattractive, in terminal, head-like, leafy clusters; corolla pinkish, very slender, trumpet-shaped, 10–15 mm long; seeds mucilaginous when wetted. $2n = 16$. Moist open areas, roadsides. BC to wOnt s to Calif, NM, Neb, Mo, Ill, introduced Alas, Yuk, Mack. Map 889

Linanthus Benth.

L. septentrionalis H.L. Mason
Slender annual, 5–30 cm high; stem erect, branched well above the base; leaves opposite, puberulent or glabrous, 5–7-palmately divided into thread-like segments, 5–20 cm long; inflorescence cymose-paniculate; flowers small, solitary, on long slender spreading pedicels; calyx-tube more or less scarious below the sinuses; calyx-lobes deltoid-subulate; corolla short-funnelform, 2–5 mm long, well exceeding the calyx, the lobes white or pale blue, the throat shorter than the tube, usually bearing a ring of hairs near the stamen insertion; filaments glabrous or with a few hairs at the base; style exserted; stigma 3-lobed; capsule with 2–4 seeds in each locule. *L. harknessii* (Curran) Greene var. *septentrionalis* (Mason) Jepson & Bailey. Dry hillsides and plains. BC to swSask s to Calif, Colo. Map 890

Microsteris Greene

M. gracilis (Hook.) Greene
Annual; stem erect, rarely decumbent, 1–4 dm high, usually well branched, glandular-puberulent, at least above; leaves entire, the lower elliptic to oblong, the upper lanceolate to linear, 2–5 cm long, all except the uppermost opposite; flowers slender, 10–14 mm long, in the axils of alternate leaf-like bracts; calyx glandular-pubescent, the slender lobes about as long as the tube, the membranous sinus of the tube soon ruptured by the enlarging capsule; corolla 8–12 mm long, the tube yellowish, the limb pink or purplish, well exserted. $2n = 14$. *Phlox gracilis* Hook. Dry to moist open ground. sBC, Alta s to Calif, NM. Introduced Yuk. Map 891 ○

Navarretia R. & P.

N. minima Nutt. / Pin-cushion Plant
Low, stiff, tufted plant, 2–6 cm high; leaves pinnatifid, with needle-like divisions; calyx-tube white-hairy and scarious between the prominent green ribs, these prolonged into the unequal, sharp-pointed lobes; corolla white, 3–7 mm long; capsule 1–3-loculed, with 1–3 seeds in each locule; seeds pale brown, angular. Moist areas on plains. sBC to sSask s to Calif, Ariz, Colo. Map 891 ●

Phlox L.

Low, caespitose, tufted or decumbent perennials; leaves firm, awl-shaped or oblong-linear, sharp-pointed, mostly opposite; flowers solitary or 2–3, sessile or short-stalked, at the ends of branches; sepals cobwebby or glandular-hairy, united below by a scarious margin; corolla salverform, the limb showy, white or bluish; stamens inserted at different levels on the corolla-tube; capsule ellipsoidal, enclosed by the calyx, 3-valved, few-seeded.

Leaves awl-shaped, less than 2 mm wide, with cobwebby hairs; corolla 8–10 mm long **P. hoodii**

Leaves linear to oblong, 2–4 mm wide, mostly without cobwebby hairs but ciliate on margins; corolla 15–18 mm long **P. alyssifolia**

P. alyssifolia Greene / Blue Phlox
Branches nearly prostrate, from a somewhat woody base; leaves mostly oblong-linear, 10–18 mm long, cuspidate, rather thick, with white, hard, ciliate margins; calyx 7–8 mm long, glandular-hairy, the teeth bristle-pointed; corolla pale bluish purple or white, 15–18 mm long. Dry gravelly slopes. BC to sSask s to Mont, Colo, SD, ND. Map 892 ●

P. hoodii Richards. / Moss Phlox
Densely tufted, with crowded branches from a somewhat woody root; leaves awl-shaped, sparsely woolly, apiculate; calyx 5–7 mm long, cobwebby, apiculate; corolla white, 8–10 mm long. $2n = 28$. Prairie grassland. BC to sMan s to Calif, Utah, Wyo; Alas, Yuk. Map 893 ●

Polemonium L.

Perennial herbs with horizontal or ascending rootstocks; stems more or less glandular-puberulent; leaves alternate, pinnate; inflorescence cymose, sometimes subcapitate; flowers showy, usually blue; calyx cup-shaped, enlarging after anthesis; corolla cup-shaped or funnelform; filaments of stamens declined, often hairy at base; style elongate; ovules 3–several in each locule; capsule 3-valved; seeds spindle-shaped, often black.

1 Corolla broadly funnelform; filaments without basal appendages; leaflets more or less verticillate and glandular-viscid **P. viscosum**

Corolla campanulate; filaments appendaged at base; leaflets not verticillate 2

2 Stem tall, solitary, 4–10 dm high; petals hairy **P. acutiflorum**

Stems low, several in a tuft, 1–3 dm high; petals glabrous **P. pulcherrimum**

P. acutiflorum Willd. / Jacob's-ladder
Stem solitary, from a horizontal rootstock, glandular-puberulent above; leaflets 15–27, lanceolate to elliptic, acute or acuminate at apex; inflorescence a strict corymbose cyme; corolla blue, rarely white; stamens included; style exserted. $2n = 18$. *P. caeruleum* L. of Ed. 1. Marshy ground and wet meadows. Eurasia: Alas, Yuk, wMack s to BC, Alta. Map 893 ○

P. pulcherrimum Hook. / Jacob's-ladder
Plants caespitose; stems rather slender, sparingly branched, somewhat glandular-puberulent; leaflets 11–23, ovate to orbicular; inflorescence cymose, the pedicels long and slender; corolla 5–8 mm long, blue with yellowish tube; stamens and style subequal to the corolla. $2n = 18$*. Dry open slopes to alpine elevations. Alas, Yuk, nwMack s to Calif, Colo. Map 894

P. viscosum Nutt. / Skunkweed
Several stems arising from a persistent rootstock; leaves mostly basal; leaflets occasionally 2-ranked, but commonly 3–5-parted, appearing verticillate, the lobes elliptic to spatulate; inflorescence subcapitate, becoming elongate; corolla blue, violet or occasionally white, 15–30 mm long, with rounded lobes. Plants viscid-pubescent and often unpleasantly sticky to handle, with a heavy odour that suggests a skunk, although the flowers are sweet-scented. $2n = 18$. Rocky alpine slopes. BC, Alta s to Ore, Nev, NM. Map 892 ○

HYDROPHYLLACEAE / Waterleaf Family

Herbs, annual or perennial; leaves alternate, rarely opposite or basal, without stipules, simple or pinnatifid; flowers perfect, regular or nearly so, mostly in cymes, sometimes solitary and axillary; sepals 5, more or less united, often appendaged in the sinuses; petals 5, united, the corolla cup-shaped or funnel-form, often appendaged within; stamens 5, inserted near the base of the corolla and alternate with the petals; pistil of 2 carpels; ovary superior, mostly 1-loculed with 2 parietal placentae, sometimes 2-loculed; style mostly 2-cleft; fruit a 2-valved capsule; seeds few–many, often reticulated or pitted.

1	Flowers solitary, axillary or terminal; petals not longer than sepals	2
	Flowers several in axillary or terminal clusters; petals usually longer than sepals	3
2	Sepals with reflexed appendages in the sinuses; leaf-lobes 3–5, entire	**Nemophila**
	Sepals lacking appendages; leaf-lobes 7–13, remotely dentate	**Ellisea**
3	Leaves reniform-orbicular, palmately lobed; stamens included	**Romanzoffia**
	Leaves not as above; stamens exserted	4
4	Inflorescence a loose, more or less globose head, well surpassed by the leaves	**Hydrophyllum**
	Inflorescence elongate, well surpassing the leaves	**Phacelia**

Ellisia L.

E. nyctelea L.
Delicate annual; stems 1–4 dm high, simple or diffusely branched, hirsute; leaves 2–6 cm long, pinnately parted into 7–13 lanceolate or oblong divisions, more or less hairy, the lower leaves opposite, the others alternate; flowers solitary on peduncles borne at or above the nodes; calyx deeply cleft, the lobes 3–5 mm long, enlarging in fruit; corolla whitish, as long as the calyx or shorter, the lobes with minute appendages within; stamens included; capsule hispid; seeds usually 4, globose, reticulated. $2n = 20$. Moist woods, stream banks. sAlta to sMan s to Mont, NM, Tex, Okla, Mich; NJ to NC. Map 895 ●

Hydrophyllum L. / Waterleaf

H. capitatum Dougl. *ex* Hook.
Perennial with a very short rootstock and slender fleshy roots; stems 1–2 dm high, pubescent; leaves mostly basal, long-petioled, the blades ovate or oval in outline, pinnately divided into 5–7 segments, these sometimes cleft or toothed, pubescent; flowers in close head-like cymes, the peduncles 1–5 cm long, definitely shorter than the leaves; calyx deeply 5-cleft, very hispid; corolla purplish blue to white, cup-shaped, the tube short, each lobe with a linear appendage within; stamens 5, exserted, bearded in the middle; style 2-cleft, exserted; capsule globose; seeds 1–4. $2n = 18^*$. Moist montane woods and open slopes. sBC to Alta s to Calif, Colo. Map 895 ○

Nemophila Nutt. / Baby-blue-eyes

N. breviflora A. Gray
Delicate annual; stems 1–2 dm long, weak, diffusely branched, with minute reflexed prickles; leaves mostly alternate, ovate in outline, pinnately divided into 3–5 oblong-lanceolate segments; flowers solitary, on axillary or terminal pedicels; calyx 5-lobed, with a linear appendage in each sinus; corolla white or purplish, usually shorter than the calyx, with 10 appendages within; stamens 5, included; capsule 3–5 mm long, with a single seed, this with an evanescent caruncle at the end. $2n = 18$. Moist montane woods and open slopes. sBC to Alta s to Calif, Colo. Map 896 ○

Phacelia Juss. / Scorpion Weed

Herbaceous perennials or annuals, mostly hirsute or hispid; leaves alternate (or opposite below); flowers often showy, mostly in 1-sided raceme-like or spike-like cymes; calyx deeply 5-cleft, without appendages; corolla white, purple or blue, 5-lobed, usually with plaits in the tube; stamens 5, often long-exserted; style 2-cleft; capsule 2–many-seeded; seeds reticulated or roughened.

1 Leaves entire, or some with 2–4 entire lobes — 2
Leaves deeply divided — 3

2 Annual; leaves linear — **P. linearis**
Perennial; leaves lanceolate to elliptic — **P. hastata**

3 Stamens scarcely exserted — **P. franklinii**
Stamens long-exserted — 4

4 Leaves incised to midrib, the segments divided into linear lobes; stamens thrice as long as corolla — **P. sericea**
Leaves incised half-way to midrib, the segments broadly lanceolate and merely toothed; stamens twice as long as corolla — **P. lyallii**

P. franklinii (R.Br.) A. Gray
Biennial or winter annual; stem erect, 2–5 dm tall, soft-hairy; lower leaves petioled, pinnately or bipinnately parted into numerous linear-oblong or triangular divisions, often toothed, hirsute; upper leaves sessile and less dissected; inflorescence congested, little elongated in age; calyx-lobes linear; corolla pale blue or almost white, 6–9 mm long; stamens slightly exserted, the filaments subtended by pairs of lamellae; seeds brownish, minutely alveolate in vertical lines, oval, angular, several in the capsule. $2n = 22$*. Dry, open, especially disturbed, areas. Alas, Yuk, swMack to sHuds Bay s to BC, Ida, Wyo, sMan; Lk Superior. Map 896 ●

P. hastata Dougl. *ex* Lehm.
Perennial or possibly biennial, with a stout taproot and caudex, the stems single or clustered from the base; stems 3–5 dm high, canescent with fine pubescence and also bristly with long stiff hairs; leaves mostly simple, or some of them with lanceolate or linear lobes near the base, 4–8 cm long, silvery, dull grey or brownish with a dense fine pubescence and long appressed coarse hairs, the veins conspicuous; inflorescence dense, hispid; flowers white, pinkish or bluish, the corolla slightly longer than the calyx-lobes; filaments usually bearded, long-exserted; ovules few, usually 4. $2n = 22, 44$. *P. heterophylla* Pursh of Ed. 1. Dry slopes and valleys. sBC, Alta s to Calif, Colo, Neb. Map 897

P. linearis (Pursh) Holz.
Annual, 1–4 dm tall; stems simple to extensively branched, puberulent and hispid; leaves mostly sessile, 2–6 cm long, linear or lanceolate, simple or some of them deeply cleft into 3–7 linear or lanceolate divisions; inflorescence at first congested, becoming elongated in fruit; sepals linear, hispid-ciliate; corolla bright blue, rarely white, about 1 cm long; filaments glabrous, scarcely exserted; seeds several, oblong, black, coarsely reticulated. $2n = 22$. Dry slopes and plains. sBC to Alta s to Calif, Utah, Wyo. Map 898

P. lyallii (A. Gray) Rydb.
Perennial, 1–2 dm high, with a stout caudex; stems sparsely hirsute or villose above; leaves oblanceolate in outline, 5–10 cm long, green, sparingly strigose, coarsely pinnatifid, the lobes broadly oblong or lanceolate; inflorescence short and dense; corolla deep blue, about 8 mm long; stamens exserted. $2n = 22$. Rocky alpine slopes. seBC, Alta s to Mont. Map 899

P. sericea (Graham) A. Gray
Perennial, 1–4 dm high, with a branching caudex; stems silky-pubescent or canescent; leaves pinnately or bipinnately parted into numerous linear-oblong lobes, silvery-silky, at least when young, the lower ones petioled; inflorescence spike-like, elongating, up to 2 dm long; corolla 5–7 mm long, violet-blue, rarely white; stamens long-exserted, the filaments glabrous. $2n = 22$*. Open woods and slopes to subalpine elevations. sBC, Alta s to Calif, Nev, Colo. Map 900

Romanzoffia Cham.

R. sitchensis Bong.
Delicate perennial herb, bulbiferous at the base, resembling some saxifrages in general appearance; stems slender, 1–3 dm high, glabrous or nearly so; leaves mostly basal, with long petioles, the blades thin, glabrous or nearly so, reniform, 1–3 cm broad, 5–7-lobed, the lobes rounded-ovate and mucronate; flowers loosely racemose; calyx-lobes oblong-linear, glabrous; corolla 6–9 mm long, white or cream; seeds numerous. $2n = 22$*. Moist rocks and ledges to alpine elevations. sAlas, BC, Alta to Ore, Ida, Mont. Map 901

BORAGINACEAE / Borage Family

Herbs, mostly rough-hairy; leaves alternate or rarely opposite, entire; flowers regular or nearly so, hypogynous, perfect, in 1-sided, circinate or scorpioid cymes, spike-like or racemose; calyx 5-parted; corolla 5-lobed, commonly salverform, often bearing 5 appendages in the throat; stamens 5, inserted on the corolla, usually included in the tube; ovary commonly 4-lobed and 4-loculed; style single; stigmas 1 or 2; fruit usually of 4 nutlets, each 1-seeded.

1 Ovary deeply 4-lobed, the style arising between the lobes; plants usually rough-hairy 2
Ovary grooved, the style arising from the top; plants glabrous **Heliotropium**

2 Nutlets with hooked prickles 3
Nutlets lacking hooked prickles 5

3 Nutlets prickly all over **Cynoglossum**
Nutlets prickly mainly on the margin 4

4 Fruiting pedicels erect or ascending; inflorescence bracteate; corolla 3–4 mm wide **Lappula**
Fruiting pedicels deflexed; inflorescence essentially ebracteate; corolla often much wider **Hackelia**

5 Leaves opposite or in whorls of 3; flowers mostly solitary (2–4), axillary **Asperugo**
Leaves alternate (the lower opposite in *Plagiobothrys);* flowers several–many, in axillary or terminal, frequently raceme-like, cymes 6

6 Flowers in leafy cymes 7
Flowers in non-leafy cymes, with or without small bracts 8

7 Style conspicuously exserted; corolla greenish white, lobes acute **Onosmodium**
Style included; corolla yellow or orange, the lobes rounded **Lithospermum**

8 Cymes bracteolate to the tip 9
Cymes bractless or only the lowest flowers with bracts 13

9 Corolla blue 10
Corolla white or dark crimson-brown 12

10 Stamens exserted; corolla at least 1 cm long or wide 11
Stamens included; corolla not as long or wide **Lycopsis**

11 Pedicels arcuate, much longer than calyx; corolla rotate, regular **Borago**
Pedicels erect or ascending, shorter than calyx; corolla tubular-funnelform, the upper lobes larger **Echium**

12 Corolla white; leaves oblanceolate to spatulate **Cryptantha**
Corolla dark crimson-brown; leaves linear to lanceolate **Nonea**

13 Flowers blue 14
Flowers white, yellow or reddish purple 15

14 Corolla tubular-funnelform, 1 cm long or more; flowers somewhat congested **Mertensia**
Corolla rotate, less than 1 cm long; flowers generally distant **Myosotis**

15 Stem mostly 5–10 dm tall; flowers nodding; corolla 12–18 mm long **Symphytum**
Stem mostly 1–4 dm tall; flowers not nodding; corolla 2–8 mm long 16

16 Corolla yellow; leaves lanceolate to oblong-elliptic **Amsinckia**
Corolla white; leaves linear to oblanceolate 17

17 Nutlets with a ventral groove or slit; plants of dry habitats **Cryptantha**
Nutlets keeled on the ventral surface, without a groove; plants of moist habitats **Plagiobothrys**

Amsinckia Lehm.

A. menziesii (Lehm.) Nels. & Macbr.
Annual herb; stem 2–5 (7) dm tall, simple or branched, hirsute; leaves 2–7 (12) cm long, lanceolate to oblong-elliptic; inflorescence of raceme-like terminal and axillary cymes; calyx 5–10 mm long in fruit, divided almost to the base; corolla yellow, 4–7 mm long, tube scarcely exserted, throat open, appendages absent; stamens included; nutlets 2.5–3.0 mm long, tuberculate and wrinkled. $2n = 16$, 26, 34. Dry open ground, railway yards. Introduced. Rare.

Asperugo L.

A. procumbens L. / Madwort
Annual herb; stem 3–12 dm long, weak, trailing, with retrorse prickles; leaves opposite or more or less whorled, 2–10 cm long, oblanceolate, thin, often deciduous; flowers axillary, mostly solitary; calyx reticulate-veined, deltate-lobed, enlarging to form a compressed 2-lipped envelope, 1–2 cm wide, around the fruit; corolla funnelform, blue, 2–3 mm wide, appendages present; stamens included; nutlets about 2.5 mm long, ovate, flattened, verrucose. $2n = 48$. Waste ground. Introduced. Rare.

Borago L. / Borage

B. officinalis L.
Annual; stems 1.5–7.0 dm tall, hispid; lower leaves petiolate, blades 5–20 cm long, lanceolate to ovate or broadly elliptic, upper leaves sessile, clasping; flowers in open bracteated cymes; calyx about 10 mm long in flower, elongating in fruit, divided almost to base; corolla rotate, 2–3 cm wide, blue, appendages present; stamens exserted, connivent; nutlets 7–10 mm long, oblong, strongly ribbed. $2n = 16$. Waste ground, row crops. Introduced. Occasional.

Cryptantha Lehm.

Herbs of dry open places; stems and leaves bristly-hairy; leaves mostly alternate, entire, linear to oblanceolate; inflorescence scorpioid and racemose or spike-like; calyx-lobes enlarging in fruit; corolla small, white or yellow, salverform, the throat with crest-like appendages; stamens included, the filaments short or absent; nutlets attached to an elongate gynobase along a median ventral groove.

1 Annual; stems often diffusely branched; corolla less than 2 mm wide — 2
Perennial; stem simple or branched from base; corolla 6–11 mm wide — 3

2 Cymes bracteate throughout — **C. minima**
Cymes with only the lowest flower(s) bracteate — **C. fendleri**

3 Lower leaves narrowly oblanceolate, 5 mm wide or less; corolla 6–8 mm wide; nutlets 2–3 mm long — **C. macounii**
Lower leaves obovate-spatulate or broadly oblanceolate, 5–8 mm wide; corolla 7–11 mm wide; nutlets 3–4 mm long — **C. nubigena**

C. fendleri (A. Gray) Greene
Annual; stem branching from near the base, or simple below and widely branching above, hispid; leaves 2–5 cm long, narrowly oblanceolate, appressed hispid; calyx 4–7 mm long and closed in fruit; pedicels falling off with fruiting calyx; corolla white, inconspicuous, about 1 mm broad; nutlets 4, or fewer, smooth, acuminate, 1.5–2 mm long. Sandy soil and dunes. sBC to sSask s to Ore, Nev, NM, Neb. Map 902

C. macounii (Eastw.) Payson
Caespitose perennial with a rather slender taproot; stems slender, 1–2 dm high, conspicuously setose with spreading bristles; leaves linear or linear-lanceolate, 2–5 cm long, bristly and strigose; inflorescence narrow, densely setose; sepals 6–7 mm long in fruit; corolla white, tube about 3 mm long, limb 6–8 mm broad; nutlets 2.5–3 mm long, somewhat rugose and ridged on dorsal surface. $2n = 18$*. C. *nubigena* var. *macounii* (Eastw.) Boivin. Dry prairies. sBC to sSask s to Mont. Map 903 ●

C. minima Rydb.
Annual; stem 1–2 dm tall, branched; leaves linear-oblanceolate, up to 6 cm long, reduced upwards; inflorescence of 1 or 2 bracteate cymes; calyx 2–2.5 mm long in flower, about 5 mm long in fruit, eventually deciduous; corolla white, 2.5–3.0 mm long; nutlets heteromorphic, one 1.6–2.0 mm long, about 1.2 mm wide, the others 1.2–1.5 mm long, 0.7–0.9 mm wide. Eroding areas. Alta, Sask s to Mont, NM, Tex, Okla. Map 904 ○

C. nubigena (Greene) Payson
Biennial, or short-lived perennial, with a taproot; stem rather stout, simple or branched from the base, 1.5–3.5 dm high, coarsely setose with spreading bristles; leaves forming a rosette at the base the first year, these spatulate to oblanceolate, or the outermost almost obovate, obtuse, 2–5 cm long; cauline leaves somewhat narrower and longer; pubescence setose-bristly and also subtomentose, pustulate; inflorescence rather narrow, somewhat glomerate, extending over upper two-thirds or more of the stem, the lower bracts conspicuous; sepals 8–10 mm long in fruit; corolla white, tube 3–4 mm long, limb about 10 mm broad; nutlets 3–4 mm long, more or less rugose-tuberculate on dorsal surface. In our area var. **celosioides** (Eastw.) Boivin (*C. bradburyana* Payson of Ed. 1). Dry prairies. BC, Alta s to Calif, Ida, Mont, Colo, Neb, ND. Map 904 ●

Cynoglossum L.

Herbs, mostly biennial or perennial; flowers in ebracteate raceme-like cymes, or panicles; calyx cleft almost to base; corolla rotate to funnelform with a short, broad tube, closed by 5 appendages; stamens included; nutlets with stout, peltate spines.

Stem leafy to the inflorescence	**C. officinale**
Stem leafless above	**C. boreale**

C. boreale Fern.
Perennial; stems 4–8 dm tall, hirsute; leaves more strigose, the basal and lower cauline petiolate, blades 7–20 cm long, 2–7 cm wide, elliptic, upper cauline leaves clasping; inflorescence racemose, elongate in fruit; sepals 2–3 mm long; corolla 5–8 mm wide, funnelform, blue (white); fruiting pedicels 5–15 mm long, recurved; nutlets 4–5 mm long. Dry woods. BC to Nfld s to Ia, Wisc, Ind, NY, Conn. Possibly introduced in our area. Map 903 ○

C. officinale L. / Hound's-tongue
Coarse biennial; stem 4–8 dm tall, soft-pubescent; lower leaves petioled, oblong to lanceolate; upper leaves sessile and clasping, lanceolate, entire, soft-hairy; racemes several, from the upper axils; corolla funnelform, red-purple, the tube about as long as the calyx; nutlets flat on upper face, somewhat margined, densely covered with barbed prickles. $2n$ = 24, 48. Waste ground and pasture. Introduced. Occasional.

Echium L.

E. vulgare L. / Viper's-bugloss; Blue Devil
Biennial with a black taproot, forming a flat rosette in the first year; stem erect, 2–8 dm tall, bristly with short and long hairs, the latter often with conspicuous swollen bases; leaves linear-lanceolate to oblong, entire, hispid; flowers conspicuous, disposed on the upper sides of short branches that elongate in fruit; corolla usually bright blue, 15–20 mm long, tubular-funnelform, irregular, unequally 5-lobed; stamens unequal, at least the longer ones exserted; style 2-cleft at apex; nutlets rounded on one side, angular and with a median ridge on the other side, pointed at top, flattened at base. $2n = 16, 32, 64$. Waste ground and roadsides. Introduced. Occasional.

Hackelia Opiz / Stick-seed

Biennial or perennial; stems stiffly erect, rough-hairy, 5–10 dm tall; leaves linear to broadly lanceolate, entire; flowers in terminal scorpioid racemes, mostly bractless or with minute bracts; pedicels recurved in fruit; calyx and corolla as in *Lappula;* nutlets attached to a pyramidal base and surpassing the style, the back of the nutlet with an upturned rim bearing barbed prickles.

1 Cauline leaves lanceolate or elliptic-oblong, pointed at both ends; corolla about 2 mm wide; marginal prickles of nutlet subulate or slightly flattened **H. americana**
Cauline leaves linear-oblong, broad at the base; corolla 4–8 mm wide; marginal prickles of nutlet very flat 2

2 Style much shorter than sepals; backs of nutlets more or less keeled down the middle, lacking spines **H. floribunda**
Style about as long as sepals; backs of nutlets not keeled but spiny **H. jessicae**

H. americana (A. Gray) Fern.
Stems 5–10 dm tall, freely branched above, the branches bearing spreading racemes; lower leaves ovate-lanceolate, tapered to a petiole-like base, usually shrivelled at flowering time; middle and upper leaves mostly lanceolate, pointed at both ends; flowers 1.5–2 mm broad; corolla white or pale blue; nutlets with a dorsal area 2–3 mm long, bearing a marginal row of somewhat flattened prickles. *H. deflexa* (Wahl.) Opiz ssp. *americana* (Gray) Hult. Moist woods. swMack to eOnt s to Wash, Ida, Colo, Neb, Ia, Wisc; Vt; eQue, NB. Map 905

H. floribunda (Lehm.) I.M. Johnston
Stems 5–10 dm tall, strigose or villose; lower leaves narrowly oblanceolate, tapering into margined petioles; middle and upper leaves oblong to linear-lanceolate, erect or ascending, attached by a broad base, strigose; flowers 5–8 mm broad; corolla blue or white; nutlets 3–5 mm long, with a dorsal area 3–4 mm long, bearing a marginal row of very flat prickles, these often confluent at the base. $2n = 24$. Moist woods and meadows, banks. BC to sOnt s to Calif, NM, Neb, ND, Minn. Map 906

H. jessicae (McGreg.) Brand
Similar to *H. floribunda*; stems simple to the very large inflorescence; lower leaves large, oblanceolate and generally long-petioled; nutlets 4–5 mm long, the dorsal area rugose and with a few short spines or bristles. Open montane woods and meadows. BC, Alta s to Calif, Utah, Wyo. Map 907

Heliotropium L. / Heliotrope

H. curassavicum L.
Perennial herb; branches spreading, often decumbent, 1–4 dm long, glabrous and usually glaucous; leaves numerous, succulent, oblanceolate to spatulate, 2–4 cm long, entire, glabrous; flowers in short, dense scorpioid spikes; corolla white, sometimes bluish, with a yellow centre, 6–8 mm broad; stigma sessile, disc-shaped, broad; fruit 2.5–3 mm long, tardily splitting into 4 nutlets. $2n = 52$. Saline flats. Alta to swMan s to Mont, NM, Okla, Fla. Map 908

Lappula Moench / Stick-tights; Beggar-ticks

Annuals; leaves linear or oblong; flowers small, borne on erect pedicels in bracted racemes; calyx-lobes only slightly enlarged in fruit; corolla blue or white, salverform; nutlets 4, erect on the elevated receptacle, armed with barbed prickles.

Marginal prickles of nutlets slender, not confluent at the base, in 2 rows **L. squarrosa**
Marginal prickles of nutlets flattened, often confluent at the base, in 1 row **L. occidentalis**

L. occidentalis (S. Wats.) Greene
Native annual, but often weed-like; similar to *L. squarrosa* but the marginal prickles of the nutlets definitely in 1 row, the prickles flattened and sometimes confluent at the base. $2n = 48$. *L. redowskii* (Hornem.) Greene of Ed. 1. Open sandy ground, roadsides, railway grades. neAsia: Alas, Yuk, swMack to sMan, s to Calif, NM, Tex, Mo. Introduced sGtLakes, NEng. Map 909

L. squarrosa (Retz.) Dumort. / Blue-bur
Annual or winter annual; stems erect, simple below, branched above, 1–5 dm tall, hairy; leaves narrowly oblanceolate to linear, reduced upwards to bracts in the inflorescence, strigose and puberulent; flowers 3–4 mm broad, blue; nutlets 3.5–4 mm long, the margins armed with 2 rows of prickles, the backs muricate. $2n = 48$. *L. echinata* Gilib. of Ed. 1. Roadsides, waste ground, fields. Introduced. Common.

Lithospermum L. / Puccoon

Perennial with thick, reddish brown roots (yielding a purplish dye); stems 2–5 dm high, leafy, stiff-hairy; leaves alternate, sessile, linear or lanceolate, hairy; flowers yellowish or orange, in leafy spike-like cymes; early conspicuous flowers may be sterile, the nutlets arising from later cleistogamous flowers; calyx 5-lobed, persistent; corolla funnelform or salverform, the lobes rounded, often toothed; nutlets ovoid, very hard, whitish, shiny and smooth.

Corolla 10 mm or more long, the tube well surpassing the calyx, bright yellow or orange; nutlets 3–4 mm long **L. incisum**

Corolla less than 10 mm long, the tube seldom much longer than the calyx, pale yellow or greenish; nutlets 4–6 mm long **L. ruderale**

L. incisum Lehm.
Stems 1–5 dm high, erect or diffusely branched from the base and somewhat decumbent, strigose or hirsute; leaves linear or linear-oblong, strigose; early flowers usually bright yellow, 1–3 cm long, the corolla-lobes fringed or toothed; later flowers smaller, often apetalous, self-fertilized and producing most of the fruit; pedicels becoming recurved; nutlets ivory-white, somewhat pitted. $2n = 24$. Dry hills and plains. BC to sOnt s to NM, Tex, Okla, Mo, Ind. Map 910

L. ruderale Lehm.
Stems 2–5 dm high, several ascending from the stout base, forming a dense clump, rough-hairy; leaves linear or lanceolate, strigose or pilose, canescent, crowded on the stems, the upper ones almost concealing the flowers; calyx about 5 mm long; corolla pale yellow or greenish, the tube 5–8 mm long, the lobes about 1.5 mm long; cleistogamous flowers not formed; nutlets white or brownish. $2n = 24$. Dry slopes and grassland. BC to swSask s to Calif, Colo. Map 911

Lycopsis L.

L. arvensis L. / Small Bugloss
Annual; stem 1–5 dm high, usually branched, very bristly-hairy; leaves alternate, mostly lanceolate, hirsute; flowers slightly irregular, blue, in dense, bracteate, raceme-like clusters; calyx as long as the corolla-tube; corolla 6–8 mm long, salverform, the tube bent near the middle, the lobes slightly unequal, the throat closed by hairy scales; nutlets erect, ovoid, angular and wrinkled. $2n = 48$. Dry fields and waste ground. Introduced. Rare.

Mertensia Roth / Lungwort; Bluebells

Perennial with 1–several stems from the crown; stems and leaves soft-hairy or glabrous; leaves alternate, entire; inflorescence loose and short, paniculate, or raceme-like; calyx small, deeply 5-cleft; corolla tubular-funnelform, commonly cup-shaped above the tube, usually blue when mature, often pink or red in the bud, sometimes with crests in the throat; style long and filiform; nutlets erect, usually wrinkled when dry, attached above the base to a convex receptacle.

1 Plant usually more than 4 dm high; cauline leaves with distinct lateral veins; stems pubescent **M. paniculata**

Plant usually less than 4 dm high; cauline leaves without distinct lateral veins; stems glabrous 2

2 Cauline leaves elliptical or ovate; corolla-tube at least twice as long as the throat and limb **M. longiflora**

Cauline leaves mostly oblong-lanceolate; corolla-tube only slightly if at all longer than the throat and limb **M. lanceolata**

M. lanceolata (Pursh) A.DC. / Prairie Mertensia

Stems 2–4 dm tall, leafy, glabrous; lower leaves petioled, the upper sessile, the blades linear-oblong to ovate-lanceolate, usually glabrous beneath and roughly hairy or pustulate above; calyx-lobes lanceolate, ciliate on the margins; corolla about 1 cm long, the tube and limb about the same length. $2n = 24$. Prairie slopes and hillsides. Alta, Mont to Sask s to NM, Colo. Map 912

M. longiflora Greene

Stems 1–3 dm tall, glabrous, erect or ascending, from a short tuberous root; lower leaves petioled, obovate or elliptic-spatulate, 2–8 cm long; upper leaves sessile, elliptic-obovate or ovate, glabrous or hirsute; inflorescence congested, elongating in fruit; calyx glabrous; corolla 15–20 mm long, with conspicuous scale-like appendages. $2n = 24$. Open woods and slopes in mountains. BC, Alta s to Calif, Ida, Mont. Map 913 ○

M. paniculata (Ait.) G. Don. / Tall Mertensia

Stems 2–8 dm tall, hirsute or villose, erect, from a woody caudex; basal leaves long-petioled, cordate-ovate to elliptic-lanceolate, pubescent on both sides; cauline leaves short-petioled to sessile, lanceolate, often acuminate, hirsute; inflorescence at first congested, later paniculate; calyx strigose; corolla 8–14 mm long, the throat with conspicuous scales. $2n = 24, 72^*$. In our area ssp. **paniculata**. Moist woods. Alas, Yuk, swMack to sHuds Bay s to Ore, Mont, Alta, Sask, Man, Minn, Ia, Mich. Map 913 ●

Myosotis L. / Forget-me-not

Herbs, annual or perennial; inflorescence of elongating, terminal, raceme-like cymes; corolla usually blue, rotate to funnelform, tube usually short, closed by 5 appendages; anthers included; style filiform; stigma capititate.

1 Calyx with straight appressed hairs **M. laxa**

Calyx with some hooked, spreading hairs 2

2 Fruiting pedicels shorter than calyx; inflorescence extending almost to base of stem **M. micrantha**

Fruiting pedicels longer than calyx; inflorescence shorter 3

3 Corolla 2–3 mm wide, lobes ascending; calyx-lobes about as long as the tube **M. arvensis**

Corolla wider, lobes spreading horizontally; calyx-lobes distinctly longer than the tube **M. alpestris**

M. alpestris Schmidt / Alpine Forget-me-not
Low perennial, 1–3 dm tall, the stems densely tufted, nearly erect, hirsute, usually with papery sheaths at the base; basal leaves spatulate to linear-oblanceolate, hirsute; upper leaves linear-lanceolate; flowers in bractless, 1-sided racemes; calyx bearing some hooked or gland-tipped hairs at the base, the lobes about 2 mm long, erect in fruit; corolla bright blue, salverform, 4–6 mm broad, the throat crested, the tube about as long as the calyx; stamens included; nutlets small, smooth, erect, attached by the base to the flat receptacle. $2n = 24^*$, 48, 72. In our area ssp. **asiatica** Vestergr. Moist alpine and subalpine meadows. Eurasia: Alas, Yuk, swMack s to BC, Ida, Wyo; SD. Map 914 •

M. arvensis (L.) Hill / Forget-me-not
Annual; stem erect, 1.5–3.0 (5.0) tall; leaves 2–5 cm long, oblong to lanceolate or oblanceolate; fruiting pedicels 5–9 mm long; calyx 3.0–4.5 mm long in fruit, the lobes as long as or slightly longer than the tube, with both straight and hooked hairs. $2n = 36$, 48, 52, 66. Fields, waste ground. Introduced. Rare.

M. laxa Lehm.
Perennial; stems erect, becoming decumbent and loosely branched; leaves 2–6 cm long, oblong-lanceolate; inflorescence racemose, up to 2 dm long in fruit, generally leafy below; pedicels divergent in fruit, longer than calyx; calyx 3–4 (8) mm long in fruit, the lobes about as long as the tube; corolla blue, 4–8 (10) mm wide. $2n = 88$. Marshy ground. Introduced. Rare.

M. micrantha Pallas *ex* Lehm.
Annual; stem 1–2 dm tall, slender, branched from the base and soon forking to form the inflorescence; leaves 8–20 mm long, oblanceolate; inflorescence racemose, with leafy bracts below; fruiting pedicels 1.0–1.5 mm long; calyx about 4 mm long in fruit, the lobes about as long as the tube, hirsute with hooked, retrorse hairs. Roadsides and waste, sandy ground. Introduced. Rare.

Nonea Medic.

N. vesicaria (L.) Reichenb.
Annual or biennial, setose hispid; stems 1.5–5.0 dm tall, erect; leaves 2–15 (20) cm long, linear to lanceolate; flowers in bracteate axillary and terminal raceme-like cymes; calyx 5–7 mm long in flower, 10–15 mm in fruit, the lobes acute, shorter than tube; corolla dark crimson-brown, about 5 mm wide, appendages present, hairy; stamens included; nutlets 3.5–4.5 mm long, ovoid, strongly ribbed. $2n = 30$. Waste ground. Introduced. Rare.

Onosmodium Michx. / False Gromwell

O. molle Michx.
Coarse perennial; stems 4–8 dm high, stout, branched above, densely pubescent with appressed and spreading hairs; leaves mostly ovate to ovate-lanceolate,

4–6 cm long, conspicuously 5–7-ribbed, entire, coarsely hirsute-strigose; flowers in leafy-bracted scorpioid racemes or spikes; calyx 7–12 mm long, the lobes narrow, densely hirsute; corolla dull white or greenish, tubular, 12–20 mm long, canescent, the lobes 3–4 mm long, triangular and erect; stamens included; style long-exserted; nutlets about 4 mm long, white or brownish, shiny, smooth or very sparingly pitted, attached by a broad base to a flat receptacle. In our area var. **occidentale** (Mack.) Johnston. (*O. occidentale* Mack. of Ed. 1.) Gravelly banks and dry open woods. Alta; seSask to sOnt s to Mont, Utah, Colo, NM, Tex, Okla, La, Ill, NY, Md. Map 914 ○

Plagiobothrys F. & M.

P. scouleri (H. & A.) Johnston
Low annual, diffusely branched at base; stems spreading or ascending, branched, 5–20 cm long, strigose; leaves linear, 1–4 cm long, strigose, the lowest ones opposite, the others alternate; flowers very small, in 1-sided racemes, leafy-bracted; calyx 2–4 mm long in fruit, the lobes surpassing the nutlets; corolla white, short-salverform, about 1.5 mm long, inconspicuous; stamens included; nutlets more or less 3-sided with dorsal and ventral keels, rounded at base, wrinkled with low anastomosing ridges, attached obliquely to the receptacle. *P. scopulorum* (Greene) Johnston of Ed. 1. Saline flats and moist sandy ground. BC to Man s to Calif, NM; Alas, Yuk. Map 915

Symphytum L.

S. officinale L. / Comfrey
Perennial herb, hispid; stems 3–12 dm tall, erect; lower leaves petiolate, blades 15–30 cm long, lanceolate to ovate, upper leaves sessile, often decurrent; flowers in elongating terminal cymes, generally nodding; calyx 5–9 mm long, deeply lobed; corolla 12–18 mm long, about 10 mm wide, tubular-campanulate, dull reddish-purple or whitish, appendages present; nutlets 5–6 mm long, smooth or wrinkled, shining. $2n = 24, 26, 40, 48, 54$. Waste ground. Introduced. Rare.

VERBENACEAE / Vervain Family

Verbena L.

V. bracteata Lag. & Rodr. / Carpet Vervain
Annual or perennial herb, usually prostrate or decumbent, rarely erect, much branched and mat-like, the branches 1–5 dm long, more or less hirsute; leaves opposite, 1–5 cm long, spatulate-obovate in general outline, 3-lobed or pinnately incised and serrate, narrowed to a short petiole, hirsute; spikes terminal, sessile, elongate, with conspicuous linear or lanceolate bracts, these much longer than the flowers; calyx 3–4 mm long, hirsute, the lobes very short; corolla light blue to purple, gamopetalous, 5-lobed, weakly 2-lipped, its tube slightly longer than the calyx; stamens 4, didynamous, on the corolla-tube; pistil of 2 united carpels,

the ovary 4-lobed; stigma 2-lobed; fruit of 4 linear nutlets, about 2 mm long, reticulate above, striate below. $2n$ = 14, 28. Sandy prairie, roadsides. sBC to sOnt s to Calif, Tex, Fla. Map 916

LABIATAE / Mint Family

Annual or perennial herbs, often with a strong odour; stems usually 4-angled; leaves opposite, simple, usually toothed, often dotted with small glands containing volatile oil; flowers perfect, hypogynous, typically zygomorphic, chiefly in cymose clusters, these often arranged in terminal spikes or racemes; calyx tubular or cup-shaped, regular or zygomorphic; corolla usually very irregular and 2-lipped; stamens 4 or 2, inserted on the corolla-tube; ovary deeply 4-lobed, but 2-carpellate; style basal or terminal, 2-cleft at the summit; fruit of 4 small nutlets in the bottom of the calyx, each nutlet with a single erect seed.

1	Stamens 2 (staminodes may be present)	2
	Stamens 4	5
2	Flowers terminal, in head-like clusters or spike-like racemes	3
	Flowers clustered in leaf-axils	4
3	Corolla 2.0–3.5 cm long, pink to purple; flowers in a dense head-like cluster	**Monarda**
	Corolla much shorter, blue-violet; flowers in slender spike-like racemes	**Salvia**
4	Plants of dry habitats, annual; corolla 2-lipped, bluish	**Hedeoma**
	Plants of moist habitats, perennial; corolla with lobes equal, white	**Lycopus**
5	Calyx with a helmet-shaped projection on the upper side	**Scutellaria**
	Calyx lacking helmet-shaped projection	6
6	Corolla more or less regular, 4-lobed; stamens all of equal length	**Mentha**
	Corolla 2-lipped; stamens didynamous	7
7	Calyx distinctly 5-toothed, the upper tooth almost as wide as the lower 4 together	**Dracocephalum**
	Calyx not as above	8
8	Flowers in axillary clusters, not forming an obvious terminal inflorescence	9
	Flowers in a terminal, spike-like inflorescence	12
9	Calyx-teeth spinescent	10
	Calyx-teeth not spinescent	11
10	Corolla about 1 cm long; leaves palmately divided	**Leonurus**
	Corolla 1.5–3.5 cm long; leaves more or less entire or serrate	**Galeopsis**
11	Flowers pedicellate; upper leaves petiolate	**Glechoma**
	Flowers sessile; upper leaves sessile, clasping	**Lamium**

12 Upper pair of stamens longer than the lower pair; calyx 15-nerved or more 13
Upper pair of stamens shorter than the lower pair; calyx 5–10-nerved 14

13 Corolla white, purple-dotted; stem pubescent **Nepeta**
Corolla blue; stem glabrous **Agastache**

14 Calyx 2-lipped, the upper 3-toothed **Prunella**
Calyx not 2-lipped, equally 5-toothed 15

15 Stem and leaves glabrous; calyx-teeth triangular, as long as wide **Physostegia**
Stem and leaves glandular, pilose; calyx-teeth lance-subulate, much longer than wide **Stachys**

Agastache Clayt.

A. foeniculum (Pursh) Ktze. / Giant Hyssop
Perennial with creeping rootstocks and erect stems, 5–10 dm tall, little branched, glabrous, but pubescent in the inflorescence; leaves short-petiolate, ovate, acute or acuminate, coarsely serrate, whitish beneath with minute appressed pubescence; flowers small, crowded in interrupted terminal spikes; calyx 15-nerved, pubescent, violet or bluish, the teeth 1–2 mm long; corolla blue, surpassing the calyx; stamens 4, exserted, the upper pair curved downward and crossing the lower pair. $2n = 18$. Grassland and thickets. sMack, BC to Ont s to BC, Colo, Neb, SD, Ill. Map 917

Dracocephalum L.

Herbs with erect stems and serrate or crenate leaves; inflorescence of small bluish flowers, short and dense or elongated and interrupted; calyx tubular, 13–15-nerved, 5-lobed, the upper lobe much wider than the others; corolla scarcely longer than the calyx, 2-lipped, the upper lip notched and slightly arched, the lower lip spreading, 3-lobed, the middle lobe notched or cleft; stamens 4, the upper pair slightly longer; nutlets ovoid, 3-sided, smooth.

Flowers in compact clusters, mostly at ends of branches; calyx hirsute; bracts with sharp spinescent teeth **D. parviflorum**
Flowers whorled, in elongate, interrupted spikes; calyx puberulent; bracts entire or merely ciliate **D. thymiflorum**

D. parviflorum Nutt. / American Dragonhead
Annual or biennial; stems solitary or clustered, simple or branched, 2–6 dm high, puberulent or nearly glabrous; leaves petiolate, lanceolate, ovate or oblong, 2–6 cm long, coarsely toothed, puberulent beneath; flowers crowded in a dense terminal, ovoid or short-cylindric spike, 2–3 cm broad; bracts foliaceous, serrate, the teeth spinulose; calyx 10–14 mm long, the upper lobe ovate and longer than the rest, all spine-tipped; corolla light blue or violet, only slightly exceeding the calyx. $2n = 14$. *Moldavica parviflora* (Nutt.) Britt. of Ed. 1. Open woods and dry, disturbed areas. Alas, Yuk, swMack to James Bay, sOnt s to Ore, Ariz, NM, Neb, Mo, Wisc, NY; Que; Nfld. Map 918

D. thymiflorum L. / Thyme-leaved Dragonhead
Perennial; stems 1–5 dm high, usually unbranched, angled, puberulent; leaves petioled, the lower ovate, crenate, the upper ones ovate-lanceolate to narrowly oblong, denticulate or entire; flowers in loose whorls, subtended by small leaves, and forming a long interrupted spike; calyx 6–8 mm long, puberulent, purple-tinged, the lobes awn-tipped, the upper lobe broader than the others; corolla blue or purplish, slightly longer than the calyx. $2n = 20$. *Moldavica thymiflora* (L.) Rydb. of Ed. 1. Waste ground, roadsides. Introduced. Occasional.

Galeopsis L.

Annual herbs; flowers in dense terminal or axillary whorls; calyx strongly ribbed, teeth unequal, spine-tipped; corolla 2-lipped, the upper laterally compressed, arching, the lower 3-lobed with 2 conical projections at the base; stamens 4, anthers pubescent; nutlets obovate.

Corolla usually pink, purple or white, 1.5–2.0 cm long, the tube scarcely exceeding the calyx **G. tetrahit**

Corolla usually yellow, about 3.0 cm long, the tube much longer than the calyx **G. speciosa**

G. speciosa Mill. / Yellow Hemp Nettle
Similar to *G. tetrahit*; differing in the characters of the key. Generally a more robust plant with the stem uniformly hispid. $2n = 16$. Fields and roadsides. Introduced. Rare.

G. tetrahit L. / Hemp Nettle
Annual weedy herb; stems 3–8 dm tall, rather coarse, swollen below the nodes, bristly-hairy; leaves petioled, ovate or lanceolate, 3–12 cm long, acuminate, rounded or cuneate at base, coarsely serrate, pubescent; flowers in dense whorls subtended by the upper leaves; calyx enlarged in fruit, its 5 spinescent lobes becoming 5–11 mm long, the calyx-tube strongly 5-ribbed; corolla 15–22 mm long, white or pink or variegated, commonly with 2 yellow spots, the upper lip arched and entire, the lower 3-cleft, spreading and with 2 protuberances at its base; anthers pubescent; nutlets obovate, 3–4 mm long, smooth. $2n = 32$. Fields, roadsides, waste ground. Introduced. Common.

Glecoma L.

G. hederacea L. / Ground Ivy; Gill-over-the-ground
Creeping perennial herb, the prostrate stems slender, 2–5 dm long, pubescent or nearly glabrous; leaves petioled, round-reniform, 1–4 cm broad, coarsely crenate, green or tinged with purple; flowers in small axillary clusters; calyx tubular, 15-nerved, 5–7 mm long, oblique; corolla purplish blue, 1–2 cm long, the upper lip shallowly 2-lobed, the larger lower lip with short lateral lobes and a broad middle lobe. $2n = 18, 24, 36$. Waste ground, roadsides, cemeteries. Introduced. Occasional.

Hedeoma Pers.

H. hispidum Pursh / Pennyroyal
Small strong-scented annual; stems simple or branched from the base, 4–16 cm high, pubescent with recurved hairs; leaves crowded, sessile or nearly so, linear or narrowly oblong, entire, glabrous or hispid-ciliate; flowers small, in axillary clusters, the upper whorls often rather crowded to form a spike; calyx 13-ribbed, somewhat 2-lipped, hairy in the throat, the calyx-teeth subulate; corolla blue, 2-lipped, about 6 mm long; fertile stamens 2; nutlets ovoid, smooth. Dry eroded slopes, prairie. Alta to Que s to Colo, Tex, Fla, NY. Map 919 ●

Lamium L.

L. amplexicaule L. / Henbit; Dead Nettle
Annual or biennial, branched from the base, the slender glabrous stems ascending or decumbent, 1–4 dm long; lower leaves long-petioled, orbicular, ovate or cordate, 1–3 cm broad, incised-crenate, pubescent above; upper leaves similar but sessile and clasping; flowers in dense whorls, subtended by the upper leaves; calyx 5-nerved, villose, 5–6 mm long, with acuminate teeth; corolla 10–16 mm long, purplish, strongly 2-lipped, the upper lip erect, oval, concave, hairy, 3–5 mm long, the middle lobe of the lower lip spotted. $2n = 18$. Fields, waste ground. Introduced. Occasional.

Leonurus L. / Motherwort

L. cardiaca L.
Perennial herb, rhizomatous; stem 5–15 dm tall, erect, branched, pubescent; leaves petiolate, blades 3–12 cm long, palmately 3–7-lobed, the upper trifid; flowers in distant axillary whorls; calyx 4–8 mm long, distinctly nerved, teeth spiny; corolla 8–12 mm long, pink or white, upper lip villous; stamens 4; nutlets obovate, hispid at the apex. $2n = 18$. Waste ground, roadsides. Introduced. Rare.

Lycopus L. / Water Horehound

Perennial herbs, resembling *Mentha* but not odorous; stems erect, mostly from stolons; leaves sessile or petioled, serrate to pinnatifid; flowers small, white or lavender, in dense axillary clusters; calyx cup-shaped, regular, 4–5-toothed; corolla nearly regular, 4-lobed, hairy in the throat; fertile stamens, 2, the second pair wanting or represented by staminodia; nutlets 3-angled, the margins thickened.

1 Calyx-lobes deltoid or broadly lanceolate, less than 1 mm long, blunt or acutish, at maturity not surpassing the nutlets **L. uniflorus**
Calyx-lobes narrowly lanceolate, 1–2 mm long or longer, with subulate tips, at maturity surpassing the nutlets 2

2 Lower and middle leaves serrate, sessile; plant tuber-bearing; nutlets with the upper outer margin truncate and often undulate **L. asper**

Lower and middle leaves incised or pinnatifid, with petioles or narrow bases; plant without tubers; nutlets with the upper outer margin well rounded and entire **L. americanus**

L. americanus Muhl.
Stoloniferous plant, without tubers; stems erect, slender, 2–8 dm tall, glabrous or sparingly hairy; leaves mostly lanceolate, 2–7 cm long, acuminate, the principal ones serrate, laciniate or pinnatifid, abruptly narrowed to a petiole or petiole-like base, the uppermost shallowly serrate to nearly entire; calyx-lobes subulate; nutlets 1–1.5 mm long, less than 1 mm broad. $2n = 22$. Marshy ground. BC to Nfld s to Calif, Tex, Fla. Map 919 ○

L. asper Greene
Plant bearing tubers; stems stiffly erect, stout, spreading-pubescent; leaves sessile, usually ascending, lanceolate to lanceolate-oblong, the larger ones with 6–12 salient teeth on each margin; calyx-lobes acuminate; nutlets about 2 mm long and about as broad. Shores and marshes. BC to sOnt s to Calif, NM, Tex, Mo, Ill, Mich; Alas; James Bay. Map 920 ●

L. uniflorus Michx.
Plant stoloniferous and often tuberiferous; stems 2–6 dm tall, slender, puberulent to glabrate; leaves lanceolate or elliptic-oblong, gradually narrowed at both ends, coarsely toothed except for the lower cuneate portion; nutlets about 1.4 mm long, with a few tubercles across the upper outer margin. $2n = 22$. Marshy ground, stream banks. Alas, swMack, BC to Nfld s to Calif, Ida, Mont, Neb, Kans, Okla, Ark, NC. Map 920 ○

Mentha L.

Perennial herbs, smelling of mint, rhizomatous; leaves cauline; calyx regular or weakly 2-lipped; corolla more or less regular, 4-lobed, the upper lobe somewhat wider, often emarginate; stamens 4, exserted; nutlets ovoid.

Flowers in axillary whorls **M. arvensis**

Flowers in terminal spikes **M. spicata**

M. arvensis L. / Wild Mint
Stem ascending or erect, 2–5 dm tall, retrorse-pubescent, especially on the angles; leaves short-petioled, oblong-lanceolate to ovate-lanceolate, rounded to cuneate at base, serrate, glandular-punctate, pubescent or glabrescent; flowers small, in dense axillary clusters, the subtending leaves scarcely reduced upwards; calyx 2–3 mm long, 10–13-nerved, pubescent, the teeth as long as the tube; corolla pinkish, purple or white, about twice the length of the calyx, nearly regular, 4–5-lobed or notched. $2n = 12, 24, 48, 64, 72, 90, 92, 96, 120, 132$. Produces a hybrid, *M.* × *gracilis* Sole, with the following species. Marshy ground and sloughs. Circumpolar: Alas, Yuk, swMack to James Bay, Nfld s to Calif, NM, Tex, Mo, WVa, Del. Map 921

M. spicata L. / Spearmint
Stem 3–9 (12) dm tall, erect, branched, glabrous; leaves more or less sessile, 2–6 (9) cm long, lanceolate to oblong-lanceolate, serrate, essentially glabrous; flowers whorled, in narrow, terminal, leafless spikes; bracts subulate-lanceolate, as long as or longer than the calyx; corolla 2–4 mm long, lilac. $2n = 36, 48$. Roadsides. Introduced. Rare.

Monarda L. / Wild Bergamot; Horse Mint

M. fistulosa L.
Erect perennial; stems 3–7 dm tall, pubescent to glabrate; leaves short-petioled or subsessile, triangular-ovate to lanceolate, more or less serrate, greyish with soft pubescence or nearly glabrous; flowers large, rose or lilac, rarely whitish, densely aggregated in a terminal head-like cluster closely surrounded by bracts; calyx tubular, 15-nerved, hairy in the throat, the teeth 1 mm long; corolla 2–3.5 cm long, pubescent, strongly 2-lipped, the upper lip narrow, bearded at the tip, the lower lip spreading, 3-lobed; stamens 2, ascending under the upper lip of the corolla, exserted; nutlets oblong, smooth. $2n = 34, 36$. In our area var. **menthifolia** (Grah.) Fern. Open woods, fields, roadsides. BC to Que s to Ariz, NM, Tex, La, Ala, Ga. Map 922

Nepeta L.

N. cataria L. / Catnip
Perennial herb, very aromatic; stems clustered, erect, branched, 3–8 dm tall, whitish-puberulent or tomentose; leaves petioled, ovate, 2–8 cm long, coarsely crenate-dentate, grey-puberulent beneath; flower clusters 2–5 cm long, terminating the main stem and its branches; calyx tubular, incurved, 15-nerved, pubescent, 6–7 mm long, the teeth subulate; corolla 8–12 mm long, whitish, dotted with purple, the tube elongate, the upper lip erect, 2-lobed, the lower lip spreading, the lateral lobes small, the median one crenulate; stamens 4, the upper pair slightly the longer. $2n = 18, 36$. Waste ground, roadsides. Introduced. Rare.

Physostegia Benth.

P. parviflora Nutt. / False Dragonhead
Smooth perennial with erect stems 3–9 dm tall; leaves sessile, the principal ones elliptic-lanceolate or oblong-lanceolate, acuminate, tapering at base, serrate with hooked or blunt teeth, the uppermost pairs usually broadly rounded at base; flowers rose-pink or purplish, large and showy, borne singly in the axils of bracts, scattered or crowded in an elongate terminal spike; secondary spikes often develop in the axils of upper leaves; calyx more or less inflated, obscurely 10-nerved, finely puberulent; corolla 8–16 mm long, funnelform with enlarged throat, 2-lipped, the upper lip erect and entire, the lower lip spreading and 3-lobed; nutlets ovoid, smooth. $2n = 38$. *Dracocephalum nuttallii* Britt. of Ed. 1. Moist woods and stream banks. swMack, BC to wOnt s to Ore, Wyo, Neb, Minn. Map 923

Prunella L.

P. vulgaris L. / Heal-all
Small perennial, with rootstock; stems decumbent or erect, 1–3 dm high, glabrous or pubescent; leaves petioled, ovate to lanceolate or oblong, cuneate or rounded at base, entire or obscurely toothed, glabrous or sparingly hairy; spikes terminal, dense, 2–5 cm long, usually subtended by the uppermost leaves; bracts of spike reniform or ovate, cuspidate and ciliate; calyx 2-lipped, the upper lip almost truncate, bearing 3 short awns; corolla purplish blue, pink or white, 10–15 mm long, 2-lipped, the upper lip arched and entire, the lower lip 3-lobed with broad median lobe; stamens 4, the filaments forked near tip, one prong bearing the anther; nutlets smooth. $2n = 28$. In our area ssp. **lanceolata** (Bart.) Hult. Moist open woods, fields. Circumpolar: Alas, BC to Nfld s to Calif, Tex, Fla. Map 924

Salvia L.

S. nemorosa L. / Wood Sage
Perennial herb; stems erect, 3–8 dm tall, branched above, densely hairy; lower leaves petioled, oblong, subcordate at base, crenate, rugose, nearly glabrous above, densely puberulent beneath; upper leaves similar, but sessile and ovate to lanceolate; flowers in whorls subtended by ovate coloured bracts, arranged in a dense spike-like raceme 1–2 dm long; calyx 2-lipped, purplish, hairy, resin-dotted; corolla bluish purple, 2-lipped, pubescent, resin-dotted; stamens 2; anther connective elongated, attached near its middle to the filament, the upper portion bearing a fertile anther, the lower portion imperfect. $2n = 14$. Roadsides and waste ground. Introduced. Rare.

Scutellaria L. / Skullcap

S. galericulata L.
Perennial herbs with rhizomes; stems 1–8 dm tall, slender, ascending or erect, simple or branched, glabrous or pubescent; leaves 2–5 cm long, oblong-lanceolate, cordate at base, crenate; flowers in axillary pairs on pedicels about 2 mm long; calyx cup-shaped, 2-lipped, the lips more or less equal, entire, the upper with a protuberance; corolla 12–20 mm long, blue, tubular, dilated at throat, 2-lipped; stamens 4, the lower pair longer; anthers hairy; nutlets papillose, elevated on a gynophore. $2n = 16, 32$. Wet meadows, stream banks. Circumpolar: Alas, Yuk, swMack to James Bay, sLab, Nfld s to Calif, Ida, Ariz, NM, Tex, Mo, Ia, Ill, Ind, WVa. Map 925

Stachys L. / Hedge Nettle

S. palustris L.
Perennial with creeping rhizomes; stems erect, simple or sparingly branched, 3–8 dm tall, villose-hirsute; leaves sessile or short-petioled, ovate to oblong, 4–8 cm long, crenate or crenate-serrate, pubescent beneath, strigose above; flowers in axillary clusters, forming an interrupted terminal spike; calyx funnelform,

the 5 teeth nearly equal, subulate; corolla rose-purple, mottled with lighter and darker spots, 10–15 mm long, strongly 2-lipped, the upper lip erect, concave, the lower lip 3-lobed; stamens 4, ascending under the upper lip; nutlets dark brown, about 2 mm long. $2n$ = 64, 96, 97, 102, 103. In our area ssp. **pilosa** (Nutt.) Epling. Wet meadows, moist open woods. Circumpolar: Alas, Yuk, swMack to James Bay, sQue, Nfld s to Ore, Nev, Ariz, NM, Kans, Ohio, NY. Map 926

SOLANACEAE / Nightshade Family

Herbs or rarely woody plants; leaves alternate; flowers perfect, mostly regular, hypogynous, gamopetalous; calyx 5-lobed, usually persistent in fruit; corolla 5-lobed, rotate to tubular, mostly plaited; stamens mostly 5 and equal, inserted on the corolla; ovary superior, usually 2-loculed; ovules many, on axile placentae; style and stigma single; fruit a berry or a capsule.

1	Shrub, often spiny	**Lycium**
	Herbs	2
2	Corolla 2–3 cm long or wide; plants not spiny	3
	Corolla smaller, if as large plants spiny	4
3	Leaves alternate; fruit a capsule, enclosed by calyx	**Hyoscyamus**
	Leaves in pairs; fruit a berry	**Chamaesaracha**
4	Corolla rotate; stamens coherent, forming a tube around the style	**Solanum**
	Corolla campanulate; stamens free	**Physalis**

Chamaesaracha A. Gray

C. grandiflora (Hook.) Fern.
Annual; stem 1.5–9.0 dm tall, erect, mostly simple, densely viscid-villous; leaves 5–20 cm long, lanceolate-ovate, more or less entire; flowers in axillary fascicles; calyx 5–10 mm long, lobes linear-attenuate, longer than tube; corolla rotate, 3–5 cm wide, white with a yellow eye; fruit a berry enclosed by the enlarged globose calyx. Waste ground and roadsides, generally sandy. Introduced. Rare.

Hyoscyamus L.

H. niger L. / Black Henbane
Biennial or annual herb, clammy-hairy, fetid and poisonous; stems stout, erect, 3–10 dm tall, from a spindle-shaped root; leaves ovate to oblong in general outline, 1–2 dm long, pubescent, irregularly toothed or lobed, the lower ones short-petioled, the upper sessile and somewhat clasping; flowers sessile, in 1-sided leafy spikes; calyx cup-shaped, 2–2.5 cm long in fruit; corolla 2–3 cm long, funnelform, oblique, yellowish with purple veins; stamens 5; capsule subglobose, enclosed in the calyx, opening by a lid near the top; seeds reniform, papillate. $2n$ = 34. Waste ground, roadsides, gardens. Introduced. Occasional.

Lycium L.

L. halimifolium Mill. / Matrimony Vine
Shrub, sparingly branched, glabrous, often spiny at the nodes, the branches slender, recurved-drooping, sometimes climbing; leaves petioled or sessile, often fascicled, lanceolate or oblanceolate, 2–5 cm long, cuneate at base; calyx 3–5-cleft, about 4 mm long; corolla short-funnelform, 1–2 cm long, dull pinkish violet, the lobes shorter than the tube; stamens exserted; fruit an ellipsoid berry, scarlet or orange-red, 1–2 cm long. $2n = 24, 48$. Waste ground. An escape from cultivation. Introduced. Rare.

Physalis L.

P. ixocarpa Brot. / Tomatillo
Annual; stem 4–10 dm tall, branched, glabrous; leaves petiolate, 3–6 cm long, ovate to rhombic, entire to irregularly dentate; calyx 5–10 mm long, teeth deltoid, shorter than tube; corolla 1.0–2.5 cm wide, yellow with 5 blackish basal spots; fruit a berry enclosed by the enlarged calyx. $2n = 24$. A garden weed. Introduced. Rare.

Solanum L. / Nightshade

Low herbs or slender vines; leaves alternate, petioled; calyx 5-cleft, regular; corolla rotate or cup-shaped, regular, plaited in the bud; stamens 5, the filaments very short; anthers large, converging around the style in cone-shaped formation, opening by apical pores or short slits; fruit a many-seeded berry.

1	Plants spiny	**S. rostratum**
	Plants not spiny	2
2	Leaves pinnatifid	**S. triflorum**
	Leaves more or less entire or dentate	3
3	Stem hirsute, with long spreading viscid hairs	**S. sarrachoides**
	Stem glabrous or puberulent, with appressed hairs	**S. nigrum**

S. nigrum L. / Black Nightshade
Erect annual, 3–6 dm high, often widely branched; leaves petioled, thin, ovate to lanceolate, acuminate, cuneate at base, the margin entire, undulate or irregularly dentate, glabrous or somewhat hairy; flowers white, in small peduncled clusters, lateral from the internodes; berry globose, green, turning black at maturity. $2n = 48, 72$. Waste ground. Introduced. Rare.

S. rostratum Dunal
Annual; stem 3–10 dm tall, branching, stellate-pubescent, spiny; leaves petiolate, 6–20 cm long, 2–7 cm wide, lobed or pinnatifid; corolla yellow, 2–3 cm wide;

stamens with one anther longer than the others; fruit enclosed by the enlarged spiny calyx. $2n = 24$. Waste ground. Introduced. Rare.

S. sarrachoides Sendt.
Similar to *S. nigrum*; differing in the character of the key and the accrescent calyx that cups the lower half of the yellowish green fruit. $2n = 24$. *S. nigrum* var. *villosum* L. of Ed. 1. Waste ground, gardens. Introduced. Rare.

S. triflorum Nutt. / Wild Tomato
Native annual, low and spreading, branched from near the base; leaves oblong or ovate-lanceolate in general outline, deeply pinnatifid or lobed, the sinuses wide and rounded, sparsely hairy; peduncles 1–3-flowered, the pedicels becoming reflexed; corolla 8–10 mm broad, white; berry globose, 8–14 mm wide, green at maturity. $2n = 24$. Disturbed sandy prairies, roadsides. BC to Que s to Calif, Tex, Okla, Mo, Ia, Minn. Map 927

SCROPHULARIACEAE / Figwort Family

Herbs, or rarely woody; leaves commonly opposite, sometimes alternate or whorled, without stipules; flowers perfect, hypogynous, more or less irregular; calyx variously toothed or cleft, persistent in fruit; corolla gamopetalous, usually 2-lipped, the upper lip often 2-lobed, the lower lip 3-lobed; stamens 2, 4 or 5, occasionally 4 fertile and 1 sterile, inserted on the corolla-tube; pistil 1, the ovary 2-loculed, with numerous ovules on axile placentae; style single; fruit a capsule, usually 2-valved, ovoid to globose, many-seeded.

1	Leaves all basal	2
	Leaves not all basal	4
2	Flowers solitary, arising from the basal rosette of leaves on long peduncles	**Limosella**
	Flowers several in a spike-like raceme	3
3	Corolla absent; stamens 2	**Besseya**
	Corolla present; stamens 4	**Pedicularis**
4	Corolla spurred	5
	Corolla not spurred	6
5	Flowers yellow	**Linaria**
	Flowers bluish purple	**Chaenorrhinum**
6	Cauline leaves alternate	7
	Cauline leaves opposite	11
7	Corolla absent; stamens 2	**Besseya**
	Corolla present; stamens 4 or 5	8
8	Stamens 5, filaments villous; corolla distinctly 5-lobed, regular	**Verbascum**
	Stamens 4, filaments not villous; corolla obscurely 5-lobed, irregular	9

9 Floral bracts coloured, large, concealing much (all) of the flower **Castilleja**
Floral bracts green or purplish brown, smaller, the flowers quite evident 10

10 Plants lacking basal leaves; cauline leaves entire, linear, the upper trifid; flowers yellow **Orthocarpus**
Plants not as above **Pedicularis**

11 Calyx 4-lobed 12
Calyx 5-lobed 16

12 Stamens 2 **Veronica**
Stamens 4 13

13 Corolla pinkish purple **Odontites**
Corolla white to yellow 14

14 Calyx about 1 cm long in flower, constricted at the neck, becoming enlarged, membranaceous, veiny, and inflated in fruit **Rhinanthus**
Calyx much shorter, not inflated in fruit 15

15 Leaves 2–10 mm long, ovate, lobed **Euphrasia**
Leaves 10–40 mm long, linear to lanceolate, more or less entire **Melampyrum**

16 Stamens 2 **Gratiola**
Stamens 4 (a staminode may be present) 17

17 Staminode present, filamentous **Penstemon**
Staminode absent or scale-like 18

18 Flowers blue; upper leaves often whorled **Collinsia**
Flowers not blue; leaves all opposite 19

19 Corolla 7–14 mm long, green, tinged reddish brown to purple **Scrophularia**
Corolla usually much longer, yellow or pinkish red **Mimulus**

Besseya Rydb. / Kitten-tails

B. wyomingensis (A. Nels.) Rydb.
Perennial herb, 1.5–3 dm high, grey-puberulent or villose throughout; basal leaves long-petioled, ovate to oblong, 2–5 cm long, crenate-serrate, subcordate to cuneate at base; cauline leaves small, sessile; flowering stalk 5–25 cm tall; flowers in a dense spike; calyx 2-lobed, asymmetrical; corolla none; stamens 2, exserted, purplish; style filiform, with small capitate stigma; capsule broadly oval, emarginate, villose. *B. cinerea* (Raf.) Pennell of Ed. 1. Open slopes and dry grassland to alpine elevations. BC, Alta, swSask s to Utah, Colo, Neb. Map 928

Castilleja Mutis *ex* L.f. / Indian Paint-brush

Perennial herbs; leaves alternate, sessile, entire or cleft; flowers in dense, terminal, leafy (bracted) spikes, the bracts often brightly coloured with red or yellow and larger than the flowers; calyx tubular, laterally flattened, 4-lobed, more deeply cleft above and below than on the sides; corolla tubular, flattened, 2-lipped; upper lip (galea) arched, elongated, entire; lower lip usually very short, 3-lobed or 3-toothed; stamens 4, in pairs, ascending under and enclosed by the galea; anther-sacs unequal, one attached at its middle, the other pendulous; style slender, usually exserted; capsule loculicidal; seeds many, reticulated.

A complex group, the species often intergrading.

1 Inflorescence pink, red, crimson or purple 2
Inflorescence yellow 5

2 Upper leaves lobed; bracts scarlet **C. hispida**
Upper leaves mostly entire; bracts not scarlet 3

3 Inflorescence pink to reddish **C. miniata**
Inflorescence crimson to purple 4

4 Leaves 2–5 mm wide; inflorescence 1–2 cm wide, generally much longer than wide **C. raupii**
Leaves generally more than 5 mm wide; inflorescence 2–4 cm wide, about as long as wide to about twice as long **C. rhexifolia**

5 Flowers 4.5–5.5 cm long **C. sessiliflora**
Flowers shorter 6

6 Upper leaves lobed 7
Upper leaves mostly entire 8

7 Stem simple; leaves ascending, strongly imbricate often completely obscuring the stem; plants villous **C. cusickii**
Stem generally branched; leaves spreading, not obscuring the stem; plants puberulent, short-hirsute **C. lutescens**

8 Plants glabrous to villous; bracts mostly entire 9
Plants puberulent, short-hirsute; bracts lobed **C. lutescens**

9 Galea usually at least two-thirds as long as corolla-tube **C. miniata**
Galea less than half as long as corolla-tube **C. occidentalis**

C. cusickii Greenm.
Stems 1–6 dm tall, simple; leaves viscid-villous, the lower linear-lanceolate, usually entire, the upper leaves broader, with 1–3 pairs of narrow, acute lobes; bracts yellow, broader than leaves, oblong, with 1 or 2 pairs of short, narrow, ascending lobes; calyx 2–3 cm long, the lobes rounded; corolla shorter or longer than calyx, the galea short, less than half the length of the corolla-tube, the lower lip prominent, one-third to nearly the length of the galea. $2n = 24$. *C. septentrionalis* Lindl. (in part) of Ed. 1. Grassland. BC, Alta s to neOre, Ida, Mont, Wyo. Map 929

C. hispida Benth.
Stems 3–5 dm tall, puberulent and also hirsute-villous, especially above; leaves hirsute-villous or glabrate, lanceolate to ovate-lanceolate, mostly 3–5-cleft above

the middle, the lobes divaricate; bracts also 3–5-cleft, usually tipped with scarlet; calyx about 2.5 cm long; corolla 2–4 cm long; the galea usually about as long as the corolla-tube, the lower lip less than one-fifth the length of the galea. $2n = 24$, 48, 96. Appears to intergrade with *C. rhexifolia*. Open woods and meadows. BC, Alta s to Ore, Ida, Mont. Map 930

C. lutescens (Greenm.) Rydb.
Stems 3–6 dm tall, usually branched above; leaves linear to linear-lanceolate, mostly entire but not infrequently with a pair of acute, linear lobes, puberulent to densely short-hirsute; bracts yellow, puberulent, broader than the leaves, oblong, the tips rounded, entire or with 3–7 narrow lobes; calyx 1.5–2.5 cm long, the lobes acute, 1–7 mm long; corolla as long as or longer than calyx, the galea short, less than half the length of the corolla-tube, the lower lip one-fifth to half the length of the galea. $2n = 48$, 96, 120. *C. septentrionalis* Lindl. (in part) of Ed. 1. Grassy slopes. BC, Alta s to neOre, Ida, Mont. Map 931

C. miniata Dougl. *ex* Hook. / Common Red Paint-brush
Stems 2–6 dm tall, glabrous or somewhat hairy; leaves narrowly lanceolate to ovate, 3-nerved, usually glabrous or sparingly pubescent, entire or sometimes shallowly 3-lobed above; bracts showy, red or yellowish, entire or variously cleft; calyx 1.5–2.5 cm long, subequally cleft above and below; corolla 2–3.5 cm long, the galea more than half as long as the corolla-tube, the lower lip at most one-fifth the length of the galea. $2n = 24$*, 48, 72, 96*, 120, c.144. Appears to intergrade with *C. rhexifolia*. Open woods and meadows. sAlas, BC to Man s to Calif, NM. Map 932

C. occidentalis Torr.
Stems 0.5–4.0 dm tall, simple or tufted, erect to somewhat decumbent at the base, glabrous to villous or viscid-villous below, villous to densely viscid-villous above; leaves linear to lanceolate, entire or sometimes with 1–2 linear lobes; bracts yellow, often becoming dark bluish-black on drying, ovate to oblong, entire or occasionally with 1–2 short rounded lobes; calyx 1.5–2.5 cm long, the lobes 1–4 mm long; corolla 1.8–3.0 cm long, the galea much shorter than the corolla-tube, the lower lip one-quarter to one-third the length of the galea. $2n = 24$, 48, 96*. Appears to intergrade with *C. rhexifolia*. The taller, more erect, less villous-viscid populations, with somewhat larger flowers, occurring at lower elevations are often treated as *C. sulphurea* Rydb., but in our area they appear to intergrade completely. The relationship to *C. elegans* (Ostenf.) Malte remains to be established. Foothills and mountains to alpine elevations. BC, Alta s to Ida, Utah, Colo, NM; SD. Map 933 •

C. raupii Pennell
Stems 3–5 dm tall, slender, green or purplish, puberulent to finely villous above; leaves linear, attenuate, entire, or sometimes with a pair of slender lateral lobes, finely pubescent; bracts violet-purple, oval to lanceolate, with a pair of short linear lateral lobes, puberulent, or the margins hirsute; calyx violet-purple, cleft and lobed, puberulent; corolla 15–18 mm long; galea 5–6 mm long, hirsute; lip

violet-purple, 2.5–3 mm long. $2n = 72$. Moist woods, bogs, shores. Alas, Yuk, Mack to James Bay s to nBC, Alta, Sask, neMan. Map 934 ●

C. rhexifolia Rydb.
Stems 1–4 dm tall, simple, glabrous to villous; leaves lanceolate to oblong, entire or the upper with a few, short, rounded lobes, veins prominent; bracts crimson to purple, ovate to oblong, entire or 2–3-lobed; calyx 1.5–2.5 cm long, lobes blunt; corolla 2.0–3.5 cm long, the galea not much more than half the length of the corolla-tube, the lower lip dark green, thickened, one-fifth to one-third the length of the galea. $2n = 24, 48^*, 96$. Appears to intergrade with *C. hispida*, *C. miniata*, and *C. occidentalis*. Subalpine and alpine slopes. BC, Alta s to neOre, Utah, Colo. Map 934 ○

C. sessiliflora Pursh
Stems 1–4 dm tall, villous-tomentose; leaves entire, linear below, broader above, with 3–5 linear, divergent lobes; bracts leaf-like, green, puberulent or lanulose; calyx 2.5–4.0 cm long, yellow, glandular-puberulent, lobes linear, acute, 5–15 mm long; corolla 4.5–5.5 cm long, yellow, pink or purplish, the galea short, about 1 cm long, the lower lip prominent, one-third to half the length of the galea. $2n = 24$. Prairies. sAlta to sMan s to Ariz, Tex, Okla, Mo, Ill, Wisc. Map 933 ○

Chaenorrhinum Reichenb.

C. minus (L.) Lange / Dwarf Snapdragon
Annual; stems 0.5–4 dm high, usually branched, glandular-pubescent; lower leaves opposite and oblong; upper leaves alternate, lanceolate or linear, obtuse, 1–2 cm long, narrowed to the base; flowers borne on short pedicels in the leaf-axils; calyx 5-cleft; corolla bluish purple with yellow in the throat, 5–7 mm long, slightly longer than the calyx, strongly 2-lipped, spurred at the base, the spur 1.5–2 mm long; capsule about 5 mm long, pubescent, opening by 2 pores; seeds ellipsoid-ovoid, truncate, roughened with longitudinal crests. $2n = 14$. Railway grades and waste ground. Introduced. Occasional.

Collinsia Nutt.

C. parviflora Lindl. / Blue-eyed Mary
Small annual; stem very slender, simple or diffusely branched, 1–3 dm high, puberulent or glabrate and sometimes glandular; leaves simple, opposite or whorled, ovate to linear-lanceolate, obtuse, mostly entire, 1–2 cm long, narrowed to a slender base; flowers pedicelled, solitary or 2–6 in a whorl, subtended by the upper leaves; calyx 5-cleft; corolla 4–6 mm long, deeply 2-lipped; upper lip 2-cleft, blue, varying to white; lower lip blue-violet, deflexed, enclosing the 4 fertile stamens; fifth stamen clavate; capsule 3–4 mm long, with 2–4 smooth seeds. $2n = 14, 28$. Moist open woods. seAlas, sYuk s to Calif, Colo; sMan; Ont; nMich; Vt. Map 935

Euphrasia L.

E. arctica Lange *ex* Rostrup / Eyebright
Slender annual, probably a root-parasite; stem 0.5–3 dm tall, simple or with ascending branches from below the middle, thinly pubescent; leaves opposite, sessile, ovate or orbicular, 2–18 mm long, palmately veined, with 3–5 teeth on each side, more or less pubescent; flowers solitary in the upper leaf-axils; calyx unequally 4-cleft, pubescent; corolla 3–4 mm long, white with purple markings and yellow eye; upper lip concave, slightly 2-lobed; lower lip spreading, the oblong lobes notched; stamens 4; capsule flattened, pubescent; seeds marked with narrow longitudinal wings. $2n = 44$. In our area var. *disjuncta* (F. & W.) Cronq., *E. disjuncta* Fern. & Wieg. of Ed. 1. Moist banks and open ground. Greenland, Iceland, Scandinavia: Alas, Yuk, swMack to Huds Bay, nQue, Nfld s to BC, Alta, nSask, sMan, Minn, Mich, Me. Map 936 ○

Gratiola L.

G. neglecta Torr.
Annual, 1–3 dm high; stems widely and diffusely branched, the upper parts of the plant glandular-puberulent and clammy; leaves opposite, mostly 2–4 cm long, linear to oblanceolate, entire or undulate-dentate, tapering to a narrow base; flowers solitary on filiform, axillary pedicels, these 1–2 cm long and widely divergent; sepals 4–6 mm long, accompanied by a pair of bractlets; corolla 8–10 mm long, pale yellow to whitish, 2-lipped, bearded in the throat; fertile stamens 2, the anther with a broad connective and transverse locules; sterile stamens minute or none; capsule 3–5 mm long, 4-valved; seeds short-cylindric, about 0.5 mm long. Wet and muddy places. sBC to Que, NS s to Calif, Tex, Ga. Map 936 ●

Limosella L.

L. aquatica L. / Mudwort
Small annual, usually growing on mud, forming new rosettes by slender runners; plant with fibrous roots, a cluster of basal leaves and a few to several short 1-flowered peduncles from the crown, glabrous throughout; leaves with a slender petiole, which in mud is short but in water elongates to 6–12 cm; leaf-blade elliptical to oblong, 2–6 mm wide, obtuse, entire, fleshy; calyx cup-shaped, 5-lobed; corolla cup-shaped, nearly regular, about 2 mm long, pink or white; stamens 4, the anther-sacs confluent; capsule longer than calyx, many-seeded. $2n = 40^*$. Mud-flats and shallow water. Circumpolar: Alas, Yuk, Mack to Huds Bay, nQue, Nfld s to Calif, NM, Neb, Mo, Minn, James Bay. Map 937 ●

Linaria Mill / Toad-flax

Perennial herbs, glabrous throughout; stems erect, often tufted, 3–10 dm tall; leaves mostly alternate, numerous, crowded on the stems, entire; flowers yellow, borne in dense terminal racemes; calyx deeply 5-cleft; corolla yellow, spurred at the base, 2-lipped; upper lip erect, 2-lobed; lower lip 3-lobed, with a prominent orange palate in the throat; stamens 4, in pairs, included; capsule ovoid, opening by apical valves.

Cauline leaves ovate-lanceolate, clasping at base; flowers 3–4 cm long **L. dalmatica**

Cauline leaves linear, narrowed to a petiole-like base; flowers 2–3 cm long **L. vulgaris**

L. dalmatica (L.) Mill.
Stems 6–10 dm tall, robust, simple or branched, glabrous and more or less glaucous; leaves mostly ovate-lanceolate, cordate-clasping at base, 3–8 cm long. $2n = 12$. Roadsides and waste ground. Introduced. Occasional.

L. vulgaris Hill / Butter-and-eggs; Toad-flax
A noxious weed, spreading by rootstocks and forming dense patches; stems 2–8 dm tall, glabrous or somewhat puberulent above; leaves very numerous, pale green, 2–5 cm long, linear or linear-lanceolate; capsule 8–12 mm long; seeds flattened, with a circular wing. $2n = 12$. Roadsides, fields, waste ground. Introduced. Common.

Melampyrum L.

M. lineare Desr. / Cow-wheat
Annual, 1–4 dm high, usually with a few slender branches; stems puberulent or nearly glabrous; leaves opposite, lanceolate or linear-lanceolate, mostly entire, the uppermost often aristate-toothed near the base; flowers solitary in the leaf-axils or in terminal leafy spikes; corolla whitish to purplish and yellow-tipped, 2-lipped, 8–12 mm long; capsule flat, 1–4-seeded; seeds 2–4 mm long, blackish with light-coloured tips when mature. $2n = 18, 36$. Sandy ground, pine woods. BC to Nfld s to Wash, Ida, Mont, Minn, Wisc, Ind, Tenn, Ga, SC. Map 938 ●

Mimulus L. / Monkey-flower

Annual or perennial herbs, often with soft weak stems; leaves opposite, commonly dentate; flowers borne in leaf-axils or in leafy racemes, usually showy; calyx tubular, angled, the lobes usually shorter than the tube; corolla 2-lipped, the lobes of the upper lip erect or reflexed, those of the lower lip spreading, the throat with two longitudinal ridges; stamens 4, didynamous; stigma usually 2-lipped; capsule 2-valved; seeds many.

1 Corolla rose-red, 3.5–5 cm long **M. lewisii**
Corolla yellow (often dotted with red), not over 4 cm long 2

2 Corolla 1.5–4 cm long; calyx-teeth quite unequal **M. guttatus**
Corolla less than 1.5 cm long; calyx-teeth equal or nearly so **M. floribundus**

M. floribundus Lindl.
Annual, 0.5–3 dm high, glandular-hairy and viscid to nearly glabrous; stems more or less branched, erect or weak and reclining; leaves ovate to ovate-lanceolate, 1–4 cm long, dentate, petiolate; flowers few, the pedicels filiform, mostly spreading at maturity; calyx 4–10 mm long, plicate-carinate, the teeth lanceolate, nearly equal; corolla slender, 7–15 mm long, yellow, marked with red in the throat; style elongated; capsule included in the inflated calyx. $2n = 32$. Moist montane banks. BC, Alta s to Calif, NM. Map 938 ○

M. guttatus DC. / Yellow Monkey-flower
Annual or perennial; glabrous or somewhat glandular-pubescent above; stems 0.5–6 dm high, stout or weak, erect or sometimes trailing; leaves 1–5 cm long, broadly oval to obovate, coarsely and unevenly sinuate-dentate, the upper ones sessile, the lower petiolate; flowers racemose or solitary, on slender pedicels; calyx 1–2 cm long, inflated in fruit; corolla mostly 2–4 cm long, bright yellow, usually spotted with red, the ridges hairy and nearly closing the throat. $2n = 28$, 30, 32, 56. Wet meadows and mountain streams. seAlas, sYuk, swMack, swSask s to Calif, NM, ND. Map 939 ○

M. lewisii Pursh / Red Monkey-flower
Perennial with rootstocks; stems 3–7 dm tall, more or less viscid-puberulent; leaves 3–7 cm long, sessile, oblong to ovate-lanceolate, rather thin, irregularly dentate to nearly entire; flowers few to several, the pedicels usually longer than the subtending leaves; calyx sharply angled, about 2 cm long; corolla rose-red, 3–5 cm long; anthers more or less bearded. $2n = 16$. Mountain streams. seAlas s to Calif, Utah, Wyo. Map 937 ○

Odontites Ludw.

O. serotina (Lam.) Dum.
Annual; stem 1–5 dm tall, generally branched, pubescent; leaves opposite, sessile, 1–4 cm long, lanceolate with 2–3 pairs of blunt teeth, pubescent; inflorescence an often secund spike-like raceme; calyx 4-lobed, about 5 mm long; corolla about 1 cm long, pinkish red, zygomorphic, upper lip straight, the lower shallowly 3-lobed; stamens 4; capsule 6–8 mm long, pubescent. $2n = 20$. Roadsides. Introduced. Rare.

Orthocarpus Nutt.

O. luteus Nutt. / Owl-clover
Annual; stems erect, simple or with a few branches, 1–4 dm tall, glandular-pubescent; leaves alternate, numerous, sessile, linear, 1–4 cm long, pubescent, usually entire; flowers small, in terminal spikes, the bracts mostly 3-cleft; calyx 6–8 mm long, 4-cleft; corolla 10–14 mm long, golden yellow, 2-lipped, pubescent; stamens 4, the two anther-sacs dissimilar; capsule 5–7 mm long; seeds about 1 mm long, ridged. $2n = 28$. Dry prairie grassland. BC to wOnt s to Calif, NM, Neb, Minn. Map 940

Pedicularis L. / Lousewort

Mostly perennial herbs, caulescent or acaulescent; leaves alternate, opposite or basal, toothed or pinnately lobed; flowers usually in terminal spikes or spike-like racemes, sometimes axillary; calyx 2–5-lobed; corolla strongly 2-lipped, the lower lip 3-lobed, the upper lip (galea) compressed on the sides, arched and often beaked at apex; stamens 4, didynamous, not exserted; capsule flattened, curved or oblique; seeds numerous, with rough surfaces.

1	Stem branched	2
	Stem simple	3
2	Flowers purple, in short spikes or solitary in leaf-axils	**P. parviflora**
	Flowers yellow, occasionally crimson-tinged, in few-flowered spikes	**P. labradorica**
3	Corolla pink to purple	4
	Corolla white, cream or yellow, sometimes crimson-tinged or the galea crimson	8
4	Galea with a long, slender, upcurved beak; corolla and often the stem crimson-purple	**P. groenlandica**
	Galea beakless	5
5	Galea with 2 small teeth just below the tip	6
	Galea lacking teeth	7
6	Stem leafy, woolly-villous	**P. arctica**
	Stem essentially leafless (0–2), glabrous	**P. sudetica**
7	Plants 2 dm tall or more; stem glabrous	**P. bracteosa**
	Plants less than 2 dm tall; stem woolly-villous	**P. lanata**
8	Leaves divided	9
	Leaves not divided, the margins finely serrate	**P. racemosa**
9	Galea with a downcurved sickle-shaped beak	**P. contorta**
	Galea beakless	10
10	Plants 2 dm tall or more	**P. bracteosa**
	Plants less than 2 dm tall	11
11	Stems leafy; inflorescence elongate; flowers yellow, the galea often crimson	**P. flammea**
	Stems leafless; inflorescence capitate; flowers white or cream	**P. capitata**

P. arctica R.Br.
Stem stout, glabrous below, with many brown scales at the base, 5–15 cm high; leaves linear-oblong, pinnatifid into several small oblong, crenate or subentire lobes; corolla rose or purple, 18–20 mm long; galea somewhat falcate, with a slender tooth on each side below the tip; lower lip much shorter than the galea; rachis and calyx glabrous to sparsely lanate; seeds smooth, brown. $2n = 16$. *P. langsdorfii* Fisch. ssp. *arctica* (R.Br.) Pennell. Moist alpine slopes. Siberia, Greenland: Alas, Yuk to Ellesmere Is s to BC, Alta, sMack, sBaffin Is. Map 939 •

P. bracteosa Benth.
Stems 3–10 dm tall, nearly glabrous; basal leaves petiolate; stem-leaves sessile, pinnately divided or lobed, the divisions variously toothed, often serrate; bracts lanceolate and attenuate, the lower ones often prominent; spike dense, 5–15 cm long; calyx-lobes linear-lanceolate, often attenuate, glandular-puberulent or villose; corolla yellowish, or tinged with purple, 15–20 mm long, the tube long and slender, the galea hooded at apex. $2n = 16$. Moist woodland and open montane slopes. BC, Alta s to Calif, Colo. Map 941

P. capitata Adams
Stem 5–15 cm high, scapiform or bearing 1–2 reduced leaves, pubescent; leaves mostly basal, pinnately parted, the narrow segments variously lobed and toothed, pubescent to glabrous; lowermost bracts prominent, leaf-like; spike short, 1–3 cm long; corolla white, often somewhat tinged with rose or purple at tip, 25–30 mm long, the galea strongly arched. $2n = 16$. Calcareous alpine slopes. Siberia, Greenland: Alas, Yuk, seVictoria Is to Ellesmere Is s to BC, Alta, nMack, sBaffin Is. Map 942

P. contorta Benth.
Stems 2–3 dm tall, glabrous, sometimes clustered; leaves glabrous, 2–10 cm long, narrowly oblong in outline, pinnately divided, the divisions linear, dentate-serrate; raceme spike-like, 3–15 cm long, the bracts deeply cleft into linear divisions; calyx often traversely wrinkled; corolla white or yellowish, 15–20 mm long, with a long coiled beak enclosed by the lower lobes. Dry grassy slopes and open woods. BC, Alta s to Calif, Wyo. Map 943 ○

P. flammea L.
Small plant with spindly roots; stems 6–10 cm high, reddish purple, sparingly leafy, glabrous or nearly so; leaves 2–6 cm long, pinnately divided, the numerous lobes oblong or oval, crenate, 3–5 mm long; spike-like raceme 2–5 cm long, glabrous to lanate; calyx as long as corolla-tube or slightly longer; corolla 10–20 mm long, about 2 mm wide, entirely yellow (forma *flavescens* Polunin) or commonly with a deep-crimson galea. $2n = 16$. Includes *P. albertae* Hult. Calcareous alpine slopes. Greenland, Iceland, Lapland: nMack to eDevon Is to Nfld s to Huds Bay, James Bay, sLab; Alta; Gaspé. Map 944

P. groenlandica Retz. / Little Red Elephant; Elephant Head
Stems 3–5 dm tall, often clustered, glabrous, sometimes reddish purple; leaves petioled, or the upper sessile, lanceolate in outline, 5–20 cm long, deeply pinnatifid, the divisions narrow and toothed; spike dense, 3–6 cm long; corolla reddish purple or pinkish, the tube short, the beak curved outward and upward and about 1 cm long (the galea resembling the head and trunk of an elephant). $2n = 16$. Boggy meadows and marshes. swGreenland: seYuk, BC to nMan, nOnt, nQue, Lab s to Calif, NM, cSask, cMan. Map 945

P. labradorica Wirsing
Probably biennial, or short-lived perennial; stem 1.5–3 dm high, branched from near the base and upward, pubescent; lower leaves pinnatifid, the upper merely toothed; flowers in short terminal spikes and solitary in the upper leaf-axils; corolla about 12 mm long, yellow or the galea purplish. $2n = 16$. Moist woodland and bogs. Asia, swGreenland: Alas, Yuk, Mack to nHuds Bay, sBaffin Is s to BC, Alta, nSask, nMan, James Bay, sLab. Map 947 ●

P. lanata Cham. & Schlecht.
Resembling *P. arctica*, but the rachis of spike and the calyx densely villous, the galea without minute lateral teeth and the lower lip of the corolla nearly as long as the galea; seeds with a loose-fitting, ashy-grey, honeycombed seed-coat. $2n = 16$. Alpine slopes. neSiberia, Greenland: Alas, Yuk to Ellesmere Is to nQue s to BC, Alta, sMack, nHuds Bay. Map 946 ●

P. parviflora J.E. Smith / Swamp Lousewort
Annual or biennial; stem glabrous, 3–6 dm tall, branched above; leaves alternate, or sometimes opposite, pinnatifid throughout; bracts deeply pinnatifid; flowers in loose spikes or solitary in the axils; calyx 2-cleft, the lobes serrate; corolla purple, 12–15 mm long, the galea rounded at apex; capsule about twice as long as the calyx. $2n = 16$. Includes *P. macrodonta* Richards. Wet meadows and bogs. sAlas, sYuk, swMack to James Bay s to BC, Alta, cSask, sMan, nOnt. Map 943 ●

P. racemosa Dougl. *ex* Benth.
Stems 2–5 dm tall, glabrous or nearly so, usually clustered; leaves mostly cauline, lanceolate or linear, not divided, but crenate-serrate on the margins; flowers rather few in an open spike-like raceme or in upper leaf-axils; corolla white, 12–15 mm long; galea curved downward and produced into a strongly incurved beak, the lower lip large. $2n = 16$. Dry, open subalpine slopes. BC, Alta s to Calif, NM. Map 946 ○

P. sudetica Willd.
Stems mostly 1–4 dm tall, arising singly or clustered, glabrous to villous; basal leaves well-developed, glabrous or rarely sparsely villous below, cauline leaves 0–2 (3); inflorescence capitate, often subtended by 1–2 leafy bracts, generally densely woolly-villous, elongating in fruit; corolla 15–24 mm long, pink to crimson-purple, galea 6–13 mm long, arching, with 2 small teeth near the tip. $2n = 16$. In our area ssp. **interioides** Hult. Bogs. More or less circumpolar: Alas, Yuk, to Ellesmere Is to sBaffin Is s to BC, Alta, sMack, Huds Bay, James Bay; Wyo; Colo. Map 947 ○

Penstemon Mitch. / Beard-tongue

Perennial herbs, rarely somewhat woody at the base; stems usually erect, often tufted and with a rosette of basal leaves; cauline leaves opposite, mostly sessile and often clasping, the upper ones progressively reduced in size; flowers showy, irregular, in terminal racemes or dense nodal clusters; calyx deeply 5-parted;

corolla tubular, gradually or abruptly dilated in the throat, 2-lipped; upper lip 2-lobed; lower lip equalling or longer than the upper, 3-lobed; fertile stamens 4, didynamous, their filaments arched; sterile stamen (staminode) present, filiform, about as long as the fertile ones, usually narrowly spatulate and bearded at the apex; style elongate, stigma capitate; capsule ovoid or conic, septicidal, with numerous angular seeds.

1	Flowers about 1 cm long, in dense terminal spikes or nodal clusters	2
	Flowers more than 1 cm long, usually at least 1.5 cm, not so crowded	3
2	Corolla yellow	**P. confertus**
	Corolla blue	**P. procerus**
3	Flowers mostly 1.5–2.5 cm long	4
	Flowers mostly 3–4 cm long	7
4	Corolla white, broadly funnelform; inflorescence glandular	**P. albidus**
	Corolla blue to lilac	5
5	Plants glabrous, glaucous; lower bracts broadly ovate to suborbicular; corolla broadly funnelform	**P. nitidus**
	Plants generally glandular-pubescent in the inflorescence, not glaucous; lower bracts lanceolate; corolla not much expanded towards the tip	6
6	Corolla blue, mostly 1.2–1.5 cm long	**P. albertinus**
	Corolla lilac, mostly 1.5–2.5 cm long	**P. gracilis**
7	Anthers glabrous; staminode exserted, densely bearded	**P. eriantherus**
	Anthers woolly; staminode included, not densely bearded	8
8	Herbs, erect; inflorescence paniculate	**P. lyallii**
	Shrubs; inflorescence racemose	9
9	Plants prostrate, mat-forming; leaves ovate to elliptic	**P. ellipticus**
	Plants bushy, branches ascending; leaves lanceolate	**P. fruticosus**

P. albertinus Greene
Stems about 2 dm high, glabrous to slightly glandular in the inflorescence; basal leaves rhombic-ovate to broadly spatulate; stem-leaves elliptic-ovate to lanceolate, subsessile, entire or sharply denticulate; calyx somewhat glandular, 3–5 mm long, the lobes acute, only slightly scarious-margined below; corolla light blue, 12–20 mm long. $2n = 16$. Includes *P. virens* Pennell of Ed. 1. Dry, open montane and subalpine slopes. seBC, Alta s to Ida, Mont. Map 948 ○

P. albidus Nutt. / White Beard-tongue
Stems stout, clustered, 1.5–4 dm high, closely puberulent; lower leaves petioled, oblong to spatulate; upper leaves sessile, lanceolate, mostly entire, more or less puberulent; calyx glandular-hairy; corolla about 2 cm long, white, or tinged with violet, glandular-hairy within, the throat widened. Leaves and flowers blackened in drying. Dry hillsides and plains. Alta to sMan s to NM, Tex, Okla, wIa, Minn. Map 948 ●

P. confertus Dougl. / Yellow Beard-tongue
Stems slender, 1–5 dm tall, glabrous, often forming small clumps; lower leaves oblanceolate, glabrous, entire; upper leaves mostly lanceolate; inflorescence interrupted; calyx 5 mm long, the lobes acuminate, scarious-margined, erose; corolla about 1 cm long, sulphur-yellow. $2n = 16, 32^*$. Meadows and moist open woods. seBC, Alta s to Ore, Mont. Map 949

P. ellipticus Coult. & Fisher
Stems mostly 1–2 dm high, forming loose mats from a procumbent woody base, the branches minutely puberulent; leaves broadly elliptical or rotund, entire or distantly toothed, the lower short-petioled; calyx densely viscid-puberulent, the lobes attenuate; corolla 3–3.5 cm long, light violet-purple, strongly widened at the throat; anthers woolly; sterile stamen short, bearded with yellowish hairs. $2n = 16^*$. Rocky banks and slopes at subalpine elevations. seBC, Alta s to Ida, Mont. Map 950

P. eriantherus Pursh
Stem stout, 1–3 dm high, leafy, pubescent, glandular above; basal leaves petioled, oblong to spatulate; stem-leaves oblong to linear, entire or toothed, greyish-puberulent, sometimes villose; corolla about 2.5 cm long, purplish, the throat strongly enlarged and bearded within; sterile stamen with long yellow hairs. Dry, open, slopes. seBC, Alta s to Ore, Colo, Neb. Map 951

P. fruticosus (Pursh) Greene
Stems mostly 1–4 dm high, forming dense clumps, woody at base; leaves linear-lanceolate, mostly 2–5 mm wide, sharply serrate to nearly entire; upper leaves of floral shoots much reduced; calyx densely glandular-hairy, the lobes slender, attenuate-caudate; corolla 3.5–4.5 cm long, lilac-purple; anthers woolly; sterile stamen very slender, yellow-bearded. $2n = 16$. In our area ssp. **scouleri** (Lindl.) Pennell & Keck. Dry rocky slopes at subalpine and alpine elevations. sBC, Alta s to Ore, Mont, Wyo. Map 952 ○

P. gracilis Nutt. / Lilac-flowered Beard-tongue
Stems slender, erect, 2–4 dm tall, glabrous or somewhat puberulent; basal leaves short-petioled, narrowly oblanceolate; stem-leaves linear-lanceolate, mostly denticulate, glabrous or puberulent; inflorescence slender and open, the pedicels glandular-puberulent; corolla slender, 18–20 mm long, lilac or whitish. $2n = 16$. Dry prairies, often sandy. BC to wOnt s to NM, Neb, Ia, Wisc. Map 952 ●

P. lyallii A. Gray
Stems 3–5 dm high, glabrous or puberulent, usually in clumps, from a woody caudex; leaves lanceolate or linear-lanceolate, attenuate, sharply denticulate to entire, grey-puberulent or glabrous; calyx glandular, the lobes lanceolate, attenuate; corolla 3–4 cm long, pale purple or light lavender; anthers woolly; sterile stamen glabrous. Includes *P. linearifolius* Coult. & Fisher. Rocky slopes. seBC, Alta s to Wash, Ida, Mont. Map 953

P. nitidus Dougl. *ex* Benth. / Smooth Blue Beard-tongue
Stems stout, 2–3 dm high, often in clumps, glabrous; leaves ovate, lanceolate or oblanceolate, 3–5 cm long, thick, entire, glabrous and glaucous; panicle narrow, with large, ovate or lanceolate, abruptly acuminate bracts; corolla 15–18 mm long, usually blue, rarely pinkish or purplish; sterile stamen bearded on the expanded apex and well back from the apex; flowering in May. $2n = 16^*$. Prairies. seBC to sMan s to Wash, Wyo, neColo, ND. Map 954

P. procerus Dougl. *ex* Grah. / Slender Blue Beard-tongue
Stems slender, 1.5–4 dm high, often decumbent at base, glabrous or puberulent; basal leaves petioled, generally oblanceolate, glabrous; stem-leaves oblanceolate or lanceolate, sessile, entire; inflorescence spike-like and compact, or interrupted below; calyx 5 mm long, the lobes cuneate or obovate, abruptly cuspidate, scarious-margined and erose; corolla deep purplish blue, rarely pinkish or white, 6–10 mm long. $2n = 16, 32$. Meadows and open woods. sYuk, BC to Sask, swMan, s to Calif, Colo. Map 955

Rhinanthus L.

R. minor L. / Yellow Rattle
Annual, with erect stem 3–6 dm tall, simple, or with a few branches above, glabrous or pubescent at the nodes and in decurrent lines; leaves opposite, sessile, lanceolate-oblong to linear, serrate or crenate; inflorescence a 1-sided, leafy-bracted, spike-like raceme, puberulent to villose; calyx membranaceous, flattened, 4-toothed, enlarging in fruit; corolla pale yellow, 1–2 cm long, zygomorphic; anthers hairy; capsule nearly orbicular, flat, 10–15 mm long; seeds large, flat, 4–6 mm long, narrow-winged. $2n = 14$. *R. crista-galli* L. of Ed. 1. Includes *R. borealis* (Sterneck) Chab., *R. groenlandicus* (Ostenf.) Chab. Meadows and open woodland. Greenland, Iceland, Europe: Alas, Yuk, swMack to Huds Bay, nQue, Nfld s to Ore, Colo, Sask, Man, Ont, NY. Map 956

Scrophularia L.

S. lanceolata Pursh
Perennial herb; stem 5–20 dm tall, squarish; leaves cauline, opposite, short-petiolate, 2–15 cm long, broadly lanceolate, serrate; inflorescence terminal, open, glandular; calyx 5-lobed; corolla 7–14 mm long, greenish purple, the tube more or less globose, upper 2 lobes fused, median lower lip deflexed; stamens 4; staminode scale-like; capsule 6–10 mm long. $2n = 96$. Waste ground. Introduced. Rare.

Verbascum L.

Mostly biennial herbs producing a rosette of leaves the first year and an erect stem the next; leaves simple, generally alternate, the basal forming a rosette; flowers in racemes, spikes or panicles; calyx equally 5-lobed; corolla rotate, more or less regular, usually yellow; stamens (4) 5, some or all the filaments villous; capsule septicidal; seeds many.

1 Hairs on the stamen-filaments violet **V. nigrum**
Hairs on the stamen-filaments white or yellow 2

2 Upper cauline leaves long decurrent **V. thapsus**
Upper cauline leaves not, or only shortly, decurrent **V. phlomoides**

V. nigrum L. / Black Mullein
Stem 5–10 dm tall, angled, stellate-pubescent; basal leaves long-petiolate, blades ovate to lanceolate, base cordate, margin crenate; cauline leaves more or less sessile; flowers in a terminal raceme sometimes with axillary racemes from the upper leaves; calyx 2.5–4.5 mm long, lobes linear; corolla 1.8–2.5 cm wide; capsule 4–5 mm long, ovoid. $2n = 30$. Roadsides. Introduced. Rare.

V. phlomoides L.
Stem 3–12 dm tall, tomentose; basal leaves petiolate, oblong-elliptical, entire or crenate, cauline leaves sessile, clasping; flowers in a simple, terminal raceme; calyx 5–12 mm long; corolla 2.0–5.5 cm wide; capsule 5–8 mm long, elliptic-ovoid. $2n = 34$. Railways and waste ground. Introduced. Rare.

V. thapsus L. / Common Mullein
Plant 10–20 dm tall; stem and leaves densely grey-tomentose with felt-like hairs; leaves alternate, elliptical to oblanceolate, 1–4 dm long, the lower ones petioled, the upper sessile, decurrent along the stem; inflorescence a spike-like raceme, 1–5 dm long, about 3 cm thick; calyx deeply 5-cleft; corolla bright yellow, 1.5–2 cm wide, 5-lobed, with a very short tube; stamens 5, exserted, all fertile, the three upper ones white-villose and short; capsule globular, tomentose; seeds marked with longitudinal ridges. $2n = 36$. Roadsides, dry open ground. Introduced. Common.

Veronica L. / Speedwell

Herbs, annuals or perennials, often in moist habitats; leaves mostly opposite, the upper sometimes alternate; flowers white to blue or purple, small, perfect, solitary to racemose or spicate; calyx deeply 4-cleft; corolla rotate, 4-lobed, only slightly irregular; stamens 2, exserted; style slender; ovary short, flattened; capsule more or less flattened, usually notched at apex, loculicidal; seeds few–many.

1 Flowers solitary in the leaf-axils 2
Flowers in axillary or terminal racemes 4

2 Pedicels shorter than sepals **V. peregrina**
Pedicels longer than sepals 3

3 Corolla 8–12 mm wide; capsule 6–10 mm wide **V. persica**
Corolla less than 8 mm wide; capsule 3–6 mm wide **V. agrestis**

4	Racemes axillary	5
	Racemes terminal	8
5	Stems hirsute, with 2 opposite rows of curled white hair; plants of dry habitats	**V. chamaedrys**
	Stems essentially glabrous; plants of moist habitats	6
6	Leaves short-petiolate	**V. americana**
	Leaves sessile	7
7	Leaves lanceolate to oblong, clasping at base	**V. catenata**
	Leaves linear, not clasping at base	**V. scutellata**
8	Leaves 5–15 cm long, conspicuously serrate	**V. longifolia**
	Leaves mostly less than 5 cm long, entire or obscurely toothed	9
9	Capsule wider than long; lower leaves short-petiolate	**V. serpyllifolia**
	Capsule longer than wide; leaves sessile	**V. alpina**

V. agrestis L.
Small annual with prostrate or ascending stems, mostly forking at base, the branches 1–3 dm long; leaves ovate to elliptic-oblong, coarsely crenate-serrate, rounded to truncate at base; corolla blue or bluish white; pedicels recurved in fruit and up to 1.5 cm long; capsule deeply notched, 3–6 mm broad; seeds about 1.4 mm long. $2n = 14, 28$. Gardens and waste ground. Introduced. Rare.

V. alpina L.
Perennial, with slender rootstock; stems erect, somewhat decumbent at base, 0.5–3 dm high, villose, glandular; leaves ovate to oval or oblong, obtuse, sessile or subsessile, entire or somewhat crenate, glabrous or pubescent; raceme short and dense, glandular-pubescent, elongating in fruit; pedicels 2–4 mm long; corolla blue, 4–5 mm wide; capsule obtuse or emarginate, glandular hairy. $2n = 18^*$, 36. In our area mostly var. **alterniflora** Fern., with the fruits somewhat distant and interrupted; also occurring, var. **unalaschensis** Cham. & Schlecht., with the fruits contiguous and not interrupted. Moist subalpine and alpine areas. Greenland, Europe: Alas, Yuk, wMack s to Calif, neUtah, Ariz, NM; eHuds Bay to Lab, Gaspé. Map 957

V. americana (Raf.) Schw. / American Brooklime
Glabrous perennial with decumbent stems, rooting at the base, 2–6 dm long; leaves ovate to oblong, short-petioled, 2–7 cm long, usually serrate with low distant teeth; racemes borne in the upper leaf-axils, the mature pedicels slender, spreading, 5–10 mm long; flowers blue to pale violet, 4–6 mm broad; capsule orbicular, not notched or slightly so. $2n = 36$. Stream banks and marshy ground. eAsia: seAlas, sYuk, swMack, cSask, sMan to James Bay, nQue, Nfld s to Calif, Ariz, NM, Tex, Okla, NC. Map 958

V. catenata Pennell / Water Speedwell
Perennial with creeping rootstocks and ascending branches, 4–8 dm long, stout and succulent; leaves sessile, with clasping base, narrowly lanceolate or oblong, entire or obscurely toothed; racemes slender, loose, 4–12 cm long, glabrescent; corolla white or pink; capsule shallowly notched, the lobes turgid and separated by deep furrows. $2n = 36$. *V. salina* Schur of Ed. 1. Marshy ground, streams, ditches. Eurasia: BC to swQue s to Calif, NM, Neb, Okla, Mo, Ohio, Pa. Map 959 ○

V. chamaedrys L. / Bird's-eye
Perennial; stems weak, prostrate to erect, 1–3 dm long, with 2 rows of generally dense, white hair; leaves 1–3 cm long, sessile or short-petiolate, ovate, base cordate, margin crenate-serrate; inflorescence axillary; pedicels 4–9 mm long; corolla about 1 cm wide, blue with a white eye; capsule obcordate, shorter than calyx. $2n = 16, 32$. Gardens and waste ground. Introduced. Rare.

V. longifolia L.
Perennial with erect stems, 3–10 dm tall; leaves opposite or in whorls of 3, short-petioled, ovate-lanceolate, 4–8 cm long, sharply serrate; racemes terminal, spike-like, erect, the axis pubescent; flowers crowded, blue; corolla pubescent in the throat; capsule 3 mm long, about half as long as the persistent style. $2n = 34, 68$. An occasional escape from cultivation. Introduced. Rare.

V. peregrina L.
Small annual, glandular-hairy; stems erect, 1–3 dm tall, simple or with ascending branches; leaves 1–2.5 cm long, narrowly oblanceolate to oblong or linear, entire or denticulate, sessile or short-petioled; flowers whitish, borne singly in the axils of the upper bract-like leaves, the pedicels about 1 mm long; corolla 2–3 mm long, somewhat shorter than the calyx; capsule flattened, notched, 3–4 mm long. $2n = 54$. In our area ssp. **xalapensis** (HBK) Pennell. Marshy ground. eAsia: BC to Que, NS s to Calif, Tex, Fla. Map 960

V. persica Poir. / Bird's-eye
Annual, much branched, the stems prostrate or ascending, 1–3 dm long, sparsely pubescent; leaves broadly ovate, 1–2 cm long, coarsely toothed, with broad petioles; corolla blue, 8–10 mm wide; pedicels often recurved in fruit, up to 2 cm long; capsule 6–10 mm wide, with a broad deep notch and a pubescent margin. $2n = 28$. Gardens and waste ground. Introduced. Occasional.

V. scutellata L. / Marsh Speedwell
Perennial with slender stems, decumbent at base and rooting at the nodes, glabrous or somewhat hairy; leaves linear to linear-lanceolate, sessile or nearly so, entire or remotely denticulate, 3–6 cm long, little reduced upward; racemes 5–10 cm long, flexuous, the pedicels filiform, becoming deflexed; corolla blue or pale violet; capsule flat, deeply notched. $2n = 18$. Swamps and shores. More or less circumpolar: Alas, Yuk, swMack to James Bay, nQue, Nfld s to Calif, Colo, ND, Ia, Va. Map 959 ●

V. serpyllifolia L.
Perennial, creeping at the base, the flowering stems ascending to erect, 1–2 dm tall, glandular-puberulent; leaves ovate to broadly oval, mostly 1–2 cm long, entire or crenulate, the lower ones petioled; raceme at first short and compact, elongating in fruit to 5–10 cm; sepals puberulent; corolla blue, 3–4 mm long; capsule obreniform, glandular-pubescent. $2n = 14$. In our area ssp. **humifusa** (Dickson) Syme. Moist montane slopes. seAlas, Yuk s to Calif, Ariz, NM; James Bay to Nfld s to Minn, Mich, NY. Introduced s. Map 961

OROBANCHACEAE / Broom-rape Family

Perennial herbs lacking chlorophyll, parasitic on the roots of other plants; stem erect, yellow, brown or purplish; leaves alternate, scale-like; flowers solitary or in spikes, racemes, panicles; calyx 4- or 5-lobed; corolla 5-lobed, generally 2-lipped; stamens 4, didynamous; ovary superior, 1-loculed with 2–4 parietal placentae; fruit a capsule; seeds many, microscopic.

Lower lip of corolla as long as or longer than the upper; stamens included; placentae 4 **Orobanche**
Lower lip of corolla much shorter than the upper; stamens exserted; placentae 2 **Boschniakia**

Boschniakia C.A. Meyer

B. rossica (Cham. & Schlecht.) Fedtsch.
Plants brown, essentially glabrous; stems 1–4 dm tall, solitary or in clusters; leaves and bracts scale-like, elliptic to ovate; inflorescence a dense terminal spike; flowers about 1 cm long; capsule 6–7 mm long. Parasitic on *Alnus crispa*. Open woodland and scrub. Asia: Alas, Yuk, wMack s to nBC, nAlta. Map 962 ○

Orobanche L.

Brownish or pinkish herbs; flowers purplish, tubular; calyx of 4 or 5 more or less united sepals; corolla more or less 2-lipped, the upper lip 2-lobed, notched or entire, the lower lip 3-lobed; stamens 4, didynamous, on the corolla-tube; style long; stigma large; capsule with 1 locule.

1 Flowers sessile or nearly so, in a dense, spike-like cluster at the top of the thick stem; each flower with 1 or 2 bracts **O. ludoviciana**
Flowers borne singly on long pedicels, without bracts below the calyx; branches from a thick, underground stem-base (caudex) 2

2 Caudex somewhat woody, usually forked, its branches bearing a few or several flowers, each on an erect pedicel **O. fasciculata**
Caudex soft, each branch bearing a single flower **O. uniflora**

O. fasciculata Nutt. / Clustered Broom-rape
Branches 3–10 cm high above the thick, hard caudex; each branch with several brownish scales and 3–10 flowers on long pedicels; corolla purplish, slightly 2-lipped, 2–3 cm long, its lobes rounded or mucronate. $2n = 48$. Parasitic on Compositae and various other plants. Prairie and other open areas. sAlas, sYuk, BC to sOnt s to Calif, NM, Tex, Okla, Ill, Ind, Mich. Map 962 ●

O. ludoviciana Nutt. / Broom-rape
Stems stout, single or clustered, 10–25 cm high, and sticky, glandular-hairy; scales ovate about 8 mm long; corolla purplish, or occasionally yellowish, 2-lipped, 1.5–2 cm long, its lobes acute. $2n = 48$. Parasitic on *Artemisia* and other Compositae. Prairies. sBC to swMan s to Calif, NM, Tex, Okla, Ind. Map 963 ●

O. uniflora L. / One-flowered Cancer-root
The soft underground caudex produces very short branches with 1–few slender pedicels 5–20 cm above ground, each pedicel 1-flowered; corolla white, purplish or yellowish, curved 1.5–2.5 cm long, its lobes oblong to obovate, minutely ciliate; calyx-lobes attenuate. $2n = 36, 48, 72$. Parasitic on various plants. Moist woods. BC to Nfld s to Calif, Tex, Fla. Map 963 ○

LENTIBULARIACEAE / Bladderwort Family

Small carnivorous herbs, in water or on moist ground; entrapping insects and other small organisms by means of minute bladders or sticky leaves; flowers irregular; calyx 2–5-cleft; corolla deeply 2-lipped, with a spur at the base; stamens 2, adnate to the base of the corolla-tube; anthers 1-loculed; ovary 1-loculed, with free-central placentation; stigma 1–2-lobed; capsule 2-valved or opening irregularly; seeds numerous, wrinkled.

Flowers bluish violet; leaves entire, in a basal rosette	**Pinguicula**
Flowers yellow; leaves generally dissected, not in a basal rosette	**Utricularia**

Pinguicula L.

Plants terrestrial, with a basal rosette of broad, entire leaves, upper surface secretory, margins inrolled; flowers solitary; calyx with the upper lip 3-lobed, the lower 2-lobed; corolla open at the throat, 2-lipped, spurred; capsule 2-valved.

Scape glandular-villous; corolla less than 10 mm long	**P. villosa**
Scape glabrous; corolla more than 15 mm long	**P. vulgaris**

P. villosa L.
Similar to *P. vulgaris* but smaller; distinguished by the characters of the key. $2n = 16$. Sphagnum bogs. More or less circumpolar: Alas, Yuk, Mack to nHuds Bay, nQue, Lab s to BC, nAlta, nSask, James Bay. Map 964 ○

P. vulgaris L. / Common Butterwort
Leaves yellowish green, elliptical to oval, 2–5 cm long, 1–2 cm wide; flower-stal 4–12 cm high; flower purplish and suggesting a violet; corolla about 1 cm wid and 1–2 cm long, with a slender, tapering, straight, nectar-producing spu 6–8 mm long; capsule 2-valved. $2n = 64$. Bogs, damp rocks and ledges, chiefl in calcareous areas. More or less circumpolar: Alas, Yuk, Mack to sBaffin Is t Nfld s to Ore, Mont, sSask, Man, Minn, Mich, NY, Vt. Map 964 ●

Utricularia L. / Bladderwort

Perennial plants of aquatic or muddy habitats; leaves alternate or whorled, enti or finely dissected and bearing bladders; flowers solitary or in few-flowere racemes; calyx 2-lipped, divided to base; corolla 2-lipped, spurred or saccat throat closed; capsule indehiscent or variously dehiscent.

1 Leaves obscure (small, simple, subterranean) **U. cornut**
Leaves evident (large, dissected)

2 Leaves and bladders borne on separate branches **U. intermedi**
Leaves and bladders borne on the same branch

3 Plants coarse; stem at least 0.5 mm thick; leaves terete, capillary; flowers 14–20 mm long; bladders abundant **U. vulgari**
Plants delicate; stem less than 0.5 mm thick; leaves flattened; flowers 4–8 mm long; bladders few **U. mino**

U. cornuta Michx. / Horned Bladderwort
Terrestrial with delicate underground stems; leaves simple, linear-filiform wit minute, marginal bladders; scapes wiry, mostly 10–25 cm tall, usually bearin 1–3 flowers; flowers 15–25 mm long, subtended by an acute bract and 2 smalle bractlets; capsule covered by the beaked calyx. Bogs and muddy shores. neAlta nSask to James Bay, Nfld s to eTex, Fla. Map 965 ○

U. intermedia Hayne / Flat-leaved Bladderwort
Stems slender bearing leaves and bladders on separate branches; leaves dissecte into delicate, thread-like but flattened segments, these minutely serrulate; blad ders few, relatively large, about 5 mm long; flowers 10–15 mm long; fruitin pedicels erect; winter-buds 3–10 mm long. $2n = 44$. Ponds and lakes. Circum polar: Alas, Yuk, swMack to Huds Bay, sLab, Nfld s to Calif, Alta, Sask, Ia Ohio, Del. Map 966

U. minor L. / Small Bladderwort
Small, very delicate, submersed plant, with thread-like stems and repeatedl forked leaves; leaf-segments very slender, somewhat flattened and not serrulate bladders often few, 2 mm long or less, borne on the leaves; flowers 4–8 mn long, pale yellow, with a very short spur and obscure palate; fruiting pedicel recurved; winter-buds rounded, lax, 1.5–5 mm long. $2n = 36$. Ponds, bogs ditches. Circumpolar: Alas, Yuk, swMack to James Bay, sLab, Nfld s to Calif Utah, Colo, Neb, Ia, Ind, NJ. Map 967

ʼulgaris L. / Common Bladderwort
ally floating near the surface of the water, the rather coarse submersed stems ı crowded foliage; flowering stalks stout, rising well above the water; leaves ed or somewhat pinnately dissected into coarse capillary segments and bear- numerous bladders 3–5 mm long; flowers 14–20 mm long; corolla yellow ı brown stripes on the palate; fruiting pedicels arched-recurving; winter-buds cm long, their minute leaves crowded and hairy. $2n$ = 36, 42, 44. In our ssp. **macrorhiza** (Le Conte) Clausen. Ponds, lakes, ditches. Circumpolar: s, Yuk, swMack to Huds Bay, nQue, Nfld s to Calif, Tex, Fla. Map 965 •

ANTAGINACEAE / Plantain Family

ıtago L. / Plantain

ıual or perennial acaulescent herbs; leaves all basal, simple; blades often minently ribbed; inflorescence a spike, usually elongate and on a long pe- ıcle; flowers small, white or greenish, regular, usually perfect, sometimes sexual; calyx-lobes 4, persistent, often scarious-margined, all alike or 2 longer; ɔlla mostly salver-shaped, constricted at the throat, 4-lobed, scarious, at least he margins; stamens 4 or 2, attached to the corolla-tube, the anthers versatile; ry superior, 2-loculed; style and stigma filiform; fruit a membranous, cir- ıscissile capsule (pyxis); seeds 1–several in each locule; seed-coat mucilagi- s when wetted.

Plants perennial; leaves linear; corolla-tube pubescent externally **P. maritima**
Plants not as above 2

Leaves linear to filiform, rarely over 5 mm wide; peduncles and leaves more or less silky-pubescent; annuals 3
Leaves lanceolate or broader, rarely less than 10 mm wide; peduncles and leaves glabrous or hairy, but not silky-pubescent; perennials 4

Leaves filiform, about 1 mm wide; stamens 2; corolla-lobes 0.5–1.0 mm long, often erect **P. elongata**
Leaves linear, over 1 mm wide; stamens 4; corolla-lobes about 2 mm long, spreading or reflexed **P. patagonica**

Leaf-blades prostrate, mostly ovate, abruptly narrowed into the petiole; seeds 5–22 to a capsule **P. major**
Leaf-blades ascending, mostly lanceolate, gradually tapering into the petiole; seeds 2–7 to a capsule 5

Plant sparingly hairy; brown or red wool in the crown **P. eriopoda**
Plant villose, with jointed hairs; not woolly in the crown **P. canescens**

:anescens Adams
ennial with stout caudex; leaves short-petioled, lanceolate, 3–5-nerved, entire, 5 cm long, villose with jointed hairs; flowering stalk villose, 1–1.5 dm long,

the spike in flower about 3 cm long, elongating in fruit; bracts and sepals rounded-ovate, with scarious margins; capsule oblong, rounded at apex; seeds 2–4, black, 1–1.5 mm long. $2n = 12^*$. *P. septata* Morris of Ed. 1. Non-alkaline grassy and gravelly slopes. eSiberia: Alas, Yuk, wMack, Banks Is, Victoria Is s to Alta, Mont. Map 968 ●

P. elongata Pursh
Small annual, leaves narrowly linear or filiform, 3–8 cm long, somewhat puberulent; flowering stalk filiform, 2–8 cm tall, the spike very slender and loosely flowered; flowers imperfectly dioecious or polygamous; bracts ovate, keeled; sepals about as long as the bracts, scarious-margined; corolla-lobes becoming erect and closed over the capsule; stamens 2; capsule 2–3 mm long, circumscissile just below the middle; seeds usually 4, about 2 mm long. $2n = 12^*, 36$. In our area ssp. **elongata**. Dry to moist alkaline prairie. BC to swMan s to Calif, Tex. Map 968 ○

P. eriopoda Torr.
Perennial with a stout crown and usually yellowish brown or red wool among the leaf-bases; leaves thick, oblanceolate to narrowly ovate, 3–7-nerved, 5–15 cm long, gradually tapering into a margined petiole; spikes on long stalks, slender, loosely flowered, 5–20 cm long; bracts broadly ovate, about as long as the scarious calyx; capsule conic, 3–4 mm long; seeds 2–4, black or brown, about 2 mm long. $2n = 24^*$. Saline or alkaline meadow and shores. Alas, Yuk, wMack to sMan s to Calif, Nev, Utah, Colo, Neb, Minn; St Lawrence R. Map 969

P. major L. / Common Plantain; Whiteman's-foot
Perennial weed; glabrous or somewhat pubescent; leaves broad, 0.5–3 dm long, the petiole long and often broad, the blade oval to ovate, 5–7-ribbed, entire to somewhat toothed; flowering stalks 1–5 dm long, the spikes variable, usually rather dense; bracts ovate; sepals about 2 mm long; stamens 4; capsule ovoid, 2–4 mm long, curcumscissile near the middle or toward the base; seeds about 1 mm long, angular. $2n = 24$. Waste ground, roadsides, lawns. Introduced. Common.

P. maritima L. / Sea-side Plantain
Perennial; plants glabrous; leaves linear to lanceolate, up to 22 cm long, 1.5 cm wide, entire or with a few short teeth, 3–5(7)-nerved; flowering stalks 1–several, 5–20 (25) cm tall; spikes 1–8 (10) cm long, generally dense; bracts 1.5–4.0 (6) mm long, broadly ovate, about as long as calyx; corolla-lobes 1.0–1.5 mm long; stamens 4; capsule ovoid or conic; seeds mostly 2–4, 1.5–3.0 mm long. $2n = 12, 18, 24$. Saline marshes. Circumpolar: sAlas s to Calif; nAlta; Huds Bay, nQue, Nfld s to NJ. Map 970 ○

P. patagonica Jacq.
Annual plant, woolly or silky throughout; leaves ascending, 3–10 cm long, linear, acute or acuminate, narrowed into a slender petiole; flowering stalks 5–20 cm tall, the spikes dense, obtuse, silvery-hairy, drying brownish; bracts linear, hid-

den or clearly projecting beyond the flowers; corolla-lobes spreading; capsule oblong, circumscissile about the middle; seeds 2, brown, concave on inner side, the minute hilum shaped like the figure 8. $2n = 20$*. In our area var. **patagonica** (*P. purshii* R. & S. of Ed. 1) and var. **spinulosa** Gray (*P. spinulosa* Dcne. of Ed. 1) with the lower bracts rigid, villous, about 1 cm long. Dry plains and eroding slopes, often in sandy areas. sBC to sMan s to Calif, Tex. Map 970 ●

RUBIACEAE / Madder Family

Annual or perennial herbs with slender, mostly 4-sided stems; leaves simple, entire, opposite or appearing whorled in verticils of 3–8; flowers small, perfect, regular, gamopetalous, epigynous; corolla 3–4-lobed, rotate or funnelform; sepals erect or obsolete; stamens as many as the lobes of the corolla and inserted on its tube; ovary inferior, 2-loculed; fruit of 2 globose 1-seeded nutlets or a dehiscent capsule.

1 Leaves opposite **Houstonia**
Leaves whorled 2

2 Flowers in an open cyme; corolla rotate **Galium**
Flowers in bracteate heads; corolla slender-funnelform **Asperula**

Asperula L. / Woodruff

A. arvensis L. / Quinsywort
Annual; stem 1.0–5.5 dm tall, branched, glabrous; leaves mostly 1–2 cm long, in whorls of 6–8, linear-lanceolate; flowers in bracteate heads about 1 cm wide; bracts long, ciliate; corolla 4-lobed, 5–7 mm long; stamens 4; fruit 2–3 mm wide. $2n = 22$. Roadsides. Introduced. Rare.

Galium L. / Bedstraw; Cleavers

Herbs with slender, square or 4-angled stems, erect or lax and spreading; leaves and accompanying leaf-like stipules in whorls of 2–8; flowers small, often white, cymose, terminating the branches or axillary; calyx obsolete; corolla 4-lobed, rarely 3-lobed, rotate; stamens 4, rarely 3; styles 2; ovary inferior; fruit of 2 rounded nutlets, each 1-seeded, indehiscent, in some species bristly.

1 Leaves in whorls of 4 2
Leaves in whorls of 6–8 4

2 Stem stiff, erect; leaves 3-nerved; ovary pubescent **G. boreale**
Stem weak, matted or ascending, leaves 1-nerved; ovary glabrous 3

3 Corolla 3-lobed; stem scabrous; leaves spreading, midrib often scabrous **G. trifidum**
Corolla 4-lobed; stem mostly glabrous; leaves often deflexed, midrib usually glabrous **G. labradoricum**

4 Leaves linear, strongly revolute; internodes puberulent, the angles not retrorsely scabrous **G. verum**

Leaves lanceolate to elliptic, not strongly revolute; internodes glabrous, the angles mostly retrorsely scabrous, at least below 5

5 Leaves in whorls of 6, elliptic, usually more than 3 mm wide, generally 3–4 times longer than wide **G. triflorum**

Leaves in whorls of 6–8, linear-lanceolate to narrowly oblanceolate, usually 2–3 mm wide, more than 4 times longer than wide 6

6 Flowers 1.0–1.5 mm wide, greenish yellow; fruits 1.5–2.8 mm long (excluding spines) **G. spurium**

Flowers 2 mm wide, white; fruits mostly 2.8–4.0 mm long (excluding spines) **G. aparine**

G. aparine L. / Cleavers
Annual; stems 3–10 dm long, lax, prostrate or spreading on bushes, retrorsely hispid, nodes usually tomentose; leaves in whorls of 6–8, oblanceolate, 1–5 cm long, the apex rounded and apiculate, margins and midribs retrorsely hispid; flowers few, in axillary clusters; corolla about 2 mm across, greenish white; fruit 3–4 mm long, covered with minute hooked bristles, rarely smooth. $2n = 42, 44, 48, 62, 66, 68$. Waste ground, fields, gardens. Introduced. Rare.

G. boreale L. / Northern Bedstraw
Erect perennial with slender rootstocks; stems erect, 3–6 dm high, branching, leafy, square, mostly smooth; leaves in whorls of 4, linear-lanceolate, 2–5 cm long, 3-nerved, obtuse; flowers faintly fragrant, numerous, in a fairly dense, repeatedly 3-forked cluster; corolla white, about 2 mm long; fruits in pairs, about 2 mm long, densely hairy. $2n = 22, 44, 55, 66$. Includes *G. septentrionale* R. & G. Woodlands, moist prairie valleys, roadsides. Circumpolar: Alas, Yuk, swMack to Huds Bay, sQue s to Calif, Tex, Mo, Ohio, Del. Map 971

G. labradoricum Wieg.
Slender perennial, mostly 1–3 dm tall; stems simple or branched above, glabrous or pubescent at the nodes; leaves mostly in whorls of 4, soon recurved or deflexed, linear-oblanceolate, 8–14 mm long, roughened on the margins; flowers few, mostly terminal and in 3's; corolla white, 4-lobed, about 2 mm across; fruit smooth, 1–2 mm long. $2n = 24$. Marshy ground, moist woods, bogs. swMack, BC to Nfld s to Minn, Ill, Ohio, Pa, NJ. Map 972

G. spurium L. / False Cleavers
Similar to *G. aparine*; differing in the characters of the key, the glabrous or only lightly pubescent nodes, and the somewhat shorter leaves, up to 4 cm long. $2n = 20$. Waste ground, fields, gardens. Introduced. Occasional.

G. trifidum L. / Small Bedstraw
Slender perennial, mostly 1–3 dm high; stems much branched, slender and weak, erect or decumbent, often matted, usually minutely retrorse-scabrous; leaves

mostly in whorls of 4, linear-spatulate, 6–14 mm long, obtuse, usually scabrous on the margins and midrib beneath; flowers terminal or axillary, the pedicels capillary and becoming arcuate; corolla greenish white, 3-lobed, the lobes obtuse; fruit smooth, 1–2 mm long. $2n = 24$. In our area ssp. **trifidum**. Marshy ground, stream banks, bogs. More or less circumpolar: Alas, Yuk, Mack to Huds Bay, sLab, Nfld s to Ariz, Tex, Okla, Ga. Map 973

G. triflorum Michx. / Sweet-scented Bedstraw
Perennial with slender rootstock; stems 3–10 dm long, spreading or reclining, simple or sparingly branched, glabrous or somewhat scabrous; leaves in whorls of 5 or 6, narrowly elliptic to oblanceolate, bristle-tipped, the margins scabrous-bristly; herbage sweet-scented in drying; inflorescences axillary from the principal leaves, also terminal, the peduncles 3-flowered or repeatedly branched and several-flowered, becoming widely divaricate; corolla whitish, the lobes minute; fruit 2–3 mm long, bearing hooked bristles. $2n = 44, 66$. Moist woods. Circumpolar: Alas, Yuk, swMack to James Bay, sLab, Nfld s to Calif, NM, Okla, Fla. Map 974

G. verum L. / Yellow Bedstraw
Perennial weed, 3–8 dm tall, from a creeping rootstock; stems erect, puberulent, at least above; leaves mostly in whorls of 8, linear, 1–3 cm long, acute, pubescent beneath, roughened above, becoming deflexed; flowers in dense clusters, from the upper axils; corolla yellow; fruit about 1 mm long, smooth. $2n = 22, 44$. Waste ground, railway grades. Introduced. Occasional.

Houstonia L. / Bluets

H. longifolia Gaertn.
Tufted perennial, 1–2.5 dm high; stems numerous, simple or branched above, often purplish, glabrous or somewhat puberulent; leaves opposite, sessile, linear-oblong, 1–3 cm long, narrowed to the base, 1-nerved, glabrous or minutely roughened on the margins, bearing small whitish or purplish stipules at the base; flowers numerous in loose or crowded cymes; sepals 4, linear-lanceolate, 1–2 mm long, in fruit often longer than the capsule; corolla purplish to white, 4-lobed, funnelform, 6–9 mm long, pubescent within; stamens 4, often exserted; style slender; stigmas 2; fruit an ovoid capsule, protruding beyond the hypanthium, dehiscent across the top. $2n = 12^*$. Open sandy woods and dunes. Alta to sQue s to Okla, Miss, Ga. Map 975

CAPRIFOLIACEAE / Honeysuckle Family

Shrubs or trailing plants with opposite exstipulate leaves; flowers perfect, regular or irregular, epigynous; corolla gamopetalous, rotate to tubular, 5-lobed and often 2-lipped; stamens 5 (4 in *Linnaea*), adnate to the corolla and alternate with its lobes; anthers versatile; ovary inferior, 2–5-loculed, with 1–several ovules in each locule; fruit succulent with 1 or a few seeds, or dry and 1-seeded.

1	Leaves pinnately compound	**Sambucus**
	Leaves simple	2
2	Plants slender, trailing; flowers nodding; stamens 4(5)	**Linnaea**
	Plants erect or twining shrubs	3
3	Leaves entire or undulate	4
	Leaves toothed or lobed	**Viburnum**
4	Flowers campanulate, more or less regular, in clusters; fruit white or purplish	**Symphoricarpos**
	Flowers funnelform or tubular, mostly irregular, mostly in pairs; fruit red, yellow or black	**Lonicera**

Linnaea L. / Twin-flower

L. borealis L.
A beautiful woodland plant with fragrant flowers; stems slender, creeping, semi-woody, often 1 m long, forming loose mats, with short erect leafy branches, 3–10 cm high, somewhat pubescent; leaves evergreen, short-petioled, broadly oval to obovate, 1–2 cm long, often crenate at apex; peduncles filiform, 3–10 cm tall, glandular-puberulent, forking into 2 pedicels at the top, each bearing a delicate nodding flower; calyx-lobes linear-lanceolate; corolla whitish, tinged with rose-purple, funnelform, 8–15 mm long, hairy inside; stamens 4, didynamous, attached near the base of corolla; ovary 3-loculed, but 1-seeded; style exserted; stigma capitate; fruit ovoid, dry, enclosed by glandular-pubescent bractlets. $2n = 32$. In our area ssp. **americana** (Forbes) Hult. Woods. Circumpolar: Alas, Yuk, Mack to Huds Bay, nQue, Nfld s to Calif, Ariz, NM, SD, Ind, WVa. Map 976

Lonicera L. / Honeysuckle

Erect shrubs or vines; leaves opposite, simple, usually entire; flowers often showy, borne in short clusters or in pairs; calyx-teeth very short; corolla tubular or funnelform, often swollen on one side at the base, 5-lobed, often distinctly zygomorphic; stamens 5; stigma capitate; ovary 2–3-loculed; fruit a several-seeded berry.

1	Stems twining or reclining; flowers conspicuous, yellow to reddish, in terminal clusters; fruit red; upper leaves connate	**L. dioica**
	Stems erect; flowers in pairs on axillary peduncles; leaves all separate	2
2	Bracts at base of flowers large and leaf-like, at least in fruit; corolla almost regular; berries of the paired flowers separate	**L. involucrata**
	Bracts small and inconspicuous; corolla 2-lipped; berries of the paired flowers more or less united	3
3	Flowers pink or white; berry red or yellow	**L. tartarica**
	Flowers yellow; berry red or black	4

4 Berries bluish-black, solitary; northern plants **L. caerulea**
Berries red, in pairs; mountain plants **L. utahensis**

L. caerulea L. / Fly Honeysuckle
Low shrub, up to 1 m high, with erect branches, the young stems generally quite hairy; leaves mostly oblong, blunt at apex, pubescent beneath, nearly sessile; flowers on short peduncles; bracts linear-subulate, longer than the ovaries; flowers yellowish; corolla 10–15 mm long, slightly gibbous at base; ovaries of 2 paired flowers united and enclosed by a cup-like structure; berry blue, edible. $2n = 18, 36$. In our area var. **villosa** (Michx.) T. & G. (*L. villosa* (Michx.) R. & S. of Ed. 1). Swamps and bogs. Circumpolar: BC to Calif; Alta to Nfld s to Minn, Wisc, Mich, Pa. Map 977 ●

L. dioica L. / Twining Honeysuckle
Twining, ascending or reclining plant, the older stems woody; leaves oblong to ovate or obovate, short-petioled or sessile, glabrous above, white-glaucous and pubescent beneath, often undulate, the upper 1–4 pairs connate; flowers in a short terminal interrupted spike; corolla yellow to reddish, 1.5–2.5 cm long, bilabiate, hairy, the tube gibbous at base; stamens and style hairy; berries red. $2n = 18$. In our area var. **glaucescens** (Rydb.) Butters. Open woods, rocky slopes. swMack, seBC to Que s to Alta, Neb, Okla, NC. Map 978

L. involucrata (Richards.) Banks / Bracted Honeysuckle
Erect or ascending shrub, 1–3 m high; leaves short-petioled, ovate to obovate, 5–15 cm long, acute or acuminate at apex, more or less pubescent; flowers in pairs, the peduncle axillary, 2–4 cm long; bracts 4 or united into 2, close under the flowers, green to deep purple, ascending, finally reflexed, 1–2 cm long; corolla yellow, 10–13 mm long, saccate at base, glandular-hairy; berries purple-black, about 8 mm in diam., unpleasant in flavour. $2n = 18$. Moist woods. BC to Que s to Calif, NM, Wisc, Mich. Map 979

L. tartarica L. / Tartarian Honeysuckle
An ornamental shrub, 1–3 m high; branches slender, smooth, becoming hollow; leaves thin, glabrous, short-petioled, ovate to oblong, 3–6 cm long; peduncles slender, about 2 cm long; subulate bracts and ovate bractlets at base of ovary; corolla bilabiate, glabrous, pink or white, 1.5–2 cm long; berries red or yellow, united at base. $2n = 18$. An escape from cultivation. Introduced. Occasional.

L. utahensis S. Wats. / Red Twin-berry
Widely branching shrub, 1–2 m high; leaves usually broadly oval, obtuse, 2–5 cm long, thin, usually glabrous, pale beneath; peduncles axillary, 1–2 cm long, 2-flowered; bractlets minute; corolla pale yellow, funnelform, gibbous at base, 1–2 cm long, hairy within; berries red, almost globular, slightly united, strongly divergent, 5–8 mm long, insipid in flavour. Includes *L. ebractulata* Rydb. Moist open woods and slopes in mountains to subalpine elevations. BC, Alta s to Calif, Utah, Wyo. Map 977 ○

Sambucus L. / Elderberry

S. racemosa L.
Shrubs, mostly 1–3 mm tall, often forming large clumps; stems with a wide pith; leaves pinnate, the leaflets 5–7, lanceolate to elliptic, acuminate, serrate, 5–17 cm long, 2–6 cm wide; inflorescence 4–10 cm long, about as wide; flowers white, numerous, crowded; calyx-lobes obsolete; corolla 3–6 mm across; stamens 5; stigmas 3; fruit a globose berry-like drupe, black or red, 5–6 mm wide. $2n = 36$, 38. In our area var. **melanocarpa** (Gray) McMinn (*S. melanocarpa* A. Gray of Ed. 1) with blackish purple fruits and glabrous leaves, somewhat pubescent when young, and var. **pubens** (Michx.) Koehne (*S. pubens* Michx. of Ed. 1) with red fruits and leaves pubescent below. Moist woods. Circumpolar: sAlas, BC to Nfld s to Calif, NM, SD, Ill, Tenn, Ga. Map 980

Symphoricarpos Duhamel / Snowberry; Buckbrush

Low bushy shrubs, freely branched; leaves simple, short-petioled, oval to ovate-oblong, obtuse, entire or occasionally wavy-toothed; flowers white or pink, terminal or axillary, mostly in short dense clusters; calyx-teeth short, persistent; corolla bell-shaped or funnelform, 4–5-lobed, bearded within; stamens 4–5, adnate to corolla; ovary 4-loculed, only 2 of the locules with fertile ovules; fruit a berry-like drupe, 2-seeded, white, greenish or becoming darkened.

Style 2–3 mm long, glabrous, included; stamens usually included **S. albus**
Style 4–8 mm long, usually hairy, exserted; stamens usually exserted **S. occidentalis**

S. albus (L.) Blake / Snowberry
Slender shrub, up to 1 m high, often dwarf, the younger parts pubescent or glabrous; leaves oval to ovate, mostly 2–4 cm long; corolla bell-shaped, 4–7 mm long, hairy within; style 2–3 mm long; glabrous, included; fruit drying white, oval to globose, 6–12 mm long, not edible, in clusters or borne singly. $2n = 36$, 54, 72. Open woodland, valley slopes. seAlas, sMack, BC to Que, NS s to Calif, Ida, Colo, Neb, Va. Map 981

S. occidentalis Hook. / Buckbrush; Wolfberry
Robust shrub, up to 1 m high, spreading freely by suckering; young twigs usually puberulent; leaves oval to ovate-oblong, mostly 3–5 cm long, thick and leathery; corolla funnelform, 5–9 mm long, densely hairy within; style 4–8 mm long, usually exserted; fruits usually in dense clusters, globose, greenish white, drying purplish, beaked by the persistent calyx; nutlets 3–4 mm long. Open woodland, prairie slopes and valleys. swMack, BC to wOnt s to Wash, Utah, NM, Okla, Mo, Mich. Map 982

Viburnum L. / Bush Cranberry; Pembina

Shrubs or small trees; leaves petiolate, simple, usually lobed or toothed; flowers white, in compound cymes; calyx 5-toothed; corolla 5-lobed, spreading, usually regular; stamens 5; style short, 3-cleft; ovary 1–3-loculed; fruit a 1-seeded drupe, red, the stone flattened; winter-buds of our species with 2 connate outer scales and often reddish.

Outer flowers of cyme similar to rest; most of the leaves not deeply 3-lobed; inflorescence branch with 1 pair of leaves **V. edule**

Outer flowers of cyme larger, sterile; most of the leaves deeply 3-lobed; inflorescence branch with 2 pairs of leaves **V. opulus**

V. edule (Michx.) Raf. / Low-bush Cranberry; Mooseberry
A branching shrub, 1–2 m high; leaves mostly 6–10 cm long, orbicular or ovate to obovate in outline, often 3-lobed at apex, palmately 3–5-nerved, coarsely toothed; petioles without stipules; cyme 1–3 cm across, 3–30-flowered; fruits ovoid, light red, 8–10 mm long, acid and juicy, strong-scented, edible. $2n = 18$. Moist woods. Alas, Yuk, wMack to James Bay, nQue, Nfld s to Ore, Ida, Colo, Minn, Pa. Map 983

V. opulus L. / High-bush Cranberry; Pembina
Shrub or small tree, 1–4 m high, with grey bark; leaves ovate, 6–12 cm long, deeply 3-lobed, the lobes long-acuminate and divergent, coarsely toothed; petioles with slender stipules at base and glands upwards; cyme 5–15 cm across, flat-topped, many-flowered; the marginal flowers neutral and 1–2 cm broad; fruits globose to ellipsoid, orange to red, 8–10 mm in diam., acid, finally juicy, edible. $2n = 18$. In our area ssp. **trilobum** (Marsh) Clausen (*V. trilobum* of Ed. 1). Moist woods. Eurasia: BC to Nfld s to Wash, Ida, Wyo, SD, Neb, Ia, Ill, Pa. Map 984

ADOXACEAE / Moschatel Family

Adoxa L.

A. moschatellina L. / Moschatel
Delicate smooth musk-scented herb, 6–18 cm tall, with scaly rhizome and ternately divided leaves; basal leaves long-petioled, 1–3-ternate, the leaflets obovate to oblong, 3-cleft; stem-leaves a single pair, smaller, ternate; flowers small, green or yellowish, 3–6 in a head-like cluster on a slender stalk, 2–10 cm long; calyx-lobes 2, 3 or 4; corolla rotate, 4–6-lobed; stamens 8 or 10, in pairs on the corolla tube; ovary 3–5-loculed; style 3–5-parted; fruit a small green dry drupe with 3–5 nutlets. $2n = 36^*, 54$. Moist woods and thickets. More or less circumpolar: Alas, Yuk, swMack to sMan s to BC, NM, Alta, Sask, Ia, Wisc; NY. Map 985

VALERIANACEAE / Valerian Family

Valeriana L.

Perennial herbs with scented rootstocks and erect stems; leaves in basal rosettes and opposite on the stems, at least the cauline ones pinnately divided; flowers small, white or pinkish, sessile in panicled cymes, perfect or unisexual, epigynous, gamopetalous; corolla 5-lobed; sepals of several bristles, inrolled and inconspicuous until fruiting; stamens 3, epipetalous, exserted; ovary inferior, 1-ovuled; fruit dry, 1-seeded, flat, 3-nerved on one face, 1-nerved on the other, crowned by a pappus-like calyx.

Corolla 1–3 mm long, the tube equalling or little longer than the lobes; basal leaves undivided; stem-leaves with 9–15 leaflets (lobes) **V. dioica**

Corolla 5–8 mm long, the tube much longer than the lobes; basal leaves mostly 3–5-divided; stem-leaves with 3–5 (7) leaflets **V. sitchensis**

V. dioica L.
Stem slender, 3–7 dm tall, glabrous or nearly so; basal leaves petiolate, the blades undivided, 2–5 cm long, elliptical, lanceolate or spatulate, glabrous, entire to repand; cauline leaves 2–4 pairs, short-petiolate to sessile, pinnate to pinnatifid with 3–7 divisions, mostly entire; inflorescence at first compact, becoming open and expanded; corolla rotate, white; achenes 3–5 mm long, glabrous. $2n = 16^*$. In our area ssp. **sylvatica** (Rich.) F.G. Mey. (*V. septentrionalis* Rydb. of Ed. 1). Wet meadows and bogs. Europe: Yuk, wMack to Huds Bay, sLab, Nfld s to Wash, Ida, Wyo, Alta, Sask, sMan, sOnt. Map 986

V. sitchensis Bong.
Stem slender to robust, 3–7 dm tall, glabrous or nearly so; basal leaves (often lacking) petiolate, commonly 3–5-divided, sometimes undivided, the segments ovate to obovate, entire to dentate; cauline leaves commonly 2–4 pairs, 3–5-pinnate or pinnatifid, the leaflets broadly ovate to lanceolate, or the terminal one sometimes obovate, the margins entire or variously toothed; corolla funnelform, white or tinged with pink or mauve; achenes 3–6 mm long. $2n = 46^*$, c.96. In our area ssp. **sitchensis** (includes *V. scouleri* Rydb.). Mountain woods and slopes to subalpine elevations. Alas, Yuk, swMack s to Calif, Ida, Mont; Ont, Que s to Mich, Ohio, NY. Map 987

DIPSACACEAE / Teasel Family

Knautia L.

K. arvensis (L.) Duby / Blue Buttons
Rough perennial herb; stems erect, sparingly branched above, hairy, 4–9 dm tall; leaves opposite, mostly pinnately lobed or pinnatifid into several lanceolate segments; flowers small, in dense involucrate heads (resembling Compositae),

borne on the ends of few long branches; involucral bracts ovate-lanceolate, 10–15 mm long; receptacle hairy but without bracts; calyx cup-shaped and mostly 8-awned; corolla lilac or pink, funnel-shaped, more or less zygomorphic, 9–12 mm long; stamens 4, separate, attached to the corolla; pistil with a slender style and an inferior ovary, closely enveloped by a 4-ribbed involucel; fruit an achene, 4–5 mm long, covered with bristly hairs. $2n = 20, 40$. Waste ground, roadsides, fields. Introduced. Occasional.

CUCURBITACEAE / Gourd Family

Echinocystis T. & G.

E. lobata (Michx.) T. & G. / Wild Cucumber; Balsam Apple
Tall climbing annual herb, often to a height of 4–8 m, with branched tendrils; stem angular, grooved, nearly glabrous; leaves alternate, petiolate, thin, 3–7-lobed, the lobes triangular and sharp-pointed, scabrous on both surfaces; plants monoecious, the flowers small, greenish or yellowish-white, epigynous; staminate flowers in long axillary panicles, the pistillate 1–2 from the same axils; calyx small, 6-lobed; corolla open, spreading, 6-lobed; stamens 2 or 3, united by their filaments; pistil with an inferior, 2-loculed ovary; style short; stigma broad and lobed; fruit green, ovoid, fleshy, 3–5 cm long, weakly spiny, opening at apex, 4-seeded; seeds about 2 cm long, flat, roughened, black or brown. $2n = 32$. Waste ground. Introduced. Rare.

CAMPANULACEAE / Bluebell Family

Campanula L.

Perennial herbs with milky juice; leaves alternate, simple, without stipules; flowers showy, usually blue or violet, perfect, regular, in racemes or panicles or solitary; sepals 5, narrow, united at the base; corolla of 5 united petals, bell-shaped or funnelform; stamens 5, attached at the very base of the corolla but free from it, widened below; ovary inferior, 3–5-loculed, with numerous ovules, an elongate style and commonly 3 stigmas; capsule short, opening by pores on the side, many-seeded; seeds smooth.

1 Calyx pubescent; flowers solitary — 2
Calyx glabrous; flowers usually several — 3

2 Leaves entire; corolla funnelform — **C. uniflora**
Leaves denticulate; corolla campanulate — **C. lasiocarpa**

3 Corolla 2–3 cm long; inflorescence secund; stem-leaves 1 cm wide or more — **C. rapunculoides**
Corolla 1–2 cm long; inflorescence not secund; stem-leaves less than 1 cm wide — **C. rotundifolia**

C. lasiocarpa Cham. / Alpine Harebell
Stems 1.5 dm high or less, glabrous or nearly so; lower leaves spatulate to obovate, 2–4 cm long, petioled; upper leaves linear to linear-lanceolate; capsule over 1 cm long, more or less villose. $2n = 34$. Alpine slopes. Japan, neAsia: Alas, Yuk, swMack s to Wash, Alta. Map 988

C. rapunculoides L. / Garden Bluebell
Spreading by slender rootstocks and forming thick roots; stems erect, 6–10 dm high; lowest leaves long-stalked, cordate-ovate; upper leaves short-stalked to sessile, ovate-lanceolate, irregularly serrate-dentate; flowers purplish, nodding, single in the axils of bracts. $2n = 68$, 102. Waste ground, an escape from cultivation. Introduced. Occasional.

C. rotundifolia L. / Bluebell; Harebell
Stems slender, 2–4 dm high; lowest leaves (early withering and seldom present on flowering stems) long-stalked, with ovate or ovate-cordate blades, mostly toothed; cauline leaves numerous, linear to linear-lanceolate, smooth; flowers usually several, on slender stalks in a loose cluster; calyx-lobes awl-shaped; corolla purplish blue, or rarely white, 15–25 mm long. $2n = 34$, 68*, 102. Dry meadows, hillsides, and open woods to alpine elevations. Circumpolar: Alas, Yuk, swMack to James Bay, nQue, Nfld s to Calif, NM, Tex, Neb, Ia, Pa, NJ. Map 989

C. uniflora L. / Alpine Harebell
Stems 1 dm high or less, glabrous; lower leaves spatulate, 2–3 cm long, entire or nearly so; upper leaves linear; capsule 1 cm long, often pubescent. $2n = 34$. Alpine slopes. neSiberia, Greenland, nEurope: Alas, Yuk to Ellesmere Is to nQue, Lab s to Mont, Colo, sMack, Huds Bay. Map 990 •

LOBELIACEAE / Lobelia Family

Herbs, rarely shrubs or small trees, with simple, alternate leaves; flowers irregular, epigynous; sepals 5, partly united; corolla of 5 united petals, zygomorphic; stamens 5, attached at the very base of the corolla, united into a tube around the style; ovary inferior; fruit a pod-like capsule with many small seeds.

Corolla-tube split to near the base on the upper side; capsule 2-loculed **Lobelia**
Corolla-tube not split; capsule 1-loculed **Downingia**

Downingia Torr.

D. laeta Greene
Small smooth annual, 5–20 cm high, simple or somewhat branched, sometimes rooting at the nodes; leaves alternate, sessile, narrow, entire, soon deciduous, smaller than the flower-bracts; flowers sessile in the axils of bracts; sepals foliaceous, nearly distinct; corolla 4–7 mm long, not longer than the calyx, white

to light blue or purplish, marked with white or yellow; filaments and anthers wholly united; two of the anthers tipped with bristles; ovary so long and narrow as to suggest a flower stalk. Slough margins and shallow water. Alta to Sask s to Calif, Utah, Wyo. Map 990 ○

Lobelia L.

Herbs, rarely shrubs; flowers solitary in leaf-axils or in terminal racemes; corolla dorsally split, usually to the base; some or all the anthers bearded at the tip; placentation axile; capsule 2-loculed.

1 Plants aquatic; leaves all basal **L. dortmanna**
Plants terrestrial; stems leafy 2

2 Flowers in a spike-like raceme; leaves lanceolate to oblanceolate, usually more than 10 mm wide **L. spicata**
Flowers in a loose raceme, on lax pedicels; leaves linear, less than 10 mm wide **L. kalmii**

L. dortmanna L.
Perennial, aquatic, glabrous; stem 1–10 dm tall, leafless except for a few scales; leaves 2–8 cm long, linear, hollow, in a basal rosette; inflorescence emergent, racemose, few-flowered; calyx about 5 mm long; corolla 10–20 mm long, lilac; anthers included; capsule 5–10 mm long, nodding. $2n = 16$. Shallow water at the margin of ponds, lakes. nEurope: sBC to Wash; Alta to Nfld s to cSask, sMan, Minn, NJ. Map 991 ▲

L. kalmii L. / Kalm's Lobelia
Biennial; stems slender, simple or a few erect branches, 1–5 dm high; basal leaves spatulate or obovate, petioled, usually pubescent; cauline leaves linear to lanceolate; raceme few-flowered, lax, often one-sided; pedicels slender, spreading, with median bracteoles; flowers 8–15 mm long; sepals 3 mm long; corolla blue with white centre, or rarely white throughout. $2n = 14$*. Calcareous bogs and meadows. swMack, BC to Nfld s to Wash, Mont, Colo, SD, Ia, Ohio, NJ. Map 991 ●

L. spicata Lam.
Perennial; stem 2–10 dm tall, simple or branched, pubescent below; leaves ascending, 3–10 cm long, lanceolate to oblanceolate or obovate, the upper sessile; raceme slender, dense, bracteolate; flowers (5) 9–12 mm long, blue to white, the lower lip of the corolla pubescent at the base. $2n = 14$. Moist prairie. Alta to Que, NS s to Mont, Neb, Tex, Okla, Ark, Ga. Map 991 ○

COMPOSITAE / Composite Family

Herbs, or rarely shrubs, diverse in habit and foliage; flower-heads, often simulating single flowers, each composed of many small individual flowers on a common receptacle, with an encircling involucre of bracts; small bracts or scales (chaff) are often present among the flowers on the receptacle; individual flowers epigynous, gamopetalous, regular or irregular, perfect or unisexual; stamens commonly 5, alternating with the corolla-lobes, epipetalous, the elongate anthers commonly united into a tube; ovary inferior, of 2 carpels, but 1-loculed and with 1 ovule; style usually 2-cleft; fruit an achene, commonly crowned with a pappus consisting of scales or hairs, these structures often regarded as a modified calyx; flowers of two main types: *tubular* or disc flowers with a straight 5-lobed corolla, and *ligulate* or ray flowers with a strap-shaped corolla; a flower-head may be *discoid* (of tubular flowers only), or *ligulate* (of ligulate flowers only), or *radiate* (with tubular flowers in the centre and ligulate flowers on the margin and projecting as *rays*).

Keys adapted from those provided by Cronquist in Davis, *Flora of Idaho*.

1 Flowers all ligulate (ray) and perfect; juice milky **Group I**
Flowers not all ligulate (ray); ligulate flowers when present pistillate or neuter; juice watery 2

2 Heads radiate 3
Heads discoid (lacking rays) 5

3 Rays yellow or orange 4
Rays white, to pink, purple or blue **Group IV**

4 Pappus wholly or partly of capillary bristles, sometimes plumose; receptacle naked **Group II**
Pappus chaffy or of firm awns, or a crown or none **Group III**

5 Pappus of capillary bristles, sometimes plumose **Group V**
Pappus of scales, or of awns, or a crown or none **Group VI**

GROUP I (Flowers all ligulate and perfect; juice milky)

1 Flowers pink, rarely purple or white; stem diffusely branched; leaves few, cauline, mostly linear or reduced to scales; plants of dry habitats 2
Flowers yellow, less commonly blue, purple, white or orange-red; stem and leaves not as above; plants of various habitats 3

2 Pappus-bristles plumose **Stephanomeria**
Pappus-bristles not plumose **Lygodesmia**

3 Pappus well developed, at least in part of capillary bristles 4
Pappus of minute scales or none 18

4 Pappus scale-like at the base, with plumose terminal bristles **Microseris**
Pappus not scale-like at the base, entirely of capillary bristles 5

5 Pappus-bristles plumose; involucre uniseriate or biseriate 6
Pappus-bristles not plumose; involucre of 2 or more series 7

6 Heads several; involucre biseriate; leaves lanceolate, coarsely toothed, bristly; branches of the pappus not interwebbed **Picris**

Heads solitary; involucre uniseriate; leaves linear, more or less entire, not bristly; branches of the pappus interwebbed **Tragopogon**

7 Plants scapose, leaves all basal; heads solitary 8

Plants with at least some cauline leaves, though they may be reduced to bracts; heads few to many, rarely solitary 10

8 Achenes 4–5-ribbed, the upper body at least spinulose or muricate; involucral bracts in 2 unequal series **Taraxacum**

Achenes 5–10-nerved, not spinulose or muricate; involucral bracts more or less equal or somewhat imbricate in several series 9

9 Leaf-margins conspicuously white villous-ciliolate **Nothocalais**

Leaf-margins not villous-ciliolate **Agoseris**

10 Achenes flattened, at least twice as wide as thick; flowers yellow or bluish 11

Achenes not flattened; flowers yellow, orange-red, purplish or white 13

11 Achenes beaked, the pappus borne on a disc at the tip **Lactuca**

Achenes beakless 12

12 Flowers yellow; pappus white **Sonchus**

Flowers bluish, rarely white or yellow; pappus brownish **Lactuca**

13 Flowers yellow 14

Flowers orange-red, purplish or white 15

14 Perennials with fibrous roots; lower stem and leaves generally long-hairy **Hieracium**

Annuals or taprooted perennials; lower stem and leaves generally glabrous, sometimes puberulent, tomentose or hispid **Crepis**

15 Flowers orange-red **Hieracium**

Flowers white or purplish 16

16 Flowers purplish **Prenanthes**

Flowers white 17

17 Lower leaves sagittate to triangular; plants essentially glabrous below the inflorescence **Prenanthes**

Lower leaves elliptic to oblanceolate; plants hairy **Hieracium**

18 Pappus absent; flowers yellow **Lapsana**

Pappus of minute scales; flowers blue or white **Cichorium**

GROUP II (Heads radiate; rays yellow or orange; pappus partly or wholly of capillary bristles; receptacle naked)

1 Leaves opposite **Arnica**

Leaves alternate or basal 2

2 Pappus single, bristles sometimes unequal but not distinctly divided into an inner and outer series of two lengths 3

Pappus double, bristles of the outer series inconspicuous, distinctly shorter than the inner 6

3 Involucral bracts uniseriate, equal, narrow, commonly with a few much shorter outer ones at the base **Senecio**

Involucral bracts either imbricate in several series, or broad and somewhat leafy, or both 4

4 Heads solitary **Haplopappus**

Heads several 5

5 Plants with a taproot; leaves often somewhat spinescent **Haplopappus**

Plants with fibrous roots; leaves never spinescent **Solidago**

6 Heads solitary; plants less than 15 cm tall **Erigeron**

Heads several; plants more than 15 cm tall **Heterotheca**

GROUP III (Heads radiate; rays yellow or orange; pappus chaffy, or of awns, or a crown or none)

1 Plants with a woody caudex; leaves linear; involucre glandular-glutinous **Gutierrezia**

Plants not as above 2

2 Plants more or less scapose, cauline leaves none, or few and reduced to bracts 3

Plants with leafy stems 4

3 Leaves long-petioled, blades triangular, up to 15 cm wide **Balsamorhiza**

Leaves not long-petioled, oblanceolate, up to 1 cm wide **Hymenoxys**

4 Involucral bracts of 2 distinct, dissimilar series 5

Involucral bracts in one or more series, all more or less the same 7

5 Receptacle naked; stem 1–2 dm tall; plants of dry habitats **Hymenoxys**

Receptacle chaffy; stem usually much taller; plants of moist habitats 6

6 Leaves divided into linear segments; pappus of 2 minute teeth or obsolete **Coreopsis**

Leaves not divided into linear segments; pappus conspicuous, of 2–4 retrorsely barbed awns **Bidens**

7 Plants glandular-hairy, heavily scented annuals; rays 1–3, inconspicuous **Madia**

Plants not as above 8

8 Receptacle naked 9

Receptacle chaffy or bristly 11

9 Leaves opposite, at least below; rays 5–6, inconspicuous **Bahia**

Leaves alternate; rays conspicuous 10

10 Involucre resinous (sticky); rays more or less entire, erect or spreading **Grindelia**

Involucre not resinous; rays distinctly 3-lobed, deflexed **Helenium**

11 Leaves opposite, at least the lower **Helianthus**

Leaves alternate 12

12 Rays crimson-purple at base; receptacle bristly **Gaillardia**
Rays entirely yellow; receptacle chaffy 13

13 Receptacle cylindrical; rays 3–7 **Ratibida**
Receptacle flat to hemispherical; rays 8–30 14

14 Leaves much divided; disc yellow **Anthemis**
Leaves entire or toothed; disc dark crimson-brown to blackish purple 15

15 Receptacle flat or nearly so; leaves ovate, often toothed **Helianthus**
Receptacle hemispherical; leaves elliptic to oblanceolate, entire **Rudbeckia**

GROUP IV (Heads radiate; rays not yellow or orange)

1 Leaves opposite; pappus chaffy; receptacle chaffy **Galinsoga**
Leaves alternate or basal; pappus and receptacle various 2

2 Basal leaves mostly 1–6 cm long, linear to oblanceolate or spatulate, forming a dense rosette; plants with a well-developed taproot; and *either* densely caespitose, seldom more than 5 cm tall, the heads sessile among the leaves, *or* with stout, simple stems, 5–15 (25) cm tall, heads solitary, involucres 2–4 cm wide, bracts ciliate to pectinate or fimbriate, rays blue and achenes densely pubescent, obovate to triangular **Townsendia**
Basal leaves, if any, and plants not as above 3

3 Pappus of capillary bristles, at least in part; receptacle naked 4
Pappus of 1 or 2 short awns, a mere crown or none 7

4 Basal leaves cordate or sagittate **Petasites**
Basal leaves not cordate or sagittate 5

5 Involucral bracts subequal or more or less imbricate, often green in part, but neither definitely leafy nor with chartaceous base and herbaceous green tip; style-branches lanceolate or broader, acute to obtuse, 0.5 mm long or less, or obsolete **Erigeron**
Involucral bracts either subequal and the outer leafy or more commonly evidently imbricate, with chartaceous base and evident green tip, sometimes chartaceous throughout; style-branches lanceolate or narrower acute or acuminate, ordinarily more than 0.5 mm long 6

6 Taproot present; leaves pinnately dissected or toothed; rays blue to bluish purple; plants of dry habitats **Machaeranthera**
Taproot absent **Aster**

7 Rays white 8
Rays pink or purple 11

8 Leaves oblanceolate, crenate-serrate to lobed; heads solitary, about 4 cm wide **Chrysanthemum**
Leaves pinnately dissected or linear with margins serrulate or pectinate; heads several, much smaller 9

9 Heads small, rays 5–12, 2–5 mm long **Achillea**
Heads larger, rays 10–25, 5–13 mm long 10

10 Receptacle naked; plants pineapple-scented or odourless **Matricaria**
Receptacle chaffy, at least towards the centre; plants ill-scented **Anthemis**

11 Rays pink **Achillea**
Rays purple **Ratibida**

GROUP V (Heads discoid; pappus capillary)

1 Receptacle densely bristly 2
Receptacle naked 6

2 Leaves with spiny margins 3
Leaves not spiny-margined 5

3 Pappus-bristles plumose **Cirsium**
Pappus-bristles naked or merely minutely barbellate 4

4 Leaves mottled; achenes 6–7 mm long **Silybum**
Leaves not mottled; achenes 3–4 mm long **Carduus**

5 Achenes attached basally to the receptacle **Saussurea**
Achenes attached laterally or obliquely to the receptacle **Centaurea**

6 Flowers all perfect 7
Flowers of some heads all pistillate or the outer flowers pistillate 13

7 Heads white to pink or purple 8
Heads yellow or orange 10

8 Leaves whorled **Eupatorium**
Leaves not whorled 9

9 Flowers white to cream; leaves triangular to ovate **Brickellia**
Flowers pink to purple; leaves linear to oblanceolate **Liatris**

10 Shrubs, or herbs with a woody caudex 11
Herbs, not woody at base 12

11 Leaves oblong-spatulate, margins spiny **Machaeranthera**
Leaves linear to filiform, not spiny **Chrysothamnus**

12 Leaves opposite **Arnica**
Leaves alternate **Senecio**

13 Basal leaves cordate or sagittate **Petasites**
Basal leaves, if any, not cordate or sagittate 14

14 Plants more or less white-woolly; involucral bracts mostly with dry, scarious, thin, white to yellowish or brownish tips 15
Plants not at all white-woolly; involucral bracts not markedly scarious at the tip 17

15 Plants perennial with taproot, or low spreading annuals; heads all with outer pistillate flowers and central perfect flowers **Gnaphalium**

Plants perennial with fibrous roots, often with rhizomes or stolons, but lacking a taproot; dioecious or nearly so, the heads of at least some plants wholly staminate or pistillate 16

16 Basal leaves soon deciduous, not markedly larger than the numerous well-developed cauline leaves; plants rhizomatous but lacking stolons; pistillate heads commonly with a few central staminate flowers **Anaphalis**

Basal leaves generally forming a persistent tuft or rosette; stem seldom very leafy; plants often stonoliferous; heads strictly dioecious **Antennaria**

17 Involucres 6–9 mm high; leaves entire **Aster**

Involucres shorter or leaves divided **Erigeron**

GROUP VI (Heads discoid; pappus not capillary)

1 Involucres lacking hooked prickles and spines 2

Involucres, at least some, armed with hooked prickles or straight spines 12

2 Basal leaves triangular-ovate to subreniform, white-woolly below; involucral bracts 4–5 **Adenocaulon**

Basal leaves either absent, or of distinctly different shape or not woolly below; involucral bracts usually more numerous 3

3 Plants annual, woolly or glandular; leaves opposite, at least below, linear-lanceolate, 1–6 mm wide, entire 4

Plants not as above 5

4 Plants woolly, not glandular, 1–15 cm tall **Psilocarphus**

Plants glandular, usually much taller **Madia**

5 Receptacle naked or merely hairy 6

Receptacle chaffy 9

6 Involucral bracts with yellowish, more or less petaloid, scarious tips; a pappus of scales present **Hymenopappus**

Involucral bracts not petaloid; pappus absent or a minute crown 7

7 Inflorescence paniculate, spicate or racemose **Artemisia**

Inflorescence corymbose 8

8 Receptacle flat; leaves punctate or villous; plants aromatic but not pineapple-scented **Tanacetum**

Receptacle conical; leaves not punctate or villous; plants pineapple-scented **Matricaria**

9 Involucres more than 6 mm high; pappus present, the awns sometimes inconspicuous; flowers perfect 10

Involucres less than 6 mm high; pappus absent; flowers imperfect 11

10 Pappus-awns evident, retrorsely barbed; involucral bracts not fused; plants of moist habitats **Bidens**

Pappus-awns inconspicuous; the larger involucral bracts fused one-third to two-thirds from the base; plants of dry habitats **Thelesperma**

11 Staminate and pistillate flowers in different heads; involucre of pistillate heads nut-like, tuberculate above **Ambrosia**

Staminate and pistillate flowers in the same head; involucres normal, not nut-like, tubercles absent **Iva**

12 Involucres, at least some, armed with hooked prickles 13

Involucres lacking hooked prickles, but at least some armed with straight spines 14

13 Heads all alike, with many perfect flowers **Arctium**

Heads of 2 kinds, staminate and pistillate; involucral bracts of the latter fused, the nut-like body covered with hooked prickles **Xanthium**

14 Heads all alike, flowers mostly perfect; corollas conspicuous, usually white or orange; involucre 10 mm high or more 15

Heads of 2 kinds, staminate and pistillate; corollas inconspicuous or absent; involucre less than 10 mm high 16

15 Involucral bracts 2–3 cm long; heads 2–3 cm wide; flowers orange **Carthamus**

Involucral bracts shorter; heads narrow; flowers usually white **Centaurea**

16 Pistillate involucres with one row of spines **Ambrosia**

Pistillate involucres with several rows of spines **Franseria**

Achillea L. / Yarrow; Milfoil

Perennial herbs with rhizomes; stems erect, 3–8 dm tall, more or less villose; leaves alternate, subentire or incised to pinnately dissected; heads radiate, relatively small, in a corymb-like inflorescence; involucral bracts imbricate in 3–4 series, dry, with hyaline or scarious margins; ray-flowers few, 5–12, pistillate and fertile, white or rarely pinkish, short and broad; disc-flowers 10–75, perfect and fertile; receptacle chaffy; achenes oblong, flattened, marginal, glabrous; pappus none.

1 Leaves mostly bipinnately dissected **A. millefolium**

Leaves pectinate or serrate 2

2 Leaves serrate (or subentire); ligules 4–5 mm long and nearly as broad **A. ptarmica**

Leaves pectinate; ligules 2–3 mm long **A. sibirica**

A. millefolium L. / Common Yarrow
Aromatic herb with horizontal rootstocks; stems mostly 3–7 dm tall; herbage sparsely to densely woolly-villose; leaves alternate on the stem, also in basal tufts, 2 or 3 times pinnately dissected into small narrow segments, the ultimate divisions sharply pointed; basal leaves petiolate, plume-like, 3–10 cm long, the cauline leaves mostly sessile and shorter; heads numerous in a dense cluster, round-topped or flattish; involucre 4–5 mm high, the bracts pale or dark-margined; rays about 5, white or rarely pinkish, 2–3 mm long; disc-flowers 10–30. $2n = 18, 27, 36, 45, 54, 72$. In our area ssp. **lanulosa** (Nutt.) Piper (*A. lanulosa* Nutt.), $2n = 36$*, characterized by pale bracts, densely woolly herbage and leaves with their ultimate divisions narrowly lanceolate or linear, and, in northern

parts, ssp. **borealis** (Bong.) Breitung (*A. nigrescens* (E. Mey.) Rydb.) with the margins of the involucral bracts dark brown or black. Widely distributed as a weed in North America is the Eurasian ssp. *millefolium*, which may be present in our area, but it is not readily distinguishable from ssp. *lanulosa*. Prairie, clearings, roadsides, waste ground. Circumpolar: Alas, Yuk to nHuds Bay, nQue, Nfld, ubiquitous s. Map 992

A. ptarmica L. / Sneezeweed
Stem 3–6 dm tall, villose or tomentose above to nearly glabrous below; leaves sessile, linear or narrowly lanceolate, acuminate, 3–10 cm long, glabrous or nearly so, serrulate to nearly entire; heads in an open cluster; involucre 2–4 mm wide and 4–5 mm high; rays usually 8–10 (more in 'double' flowers), white, 4–5 mm long; disc-flowers numerous. $2n = 18$. Roadsides and waste ground. Introduced. Rare.

A. sibirica Ledeb.
Stem 5–8 dm tall from a slender rootstock, sparsely villose to glabrate; leaves sessile or nearly so, linear or linear-lanceolate, 5–10 cm long, acuminate, pectinately pinnatifid or incised, the divisions sharply toothed, villose or glabrate; heads several or many in a crowded cluster; involucre 4–7 mm wide and 4–5 mm high; rays 6–12, white, 1–2 mm long; disc flowers 25–30. $2n = 18, 36^*$. Moist thickets and forests, often on stream banks. eAsia: Alas, Yuk, wMack to Man s to cBC, Alta, Sask, ND; Gaspé. Map 993 ●

Adenocaulon Hook.

A. bicolor Hook. / Pathfinder; Trail-plant
Perennial herb with rootstocks, 3–9 dm tall, the stem white-woolly to glabrate; leaves alternate, mostly near the base; petioles winged; blades thin, large, triangular to ovate, the base somewhat cordate, entire to sinuate-dentate or shallowly lobed, green and glabrate above, densely white-tomentose beneath; inflorescence paniculate, the peduncles beset with very sticky hairs; heads discoid, about 3 mm long, the 4 or 5 involucral bracts ovate, acute, spreading or reflexed in fruit; flowers whitish, tubular, the outer ones pistillate, the inner ones functionally staminate; receptacle naked; achenes 4–6, clavate, 5–8 mm long, stipitate-glandular above; pappus none. $2n = 46$. Moist montane woods. sBC, Alta s to Calif, Ida, Mont; SD; sOnt, nMinn, nMich. Map 993 ○

Agoseris Raf. / False Dandelion

Perennial scapose herbs with taproot; leaves basal-tufted, elongated, entire to pinnatifid; heads solitary on long scapes, yellow or burnt orange (drying to pink or purple); involucre cup-shaped or oblong, its bracts usually imbricated in several series, the outer shorter and broader; receptacle flat, naked; flowers all ligulate; achenes oblong or linear, not flattened, about 10-ribbed, narrowed above to a beak when mature; pappus of numerous white capillary bristles.

Flowers burnt orange, often drying purple or pinkish; beak of achene slender, more than half as long as the body, obscurely or scarcely nerved **A. aurantiaca**

Flowers yellow, often drying pinkish; beak of achene stout, up to about half as long as the body, clearly nerved throughout **A. glauca**

A. aurantiaca (Hook.) Greene

Leaves oblanceolate, linear-oblong or linear, entire or irregularly toothed to pinnatifid, glabrous or somewhat pubescent; scapes 1–6 dm tall, glabrous to villose, usually conspicuously white-woolly under the head; involucral bracts narrow, acuminate or the outer blunt, often purple-spotted, especially along the midrib; flowers burnt orange, drying to deep pink or purple; achene beak obscurely or scarcely striate, longer than the body or more than half as long. $2n$ = 36*, 72. Mountain woodlands and meadows to alpine elevations. sYuk, swMack s to Calif, NM; Que. Map 994

A. glauca (Pursh) Raf.

Leaves lanceolate, oblanceolate or linear, mostly 1–2 dm long, entire or toothed to laciniate-pinnatifid, glabrous or pubescent, often glaucous; scapes mostly 1–4 dm tall, glabrous to woolly; involucral bracts usually lanceolate to linear-lanceolate, acute to acuminate, sometimes purple-spotted; rays yellow when fresh; achene with a stout striate beak up to half the length of the body, rarely nearly beakless. $2n$ = 18*, 36*. In our area mostly var. **glauca**, essentially glabrous, more or less glaucous, and at higher elevations in the mountains the densely pubescent var. **dasycephala** (T. & G.) Jepson. Prairies, meadows, roadsides, and open mountain slopes to alpine elevations. seAlas, sYuk to cMan s to Calif, Ariz, Colo, SD, Minn; nwMack; sHuds Bay. Map 995

Ambrosia L. / Ragweed

Annual or perennial herbs with leafy stems; leaves opposite or alternate, mostly lobed or dissected; heads monoecious or rarely dioecious; staminate heads in terminal spikes or narrow racemes; involucre saucer-shaped or hemispherical, of 5–12 partly united bracts; receptacle flat, its bracts filiform; filaments united but the anthers almost separate; pistillate heads below the staminate in the axils of the leaves or bracts, 1-flowered, enclosed by a nut-like involucre, this ellipsoid or obovoid and usually with a single series of tubercles or short spines near apex; corolla lacking; achene ovoid or obovoid; pappus none.

1 Leaves palmately 3–5-lobed or undivided; body of fruit about 5 mm long **A. trifida**

Leaves once or twice pinnatifid; body of fruit about 3 mm long 2

2 Annual; leaves petiolate, rather thin, the lower usually bipinnatifid **A. artemisiifolia**

Perennial, with creeping rootstocks; leaves short-petiolate or sessile, rather thick and harsh, mostly once-pinnatifid **A. psilostachya**

A. artemisiifolia L. / Common Ragweed
Annual weed; stems erect, mostly 3–10 dm tall, usually bushy-branched, variously hairy or nearly glabrous; leaves opposite or alternate, short-stalked, once or, more often, twice pinnatifid into narrow segments, smoothish above; staminate heads numerous, their involucres about 3 mm wide, saucer-shaped, glabrous or hairy, indistinctly radiate-veined; pistillate heads in small clusters in the upper leaf-axils; fruit 3–5 mm long, with a terminal beak surrounded by a ring of short spines; flowering in August and early September. $2n = 36^*$. In our area var. **elatior** (L.) Desc. Drying sloughs, roadsides, waste ground. s3C to Nfld s to Calif, Tex, Fla. Map 996

A. psilostachya DC. / Perennial Ragweed
Perennial, with horizontal rootstocks; stems 3–8 dm high, somewhat hairy and harsh to touch; leaves opposite above, alternate below, nearly or quite sessile, usually only once pinnatifid but the lobes sometimes incised, scabrous above, somewhat strigose, rather thick; staminate involucre pubescent; pistillate heads 1–3 in the axils of upper leaves; fruit with conical beak not over 0.5 mm long, unarmed or with 4 short blunt tubercles. $2n = 36, 54, 72, 108, 144$. In our area var. **coronopifolia** (T. & G.) Farw. Railway tracks. Introduced. Rare.

A. trifida L. / Great Ragweed
Annual weed, usually stout and tall, often over 1 m high, scabrous to nearly glabrous; leaves opposite, petiolate, mostly ovate in outline, palmately 3–5-lobed or sometimes lobeless, serrate, more or less scabrous; staminate involucres with 3 strong ribs on one side; pistillate heads in axillary clusters; fruit with a conical beak 2–4 mm long and several sharp tubercles terminating the ridges. $2n = 24$. Waste ground, usually moist. Introduced. Occasional.

Anaphalis DC. / Pearly Everlasting

A. margaritacea (L.) Benth. & Hook.
Perennial herb with rhizomes; stems erect, sometimes in tufts, 2–6 dm tall, very leafy, loosely white-woolly, the pubescence sometimes tawny in age; leaves alternate, sessile, entire, 3–10 cm long, lanceolate to linear, commonly acuminate, the margins often revolute, white-tomentose beneath, usually glabrate and green above; heads several to numerous, discoid, many-flowered, crowded in a short broad inflorescence; involucre 5–7 mm high, the bracts imbricate in several series, scarious, pearly white, sometimes with a small dark spot, the outer spreading when dry; receptacle naked; plants dioecious or nearly so, the pistillate heads often with a few staminate flowers in the centre; flowers yellow; pappus of capillary bristles, not thickened at apex as in staminate flowers of *Antennaria*, and not united at base as in pistillate flowers of *Antennaria*; achenes papillate; heads last all summer. $2n = 14, 28$. Open woods in foothills and mountains. Asia: sAlas, swMack to James Bay, sLab, Nfld s to Calif, Ariz, NM, SD, Sask, Man, Kans, NC. Map 997

Antennaria Gaertn. / Pussy-toes; Everlasting

Perennial woolly or silky herbs, generally mat-forming; leaves alternate and basal, mostly entire; dioecious; heads discoid, small, many-flowered, solitary to several or many in a usually compact inflorescence; involucral bracts scarious at tips, white or coloured, imbricated; receptacle naked; staminate flowers with tubular corolla, undivided style, usually a scanty pappus and the bristles often clavate; pistillate flowers with filiform corolla, a 2-cleft style, and copious pappus of capillary bristles united at the base and falling off together; achenes terete or flattish. Staminate plants are rare in some species; many of the species produce seeds without fertilization. A difficult group and the status of species is problematical.

1 Plants with well-developed stem-leaves; basal leaves, if present, upright around the stem; leaves mostly linear-lanceolate 2
Plants with reduced stem-leaves; basal leaves in well-developed appressed rosettes; leaves usually ovate to spatulate 6

2 Plants low, densely caespitose; flowering stems 2–5 cm high, scarcely exceeding the basal leaves; heads solitary **A. dimorpha**
Plants taller; not densely caespitose; heads several 3

3 Involucres 4–5 mm high; bracts scarious throughout, glabrous **A. luzuloides**
Involucres 6–10 mm high; bracts scarious only at the tip, woolly at the base 4

4 Plants 10–20 cm tall; upper stem leaves half as long as the lower **A. lanata**
Plants 20–50 cm tall; upper stem leaves much shorter than the lower 5

5 Bracts brownish black or white only at the tip **A. pulcherrima**
Bracts mostly white, with only a brownish black spot at the base **A. anaphaloides**

6 Heads on long slender peduncles in loose racemes; bracts glabrous or nearly so; leaves glabrous on the upper surface **A. racemosa**
Heads on short peduncles, in corymbs, often congested, bracts woolly at base; leaves glabrous or usually tomentose 7

7 Leaves generally about 1 cm wide, usually green and glabrous above; stolons often leafless, with a terminal rosette **A. neglecta**
Leaves generally narrower, usually greyish and tomentose above and below; stolons leafy 8

8 Upper portion of bracts white, cream or pink 9
Upper portion of bracts greenish, brown or black 12

9 Bracts with pink tips **A. rosea**
Bracts with white or cream tips 10

10 Bracts with a dark spot at base; lower leaves narrowly oblanceolate **A. corymbosa**
Bracts lacking dark spot at base; lower leaves usually broader 11

11 Heads large, the involucre 8–13 mm high; stem-leaves 5–7 **A. aprica**
Heads smaller, the involucre 4–7 mm high; stem-leaves 8–12 **A. parvifolia**

12 Heads solitary; plants greenish **A. monocephala**
Heads several (3–5); plants grey or whitish, rarely greenish 13

13 Bracts blackish, acute **A. alpina**
Bracts pale brown, rounded **A. umbrinella**

A. alpina (L.) Gaertn.
Mat-forming perennial, the flowering stems 5–12 cm high, loosely tomentose; lower leaves oblanceolate, white-tomentose on both sides, or glabrate above; cauline leaves few, mostly linear; heads 3–7, in a dense cluster; involucral bracts very dark greenish black or brown, sometimes with sordid or white tips. $2n = 56$, 63, 70, 84. Includes *A. media* Greene, *A. canescens* (Lange) Malte, *A. ungavensis* (Fern.) Malte. Circumpolar: Alas, Yuk to Ellesmere Is, Lab s to Calif, Colo, sMack, Huds Bay. Map 998

A. anaphaloides Rydb.
Similar to *A. pulcherrima*; distinguished by the character of the key and the somewhat smaller involucre, 5–7 mm high. $2n = 28$.* Open woods and meadows. BC, Alta, swSask s to Ore, Nev, Utah, Colo. Map 999

A. aprica Greene
Similar to *A. parvifolia* but with shorter, stouter stems, larger heads and somewhat larger leaves; stems mostly less than 15 cm tall; basal leaves 1–3 cm long, with very thick white tomentum; involucre of pistillate head 7–11 mm high, the bracts whitish or sometimes with a pinkish tinge, usually blunt, sometimes acutish. $2n =$ c.84, 104*. *A. parvifolia* Nutt. of Ed. 1. Dry prairies and open areas. BC to Man s to Wash, Nev, Ariz, Colo, Neb, Minn. Map 1000

A. corymbosa E. Nels.
Perennial with leafy stolons, forming loose mats; flowering stems slender, 1–3 dm tall, tomentose; lower leaves narrowly oblanceolate, 2–4 cm long, tapering to a narrow base, thinly but persistently tomentose; cauline leaves linear; heads several, peduncled, in a compact cluster; involucres 4–5 mm high, the bracts each with a dark brown or blackish spot below the whitish terminal portion. Moist open woods. Alta, Sask s to Mont, Ida, Ore, Calif, Utah, Colo. Map 1001 ○

A. dimorpha (Nutt.) T. & G.
Low plant with much-branched caudex, forming small compact mats but not stoloniferous; flowering stems very leafy, 2–5 cm high, terminated by solitary heads; leaves oblanceolate, spatulate or linear, narrowed to a short petiole, silky-tomentose on both sides; involucre of staminate head 5–7 mm high, the bracts blackish green or brown but with hyaline tips; involucre of pistillate head 10–15 mm high, the bracts narrow, brownish, and with hyaline acuminate tips; staminate pappus-bristles barbellate upwards, not clavate. Dry, open areas. sBC to sSask s to Calif, Colo, Neb. Map 1001 ●

A. lanata (Hook.) Greene
Similar to *A. pulcherrima*, but less robust; stems mostly 1–2 dm tall; basal leaves 3–10 cm long, the main veins more or less prominent; involucre tomentose toward the base, the bracts conspicuously brown or greenish black below and generally paler or even white at the tips, commonly acute. $2n = 28^*$. Alpine and subalpine meadows. BC, Alta s to Ore, Ida, Wyo. Map 1002

A. luzuloides T. & G.
Plant with a rather woody branched caudex and stems 1–5 dm tall, grey-tomentose; lower leaves erect, usually with conspicuous nerves, narrowly oblanceolate, 3–8 mm long, thinly tomentose, narrowed to a petiole or subsessile; cauline leaves reduced upwards, mostly linear; heads several or numerous, small, the involucres 4–5 mm high, glabrous nearly or quite to the base, scarious, pale greenish brown with whitish tips; pappus-bristles of staminate heads strongly clavate and somewhat serrulate. Gravelly montane slopes. BC, Alta s to Calif, Colo. Map 1003 ○

A. monocephala DC.
Plants mat-forming, with short stolons; stems mostly 5–10 cm tall, tomentose to floccose; basal leaves 5–10 (20) mm long, oblanceolate to spatulate, mucronate, glabrous to appressed white tomentose below, green and glabrous to thinly pubescent above; head solitary, rarely 1–2 smaller laterals; involucres 4–10 mm high; bracts green at base, the middle brownish or black, the attenuate tips blackish; achenes about 1–4 mm long, glabrous or papillose. $2n = 56, 63$, c.70, c.80. Includes *A. angustata* Greene, *A. glabrata* (J. Vahl) Greene. Alpine slopes and ledges. eSiberia, wGreenland: Alas, Yuk to sBaffin Is, Lab s to BC, Alta, sMack, Huds Bay. Map 1003 ●

A. neglecta Greene
Mat-forming perennial, the stolons short and somewhat leafy or elongated and nearly naked; stems 1–4 dm tall, with slender leaves; basal leaves spatulate, oblanceolate, obovate or nearly elliptical, sometimes mucronate, densely and persistently white-tomentose beneath, thinly tomentose to glabrous and green above, 1-nerved or somewhat 3-nerved; heads several in a dense or somewhat open cluster; pistillate involucre 6–9 mm high, the bracts narrow and long-attenuate with scarious whitish tips; staminate involucre smaller than the pistillate, the bracts broader and with conspicuous white tips. $2n = 28, 56^*, 70^*$, c.84, 98^*, 112^*. Includes *A. howellii* Greene, *A. neodioica* Greene, *A. campestris* Rydb. Grassland and open woods. sYuk, swMack to sHuds Bay, Lk Superior, Nfld s to Calif, Ariz, Colo, Okla, Mo, Va. Map 1004

A. parvifolia Nutt.
Low, mat-forming plant with creeping leafy stolons; flowering stems slender, mostly 10–25 cm tall; basal leaves small, commonly 1–2 cm long, oblanceolate to spatulate, tapering to a narrow base, closely white or grey tomentose on both sides; cauline leaves scattered, linear; heads several in a narrow cyme, often capitate; involucre of pistillate head 4–7 mm high, the bracts whitish, obtuse or acute. $2n = 28^*$. *A. nitida* Greene of Ed. 1. Includes *A. microphylla* Rydb. Prairies

and other dry open areas. Alas, Yuk, swMack to James Bay s to Calif, NM, ND. Map 1005

A. pulcherrima (Hook.) Greene / Showy Everlasting
Tall perennial, not mat-forming, but producing erect basal tufts of leaves; stems 2–5 dm tall, silky-woolly; basal leaves narrowly oblanceolate, 5–15 cm long, prominently 3-ribbed, loosely tomentose; cauline leaves lanceolate to linear, the upper ones small; heads several or numerous in a compact or fairly broad inflorescence; involucre densely pubescent below, the bracts 7–9 mm high, with white or tawny tips and commonly with a small dark spot at the base, usually obtuse. $2n$ = 56*, 63. Open woods, moist grassy areas, and flats. Alas, Yuk, sMack to Huds Bay s to Wash, Utah, Colo, Sask, Man, sJames Bay; sLab, Nfld. Map 1006

A. racemosa Hook.
Perennial with long, slender, leafy stolons; stems 1–5 dm tall, slender, sparsely leafy, tomentose below, glandular-puberulent above; lower leaves short-petiolate, the blades oval to ovate or obovate, 2–6 cm long, thin, green and glabrous or glabrate above, persistently tomentose beneath; cauline leaves narrow, sessile; heads on slender peduncles in an open inflorescence, usually racemiform; involucres with little or no tomentum, the pistillate 6–8 mm high, the staminate 4–5 mm high. $2n$ = 28*. Montane woods. BC, Alta s to Ore, Ida, Wyo. Map 1007

A. rosea Greene
Similar to *A. parvifolia* but with pinkish involucral bracts, rather than white; also, not so strongly mat-forming, the basal rosettes more ascending and the leaves generally somewhat larger and more pointed. Staminate plants unknown in our region. $2n$ = 56*, 126. Includes *A. concinna* E. Nels., *A. oxyphylla* Greene. Prairies and dry open areas, often with *A. parvifolia*. Alas, Yuk, Mack to Huds Bay, nLab s to Calif, NM, Sask, Man. Map 1008

A. umbrinella Rydb.
Stems 8–15 cm tall; basal leaves narrowly oblanceolate or spatulate, abruptly acute, densely white-tomentose on both sides; heads mostly 3–8, clustered; involucre 5–6 mm high, the bracts brown to yellow-brown or dirty green at base, the scarious parts varying from brownish to dirty white, obtuse or sometimes acutish at apex. $2n$ = 56*. Includes *A. sansonii* Greene, *S. sedoides* Greene, *A. aizoides* Greene, *A. isolepis* Greene. Gravelly slopes. Alas, Yuk, Mack to Huds Bay, nQue s to Calif, Ariz, Colo. Map 1009

Anthemis L. / Chamomile
Annual or perennial herbs, often strong-scented, with alternate, pinnatifid leaves, and radiate heads terminating the branches, the rays white or yellow; involucral bracts commonly multiseriate, sometimes subequal, dry, the margins hyaline or somewhat scarious; receptacle conical, more or less chaffy; ray-flowers pistillate or neutral; disc-flowers numerous, perfect; anthers entire at base; achenes 4–5-angled or terete, sometimes flattish; pappus a short crown or none.

Rays white; annual; ill-scented **A. cotula**
Rays yellow; short-lived perennial **A. tinctoria**

A. cotula L. / Mayweed; Dog-fennel
Annual, more or less branched, 1–5 dm tall, glabrous or sparsely pubescent, ill-smelling; leaves 2–3-pinnately dissected, the segments narrow; ray-flowers sterile, the ligules white, 6–10 mm long; scales of receptacle awl-shaped, subtending the central disc-flowers only; achenes ribbed and rough-tuberculate; pappus none. $2n = 18$. Waste ground. Introduced. Rare.

A. tinctoria L. / Yellow Chamomile
Short-lived perennial, 3–7 dm tall, whitish puberulent or villose at least above; leaves pinnately divided, with winged rachis, the segments deeply toothed, more or less villose; heads on long peduncles; involucre villose-tomentose; ligules yellow, 7–15 mm long; scales of receptacle lanceolate with firm yellow tips; achenes striate; pappus a short crown. $2n = 18$. Waste ground. Introduced. Occasional.

Arctium L. / Burdock

Large biennial herbs with broad alternate leaves and several or many flowering heads; leaves petiolate, the blades ovate to oblong, often cordate, up to 5 dm long and 4 dm broad, entire or toothed, tomentose or glabrate beneath, almost glabrous above; heads discoid, nearly globose, 1.5–3 cm thick, the involucral bracts multiseriate, narrow, often with a stiff hooked tip; receptacle flat, bristly; flowers all tubular and perfect; corolla pink or purplish; pappus of many short bristles, deciduous; achenes oblong, 3–5-angled, glabrous, often dark-mottled.

1 Involucre densely arachnoid-tomentose; corolla glandular **A. tomentosum**
Involucre glabrous; corolla not glandular 2

2 Heads 2.0–2.5 cm wide, racemose or in racemose clusters, the lower sessile or on short peduncles; inner bracts shorter than corollas **A. minus**
Heads 3.0–4.5 cm wide, corymbose, the lower on long peduncles; inner bracts as long as or longer than the corollas **A. lappa**

A. lappa L. / Great Burdock
Coarse, branching plant, 15–30 dm tall; leaves and heads as described above; outer bracts spreading or reflexed, hooked; inner bracts ascending, spine-tipped; achenes 6–7 mm long. $2n = 36$. Waste ground. Introduced. Rare.

A. minus (Hill) Bernh. / Common Burdock
Coarse, branching plant, mostly 8–12 dm tall; leaves and heads as described above; outer bracts lanceolate, hooked; inner bracts tapered to a more or less straight spiny tip; achenes 5–6 mm long. $2n = 36$. Waste ground, fields. Introduced. Occasional.

A. tomentosum Mill. / Woolly Burdock
Coarse, branching plant, mostly 8–12 dm tall; inflorescence corymbose; heads 2.0–2.5 cm wide; outer bracts spreading or reflexed; inner bracts ascending, broader, as long as or longer than the corollas; achenes 5.0–6.5 mm long. $2n = 36$. Waste ground, roadsides. Introduced. Occasional.

Arnica L.

Perennial herbs with rootstocks, erect stems, mostly opposite leaves and rather large heads of yellowish flowers; leaves simple, entire or toothed; heads solitary or several, many-flowered, conspicuously radiate (rays lacking in one species); involucral bracts about equal in length, in 1–2 series; receptacle naked; ray-flowers pistillate, yellow or orange; disc-flowers perfect, yellow; style-branches flattened, truncate, penicellate; achenes slender, 5–10-nerved; pappus of numerous capillary bristles, white or brownish, barbellate to subplumose.

1 Cauline leaves mostly 5–10 pairs; pappus straw-coloured or brownish, barbellate 2
Cauline leaves mostly 2–4 pairs, not including those (in any) of the basal cluster; pappus often white or subplumose 4

2 Involucral bracts obtuse or merely acutish, bearing a tuft of long hairs at or just within the tip **A. chamissonis**
Involucral bracts definitely acute, the tip not markedly more hairy than the body 3

3 Leaves more or less toothed; rhizome usually elongated; stems not much tufted **A. amplexicaulis**
Leaves entire or nearly so; rhizome usually a short branched caudex; stems tufted **A. longifolia**

4 Rays lacking; heads nodding in the bud **A. parryi**
Rays present; heads erect or nearly so in the bud 5

5 Pappus straw-coloured or brownish, subplumose; tufts of basal leaves not evident on the flowering stems 6
Pappus white, scabrous or barbellate, not subplumose; basal leaves often conspicuous 7

6 Heads narrow, more or less turbinate, generally several; middle stem-leaves generally largest **A. diversifolia**
Heads broader, nearly hemispherical, few or solitary; lowest stem-leaves generally largest **A. mollis**

7 Basal leaves cordate or broadly ovate, usually long-stalked 8
Basal leaves linear to lanceolate or oblanceolate, usually short-stalked or stalkless (except *A. lonchophylla*) 10

8 Plants 1–3 dm tall, tufted; heads 3–9; involucre about 1 cm wide, bracts 7–9 mm long; leaves usually 2.5 cm wide or less **A. gracilis**
Plants 2–6 dm tall, stems solitary or few; heads 1–3; involucre more than 1 cm wide, bracts 10–18 mm long; leaves usually more than 2.5 cm wide 9

9 Lower leaves usually cordate; base of involucre white-pilose; achenes hairy and/or glandular **A. cordifoli**
Lower leaves seldom cordate; base of involucre sparingly hairy; achenes glabrous, at least at base **A. latifoli**

10 Plants of alpine or subalpine habitats 1
Plants of montane or lowland habitats 1

11 Involucres 9–12 mm high; rays with teeth less than 0.5 mm long; cauline leaves often remotely toothed; bracts greenish at apex **A. rydbergi**
Involucres 10–15 mm high; rays with longer teeth; cauline leaves entire; bracts often purplish at apex 1

12 Heads erect; leaves acute; involucral bracts villose and glandular; achenes densely hairy **A. angustifoli**
Heads nodding, becoming erect; leaves generally obtuse; bracts glabrous, at least above the middle; achenes glabrous below **A. louisean**

13 Leaves generally dentate or denticulate; heads 1–5, usually 3, turbinate-campanulate **A. lonchophyll**
Leaves mostly entire; heads usually solitary, sometimes 3, hemispherical or nearly so 1

14 Base of stem with tuft of brown hair; disc-corollas with spreading glandular and glandless hairs; leaves lanceolate to oblanceolate, about 10 mm wide **A. fulgen**
Base of stem lacking brown hair, sometimes with whitish hairs; disc-corollas stipitate-glandular only; leaves generally narrower **A. sorori**

A. amplexicaulis Nutt.
Perennial with coarse rhizomes; stems 3–8 dm tall, leafy, hairy and glandular or nearly glabrous; cauline leaves mostly 5–10 pairs, not much reduced upwards, elliptic or ovate-lanceolate, sessile or the lower ones short-stalked, more or less toothed; heads usually several, broadly campanulate; involucre 9–15 mm high, the bracts narrow and acute or acuminate; rays 8–14, pale yellow, about 14 mm long; pappus light brown, subplumose. $2n$ = 38, 57*, 76*. In our area ssp. **amplexicaulis**. Moist montane woods. seAlas, sYuk s to Calif, Mont. Map 1010

A. angustifolia M. Vahl
Low herb with simple stem, 1–3 dm high, sparsely to densely villose, often minutely glandular; basal leaves mostly oblanceolate, with short margined petioles, usually entire, acute or acuminate, sparsely to densely villose and obscurely glandular; cauline leaves 1–3 pairs, lanceolate to linear, sessile, acute or acuminate; heads usually solitary; involucral bracts lanceolate, white-villose, densely so at base, often purplish; rays about 2 cm long, sharply and conspicuously toothed; achenes white-hirsute; pappus white, barbellate. $2n$ = 38, 57*, 76*. *A. alpina* (L.) Olin of Ed. 1. In our area ssp. **tomentosa** (Macoun) Douglas & Douglas. Alpine slopes. Circumpolar: Alas, Yuk to Ellesmere Is to Nfld s to BC, Mont, Alta, sMack, Huds Bay, nQue. Map 1011

A. chamissonis Less.
Perennial with long rhizomes; stems solitary, 2–8 dm tall, pubescent, sometimes densely so, glandular above; cauline leaves mostly 5–8 pairs, not markedly reduced upwards, lanceolate to oblanceolate, sessile, or the lowermost connate-petiolate, entire or denticulate; heads commonly several, broadly campanulate; involucre 8–12 mm high, the bracts obtuse to acute, with a tuft of long hairs at apex; rays pale yellow, 15–20 mm long; pappus mostly straw-coloured, barbellate. $2n = 38, 57$. In our area ssp. **chamissonis** with dentate leaves, involucral hairs with prominent septa and a subplumose, tawny pappus, and ssp. **foliosa** (Nutt.) Maguire, as described above. Moist meadows and thickets. sAlas, sYuk, sMack to James Bay s to Calif, NM, sSask, sMan, nOnt. Map 1012

A. cordifolia Hook.
Perennial with creeping rootstocks, producing flowering shoots, 2–6 dm tall, also short shoots with clusters of basal leaves; herbage glandular-puberulent to villose; cauline leaves mostly 2–4 pairs, the lower cauline and the basal leaves long-petioled, more or less cordate, obtuse, dentate or entire; upper leaves small, sessile; heads 1–3, rarely more; involucre about 15 mm high, with a sparse or dense white pubescence especially towards the base, and usually also glandular; rays 2–3 cm long; achenes hirsute; pappus white or whitish, barbellate. $2n = 38$, 57*, 76*, 95. Woodland. sYuk, swMack s to Calif, NM, SD; nMich. Map 1013

A. diversifolia Greene
Stems 2–4 dm tall from the rhizome, glandular-puberulent to nearly glabrous; cauline leaves mostly 3–4 pairs, the middle ones usually the largest, sessile or with short winged stalks, ovate or deltoid to elliptic-lanceolate, irregularly toothed; heads generally several, narrow, more or less turbinate; involucre 10–14 mm high, the bracts acute or acuminate, glandular-puberulent; rays about 13; pappus straw-coloured or light brown, subplumose. $2n =$ 57*, 76*. Rocky slopes to subalpine elevations. seAlas s to Calif, Utah. Map 1014

A. fulgens Pursh
Stem stout, glandular and pubescent, especially above, often viscid, 2–6 dm tall, from a short scaly rhizome; basal leaves and persistent leaf-bases with axillary tufts of brown hair; leaves entire or nearly so, pubescent and glandular, the basal ones petiolate, oblanceolate to narrowly oblong, the cauline ones in 2–4 pairs, short-stalked or sessile, lanceolate or linear, small and distant; heads solitary, occasionally 3, broadly hemispherical; involucre 10–15 mm high, glandular and hairy, the bracts with their tips rather blunt and hairy within; achenes hairy; pappus white to somewhat brownish, barbellate. $2n = 38, 57$. Grassland. BC to sMan s to Calif, Colo, SD, ND. Map 1015

A. gracilis Rydb.
Small, tufted plants, the stem slender, 1–3 dm high, from a branched caudex; herbage glabrous or glandular and sometimes sparsely hairy above; basal leaves mostly on separate short shoots, petiolate, ovate to broadly lanceolate, more or less toothed; cauline leaves mostly 2–3 pairs, petiolate or sessile, ovate to ovate-

lanceolate; heads mostly 3–7, small, turbinate; involucre 8–12 mm high, glandular, often also hairy; rays usually 8–11, about 15 mm long; pappus whitish, barbellate. $2n$ = 38*, c.76. *A. latifolia* ssp. *gracilis* (Rydb.) Ediger & Barkley. Open, rocky, subalpine slopes. BC, Alta s to Ore, Utah, Wyo. Map 1016

A. latifolia Bong.
Perennial with creeping rootstocks, producing flowering shoots 2–6 dm tall, also occasional short shoots with clusters of leaves; herbage glandular, hairy to glabrous; basal leaves petiolate, ovate, occasionally cordate; cauline leaves mostly 2–4 pairs, elliptic-lanceolate, more or less toothed, the middle ones as large as or larger than those below; heads 1–3, occasionally more; involucre mostly 12–15 mm high, somewhat glandular and sometimes sparsely hairy; achenes glabrous, at least at base, often glandular above; pappus white, barbellate. $2n$ = 38*, 57, 76*, 112*. Moist montane woods and meadows. Alas, Yuk, swMack s to Calif, Colo. Map 1017

A. lonchophylla Greene
Stems mostly solitary from a branching rhizome, 2–4 dm tall, glandular-puberulent, often also with spreading hairs; lower leaves petiolate, elliptic to lanceolate, glandular-puberulent to glabrate, denticulate or dentate- serrate; cauline leaves mostly sessile, 2–3 pairs, reduced upwards; heads 1–9, turbinate or campanulate, the bracts glandular and more or less spreading-hairy; pappus white, barbellate. $2n$ = 57*, 76*. In our area ssp. **lonchophylla**. Dry, open montane slopes. Alas, Yuk, swMack to Huds Bay s to BC, Alta, Sask, Man, cOnt; eQue, Nfld; SD; Minn. Map 1018

A. longifolia D.C. Eat.
Flowering stems several, also sterile shoots from a branched caudex; taller stems 3–6 dm high, more or less puberulent; leaves mostly 5–7 pairs, not much reduced upwards, sessile or the lower ones connate-petiolate, narrowly lanceolate to elliptical, entire or nearly so; heads several campanulate; involucre 7–10 mm high, the bracts acute or acuminate, glandular-puberulent; rays 8–11; pappus commonly straw-coloured and barbellate. $2n$ = 57, 76. Open rocky subalpine slopes. BC, Alta s to Calif, Colo. Map 1019

A. louiseana Farr
Flowering stems 1–several from a scaly rhizome or caudex, 0.5–2.5 dm tall, sparsely hairy; basal leaves with short margined petioles, oblong to ovate-lanceolate, usually serrate, glandular-puberulent to glabrate, obtuse or abruptly pointed; cauline leaves 1–2 pairs or none, the upper reduced; heads 1–3; involucre 1–1.5 cm high, the bracts glabrous except the minutely pilose base; rays 1–2 cm long; achenes glabrous or sparsely hairy above the middle. $2n$ = 38, 57*, 76*. Includes *A. griscomii* Fern. In our area ssp. **louiseana**. Alpine slopes and ledges. neSiberia: Alas, Yuk, wMack s to nBC; Alta; eQue, Nfld. Map 1020

A. mollis Hook.
Stems 2–6 dm tall, from a branched rhizome or caudex, pubescent and glandular; leaves in 3–4 pairs, all sessile or the lower short-stalked, the lower usually the largest, narrowly ovate or elliptic-lanceolate to oblanceolate, irregularly denticulate or rarely dentate; heads solitary or few, hemispherical or campanulate; involucre about 12 mm high, the bracts acuminate, hairy at base, more glandular towards apex; rays 14–18; pappus light brown, subplumose. $2n$ = 38, 57, 76*, 95, 114, 133, 152. Moist subalpine areas. Alas, Yuk, swMack; BC, Alta s to Calif, Utah, Colo. Map 1021

A. parryi A. Gray
Stems solitary, 2–5 dm high, from a slender rhizome, mostly woolly-villose near the base, glandular at least above; lower leaves lanceolate, cuneate to a short petiole, entire to denticulate, more or less villose and glandular; upper leaves 2–3 pairs, markedly reduced; heads generally several, turbinate-campanulate, the bracts puberulent and glandular; rays lacking; achenes glabrous to hairy or glandular; pappus straw-coloured to light brown, barbellate to subplumose. $2n$ = 38, 57, 76. Open montane woods. BC, Alta s to Calif, Colo. Map 1022

A. rydbergii Greene
Stems often clustered, 1–3 dm tall, from a scaly rhizome, more or less pilose and glandular-puberulent; basal leaves short-petiolate, commonly oblanceolate, frequently formed on separate short shoots; cauline leaves 2–4 pairs, sessile or the lower with short winged petioles, lanceolate to ovate, toothed or entire, glandular-hairy or pilose to glabrate; heads 1–3, turbinate-campanulate; involucre 9–12 mm high, the bracts glandular and pilose to nearly glabrous; achenes villose; pappus white, barbellate. $2n$ = 38*, 76. Turfy slopes and shale at alpine elevations. BC, Alta s to Calif, Utah, Colo. Map 1023

A. sororia Greene
Resembling *A. fulgens* (and possibly not well separated from that species), but generally more slender with narrower leaves, also lacking tufts of brown hair near the base of the stem, and the disc-corollas without spreading glandless hairs; pappus white or nearly so. $2n$ = 38. Prairies. BC, Alta s to Calif, Utah, Wyo. Map 1024

Artemisia L. / Wormwood; Sagebrush

Herbs or shrubby plants, usually aromatic, with alternate leaves and small discoid heads arranged in spicate-racemose or panicled clusters; leaves entire or dissected; involucre imbricate, the bracts dry and more or less scarious; receptacle naked or beset with long hairs; flowers all tubular, sometimes all perfect, sometimes the marginal flowers pistillate and the central flowers of the head either fertile or sterile; anthers united, not caudate at base, the tips lanceolate or subulate; style more or less exserted; achenes usually glabrous; pappus none.

1	Stem-leaves entire to coarsely lobed or toothed	2
	Stem-leaves deeply and narrowly dissected	7

2 Plants shrubby, decidedly woody at the base, flowers all alike, perfect and fertile 3
Plants herbaceous, flowers of two kinds, the marginal female only, the central perfect but sometimes with the ovary aborted 4

3 Leaves cuneate, 3-lobed **A. tridentata**
Leaves oblanceolate to linear, entire **A. cana**

4 Leaves green, mostly entire, essentially glabrous **A. dracunculus**
Leaves white tomentose at least below, mostly (except *A. longifolia*) coarsely toothed or lobed 5

5 Leaves densely white tomentose above and below **A. ludoviciana**
Leaves green to greenish-grey above 6

6 Leaves mostly entire, linear-lanceolate, 2–5 mm wide, margins strongly revolute **A. longifolia**
Leaves, some at least, coarsely lobed or toothed, wider, not strongly revolute **A. tilesii**

7 Leaf-divisions narrow-filiform, about 1 mm wide 8
Leaf-divisions wider, mostly 2–3 mm wide 12

8 Stems woody, 5–20 dm tall **A. abrotanum**
Stems herbaceous, mostly less than 5 dm tall 9

9 Plants alpine; heads 6–10 mm wide; bracts with a broad brownish black scarious margin; central flowers fertile; corolla yellow; receptacular hairs absent **A. furcata**
Plants not as above 10

10 Plants densely white tomentose throughout including involucres; central flowers fertile; receptacular hairs present between the disc-florets **A. frigida**
Plants green or grey-green, not densely tomentose throughout; central flowers sterile; receptacular hairs absent 11

11 Heads 2–4 mm wide; stems 25–60 cm tall; lowland **A. campestris**
Heads 4–6 mm wide; corolla reddish; stems 10–25 cm tall; alpine **A. borealis**

12 Margin of leaves serrate **A. biennis**
Margin of leaves entire 13

13 Leaves green, essentially glabrous; heads large, 6–10 mm across, nodding **A. norvegica**
Leaves greyish green, tomentose at least below, heads smaller, erect 14

14 Heads 2–3 mm wide; involucral bracts 1–2 mm long; receptacular hairs present between the disc-florets; leaf-lobes rounded **A. absinthium**
Heads 3–5 mm wide; involucral bracts 2–3 mm long; receptacular hairs absent; leaf-lobes acute **A. michauxiana**

A. abrotanum L. / Southernwood
Perennial, more or less shrubby; stems 5–20 dm tall, branched with striate, glabrous or puberulent twigs; leaves mostly 3–6 cm long, 2 or 3 times pinnatifid, the segments linear to filiform, green and glabrous above, thinly tomentose

below; inflorescence paniculate; heads 3–4 mm long, more or less globose; involucre more or less pubescent; flowers all fertile; corolla yellowish; achenes glabrous, angled. $2n = 18$. An escape from cultivation. Roadsides and waste ground. Introduced. Occasional.

A. absinthium L. / Wormwood; Absinthe
Aromatic herb, sometimes woody at base, 4–10 dm tall; stem finely canescent or glabrate; leaves silky-canescent, the lower long-petioled and 2–3 times pinnately divided into oblong or lanceolate segments; upper leaves smaller, with fewer and narrower divisions; heads nodding, numerous, in a large panicle; involucre 2–3 mm high, the bracts silky-canescent; receptacle long-hairy. $2n = 18$. Roadsides and waste ground. An escape from cultivation. Introduced. Occasional.

A. biennis Willd. / Biennial Sagewort
Annual or biennial with a taproot; stem 3–10 dm high, glabrous, often reddish; leaves glabrous, twice pinnately cleft, or the upper pinnatifid, the lanceolate or linear divisions commonly sharply toothed; heads numerous in dense spikes, these arranged in a leafy panicle; involucre 2–3 mm high, glabrous. $2n = 18$. Moist ground, shores, disturbed areas. Widely adventive in nAm. Alas, swMack; BC to Nfld s to Calif, Colo, Kans, Mo, Ohio, Pa, NJ. Map 1025

A. borealis Pallas
Similar to *A. campestris;* distinguished by the characters of the key. $2n = 18^*$, 36. Rocky alpine slopes. Circumpolar: Alas, Yuk, Banks Is, Victoria Is to sBaffin Is, Nfld s to nWash, Colo, sMack, James Bay, Gaspé. Map 1026

A. campestris L.
Herb, biennial or perennial, with a prominent crown and a taproot; stems 3–6 dm tall, single or several from the crown, glabrous or somewhat hairy; basal leaves crowded, petiolate, 2–3-pinnately divided into linear segments, glabrous or silky-canescent; upper leaves progressively reduced, often ternate; heads numerous in a narrow panicle; involucre 2–3 mm high and 2–4 mm wide, glabrous or sparingly short-hairy; receptacle glabrous. $2n = 18^*$, 36. A complex in which several species are often recognized, *A. canadensis* Michx., *A. caudata* Michx., *A. camporum* Rydb., *A. forwardii* S. Wats., *A. pacifica* Nutt. In our area the almost glabrous, biennial, ssp. **caudata** (Michx.) H. & C. predominates. Prairies, shores, and open sandy woods. Alas, Yuk, swMack to James Bay, Nfld s to Ore, Ariz, NM, Tex, Okla, Ia, Mich, Vt, Me. Map 1027

A. cana Pursh / Sagebrush
Branching shrub, more or less erect, often twisted and gnarled, 3–15 dm high; young twigs canescent or greyish-tomentose; leaves linear or linear-lanceolate, 1–4 cm long, generally acute, entire or, rarely, with 1 or 2 lobes or teeth, silvery-canescent; heads mostly in small glomerules arranged in a narrow leafy panicle; involucre 3–5 mm high, the bracts canescent or tomentose; flowers yellow,

5–20 in the head, receptacle glabrous; achenes glanduliferous. $2n = 18, 36$. Dry hills and prairies. Alta to Man s to Wash, Calif, Utah, NM, Kans. Map 1028

A. dracunculus L. / Dragonwort
Perennial herb, nearly odourless to strongly fragrant, with a stout rhizome or woody caudex; stems 5–10 dm tall, glabrous or occasionally hairy; leaves mostly linear, 2–8 cm long, entire or rarely somewhat cleft, glabrous to silky-canescent, sessile; heads many, in a leafy cluster; involucre 2–3 mm high, glabrous or sparingly hairy; receptacle glabrous. $2n = 18, 36, 54, 72, 90$. Dry hills and prairies. BC to Man s to Calif, Tex, Mo. Introduced Alas, Yuk and e. Map 1029

A. frigida Willd. / Pasture Sagewort
Silvery-grey perennial with a woody crown and root, often mat-forming; stems 1–4 dm tall, branching from the base, the young twigs cancescent or tomentose; basal leaves crowded, twice-pinnatifid into linear or linear-filiform divisions, silvery-canescent but often turning brown in age; upper leaves less dissected; heads yellowish, numerous in a narrow leafy cluster; involucre 2–3.5 mm high, hairy; receptacle densely villose; achenes glabrous. $2n = 18$. Prairies and open areas to alpine elevations. Asia: Alas, Yuk, swMack to cMan s to BC, Ida, Ariz, NM, Tex, Ia, Minn. Introduced e. Map 1030

A. furcata Bieb.
Perennial, commonly with a branched caudex; stems mostly 10–20 cm tall, canescent to sericeous; basal leaves 2–5 (10) cm long, the short blades trifid or generally palmately divided 2–3 times into linear or narrowly elliptic segments, sparsely to densely, coarsely sericeous above and below; inflorescence racemose or spicate; heads mostly 5–12; pressed involucres 6–10 mm wide; bracts 3–6 mm long, ovate to broadly elliptic, the tips generally erose, occasionally fimbriate, backs villose; corolla glandular; achenes glabrous. $2n = 18, 36, 54, 90$. Rocky alpine slopes. eAsia: Alas, Yuk, nwMack; Alta; Wash. Map 1031 ○

A. longifolia Nutt.
Perennial with a suffruticose base or woody caudex; stems clustered, 3–12 dm high, closely white-tomentose; leaves linear or linear-lanceolate, long-attenuate, grey-puberulent above, white-tomentose beneath, entire or the lowest rarely toothed, revolute-margined; heads erect in a narrow panicle; involucre 4–5 mm high, tomentose, the bracts ovate or oval. $2n = 36$. *A. vulgaris* L. ssp. *longifolia* H. & C. Eroded slopes and dry saline plains. Alta to Man s to Mont, Wyo, Colo, SD. Map 1031 ●

A. ludoviciana Nutt. / Prairie Sagewort
Perennial herb with creeping rootstocks, often producing colonies; stems 3–6 dm tall, more or less tomentose, rarely glabrate; leaves densely white-tomentose on both sides or becoming glabrate and green above, sometimes yellowish; lower leaves oblanceolate, narrowly elliptic or oblong-lanceolate, entire or somewhat toothed; upper leaves lanceolate to linear; heads numerous, often dense, in a narrow panicle; involucre 3–4 mm high, usually tomentose. $2n = 18, 36, 54$. *A. vulgaris* L. ssp. *ludoviciana* H. & C. A highly variable species, in our

area mostly var. **ludoviciana** with lanceolate leaves about 10 mm wide and var. **gnaphalodes** (Nutt.) T. & G. (*A. pabularis* (Nels.) Rydb.) with linear leaves 3–5 mm wide, often conduplicate. Prairies and dry open areas. BC to Ont s to Calif, Tex, Ark, introduced e. Map 1032

A. michauxiana Bess.
Perennial with a branched caudex, often woody, and elongated rhizomes; stems often several, 2–4 dm high; leaves white-tomentose beneath, at least when young, generally green above; lower leaves bipinnately dissected into short linear or lanceolate divisions, these often toothed; upper leaves reduced; heads in an elongate narrow inflorescence; involucre 3.5–4 mm high, glabrous or nearly so. $2n = 18^*$. *A. vulgaris* ssp. *discolor* (Dougl.) H. & C. Rocky subalpine slopes. sYuk, BC, Alta s to Calif, Utah, Wyo. Map 1033

A. norvegica Fries
Low perennial with a stout rhizome or caespitose caudex; stem 1–4 dm high, villose to nearly glabrous, sometimes ascending at base; lower leaves petiolate, twice-pinnately dissected into linear or lanceolate divisions, villose to glabrate; upper leaves smaller, less dissected, often sessile; heads large, nodding, not numerous; involucre 4–5 mm high, the bracts with blackish hyaline margins. $2n = 18, 36^*$. Includes *A. arctica* Less. In our area ssp. **saxatilis** (Bess.) H. & C. Subalpine and alpine slopes. Eurasia: Alas, Yuk, wMack s to Calif, Colo. Map 1034 ●

A. tilesii Ledeb.
Tall perennial with a strong rootstock, sometimes woody at the base; stem 6–12 dm tall, finely tomentose; leaves mostly cauline, 5–20 cm long, oblanceolate, elliptic or lanceolate in outline, entire or sparingly and sharply lobed or toothed, often long-attenuate, densely white-tomentose beneath, glabrate and green above; heads numerous, erect, in an elongated pyramidal panicle; involucre 4–5 mm high, 2.5–3 mm wide, the bracts tomentose, with blunt tips; corolla yellow; receptacle and achenes glabrous. $2n = 18, 36, 54$. *A. herriotii* Rydb. of Ed. 1. A highly variable species, in our area ssp. **elatior** (T. & G.) Hult. Open woods and river flats. nSiberia: Alas, Yuk, Mack to nHuds Bay s to Ore, Mont, Alta, sMack, James Bay. Map 1035

A. tridentata Nutt. / Sagebrush
Branching aromatic shrub, 1–20 dm high, the bark often shreddy; young branches and leaves silvery-canescent; principal leaves cuneate, 1–2 cm long, mostly 3-toothed or cleft at the apex, the upper ones linear-cuneate and entire; heads paniculate; involucre 3–5 mm high, 2 mm wide; flowers all perfect. $2n = 18, 36$. Dry montane slopes. BC, Alta, s to Calif, NM, ND. Map 1034 ○

Aster L.

Perennial or occasionally annual herbs with simple, alternate, entire or variously toothed leaves, and solitary to, more often, several or numerous heads, these usually radiate; involucral bracts in 2 or more series, commonly imbricate, usually more or less herbaceous at the tip and chartaceous below; receptacle naked, flat or convex; rays pistillate and fertile, blue or violet to pink or white; disc-flowers perfect and fertile, red or purple to yellow or whitish; anthers entire at base; style-branches flattened, with mostly slender, acute or acuminate, hairy appendages; achenes several-nerved; pappus of numerous capillary bristles, sometimes with an additional short outer series.

1 Rays inconspicuous or absent; annual **A. brachyactis**
Rays conspicuous; perennial 2

2 Pappus double, the inner series of long bristles, the outer of bristles about 1 mm long; inflorescence corymbose; ligules white **A. umbellatus**
Pappus single 3

3 Rays 8–15, mostly 13, about 2 cm long, white or pinkish; involucral bracts chartaceous, keeled, often crimson-edged; leaves lanceolate to ovate, sessile, rather crowded; stem strongly ribbed **A. engelmannii**
Rays mostly more than 15; involucral bracts generally herbaceous, at least above, usually not keeled 4

4 Involucre and usually the peduncles glandular 5
Involucre and peduncles not glandular 8

5 Leaves generally less than 5 mm wide, entire 6
Leaves more than 10 mm wide, usually toothed 7

6 Leaves 2–4 cm long, about 5 mm wide, glandular-puberulent **A. campestris**
Leaves 4–8 cm long, 2–3 mm wide, glabrous **A. pauciflorus**

7 Leaves coarsely serrate, mostly 3–5 cm wide **A. conspicuus**
Leaves distantly serrulate or sometimes entire, mostly 1–2 cm wide **A. modestus**

8 Plants 50 cm tall, or more; stem stout, usually crimson-purple, hispid; leaves oblong to lanceolate, auriculate-clasping, entire or distantly serrulate **A. puniceus**
Plants not as above 9

9 Outer bracts (at least some) foliaceous, equalling or surpassing the inner 10
Outer bracts not foliaceous, not surpassing the inner 11

10 Basal and lower stem-leaves sessile, entire; inflorescence long, narrow and leafy, with many heads **A. eatonii**
Basal and lower stem-leaves petiolate or narrowed to a petiolar base, often serrate or serrulate; inflorescence with 1 or few heads **A. subspicatus**

11 Bracts mostly glabrous except the ciliate margins 12
Bracts densely pubescent 17

12 Bracts oblanceolate, the outer obtuse, appressed, well imbricate; stems mostly 10–30 cm tall, pubescent; leaves oblanceolate, generally less than 10 mm wide **A. ascendens**
Bracts linear, lanceolate or oblong, all acute 13

13 Lower leaves long-petioled, blades ovate to lanceolate, often cordate at base, serrate to serrulate 14

Lower leaves linear, lanceolate or oblanceolate, mostly sessile or nearly so 15

14 Stem pubescent, often in lines; lower leaves ovate, frequently cordate **A. ciliolatus**

Stem glabrous, except in the inflorescence; leaves lanceolate **A. × maccallae**

15 Plants glabrous throughout; lower leaves wing-petioled, the upper auriculate-clasping **A. laevis**

Plants pubescent on the stem and branches in lines decurrent from the leaf-bases 16

16 Leaves mostly 2–5 mm wide; heads few, generally less than 10, sometimes solitary **A. borealis**

Leaves usually wider; heads generally more than 10 **A. hesperius**

17 Bracts bristle-tipped 18

Bracts not bristle-tipped 19

18 Heads small, usually secund on arching branches; involucre 2.5–4.5 mm high; bracts strongly imbricate, the outer 2.0–3.5 mm long; rays 10–20, mostly less than 6.5 mm long **A. ericoides**

Heads larger, usually not secund; involucre 5–8 mm high; outer bracts 3–7 mm long; rays 22–35, 7 mm long or longer **A. falcatus**

19 Heads solitary; basal leaves entire, oblanceolate or spatulate, the cauline leaves reduced above; achene 2-nerved **A. alpinus**

Heads solitary or few; basal leaves mostly serrulate, linear-lanceolate, the cauline leaves similar, well developed; achene several-nerved **A. sibiricus**

A. alpinus L.

Plant with a solitary stem from an enlarged caudex and crown, less than 3 dm high, rather stout, pilose or short-woolly; basal leaves numerous, oblanceolate, 1–5 cm long, rounded at apex, hirsute to silky-hairy; upper leaves, linear, reduced; heads solitary; involucral bracts loosely erect, broadly linear, silky and short-woolly, in 2 or 3 series but nearly of the same length; rays violet or whitish, about 1 cm long. $2n = 18, 27, 36$. In our area ssp. **vierhapperi** Onno. Foothills prairie to dry alpine slopes. Eurasia: Alas, Yuk, wMack s to Colo. Map 1036

A. ascendens Lindl.

Fibrous-rooted perennial with a caudex or horizontal rootstock; stems 2–6 dm tall, sometimes decumbent at base, sparsely to densely pubescent; leaves firm, entire, the lower oblanceolate and petiolate, the upper ones linear to lanceolate and sessile by a more or less clasping base; heads few–many; involucre 5–8 mm high, the bracts appressed, well imbricate in 3–4 series, glabrous to conspicuously ciliate on the margins, rarely somewhat hairy on the back; outer bracts mostly oblanceolate and the apex obtuse or rounded, occasionally cuspidate; inner bracts narrower and acute or acuminate; rays 5–10 mm long, blue or violet to pink, rarely white. $2n = 16, 32, 64$*. Dry, open areas, often in disturbed ground. seBC to Sask s to Calif, NM. Map 1037

A. borealis (T. & G.) Prov.
Plant with long creeping rhizome seldom over 2 mm thick; stem slender, 1–8 dm tall, puberulent in lines from the upper leaf-bases, glabrous below; leaves narrowly linear or linear-lanceolate, 4–8 cm long, commonly 2–5 mm wide, sessile and usually slightly clasping, entire or subentire, glabrous or scabro-ciliate; heads commonly few, sometimes solitary; involucre 5–7 mm high, the bracts slender, mostly acute, more or less imbricate; rays white, light rose or pale blue. $2n = 16^*$, 32, 48, 64. *A. junciformis* Rydb. of Ed. 1. Marshy ground and bogs. Alas, Yuk, wMack to Huds Bay, Que, NS s to Ida, Colo, Minn, NJ. Map 1038

A. brachyactis Blake / Rayless Aster
Annual with a taproot; stem 1–6 dm high, simple or much branched, glabrous to sparingly puberulent above; leaves 3–10 cm long, linear, mostly sessile, entire, acute, more or less ciliate on the margins; heads numerous, paniculate; involucre 5–8 mm high, the bracts in 2–3 series but almost equal in length, linear, acuminate, the midrib prominent; rays lacking or, if present, shorter than the pappus; disc-corollas surpassed by the copious white pappus; achenes several-nerved, appressed-hairy. $2n = 14$. *Brachyactis angusta* (Lindl.) Britt. Moist, usually saline ground. Yuk, swMack to Man s to Wash, Utah, Colo, Kans, Mo, Minn, introduced e. Map 1039

A. campestris Nutt.
Stems erect from creeping rhizomes, 2–5 dm tall, minutely glandular, at least above, otherwise puberulent or nearly glabrous, often brownish or purplish; leaves linear-oblong, 3–5 cm long, firm, entire, minutely glandular, mostly sessile and slightly clasping; heads several or many; involucral bracts glandular-puberulent, in about 3 series, but nearly equal in length, linear-lanceolate, somewhat acuminate; rays 20–40, violet or purplish; pappus bristles many, sordid. $2n = 10$. *Virgulus campestris* (Nutt.) Reveal & Keener. Dry open areas. BC, Alta s to Calif, Utah. Map 1040

A. ciliolatus Lindl. / Lindley's Aster
Perennial, often with long rhizomes; stems 2–10 dm tall, glabrous or pubescent, remotely leafy; basal and lower cauline leaves commonly with long winged petioles, the blades broadly ovate to lanceolate, rounded or cordate at base, acuminate, more or less serrate, glabrous, or pubescent, especially on the veins beneath; upper leaves lanceolate, sessile; heads usually rather few, in an open inflorescence, involucre 5–8 mm high, the bracts mostly linear, attenuate, with green midrib and scarious margins, not strikingly white-coriaceous below as are those of *A. laevis*; ligules pale blue; achenes glabrous or nearly so. $2n = 48^*$, 72. Woodlands and clearings. BC to Que, NS s to BC, Mont, Wyo, SD, Minn, NY. Map 1041

A. conspicuus Lindl. / Showy Aster
Stems from a strong rootstock, 3–9 dm tall, stout, sometimes zig-zag, scabrous-puberulent, glandular above; leaves large, firm, sharply and coarsely toothed, scabrous above, pubescent beneath; lower leaves mostly obovate, soon deci-

duous, the others ovate to elliptic and sessile; inflorescence corymb-like; involucre broadly cup-shaped, 1 cm high, densely glandular, the bracts well imbricate, with a broad chartaceous base and green, spreading or recurved, pointed tip; rays violet or pale blue, 10–15 mm long. $2n$ = 108*, 126*. Woodlands and clearings. BC to Sask s to Ore, Ida, Wyo. Map 1042

A. eatonii (A. Gray) Howell
Perennial with rootstocks; stems 3–8 dm tall, often tinged with red, uniformly puberulent above; leaves glabrous or scabrous, usually entire, mostly sessile, linear-lanceolate, 6–12 cm long, usually over 7 times as long as wide; heads in a long, narrow, leafy cluster; involucre 5–9 mm high, the bracts loose or squarrose, some of them leafy and wholly green; rays mostly white or pink; achenes hairy. $2n$ = 16. Moist montane woodland, stream banks. BC, Alta, swSask s to Calif, NM. Map 1043

A. engelmannii (D.C. Eat.) A. Gray
Stout perennial with a caudex or thick rhizome; stem 3–12 dm tall, very leafy, nearly glabrous or slightly glandular; leaves sessile, ovate-lanceolate to narrowly elliptic, pointed, entire 5–10 cm long, the lower smaller and soon deciduous; inflorescence corymb-like, short; involucre 8–12 mm high, the bracts imbricate, chartaceous below, somewhat keeled, the outer ones acuminate, the inner often purplish; rays 9–15, widely spaced, white or pinkish, about 2 cm long. $2n$ = 18. Open montane woods. BC, Alta s to Wash, Nev, Colo. Map 1044

A. ericoides L. / Tufted White Prairie Aster
Stems clustered from a short rootstock or caudex, 3–8 dm tall, rough-hairy, with ascending or recurved branches above; leaves numerous, firm, linear or linear-oblong, often spinulose at apex, hirsute-strigose to glabrate; heads numerous, often secund on ascending-recurved branches; involucre 3–4 mm high, occasionally higher, the bracts in about 3 series, the outer ones oblanceolate, loose or squarrose, with spinulose tips; rays white or rarely pinkish, mostly 3–5 mm long. $2n$ = 10*, 20*. *A. pansus* (Blake) Cronq. of Ed. 1. *Virgulus ericoides* (L.) Reveal & Keener. In our area ssp. **pansus** (Blake) A.G. Jones. Dry open areas, prairies. BC to sQue, Me s to Wash, Ore, Utah, Ariz, Tex, Ark, Ill, nVa. Map 1045

A. falcatus Lindl. / Creeping White Prairie Aster
Stems from elongate rhizomes or in caespitose clusters, pubescent with appressed or spreading hairs; heads solitary or few at the ends of the branches; leaves linear, entire, strigose, the apex often spinulose; involucre 5–8 mm high, the bracts pubescent, commonly with squarrose and spinulose tips, the outer bracts nearly as long as the inner; rays usually white, mostly 25–30, 7–9 mm long. $2n$ = 20, 30*, 40. *Virgulus falcatus* (Lindl.) Reveal & Keener. In our area ssp. **falcatus**, caespitose, sparsely pubescent, heads on slender few-bracted peduncles, the involucral bracts herbaceous, not strongly squarrose, and ssp. **commutatus** (T. & G.) A.G. Jones, rhizomatous and mat-forming, usually copiously pubescent, heads on stout many-bracted peduncles, the involucral bracts squar-

rose with an indurate basal portion. Dry open areas, prairies. neAlas, Yuk, swMack to cMan s to BC, Mont, Utah, Ariz, NM, Tex, Okla, Kans, Ia, Minn. Map 1046

A. hesperius A. Gray / Western Willow Aster
Plant with creeping rhizome; stems 5–12 dm tall, usually branching above, pubescent in lines decurrent from the leaf-bases, commonly glabrous below; leaves linear to broadly lanceolate, 5–15 cm long, entire or sometimes toothed, glabrous or somewhat scabrous, sessile or the lower petiolate and deciduous; heads usually many in a leafy cluster; involucre 5–8 mm high, the bracts linear-lanceolate, imbricate or nearly equal; rays 6–14 mm long, white, pink or blue. $2n$ = c.48, 64*. Stream banks, ditches, moist ground. BC to Man s to Calif, Nev, NM, Tex, Mo, Wisc. Map 1047

A. laevis L. / Smooth Aster
Perennial with rootstocks; stems mostly 4–10 dm tall, stout, glabrous, more or less glaucous; leaves glabrous, often glaucous, thick, ovate to lanceolate or oblong, entire or toothed, the lower ones narrowed into winged petioles, the upper sessile and more or less clasping, often auriculate; heads few or numerous in panicles; involucre 6–8 mm high, the bracts imbricate, firm, white-chartaceous at base, the tips sharply pointed and green; rays blue or purple; pappus tawny; achenes glabrous or nearly so. $2n$ = 48*. Open, generally dry areas. BC to Ont, sQue s to Wash, NM, Kans, Mo, Ga, Conn. Map 1048

A. × maccallae Rydb.
Similar to *A. ciliolatus;* distinguished by the characters of the key. $2n$ = 48*. A hybrid between *A. ciliolatus* and *A. subspicatus*. Moist woods. BC, Alta. Map 1049 ○

A. modestus Lindl.
Perennial with creeping rhizomes; stem 3–8 dm tall, slender, very leafy, densely stipitate-glandular, at least above, and often spreading-hairy as well; leaves mostly 4–8 cm long, lanceolate, acuminate, thin, sessile and more or less auriculate-clasping, entire or remotely toothed, glabrous or hairy; heads rather few in a short leafy cluster; involucre 7–11 mm high, the bracts slender, acuminate, about equal, glandular, the inner often purplish; rays purple or violet. $2n$ = 18. Moist woods and boggy ground. seAlas, sYuk to Sask, seMan, Ont s to Ore, Ida, Mont, Minn, Mich. Map 1050

A. pauciflorus Nutt.
Stems from slender rootstocks, 2–5 dm tall, glabrous below, glandular only near the heads; leaves rather fleshy, entire, glabrous, the lower oblanceolate and petiolate, the upper ones linear; heads few, borne singly on the branches; involucre 5–8 mm high, the bracts rather fleshy, linear-lanceolate, glandular, rather loose and unequal in 2 or 3 series; rays blue to whitish, 4–6 mm long. $2n$ = 18. Alkaline flats. sMack, Alta to swMan s to Ariz, NM, ND. Map 1051 ●

A. puniceus L. / Purple-stemmed Aster
Perennial with a stout caudex or rhizome; stem stout, 5–15 dm tall, often reddish purple, mostly spreading-hairy, simple or much branched above; leaves mostly sessile, lanceolate to oblong, 6–16 cm long, auriculate-clasping, distantly serrate or occasionally entire, scabrous to glabrous above, often hairy along the midrib beneath; heads few or numerous in a leafy cluster; involucre 6–12 mm high, the bracts slender, loose, scarcely imbricate, some of the outer commonly foliaceous, but narrow; rays mostly 30–60, blue, occasionally rose or white, 8–16 mm long. $2n = 16^*$. Marshy ground. Alta to Nfld s to SD, Kans, Ia, Ill, Ala, Ga. Map 1049 ●

A. sibiricus L.
Stems arising singly or in clusters from a creeping rhizome, often forming loose mats and loosely ascending, 1–4 dm high, often purplish, puberulent; leaves firm, oblanceolate to oblong-elliptic, 2–7 cm long, crenulate-serrate or entire, glabrous above, often somewhat hairy beneath, the lower short-stalked; heads solitary to several; involucre 6–9 mm high, villose-puberulent, the bracts broadly linear, mostly erect, purplish; rays purple or blue, 8–12 mm long. $2n = 18^*$. Shores, river flats, and open areas to alpine elevations. Eurasia: Alas, Yuk, wMack s to BC, Ore, Ida, Wyo. Map 1052

A. subspicatus Nees
Stems 1–10 dm high, from rootstocks, glabrous or somewhat pubescent; leaves entire or slightly toothed, glabrous to pubescent, the lower petiolate, obovate or oblanceolate, often deciduous; middle leaves sessile, lanceolate to oblong-ovate, 5–12 cm long and 1–4 cm wide, mostly less than 7 times as long as wide; heads commonly few in a corymb-like cluster, the upper leaves often grading into the involucral bracts; outer bracts equalling or surpassing the inner ones, sometimes foliaceous; rays 1–1.5 cm long, rose to purple. $2n = 16, 48^*$, c.72, 80. A variable complex, not taxonomically clear. Includes *A. foliaceus* Lindl. of Ed. 1. Moist woodlands, stream banks to subalpine elevations. sAlas, BC, Alta s to Calif, NM. Map 1053

A. umbellatus Mill.
Stems from creeping rhizomes, 3–20 dm tall, leafy to the top, puberulent; leaves lanceolate to ovate, tapering to base and apex, short-stalked or subsessile, entire, smooth to scabrous above, puberulent beneath; heads mostly numerous in a flattish-topped corymb; involucre 3–5 mm high, the bracts relatively few, pubescent on the back, appressed-ascending; rays 4–7, white, wide-spreading; pappus double, the inner of firm bristles, of which some are often thickened toward the tips, the outer of short bristles less than 1 mm long. $2n = 18$. Moist woodland and swampy ground. Alta to Nfld s to Neb, Ia, Mich, Ky, Ga. Map 1051 ○

Bahia Lag.

B. oppositifolia (Nutt.) DC.
Perennial herb, more or less woody at base, with a slender tough root; stems 1–2 dm high, tufted, widely branched below, grey-puberulent; leaves nearly all opposite, petiolate, palmately 3–5-parted into linear divisions, canescent-strigose and impressed-punctate; heads terminating the upper branches, yellow, radiate, 7–10 mm wide; involucre 5–7 mm high, canescent, the bracts in 2 series, outer more or less keeled, inner somewhat scarious-margined; receptacle flat, naked; ray-flowers pistillate, the rays yellow, few, 2–4 mm long; disc-flowers perfect, yellow; achenes 4 mm long, slender, 4-angled, glandular; pappus of a few blunt scarious scales. *Picradeniopsis oppositifolia* (Nutt.) Rydb. Roadsides. Introduced. Rare.

Balsamorhiza Nutt.

B. sagittata (Pursh) Nutt. / Balsam-root
Perennial herb with a short caudex, large basal leaves and a stout tap-root; leaves long-petiolate, the blades triangular-oblong to cordate-sagittate, up to 3 dm long, entire, grey-tomentose to silvery and velvety-puberulent; flowering stem scapiform from the caudex, bearing a few reduced leaves, woolly-tomentose at least above; heads large, usually solitary, the rays yellow, 2–4 cm long; involucre 2–4-seriate, the bracts ovate to lanceolate, acuminate to attenuate, woolly-tomentose; receptacle chaffy, its scales clasping the achenes; disc-flowers numerous, light yellow; achenes 7–8 mm long, glabrous, more or less 4-sided; pappus none. $2n = 38$. Grassland and open montane woods. BC, Alta s to Calif, Colo, SD. Map 1054

Bidens L. / Beggar-ticks

Annual herbs with opposite leaves and mostly yellow flowers; leaves simple or dissected; heads radiate or discoid; involucre double, the outer commonly large and foliaceous, the inner membranous or scarious-margined and often striate; receptacle flat, chaffy; ray-flowers, when present, 3–8, neutral, yellow or whitish; disc-flowers perfect; anthers not caudate at base; achenes flattened or somewhat 4-sided, beakless; pappus mostly of 2–4 awns or teeth, retrorsely barbed.

1 Rays generally conspicuous; leaves mostly sessile, not deeply lobed or pinnate **B. cernua**
Rays inconspicuous or absent; leaves petiolate, mostly deeply 3-lobed or pinnate 2

2 Leaves pinnate, the leaflets usually distinctly stalked **B. frondosa**
Leaves simple, deeply incised or 3-lobed **B. tripartita**

B. cernua L. / Nodding Beggar-ticks
Stem simple or freely branched, 2–10 dm high, glabrous, sometimes with a decumbent rooting base; leaves sessile, often connate, linear-lanceolate, 5–15

cm long, often coarsely serrate; outer involucral bracts 5–8, usually leafy, unequal, often reflexed; inner bracts membranous, yellow and often red-streaked; rays 6–8, yellow, up to 1.5 cm long, or wanting; achenes 5–7 mm long, cuneate, the marginal nerves and the summit cartilaginous, the 4 awns (rarely 2) 2–3 mm long, retrorsely barbed. $2n = 24, 48$. Marshy ground, ponds, ditches. BC to Que, NS s to Calif, NM, Tex, Okla, Ark, NC. Map 1055

B. frondosa L. / Tall Beggar-ticks
Stem coarse, branched, 3–15 dm high, commonly reddish, furrowed, glabrous or sparsely hairy; leaves petiolate, mostly 3–5-pinnately compound, the leaflets often stalked, lanceolate, coarsely serrate, puberulent or glabrous; outer involucral bracts 10–16, 1–2 cm high, rather leafy, ciliate on the margins and often hairy on the back; inner bracts 7–9 mm long; rays inconspicuous, less than 3.5 mm long, or wanting; disc-corollas pale yellow; achenes 6–12 mm long, flat, with 2 barbed awns, 3–4 mm long. $2n = 48$. *B. vulgata* Greene of Ed. 1. Moist ground, ditches. BC to Nfld s to Calif, Tex, Ark, La, Va. Map 1056 ●

B. tripartita L.
Stem simple or branched, 2–8 dm high, glabrous; leaves undivided or more often lobed at base, the blades elliptic-lanceolate, serrate, tapering to short, wing-margined petioles; outer involucral bracts 6–10, leafy, conspicuous, erect or slightly spreading; inner bracts short, membranous; ligules usually wanting; achenes 5–10 mm long, flat, usually with 3 awns, 4–6 mm long and barbed. $2n = 48$. *B. comosa* (Gray) Wieg. of Ed. 1. Moist ground, river flats. Introduced. Rare.

Brickellia Ell.

B. grandiflora (Hook.) Nutt.
Herbaceous perennial or somewhat woody at the base; stems 2–8 dm tall, often branched above, puberulent to glabrate; leaves mostly petiolate, simple, opposite or alternate; the blades deltoid-ovate to lanceolate, 2–10 cm long, truncate to cordate or broadly cuneate at base, toothed, hairy to nearly glabrous, obscurely glandular-punctate; heads discoid, paniculate-cymose, mostly in clusters, drooping; involucre 7–11 mm high, the bracts striate, imbricate in several series, the outer slender-tipped, puberulent; receptacle flat, naked; flowers all tubular and perfect, greenish yellow or cream-coloured; achenes 10-ribbed, about 5 mm long, the base callous; pappus of numerous whitish bristles. $2n = 18$. Eroded slopes and rocky montane banks. seBC, Alta s to Calif, Colo, Tex, Ark, Neb. Map 1056 ○

Carduus L. / Thistle

C. nutans L.
Biennial or rarely annual herb, 3–25 dm tall; stem winged; leaves glabrous, or long-villous chiefly along the midrib and main veins beneath, deeply lobed, up to 40 cm long and 15 cm wide; heads mostly solitary and nodding at the ends

of the branches, usually large, the disc 4–8 cm wide when pressed, sometimes smaller and scarcely more than 1.5 cm; middle and outer involucral bracts 2–8 mm broad with a long, flat, spreading or reflexed, spine-pointed tip, the midrib very prominent; inner bracts narrower and softer, scarcely spiny, often purplish; flowers purple; pappus minutely barbellate. $2n = 16$. Roadsides and waste ground. Introduced. Occasional.

Carthamus L.

C. tinctorius L. / Safflower
Annual; stem mostly 1–6 dm tall, simple or sparingly branched, glabrous; leaves ovate to ovate-lanceolate, the lower short-petiolate, the upper sessile, margin entire or spinose-dentate; inflorescence corymbose; heads few; involucral bracts imbricate, the outer 2–3 cm long, leaf-like, spinescent; flowers orange or reddish, all tubular; achenes white, shiny; pappus absent or of narrow scales. $2n = 24$. An escape from cultivation. Introduced. Rare.

Centaurea L.

Herbs; leaves alternate or basal; receptacle more or less flat, bristly; involucral bracts imbricate, spine-tipped or with a scarious or hyaline, erose to pectinate, terminal appendage; heads discoid; flowers all tubular and perfect or more often the marginal ones sterile with enlarged, irregular corollas; stamens with appendaged anthers; pappus of several series of graduated bristles or scales, often reduced or wanting.

1	Involucral bracts, at least some, spine-tipped	2
	Involucral bracts not spine-tipped	3
2	Spines more than 10 mm long	**C. solstitialis**
	Spines less than 10 mm long	**C. diffusa**
3	Leaves pinnatifid	**C. maculosa**
	Leaves entire, toothed or lobed	4
4	Pappus 6–11 mm long; plants perennial, with deep creeping roots	**C. repens**
	Pappus 1–3 mm long; plants annual, lacking deep creeping roots	**C. cyanus**

C. cyanus L. / Cornflower; Bachelor's Button
Annual or winter-annual, 2–12 dm tall; herbage usually loosely white-tomentose when young, the lower surfaces of the leaves often persistently so; leaves up to 13 cm long and nearly 1 cm wide (excluding the lobes), narrow, often linear, entire, or the lower ones toothed or with a few narrow lobes; heads terminating the branches; involucre mostly 11–16 mm high, the bracts more or less striate, with a relatively narrow, often darkened, pectinate or lacerate fringe near the tip; flowers mostly blue, sometimes purple, pink or white, the marginal ones with enlarged, irregular corollas. $2n = 24$. Roadsides and waste ground. An escape from cultivation. Introduced. Rare.

C. diffusa Lam.
Diffusely branched annual or biennial, 1–6 dm tall, sparsely scabrous-puberulent under the thin and deciduous arachnoid tomentum; leaves small, more or less pinnatifid, the lower ones deciduous, the reduced ones of the inflorescence mostly entire; heads numerous, narrow; involucre 8–10 mm high, the middle and outer bracts coarsely pectinate or weakly spinose-ciliate, and tipped with a definite slender spine mostly 1.5–4 mm long; flowers few, creamy or sometimes purplish, the marginal ones not enlarged; pappus wanting. $2n = 18$. Roadsides and railway tracks. Introduced. Rare.

C. maculosa Lam.
Biennial or short-lived perennial, tending to be taprooted, mostly 3–15 dm tall; with a thin and loose, soon deciduous tomentum, also sparsely scabrous-puberulent; leaves obscurely to evidently glandular-punctate, pinnatifid, with narrow lobes, or the reduced ones of the inflorescence entire; heads terminating the numerous branches; constricted upwards in life; involucre mostly 10–13 mm high, its bracts striate, the middle and outer ones with short, dark pectinate tips; flowers pink-purple or rarely white, the marginal ones enlarged; pappus up to 2 mm long, or rarely wanting. $2n = 36$. Roadsides and waste ground. Introduced. Rare.

C. repens L. / Russian Knapweed
Perennial herb with deep-seated, black, creeping roots; stems erect, branched near the base, 4–8 dm high, tomentose to nearly glabrous; leaves alternate, linear to lanceolate, more or less clearly glandular-punctate, often hairy, the lower cauline ones up to 6 cm long, often pinnately lobed, the others smaller, entire or somewhat toothed; heads terminating leafy-bracted branches, discoid; involucre narrowly ovoid, whitish green, 9–12 mm high, the bracts imbricate in several series, the outer rounded-ovate with nearly entire hyaline tips, the inner ones narrower and with plumose-hairy tips; receptacle densely bristly; flowers all tubular, pale pink to purple; achenes 2–3 mm long, often whitish, obovoid, flattened; pappus capillary, deciduous. $2n = 26$. Roadsides, fields, waste ground. Introduced. Occasional.

C. solstitialis L / Yellow Star Thistle
Annual or biennial, 2–10 dm tall, with erect, branching, winged stems; lower leaves lyrate, upper leaves entire, linear, decurrent; heads solitary on branch tips, about 20 mm high; involucre globose, about 13 mm high, the lower outer bracts coriaceous, with several small spines apically, the upper outer bracts with a single slender apical spine 10–30 mm long; flowers yellow; pappus 2–5 mm long. $2n = 16$. Gardens, where probably introduced with seed. Introduced. Rare.

Chrysanthemum L.

C. leucanthemum L. / Ox-eye Daisy
Perennial herb with short rootstocks and usually growing in patches; stems erect, 2–8 dm tall, simple or forked near the top, glabrous or sparsely hairy;

leaves glabrous or somewhat hairy, the basal ones petiolate, oblanceolate or spatulate, 4–15 cm long, coarsely toothed or pinnatifid; cauline leaves petiolate to sessile, oblanceolate to linear, toothed to somewhat incised, often lacerate near the base; heads terminal on long peduncles, solitary, about 4 cm broad, radiate; disc yellow; rays white; involucral bracts imbricate, narrow, with a brown submarginal strip, becoming scarious at the margin and tip; receptacle naked; disc-flowers perfect; ray-flowers pistillate, white, 10–15 mm long; achenes terete, about 10-ribbed; pappus none. $2n = 18, 36, 54, 72, 90$. Roadsides, fields, waste ground. Introduced. Common.

Chrysothamnus Nutt.

C. nauseosus (Pall.) Britt. / Rabbit-brush
Shrub, commonly 2–6 dm high, widely branching, sometimes woody only at a very short base and producing numerous annual shoots; stems leafy, covered with a close white felt-like tomentum; leaves alternate, linear to filiform, sessile or appearing so, entire, tomentose to glabrate; heads in terminal rounded cymes, discoid, mostly 5-flowered; involucral bracts of various lengths and in more or less distinct vertical ranks, commonly hairy; receptacle small, naked; flowers yellow, all tubular and perfect; corollas mostly 6–10 mm long; achenes slender, about 5 mm long, hairy; pappus of numerous capillary bristles. $2n = 18$. Dry open areas. sBC to sSask s to Calif, Ariz, NM, Tex. Map 1057

Cichorium L.

Herbs; leaves alternate, entire or lobed; heads showy; involucral bracts biseriate, the inner longer; flowers all ligulate; pappus of short obtuse scales.

Leaves glabrous; involucre glabrous, longer than the rays **C. endivia**
Leaves hairy, at least the lower below; involucre glandular-hairy, shorter than the rays **C. intybus**

C. endivia L. / Endive
Similar to *C. intybus* but annual or biennial; distinguished by the characters of the key, the strongly thickened peduncles of the terminal heads and the longer pappus scales, one-sixth to half as long as the achene. $2n = 18$. Waste ground. An escape from cultivation. Introduced. Rare.

C. intybus L. / Chicory
Perennial with a deep taproot; stem 3–10 dm tall, stiff, widely branching, more or less hispid; lower leaves petiolate, oblanceolate, 7–15 cm long, pinnatifid or merely toothed; upper leaves sessile and clasping, small, lanceolate to oblong, often entire; heads blue, 2–3 cm wide, borne 2–3 together in scattered sessile clusters on the elongated branches; involucre cylindrical, the bracts in 2 series, the outer somewhat spreading and glandular-ciliate, the inner erect; flowers bright blue, occasionally pink or white; achenes 2–3 mm long, glabrous, striate, beakless; pappus of 1–3 series of small scales. $2n = 18$. Roadsides, waste ground. Introduced. Occasional.

Cirsium Mill. / Thistle

Herbs, biennial or perennial, with sessile alternate leaves, often pinnatifid, prickly; heads solitary to numerous, mostly large, many-flowered, cup-shaped to hemispherical or nearly globose, discoid; involucral bracts imbricate in several rows, usually spine-tipped; receptacle densely clothed with soft bristles; flowers all tubular and perfect, rarely unisexual by abortion; corolla purple, red, creamish or white, the tube slender and with long narrow lobes; styles with a thickened hairy ring; achenes oblong, flattish, glabrous; pappus of numerous bristles, more or less plumose, united into a ring at the base, deciduous.

1 Heads small, less than 2.5 cm high, involucres 1–2 cm high **C. arvense**
Heads larger, more than 2.5 cm high, involucres more than 2 cm high 2

2 Upper surface of leaves scabrous with appressed spines **C. vulgare**
Upper surface of leaves not scabrous 3

3 Pappus of mature fruit longer than corolla **C. foliosum**
Pappus of mature fruit shorter than corolla 4

4 Heads 5–7 cm high **C. drummondii**
Heads smaller, less than 5 cm high 5

5 Involucre densely arachnoid, outer bracts up to 2 mm wide at base **C. hookerianum**
Involucre not densely arachnoid, outer bracts more than 2 mm wide at base 6

6 Outer bracts eglandular, more or less equal; cauline leaves subentire to shallowly lobed **C. scariosum**
Outer bracts glandular, progressively longer from outer to inner; cauline leaves deeply pinnatifid lobed 7

7 Corollas deep purple (rarely white); achenes up to 5 mm long; leaf-lobes generally less than 7 mm wide **C. flodmanii**
Corollas light purple, achenes more than 5 mm long; leaf-lobes generally more than 7 mm wide **C. undulatum**

C. arvense (L.) Scop. / Creeping Thistle; Canada Thistle
Perennial with deep-seated, spreading, sprouting roots, generally in patches; stems 3–10 dm tall, very leafy, glabrous or nearly so; leaves glabrous, or somewhat tomentose beneath, sinuate-pinnatifid, usually very prickly; plants dioecious, the heads unisexual; involucres 1–2 cm high, the bracts spineless or the outer ones with weak spines; flowers pinkish-purple, or white; pappus of pistillate heads surpassing the corollas, that of the staminate heads shorter; achenes about 4 mm long. $2n = 34^*$. Fields, roadsides, waste ground. Introduced. Common.

C. drummondii T. & G. / Drummond's Thistle
Biennial; stem mostly 15–30 (50) cm tall, unbranched, tomentose; leaves narrowly oblanceolate, cut one-half to three-quarters to the midrib, the lobes oblong-triangular; upper cauline leaves numerous, ascending, enclosing the subsessile

cluster of 1–(3–5)–9 heads; involucre 3.0–3.5 cm high; outer bracts ovate, 5–8 mm wide at base, glandular, the slender spine 2–3 mm long; flowers rose-purple; anthers 7–9 mm long; pappus yellowish, shorter than the corolla; achenes 4.0–5.5 mm long, straw-coloured, with a light-yellow apical rim. $2n = 34$*. Open deciduous woods, clearings, meadows. wMack, eBC to nOnt s to Wyo, SD. Map 1058

C. **flodmanii** (Rydb.) Arthur / Flodman's Thistle
Slender plants, 3–8 dm tall, sometimes spreading by slender creeping roots; leaves of first-year rosettes oblanceolate, coarsely pinnatifid to merely spinose-ciliate, densely white-tomentose beneath, grey-tomentose above; flowering stems slender, cottony-tomentose; cauline leaves deeply pinnatifid, the lobes narrowly lanceolate; heads one or several; involucre 2–3 cm high, the outer bracts glutinous on the back and with spreading prickles 2–4 mm long; flowers deep purple, rarely creamish white; achenes 3–5.5 mm long with a conspicuous apical yellow band 0.5 mm wide. $2n = 22$*. Flowers July–September. Moist grassland, roadsides. Alta to Man s to Utah, Colo, Neb, Ia. Map 1059

C. **foliosum** (Hook.) DC.
Biennial; stem 2.5–7.0 dm tall, unbranched, fleshy, ribbed, pilose; leaves numerous, ascending, exceeding the heads, linear-oblanceolate, the larger 10–18 cm long, 2 cm wide, shallowly lobed, upper surface sparsely pubescent with long, multicellular hairs, lower surface tomentose to villous; heads 5 or more in a compact globose cluster; involucre 2 cm high; bracts usually subequal, the outer lanceolate to ovate-lanceolate, 2–4 mm wide at base, membranous, tapering to a slender spine 2–3 mm long, eglandular; flowers white, rarely pale pink; anthers 4.5–5.5 mm long; pappus longer than corolla by 1–10 mm; achenes about 4.5 mm long, light brown with darker striae and a narrow yellow apical rim. $2n = 34$*. Moist meadows, deciduous woods, and clearings. sYuk, sMack, s to BC, Mont, Wyo. Map 1060

C. **hookerianum** Nutt. / White Thistle
Biennial; stem 3–8 dm tall, thick and succulent below, tapering and slender above, arachnoid-villose to tomentose; leaves linear-oblong to oblanceolate, lobed or often merely toothed, arachnoid-villose to tomentose on both sides, or glabrate above; heads in dense or elongate clusters, few to numerous; involucre 2.0–2.5 cm high, the bracts generally arachnoid-tomentose, narrow, rather loose, mostly spine-tipped; flowers creamish or dirty white, the corollas unequally cleft. $2n = 34$*. Open woods and meadows. BC, Alta to nWash, nIda, nMont. Map 1061

C. **scariosum** Nutt.
Biennial; stem 2–10 dm tall, mostly unbranched, fleshy, ribbed, pubescent; lower cauline leaves narrowly obovate, tapering at base, shallowly lobed to incised half-way to midrib, glabrous above, woolly below; upper cauline leaves smaller, narrowly elliptic to more or less linear, less deeply lobed; heads 5–15, terminal, the upper leaves sometimes with axillary heads; involucre 2–3 cm high; bracts subequal or the inner progressively longer, the outer 2.5–5.0 mm wide at base,

ovate-lanceolate, acuminate, eglandular, the slender spine 2–4 mm long; flowers white to pink or reddish purple; anthers 6–10 mm long; pappus 4–5 mm shorter than corolla; achenes 5.5–6.5 mm long, light or dark brown, with a narrow, yellowish apical band. $2n = 34$*. Moist meadows, stream banks. seBC, Alta s to Wash, Ida, Utah, Wyo; Que (Mingan Is). Map 1062

C. undulatum (Nutt.) Spreng.
Similar to *C. flodmanii*, but more robust and not so readily spreading by creeping roots; stems 3–12 dm tall, densely white-tomentose; leaves coarsely toothed to pinnatifid, the lobes broadly triangular or ovate, occasionally oblong, closely white-tomentose beneath, thinly tomentose or glabrate above; heads solitary or several; involucre mostly 2.5–3 cm high, the outer bracts with a glutinous ridge on the back, often tomentose marginally, and with spreading prickles 3–5 mm long; flowers usually pink-purple; achenes mostly 5–7 mm long, the apical band, if present, very narrow. $2n = 26$*. Dry open areas. BC to swMan s to Ore, Nev, Ariz, NM, Tex, Mo. Map 1063

C. vulgare (Savi) Ten. / Bull Thistle
Biennial, 6–12 dm tall, the stem conspicuously spiny-winged by the decurrent leaf-bases, hirsute and sometimes arachnoid; leaves pinnatifid, the lobes sometimes toothed, very spiny, covered with small appressed spines on upper surface, thinly hairy beneath; heads usually several; involucre 2.5–4 cm high, the bracts with slender spines; flowers rose-purple, rarely white. $2n = 68$. Fields, waste ground, roadsides. Introduced. Common.

Coreopsis L.

C. tinctoria Nutt. / Common Tickseed
Glabrous annual herb with an erect branching stem, 4–10 dm tall; leaves mostly opposite, short-petiolate or nearly sessile, mostly 5–10 cm long, once or twice pinnatifid, the divisions linear or narrowly lanceolate; heads numerous, radiate, on long peduncles; involucre about 7 mm high, 10–15 mm wide, the bracts in 2 series, all more or less united at base; outer bracts about 2 mm long; inner bracts 5–8 mm long, orange or brown, with scarious margins; receptacle flat or slightly convex, chaffy; ray-flowers about 8, neutral, the ligules 8–12 mm long, yellow to orange, with or without a purple base; disc-flowers fertile, red-purple; achenes linear-oblong or somewhat cuneate, incurved, 1–4 mm long; pappus of 2 minute awns or none. $2n = 26$. Clay flats and slough margins. Alta to Man s to Calif, Tex, La. Map 1064

Crepis L. / Hawksbeard

Annual, biennial, or perennial herbs with milky juice; perennial species with a woody caudex and leafy crown; leaves alternate or all basal, the cauline mostly reduced, the basal entire, toothed or pinnatifid; heads 1–many, on branched or scapiform stems; flowers all ligulate and perfect, yellow; involucre cylindric or cup-shaped, the principal bracts in 1–2 series, the reduced outer ones few or many; receptacle flat or nearly so, naked or short-ciliate; achenes terete or nearly so, 10–20-ribbed, beakless or sometimes narrowed into a short beak; pappus of numerous white or whitish capillary bristles.

1 Plants annual; cauline leaves auriculate 2
Plants perennial, with a taproot; cauline leaves, if present, not auriculate 3

2 Basal rosette present at flowering; stems several; achenes less than 2.5 mm long **C. capillaris**
Basal rosette absent at flowering; stems solitary; achenes more than 2.5 mm long **C. tectorum**

3 Plants 2–6 cm high; alpine **C. nana**
Plants taller; not alpine 4

4 Leaves entire or dentate; stems glabrous 5
Leaves pinnatifid; stems tomentose or puberulent 6

5 Plants 10–30 cm tall; basal leaves glabrous **C. elegans**
Plants 20–70 cm tall; basal leaves coarsely hirsute **C. runcinata**

6 Involucre 5–10 mm wide at middle, more or less glandular hairy; plants mostly less than 35 cm tall **C. occidentalis**
Involucre less than 5 mm wide at middle, not glandular hairy; plants often more than 35 cm tall 7

7 Leaf-segments linear to narrowly lanceolate, usually entire; achenes greenish **C. atrabarba**
Leaf-segments lanceolate, usually toothed; achenes yellowish or brown **C. intermedia**

C. atrabarba Heller
Perennial herb, 1.5–7 dm tall with 1 or 2 stems from a caudex, leaves deeply pinnatifid, the lobes linear to lanceolate and entire or somewhat toothed, tomentose to glabrate; uppermost leaves linear and entire; heads 3–30, mostly 10–35-flowered; involucre 8–14 mm high and 3–5 mm wide, grey-tomentulose, sometimes with glandless black setae; outer bracts 5–10, less than half as long as the inner ones; achenes usually greenish, more or less attenuate at apex. $2n$ = 22, 33, 44, 55, 88. Dry, open mountain slopes. BC, Alta; swSask, s to Nev, Colo. Map 1065

C. capillaris (L.) Wallr.
Annual, occasionally biennial; stem 1–9 dm tall, generally branched; basal leaves variable, 3–30 cm long, mostly 1–4 cm wide, petiolate, lanceolate or oblanceolate, denticulate to runcinate-pinnatifid or even bipinnately divided; cauline leaves reduced upwards, sessile; heads numerous, corymbose; involucre 5–8 mm high; bracts tomentose, often with intermixed black glandular bristles, the outer linear, shorter than the inner; achenes 1.5–2.5 mm long, pale brown. $2n$ = 6. Roadsides, waste ground. Introduced. Rare.

C. elegans Hook.
Perennial, 1.5–2.5 dm high and as broad or broader; stems glabrous, several or numerous, stiffly erect, branched from the base, forming a dense clump topped by a mass of flower-heads; lower leaves spatulate, entire or coarsely dentate, narrowed into a petiole with a clasping base, glaucous and purplish; upper leaves linear, petiolate or sessile; heads erect, small, 6–10-flowered; involucre 8–10 mm high, 2–3 mm wide, the outer bracts 7–8, very short, the inner bracts 8–10, oblong, in 2 ranks; achenes golden brown, very slender, attenuate into a delicate beak with expanded pappus disc. $2n = 14$. Riverbanks, sand-bars, mountain slopes. Alas, Yuk, wMack s to BC, Mont, Wyo. Map 1066

C. intermedia A. Gray
Perennial, 3–7 dm tall with 1 or 2 stems from the caudex; leaves pinnatifid, the lobes lanceolate, entire or sometimes toothed, densely to sparsely grey-tomentulose; heads 10–60, mostly 7–12-flowered, involucre 10–16 mm high and 4–5 mm wide, more or less grey-tomentulose, occasionally with a few setae near apex; outer bracts 6–8, less than half as long as the inner ones; achenes yellow or brown, narrowed above. $2n = 33, 44, 55, 88$. Assigned here, largely for convenience, are polyploid apomicts that are intermediate between C. *atrabarba*, C. *occidentalis*, and several other species of this complex. Dry open areas. BC, Alta s to Wash, Calif, Wyo, Colo. Map 1067

C. nana Richards.
Perennial, 2–8 cm high, with a taproot, the caudex short or somewhat elongated; leaves mostly basal, narrowly spatulate to obovate, entire to lyrate-pinnatifid, glabrous or nearly so, glaucous, often purplish; heads several to numerous, borne among the leaves on short stalks; involucre 8–13 mm high, the outer bracts 5–8, unequal, the inner ones 10, oblong; achenes golden brown, columnar, slightly constricted above or with a short slender beak. $2n = 14$*. Rocky alpine slopes. Asia: Alas, Yuk, Prince Patrick Is to nBaffin Is to Nfld s to Calif, Utah, sMack, nHuds Bay. Map 1068

C. occidentalis Nutt.
Perennial, 1–4 dm tall, with a taproot and 1–3 stems from a caudex; stems and foliage grey-tomentulose, or glabrate in age, often glandular-hairy above; lower leaves runcinately toothed to deeply pinnatifid with toothed lobes; heads mostly 12–20-flowered; involucre 11–19 mm high, 5–10 mm wide at middle, with gland-tipped hairs; achenes fusiform, brownish. $2n = 22, 33, 44, 55, 66, 77, 88$. Dry slopes. BC, Alta s to Calif, NM, SD. Map 1069

C. runcinata (James) T. & G.
Perennial, 2–7 dm tall, with one or more strong roots and 1–3 nearly leafless stems; herbage glabrous or somewhat hispid, not at all canescent or tomentose; basal leaves forming a conspicuous rosette, oblanceolate to elliptical or obovate, entire to runcinate-toothed or pinnatifid, sessile or wing-petioled; upper leaves much reduced, mostly lanceolate or linear; heads mostly 3–15 and 20–50-flowered; involucre 9–16 mm high, glabrous or hispid, sometimes with glandular

hairs, the inner bracts mostly 10–15; achenes brownish, somewhat contracted at apex or short-beaked. $2n = 22$. Moist open areas, especially saline meadows. BC to Man s to Calif, NM, Tex, Neb, Minn. Map 1070

C. tectorum L. / Annual Hawksbeard
Annual, sometimes persisting over winter, mostly 1–10 dm tall, with a branched leafy stem, glabrous or pubescent; lower leaves petiolate, the blades lanceolate or oblanceolate, denticulate or pinnatifid; middle and upper cauline leaves sessile, auriculate, mostly linear, the uppermost with revolute margins or bract-like; heads numerous or five, 30–70-flowered; involucre 6–9 mm high; outer bracts subulate, much shorter than the 12–15 inner ones, these puberulent and sometimes also glandular-hispid; receptacle inconspicuously ciliate; achenes mostly 3–4 mm long, dark purplish-brown, fusiform, attenuate but scarcely beaked, 10-ribbed, the ribs scabrous. $2n = 8$*. Roadsides, waste ground. Introduced. Common.

Erigeron L. / Fleabane; Wild Daisy

Herbs, commonly perennial, sometimes or biennial; stems leafy or scapose; leaves alternate, rarely all basal; heads generally few or solitary, usually borne on naked or sub-naked peduncles; involucral bracts narrow, equal or imbricate, often greenish but not distinctly foliaceous, usually with thin or scarious margins and tips; receptacle flat or nearly so, naked; pistillate marginal flowers more or less numerous, usually with evident and often narrow ligules, or rarely absent; ligules white, pink, blue or purple, rarely yellow; style-appendages short, up to 0.5 mm long, rarely obsolete; pappus of a few to rather numerous capillary bristles, often with some much shorter outer scales (squamellae) or setae; achenes commonly 2-nerved, occasionally up to 14-nerved.

1 Plants producing leafy stolons **E. flagellaris**
Plants not producing leafy stolons 2

2 Leaves dissected, or at least some trifid or lobed at the apex 3
Leaves more or less entire, not dissected, trifid or lobed at the apex 7

3 Leaves 2– 4 times ternately dissected **E. compositus**
Leaves trifid, or at least some toothed or lobed at the apex 4

4 Leaves trifid 5
Leaves not trifid, some irregularly toothed or lobed at the apex, others more or less entire 6

5 Leaf-segments less than 1 mm wide, the basal segment not obviously dilated apically; involucre hirsute but not densely villose **E. compositus**
Leaf-segments more than 1 mm wide, the basal segment obviously dilated apically; involucre densely villose **E. trifidus**

6 Heads 1–2 cm wide **E. purpuratus**
Heads 3–5 cm wide **E. lanatus**

7 Rays absent, inconspicuous or short, not more than 5 mm long — 8
Rays present, longer than 5 mm — 12

8 Involucre 4–5 mm high — 9
Involucre at least 6 mm high — 10

9 Heads numerous (20+); involucre not glandular; the bracts generally barred horizontally on the median stripe — **E. canadensis**
Heads few (10±); involucre glandular, the bracts not barred on the median stripe — **E. annuus**

10 Involucre deep purple; densely villous, the hairs with purple septa — **E. humilis**
Involucre not as above — 11

11 Involucre glandular; peduncles arcuate; basal rosette generally present — **E. acris**
Involucre not glandular; peduncles straight; basal rosette often not present — 13

12 Infloresence corymbose, head occasionally solitary; rayless pistillate flowers present between the ray and disc flowers — **E. elatus**
Inflorescence racemose, head occasionally solitary, the lower peduncles often shorter than the subtending leaf; rayless pistillate flowers absent — **E. lonchophyllus**

13 Rays yellow — **E. aureus**
Rays not yellow — 14

14 Plants generally less than 10 cm tall; head solitary; involucres woolly-villose — 15
Plants usually taller; heads usually more than one; involucres not woolly-villose — 20

15 Leaves linear or narrowly oblanceolate — 16
Leaves not linear, mostly spatulate, oblanceolate or obovate — 17

16 Leaves less than 2 cm long; plants with a branching caudex — **E. radicatus**
Leaves more than 2 cm long; plants tufted, not branched at base — **E. ochroleucus**

17 Heads 3–5 cm wide; involucre 10 mm high or more — 18
Heads 2.5 cm wide or less; involucre less than 10 mm high — 19

18 Rays white; cauline leaves absent — **E. lanatus**
Rays blue or pale purple; cauline leaves present — **E. grandiflorus**

19 Involucre deep purple, densely villose, the hairs with purple septa; rays 3–6 mm long — **E. humilis**
Involucre not deep purple; rays about 10 mm long — **E. × arthurii**

20 Basal leaves lobed or toothed, occasionally entire; cauline leaves auriculate-clasping — **E. philadelphicus**
Basal and cauline leaves not as above — 21

21 Leaves 1–3 mm wide, linear to oblanceolate, not much reduced in size upwards — 22
Leaves more than 3 mm wide, lanceolate to oblanceolate or spatulate, often markedly reduced upwards — 24

22 Involucre glabrous; leaves essentially glabrous — **E. hyssopifolius**
Involucre pubescent, generally glandular; leaves coarsely pubescent, with appressed or spreading white hairs — 23

23 Stem 10–15 cm tall, sparingly branched; heads mostly 1–2; rays drying white or yellowish **E. pumilus**
Stem 15–40 cm tall, often diffusely branched; heads few to many; rays drying pink or purplish **E. divergens**

24 Leaves with depressed, usually pubescent veins; rays about 2 mm wide **E. peregrinus**
Leaves lacking depressed, pubescent veins; rays usually less than 1 mm wide 25

25 Stem leafiest about the middle; cauline leaves mostly 1–2 cm wide **E. speciosus**
Stem leafiest basally; cauline leaves less than 1 cm wide 26

26 Leaves essentially glabrous above and below **E. glabellus**
Leaves pubescent 27

27 Plants generally green; cauline leaves markedly reduced in size upwards; basal leaves not 3-nerved; rays usually drying pink or purplish **E. glabellus**
Plants generally greyish to dark olive-green; cauline leaves often not markedly reduced in size upwards; basal leaves frequently 3-nerved; rays usually drying white or yellowish **E. caespitosus**

E. acris L.
Biennial or perennial herb with short, simple or slightly branched caudex; stem 0.5–8 dm tall, glabrous to somewhat hirsute, often also glandular, especially under the heads; leaves glabrous to more or less hirsute and sometimes glandular; basal leaves oblanceolate to spatulate, entire to remotely serrulate, up to 10 cm long; cauline leaves usually ample, narrowly lanceolate to ovate; heads solitary–many; involucres 5–12 mm high, glandular or hirsute, or both, the bracts generally subequal, green or purplish, the inner long-attenuate, often almost caudate; pistillate flowers of 2 kinds, the outer with short rays, white or pink, the inner ones without rays; disc-flowers perfect, with slender tubular corollas; achenes 2-nerved, sparsely hairy; pappus of 25–35 bristles, sometimes with a few short outer setae. $2n = 18$*. Includes *E. scotteri* Boivin. In our area ssp. **debilis** (A. Gray) Piper, with a somewhat glandular involucre, the plants 3–15 cm tall, mostly with solitary heads, and ssp. **politus** (E. Fries) Schinz & Keller (var. *asteroides* (Andrz. *ex* Bess.) DC. of Ed. 1) with a densely glandular involucre, the plants taller, with several–many heads. Moist woods, clearings, shores and moist rocky slopes to alpine elevations. Circumpolar: Alas, Yuk, Mack to Huds Bay, sLab s to Calif, Utah, Colo, Sask, Man, Minn, Mich, Me. Map 1071

E. annuus (L.) Pers.
Annual or rarely biennial; stems 3–7 dm tall, finely strigose to hirsute, sometimes almost glabrous; foliage appearing scanty; basal leaves mostly oblanceolate to elliptic, tapering to the petioles, entire or toothed; cauline leaves rather few, linear to lanceolate; heads several–many in an open, rather flat-topped cluster; involucre 2–5 mm high, finely and obscurely glandular and more or less hairy; rays 50–100 (sometimes lacking), up to 6 mm long, white or sometimes pink or blue. $2n = 27$*, 36, 54. Includes *E. strigosus* Muhl. of Ed. 1. In our area mostly ssp. **septentrionalis** (Fern. & Wieg.) Wagenitz. Dry open areas, fields, disturbed ground. BC to Nfld, generally distributed s. Map 1072

E. × arthurii Boivin
Perennial; stems 5–10 cm tall, several from a basal rosette, simple; leaves petiolate, 2–5 cm long, 3–10 mm wide, oblanceolate to spatulate, coarsely pubescent, somewhat glandular below; cauline leaves 1–4, smaller, linear; heads solitary; involucre 6–7 mm high, more or less glutinous; bracts pubescent, margin glandular-serrulate; rays 8–10 mm long, white, drying pink or lilac; pappus 4–5 mm long. A hybrid between *E. acris* and *E. aureus*. Rocky alpine slopes. BC, Alta. Map 1073 ●

E. aureus Greene
Low perennial with a branched caudex; stem 2–15 cm high, more or less villose with spreading or sometimes appressed hairs; basal leaves with broad obovate or elliptical blades, often abruptly narrowed to the petiole, hirsute or hirsute-villose with appressed or loose hairs; cauline leaves few, small, narrow; heads solitary, the disc 7–16 mm wide; involucre 5–8 mm high, sparsely to densely woolly-villose with multicellular hairs, the cross-walls purple-tinged or clear; involucral bracts loose, green with purple tips, sometimes purplish throughout; ligules 25–70, yellow, 6–9 mm long; pappus double, the inner of coarse bristles, the outer of setae or narrow scales; achenes 2-nerved, hairy. $2n = 18$*. Rocky alpine slopes. BC, Alta s to Wash. Map 1074 ○

E. caespitosus Nutt.
Perennial with strong taproot and stout usually branched caudex; stems several, curved at base, then erect, 1–2.5 dm high, densely canescent or minutely hirsute with short spreading hairs; leaves short-hirsute, the basal ones oblanceolate or narrowly spatulate, tapering to a petiole, entire, more or less 3-nerved; cauline leaves several, reduced upward in size and distribution, oblong-ovate to linear; heads few or solitary; disc 9–18 mm wide; involucres 4–7 mm high, more or less pubescent with white hairs and also glandular, the bracts slightly to strongly imbricate, appressed, firm, thickened on the back; ligules 30–100, white, blue or pinkish, 5–15 mm long, 1–2.5 mm wide. $2n = 18, 36$*. Dry, open areas. BC to Man s to Wash, Utah, Ariz, NM, Neb, ND; Alas, Yuk. Map 1075

E. canadensis L. / Horseweed
Coarse annual; stems erect, 2–15 dm tall, leafy, spreading-hirsute to nearly glabrous; leaves oblanceolate to linear, more or less hirsute, coarsely ciliate near the base, entire or incised; lower leaves petioled, 2–10 cm long, the upper ones sessile, reduced in size; heads very numerous in an elongated panicle, or few on stunted plants; involucres 2–4 mm high, the bracts in 2–3 series, narrow, glabrous or slightly hirsute; ligules numerous, but shorter than the pappus, usually white; central perfect flowers few; pappus usually of one series of capillary bristles. $2n = 18$. Disturbed ground, fields, roadsides. sMack, BC to Nfld s to Calif, Tex, Fla. Map 1076 ●

E. compositus Pursh
Perennial with simple or branched stout caudex; stems 5–15 cm tall, scape-like; leaves crowded at the base, ternately lobed or dissected, the divisions linear or

narrowly oblong, more or less glandular and with at least a few spreading hairs cauline leaves few and reduced, mostly entire; heads solitary; involucres 5–10 mm high, more or less glandular and hirsute, the bracts unequal, purplish at least at the tips; rays white, pink or blue, up to 12 mm long but usually quite short and sometimes absent; pappus simple, of 12–20 coarse bristles. $2n$ = 18, 54*. In our area var. **glabratus** Macoun with leaves 2–3 times ternate and var. **discoideus** A. Gray with trifid leaves. Dry open areas to alpine elevations. Siberia (Wrangel Is), Greenland: Alas, Yuk to Ellesmere Is to Baffin Is s to Calif, Ariz, Colo, SD, Sask, sMack; eQue, Nfld. Map 1077

E. divergens T. & G.
Taprooted biennial or perennial; stem 1–4 (7) dm tall, often diffusely branched, pubescent, somewhat glandular; leaves 1–4 cm long, 1–3 mm wide,oblanceolate to linear, the lower deciduous, coarsely pubescent above and below; heads several–many; involucre 4–5 mm high; rays 5–6 mm long, numerous, blue, pink or white. $2n$ = 27, 36, 54, 72. Dry, open montane areas. BC, Alta s to Calif, NM, Neb. Map 1078 ○

E. elatus Hook.
Similar to *E. acris* and *E. lonchophyllus;* distinguished by the characters of the key. $2n$ = 36*. *E. acris* L. var. *elatus* (Hook.) Cronq. of Ed. 1. Marshy ground. Alas, Yuk, wMack to Huds Bay, Nfld s to BC, Alta, nSask, sMan, James Bay, sQue. Map 1079

E. flagellaris A. Gray
Biennial or short-lived perennial with a taproot; flowering stems erect, 5–40 cm tall, sparsely leafy or nearly naked; some of the main stems, or branches of the flowering stems, becoming trailing, sparsely leafy stolons, these sometimes rooting at the tips; stems and leaves somewhat hirsute or strigose; basal leaves petioled, oblanceolate, entire; cauline leaves linear to narrowly oblanceolate; heads solitary; involucres 3.5–5 mm high, finely glandular and more or less hirsute; ligules 5–10 mm long, white, occasionally pink or blue; pappus double. $2n$ = 27, 36, 45. Dry open woods. BC, Alta; Wyo, SD s to Nev, Ariz, Tex. Map 1073 ○

E. glabellus Nutt.
Perennial or biennial with simple or slightly branched caudex and a fibrous root system; stem erect or occasionally curved at base, 1–5 dm high, with appressed or spreading hairs; leaves more or less hirsute or strigose; basal leaves petiolate, oblanceolate, entire or irregularly toothed; cauline leaves lanceolate or linear, reduced in size upwards; heads 1–10; involucres 5–9 mm high, hirsute or strigose, the bracts linear, often with an evident brown midvein; rays 125–175, 8–15 mm long, blue or pink, occasionally white. $2n$ = 18*, 27, 36. Includes *E. formosissimus* Greene. In our area ssp. **glabellus**, with short, appressed pubescence, the leaves essentially glabrous, and ssp. **pubescens** (Hook.) Cronq. (*E. asper* Nutt.) with coarse, spreading, septate hairs throughout. Open woods,

meadows, prairies. Alas, Yuk, wMack to sMan s to BC, Ida, Utah, Colo, SD, ND, Wisc. Map 1080

E. grandiflorus Hook.
Perennial herb, 4–25 cm high, the stem decumbent at base, somewhat pilose; basal leaves oblanceolate, tapering to the petiole, hirsute-pilose; cauline leaves several, ovate to narrowly lanceolate; head solitary; involucre 8–10 mm high, densely villose or pilose with flattened multicellular hairs; cross-walls of the hairs clear or occasionally the basal ones purplish; involucral bracts about equal, loose, some often reflexed above the middle, the tips often purplish and glandular; rays 100–125, blue, 10–15 mm long, 1–2 mm wide; pappus double, the inner of 15–20 bristles, the outer of a few inconspicuous setae. $2n$ = 18, 36*. Alpine slopes. Alas, Yuk, Banks Is, wMack s to BC, Alta. Map 1078 ●

E. humilis Grah.
Perennial with short caudex and taproot and usually some fibrous roots; stem more or less erect, generally lax, 2–25 cm high, commonly villose with spreading hairs, those near the top of the stem usually with purple cross-walls; leaves villose when young, the lower ones becoming glabrate; basal leaves mostly oblanceolate, up to 8 cm long; cauline leaves mostly linear or linear-lanceolate; heads solitary; involucre 6–9 mm high, woolly-villose, the hairs commonly with dark cross-walls, the bracts loose and attenuate, usually blackish purple, rarely greenish; ligules 50–150, white to purplish, 3–6 mm long. $2n$ = 36*. Moist alpine slopes. eSiberia, Greenland, Norway: Alas, Yuk, Banks Is to sBaffin Is, nQue s to BC, Mont, sMack, James Bay. Map 1081

E. hyssopifolius Michx.
Perennial with a very short underground caudex and many fibrous roots; stems several from the caudex, slender, mostly 1–3 dm high, usually sparsely hirsute-strigose, sometimes villose or nearly glabrous; leaves crowded, longer than the internodes, linear to oblong-lanceolate, up to 3 cm long, thin and lax, subglabrous, with rather distantly ciliate margins; short leafy shoots often arising in the leaf-axils; heads solitary, rarely as many as 5, borne on long peduncles; involucre 4–6 mm high, the bracts thin, lax, acuminate to attenuate, with a few somewhat viscid hairs; ligules 20–50, white, or sometimes pink or purple, 4–8 mm long. $2n$ = 18. Shores, banks, ledges. Yuk, swMack to James Bay, Nfld s to nBC, nAlta, sMan, Ont, Mich, NY, Vt. Map 1082 ○

E. lanatus Hook.
Perennial with a slender branching caudex; stem scapose, about 5 cm tall or less, loosely villose with long hairs, especially upwards; leaves basal, oblanceolate, up to 3 cm long, often 3-toothed at apex, loosely woolly to pilose; head solitary; involucre 9–15 mm high, more or less woolly, the cross-walls of the hairs clear or some of them obscurely purplish; involucral bracts strongly purplish, especially upwards; ligules 30–80, white, sometimes blue or pink, 8–11 mm long, 1.4–2 mm wide; pappus bristles 25–35, white, often somewhat twisted. $2n$ = 36*. Rocky alpine slopes. BC, Alta s to Mont; Colo. Map 1076 ○

E. lonchophyllus Hook.
Biennial or short-lived perennial with a weak root; stem erect, usually 1–5 dm tall, more or less hirsute; leaves hirsute to glabrous, often conspicuously ciliate on the margins; basal leaves oblanceolate to spatulate, up to 15 cm long, entire; cauline leaves commonly linear, little reduced upwards; heads several or many, occasionally solitary, borne on nearly erect peduncles, often equalled or surpassed by the subtending leaves; involucre 4–9 mm high, hirsute, not glandular, the bracts usually somewhat imbricated, thin, light green, commonly purplish near the acuminate tips; rays many, 2–3 mm long, white or sometimes pinkish, only slightly surpassing the disc; pappus of 20–30 long bristles, sometimes with a few inconspicuous outer setae. $2n = 18$. Moist ground, meadows, shores. Asia: Alas, Yuk, wMack s to Calif, Utah, NM, ND; James Bay; eQue. Map 1083

E. ochroleucus Nutt.
Perennial with a taproot and a crown or short branched caudex; stem 1–40 cm high, commonly hirsute-strigose, sometimes with fine lax pubescence; leaves strigose or villose-strigose to nearly glabrous; basal leaves linear to oblanceolate, 1–12 cm long, their bases somewhat enlarged and membranous, whitish or purplish; cauline leaves linear, reduced upwards, sometimes few and bract-like; heads solitary or few; involucre 4–8 mm high, more or less densely hirsute-villose, often also somewhat viscid, the hairs sometimes with the lower cross-walls purplish; involucral bracts relatively broad, mostly dark yellowish-green, sometimes with brown midrib, often with purple tips; ligules 20–80, blue, purple or white, 4–12 mm long; pappus double, the inner of 10–20 fragile bristles, the outer of thick setae or squamellae, usually fairly conspicuous. $2n = 18$. Dry, open montane slopes. seBC, Alta s to Wyo, nwNeb. Map 1082 •

E. peregrinus (Pursh) Greene
Perennial with a rhizome or short caudex; stem 1–7 dm high, glabrous to more or less villose especially just below the heads, leafy or subscapose; basal leaves oblanceolate to spatulate, petiolate; cauline leaves ovate to linear, reduced above, often somewhat clasping, glabrous or ciliate on the margins, sometimes villose; heads usually solitary; involucre 7–11 mm high, the bracts linear, attenuate, densely glandular on the back, rarely with a few long hairs; rays 30–80, 8–25 mm long, 2–4 mm wide, deep rose-purple, sometimes pale or even white; achenes 4–7-nerved (usually 5); pappus single, of 20–30 long bristles, or sometimes with a few outer setae. $2n = 18^*$. In our area ssp. **callianthemus** (Greene) Cronq. Moist banks, meadows, and open montane woods. sAlas, sYuk s to Calif, Utah, NM. Map 1084

E. philadelphicus L.
Biennial or short-lived perennial with a simple caudex, occasionally blooming the first season; stems and leaves more or less pubescent with spreading hairs, sometimes nearly glabrous; stems 2–7 dm tall; basal leaves oblanceolate, rarely obovate, crenate or lobed, short-petioled; upper leaves sessile and clasping at base, crenate, dentate or nearly entire; heads 1–many in an open cluster; involucre 4–6 mm long, the bracts with more or less hairy brownish midvein and broad

hyaline margins; rays mostly over 150, 5–10 mm long, deep pink or purplish to white; achenes 2-nerved, sparsely hairy; pappus single. $2n$ = 18*. Moist open woods, clearings, disturbed areas. swMack, BC to Nfld s to Calif, Tex, Fla; nYuk. Map 1074 ●

E. pumilus Nutt.
Perennial with caudex and taproot; stem 5–40 cm tall, hirsute with spreading hairs, often also glandular or viscid, especially under the heads; leaves entire, more or less hirsute; basal leaves oblanceolate, sometimes narrowly so; cauline leaves more or less reduced in size and distribution; heads 1–several; involucre 4–7 mm high, more or less spreading-hirsute or somewhat glandular, the bracts narrow, green, with brown midrib; ligules 50–100, white, rose or purple, 5–15 mm long; pappus double, inner of 7–27 long bristles, outer of inconspicuous setae or squamellae. $2n$ = 18, 36. Dry hills, prairies. sBC to sSask s to Calif, NM, Kans. Map 1085

E. purpuratus Greene
Perennial with slenderly branching caudex; leaves sparsely to moderately villose and obscurely viscid, oblanceolate to spatulate, up to 2.5 cm long, some of them 3-lobed at apex, with broad rounded lobes; peduncles scapiform, 3 cm long or less, loosely villose and glandular; heads solitary; involucre 6–8 mm high, viscid-villose with long slender flattened hairs; involucral bracts few, thin, green, sometimes purplish; ligules 60 or fewer, white or pinkish, up to 5 mm long. $2n$ = 18*. In our area ssp. **pallens** (Cronq.) G. Douglas (*E. pallens* Cronq. of Ed. 1). Rocky alpine slopes. Alas, Yuk, wMack, nBC; sBC, Alta. Map 1086 ●

E. radicatus Hook.
Perennial with branching caudex and a taproot; stem 5 cm high or less, finely hirsute or villose, especially upwards; leaves all or nearly all in a basal cluster, oblanceolate or linear-oblanceolate, glabrous to finely pubescent, the margins more or less ciliate; cauline leaves, when present, 2 or 3, small, linear, pubescent; head solitary; involucre about 5 mm high, viscid and finely short-villose, the bracts narrowly lanceolate to oblong, greenish; ligules 20–50, white, 5–8 mm long; pappus double, the inner of 6–12 fragile bristles, the outer of fine squamellae or setae. $2n$ = 36*. Open slopes to alpine elevations. Alta; swSask. Map 1086 ○

E. speciosus (Lindl.) DC.
Perennial with more or less branched caudex; stem leafy, 1.5–8 dm tall, glabrous or hairy, sometimes glandular in the inflorescence; leaves commonly 3-nerved, entire, the margins ciliate, glabrous or pubescent; lower leaves oblanceolate to broadly spatulate, petioled; upper leaves lanceolate to broadly ovate, not markedly reduced; heads 1–21; involucre 6–9 mm high, glandular, or with a few long hairs, the bracts equal, narrow, with loose tips; rays 75–150, blue or rarely whitish, 7–18 mm long. $2n$ = 18*, 36. Includes *E. subtrinervis* Rydb. of Ed. 1. Moist open woods and montane meadows. sBC, Alta s to Calif, Ariz, NM, Neb, SD. Map 1087 ●

E. trifidus Hook.

Densely caespitose perennial from a compact multi-branched caudex; stem 4–8 (10) cm high, scapose or with a single median bract, sparsely to densely glandular-pubescent and generally hirsute with long, spreading, white, septate hairs, often densely lanate below the head; leaves basal, 1–2 (3) cm long, mostly trifid, occasionally entire, rarely 2-lobed or with a lobe bifid or toothed, coarsely ciliate, sparsely glandular-hirsute above and below, petiole (basal segment) mostly 1–2 mm wide, usually markedly dilated apically, the lobes mostly 3–5 (8) mm long, generally lanceolate to narrowly ovate or oblong; heads solitary; involucre 8–10 mm high; bracts 25–30, crimson or crimson-tipped, 5–12 mm long, narrowly lanceolate, sparsely to densely glandular-hirsute; rays 20–30 (40), 10–15 mm long, linear to more often spatulate or oblanceolate, mostly white, sometimes pink (drying purple); achenes 2.0–2.5 mm long; pappus-bristles 15–20, 3–4 mm long. $2n = 45$*. Probably a hybrid between *E. compositus* and *E. lanatus*. Alpine slopes. BC?, Alta. Map 1087 ○

Eupatorium L.

E. maculatum L. / Spotted Joe-Pye Weed

Perennial; stem stout, mostly 6–15 dm tall, purple or purple-spotted; leaves 6–20 cm long, 2–7 cm wide, lanceolate to ovate, petiolate, sharply serrate, pubescent below, in whorls of 3–6; inflorescence corymbose; heads many; involucre 6–9 mm high, bracts imbricate, often purplish; flowers pink to purple, all tubular; achenes 3.0–4.5 mm long, glandular-dotted; pappus of capillary bristles. $2n = 20$. Marshy ground and moist open woods. BC to Nfld s to Utah, NM, Neb, Ia, Ohio, NC, Md. Map 1088 ○

Franseria Cav.

F. acanthicarpa (Hook.) Coville / Bur Ragweed

Annual herb, resembling *Ambrosia*; stems 2–6 dm high, diffusely branched, strigose and hirsute with white hairs; leaves mostly alternate, petiolate or the upper sessile, deeply bipinnatifid, hispid-strigose on both sides or glabrate above; plants monoecious; staminate and pistillate flowers in different heads; staminate heads nodding, in long terminal clusters, without subtending leaves; involucres with united oval bracts; pistillate heads solitary or clustered in upper leaf-axils, 1-flowered and 1-beaked; involucral bracts united, enclosing the pistil and armed over the surface with 8–15 divergent spines; pappus none; fruit bur-like, reticulate-ridged and very spiny. Sand-dunes. BC to Sask s to Calif, NM, Tex. Map 1088 ●

Gaillardia Foug.

G. aristata Pursh / Gaillardia

Perennial herb with a slender taproot; stems 1–several, 3–6 dm tall, simple or sparingly branched, leafy, more or less hirsute or villose; leaves alternate, mostly oblanceolate or lanceolate in outline, the lower petiolate, upper sessile, greyish-puberulent, entire or coarsely toothed or pinnatifid; heads large and showy, radiate, solitary or few, long-pedunculate; involucre 2–3 cm wide, the bracts in 2 or 3 series, long-attenuate, hairy, often reflexed; disc purplish, 1.5–4 cm wide; rays yellow or purplish at the base, 1–3 cm long, broad, cuneate, 3-cleft; disc-corollas densely villose above; style-branches flattened and appendaged; receptacle bristly with setae longer than the achenes, the latter obpyramidal, with long stiff hairs; pappus of 5–10 awned scales, longer than the achenes. $2n = 34^*$, 68^*. Prairie grassland, dry open areas, roadsides. BC to Man to Ore, Utah, Colo, SD. Map 1089

Galinsoga Ruiz. & Pav.

G. ciliata (Raf.) Blake
Annual; stem 2–7 dm tall, freely branched, pubescent with spreading hairs; leaves opposite, the lower 2–7 cm long, 1–4 cm wide, ovate, serrate-dentate, petiolate, the upper smaller, ovate-lanceolate, sessile; inflorescence cymose; heads radiate, 3–5 mm wide, the rays usually 5, scarcely longer than the involucre; bracts few, in 1 or 2 series, ovate, glandular-villous; flowers white; achenes about 1 mm long; pappus of scales, the longest tapering to a distinct awn. $2n = 32$. Waste ground. Introduced. Rare.

Gnaphalium L. / Cudweed

Annual or perennial herbs, more or less woolly throughout; stems leafy, erect or diffuse; leaves alternate, sessile, entire; heads of small flowers in cymose clusters or glomerules; flowers all tubular, yellow or whitish, the outer pistillate and very slender, the central perfect; involucral bracts dry and scarious, white or coloured; receptacle flat and naked; anthers tailed; achenes small, nerveless; pappus of capillary bristles, deciduous.

1	Plants 2–7 dm tall; stem simple below the inflorescence; heads not subtended by conspicuous leafy bracts; involucre 4–7 mm high	2
	Plants usually shorter; stem diffusely branched; heads subtended by conspicuous leafy bracts; involucre 2–4 mm high	3
2	Stem and upper leaf-surface glandular	**G. viscosum**
	Stem and leaves not glandular	**G. microcephalum**
3	Leaves oblanceolate; involucral bracts whitish at tip; plants floccose	**G. palustre**
	Leaves linear to narrowly oblanceolate; involucral bracts discoloured at the tip; plants finely tomentose	**G. uliginosum**

G. microcephalum Nutt.
Short-lived perennial herb with a taproot, mostly 2–7 dm tall, the stems and leaves closely white-tomentose, simple or moderately branched; leaves numerous, broadly linear or the lower oblanceolate, 3–10 cm long, somewhat decurrent at the base; heads generally numerous in small glomerules, commonly forming a broad open inflorescence; involucre 4–7 mm high, the numerous bracts white

to tan, woolly only at the base or not at all. $2n = 28$. Dry open areas. BC, Alta s to Calif, Ida, Colo. Map 1090 ○

G. palustre Nutt. / Marsh or Low Cudweed
Low spreading annual, loosely floccose-tomentose; stems at first erect, later diffusely branched from the base and spreading, 5–20 cm high; leaves oblanceolate or oblong, often broadly so, 1–3 cm long; heads in leafy-bracted clusters; involucre woolly below, 3–4 mm high, the bracts usually with white tips. $2n = 14$. Slough margins, moist open areas. BC to Sask s to Calif, NM, Neb. Map 1091

G. uliginosum L.
Similar to *G. palustre*; leaves narrowly oblanceolate or linear; pubescence generally more appressed and not so floccose; involucral bracts in mature heads usually greenish or brownish to the tip. $2n = 14$. Waste ground, fields. Introduced. Occasional.

G. viscosum HBK
Annual or biennial; stem 3–9 dm tall, mostly simple; leaves 4–10 cm long, 5–20 mm wide, linear to oblanceolate, sessile, decurrent, the upper and often the lower surface glandular-pubescent; inflorescence branched, many-headed; involucres 5–7 mm high; bracts acute, yellowish to white, woolly at base. Open woods. BC, Alta s to Ore; Ont to NS s to Tenn, Pa. Map 1090 ●

Grindelia Willd.

G. squarrosa (Pursh) Dunal / Gumweed
Biennial or perennial herb with a taproot, glandular-punctate leaves and gummy heads; stems erect, often widely branched, 3–6 dm high, glabrous, often purplish; leaves alternate, narrowly oblong or oblanceolate, stiff, glabrous, commonly glandular-dotted, serrulate to coarsely toothed or entire, the lower leaves petiolate, the upper sessile and clasping; heads yellow, radiate, 2–3 cm broad, rather numerous, often forming flat-topped clusters; involucre very resinous, the bracts numerous, linear, their terete tips projecting and mostly reflexed; receptacle flat, naked; ray-flowers bright yellow, pistillate and fertile; disc-flowers perfect, their corollas yellow; style-branches narrow, flattened, with slender appendages; achenes 4–5-ribbed; pappus of a few awns, soon deciduous. $2n = 12^*$, 24. Dry prairie, especially somewhat saline flats and denuded areas. sMack, BC to Man s to Calif, NM, Tex, Okla, Mo, Minn. Map 1092

Gutierrezia Lag.

G. sarothrae (Pursh) Britt. & Rusby / Broomweed
Perennial with a caespitose woody caudex; stems numerous, erect, slender, brittle, 2–4 dm high, mostly scabrous-puberulent; leaves alternate, linear, 1–3 mm wide, entire, obscurely puberulent and often minutely glandular-punctate; heads small, radiate, yellow, usually many, clustered in terminal, flat-topped corymbs; involucre glandular-glutinous, the bracts leathery; receptacle

naked; ray-flowers few, pistillate, fertile, the ligules yellow, 2–3 mm long; disc-flowers perfect, the corollas yellow; achenes 5-angled; pappus of several narrow scales. $2n = 8^*$, 16. Prairies. Alta to Man s to seWash, eOre, Calif, NM, Tex, Okla, Neb. Map 1093

Haplopappus Cass.

Perennial herbs, sometimes woody at base, mostly with taproots; leaves alternate and entire, toothed, or pinnatifid; heads variously disposed, sometimes solitary; involucres seldom less than 6 mm high, the bracts usually imbricated; receptacle alveolate but not truly chaffy; rays usually present, yellow; disc-flowers several–many; style-branches flattened, with externally hairy appendages; pappus of capillary bristles, white or sordid or reddish.

1	Leaves mostly pinnatifid, the lower ones with the segments toothed or lobed	**H. spinulosus**
	Leaves obscurely toothed or entire	2
2	Plants 5–10 cm tall, glandular-hairy	**H. lyallii**
	Plants taller, not glandular-hairy	3
3	Involucre and often the lower leaves densely lanate to floccose	**H. uniflorus**
	Involucre and leaves not densely lanate or floccose	**H. lanceolatus**

H. lanceolatus (Hook.) T. & G.
Stems 1–several, ascending and upwardly curving from a stout fibrous crown and strong taproot, 2–4 dm tall, glabrous or nearly so; basal leaves tufted, petiolate, 5–20 cm long, linear to lanceolate, glabrous to tomentulose, entire to spiny-denticulate; cauline leaves few, small, sessile; heads 1–many; involucres 5–10 mm high, the bracts lanceolate, whitish at base and green at apex; rays 15–45; achenes 3–4 mm long, densely hairy; pappus sordid. $2n = 24$, 36. Moist, saline meadows. BC to Sask s to Calif, Nev, Utah, Colo, wNeb. Map 1094 ●

H. lyallii A. Gray
Perennial with a branched caudex and commonly with creeping rhizomes as well; stems several, 5–15 cm tall, leafy, puberulent and often glandular; leaves obovate-spatulate to oblanceolate, 2–7 cm long, usually entire, 1-nerved, glandular-puberulent, tapering to a short margined petiole; heads solitary; involucres about 1 cm high, the bracts nearly equal, narrowly lanceolate, glandular-puberulent, the outermost often oblong and foliaceous; achenes sparsely hairy or glabrous; pappus whitish. $2n = 18^*$. Dry alpine slopes. BC, Alta s to Ore, Nev, Colo. Map 1094 ○

H. spinulosus (Pursh) DC.
Perennial with woody crown and root; stems several, erect or spreading, 1–4 dm high, branched, tomentose to glabrate; leaves grey or bluish green, commonly tomentose, mostly oblong in outline and pinnatifid, the linear divisions often with bristle-pointed lobes or teeth; heads terminating the several branches; in-

volucres hemispherical, 5–8 mm high, the bracts linear, acute or bristle-tipped, imbricated in 4–6 series, usually spreading or reflexed in age; rays 15–50; achenes 2–2.5 mm long, appressed-pubescent, pappus sordid. $2n = 8, 12, 16$. Prairies, roadsides. Alta to Man s to Calif, NM, Tex, Okla, Minn. Map 1095 ●

H. uniflorus (Hook.) T. & G.
Similar to *H. lanceolatus;* distinguished by the characters of the key; generally smaller plants, the basal leaves shorter, the heads mostly solitary. *H. lanceolatus* var. *sublanatus* Cody. Dry to moist open slopes and banks, roadsides. sMack, Alta. Map 1095 ○

Helenium L.

H. autumnale L. / Sneezeweed
Perennial herb with caudex and fibrous roots; stem stout, 3–10 dm tall, glabrous to finely pubescent, very leafy, marked by decurrent wings from the leaf-bases; leaves numerous, linear-lanceolate or elliptical to narrowly ovate, shallowly toothed to nearly entire, sessile or with a very short petiole, extending down the stem as narrow wings; heads radiate, yellowish, several–many in a leafy inflorescence, hemispherical or subglobose; disc yellow, 8–20 mm wide, the disc-flowers numerous, perfect, the corolla-lobes glandular-hairy; rays 10–20, yellow or orange, often deflexed, cuneate, 3-lobed, usually pistillate; involucral bracts in 2–3 series, narrow, acuminate, strigose-puberulent, soon deflexed; receptacle naked; achenes 4–5-angled with intermediate ribs, hairy; pappus of several minute awn-tipped scales, usually brownish. $2n = 32$. In our area var. **montanum** (Nutt.) Fern. with rays 10–15 mm long and var. **grandiflorum** (Nutt.) T. & G. with rays 15–25 mm long. Moist meadows, banks. sMack, BC to Que s to Ariz, Tex, Fla. Map 1096

Helianthus L. / Sunflower

Coarse, annual or perennial herbs with leafy stems; leaves simple, entire or toothed, the lowermost opposite, the upper ones sometimes alternate; heads large, radiate, solitary or few; involucral bracts imbricated in several series; receptacle flat to low-conical, chaffy, the scales clasping the achenes; ray-flowers large, yellow, neutral; disc-flowers numerous, yellow or purplish, fertile; achenes somewhat 4-sided and compressed; pappus of small scales or awns, soon falling off.

1 Annuals; leaves mostly alternate; disc purple or dark brown; receptacle flat or nearly so — 2
  Perennials; leaves mostly opposite, disc various; receptacle convex or low-conical — 3

2 Leaves commonly ovate, the lowest often cordate; pubescence rough; involucral bracts abruptly attenuate, with conspicuous marginal hairs — **H. annuus**
  Leaves lanceolate to ovate, mostly cuneate at base; pubescence soft; involucral bracts gradually attenuate, without long marginal hairs — **H. couplandii**

3 Disc purple, red or dark brown; involucral bracts broad, firm, appressed; larger leaves somewhat rhombic **H. subrhomboideus**

Disc yellow or light brown; involucral bracts narrow, mostly with loose acuminate tips; leaves generally lanceolate 4

4 Stem conspicuously pubescent below the inflorescence with short white hairs; leaves generally folded lengthwise; involucral bracts covered with short white hairs, sometimes ciliate towards the base **H. maximilianii**

Stem glabrous or somewhat scabrous below the inflorescence, sometimes glaucous; leaves not folded lengthwise; involucral bracts usually scabrous and conspicuously hairy on the margins **H. nuttallii**

H. annuus L. / Common Annual Sunflower
Annual, mostly 6–12 dm tall, the stem rough-hispid, often mottled; leaves chiefly alternate, petiolate, ovate, obtuse, mostly toothed, subcordate, scabrous-hairy above, hispid beneath; heads large, the disc 2–4 cm wide, usually reddish purple; involucral bracts ovate or lanceolate, abruptly attenuate, densely hispid, long-hairy on the margins; scales (chaff) of the disc with 3 awns, the tips usually short-hairy. $2n = 34$. In our area ssp. **lenticularis** (Lindl.) Cockerell. The cultivated sunflower, a common escape along roadsides, in waste ground, is closely related to our native taxon. Eroded slopes. BC to NS, generally distributed s. Map 1097

H. couplandii Boivin / Annual Sunflower
Similar to *H. annuus*, but generally more slender and not so rough-hairy; leaves narrower, more often entire and rarely cordate; heads smaller, the disc mostly 1–2 cm wide; involucral bracts generally tapering and rather uniformly short-hairy; disc-bracts with the middle awn conspicuously white-pubescent at the tip. $2n = 34$. *H. petiolaris* Nutt. of Ed. 1. Eroded slopes, sandy areas. BC to Man s to Ariz, NM, Tex, Okla, Mo, Minn. Map 1098

H. maximilianii Schrad.
Perennial with stout rootstock and thick woody roots; stems 5–25 dm high, stout, solitary or tufted, rough-hairy with whitish pubescence, especially above; leaves usually alternate, linear-elliptic to lanceolate, generally conduplicate, 5–15 cm long, entire or serrate, firm, narrowed to a short winged petiole, scabrous-hairy on both sides; heads usually racemose; disc 2–3 cm wide; involucral bracts linear-lanceolate, attenuate, densely appressed-pubescent and somewhat short-ciliate on the margins; achenes glabrous. $2n = 34$. Roadsides, waste ground. Introduced. Occasional.

H. nuttallii T. & G. / Common Tall Sunflower
Perennial with slender rootstocks and thickened fleshy roots; stems 5–20 dm tall, glabrous or somewhat scabrous-hispid with scattered hairs; leaves 5–15 cm long, mostly lanceolate, entire to denticulate, narrowed to a short winged petiole, scabrous-hairy on both sides, especially beneath; lower leaves opposite, upper alternate, or occasionally all opposite; heads solitary or several; disc 1.5–3 cm

wide; involucral bracts linear-lanceolate, loose, acuminate, generally ciliate on the margins, otherwise only moderately hairy to glabrate; achenes glabrous. $2n = 34$. Moist meadows. BC to Sask s to Ida, neOre, Nev, Ariz, NM. Map 1099

H. subrhomboideus Rydb. / Rhombic-leaved Sunflower
Perennial with stout and often branched rootstock; stems stiff, erect, mostly 4–12 dm tall, densely scabrous-hispid, rarely nearly smooth, often tinged with red or purple; leaves 5–12 cm long, mostly opposite, 3-nerved, obliquely 4-sided, the lower spatulate, the upper ovate or lanceolate, cuneate to a short petiole, entire or somewhat toothed, thick, rigid, coriaceous, scabrous-hispid on both sides; heads solitary or a few on long, nearly naked peduncles; disc 1–2 cm wide, brown-purple or reddish; involucral bracts ovate or ovate-lanceolate, erect and closely appressed, deep green, usually shining, glabrous or obscurely pubescent, short-ciliate on margin; pappus on disc-achenes of 2 awn-like scales and accessory shorter scales; achenes more or less hairy at apex. $2n = 102$. *H. laetiflorus* Pers. var. *subrhomboideus* (Rydb.) Fern. of Ed. 1. Prairies, roadsides. BC to Que s to NM, Tex, Neb, Ill. Map 1100

Heterotheca Cass.

H. villosa (Pursh) Shinners / Golden Aster
Perennial herb with several spreading or ascending stems, mostly 2–5 dm tall, generally quite hairy and greyish green; leaves rather numerous, alternate, sessile or short-petiolate, oblong to oblanceolate, 2–5 cm long, entire or denticulate, strigose-hirsute, sometimes obscurely glandular; heads solitary to several near the ends of branches, leafy-bracted, radiate, yellow; involucre 5–10 mm high, the bracts more or less imbricate, strigose to hirsute and sometimes glandular; receptacle naked; ray-flowers golden yellow; pistillate, the rays 6–10 mm long; disc-flowers perfect; achenes flattened, hairy; pappus double, the outer of very short bristles, the inner of long capillary bristles. $2n = 18^*, 36^*$. *Chrysopsis villosa* (Pursh) Nutt. of Ed. 1. In our area var. **villosa** with leaf pubescence appressed, seldom at all glandular, and var. **hispida** (Hook.) Harms with leaf pubescence spreading, intermixed with glands. Dry, open, often sandy, areas. BC to Sask s to Calif, Tex, Ill, ND. Map 1101

Hieracium L. / Hawkweed

Perennial herbs with caudex or elongated rhizome and fibrous roots; juice milky; leaves alternate or all basal, entire or toothed; heads usually panicled or corymbed; involucres mostly hemispherical, the bracts more or less imbricate; flowers all ligulate and perfect, yellow to orange or white; achenes slender, mostly narrowed towards the base and truncate at summit, ribbed and grooved; pappus of capillary rough bristles, usually brownish or sordid, occasionally whitish.

1 Flowers white — **H. albiflorum**
Flowers yellow or orange — 2

2	Leaves essentially basal	3
	Leaves mostly borne on the stem	4
3	Leaves essentially glabrous; flowers yellow	**H. triste**
	Leaves densely hirsute; flowers orange	**H. aurantiacum**
4	Stem and leaves hispid or densely hirsute with long spreading hairs; larger leaves arising from the lower third of the stem and markedly reduced above the middle	**H. cynoglossoides**
	Stem and leaves not densely hirsute, both glabrous or the stem sparsely pubescent with white appressed hairs, or the leaves with short, conical projections along the margin; stems leafy about the middle, the leaves often not much reduced above	**H. umbellatum**

H. albiflorum Hook. / White Hawkweed
Stems slender, 3–8 dm tall, loosely long-hairy, often glabrous upwards; lower leaves petiolate, oblong to oblanceolate, more or less hairy, especially on the midrib below, entire or denticulate; upper leaves much smaller and distant, sessile, lanceolate, sparsely hairy; heads on slender peduncles, usually numerous; involucres 6–10 mm high, blackish green, glabrous to sparsely hairy, often glandular-puberulent at the base; rays white or cream-coloured. $2n = 18$. Open woods. seAlas, swMack s to Calif, Colo. Map 1102

H. aurantiacum L. / Orange Hawkweed; Devil's Paint-brush
Coarse perennial, spreading by long rhizomes and stolons; stems 2–7 dm tall, long-hirsute, tomentose and glandular above, naked or with 1 or 2 reduced leaves; basal leaves in rosettes, the blades oblanceolate to narrowly elliptical, tapering into winged petioles, long-hairy on both sides or nearly glabrous above; heads several in a compact corymb; involucres 6–8 mm high, with dark glandular and long-hispid pubescence; flowers reddish orange, becoming deeper red on drying. $2n = 18, 27, 36, 45, 54, 63, 72$. Fields and roadsides. Introduced. Occasional.

H. cynoglossoides Arv.-Touv. / Woolly Hawkweed
Stems mostly solitary, 2–10 dm tall, more or less densely hairy with long white or yellowish bristles; basal and lower cauline leaves similar, 8–20 cm long, linear-lanceolate, entire or obscurely denticulate, tapering into a winged petiole; upper leaves much reduced and more distant, lanceolate, sessile; heads few–many, sometimes crowded; involucres 7–12 mm high with numerous short blackish hairs, mostly gland-tipped, or sparingly to densely hairy with long, glandless, black-based, white (in age, yellow) bristles, obscurely stellate, and inconspicuously to not at all glandular; ligules 7–14 mm long. $2n = 18$. Includes *H. albertinum* Farr of Ed. 1. Open woods and montane slopes. BC, Alta s to Ore, Utah, Wyo. Map 1103

H. triste Willd.
Stems slender, usually 1–3 dm tall, finely stellate-puberulent; leaves mostly basal, long-petiolate, the blades oblanceolate, entire or callose-denticulate, glabrous or

inconspicuously short-pubescent; heads usually few; involucres and pedicels dark with dense grey pubescence and long black hairs, often glandular; flowers pale yellow. $2n = 18$. In our area ssp. **gracile** (Hook.) Calder & Taylor. *H. gracile* Hook. of Ed. 1. Subalpine meadows. Alas, Yuk, swMack s to Calif, NM. Map 1104

H. umbellatum L. / Narrow-leaved Hawkweed
Stem erect or ascending, leafy, 2–10 dm high, more or less hairy near base or glabrous; leaves lanceolate or narrowly oblong to ovate, puberulent or long-hairy to nearly glabrous or scabrous, dentate to nearly entire; upper leaves sessile, rounded to subcordate at base and somewhat clasping; heads 1–many, on arched-ascending pedicles; involucres smoky-green to blackish, 8–12 mm high, the bracts appressed, glabrous or obscurely hairy, often with a few glands; achenes about 3 mm long. $2n = 18, 27^*$. Includes *H. canadense* Michx. of Ed. 1, *H. scabriusculum* Schwein. Open woods, clearings, meadows, disturbed ground. Circumpolar: Yuk, swMack to Huds Bay, Nfld s to Ore, Ida, Colo, Neb, ND, Ia, Mich, NJ. Map 1105

Hymenopappus L'Her.

H. filifolius Hook.
Perennial herb with a stout crown and taproot; stems 2–4 dm tall, thinly floccose-tomentose to glabrate; leaves alternate, white-tomentose when young, becoming glabrate, pinnatifid into linear or filiform segments; lower leaves petiolate and elongated, the upper few and much smaller; heads few–many, in a corymb-like cluster; involucres 5–7 mm high, the bracts appressed, their tips broad, obtuse, scarious and yellowish; receptacle small, naked; ray-flowers none; disc-flowers tubular, yellowish, perfect and fertile; achenes obpyramidal, 4–5-angled, 15–20-nerved, silky-villose; pappus of minute scales, usually less than 1 mm long, almost or completely hidden by the long achene hairs. $2n = 34, 68$. Dry hills and plains. Alta, Sask s to Calif, Tex, Neb. Map 1106

Hymenoxys Cass.

Caespitose perennial herbs with stout branching caudex and strong taproot; stems simple or branched; leaves alternate, simple or pinnatifid, hairy or glabrate, usually glandular-punctate, commonly in basal rosettes; heads radiate, solitary on the stem or terminating the branches of a flat-topped cyme; involucres cup-shaped or hemispherical, the bracts in 2 or 3 series; receptacle naked; ray-flowers pistillate, fertile, the corollas showy, 3-toothed, bright yellow, becoming whitish and sharply reflexed in age; disc-flowers numerous, perfect, fertile; achenes obpyramidal, or turbinate, 5-angled, hairy; pappus of 5–8 hyaline scales.

Leaves all basal, entire; heads solitary on unbranched stalks; involucral bracts similar, in 2 or 3 indefinite series, none united **H. acaulis**

Leaves both basal and cauline; 3–7-divided into linear segments; heads 1–3(5), in flat-topped corymbs; involucral bracts dissimilar; in 2 definite series, the outer ones partly united **H. richardsonii**

H. acaulis (Pursh) Parker / Butte Marigold
Perennial herb with a thick crown; leaves tufted, all basal, linear-oblanceolate, 2–8 cm long, entire, silky-hairy, often silvery; heads solitary on scape-like stalks, involucres 5–8 mm high, silky-villose, the bracts linear to oblong; ligules usually present, 6–15 mm long, 3-toothed, yellow with orange nerves; disc-flowers with 5 short pubescent teeth; pappus-scales 5–7, often abruptly aristate. $2n = 28$, 30, 56, 60, 70. Eroded hillsides. Alta, swSask s to Ida, Nev, seCalif, NM, Tex, Kans. Map 1107

H. richardsonii (Hook.) Cockerell / Colorado Rubber-plant
Perennial herb with a branched and somewhat woody caudex, the latter usually covered with dead leaf-bases and woolly tufts; stems 1–3 dm high, branched in the inflorescence, somewhat hairy; leaves tufted, mostly basal, some cauline, divided into 3–7 linear segments, fleshy, hairy or glabrate, glandular-punctate with sunken glands; heads usually a few in a flat-topped cyme, rarely solitary; outer involucral bracts united near base, the inner bracts distinct to base and their upper margins erose or fimbriate; ligules bright yellow, 3-toothed, 7–10 mm long; pappus scales usually 5 or 6, acuminate or short-awned. $2n = 28$, 30*. Prairies, dry rocky hillsides. Alta, Sask s to Ariz, NM, Tex. Map 1108

Iva L. / Marsh Elder

Annual or perennial herbs with leafy stems; leaves opposite, at least below; heads small, greenish, discoid, solitary in the upper leaf-axils or in dense panicles; staminate and pistillate flowers in the same head, the pistillate few and marginal; involucres cup-shaped or hemispherical, of mostly 4–5 separate or partly united bracts; staminate flowers with undivided style; filaments united, anthers scarcely so; achenes obovoid, somewhat flattened; pappus none.

Perennial, with spreading roots; leaves small, sessile, heads axillary — **I. axillaris**
Robust annual; leaves broad, long-petiolate; heads in a panicle — **I. xanthifolia**

I. axillaris Pursh / Poverty-weed
Deep-rooted perennial, spreading in the soil and often forming dense patches; stems 1–5 dm tall, often not branched, sparsely hairy or glabrate; leaves sessile or nearly so, narrowly ovate to oblong or linear, 1–3 cm long, thick, entire, pubescent to glabrate, the lower opposite, those of the inflorescence alternate; heads on short axillary peduncles, nodding; involucres 3–4 mm high, the 4–5 bracts united at least at the base; staminate flowers 12–20; pistillate flowers 5–8, with evident corolla and not enclosed by scales; achenes about 3 mm long. $2n = 36$*, 54. In our area ssp. **axillaris**. Saline clay flats, banks, eroded slopes. BC to Man s to Calif, NM, Tex. Map 1109

I. xanthifolia Nutt. / False Ragweed; Burweed Marsh Elder
Coarse annual; stems stout, 5–20 dm tall, pubescent to glabrous; leaves mostly opposite, long-petiolate, the blades usually ovate, 5–20 cm long, cordate or broadly cuneate at base, coarsely and often doubly serrate, scabrous above,

puberulent beneath; juvenile and lower leaves often 3-lobed and variously incised; heads not subtended by leaves, clustered in dense terminal and axillary panicles; involucres 1.5–3 mm high, with 5 outer bracts and 5 inner scale-like bracts partly enclosing the achenes; corolla of pistillate flowers scarcely evident; achenes 2 mm long. $2n = 36$. Moist alluvial ground, roadsides, ditches, irrigation canals. Alta, Sask s to Wash, Ariz, NM, Tex, Mo, Ia, Minn. Introduced elsewhere. Map 1110

Lactuca L. / Lettuce

Annual, biennial or perennial herbs with milky juice; leaves alternate, often pinnatifid; heads paniculate, involucres narrowly conical or cylindrical in fruit, the bracts imbricated; flowers all ligulate, blue or yellow; achenes flattened, winged or strongly nerved laterally, with 1 or several nerves on each face, beaked or sometimes beakless, expanded at the apex as a disc where the pappus is attached; pappus of soft capillary bristles which fall separately. Cultivated lettuce is *L. sativa* L.

1 Flowers yellow; leaves strongly spiny on lower side of midrib and prickly on margin; achene with slender beak **L. serriola**
  Flowers blue or purplish; leaves not spiny 2

2 Annual or biennial; leaves commonly 10–40 cm long; involucre 10–14 mm high in fruit; achene not beaked **L. biennis**
  Perennial; leaves mostly 5–18 cm long; involucre 15–20 mm high in fruit; achene with rather stout beak **L. pulchella**

L. biennis (Moench) Fern.
Robust annual or biennial, with smooth stems, 5–20 dm tall, unbranched except at the top; leaves commonly 1–4 dm long, usually deeply pinnatifid, the lobes sharply toothed, glabrous or hairy on the main veins beneath; heads in a large panicle; flowers bluish to white, rarely yellowish; involucres 10–14 mm high in fruit; achenes 4–5 mm long, several-nerved on each face, scarcely beaked; pappus light brown. $2n = 34^*$. Moist open woods, clearings. sAlas, sYuk to sJames Bay, Nfld s to Calif, Colo, SD, Minn, SC. Map 1111

L. pulchella (Pursh) DC. / Common Blue Lettuce
Deep-rooted perennial, spreading underground and forming patches; stems mostly 3–6 dm tall, somewhat branched above; leaves glabrous, often somewhat glaucous, linear-lanceolate to oblong, 5–18 cm long, entire or somewhat runcinate-pinnatifid; heads few and showy, the flowers blue or blue-purple; achenes thin, marginless, 4–7 mm long, several-nerved on each face, the beak short and rather stout, pappus white. $2n = 18^*, 34$. *L. tartarica* (L.) C.A. Mey. ssp. *pulchella* Pursh (Stebbins). Moist meadows, woods, clearings. swMack to James Bay s to Calif, NM, Okla, Mo, Wisc, seOnt. Map 1112

L. serriola L. / Prickly Lettuce
Annual or winter annual (possibly biennial) with prickly leaves; stems 3–15 dm high, glabrous or hispid below; leaves oblong or oblong-lanceolate, sagittate-clasping, commonly twisted at base to lie in a vertical plane, commonly lobeless, sometimes pinnatifid, prickly-spiny on the midrib beneath and prickly on the margins; heads numerous, in loose pyramidal panicle with spreading branches; involucres about 10 mm high; flowers yellow, or turning bluish on drying; achenes about 3 mm long, several-nerved on each face, hairy above, tapering to a long slender beak; pappus white. $2n = 18$. Roadsides, waste ground. Introduced. Occasional.

Lapsana L.

L. communis L. / Nipplewort
Annual; stem 1–9 dm tall, branched, generally pubescent below, glabrous above; leaves petiolate, blade 2.5–10.0 cm long, 2–7 cm wide, ovate to suborbicular, toothed or lobed, occasionally lyrate basally, progressively less petiolate and narrowed upwards; inflorescence corymbose; heads 10–20, involucre 5–8 mm high; bracts subequal; flowers yellow, all ligulate, achenes 3–5 mm long, curved, glabrous; pappus absent. $2n = 12, 14, 16$. Waste ground, gardens. Introduced. Rare.

Liatris Schreb. / Blazing Star

Perennial herbs with stiff leafy stems from a caudex or globose corm; leaves alternate, narrow, entire, rather thick, rough, more or less punctate; heads in spikes or narrow racemes, discoid, with rose-purple flowers, rarely whitish; involucres oblong to hemispherical, the bracts imbricate in several series, often with purplish scarious margins; ray-flowers none; receptacle naked; achenes cylindrical but tapered to the base, 10-ribbed, hairy; pappus of 1 or 2 series of plumose or barbellate bristles.

Lower leaves 10–20 mm wide; heads subglobose, 1–3 cm wide; pappus bristles barbellate **L. ligulistylis**
Lower leaves 2–4 mm wide; heads cylindrical, narrower; pappus bristles plumose **L. punctata**

L. ligulistylis (A. Nels.) K. Schum.
Stems stiffly erect, glabrate below, hairy above, 2–8 dm tall, from a stout hard corm; lower leaves linear-oblanceolate, 8–15 cm long, bright green, somewhat hirsute, especially on the margins, usually with a narrowly winged petiole; upper leaves similar but much smaller, linear and sessile; heads large, few–several, in a raceme; involucres 1–2 cm high, 1–3 cm wide, the bracts elliptical to oblong, green with purplish erose tips; flowers purple to rose; achenes 5–6 mm long, pubescent; pappus purplish. $2n = 20^*$. Sandy woods. Alta to Man s to NM, SD, Mo. Map 1113

L. punctata Hook.
Stems glabrous or nearly so, 2–6 dm high, from a slender or very thick caudex; leaves numerous, mostly linear, 2–4 mm wide, rather rigid, conspicuously punctate, ciliate on the thickened whitish margins, otherwise glabrous; heads sessile and rather crowded in spikes; involucres narrowly cup-shaped, the bracts cuspidate or acuminate, punctate; flowers purple or rose; achenes 6–7 mm long, hairy. $2n = 20, 40$. Dry plains and hills. Alta to Man s to Mont, NM, Tex, Okla, Ark, Minn, Mich. Map 1114

Lygodesmia D. Don / Skeleton-weed

Annual or perennial herbs with milky juice; stems rigid and tough, stiffly erect or spreading, branching, often nearly leafless, striate, glabrous, often glaucous; leaves usually alternate, very slender, commonly reduced to mere scales; heads terminal on the branches; involucres cylindrical, with a few bracts and several small basal bractlets; flowers 3–10, commonly 5, in the head, all ligulate, pink or rose, rarely white; receptacle flat, naked; achenes slender, 4–8-ribbed; pappus of many soft bristles.

Perennial; leaves seldom over 5 cm long, usually scale-like	**L. juncea**
Annual; leaves 5–20 cm long	**L. rostrata**

L. juncea (Pursh) D. Don / Skeleton-weed; Prairie Pink
Perennial herb with long rootstocks; stems much branched from near the base, 1–4 dm tall, greenish, usually nearly bare, sometimes with clusters of rounded insect galls; lower leaves linear, seldom over 5 cm long, the upper leaves very small, scale-like or subulate; heads solitary at ends of branches; involucres 10–14 mm high, with about 5 linear bracts; flowers mostly 5; achenes 5–7 mm long; pappus pale brownish. $2n = 18$. Dry sandy prairies. seBC to Man s to Ariz, NM, Tex, Ark. Map 1115

L. rostrata A. Gray / Annual Skeleton-weed
Annual plant; stems 3–10 dm tall, erect, with ascending branches; leaves mostly 5–20 cm long, narrowly linear and entire, the uppermost small and subulate; heads spread along the branches; involucres 10–16 mm high, with 7–9 principal bracts; flowers mostly 7–9; achenes 8–10 mm long; pappus whitish. $2n = 12$. Sandy banks and dunes. Alta to Man s to Colo, Tex, Okla, Ia. Map 1116

Machaeranthera Nees

Herbs; annual or perennial; leaves alternate, entire to bi- or tri-pinnate; heads discoid or radiate; involucre in several series; bracts linear to lanceolate or subulate, the midrib green towards the tip; ray-flowers pistillate, fertile; disc-flowers perfect, fertile, yellowish; pappus of capillary bristles usually present. *Aster, Haplopappus*, in part, of Ed. 1.

1	Leaves pinnately dissected	**M. tanacetifolia**
	Leaves entire or toothed	2
2	Rays present	**M. canescens**
	Rays absent	**M. grindelioides**

M. canescens (Pursh) A. Gray
Biennial, with a taproot, 1–5 dm tall, usually well branched, cinereous-puberulent; leaves entire to slightly toothed, the teeth and apex spinulose; lower leaves oblanceolate or spatulate, sometimes deciduous; upper leaves often linear; heads commonly many, in an open panicle; involucre 6–10 mm high, canescent, the bracts well imbricate, the short tips spreading or reflexed; rays bright bluish purple; pappus bristles stiff, unequal. $2n = 8^*$. *Aster canescens* Pursh of Ed. 1. Dry open areas. sBC to Sask s to Calif, Ariz, Colo. Map 1117 ●

M. grindelioides (Nutt.) Shinners
Low perennial with a woody caudex, often caespitose; stems 1–3 dm tall, stiff, erect, tomentose or nearly glabrous; leaves sessile, oblong-lanceolate to spatulate, 1–3 cm long, thick, spiny-toothed, greyish green, hairy or nearly glabrous; heads solitary or several in a cyme, cup-shaped; rays none or very short; involucres 7–10 mm high, the bracts usually yellowish and green-tipped, more or less puberulent and glandular; achenes densely white-hairy; pappus reddish brown to tan. $2n = 8, 16$. *Haploppapus nuttallii* T. & G. of Ed. 1. Dry eroded slopes. Alta, Sask s to Nev, Ariz, NM, Mont. Map 1118

M. tanacetifolia (HBK) Nees
Annual, with taproot; stem 5–30 cm tall, branched, glandular hairy; leaves 2–10 cm long, bi- or tri-pinnatifid; inflorescence somewhat corymbose; heads few, large; involucre 6–10 mm high; bracts imbricate, tips green; rays 12–36, 10–16 mm long, whitish or blue; achenes 2–3 mm long, appressed hairy. $2n = 8$. *Aster tanacetifolius* HBK. Dry open areas. Alta s to Ariz, NM, Tex, Kans. Map 1117 ○

Madia Mol.

M. glomerata Hook. / Tarweed
Slender annual, strong-scented, hairy and very sticky on flowering parts; stems 1–6 dm tall, simple or branched above, leafy, hirsute and glandular, especially above; leaves alternate, mostly linear, sessile, 2–7 cm long, entire; heads numerous, in compact clusters near the ends of branches; involucres 6–8 mm long, fusiform, angled by the narrow backs of the bracts, glandular; ray-flowers 1–5, pistillate, the rays inconspicuous, about 1 mm long, 3-lobed, yellowish; disc-flowers several, perfect, yellowish, enclosed by a series of scales; achenes black, 5–8 mm long, those of the ray-flowers somewhat curved; pappus none. $2n = 28$. Dry open areas, coulee slopes, roadsides. BC to Man s to Calif, Ariz, Colo, ND, Minn. Introduced elsewhere. Map 1119

Matricaria L. / Wild Chamomile

Annual or biennial, glabrous herbs with alternate pinnatifid or pinnately dissected leaves; heads terminating the branches or in a corymb-like cluster, discoid or ligulate with white rays; involucre saucer-shaped or hemispherical, the bracts 2–3-seriate, obtuse, scarious-margined; receptacle conic, globose or hemispherical, naked; disc-flowers yellow, perfect, fertile; ray-flowers, if present, pistillate and fertile, the ligules white; achenes 3–5-ribbed, wingless; pappus a short crown or none.

1 Heads discoid; disc-corollas 4-lobed — **M. matricarioides**
Heads with white rays; disc-corollas 5-lobed — 2

2 Plants fragrant; receptacle conical; pappus none — **M. recutita**
Plant odourless; receptacle hemispherical; pappus a short crown — **M. perforata**

M. matricarioides (Less.) Porter / Pineapple-weed
Branching leafy annual, pineapple-scented, with glabrous stems, 1–4 dm high, leaves 1–5 cm long, 1–3-pinnatifid, the ultimate divisions short, linear or filiform; heads green, conical, rayless, 5–10 mm broad, short-peduncled; involucral bracts oval, with broad scarious margins; receptacle conical, pointed; achenes slightly angled; pappus an obscure crown. $2n = 18$. *Chamomilla suaveolens* (Pursh) Rydb. Waste ground, roadsides. Introduced. Common.

M. perforata Mérat / Scentless Chamomile
Annual or biennial, 2–10 dm tall, with ascending branches, glabrous or nearly so; leaves 2–8 cm long, bipinnatifid, the ultimate segments mostly linear or filiform; heads 3–4 cm broad, the white rays 8–16 mm long; achenes 2 mm long, strongly 3-ribbed; pappus a short crown or border. $2n = 18, 36$. *M. maritima* L. of Ed. 1. Waste ground, roadsides. Introduced. Occasional.

M. recutita L. / Wild Chamomile
Branching annual, pineapple-scented, with slender glabrous stems, 2–6 dm high; leaves usually bipinnatifid, the ultimate divisions linear or filiform; heads about 2 cm broad, the white rays 4–10 mm long; achenes 1 mm long, faintly 3–5-ribbed; pappus obsolete. $2n = 18$. *M. chamomilla* L. of Ed. 1. (*Chamomilla recutita* (L.) Rausch.) Waste ground, roadsides. Introduced. Occasional.

Microseris D. Don

M. nutans (Hook.) Sch.-Bip.
Perennial herb, more or less caulescent, 1–5 dm high, glabrous or minutely puberulent, with one or more thickened fleshy roots; main stem usually with a few branches, leafy chiefly near the base, erect or somewhat decumbent; leaves alternate, linear, entire or laciniate with slender teeth or divergent lobes; heads ligulate, usually solitary on the ends of slender branches; involucre usually granular, 1–2 cm high, of 8–10 lanceolate acuminate bracts and a few subtending

short bractlets; flowers yellow, all ligulate and perfect; achenes tapering above; pappus white, of hairy bristles, these dilated and scale-like at the base. $2n = 18$*. Open montane woods, grassy slopes. BC, Alta s to Calif, Utah, Colo. Map 1120 ○

Nothocalais Greene

N. cuspidata (Pursh) Greene
Perennial, from a stout taproot; stem 1–3 dm tall, scapose; leaves 7–30 cm long, basal, linear; margin villose-ciliate, often crisped; heads solitary, erect; involucre 17–25 mm high; bracts subequal; flowers yellow, all ligulate; achenes 8–10 mm long, beakless; pappus of long capillary bristles intermixed with attenuate scales, hardly broader than the bristles. $2n = 18$*. *Microseris cuspidata* (Pursh) Schultz-Bip. Prairies. Alta to sMan s to Colo, NM, Tex, Okla, Mo. Map 1120 ●

Petasites Mill. / Sweet Coltsfoot

Perennial herbs with thick creeping rhizomes and large basal leaves, more or less woolly-tomentose; leaves petioled, all basal and from the rhizome, other than the large sheathing bracts on the flowering scape; the latter 1–8 dm tall, arising in early spring before the leaves, soon fruiting and dying down; heads several to numerous, racemose or corymbose, somewhat dioecious; rays small or lacking; involucre cup-shaped, the bracts in one series; flowers whitish; fertile plants with chiefly pistillate flowers, tubular or ligulate; substerile plants with numerous staminate flowers in the disc and a single row of pistillate ray-flowers; receptacle flat, naked; achenes linear, 5–10-ribbed; pappus of soft white bristles.

1 Leaves nearly entire or merely dentate; pappus generally more than 15 mm long **P. sagittatus**
Leaves deeply lobed; pappus 15 mm long or less 2

2 Leaves divided almost to base into 5–7 regular, elliptical, entire or shallowly dentate lobes **P. palmatus**
Leaves divided half-way to midrib at most, lobes often deeply incised 3

3 Leaves usually deeply cordate, generally longer than wide; lobes 5–7, widest at the base, the edges convex-curved, not overlapping **P. vitifolius**
Leaves shallowly cordate, often about as long as wide; lobes 3–5, usually widest about the middle, the edges concave-curved, usually overlapping, the blade appearing fenestrate with elliptical openings **P. nivalis**

P. nivalis Greene
Stems mostly 1–5 dm tall; leaves generally 3–15 cm long, about as wide or wider, glabrous above, glabrous to sparsely woolly-floccose below; cauline bracts mostly 2–4 (6), ovate, inflated, often spathe-like, usually completely ensheathing stem at base; heads generally 1–12; involucre 7–10 mm high. Includes *P. hyperboreus* Rydb. *sensu* Porsild, Hultén. Moist subalpine to alpine banks. eSiberia: Alas, Yuk, swMack s to Wash, BC, Alta. Map 1121

P. palmatus (Ait.) A. Gray. / Palmate-leaved Coltsfoot
Petioles slender; leaf-blades 5–20 cm wide, reniform to nearly orbicular, deeply divided into 5–7 sharply toothed and cleft lobes, green and glabrate above, thinly white-tomentose beneath; mature fruiting heads, including pappus, 1–2 cm high; achenes about 2 mm long. $2n = 60$*. *P. frigidus* (L.) Fries var. *palmatus* (Ait.) Cronq. Moist woods and swamps. eAsia: Yuk, wMack to Huds Bay, nQue, Nfld s to Calif, BC, Alta, Sask, Man, Minn, Mich, Mass. Map 1122

P. sagittatus (Pursh) A. Gray / Arrow-leaved Coltsfoot
Leaves large, the blades 1–3 dm long and 1–2 dm broad, deltoid-oblong to reniform-sagittate, the margin dentate or occasionally nearly entire, sparingly woolly above, densely white-tomentose beneath; mature fruiting heads, including pappus, 2–3.5 cm long; achenes 3–3.5 mm long. $2n = 60$. Wet meadows, bogs. Alas, Yuk, Mack to Huds Bay, nQue s to Wash, Ida, Colo, SD, ND, Minn, Wisc, cOnt. Map 1123

P. vitifolius Greene
Somewhat similar to *P. nivalis* but generally larger plants; distinguished by the characters of the key, the narrower, generally more numerous cauline bracts, and the greater number of heads. $2n = 60$. Probably a hybrid between *P. palmatus* and *P. sagittatus*. *P. frigidus* (L.) Fries var. *nivalis* (Greene) Cronq. Wet woods, bogs. Yuk, swMack to Huds Bay, nQue, Lab s to Ore, Alta, Sask, Man, Minn, cOnt, cQue. Map 1124

Picris L.

P. echioides L. / Bristly Ox-tongue
Plants annual or biennial, hispid; stem 3–9 dm tall, branched; leaves alternate to 25 cm long, 7 cm wide, oblanceolate to lanceolate or oblong, coarsely toothed or sinuate, short-stalked or sessile; inflorescence more or less corymbose; involucre 10–20 mm high, biseriate; outer bracts more or less ovate, foliaceous, the inner narrower, spine-tipped; flowers yellow, all ligulate; achenes beaked, the bodies 2–4 mm long, dimorphic, the inner straight, glabrous, the outer curved, pubescent; pappus plumose. $2n = 10$. Waste ground. Introduced. Rare.

Prenanthes L. / White Lettuce

Perennial herbs with more or less thickened roots, erect leafy stems and milky juice; leaves alternate, glabrous, the lower petioled, the upper often sessile and clasping; heads often nodding, in small clusters on an elongated inflorescence; involucre cylindrical, of narrow bracts subtended by a few small bractlets; receptacle flat, naked; flowers all ligulate and perfect; achenes elongate, ribbed-striate; pappus of numerous, pale or brown bristles.

1 Lower leaves oblanceolate to elliptic; involucre coarsely pubescent with hairs 1–2 mm long **P. racemosa**
Lower leaves cordate to sagittate; involucre glabrous or sparsely glandular-pubescent 2

2 Inflorescence narrow, elongate, somewhat spiciform **P. sagittata**
Inflorescence broader, more or less corymbose **P. alata**

P. alata (Hook.) D. Dietr.
Similar to *P. sagittata*; distinguished by the character of the key and the glandular-pubescent involucre. $2n = 16$. Moist wood margins. sAlas, BC s to Ore, Alta. Map 1125 ○

P. racemosa Michx.
Plants glabrous, except in the inflorescence, glaucous; stem 5–12 dm high, stout, often purplish; lowermost leaves tapering into winged petioles, the upper ones sessile, partly clasping; leaf-blades oblanceolate to obovate to oval or broadly lanceolate, dentate to entire; inflorescence dense, spike-like, the tip recurved in the bud-stage, becoming erect as the heads enlarge; involucral bracts long-hirsute; flowers pink or purplish; pappus sordid or yellowish. Moist meadows and thin thickets. eBC to NS s to Mont, Colo, Neb, Mo, Ia, Ohio, NY, NJ. Map 1125 ●

P. sagittata (A. Gray) A. Nels.
Stem 3–7 dm tall; leaves mostly with narrowly winged petioles, the blades narrowly triangular or oblong-lanceolate and sagittate or hastate at the base, the margins irregularly toothed to entire; heads in a narrow panicle, not nodding in the flowering state, spreading or somewhat reflexed in fruit; bractlets at the base of the involucre linear; flowers white; pappus sordid. $2n = 16$. Moist woods and thickets. Alta s to Ida, Mont. Map 1126

Psilocarphus Nutt.

P. elatior A. Gray / Woolly-heads
Low woolly annual, 3–15 cm high, the stems generally branched, erect or sometimes decumbent; leaves mostly opposite, sessile, entire, linear to linear-oblong, 5–18 mm long, loosely white-tomentose; heads small, discoid, many-flowered in terminal compact clusters and in the forks of the branches, surrounded by the upper leaves and with no true involucre; receptacle nearly globular; outer flowers pistillate and fertile, with filiform corollas, each loosely enveloped by a saccate woolly bract which bears a small hyaline appendage below the summit; central flowers few, bractless, functionally staminate, with tubular corollas; pappus none; achene cylindric, brown, smooth, loose in the bladder-like bract. Dried beds of vernal pools. sBC, Alta s to neOre, Ida. Map 1127

Ratibida Raf.

R. columnifera (Nutt.) Wooton & Standl. / Prairie Cone-flower
Perennial herb with a taproot and short caudex; stems 3–5 dm tall, stiff, leafy, more or less branched, grooved, coarsely hairy; leaves alternate, pinnately divided into 5–9 linear or oblong lobes, these sometimes 2–3-cleft, strigose or hirsute; heads several, radiate, on long peduncles; rays usually yellow, rarely purplish, spreading or reflexed; disc columnar, 1–4 cm high, grey to yellow,

becoming purple-brown; disc-flowers perfect, the corolla almost without a tube; involucral bracts slender, hairy; disc-bracts (chaff) with white-pubescent tips, embracing the achenes and deciduous with them; achenes grey-black, flattened, 2 mm long, with a crown of minute greyish scales and often 1 or 2 small teeth (pappus). $2n$ = 26, 27, 28. Prairies, roadsides. BC to Man s to Calif, Tex, Ark. Introduced elsewhere. Map 1128

Rudbeckia L.

R. hirta L. / Black-eyed Susan
Perennial herb, 3–10 dm tall, more or less rough-hairy throughout; stems simple or sparingly branched, hirsute, often purplish; lower leaves long-petioled, oblanceolate to oblong-lanceolate, entire or somewhat toothed, the others mostly linear-lanceolate; leaf-surfaces strigose to hirsute, the hairs often with green bulbous bases; heads radiate, showy, terminal on the stems or branches, usually long-peduncled; rays orange-yellow, mostly 2–4 cm long; disc 1–2 cm wide, generally brown or deep purple; involucres with about 2 rows of foliaceous bracts, spreading, often becoming reflexed; receptacle enlarged, conical, chaffy, its bracts acute and hispid; corolla-tube about one-fifth as long as the throat; pappus none; style-branches with hairy tips; achenes 4-angled, glabrous. $2n$ = 38. *R. serotina* Nutt. of Ed. 1. Fields, roadsides, waste ground. Introduced. Occasional.

Saussurea DC.

Perennial herbs; leaves alternate, entire to pinnatifid; heads solitary or clustered; flowers perfect; corollas tubular, 5-lobed, bluish or purple; receptacle plane or convex, naked or chaffy with scales or bristles; achenes oblong, angled; pappus-bristles in 1 or 2 series or absent.

1 Plants 3–12 dm tall; leaves triangular-ovate to cordate **S. americana**
Plants less than 3 dm tall; leaves linear-lanceolate to elliptic 2

2 Outer bracts of the involucre about as long as the inner; receptacle naked; alpine **S. nuda**
Outer bracts of the involucre shorter than the inner; receptacle chaffy; lowland **S. glomerata**

S. americana D.C. Eat. / American Saw-wort
Stem 3–10 dm tall, simple below, branched above; lower leaves up to 15 cm long, 8 cm wide, toothed, arachnoid below, the upper leaves becoming more or less lanceolate and sessile; inflorescence congested or corymbose; involucre 10–14 mm high, the inner bracts progressively longer; flowers purple; corolla 11–12 mm long; achenes 4–6 mm long, light brown. Moist meadows and moist open slopes. seAlas, sYuk, nBC; sBC, Alta s to Wash, neOre, nIda. Map 1129 ○

S. glomerata Poir.
Stem 2–3 dm tall, slender, branched; leaves up to 12 cm long, 2 cm wide, narrowly elliptic, entire or lobed, glandular-punctate below; inflorescence corymbose; heads

several; involucre about 15 mm high; bracts membranous, the inner with dilated, erose tips; flowers lilac; corolla about 17 mm long. $2n = 26$. Farmyards. Introduced, probably an escape from cultivation. Rare.

S. nuda Ledeb.
Stems 1–2 dm high, loosely arachnoid-hairy to glabrate; leaves lanceolate, acute to acuminate, lower petioled, upper sessile, sinuate-dentate to entire, arachnoid-hairy to glabrate, sometimes obscurely glandular-dotted; heads crowded on the end of the stem; involucres about 10 mm high and broad, arachnoid-hairy, the bracts in 4–5 series, acuminate; flowers purple; pappus of short outer bristles, 1–2 mm long, the inner ones much longer. $2n = 26$*. *S. densa* (Hook.) Rydb. of Ed. 1. In our area var. **densa** (Hook.) Hult. Rocky alpine slopes, eAsia: Alas; BC, Alta. Map 1129 ○

Senecio L. / Groundsel; Ragwort

Perennial herbs, or rarely annuals; leaves alternate, entire, toothed or lobed; heads solitary to numerous, radiate or sometimes discoid; involucres mostly uniseriate, the bracts linear and connivent-erect, often with a few bractlets at the base; rays pistillate and fertile, yellow to orange, occasionally reddish; disc-flowers perfect and fertile, usually yellow; achenes slender, 5–10-ribbed; pappus of numerous soft white bristles.

1 Plants annual, of marshy or aquatic habitats, densely villous throughout; tips of the involucral bracts not bearded or black **S. congestus**
Plants not as above 2

2 Involucral bracts black-tipped, sometimes only the outer 3
Involucral bracts not black-tipped 9

3 Heads rayless or rays minute, the heads appearing rayless; plants annual, 1–4 dm tall 4
Heads with rays; plants perennial 5

4 Plants glandular-viscid **S. viscosus**
Plants not glandular-viscid **S. vulgaris**

5 Stem-leaves not reduced in size upwards, largest leaves often about the middle 6
Stem-leaves reduced in size upwards, largest at the base 7

6 Leaves triangular, toothed **S. triangularis**
Leaves not triangular, blades deeply lobed or pinnatifid **S. eremophilus**

7 Plants glabrous (at least by flowering); leaves irregularly toothed **S. foetidus**
Plants pubescent, sometimes sparsely; leaves denticulate to entire 8

8 Plants rhizomatous; lower leaves mostly linear to lanceolate, not distinctly petiolate **S. lugens**
Plants with an abruptly shortened, button-like caudex; lower leaves mostly lanceolate to ovate or suborbicular, usually distinctly petiolate **S. integerrimus**

9 Leaves mostly cauline, not reduced in size upwards 10
Leaves mostly basal, reduced in size upwards 13

10 Leaves pinnatifid 11
Leaves not pinnatifid, mostly more or less entire, toothed or shallowly lobed 12

11 Plants glandular-viscid **S. viscosus**
Plants not glandular-viscid **S. eremophilus**

12 Stem 3–15 dm tall, unbranched; leaves triangular **S. triangularis**
Stem 1.0–1.5 dm tall, freely branched from base; leaves not triangular **S. fremontii**

13 Heads rayless 14
Heads with rays 15

14 Involucre generally greenish, heads many; leaves thin; lowland plants **S. indecorus**
Involucre generally crimson-purple, heads mostly 1–6, leaves thick, firm; alpine or subalpine plants **S. pauciflorus**

15 Heads solitary, rarely 2 or 3 16
Heads several 19

16 Plants silvery-grey to dark olive-green; basal leaves oblanceolate to elliptic, more or less entire, densely tomentose below; involucre 6–8 mm high **S. canus**
Plants not as above 17

17 Plants 2–5 dm tall, leaves linear, the lower 10–20 cm long; heads 3–5 cm wide **S. megacephalus**
Plants 0.5–1.5 dm tall; leaves not linear, shorter; heads 2–3 cm wide 18

18 Plants glabrous throughout **S. cymbalarioides**
Plants floccose, at least in the axils of the basal leaves **S. conterminus**

19 Plants silvery-grey to dark olive-green; basal leaves oblanceolate to elliptic, more or less entire, densely tomentose below; involucre 6–8 mm high **S. canus**
Plants not as above 20

20 Basal leaves or some of them cordate or subcordate, conspicuously and sharply toothed **S. pseudaureus**
Basal leaves not cordate or subcordate, mostly cuneate, margins crenulate, serrate or more or less entire 21

21 Leaves relatively thick and firm, elliptic to subrotund, often cuneate, crenate to lobulate, apex truncate often with 2 or 3 coarse teeth **S. streptanthifolius**
Leaves relatively thin, elliptic to oblanceolate, crenate-serrate, apex rounded **S. pauperculus**

S. canus Hook. / Prairie Groundsel
Perennial herb with 1 or more stems from the caudex, 1–4 dm tall, rarely taller; white-tomentose throughout, except the upper leaf-surfaces sometimes glabrate; basal leaves more or less tufted, 1–8 cm long, rather short-petioled, broadly elliptic to oblanceolate, thick, obtuse, entire to dentate; cauline leaves progressively smaller upwards, sessile and auricled, oblong to lanceolate in outline,

lyrate-pinnatifid to entire; heads several; involucres 4–8 mm high; rays bright yellow. *Packera cana* (Hook.) Weber & Löve. $2n = 46, 92, 138$. Dry open, often rocky areas, occasionally to alpine elevations. BC to Sask s to Calif, Colo, Neb. Map 1130

S. congestus (R.Br.) DC. / Marsh Ragwort
Annual or biennial, with stout soft hollow stems, generally villose, 2–8 dm tall, unbranched except in the inflorescence; leaves lanceolate to linear, ascending, entire or toothed or shallowly pinnatifid, the upper more or less clasping; heads several or many in an open or congested cluster, the peduncles and bracts pubescent; involucres about 8 mm high; rays pale yellow, short; pappus copious. $2n = 48$. Marshes, ponds, ditches. Circumpolar: Alas, Yuk to Prince Patrick Is to nBaffin Is to sLab, eQue s to BC, Alta, Sask, Man, ND, Ia, Wisc, nOnt. James Bay. Map 1131 •

S. conterminus Greenm.
Rhizomatous perennial; stems 0.5–1.2 (1.5) dm tall, usually single, more or less glabrous to densely tomentose, the axils of basal leaves floccose; basal leaves 1–6 (8) cm long, fleshy, blade ovate-orbicular to oblong-spatulate, usually shorter than petiole, crenate-serrate to irregularly or pinnately lobed, often tomentose or floccose; cauline leaves 0–4, reduced upwards, lanceolate to oblanceolate, irregularly lobed; heads solitary, occasionally 2 or 3; involucre 8–12 mm high. $2n = c.184^{*}$. *S. resedifolius* Less. of Ed. 1. Rocky alpine slopes. Alta, BC s to nMont. Map 1132

S. cymbalaroides Buek
Rhizomatous perennial, glabrous throughout; stems mostly 0.5–1.5 dm tall, usually single; basal leaves 2–4 (6) cm long, blade broadly ovate to orbicular, much shorter than the petiole, margin crenate to more or less entire; cauline leaves 2–3, reduced upwards, lanceolate to oblanceolate, pinnately lobed; heads solitary, rarely 2; involucre 6–8 mm high; bracts green or crimson. $2n = 46^{*}$. *S. subnudus* DC. of Ed. 1. Moist alpine and subalpine meadows. sBC, Alta s to Calif, Ida, Wyo. Map 1131 ○

S. eremophilus Richards.
Stout perennial, 5–10 dm tall, freely branching, leafy up to the inflorescence; leaves glabrous or sparsely hairy, mostly 1–2 dm long, the lower petioled, the upper sessile; leaf-blades deeply pinnatifid, the lobes usually incised or sharply toothed; heads numerous in large corymbose clusters; involucres 6–8 mm high, the bracts linear-lanceolate, often black-tipped; rays 7–9. $2n = 40$. Moist open woods, roadsides. sMack, BC to wOnt s to BC, Ariz, NM, SD, ND. Map 1133 •

S. foetidus Howell
Robust perennial, the stems 5–9 dm tall, nearly glabrous at maturity; lower leaves long-petioled, the blades mostly oblanceolate, rather thick, entire, sharply dentate or denticulate; upper leaves sessile, much reduced, lanceolate, denticulate; inflorescence at first congested, becoming open and more or less flat-

topped, the lower peduncles markedly elongated; involucral bracts 6–9 mm long, black-tipped; rays few, light yellow, about 7 mm long. $2n = 40$. *S. hydrophiloides* Rydb. of Ed. 1. In our area var. **hydrophiloides** (Rydb.) Barkley. Moist montane meadows. BC, Alta s to Calif, Ida, Mont. Map 1134

S. fremontii T. & G.
Glabrous perennial with a branching caudex; numerous closely crowded stems from the caudex, somewhat decumbent at base, extend 10–15 cm high; leaves 2–3 cm long, somewhat succulent, obovate to spatulate, generally toothed or lobed, the lower with winged petioles, the upper sessile; heads terminating the branches; involucres 8–10 mm high; rays about 8 mm long, bright yellow. $2n = 40$*, 80. Subalpine and alpine talus slopes. BC, Alta s to Calif, Utah, Colo. Map 1133 ○

S. indecorus Greene
Perennial, 2–9 dm tall from a short caudex; stems and leaves glabrous or soon glabrate, occasionally with sparse tomentum in the axils; basal leaves slender-petioled, oblong, elliptic or ovate, cuneate or subcordate at base, serrate or incised; cauline leaves petioled to sessile, incised-pinnatifid, the lobes often toothed; heads mostly 6–40, generally discoid, the flowers pale yellow; involucres 7–10 mm high, the bracts often purple-tipped; corolla-tube 3.6–6 mm long; achenes 2–3 mm long. $2n = 176–200$. *Packera indecora* (Greene) Löve & Löve. Moist open woods, wet meadows. Alas, Yuk, swMack to Huds Bay, nQue, Lab s to Wash, Wyo, Sask, eOnt, James Bay, eQue. Map 1135

S. integerrimus Nutt.
Robust perennial with a short caudex, the stems solitary, 3–7 dm tall; stems and leaves hirsute to arachnoid-villose when young, often becoming somewhat glabrate in age; basal leaves oblanceolate, elliptic, ovate or subrotund, entire to irregularly toothed; cauline leaves progressively reduced and becoming sessile upwards; inflorescence usually large, congested to open, of 5–35 heads, the peduncle of the terminal head shorter and thicker than the others; involucral bracts 5–9 mm long, black-tipped; rays yellow, 6–15 mm long. $2n = 40, 80$. In our area mainly var. **exaltatus** (Nutt.) Cronq., which in the mountains is not easily distinguished from *S. lugens*. Dry to moist plains and open woods to subalpine elevations. BC to Man s to Calif, Colo, Neb, Minn. Map 1136

S. lugens Richards.
Tall perennial with a short rhizome, stems solitary, 2–6 dm high, slender, leafy near base, naked except for a few reduced leaves upward; stems and leaves loosely arachnoid or floccose-woolly when young, becoming glabrate but usually some of the woolly hairs persisting at maturity, often glaucescent; lower leaves narrowly oblong-oblanceolate, long-tapering at base into margined petioles, the blades denticulate to nearly entire; cauline leaves much reduced, linear-lanceolate to linear, sessile; inflorescence with mostly 5–15 heads, the central one often short-peduncled; bracts mostly 6–8 mm long, black-tipped. $2n = 40, 80$*. See note under previous species. Moist subalpine and alpine slopes. Alas, Yuk, wMack s to BC, Mont, Wyo; Wash. Map 1137

S. megacephalus Nutt.
Stout perennial, more or less tomentose, with a solitary stem, 2–5 dm tall; leaves oblanceolate, oblong or linear, rather thick, entire or denticulate, the lower ones 1–2 dm long and petioled, the upper ones greatly reduced and sessile; heads solitary or 2–4, large, the involucres 12–16 mm high; bracts slender, acuminate, the outer lower ones often surpassing the inner; rays 15–20 mm long, yellow or orange. $2n = 40$. Rocky alpine and subalpine slopes. BC, Alta s to Ida, Mont. Map 1138 ○

S. pauciflorus Pursh
Perennial with rather stout stems, 1–5 dm high, and thick fleshy leaves, glabrous or nearly so; basal leaves long-petioled, elliptic or reniform, cuneate to cordate at base, coarsely dentate; median and upper leaves sessile, lanceolate to oblanceolate, mostly pinnatifid with blunt lobes and coarse teeth; inflorescence with 1–6 heads, rarely more; heads discoid or radiate; involucres usually purple; corolla-tube 3–4 mm long, the lobes orange to red; achenes dark red-brown, plump, 3–3.5 mm long. $2n = 176–184$. *Packera pauciflora* (Pursh) Löve & Löve. Moist alpine and subalpine meadows. Alas, Yuk, swMack s to Wash, Wyo; Calif; Man, James Bay, nQue, Lab, Nfld s to Gaspé. Map 1138 ●

S. pauperculus Michx.
Perennial with slender stems, 1–6 dm tall, glabrous or glabrescent, often with tufts of woolly tomentum; basal leaves more or less tufted, slender-petioled, the blades oblanceolate, spatulate or oblong-elliptic, crenate to serrate or nearly entire, cuneate at base; cauline leaves more or less pinnatifid, greatly reduced upwards and sessile; heads several, mostly long-pedicelled; involucres 6–9 mm high; flowers deep yellow, the rays usually showy. $2n = 44^*, 46, 88^*, 92^*$. *Packera paupercula* (Michx.) Löve & Löve. A complex in which two or three species may eventually be recognized. Moist meadows, shores, open woods, roadsides. Alas, Yuk, sMack to Huds Bay, nQue, Nfld s to BC, Ore, NM, SD, Mo, Tenn, Ga, NC. Map 1139

S. pseudaureus Rydb.
Perennial with stems 3–7 dm tall from a short caudex, sparsely tomentose at first, soon essentially glabrous; basal leaves long-petiolate, the blades thin, mostly ovate, serrate, cordate to truncate at base; cauline leaves reduced upwards, becoming sessile and clasping, mostly incised-pinnatifid at least at their bases, otherwise sharply toothed; heads several–many; involucres 5–8 mm high; ligules bright yellow. $2n = 46$. *Packera pseudaurea* (Rydb.) Weber & Löve. Moist meadows, stream banks, thickets. BC, Alta s to Calif, Nev, NM; sMan s to ND, SD, Kans, Mo, Minn. Map 1140

S. streptanthifolius Greene
Stems nearly glabrous, 1–5 dm tall, from a caudex or short rhizome; basal and lower cauline leaves long-petioled, thickish and somewhat fleshy, the blades obovate to spatulate, entire below and toothed at or near the apex, cuneate at base, glabrous; upper cauline leaves reduced, sessile, commonly incised and

toothed near their bases; heads several or numerous; involucres 5–7 mm high; rays 6–8 mm long, bright yellow. $2n = 46^*$. *S. cymbalarioides* Nutt. of Ed. 1. *Packera streptanthifolia* (Greene) Weber & Löve. Dry to moist open areas, open woods. Yuk, swMack s to Calif, NM. Map 1141

S. triangularis Hook.
Coarse perennial, 3–15 dm tall, the stems erect, often clustered, very leafy, glabrous; leaves petioled, except the uppermost, the blades thin and shining, elongate-triangular or lanceolate, 5–15 cm long, coarsely and sharply dentate; heads few or many, in an open, often flat-topped inflorescence; involucres 6–10 mm high; rays yellow, about 1 cm long. $2n = 40^*$, 80. Moist banks and streams to subalpine elevations. Alas, Yuk, swMack s to Calif, NM. Map 1142

S. viscosus L.
Annual plant, densely glandular-viscid throughout; stems mostly 1–4 dm tall, simple or branched, leafy; leaves 3–12 cm long, 1–5 cm wide, pinnately lobed, the lower petiolate, the upper sessile; heads several–many; involucre 6–8 mm high; ray florets present but inconspicuous; achenes glabrous. $2n = 40$. Waste ground. Introduced. Rare.

S. vulgaris L. / Common Groundsel
Annual plant, 1–5 dm tall, the stems leafy and glabrous or sparingly hairy; leaves undulate or deeply pinnatifid with oblong or rounded, often incised lobes; lower leaves tapering to the petiole, the upper sessile and more or less clasping; heads several or many, discoid; involucres 5–8 mm high, the bracts glabrous, the bractlets minute and black-tipped; pappus conspicuous, often longer than the corollas. $2n = 20$, 40. Waste ground, gardens. Introduced. Common.

Silybum Vaill. *ex* Adans.

S. marianum (L.) Gaertn. / Milk Thistle
Annual or biennial, essentially glabrous; stem 6–15 dm tall, branched, lightly arachnoid; leaves alternate, the lower up to 60 cm long, 30 cm wide, pinnately lobed with spiny margins, petiolate, the upper progressively smaller, less lobed, sessile or clasping; heads solitary at the ends of branches; large; involucral bracts imbricate, in several series, the outer about 4 cm long, 1 cm wide, ovate-lanceolate, spiny; flowers purple, all tubular; achenes 6–7 mm long; pappus of slender, unequal bristles. $2n = 34$. Waste ground, gardens. Introduced. Rare.

Solidago L. / Goldenrod

Perennial herbs with rhizome or caudex and fibrous roots; stems mostly slender and erect; leaves alternate, entire or toothed, the cauline leaves sessile or nearly sessile; basal vegetative shoots with crowded leaves often present; inflorescence usually large, composed of numerous small heads with yellow flowers; involucres 3–6 mm high, the bracts narrow, more or less imbricate in several series; ray-flowers pistillate and fertile, the rays yellow, short, 2–5 mm long; disc-flowers perfect and fertile, yellow; style-branches flattened, with lanceolate hairy appendages; achenes several-ribbed, nearly terete; pappus simple, of capillary bristles, usually white. Blooming chiefly in late summer and autumn.

1 Petiole of the basal leaves conspicuously ciliate, otherwise leaves essentially glabrous; plants mostly 0.5–2.0 (3.0) dm tall **S. multiradiata**

Petiole of the basal leaves (if present) not as above; plants generally taller 2

2 Heads corymbose, the inflorescence rounded or flat-topped 3

Heads not corymbose, usually racemose, the inflorescence cylindrical to narrowly oval, or paniculate, with spreading or arching branches, the heads often secund 4

3 Involucral bracts striate; stem and leaves pubescent **S. rigida**

Involucral bracts not striate; stem and leaves glabrous **S. graminifolia**

4 Leaves glabrous; stems mostly glabrous below the inflorescence 5

Leaves pubescent or puberulent; stems mostly pubescent or puberulent 7

5 Heads glutinous **S. spathulata**

Heads not glutinous 6

6 Lower stem-leaves well developed, narrowly oblanceolate, a basal rosette often also present; stem-leaves reduced upwards; plants mostly 2–4 dm tall **S. missouriensis**

Lower stem-leaves and basal rosette absent, or reduced and soon withered; stem leafiest from about the middle, the leaves lanceolate, only gradually or scarcely reduced upwards; plants mostly 5–15 dm tall **S. gigantea**

7 Plants mostly 1–4 dm tall; lower stem-leaves well developed, oblanceolate, a basal rosette often also present; stem-leaves linear to linear-lanceolate, sparse, often distant, reduced upwards **S. nemoralis**

Plants not as above 8

8 Leaves elliptic to obovate, firm, densely hirsute above and below, grey-green; middle stem-leaves mostly 2–3 times longer than wide; plants usually 1–4 dm tall **S. mollis**

Leaves lanceolate, sometimes glabrous above; middle stem-leaves mostly more than 4 times longer than wide; plants usually 4–12 dm tall **S. canadensis**

S. canadensis L. / Canada Goldenrod
Usually robust plants; stems 3–9 (12) dm tall, leafy, pubescent at least above; leaves crowded, mostly 4–7 (12) cm long, lanceolate, acuminate, serrate, densely short-hairy above and below; heads in a narrow or broadly pyramidal, panicle, the branches usually ascending, often secund, densely pilose; involucres 3–5 mm high, the bracts linear to linear-lanceolate, acuminate to acute; achenes pubescent. $2n$ = 18, 36, 54. Includes *S. lepida* DC. and *S. pruinosa* Greene of Ed. 1. In our area mostly var. **gilvocanescens** Rydb. with pubescence as above, also var. **canadensis**, the leaves puberulent below, puberulent or glabrous above. Open woods, moist meadows. Alas, Yuk, swMack to Huds Bay, Nfld s to Calif, NM, Tex, Fla. Map 1143

S. gigantea Ait.
Tall robust plants; stems commonly clustered, sometimes solitary, 5–20 dm tall, erect, glabrous, often glaucous up to the inflorescence; leaves numerous, lanceolate, cuneate, acuminate, sharp-serrate above the middle, glabrous or sometimes hairy on the veins below; heads in a broad pyramidal panicle, the lower branches often elongated and recurved; involucres 3–5 mm high, the bracts linear or linear-lanceolate; achenes pubescent or nearly glabrous. $2n = 18^*, 36, 54^*$. In our area var. **serotina** (Ait.) Cronq. (includes *S. shinnersii* (Beaudry) Beaudry). Thickets, moist open woods, alluvial flats, shores. swMack, BC to NS s to Ore, Utah, NM, Tex, Okla, Ark, Ga. Map 1144

S. graminifolia (L.) Salisb.
Stems 3–7 dm tall, leafy, striate, glabrous or nearly so; leaves linear to lanceolate, entire, glabrous, more or less punctate, 3-nerved, or the broader ones 5-nerved, the lateral veins obscure; heads in small compact clusters at the ends of the branches, forming a flat-topped or corymbose inflorescence; involucres 3–5 mm high, more or less glutinous, the bracts obtuse or rounded, yellowish. $2n = 18$. In our area var. **major** (Michx.) Fern. Meadows and shores. swMack, BC to Nfld s to NM, Okla, Mo, Ky, NC. Map 1145

S. missouriensis Nutt.
Stems solitary, 1–6 dm tall, glabrous or sparsely puberulent, abundantly produced from creeping rhizomes; basal leaves narrowly oblanceolate, petioled, mostly triple-veined, glabrous except for scabrous-ciliate margins, entire or somewhat serrate; upper leaves smaller, lanceolate to linear, mostly sessile; heads mostly secund, in a broad rounded or flat-topped panicle; involucres 3–5 mm high, the bracts linear-oblong, blunt or acutish; achenes hairy to nearly glabrous. $2n = 18^*, 36^*$. Grassland and open woods. BC to Ont s to Wash, Ariz, NM, Tex, Okla, Ark, Tenn. Map 1146

S. mollis Bartl.
Stems stout, stiffly erect, grey-hairy, 2–5 dm high, solitary or clustered; lower leaves ovate to elliptic or oblanceolate, rigid, velvety-pubescent, entire to toothed, conspicuously 3-nerved, petioled; upper leaves narrower, much smaller, sessile; heads scarcely secund, compactly arranged in a cylindrical or ovoid panicle; involucres 3–6 mm high, the bracts ovate or oblong, acute or obtuse, yellowish, ligules bright yellow. $2n = 54$. Dry grassland and roadsides. Alta to swMan s to Mont, NM, Tex, Okla, Ia, Minn. Map 1147

S. multiradiata Ait.
Stems slender, 0.5–5 dm tall, commonly tufted from a branching caudex, loosely pubescent at least in the inflorescence, often reddish-tinged; basal and lower cauline leaves oblanceolate, 2–10 cm long, toothed, glabrous, the petiolar base ciliate-margined; upper leaves few and reduced; heads in a loose or dense corymb-like inflorescence, with rather long peduncles, the pedicels white-villose; involucres 4–6 mm high, the bracts linear to lanceolate, mostly acuminate, usually ciliolate. $2n = 18, 36^*$. Moist open areas to alpine elevations. eAsia: Alas, Yuk,

Mack, Banks Is, Victoria Is to nHuds Bay, sBaffin Is, Nfld s to Calif, NM, nSask, nMan, James Bay, Gaspé. Map 1148

S. nemoralis Ait.
Stems grey-puberulent, 2–6 dm high, solitary or tufted from a thick caudex; leaves narrowly lanceolate or linear-oblanceolate, nearly entire or somewhat crenate, puberulent, the lower ones long-petioled, the upper reduced and nearly sessile; heads mostly crowded in a narrow panicle or on spreading to recurved branches of a broad inflorescence, often one-sided; involucre 4–6 mm high, the bracts linear-oblong, obtuse, rounded or acute, often with more than one nerve. $2n = 18, 36^*$. In our area var. **longipetiolata** (Mack. & Bush) Palmer & Stayerm. (var. *decemflora* (DC.) Fern. of Ed. 1). Dry open pine woods and prairies. BC to Que, NS s to Ariz, Tex, Ga. Map 1149

S. rigida L. / Stiff Goldenrod
Stems 3–8 dm tall, coarse and stiff, puberulent, from a stout caudex; leaves oblong to elliptic, thick and firm, nearly entire, canescent on both sides, 1-nerved but the lateral nerves sometimes evident; basal leaves long-petioled, the upper ones sessile; heads relatively large in rather dense corymb-like cymes; involucres 5–7 mm high, the bracts oblong, obtuse, several-nerved, greenish; achenes usually with loose hairs near the top. $2n = 18^*, 36$. In our area var. **humilis** Porter (*S. parvirigida* Beaudry). Dry open areas, especially in sandy soil. Alta to Ont s to NM, Tex, La, Ga. Map 1150

S. spathulata DC.
Stems solitary or tufted, 1–4 dm high, decumbent at base, often reddish, glabrous or sparingly pubescent above; basal leaves spatulate-obovate or oblanceolate, rounded at apex, more or less crenate near tip, 1-nerved, petioled; upper leaves similar but smaller; heads racemose in a dense, often interrupted, inflorescence, cylindrical to ovoid in outline, not at all secund; involucres 4–6 mm high; bracts linear-oblong to oblong, obtuse; achenes about 2 mm long, hairy. $2n = 18^*, 36$. *S. decumbens* Greene of Ed. 1. Grassland and open woods to alpine elevations. Alas, Yuk, wMack to Man s to Calif, Ariz, NM. Map 1151

Sonchus L. / Sow Thistle

Annual or perennial herbs with milky juice; leaves alternate, pinnatifid or entire, mostly clasping the stem, often prickly-margined; heads solitary, or several in an irregular corymbose inflorescence; flowers all ligulate and perfect, yellow, numerous in the head; involucral bracts imbricate in several series, often basally thickened in age; achenes flattened, longitudinally ribbed, glabrous, beakless; pappus of numerous soft white capillary bristles.

1 Perennials, spreading by long horizontal roots; heads 3–5 cm wide in flower; fruiting involucre 14–24 mm high 2
Annuals; heads 1.5–2.5 cm wide; fruiting involucre 9–13 mm high, glabrous 3

2 Involucre and peduncles with coarse gland-tipped hairs **S. arvensis**
Involucre and peduncles glabrous or nearly so **S. uliginosus**

3 Basal auricles of leaves mostly acute; achenes striate and transversely wrinkled **S. oleraceus**
Basal auricles of leaves rounded; achenes 3-ribbed on each side, not wrinkled **S. asper**

S. arvensis L. / Perennial Sow Thistle
Perennial herb, spreading by long horizontal roots; stems 4–20 dm tall, glabrous below the inflorescence; leaves prickly-margined, the lower and middle ones usually pinnately-lobed or pinnatifid, 6–40 cm long, the upper ones less lobed and smaller, often markedly auriculate; heads several in an open corymbose cluster; fruiting involucres hemispherical to cup-shaped, 14–24 mm high; involucre and peduncles with coarse spreading gland-tipped hairs; achenes 2.5–3.5 mm long, with 5 or more longitudinal ribs on each face, strongly rugulose. $2n = 54$. Waste ground, fields, gardens. Introduced. Common.

S. asper (L.) Hill / Annual Sow Thistle
Annual, with a short taproot, 1–10 dm tall, nearly glabrous; leaves 6–30 cm long, pinnatifid or frequently obovate and lobeless, with rounded auricles, spinulose-dentate; heads few–several in a corymbose cluster; fruiting involucres 9–13 mm high; achenes 2–3 mm long, with usually 3 longitudinal ribs on each face, not rugulose. $2n = 18$. Waste ground, fields, roadsides. Introduced. Common.

S. oleraceus L. / Annual Sow Thistle
Similar to *S. asper*, commonly less prickly, the leaves markedly pinnatifid only occasionally merely toothed; the achenes transversely rugulose and rather obscurely 3–5-ribbed on each face. $2n = 16, 18, 32, 36$. Waste ground, gardens, roadsides. Introduced. Occasional.

S. uliginosus Bieb. / Perennial Sow Thistle
Similar to *S. arvensis*, but lacking gland-tipped hairs; leaves somewhat less lobed; involucres narrower and paler; involucral bracts glabrous, white-margined and the outer ones more strongly carinate-margined in age; growing somewhat taller and stiffer under comparable conditions. $2n = 36$*. Fields, waste ground, roadsides. Common.

Stephanomeria Nutt.

S. runcinata Nutt.
Perennial; stems 1–2 dm tall, usually several, generally branched; leaves alternate, 1–7 cm long, 4–15 mm wide, linear to linear-lanceolate, runcinate-pinnatifid, upper leaves reduced; heads all ligulate, mostly 5-flowered, pink or white; involucre 9–12 mm high; achenes about 5 mm long; pappus white, plumose. $2n = 16$. *S. tenuifolia* (Torr.) Hall of Ed. 1. Dry hills and plains. Alta, Sask s to Mont, Colo, Neb. Map 1152

Tanacetum L. / Tansy

Perennial herbs with alternate, 1–3-pinnately dissected leaves; heads few or many in a corymbiform inflorescence; heads many-flowered, discoid or nearly so; involucral bracts imbricate, dry, the tips and margins commonly more or less scarious; receptacle convex or flat, naked; marginal flowers pistillate, more or less radiate, rarely wanting; disc-flowers perfect, with yellow tubular corollas; achenes angled or ribbed, commonly glandular; pappus a short crown, or none.

Leaves villose, at least when young, scarcely punctate; heads few and large, the disc 1–2 cm wide **T. huronense**

Leaves glabrous or nearly so, punctate; heads numerous and smaller, the disc 0.5–1 cm wide **T. vulgare**

T. huronense Nutt.
Plant with slender rhizomes; stems 1–8 dm tall; leaves sessile to short-petiolate, more or less villose, 2–3-pinnatifid, the ultimate segments narrow; primary rachis scarcely or not at all winged; heads 1–15; disc 1–2 cm wide; rays often evident, sometimes inconspicuous. $2n = 54$. *Chrysanthemum bipinnatum* L. ssp. *huronense* (Nutt.) Hult. Shores, sand-dunes, gravel bars. Alas, Yuk, nwMack s to wBC; neAlta, nwSask; Huds Bay; Gt Lakes; seQue, Me. Map 1153 •

T. vulgare L. / Common Tansy
Coarse aromatic herb with a stout rhizome; stem erect, 4–18 dm tall; leaves numerous, 1–2 dm long, sessile or short-petiolate, glabrous, punctate, pinnatifid, the pinnae again pinnatifid or deeply lobed, the pinnules often toothed; rachis broadly winged; heads 20–200, in a flat-topped corymb; disc 0.5–1 cm wide; rays inconspicuous or none. $2n = 18$. Waste ground, roadsides. Introduced. Common.

Taraxacum Weber / Dandelion

Acaulescent perennials with fleshy taproots and milky juice; leaves in a rosette, pinnatifid, toothed or rarely entire; heads solitary on a hollow scape, this elongating in fruit; involucral bracts in 2 series, the outer generally much shorter than the inner and often reflexed; receptacle naked; flowers all ligulate, yellow; achenes columnar to spindle-shaped, longitudinally ribbed, muricate at least above, prolonged into a slender beak; pappus of numerous capillary bristles, white or brownish.

1 Outer involucral bracts appressed or ascending **T. ceratophorum**

Outer involucral bracts reflexed 2

2 Leaves divided almost to midrib, terminal segment small; achenes reddish **T. laevigatum**

Leaves more shallowly lobed, terminal segment large; achenes olivaceous to brown **T. officinale**

T. ceratophorum (Ledeb.) DC.
Scapes 2–25 cm tall; leaves 2–20 cm long, oblanceolate, sinuate-dentate to pinnatifid, glabrous or slightly hairy; petiolar bases often broadly winged; involucres 1.0–2.5 cm high; outer bracts appressed or somewhat spreading; inner bracts corniculate or not; body of achenes, 3–4 mm long, yellowish to brownish or black, muricate to near the base. $2n = 24, 32, 40$. Includes *T. eriophorum* Rydb., and *T. lyratum* (Ledeb.) DC. of Ed. 1. Open areas to alpine elevations. Circumpolar: Alas, Yuk, to Prince Patrick Is, nBaffin Is, Nfld s to Calif, NM, nSask, Huds Bay, nQue. Map 1154

T. laevigatum (Willd.) DC. / Red-seeded Dandelion
Scapes 5–30 cm tall; leaves generally very deeply dissected for their whole length, the lobes narrow, the terminal one seldom much larger than the lateral ones, involucres mostly 1–2 cm high, the inner bracts often somewhat corniculate, the outer bracts appressed to reflexed; body of achene reddish at maturity, usually muricate only above the middle, somewhat wrinkled below; pappus sordid. $2n = 16, 24, 32$. Waste ground, roadsides, lawns. Introduced. Common.

T. officinale Weber / Common Dandelion
Robust plant, the scapes 5–40 cm tall; leaves 5–40 cm long, oblanceolate, commonly sparsely hairy beneath and on the midrib, sometimes completely glabrous, more or less runcinate-pinnatifid or lobed, the terminal lobe usually larger than the others; petiolar base slender and obscurely winged; heads usually large; involucres 1.5–2.5 cm high; outer bracts reflexed; inner bracts slightly, if at all, corniculate; body of achene 3–4 mm long, pale grey to olive-brown, muricate above or sometimes to near the base; pappus white. $2n = 16, 24, 32, 40$. Waste ground, roadsides, lawns. Introduced. Common.

Thelesperma Less. / Green-thread

T. marginatum Rydb.
Perennial, essentially glabrous; stem 1–2 dm tall; lower leaves opposite, 1.5–5.0 cm long, irregularly pinnatifid, the lobes 1.5–3.0 mm wide, upper leaves linear, reduced; heads 1–3, discoid; involucre 6–9 mm high; bracts in 2 series, the outer herbaceous, the inner membranous with scarious margins, united in the lower one-third to two-thirds; flowers yellow; pappus of 2 inconspicuous awns. Dry eroded hills. Alta s to Wyo. Map 1153 ○

Townsendia Hook.

Caulescent or acaulescent herbs with taproots; leaves alternate or basal, entire, strigose; heads solitary or few, medium to large, sessile or peduncled; involucral bracts appressed, somewhat imbricate, lanceolate, the margins white-scarious; receptacle flat, naked; ray-flowers pistillate, blue or purplish to pinkish or white; disc-flowers perfect, yellow; style-branches flattened, with externally hairy appendages; achenes of disc-flowers flattened, 2-nerved, usually pubescent with forked or barbed hairs; pappus of disc-flowers a single series of rigid scabrous bristles, that of the ray-flowers similar or composed of small scales.

1	Stems present, mostly 1–2 dm tall	**T. parryi**
	Stems absent, plants cushion-like	2
2	Involucres densely woolly-villous; leaves spatulate	**T. condensata**
	Involucres not densely woolly-villous; leaves linear to linear-lanceolate or oblanceolate	3
3	Corolla of disc-flowers about 5 mm long; involucre about 15 mm wide	**T. hookeri**
	Corolla of disc-flowers 8–10 mm long; involucre 20–25 mm wide	**T. exscapa**

T. condensata D.C. Eat.
Low caespitose perennial; leaves 1.2–3.5 cm long, 2–4 mm wide, spatulate, with a short abruptly expanded blade; heads sessile among the leaves; involucre 8–18 mm high, 10–40 mm wide; bracts lanceolate to linear, acuminate, ciliate, villous; rays 8–16 mm long, white, pink or lavender. $2n = 18$, c.36. Exposed alpine ridges. Alta s to Ida, wMont, Wyo; ecCalif. Map 1155 ○

T. exscapa (Richards.) Porter
Low caespitose perennial, the stout caudex more or less branched; leaves mostly 2–5 cm long, 2–4 mm wide, linear or linear-oblanceolate, crowded, canescent-strigose; heads sessile among the erect leaves; involucre 12–20 mm high, its bracts linear-lanceolate, the margins scarious and lacerate-ciliate; rays white to pinkish, 12–22 mm long; pappus of the disc-flowers conspicuously longer than the disc-corollas. $2n = 18$. *T. sericea* Hook. of Ed. 1. Dry hillsides and prairies. seBC to swMan s to Nev, Ariz, NM, Tex, Okla. Map 1155 ●

T. hookeri Beaman
Similar to *T. exscapa* but noticeably smaller plants; distinguished by the characters of the key; the shorter involucre, 9–15 mm high; the shorter rays, 8–14 mm long; and the pappus of the disc-flowers which barely exceeds the disc-corollas. $2n = 18$. Dry prairie slopes and eroding hills. ecBC to swSask s to Utah, Colo, SD. Map 1156

T. parryi D.C. Eat.
Stout perennial, 0.5–2.5 dm tall, with a crown of basal leaves and 1–several flowering stems, each with a single large head; stems appressed-pubescent; basal leaves crowded, spatulate to broadly oblanceolate, 2–6 cm long, strigose beneath, glabrous above; cauline leaves few and small; involucre 9–16 mm high, its bracts narrowly lanceolate, acuminate, scarious and erose on the margins; rays violet or bluish purple, 1–2 cm long; pappus of long bristles. $2n = 18, 36^*$. Dry hills, banks, roadsides. seBC, Alta s to Ida, Wyo; neOre. Map 1157

Tragopogon L.

Herbs; leaves linear, clasping; heads large, solitary or few; bracts uniseriate; flowers all ligulate; achene beaked; pappus single, of mostly plumose bristles.

1 Flowers purple **T. porrifolius**
Flowers yellow 2

2 Pedunde enlarged below the head; involucral bracts generally 13; achenes 25–35 mm long **T. dubius**
Peduncle not enlarged below the head; involucral bracts generally 8; achenes 15–25 mm long **T. pratensis**

T. dubius Scop. / Goat's-beard
Stout glabrous latex-bearing herb, 4–10 dm tall, with a taproot; leaves alternate, 5–30 cm long, linear, entire, with a clasping base; heads solitary on slender peduncles, these enlarged below the head; bracts equal, lanceolate-attenuate, surpassing the flowers, elongating to 4–7 cm in fruit; achenes 25–35 mm long, linear-fusiform, 5–10-ribbed, long-beaked, these forming a large conspicuous fruiting-head; pappus of plumose bristles, united at the base, the plume-branches interwebbed. $2n = 12$. Roadsides, waste ground. Introduced. Common.

T. porrifolius L.
Perennial, glabrous; stem 4–10 dm tall; leaves up to 30 cm long; involucral bracts 5–11, mostly 8, 2.5–4.0 cm long in flower, becoming 4–7 cm long in fruit; rays purple, shorter than bracts; achenes 25–40 mm long. $2n = 12$. Roadsides. Introduced. Rare.

T. pratensis L.
Perennial, glabrous; stem 1.5–8.0 dm tall; leaves to 30 cm long, less than 2 cm wide, often narrower; involucral bracts usually 8, 1.2–2.4 cm long in flower, becoming 1.8–3.8 cm long in fruit, equalling or shorter than rays; achenes 12–24 mm long. $2n = 12$. Roadsides, waste ground. Introduced. Rare.

Xanthium L.

X. strumarium L. / Cocklebur
Coarse annual herb, 2–12 dm high; stem simple or branched, more or less scabrous-hairy, sometimes purple-spotted; leaves alternate, slender-petioled, ovate to reniform, generally cordate or deltoid at base, coarsely toothed or undulate, sometimes 3–5-lobed; staminate and pistillate flowers in different heads, borne in axillary clusters; staminate uppermost, many-flowered, with an involucre of separate bracts and a cylindric chaffy receptacle; corollas discoid; filaments monadelphous and anthers free; pistillate heads 2-flowered, the involucral bracts all fused, forming a bur-like body, 2-beaked at apex and bearing stiff hooked bristles; corolla lacking; fruit bur-like, 1–3.5 cm long, covered with hooked spines, containing 2 oblong flattened achenes. $2n = 36$. Moist ground, shores, waste ground, fields. BC to Que s to Calif, NM, Tex, Okla, Ark, Ill, Pa, Mass. Map 1158

Geographical Abbreviations

Abbreviation	Name	Abbreviation	Name
Ala	Alabama	Mo	Missouri
Alas	Alaska	Mont	Montana
Alta	Alberta	NAm	North America
Apal	Apalachia	NB	New Brunswick
Arct Arch	Canadian Arctic Archipelago	NC	North Carolina
Ariz	Arizona	ND	North Dakota
Ark	Arkansas	Neb	Nebraska
BC	British Columbia	NEng	New England
Calif	California	Nev	Nevada
Can	Canada	Nfld	Newfoundland
Colo	Colorado	NH	New Hampshire
Conn	Connecticut	NJ	New Jersey
Daks	Dakotas	NM	New Mexico
DC	District of Columbia	NS	Nova Scotia
Del	Delaware	NWT	Northwest Territories
Fla	Florida	NY	New York
Ga	Georgia	Okla	Oklahoma
Gt Lakes	Great Lakes	Ont	Ontario
Huds Bay	Hudson Bay	Ore	Oregon
Ia	Iowa	Pa	Pennsylvania
Ida	Idaho	Que	Quebec
Ill	Illinois	RI	Rhode Island
Ind	Indiana	SAm	South America
Kans	Kansas	Sask	Saskatchewan
Ky	Kentucky	SC	South Carolina
La	Louisiana	SD	South Dakota
Lab	Labrador	Tenn	Tennessee
Mack	District of Mackenzie	Tex	Texas
Man	Manitoba	USA	United States
Mass	Massachusetts	VA	Virginia
Md	Maryland	Vt	Vermont
Me	Maine	Wash	Washington
Mex	Mexico	Wisc	Wisconsin
Mich	Michigan	WVa	West Virginia
Minn	Minnesota	Wyo	Wyoming
Miss	Mississippi	Yuk	Yukon

Glossary

acaulescent. Stemless or apparently so, or with the stem below the surface of the ground.

achene. A small, dry 1-seeded indehiscent fruit, distinguished from a nutlet by its relatively thin wall.

actinomorphic (flower). Regular; radially symmetrical.

acuminate. Gradually tapering to a point.

acute. Sharply pointed, but not drawn out.

adnate. Having one organ united wholly or in part with another organ, as stipules adnate to the petiole.

adventitious. Applied to roots, buds, etc., which develop in an unusual or irregular position.

agamospecies. A species in which seeds are usually produced asexually. Such species are often members of a complex and usually difficult taxonomically.

agamospermy. Production of seeds by asexual processes.

alpine. The cold, treeless zone on mountains, usually above 2,275 m (6,500 ft) in our area.

alternate. Not opposite each other.

alveolate. Honeycomb-like.

ament. See *catkin.*

anastomosing. Connecting by cross-veins, forming a network.

androgynous (inflorescence in *Carex*). Composed of both staminate and pistillate flowers, the staminate at the apex.

annual. Of only one year's growth or duration; a plant that completes its life-cycle in one season. *Winter annual,* a plant from autumn-sown seed which blooms and fruits the next year.

annular. In the form of a ring.

annulus. A ring of thick-walled cells partly surrounding the spore-case (in most ferns).

anterior (in the flower). The side away from the axis and towards the bract.

anther. The pollen-bearing part of the stamen; composed usually of two pollen-sacs.

anthesis. The expansion or time of expansion of a flower.

apetalous. Without petals.

aphyllopodic. Having the lowest leaves reduced to scales so that the first foliage leaves are well above the plant base. See *phyllopodic.*

apical. Relating to the apex or tip.

apiculate. Ending in an abrupt short point.

appressed. Lying flat and close against.

aquatic. Living in water; describing plants floating at the surface or wholly submerged.

arachnoid. Cobwebby.

arcuate. Curved into an arc of a circle.

aristate. Having a bristle or awn, usually terminal in position.

articulate. Jointed.

ascending. Upcurved; growing obliquely upward.

attenuate. Tapering to a very narrow or slender structure.

auricle. An ear-shaped appendage or lobe.

auriculate. With auricles.

awl-shaped. Tapering from the base to a slender rigid point.

awn. A bristle-like appendage.

awned. Having awns.

axil. The upper angle formed by a leaf or branch with the stem.

axile, axial. Refers to the axis of an organ; applied to placentae in the ovary.

axillary. Located in or arising from an axil.

barbed. With rigid points or short bristles, often reflexed.
barbellate (pappus). With minute stiff hairs.
basal. Located at the base of a plant or of an organ of a plant.
basifixed. Attached by the base; as an anther upon the filament.
beaked. Ending in a prolonged tip.
bearded. With long or stiff hairs.
berry. In a strict sense, a fleshy or pulpy fruit formed from a single ovary and containing one to many seeds; loosely, any juicy of pulpy fruit.
bidentate. Having two teeth.
biennial. Of two years' duration.
bilabiate. Two-lipped, applied to calyx or corolla.
bipinnate. Twice or double pinnate.
bisexual (flowers). With both stamens and ovaries; perfect.
blade. The flat expanded part of a leaf.
bloom. A whitish powdery covering, often of a waxy nature.
bract. A scale or leaf, usually small, subtending a flower or belonging to a flower-cluster.
bracteate. With bracts.
bractlet, bracteole. A secondary bract, as one upon a flower stalk; a very small bract.
bud. The undeveloped state of a stem, branch, or flower.
bud-scales. Modified leaves covering a bud.
bulb. A short, thick bud, usually underground, bearing fleshy scales or coats that are stored with reserve food.
bulblet. A small bulb, especially when borne on the stem, sometimes in place of flowers.
bulbose. Bulb-like in form.

caducous. Falling very early.
caespitose. Growing in dense tufts.
callus. A small, hard protuberance.
calyculate. With bracts around the calyx or involucre, suggestive of an outer calyx.
calyx. The outer part of the perianth, usually green; the set of sepals.
campanulate. Bell-shaped.
canaliculate. Channelled; longitudinally grooved.
canescent. Pale or grey, because of a close, fine pubescence.
capillary. Hair-like.
capitate. Shaped like a head; forming a dense cluster.
capsule. A dry dehiscent fruit formed from a compound ovary. Compare *legume* and *follicle.*
carpel. The basic unit of the gynoecium, consisting of the ovary, style, and stigma. Where only a single carpel is present carpel and gynoecium are synonymous, as in our Leguminosae, in *Actaea* and *Prunus.* See *gynoecium* and *pistil.*
carpellate. Having carpels.
carpophore. A stalk-like elongation of the receptacle; as in *Silene.*
cartilaginous. Firm and tough.
caryopsis. A grain, as of grasses; a seed-like fruit with the thin pericarp adherent to the enclosed seed.
catkin. An ament; a scaly spike or raceme, usually unisexual bearing many small naked or apetalous flowers.
caudate. With a tail-like terminal appendage.
caudex. The persistent thickened stem-base of a perennial herb.
cauline. Pertaining to the stem.
chaff. Dry, thin scales (flower-bracts of the Compositae).
chasmogamous. With flowers opening, pollination normal. Opposite of cleistogamous.
chromosome number. The number of chromosomes occurring in the somatic nucleus of an organism. Given in the text as follows: $2n =$ a number.
ciliate. With marginal hairs.
cinereous. Ash-coloured; light grey.
circinate. Coiled into a ring or partially so.
circumpolar. Occurring more or less continuously around the globe in the north temperate zone.
circumscissile. Transversely dehiscent the top falling off as a lid.
cladophyll. A branch, usually flattened, that functions as a leaf, the true leaves

generally being reduced to scales.

clasping (leaf). With the base partly or completely surrounding the stem.

clavate. Club-shaped; gradually thickened upward.

claw. The narrowed base or stalk of a petal (in some flowers).

cleft. Deeply lobed (to about the middle or deeper). Compare *parted*, meaning more deeply cut.

cleistogamous. Fertilized without the opening of the flower.

coma. A tuft of soft hairs, especially at the tip of a seed.

commissure. Face by which two carpels adhere (Umbelliferae).

compound (leaf). Composed of two or more leaflets.

– (ovary). Developed from two or more carpels, as indicated by two or more locules, or placentae, or styles, or stigmas.

conduplicate. Folded together lengthwise.

conical. Cone-shaped; largest at the base and tapering to the apex.

connate. Said of similar organs more or less united.

connective. The portion of a stamen which connects the two lobes of an anther.

connivent. Converging; coming into contact.

contorted. Bent or twisted on itself.

convolute. Rolled up longitudinally.

cordate. Heart-shaped, with the point upward; commonly applied to the base of a leaf.

coriaceous. Of leathery texture.

corm. The enlarged fleshy base of a stem; a solid but bulb-like underground stem.

corolla. The inner whorl of the perianth, composed of petals.

corymb. A flat-topped or round-topped flower-cluster in which the axis is shortened and the stalks of the lower flowers lengthened, the marginal flowers blooming first. See *cyme.*

corymbose. Arranged in a corymb.

cotyledon. A leaf in the embryo within a seed.

coulee. A dry prairie valley.

creeping. Growing along or near the surface of the ground and rooting.

crenate. With rounded teeth; scalloped.

crenulate. Finely crenate.

crested. Bearing a ridge or projection on the top of a petal or other organ.

crisped. Irregularly curled along the margin.

crown. An inner appendage to a petal, or to the throat of a corolla.

cruciform. Cross-shaped; e.g. corolla of Cruciferae.

cucullate. Hooded, or resembling a hood.

culm. The stem of a grass of a sedge.

cuneate. Wedge-shaped.

cuspidate. Ending in a sharp stiff point (a cusp).

cyme. A flat-topped or convex flower-cluster, the central flowers of which bloom first. See *corymb.*

cymose. Arranged in a cyme.

deciduous. Falling away at the end of the growing season; not evergreen.

declined. Directed obliquely.

decumbent (stem). Reclining, but with the tip ascending.

decurrent (leaf). Extending downward; the blade prolonged downward as two wings along the petiole or the stem.

decussate. Opposite, with successive pairs at right angles to each other.

deflexed. Bent abruptly downward.

dehiscence. The act or method of opening, as in fruit, anthers.

dehiscent. Splitting open along regular lines at maturity.

deltoid. Broadly triangular; like the Greek letter Δ.

dendritic. Branched in tree-like form, as in some hairs.

dentate. Toothed, with the teeth pointed and directed outward. Compare *crenate* and *serrate.*

denticulate. Minutely dentate.

diadelphous (stamens). Arranged in two, often unequal, sets.

dichotomous. Forking more or less regularly into two branches of about equal size.

didynamous. With two stamens longer than the other two; in two pairs.

diffuse. Widely or loosely spreading.

digitate. Compound, with the leaflets borne finger-like on the petiole.

dioecious. Unisexual, with male and female flowers or cones on separate plants.
diploid. An organism with two full chromosome complements in each somatic nucleus. The number of chromosomes in a single full complement is said to be the basic number.
disc. An enlargement of the receptacle in the centre of the flower of various plants. In Compositae the central part of the head, composed of tubular flower.
discoid. Resembling a disc.
discoid head. In Compositae, one without ray-flowers.
dissected. Cleft, or divided into narrow segments.
distichous. Arranged in two vertical rows.
distinct. Separate; not united with parts of the same series.
divaricate. Spreading; widely divergent.
divided. Cleft to the midrib or to the base.
dorsal. Upon or pertaining to the back (outer surface) of an organ; abaxial.
drupe. A fleshy or pulpy fruit in which the inner portion is hard or stony; a stone-fruit, like a plum.
drupelet. A little drupe, such as the individual units of a raspberry.

ecotype. The product arising from a genotypic response of a population to a particular habitat.
ellipsoid. Solid but with an elliptical outline.
emarginate. With a shallow notch at the apex.
embryo. The immature plant within the seed.
endosperm. The reserve food stored around the embryo.
entire. With smooth margins, not toothed, cleft, or lobed.
epicalyx. A series of sepal-like appendages associated with the calyx, often alternating with the sepals.
epigynous. Applied to a flower whose ovary is inferior, the sepals and petals being borne at the summit of the ovary.
erose. With jagged margin.
evergreen. Applied to plants whose leaves remain green through the winter.
excurrent. Running through apex and beyond as a short point (mucro).
exfoliating. Peeling off in shreds, thin layers, etc.
exserted. Extending beyond the surrounding organs, as stamens from a corolla; not included.
exstipulate. Lacking stipules.
extrorse. Facing outward.

fascicle. A close bundle or cluster.
fenestrate. With holes; mainly applied to leaves.
fertile. Productive; capable of producing viable seeds or spores.
fibrillose. Furnished with fibre-like structures.
filament. The stalk of a stamen below the anther.
filiform. Long, slender, and terete; thread-shaped.
fimbriate. Fringed.
fistulose. Hollow.
flaccid. Lax and weak; flabby.
fleshy. Succulent, thick and juicy.
flexuous. Zig-zag; curved alternately in opposite directions.
floccose. With tufts of soft woolly hairs.
floret. A small flower such as one of a grass; one of a dense cluster.
follicle. A pod-like fruit which opens along one side.
frond. The expanded leaf-like portion of a fern.
fruit. The seed-bearing product of a plant; the ripened ovary often with other flower parts.
fugacious. Falling or withering early.
funiculus. The stalk of an ovule or seed.
funnelform. Funnel-shaped; expanding gradually upwards.
fusiform. Spindle-shaped; narrowed towards each end from an enlarged middle.

galea. A hood-like part of a perianth or corolla; the upper lip of some bilabiate corollas.

gamopetalous. With united petals.
gemma (pl. **gemmae**). An often bud-like asexual propagule, usually borne on the stem in the leaf-axils.
geniculate. Bent abruptly like a knee.
genotype. The genetic constitution of an organism.
gibbous. Enlarged or swollen on one side.
glabrate. Nearly glabrous, or becoming glabrous.
glabrous. Without hairs.
gland. A secreting structure; in plants usually producing nectar or volatile oil.
glandular. Bearing glands.
glaucescent. Somewhat glaucous.
glaucous. Covered with a 'bloom'; blue-white or blue-grey.
globose, globular. Spherical or nearly so.
glomerulate. Collected into heads.
glume. A chaff-like bract; specifically one of the two bracts at the base of the grass spikelet.
glutinous. Covered with a sticky exudate.
grain. The fruit of grasses; a caryopsis.
gynaecandrous. In *Carex,* having a spike with the upper flowers pistillate and the lower ones staminate.
gynobase. Enlargement of the receptacle on which the gynoecium rests.
gynoecium. The female reproductive structure *in toto*. It may consist of one or more free carpels or two or more fused carpels. See *carpel* and *pistil.*
gynophore. The elongate stalk (stipe) of an ovary.

habit. The general appearance of a plant, or its mode of growth.
habitat. A plant's natural place of occurrence.
hastate. Like an arrowhead but with the basal lobes diverging.
head. A dense cluster of flowers or fruits on a very short axis or receptacle.
herb. A plant with no persistent woody stem above the ground.
herbaceous. Pertaining to an herb; leaf-like in colour or texture.
hirsute. With rather coarse, spreading hairs.
hispid. With rigid or bristly hairs.
hyaline. Thin and translucent; nearly transparent.
hybrid. A cross between two species.
hypanthium. A saucer-shape, cup-shaped or tubular receptacle, often surrounding or enclosing the ovaries, and bearing the sepals and petals and often the stamens on its margin.
hypogynous. Having the corolla or other flower parts inserted below the ovary and free from it.

imbricate. Overlapping, like shingles on a roof.
immersed. Growing wholly under water.
imperfect (flowers). Lacking either stamens or ovaries.
incised. Cut sharply and irregularly, more or less deeply.
included. Not protruding from the surrounding structures; not exserted.
indehiscent. Not splitting open.
indurate. Hardened.
indusium. In ferns, a membrane covering the sorus.
inferior (ovary). With the other floral parts (sepals, petals, and stamens) attached to its top or on a hypanthium that projects above it. A flower with an inferior ovary is said to be epigynous.
inflexed. Bent inward; incurved.
inflorescence. Mode of arrangement of flowers in a cluster; flower-cluster.
internode. The portion of a stem between two nodes.
introduced. Not native.
introrse. Facing inward.
involucral. Pertaining to an involucre.
involucrate. With an involucre.
involucre. A set of bracts surrounding a flower-cluster or a single flower.
involute. Rolled inward.
irregular (flower). Having one or more organs of the same series unlike or unequal. See *zygomorphic.*

keel. The two lower united petals of the flower of the pea type; a sharp ridge, as on bracts or sepals.
keeled. Ridged, like the keel of a boat.
kranz type venation. Venation appearing under low magnification ($\times 10$) as a

conspicuous dark green reticulum when the leaf surface is lightly scraped with a sharp blade.

labiate. Provided with a lip-like organ.
lacerate. Irregularly cleft, as if torn.
laciniate. Slashed; cut into narrow lobes.
lamina. A plate or blade; the blade of a leaf or a petal.
lanceolate. Lance-shaped; several times longer than broad, widening above the base and tapering to the apex.
lateral. Belonging to or borne on the sides.
latex. The milky juice of certain plants.
leaflet. A single division of a compound leaf.
legume. A dry dehiscent fruit derived from a single carpel and usually splitting along two lines; also used as a name for plants of the Pea Family.
lemma. The lower of two bracts enclosing the flower in Gramineae. See *palea.*
lenticels. Small dots or spots on the bark of young twigs in many shrubs and trees.
lenticular. Shaped like a biconvex lens.
ligulate. Provided with or resembling a ligule.
ligule. The strap-shaped corolla of ray-flowers of numerous Compositae; a small thin projection from the top of the leaf-sheath, as in many grasses.
limb. The expanded portion of a petal, or sepal or gamopetalous corolla.
linear. Long and narrow, with parallel margins.
lip. Either of the divisions of a bilabiate corolla or calyx; the characteristic upper (apparently lower) petal in orchids.
locular, loculed. Chambered; having locules.
locule. One of the cavities or compartments of an ovary or anther.
loculicidal. Splitting along the middle of the back of each locule.
lobe. Any segment of an organ, especially if rounded.
lobed. Provided with a lobe or lobes.
loment. A jointed legume, usually constricted between the seeds.
lyrate. Pinnatifid, with the terminal lobe or segment considerably larger than the others; lyre-shaped.

maculate. Spotted.
malpighian (hairs). Attached by their middle.
marcescent. Withering, but the remains persisting.
membranaceous, membranous. Thin and somewhat translucent.
midrib, midvein. The central vein, or rib, of a leaf or other organ.
monadelphous (stamens). United by their filaments into one set.
moniliform. Like a string of beads.
monoecious. Having separate male and female flowers or cones on the same plant.
montane. Of the lower mountain slopes.
mucronate. Tipped with a short, sharp abrupt point.
muricate. Roughened with short, hard points.
muskeg. Wooded peatland. Predominant species being *Picea mariana* and to a lesser extent *Larix laricina. Sphagnum* species and ericaceous shrubs, e.g. *Andromeda polifolia,* generally in association.
mycorrhizal. Having an associated (symbiotic) fungus.

nectariferous. Having a nectary; producing nectar.
nectary. A gland where nectar is secreted.
nerve. A prominent line or vein in a leaf or other organ.
node. A point on a stem from which leaves or branches arise.
nodulose. With minute knobs or knots.
nut. A dry, 1-seeded, hard-walled, indehiscent fruit.
nutlet. A small nut.

obcordate. Inverted-heart-shaped, with the point basal.
oblanceolate. Lanceolate with the broadest part towards the apex.
oblique. Slanting, or with unequal sides.
oblong. Longer than broad and with nearly parallel sides.
obovate. Ovate, with the broadest part

towards the apex.
obtuse. Blunt or rounded at the end.
ochroleucous. Yellowish white.
ocreae. A sheathing stipule.
operculate. With a cap or lid.
opposite. Situated on opposite sides of an axis, as leaves, when there are two at one node; situated directly in front of another organ, as stamens opposite the petals.
orbicular. Approximately circular.
oval. Broadly elliptical.
ovary. The part of the carpel or pistil containing the ovules.
ovate. Egg-shaped in outline with the broader end basal.
ovoid. Shaped like a hen's egg.
ovule. An organ within the ovary which after further development becomes a seed.

palea. The upper of two bracts enclosing the flower in Gramineae. See *lemma*.
palmate (leaf). Radiately lobed or divided, with three or more veins arising from one point.
palmately. In palmate manner.
panicle. A compound or branched flower-cluster, the lower branches longest and flowering first; a compound raceme.
paniculate. Arranged in a panicle.
papilionaceous (corolla). Butterfly-shaped.
papillose. Covered with papillae, which are minute, blunt, rounded projections.
pappus. The hairs, bristles, scales, or other structures which are borne upon the achene of many species of Compositae.
parasite. An organism which derives food and nutrients from another living organism.
parietal. Borne on or pertaining to the wall of an ovary or capsule.
parted. Deeply cut or lobed, nearly to the base.
pectinate. Pinnatifid with narrow, closely set segments; comb-like.
pedate. Describes a palmately lobed leaf in which the two lateral lobes are divided or cleft.
pedicel. The stalk of a single flower in a cluster.
peduncle. A primary flower-stalk, supporting either a solitary flower or a cluster.
peltate. Describes a flat organ attached to its stalk in from the margin.
pendent. Hanging down.
penicellate. Shaped like a little brush.
perennial. A plant, or part of a plant, which persists for more than two seasons.
perfect (flower). With both stamens and ovaries present.
perfoliate. Applied to a sessile leaf with basal portion continuous around the stem.
perianth. The outer, sterile appendages of a flower, consisting of the sepals and petals (biseriate) or the sepals only (uniseriate).
pericarp. The wall of the fruit.
perigynium. The sac-like membrane enclosing the ovary in *Carex*.
perigynous. Borne around the ovary, on the edge of a hypanthium.
persistent. Describes organs that remain attached after maturity, as during the winter.
petal. A unit or part of the corolla.
petiolate, petioled. With a petiole.
petiole. Stalk of a leaf.
petiolulate. Having a petiolule.
petiolule. Stalk of a leaflet.
phyllopodic. Having the lowest leaves with well-developed blades, not reduced and scale-like. See *aphyllopodic*.
pilose. With soft spreading hairs.
pinna (pl. **pinnae**). One of the primary divisions of a pinnate or compoundly pinnate leaf or frond.
pinnate (leaf). Compound, with the leaflets arranged on each side of a common axis.
pinnatifid. Pinnately cleft.
pinnule. A secondary pinna; one of the pinnately arranged divisions of a pinna.
pistil. A gynoecium in which two or more carpels fuse to form the ovary; such fusion may also include the style and stigmas. The compound nature of the pistil and the number of carpels involved may be evident from the number of locules in the ovary, the number of placentae, the number of styles, or the divisions of the stigma, but this is

not invariably the case. In plants with pistils, pistil and gynoecium are synonymous. See *carpel* and *gynoecium.*

pistillate. With pistils but without stamens.

pitted. With small depressions or pits.

placenta. An ovule-bearing surface (within an ovary).

plicate. Folded like a fan; plaited.

plumose. Having fine hairs on each side, like the plume of a feather.

pod. General term for any dry fruit which opens to release its seeds.

pollen. The powdery contents of an anther; composed of minute grains which produce pollen-tubes in the ovary.

pollinium (pl. **pollinia**). A mass of waxy pollen, as in Orchidaceae and Asclepiadaceae.

polygamous. Having some perfect and some unisexual flowers.

polypetalous. Having separate petals.

polyploid. An organism with three or more full chromosome complements in each somatic nucleus; i.e. the chromosome number is three or more times the basic number. See *diploid.*

polysepalous. Having separate sepals.

pome. A fleshy fruit of the apple type.

posterior. On the side towards the axis. See *anterior.*

prickle. A spiny outgrowth from the surface of a plant.

procumbent. Lying on the ground or trailing, but not rooting at the nodes.

prostrate. Flat on the ground.

puberulent. Minutely pubescent.

pubescent. Covered with hairs, especially if short, soft, and down-like.

punctate. Marked with translucent or coloured dots or pits.

pustulate. With blister-like elevations.

pyxis. A capsule with circumscissile dehiscence, the upper portion coming off like a lid.

raceme. An inflorescence of pedicelled flowers upon a common, elongated axis (rachis).

racemose. In racemes; or resembling a raceme.

rachilla. Axis of a spikelet (as in Gramineae).

rachis. The central axis of a raceme or spike of flowers, or of a compound leaf.

radiate. Furnished with ray-flowers.

radical. Belonging to the root, or apparently coming from the root.

ray. A strap-like marginal flower of many Compositae; the branch of an umbel or similar inflorescence.

receptacle. The end of the flower-stalk, bearing the floral organs; or, in Compositae, bearing the flowers.

recurved. Curved downward or backward.

reflexed. Abruptly turned or bent backward.

regular. Having the members of each part alike in size and shape. See *actinomorphic.*

reniform. Kidney-shaped.

repand. With a slightly uneven, wavy margin.

reticulate. In the form of a network; net-veined.

retrorse. Directed back or downward.

retuse. With a shallow notch in a rounded or blunt apex.

revolute. Rolled downward from the margins or apex.

rhizome. A more or less elongated, usually underground, horizontal or ascending stem; a rootstock.

rhombic. Obliquely four-sided.

rib. A primary or prominent vein of a leaf.

ribbed. With prominent veins.

rootstock. See *rhizome.*

rosette. A cluster of leaves or other organs, often basal in position.

rostrate. Having a beak.

rotate (corolla). Wheel-shaped, flat and circular in outline.

rugose. Having a wrinkled surface.

runner. An elongate, slender, prostrate branch, rooting at the nodes or tip; a very slender stolon.

saccate. Sac-like or dilated.

sagittate. Shaped like an arrowhead, with the lobes directed downward.

salverform. With the slender corolla-tube abruptly expanded into a flat limb.

samara. A dry, indehiscent, winged fruit.

saprophyte. A plant which derives its

food from non-living organic matter.
scabrous. Rough to the touch, usually because of short, stiff hairs.
scale. A term applied to various kinds of small, flat structures, such as thin scarious bracts.
scape. A leafless flowering stalk arising from the ground or from a short basal axis.
scarious. Thin, dry, and translucent, not green.
schizocarp. A fruit that splits into its separate carpels at maturity.
scorpioid. Coiled up in the bud, unrolling in growth.
scurfy. Covered with small scales.
secund. Arranged or bent to lie along one side of an axis only.
segment. One of the parts of a leaf or other organ that is cleft or divided.
sepal. One of the parts or lobes of a calyx.
septate. Divided by partitions.
septicidal (capsule). Dehiscing along the partitions and between the locules.
septum. A partition within an organ.
serrate. Toothed, the teeth sharp and pointing forward.
serrulate. Finely serrate.
sessile. Without a stalk.
setose. Beset with bristles.
sheath. The part of a leaf or leaf-base which clasps or encloses the stem.
sheathing. Enclosing as by a sheath.
shrub. A woody plant, smaller than a tree, usually with several stems.
silicle. A short silique.
silique. The characteristic pod-like capsule of the Cruciferae.
simple (leaf). With the blade all in one piece; not divided into leaflets.
sinuate. With a strongly wavy margin.
sinus. The cleft or space between two lobes.
smooth. Without roughness; not pubescent.
sobole. A basal shoot or sucker arising from a stem or rhizome. Where present usually produced in some abundance.
sorus (pl. **sori**). A cluster of sporangia.
spadix. A spike with a thick or fleshy axis.
spathe. A large bract enclosing a flower-cluster.
spatulate. Spatula-shaped; gradually narrowed from a rounded summit.
spicate. Arranged in or resembling a spike.
spike. An elongated inflorescence with sessile or nearly sessile flowers.
spikelet. A small or secondary spike; the characteristic unit of the flower-cluster of Gramineae.
spine. A sharp rigid outgrowth from the stem; a thorn.
spinose. With spines or spine-like.
spinulose. Minutely spinose.
sporangium (pl. **sporangia**). A spore-sac, especially in ferns, in which the spores are produced.
sporophyll. A leaf bearing a sporangium or sporangia.
spur. A hollow, slender extension from the corolla or calyx.
squarrose. Spreading or recurved at the tip.
stamen. The pollen-bearing organ of a flower.
staminate (flowers). With stamens but lacking ovaries.
staminode. A sterile stamen.
standard. The large upper petal in a typical flower of the Leguminosae.
stellate. Star-shaped.
sterile. Without spores or without seed.
stigma. The receptive surface at the top of the style to which the pollen grains adhere.
stipe. The stalk-like support of an ovary; the leaf-stalk of a fern.
stipitate. Having a stipe.
stipulate. With stipules.
stipule. An appendage at or near the base of a leaf petiole.
stolon. A horizontal branch from the base of a plant and taking root and producing new plants at the apex or at the nodes.
stoloniferous. Producing stolons.
striate. Marked with fine longitudinal lines (grooves or ridges).
strict. Close or narrow and upright.
strigillose. Diminutive of strigose.
strigose. With appressed straight stiff hairs.

style. The structure connecting the ovary and stigma.
stylopodium. A disc-like expansion at the base of a style, as in Umbelliferae.
subalpine. The mainly forested zone between the montane and alpine zones.
subulate. Awl-shaped.
succulent. Soft and juicy.
sucker. A shoot arising from the underground part of a plant.
suffruticose. Somewhat shrubby.
superior (ovary). Borne above the insertion of other flower parts and free from these parts.
suture. A line of splitting or opening.
sympetalous. Having the petals more or less united; gamopetalous.

taproot. A prominent vertical root from the main axis of the plant.
taxon (pl. **taxa**). A taxonomic group (groups) of any rank.
tendril. A slender clasping or twining portion of a leaf or stem.
terete. Circular in cross-section.
ternate. Divided into three segments, or arranged in threes.
throat. Refers to the junction of the tube and limb of a gamopetalous corolla.
thyrsoid. Forming a contracted and compact compound flower-cluster.
tomentose. Densely pubescent with matted hairs.
tomentum. A covering of matted, woolly hairs.
torulose. Cylindric, with swollen portions at intervals.
trifoliolate. With three leaflets.
truncate. With the base or apex transversely straight, or nearly so, as if cut off.
tuber. A thick, short, underground branch or part of a branch, having several buds or 'eyes'.
tubercle. A small, usually rounded, projection from the surface of a leaf, seed, etc.; the persistent base of the style in some Cyperaceae.
tuberous. Resembling a tuber.
tubular (corolla). Prolonged into a tube, without much expansion at the summit.
turion. A small bulb-like offset, as found in some species of *Epilobium*.
twining. Winding spirally about a support.

umbel. A flower-cluster with all the pedicels arising from the same level. In a compound umbel, the branches are again umbellately branched at their tips.
umbellate. In or like an umbel.
umbonate. Bearing a stout protuberance.
umbraculate. Umbrella-shaped.
unisexual (flowers). With stamens or ovaries, but not both; imperfect.
urceolate. Pitcher-like; urn-shaped.
utricle. A small, thin-walled, 1-seeded, more or less inflated fruit.

valvate. Meeting along the margins only and not overlapping. Compare *imbricate.*
valve. One of the portions of the wall of a pod or capsule into which it separates at dehiscence.
vein. Thread of fibrovascular tissue in a leaf, petal, or other flat organ.
veinlet. A small vein.
venation. Arrangement of veins.
ventral. Being on the side next the axis.
vernation. The arrangement of leaves in the bud.
verrucose. Warty; covered with low rounded protuberances.
versatile (anther). Attached at or near the middle to the filament.
verticil. A whorl.
verticillate. Arranged in whorls.
villose, villous. Bearing long, soft (not matted) hairs. Compare *tomentose.*
viscid. Sticky; glutinous.

whorl. A group of three or more similar organs, as leaves, arising from a node; a verticil.
wing. Any membranous expansion; one of the two lateral petals of the flower of the leguminosae.
woolly. Covered with long and matted or tangled hairs.

xerophyte. Plant of a very dry habitat.

zygomorphic. Applied to flowers capable of being bisected into similar halves along only one plane; bilaterally symmetrical.

Bibliography

Barkley, T.M. (Editor). *Atlas of the Flora of the Great Plains.* Ames: The Iowa State University Press, 1977

Boivin, B. *Flora of the Prairie Provinces.* Provancheria, vols. 2–5. Quebec: Université de Laval, 1967–79

Calder, J.A., and Taylor, R.L. *Flora of the Queen Charlotte Islands, Part 1.* Canada Department of Agriculture, Research Branch Monograph No. 4, 1968

Clapham, A.R., et al. *Flora of the British Isles,* 2nd ed. Cambridge University Press, 1962

Correll, D.S., and Johnstone, M.C. *Manual of the Vascular Plants of Texas.* Renner: Texas Research Foundation, 1970

Cronquist, A., et al. *Intermountain Flora,* vol. 6. New York: Columbia University Press, 1977

Federov, A.A. (Editor). *Chromosome Numbers of Flowering Plants.* Leningrad: Academy of Sciences of the U.S.S.R., Komarov Botanical Institute, 1969

Goldblatt, P. (Editor). *Index to Plant Chromosome Numbers 1975–1978.* Monographs in Systematic Botany, vol. 5. St. Louis: Missouri Botanical Garden, 1981

Hitchcock, C.L., et al. *Vascular Plants of the Pacific Northwest,* vols. 1–5. Seattle: University of Washington Press, 1950–69

Hultén, E. *The Amphi-Atlantic Plants.* Stockholm: Almquist & Wiksell, 1958

– *The Circumpolar Plants,* vol. 1. Stockholm: Almquist & Wiksell, 1962

– *Flora of Alaska and Neighboring Territories.* Stanford University Press, 1968

– *The Circumpolar Plants,* vol. 2. Stockholm: Almquist & Wiksell, 1971

Löve, A., and Löve, D. *Cytotaxonomical Atlas of the Arctic Flora.* Va Duz: J. Cramer, 1975

Moore, R.J. (Editor). *Index to Plant Chromosome Numbers 1967–1971.* Regnum Vegetabile, vol. 90. Utrecht: Oosthoek's Uitgeversmaatshappij, 1973

Porsild, A.E., and Cody W.J. *Vascular Plants of Continental Northwest Territories, Canada.* Ottawa: National Museums of Canada, 1980

Scoggan, H.J. *The Flora of Canada,* vols. 1–4. Ottawa: National Museums of Canada, 1978–9

Taylor, R.L., and MacBryde, B. *Vascular Plants of British Columbia.* Vancouver: University of British Columbia Press, 1977

Tutin, T.G., et al. (Editors). *Flora Europaea,* vols. 1–5. Cambridge University Press, 1964–80

Welsh, S. *Flora of Alaska and Adjacent Parts of Canada.* Provo: Brigham Young University Press, 1974

DISTRIBUTION MAPS

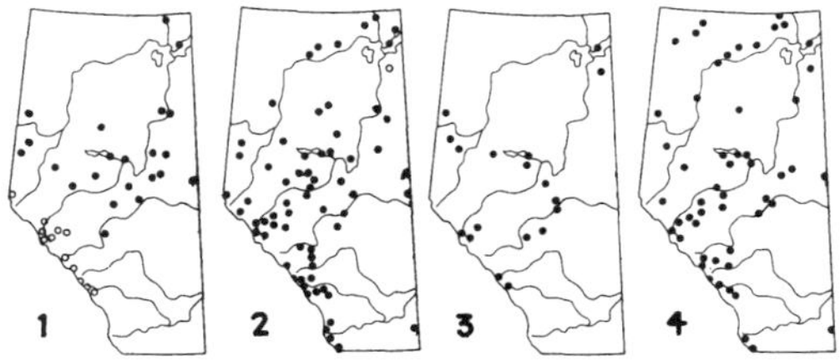

1 ○ *Lycopodium alpinum*
● *Lycopodium obscurum*
2 ● *Lycopodium annotinum*
○ *Lycopodium inundatum*
3 *Lycopodium clavatum*
4 *Lycopodium complanatum*

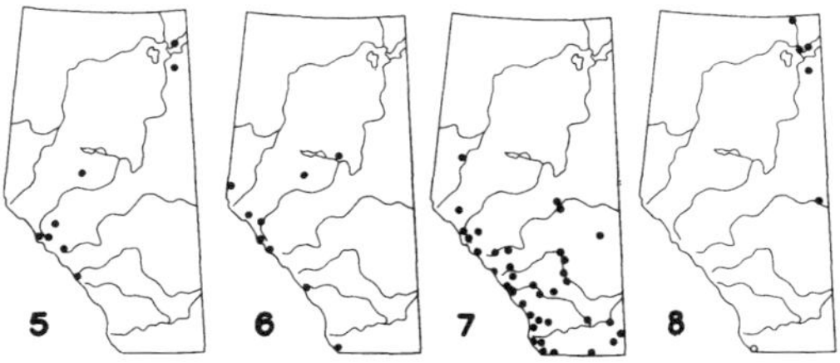

5 *Lycopodium selago*
6 *Lycopodium sitchense*
7 *Selaginella densa*
8 ● *Selaginella rupestris*
○ *Selaginella wallacei*

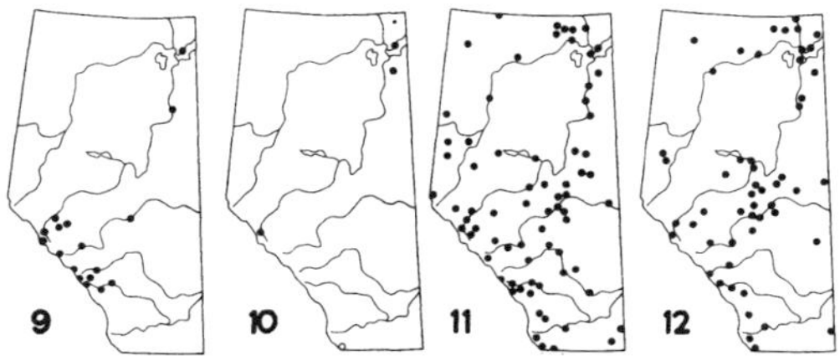

9 *Selaginella selaginoides*
10 ○ *Isoetes bolanderi*
● *Isoetes echinospora*
11 *Equisetum arvense*
12 *Equisetum fluviatile*

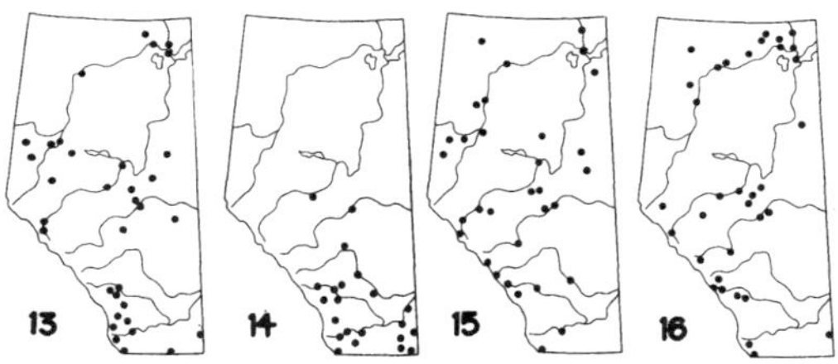

13 *Equisetum hyemale*
14 *Equisetum laevigatum*
15 *Equisetum palustre*
16 *Equisetum pratense*

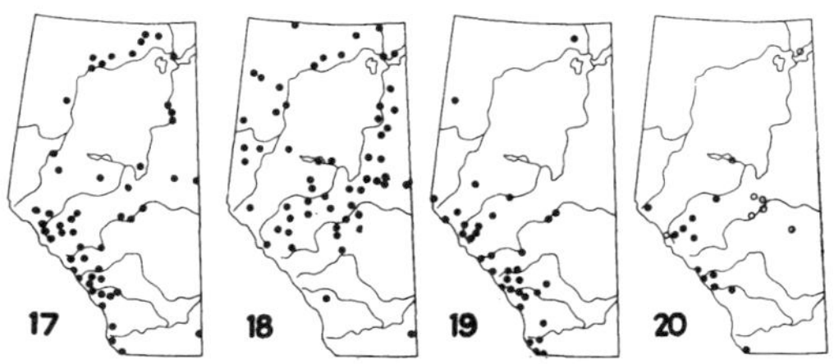

17 *Equisetum scirpoides*
18 *Equisetum sylvaticum*
19 *Equisetum variegatum*
20 ● *Botrychium boreale*
○ *Botrychium multifidum*

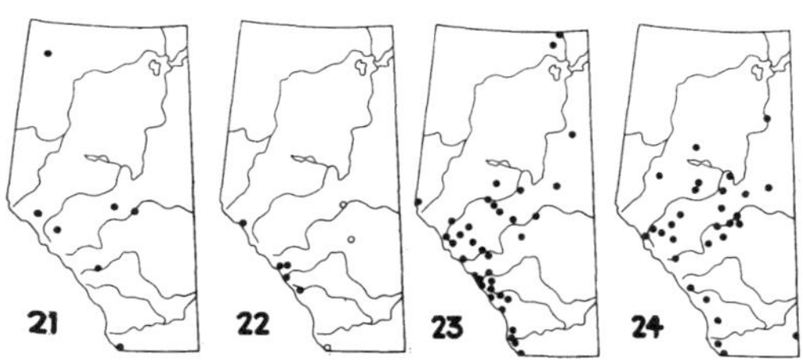

21 *Botrychium dusenii*
22 ● *Botrychium lanceolatum*
○ *Botrychium simplex*
23 *Botrychium lunaria*
24 *Botrychium virginianum*

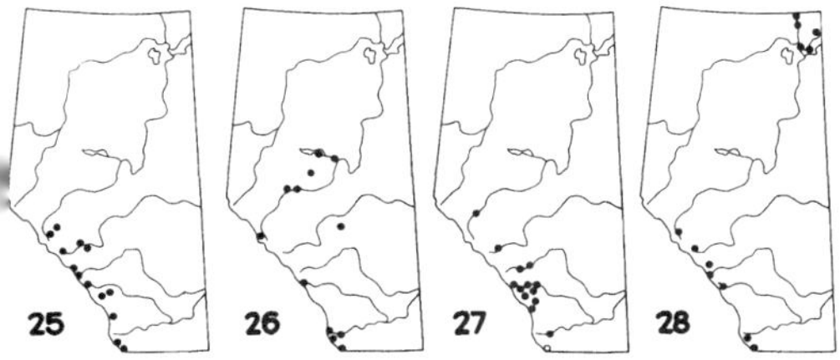

25 *Asplenium viride*
26 *Athyrium filix-femina*
27 ● *Cheilanthes feei*
○ *Cheilanthes gracillima*
28 *Cryptogramma acrostichoides*

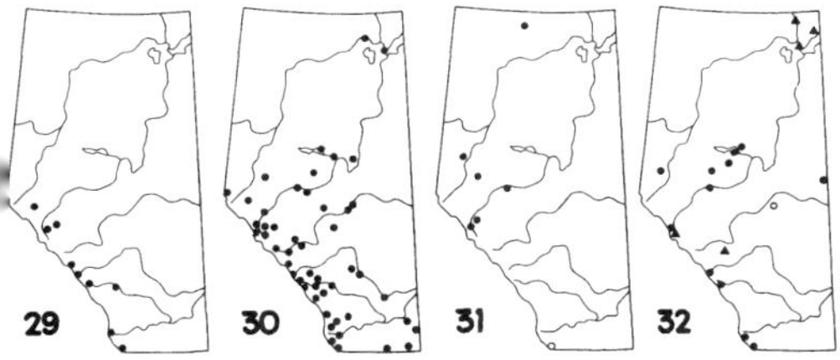

29 *Cryptogramma stelleri*
30 *Cystopteris fragilis*
31 ● *Cystopteris montana*
○ *Adiantum pedatum*
32 ● *Dryopteris assimilis*
○ *Dryopteris cristata*
▲ *Dryopteris fragrans*

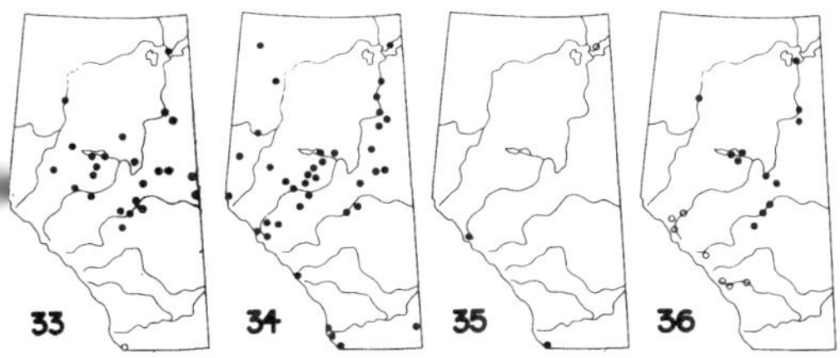

33 ● *Dryopteris carthusiana*
○ *Dryopteris filix-mas*
34 *Gymnocarpium dryopteris*
35 ○ *Gymnocarpium jessoense*
● *Athyrium distentifolium*
36 ● *Matteuccia struthiopteris*
○ *Pellaea glabella*

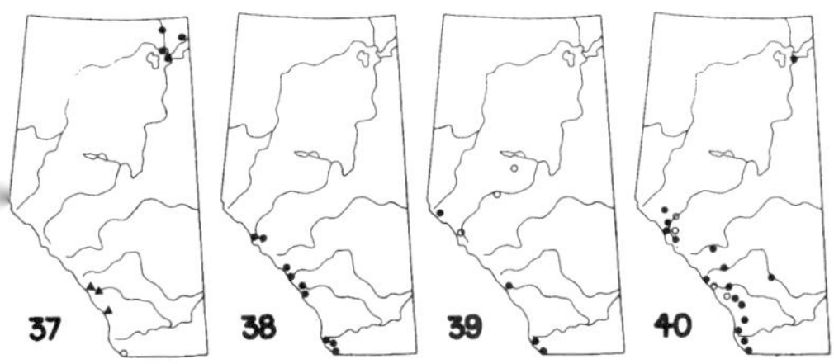

37 ▲ *Pellaea atropurpurea*
○ *Polypodium hesperium*
● *Polypodium virginianum*
38 *Polystichum lonchitis*
39 ● *Pteridium aquilinum*
○ *Phegopteris connectilis*
40 ○ *Woodsia glabella*
● *Woodsia oregana*

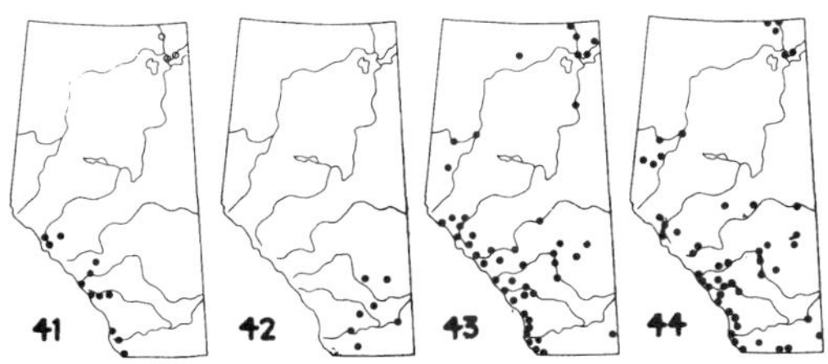

41 ○ *Woodsia ilvensis*
● *Woodsia scopulina*
42 *Marsilea vestita*
43 *Juniperus communis*
44 *Juniperus horizontalis*

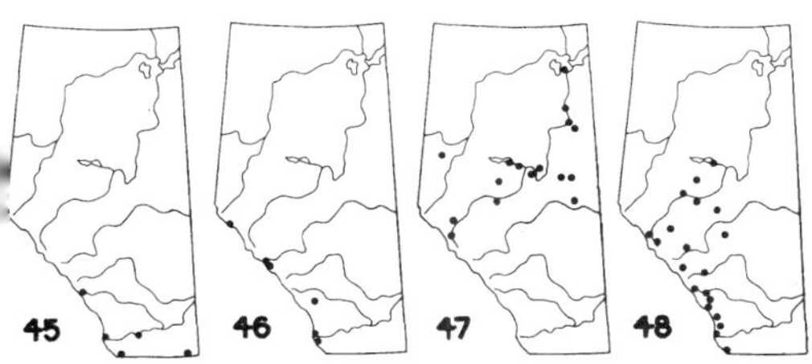

45 *Juniperus scopulorum*
46 *Thuja plicata*
47 *Abies balsamea*
48 *Abies lasiocarpa*

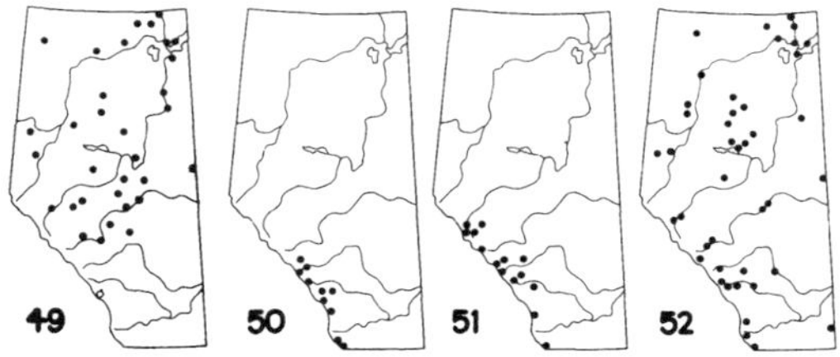

49 ● *Larix laricina*
○ *Larix occidentalis*
50 *Larix lyallii*
51 *Picea engelmannii*
52 *Picea glauca*

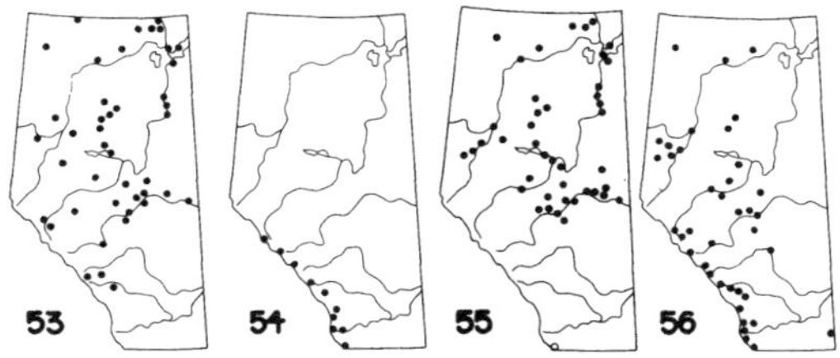

53 *Picea mariana*
54 *Pinus albicaulis*
55 ● *Pinus banksiana*
○ *Pinus monticola*
56 *Pinus contorta*

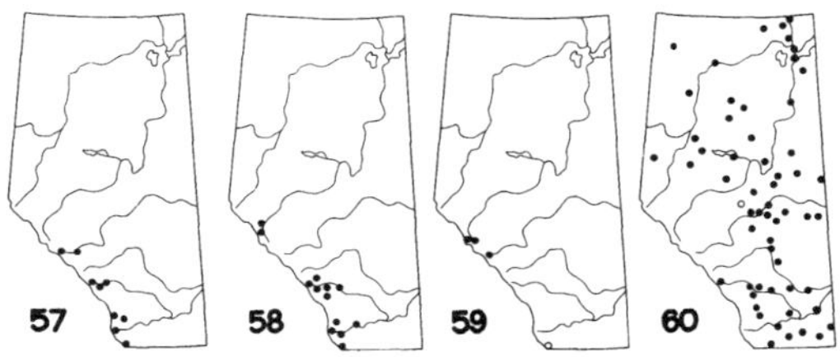

57 *Pinus flexilis*
58 *Pseudotsuga menziesii*
59 ○ *Taxus brevifolia*
● *Tsuga heterophylla*
60 ● *Typha latifolia*
○ *Sparganium glomeratum*

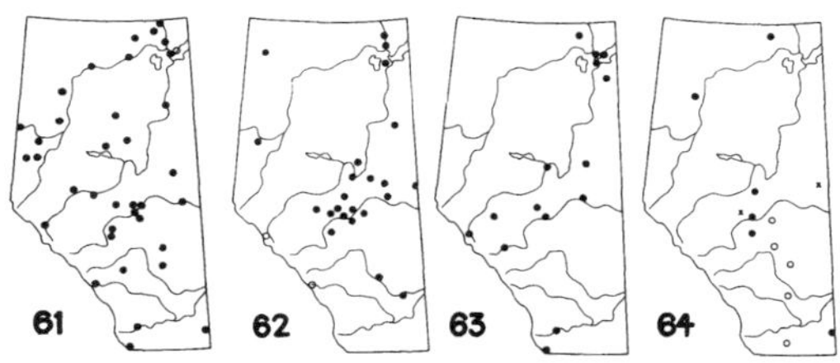

61 ● *Sparganium angustifolium*
○ *Sparganium fluctuans*
62 ● *Sparganium eurycarpum*
○ *Sparganium hyperboreum*
63 *Sparganium minimum*
64 ● *Najas flexilis*
○ *Ruppia maritima* × fossil

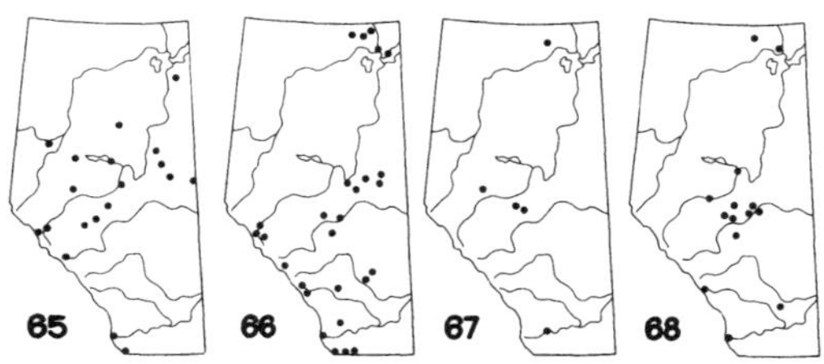

65 *Potamogeton alpinus*
66 *Potamogeton filiformis*
67 *Potamogeton foliosus*
68 *Potamogeton friesii*

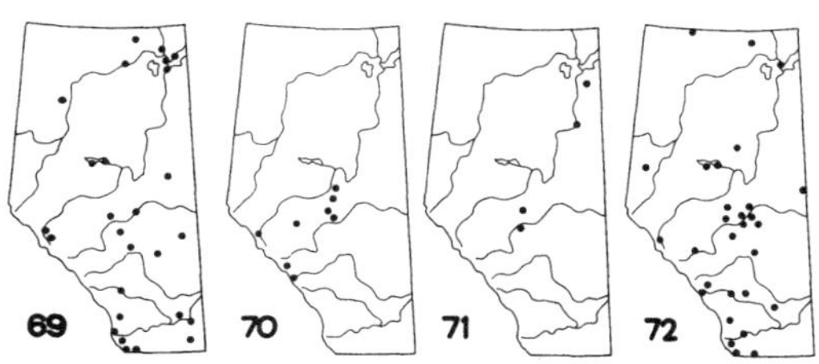

69 *Potamogeton gramineus*
70 *Potamogeton natans*
71 *Potamogeton obtusifolius*
72 *Potamogeton pectinatus*

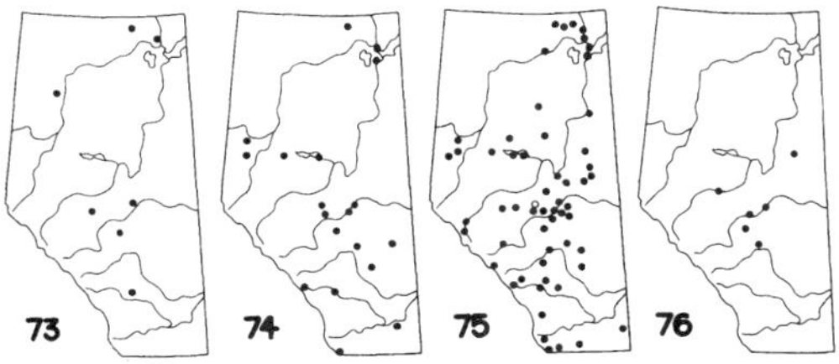

73 *Potamogeton praelongus*
74 *Potamogeton pusillus*
75 ● *Potamogeton richardsonii*
○ *Potamogeton robbinsii*
76 *Potamogeton strictifolius*

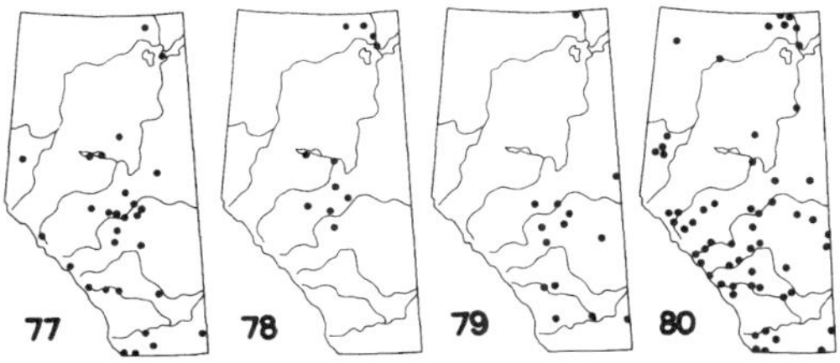

77 *Potamogeton vaginatus*
78 *Potamogeton zosteriformis*
79 *Zannichellia palustris*
80 *Triglochin maritima*

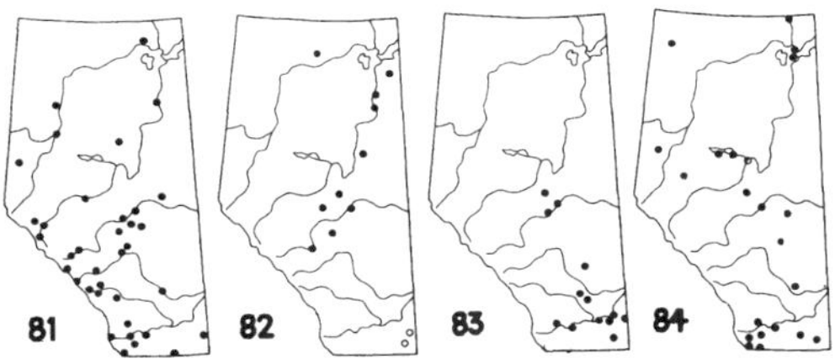

81 *Triglochin palustris*
82 ○ *Lilaea scilloides*
● *Scheuchzeria palustris*
83 *Alisma gramineum*
84 ● *Alisma plantago-aquatica*
○ *Sagittaria latifolia*

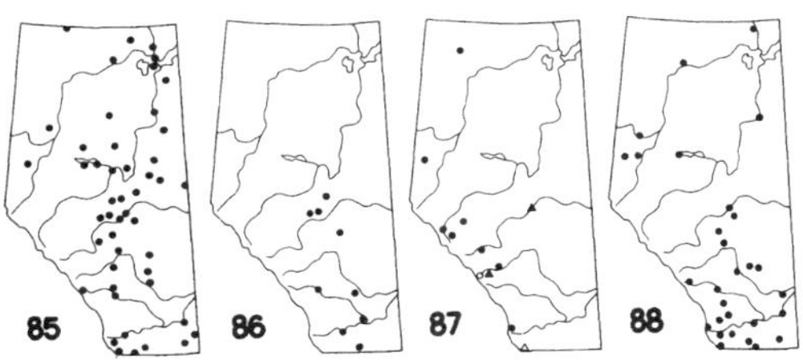

85 *Sagittaria cuneata*
86 *Elodea longivaginata*
87 ● × *Agroelymus hirtiflorus*
○ × *Agroelymus mossii*
▲ × *Agroelymus turneri*
△ *Agropyron* × *brevifolium*
88 × *Agrohordeum macounii*

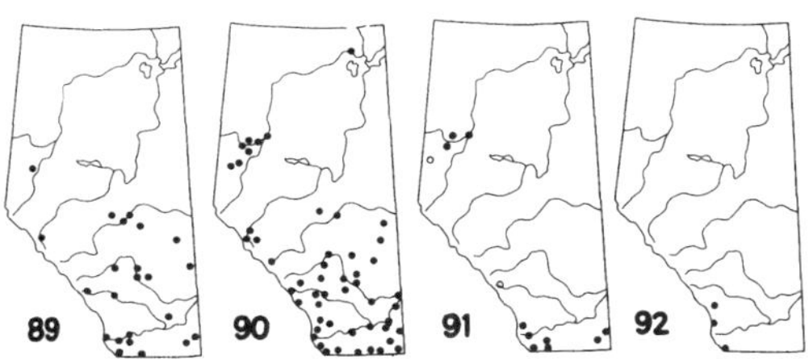

89 *Agropyron albicans*
90 *Agropyron dasystachyum*
91 ● *Agropyron* × *pseudorepens*
○ × *Agroelymus bowdenii*
92 *Agropyron scribneri*

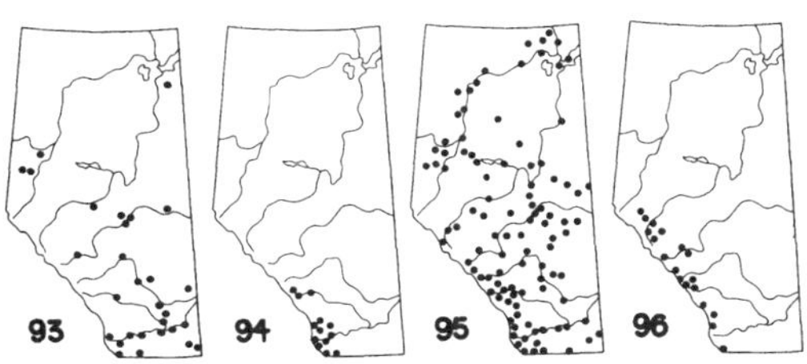

93 *Agropyron smithii*
94 *Agropyron spicatum*
95 *Agropyron trachycaulum*
96 *Agropyron violaceum*

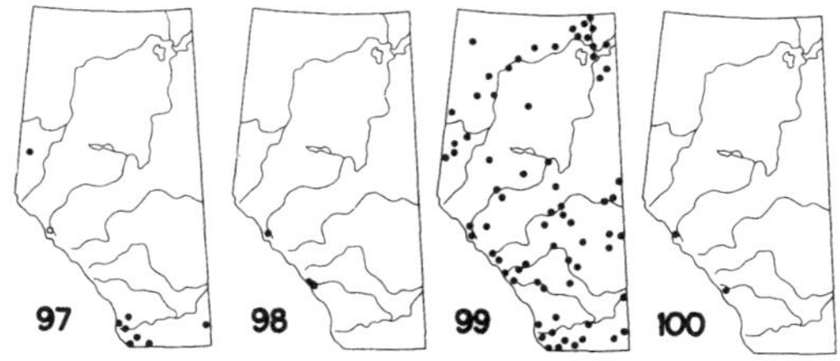

97 ● *Agrostis exarata*
○ *Agrostis mertensii*
98 *Agrostis humilis*
99 *Agrostis scabra*
100 *Agrostis thurberiana*

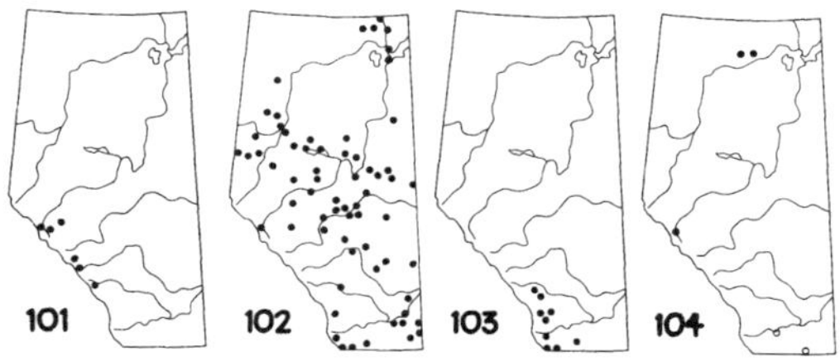

101 *Agrostis variabilis*
102 *Alopecurus aequalis*
103 *Alopecurus occidentalis*
104 ● *Arctagrostis arundinacea*
○ *Aristida longiseta*

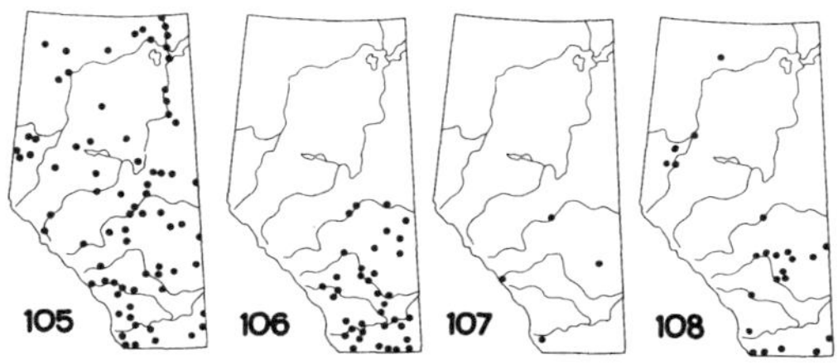

105 *Beckmannia syzigachne*
106 *Bouteloua gracilis*
107 *Bromus altissimus*
108 *Bromus anomalus*

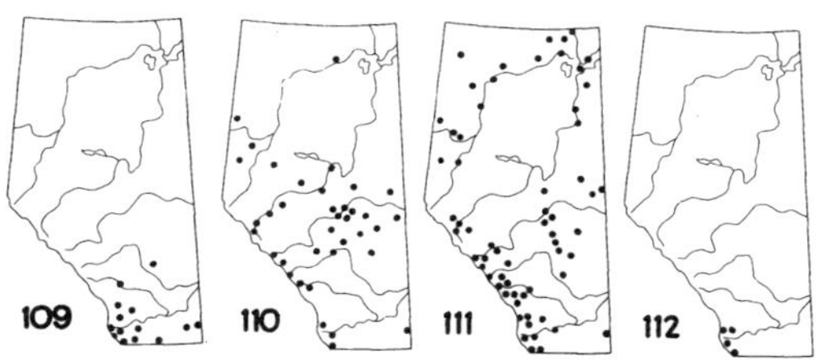

109 *Bromus carinatus*
110 *Bromus ciliatus*
111 *Bromus inermis*
112 *Bromus vulgaris*

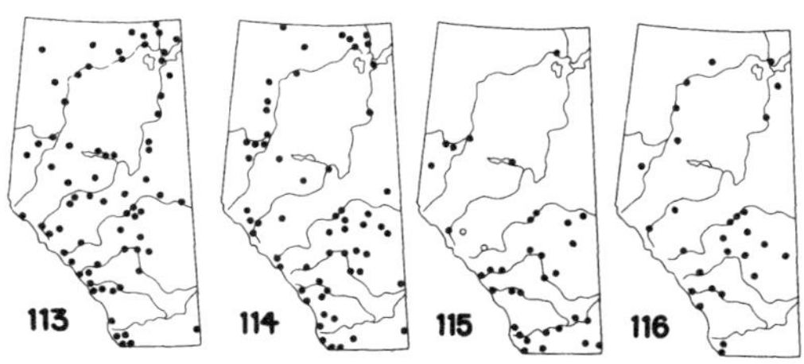

113 *Calamagrostis canadensis*
114 *Calamagrostis inexpansa*
115 ○ *Calamagrostis lapponica*
● *Calamagrostis montanensis*
116 *Calamagrostis stricta*

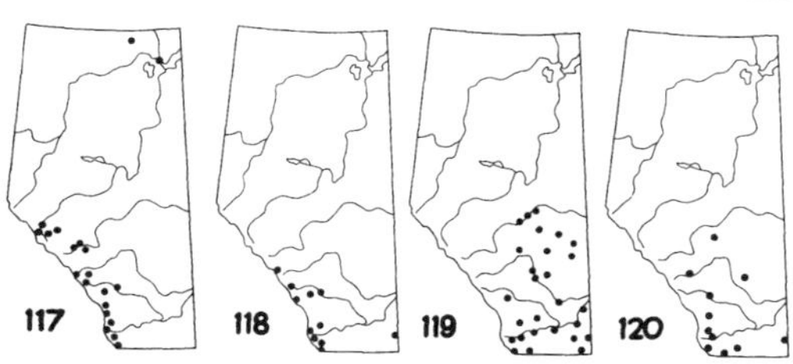

117 *Calamagrostis purpurascens*
118 *Calamagrostis rubescens*
119 *Calamovilfa longifolia*
120 *Catabrosa aquatica*

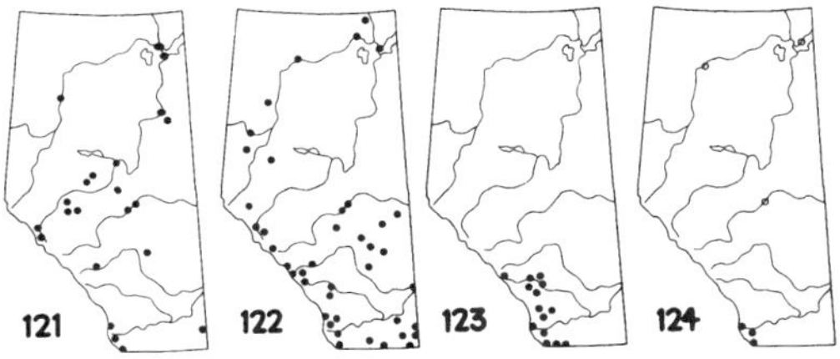

121 *Cinna latifolia*
122 *Danthonia californica*
123 *Danthonia parryi*
124 ○ *Danthonia spicata*
● *Danthonia unispicata*

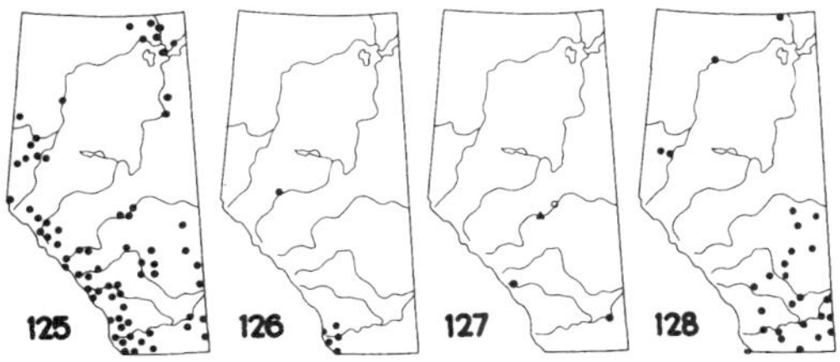

125 *Deschampsia cespitosa*
126 *Deschampsia elongata*
127 ● *Dicanthelium acuminatum*
▲ *Dicanthelium leibergii*
○ *Dicanthelium oligosanthes*
128 *Distichlis stricta*

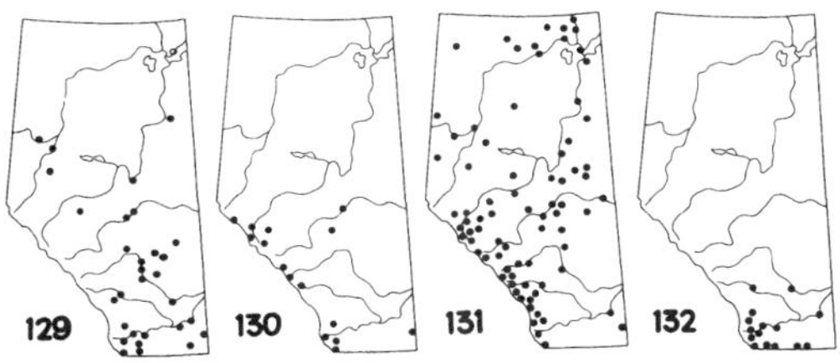

129 ● *Elymus canadensis*
○ *Elymus mollis*
130 *Elymus glaucus*
131 *Elymus innovatus*
132 *Elymus piperi*

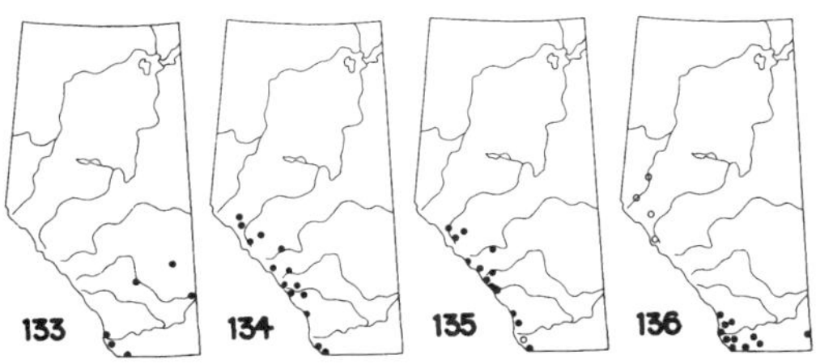

133 *Elymus virginicus*
134 *Festuca baffinensis*
135 ● *Festuca brachyphylla*
○ *Festuca subulata*
136 ● *Festuca idahoensis*
○ *Festuca altaica*

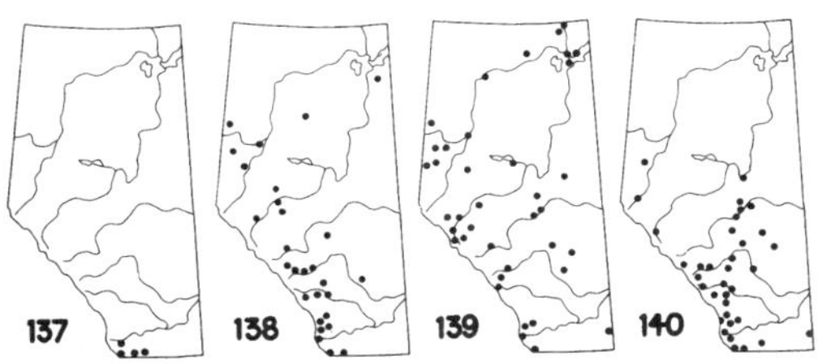

137 *Festuca occidentalis*
138 *Festuca rubra*
139 *Festuca saximontana*
140 *Festuca scabrella*

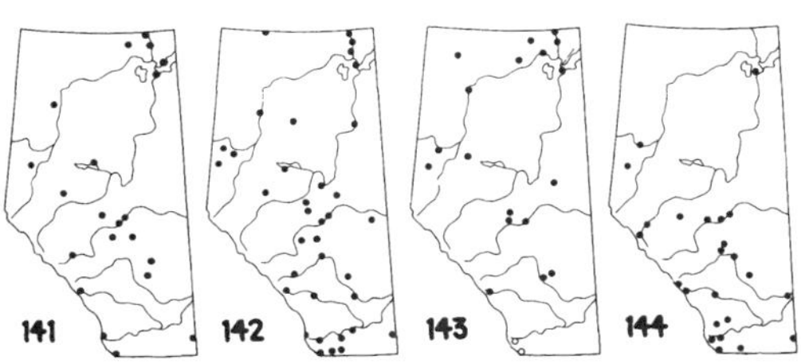

141 *Glyceria borealis*
142 *Glyceria grandis*
143 ○ *Glyceria elata*
● *Glyceria pulchella*
144 *Glyceria striata*

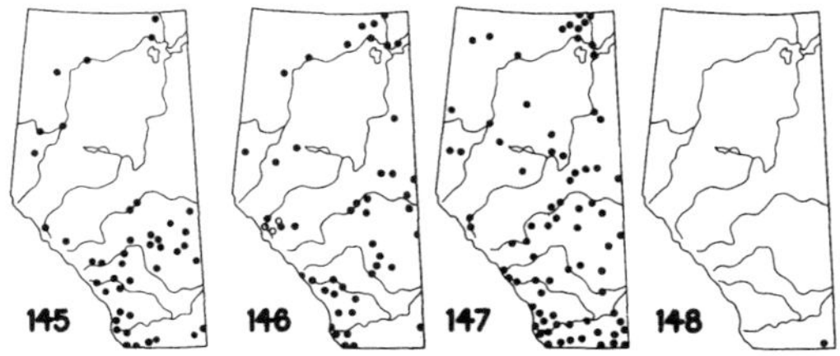

145 *Helictotrichon hookeri*
146 ○ *Hierochloe alpina*
● *Hierochloe odorata*
147 *Hordeum jubatum*
148 *Hordeum pusillum*

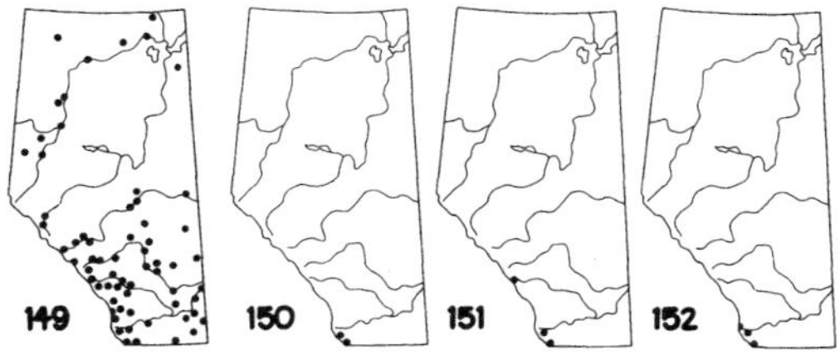

149 *Koeleria macrantha*
150 *Melica smithii*
151 *Melica spectabilis*
152 *Melica subulata*

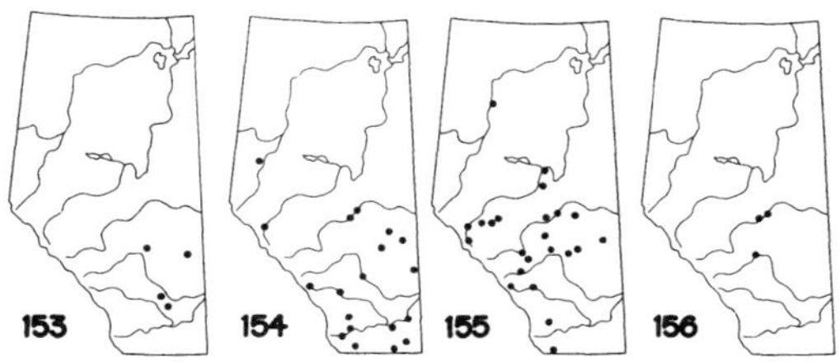

153 *Muhlenbergia asperifolia*
154 *Muhlenbergia cuspidata*
155 *Muhlenbergia glomerata*
156 *Muhlenbergia racemosa*

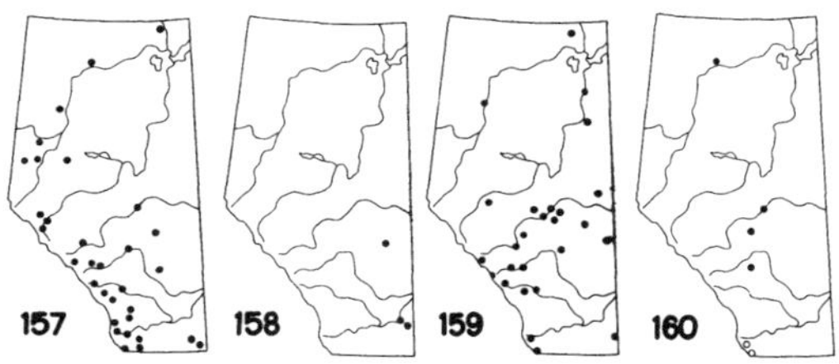

157 *Muhlenbergia richardsonis*
158 *Munroa squarrosa*
159 *Oryzopsis asperifolia*
160 ● *Oryzopsis canadensis*
○ *Oryzopsis exigua*

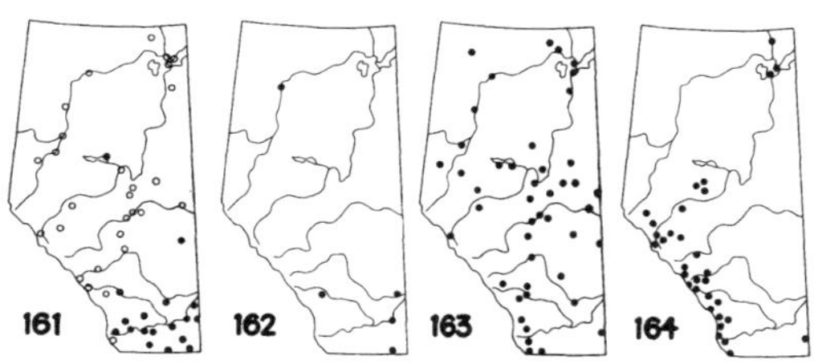

161 ● *Oryzopsis hymenoides*
○ *Oryzopsis pungens*
162 *Oryzopsis micrantha*
163 *Phalaris arundinacea*
164 *Phleum commutatum*

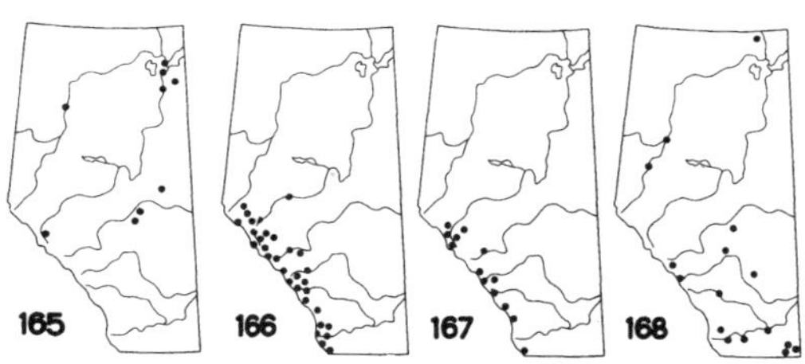

165 *Phragmites australis*
166 *Poa alpina*
167 *Poa arctica*
168 *Poa arida*

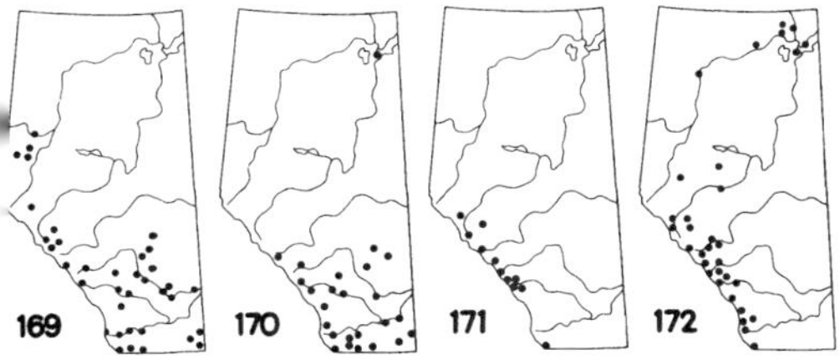

169 *Poa canbyi*
170 *Poa cusickii*
171 *Poa epilis*
172 *Poa glauca*

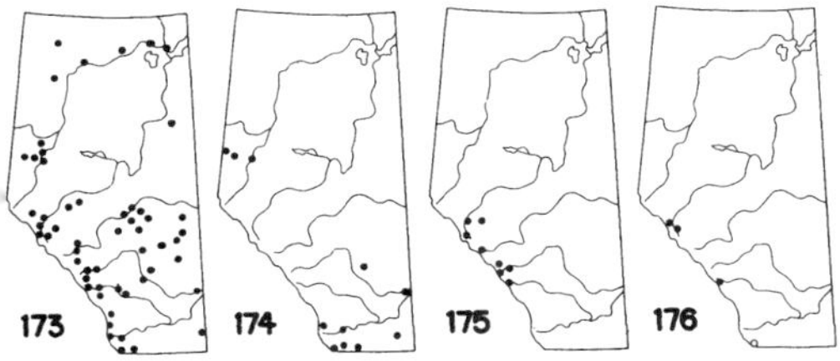

173 *Poa interior*
174 *Poa juncifolia*
175 *Poa leptocoma*
176 ● *Poa lettermanii*
○ *Poa gracillima*

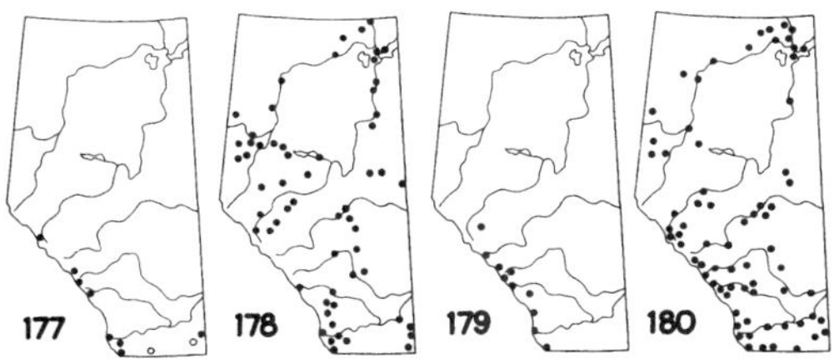

177 ● *Poa nervosa*
○ *Poa nevadensis*
178 *Poa palustris*
179 *Poa pattersonii*
180 *Poa pratensis*

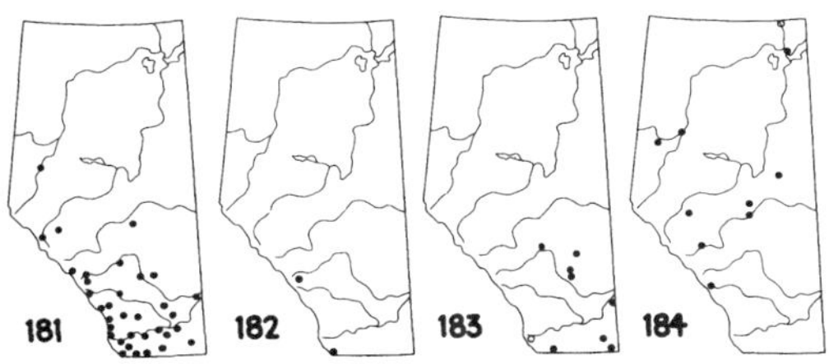

181 *Poa sandbergii*
182 *Poa stenantha*
183 ● *Puccinellia cusickii*
○ *Puccinellia pauciflora*
184 ● *Puccinellia distans*
○ *Puccinellia hauptiana*

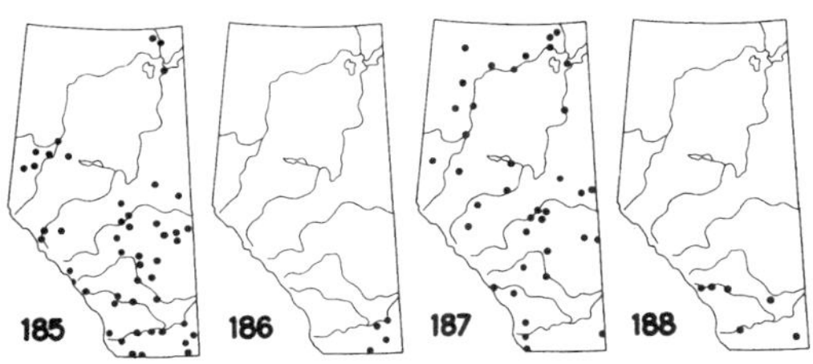

185 *Puccinellia nuttalliana*
186 *Schedonnardus paniculatus*
187 *Schizachne purpurascens*
188 *Schizachyrium scoparium*

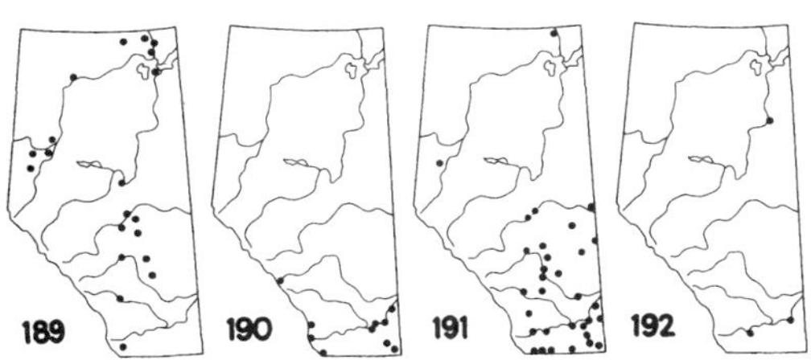

189 *Scolochloa festucacea*
190 *Sitanion hystrix*
191 *Spartina gracilis*
192 *Spartina pectinata*

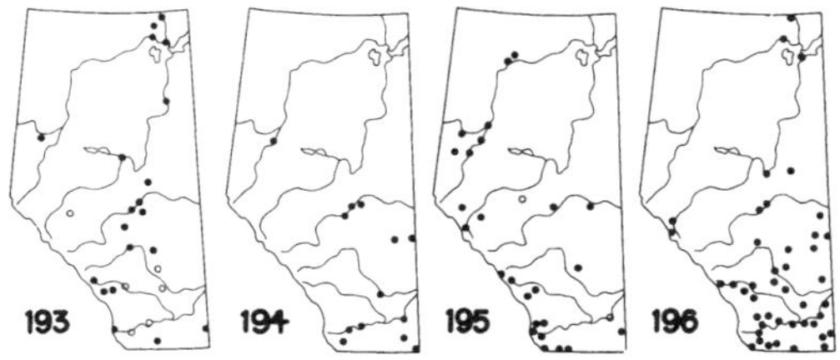

193 ● *Sphenopholis intermedia*
○ *Sphenopholis obtusata*
194 *Sporobolus cryptandrus*
195 ○ *Sporobolus neglectus*
● *Stipa columbiana*
196 *Stipa comata*

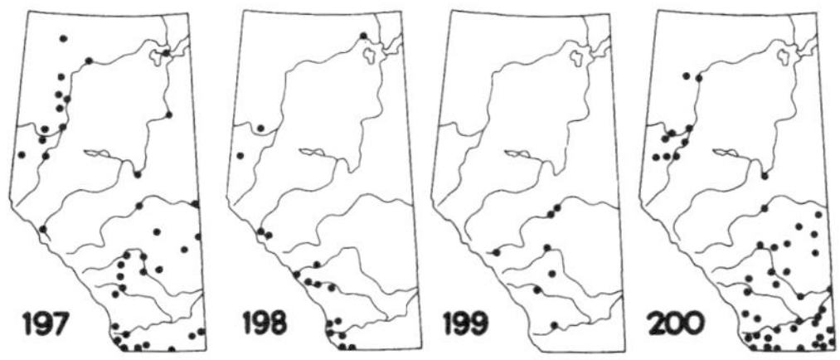

197 *Stipa curtiseta*
198 *Stipa richardsonii*
199 *Stipa spartea*
200 *Stipa viridula*

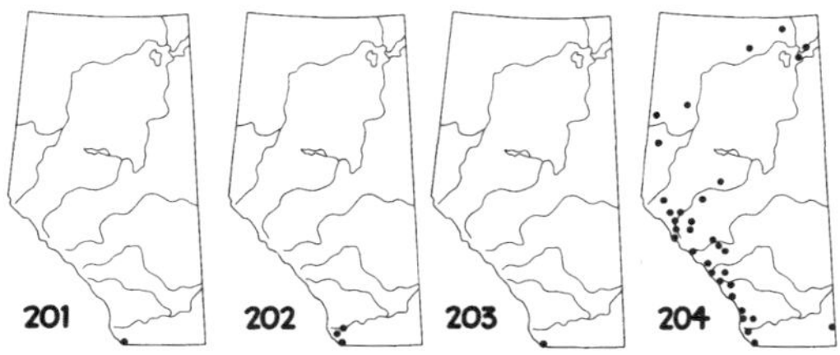

201 *Trisetum canescens*
202 *Trisetum cernuum*
203 *Trisetum montanum*
204 *Trisetum spicatum*

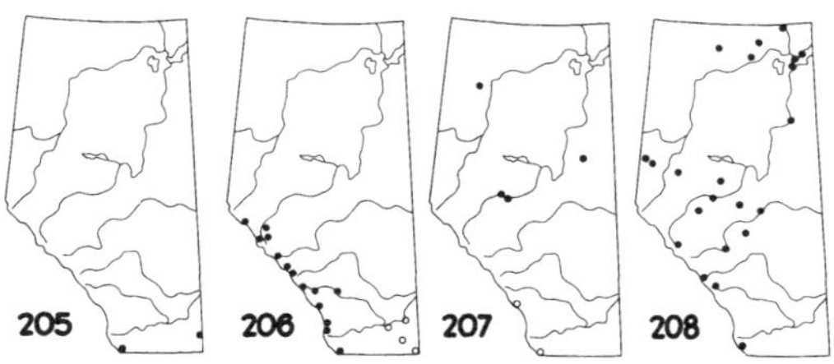

205 *Trisetum wolfii*
206 ● *Vahlodea atropurpurea*
○ *Vulpia octoflora*
207 ● *Carex adusta*
○ *Carex aperta*
208 *Carex aenea*

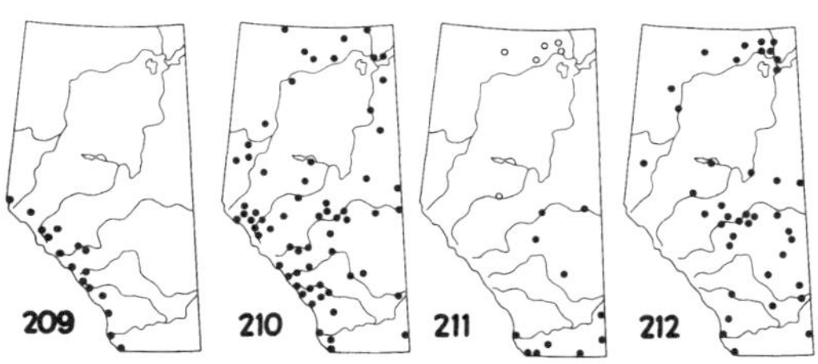

209 *Carex albo-nigra*
210 *Carex aquatilis*
211 ○ *Carex arcta*
● *Carex athrostachya*
212 *Carex atherodes*

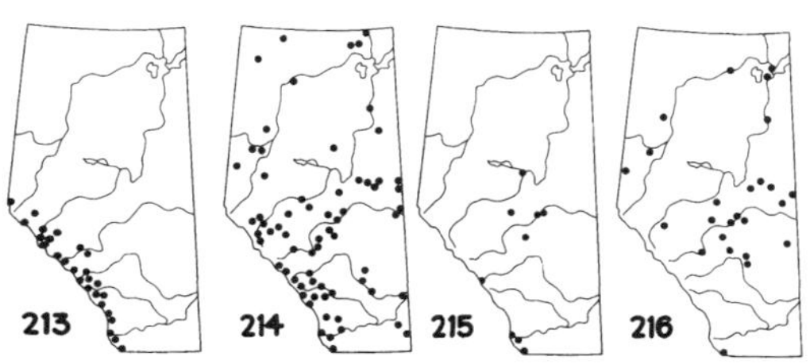

213 *Carex atrosquama*
214 *Carex aurea*
215 *Carex backii*
216 *Carex bebbii*

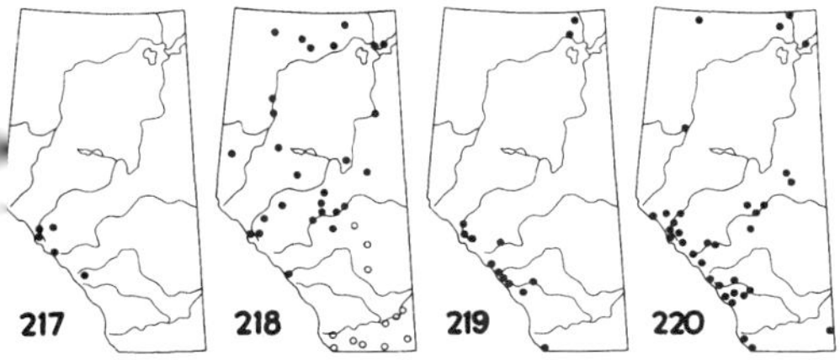

217 *Carex bipartita*
218 ○ *Carex brevior*
● *Carex brunnescens*
219 *Carex buxbaumii*
220 *Carex capillaris*

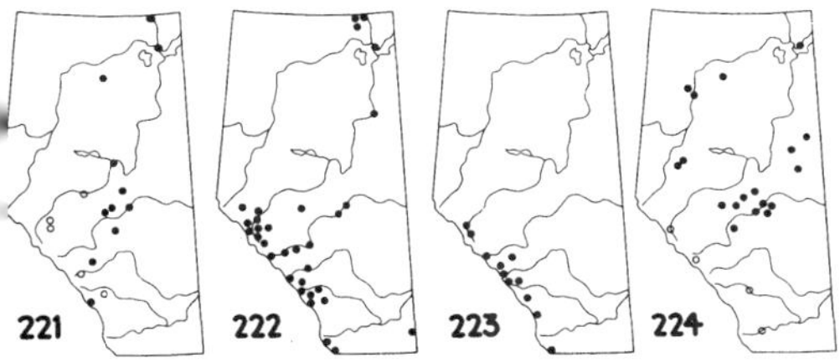

221 ○ *Carex capitata*
● *Carex chordorrhiza*
222 *Carex concinna*
223 *Carex concinnoides*
224 ○ *Carex crawei*
● *Carex crawfordii*

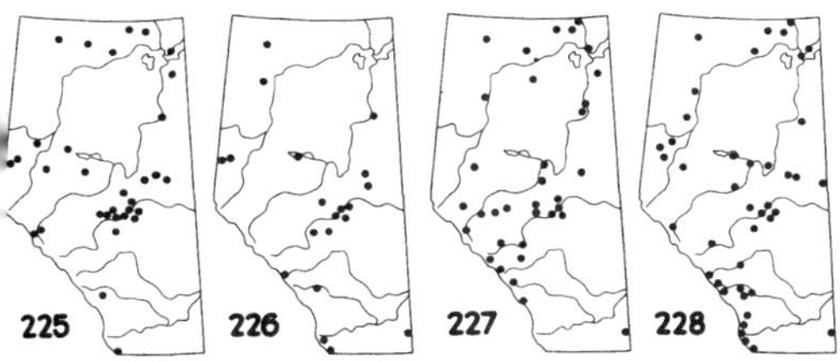

225 *Carex curta*
226 *Carex deweyana*
227 *Carex diandra*
228 *Carex disperma*

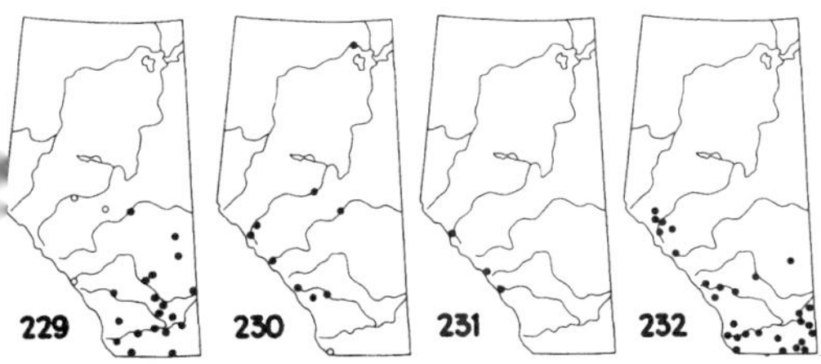

229 ○ *Carex deflexa*
● *Carex douglasii*
230 ● *Carex eburnea*
○ *Carex epapillosa*
231 *Carex enanderi*
232 *Carex filifolia*

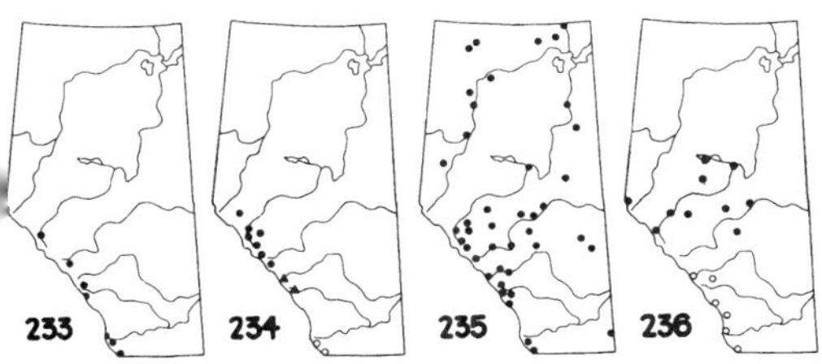

233 *Carex flava*
234 ● *Carex franklinii*
○ *Carex geyeri*
▲ *Carex glacialis*
235 *Carex gynocrates*
236 ○ *Carex haydeniana*
● *Carex heleonastes*

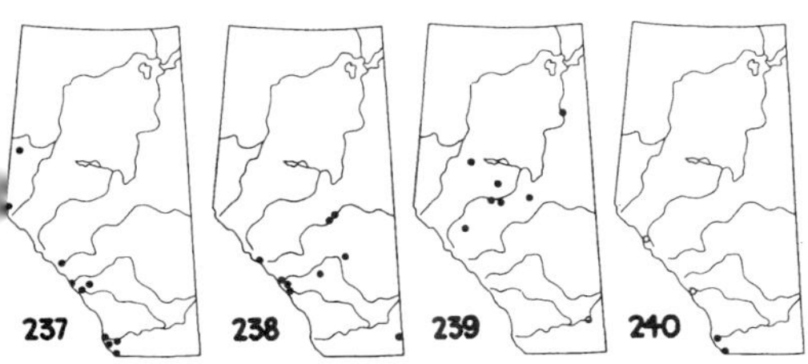

237 *Carex hoodii*
238 *Carex hookerana*
239 ● *Carex houghtoniana*
○ *Carex hystricina*
240 ○ *Carex illota*
● *Carex kelloggii*

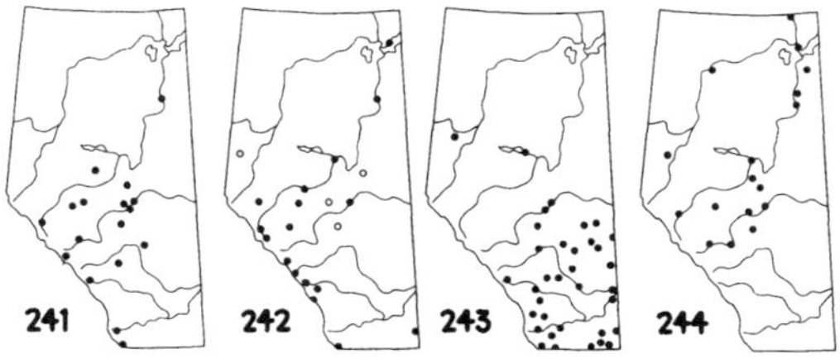

241 *Carex interior*
242 ○ *Carex lacustris*
● *Carex leptalea*
243 *Carex lanuginosa*
244 *Carex lasiocarpa*

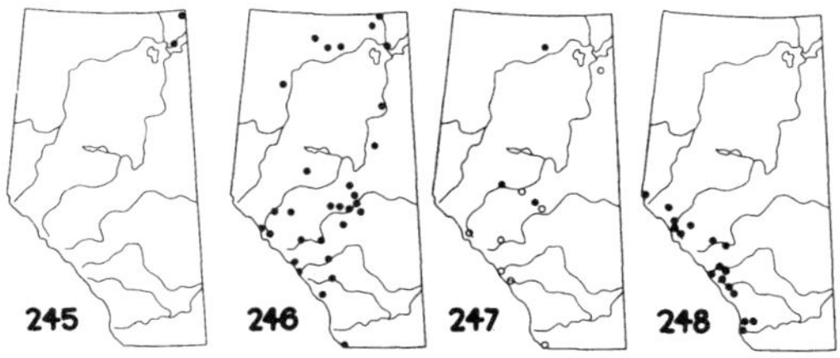

245 *Carex lenticularis*
246 *Carex limosa*
247 ○ *Carex livida*
● *Carex loliacea*
248 *Carex macloviana*

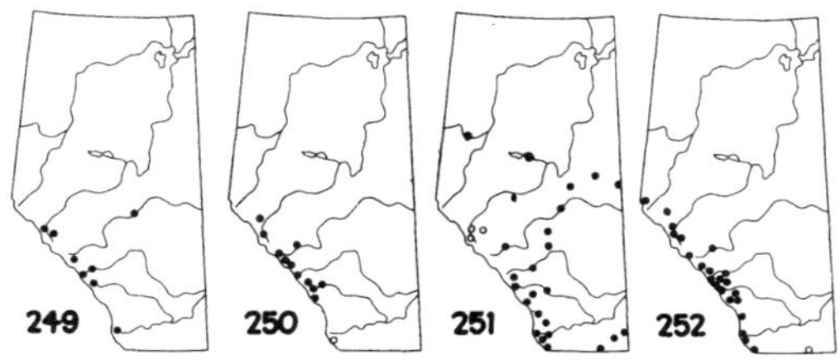

249 *Carex maritima*
250 ○ *Carex mertensii*
● *Carex microglochin*
251 ● *Carex microptera*
○ *Carex misandra*
252 ● *Carex nardina*
○ *Carex nebraskensis*

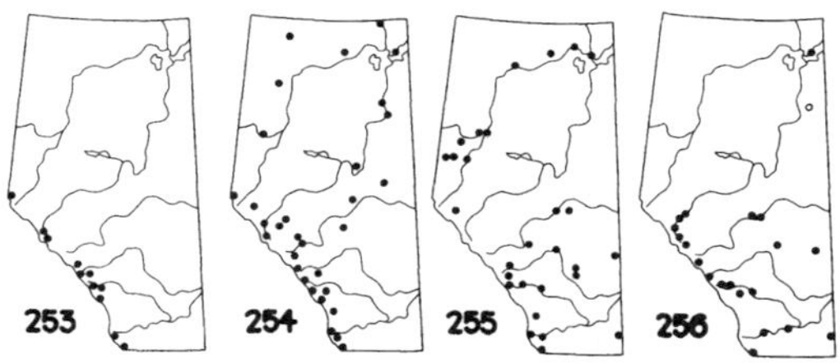

253 *Carex nigricans*
254 *Carex norvegica*
255 *Carex obtusata*
256 ● *Carex viridula*
○ *Carex oligosperma*

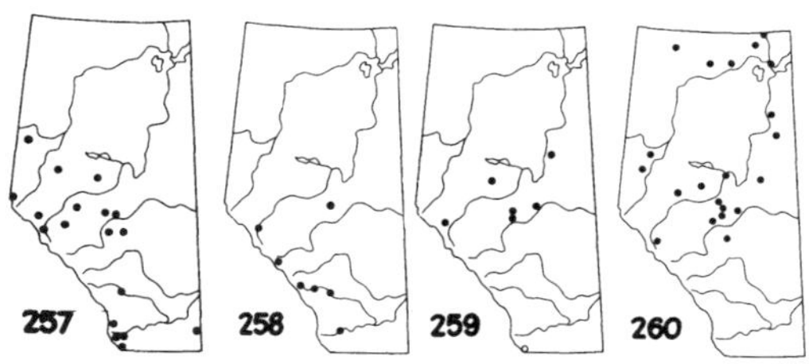

257 *Carex pachystachya*
258 *Carex parryana*
259 ● *Carex pauciflora*
○ *Carex paysonis*
260 *Carex paupercula*

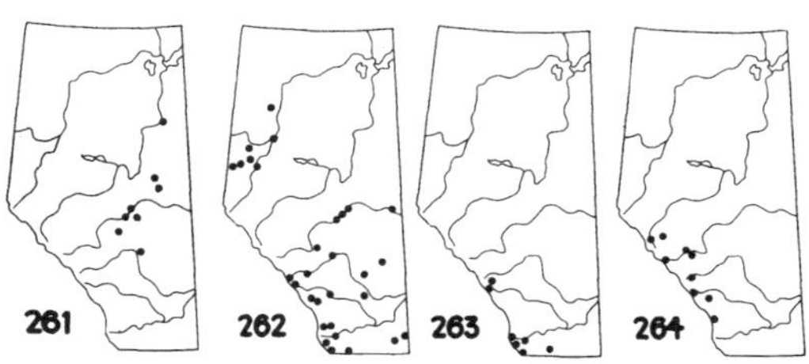

261 *Carex peckii*
262 *Carex pensylvanica*
263 *Carex petasata*
264 *Carex petricosa*

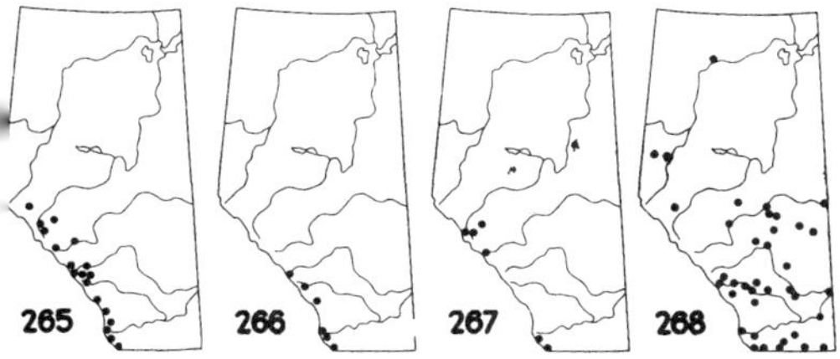

265 *Carex phaeocephala*
266 *Carex platylepis*
267 *Carex podocarpa*
268 *Carex praegracilis*

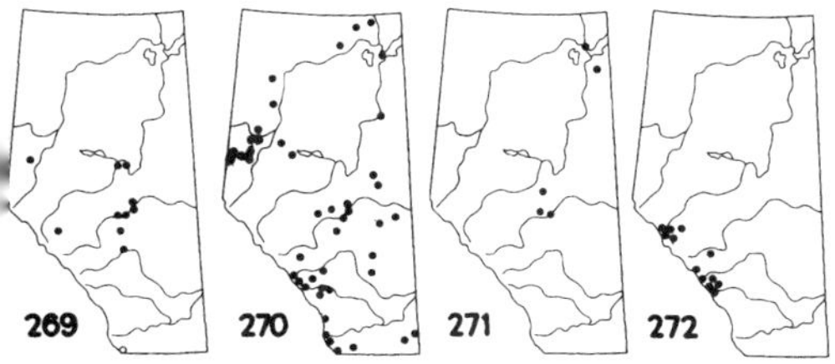

269 ● *Carex prairea*
○ *Carex preslii*
270 *Carex praticola*
271 *Carex pseudo-cyperus*
272 *Carex pyrenaica*

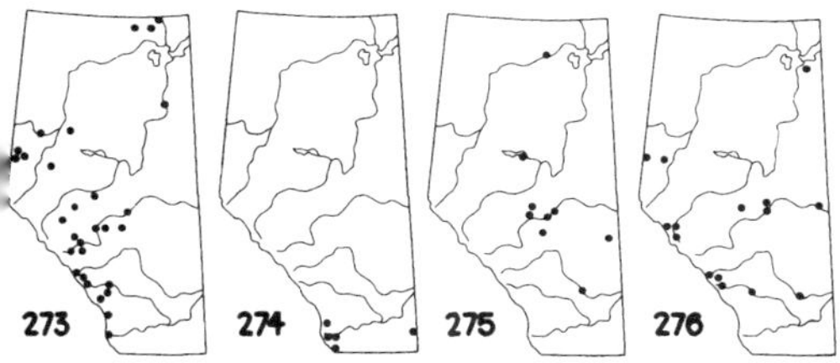

273 *Carex raymondii*
274 *Carex raynoldsii*
275 *Carex retrorsa*
276 *Carex richardsonii*

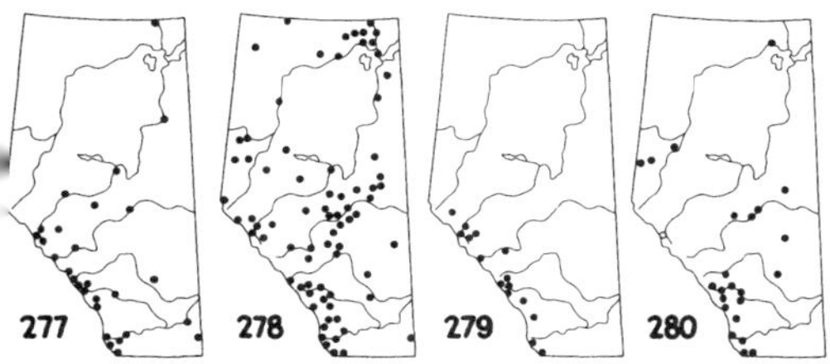

277 *Carex rossii*
278 *Carex rostrata*
279 *Carex rupestris*
280 ● *Carex sartwellii*
○ *Carex scoparia*

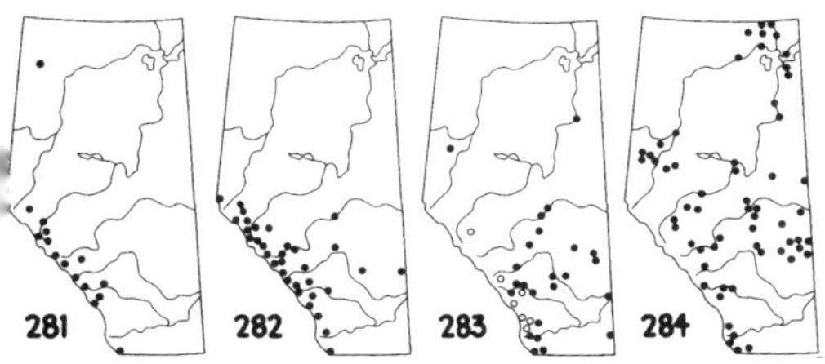

281 *Carex saxatilis*
282 *Carex scirpoidea*
283 ○ *Carex scopulorum*
● *Carex sprengelii*
284 *Carex siccata*

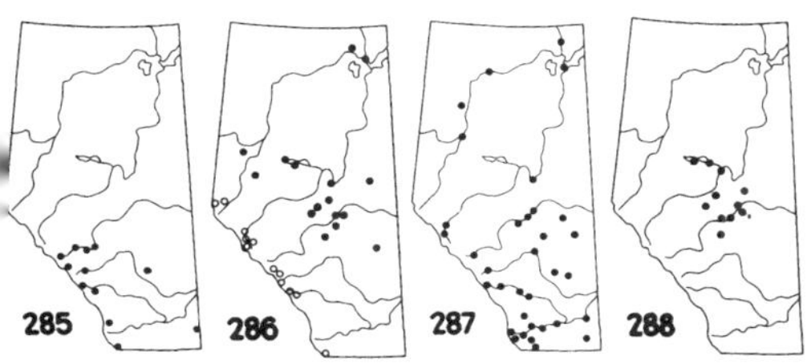

285 *Carex simulata*
286 ○ *Carex spectabilis*
● *Carex sychnocephala*
287 *Carex stenophylla*
288 *Carex stipata*

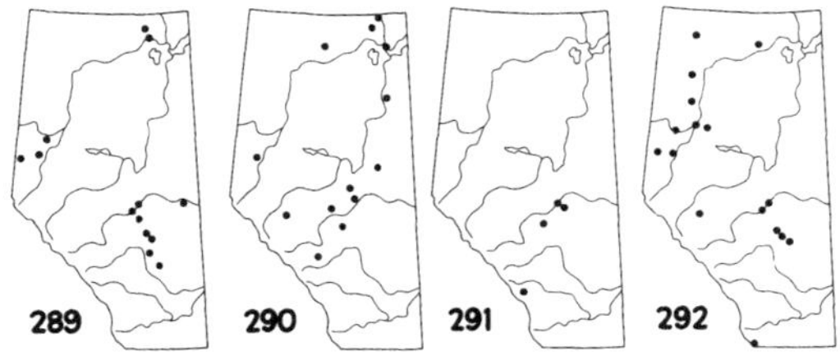

289 Carex tenera
290 Carex tenuiflora
291 Carex tincta
292 Carex torreyi

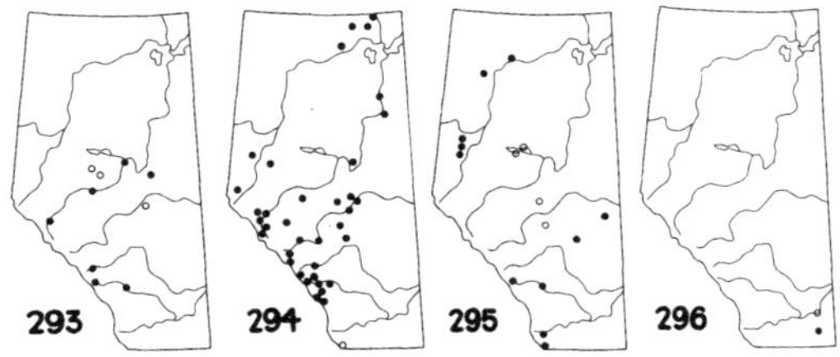

293 ○ *Carex trisperma*
● *Carex umbellata*
294 ● *Carex vaginata*
○ *Carex vesicaria*
295 ○ *Carex vulpinoidea*
● *Carex xerantica*
296 ○ *Cyperus squarrosus*
● *Cyperus schweinitzii*

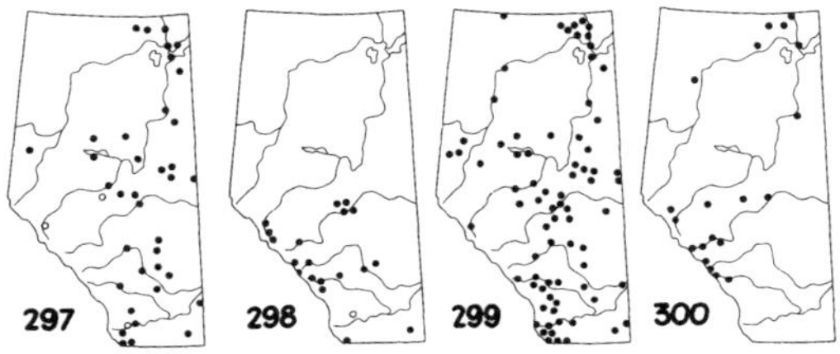

297 ● *Eleocharis acicularis*
○ *Eleocharis compressa*
298 ● *Eleocharis quinqueflora*
○ *Eleocharis ovata*
299 *Eleocharis palustris*
300 *Eriophorum brachyantherum*

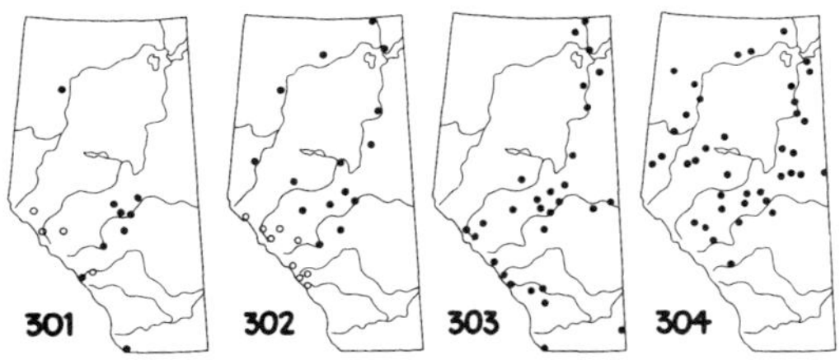

301 ○ *Eriophorum callitrix*
● *Eriophorum chamissonis*
302 ● *Eriophorum gracile*
○ *Eriophorum scheuchzeri*
303 *Eriophorum polystachion*
304 *Eriophorum vaginatum*

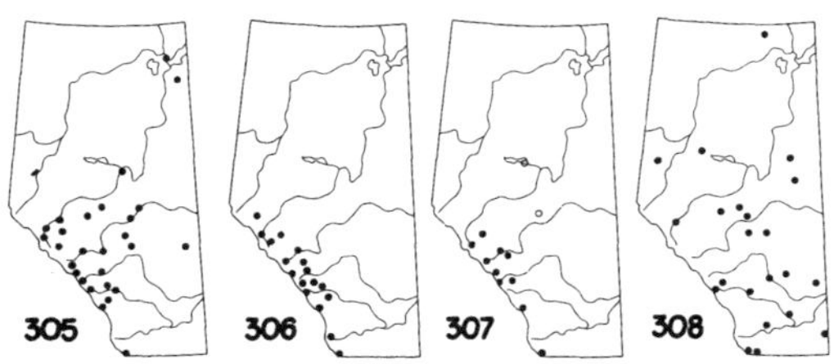

305 *Eriophorum viridi-carinatum*
306 *Kobresia myosuroides*
307 ● *Kobresia simpliciuscula*
○ *Rhynchospora capillacea*
308 *Scirpus acutus*

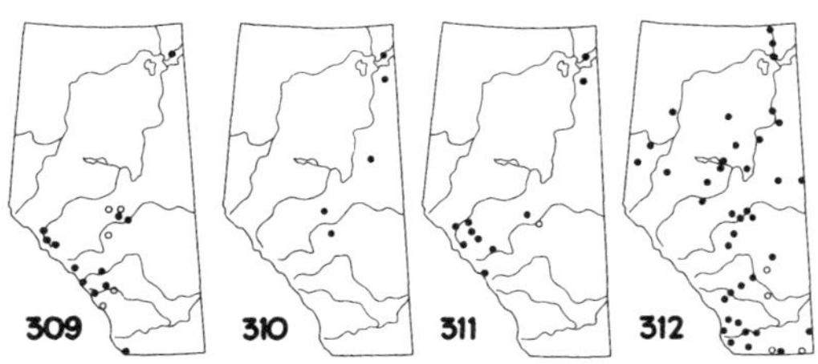

309 ● *Scirpus cespitosus*
○ *Scirpus clintonii*
310 *Scirpus cyperinus*
311 ○ *Scirpus fluviatilis*
● *Scirpus hudsonianus*
312 ● *Scirpus microcarpus*
Scirpus nevadensis

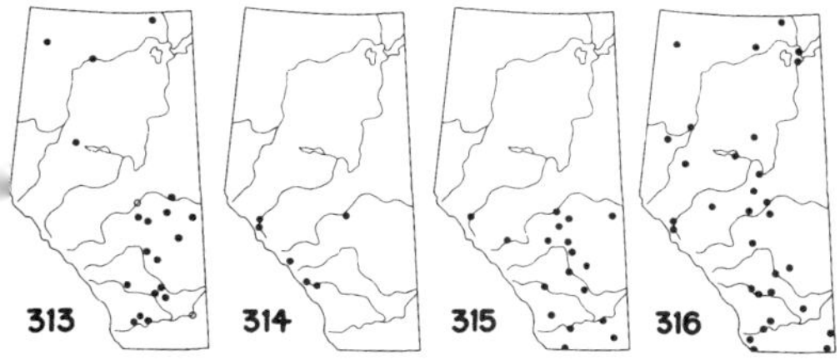

313 ○ *Scirpus pallidus*
● *Scirpus paludosus*
314 *Scirpus pumilus*
315 *Scirpus pungens*
316 *Scirpus validus*

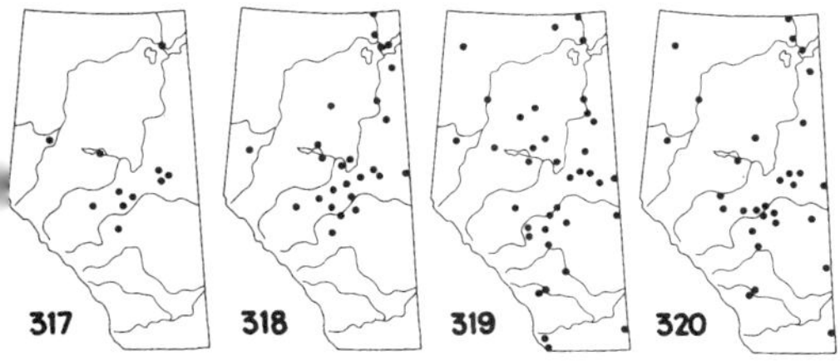

317 *Acorus americanus*
318 *Calla palustris*
319 *Lemna minor*
320 *Lemna trisulca*

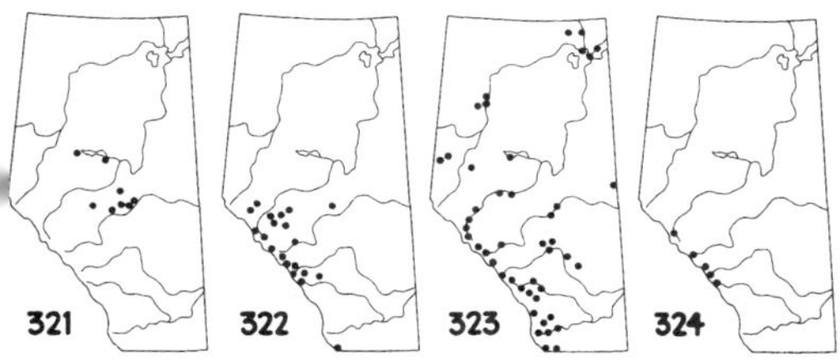

321 *Spirodela polyrhiza*
322 *Juncus albescens*
323 *Juncus alpinoarticulatus*
324 *Juncus biglumis*

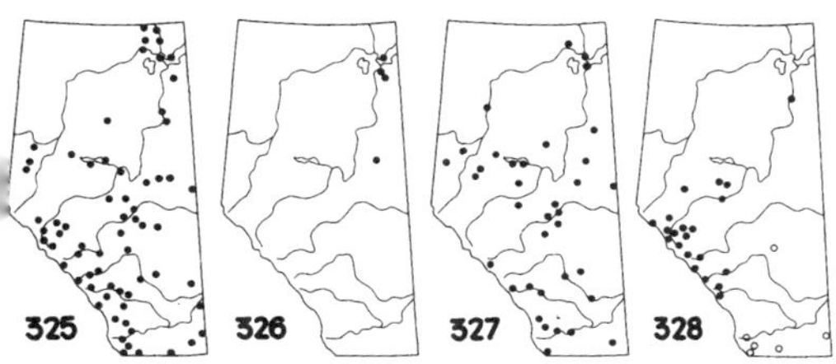

325 *Juncus balticus*
326 *Juncus brevicaudatus*
327 *Juncus bufonius*
328 ● *Juncus castaneus*
○ *Juncus confusus*

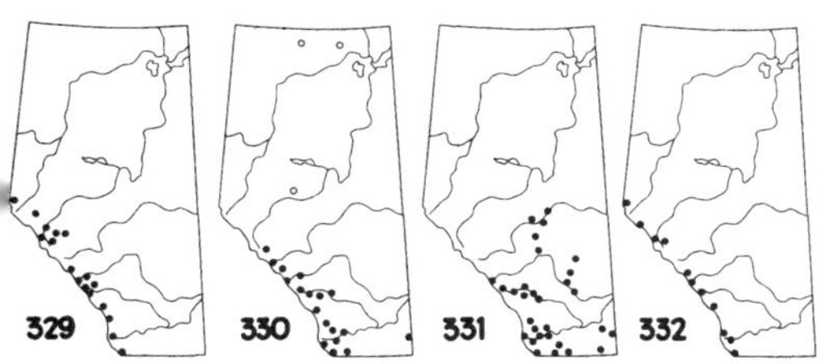

329 *Juncus drummondii*
330 ● *Juncus ensifolius*
○ *Juncus filiformis*
331 *Juncus longistylis*
332 *Juncus mertensianus*

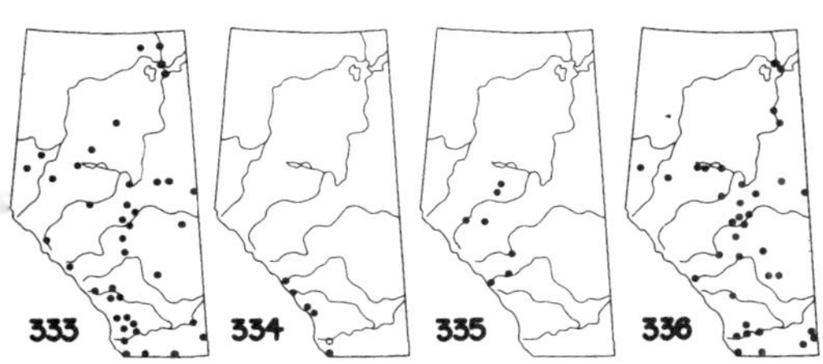

333 *Juncus nodosus*
334 ● *Juncus parryi*
○ *Juncus nevadensis*
335 *Juncus stygius*
336 *Juncus tenuis*

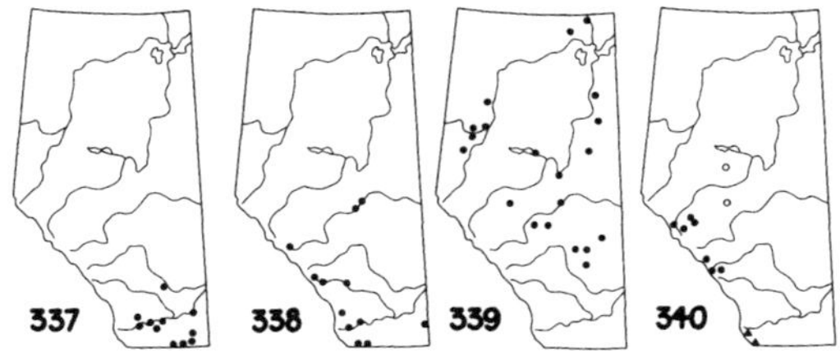

337 *Juncus torreyi*
338 *Juncus tracyi*
339 *Juncus vaseyi*
340 ○ *Luzula acuminata*
● *Luzula arcuata*
▲ *Luzula hitchcockii*

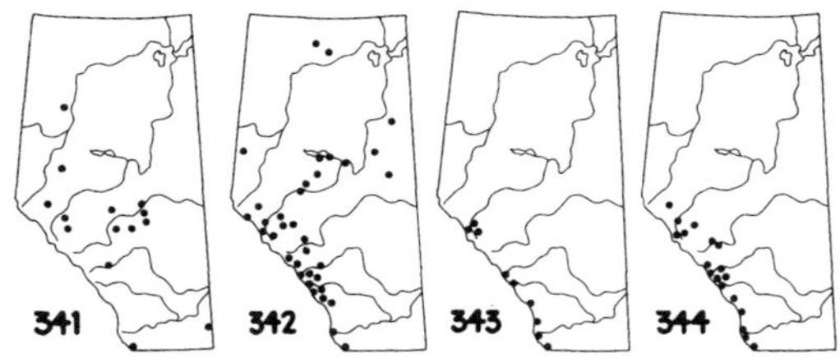

341 *Luzula multiflora*
342 *Luzula parviflora*
343 *Luzula piperi*
344 *Luzula spicata*

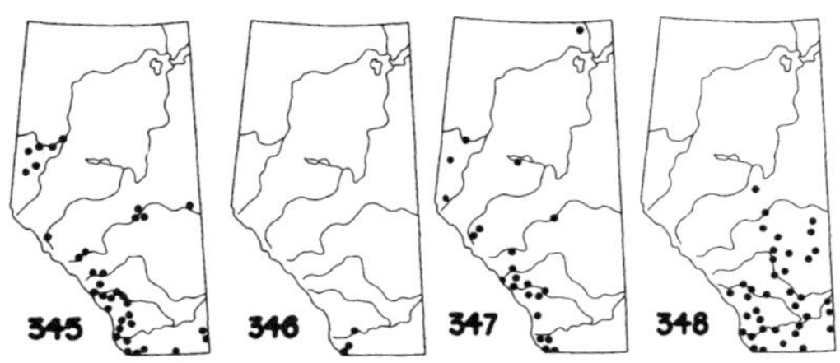

345 *Allium cernuum*
346 *Allium geyeri*
347 *Allium schoenoprasum*
348 *Allium textile*

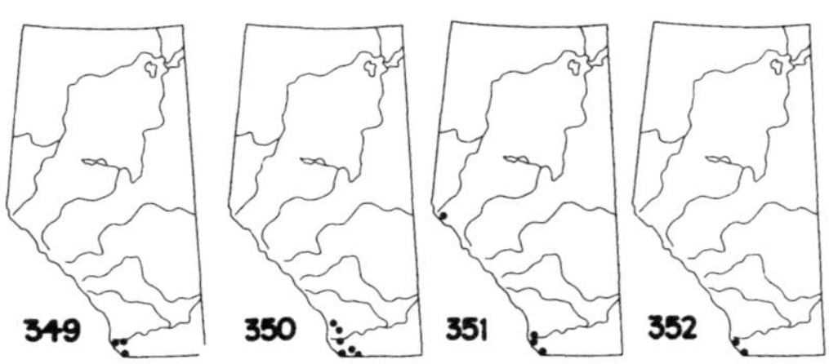

349 *Calochortus apiculatus*
350 *Camassia quamash*
351 *Clintonia uniflora*
352 *Disporum hookeri*

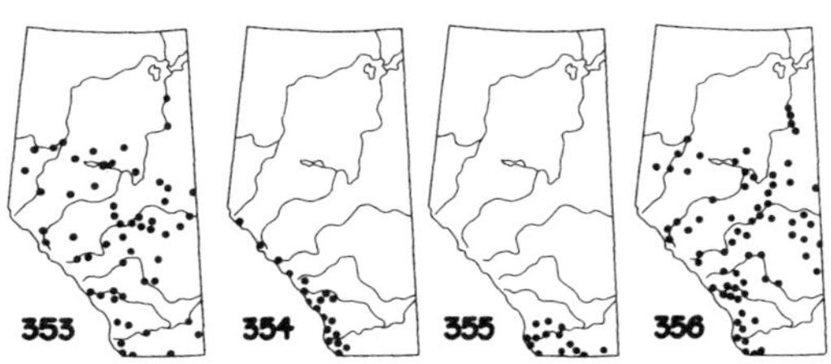

353 *Disporum trachycarpum*
354 *Erythronium grandiflorum*
355 *Fritillaria pudica*
356 *Lilium philadelphicum*

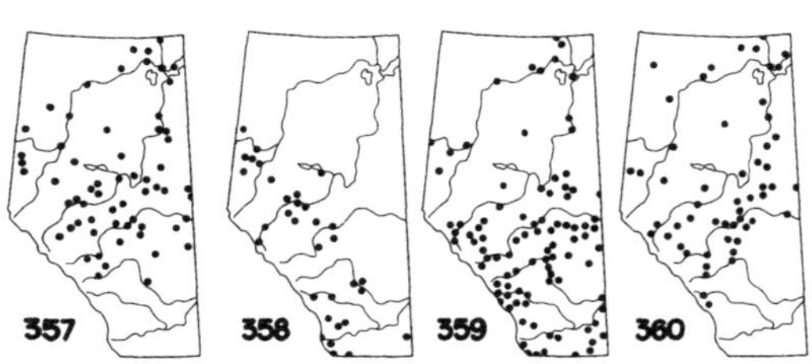

357 *Maianthemum canadense*
358 *Smilacina racemosa*
359 *Smilacina stellata*
360 *Smilacina trifolia*

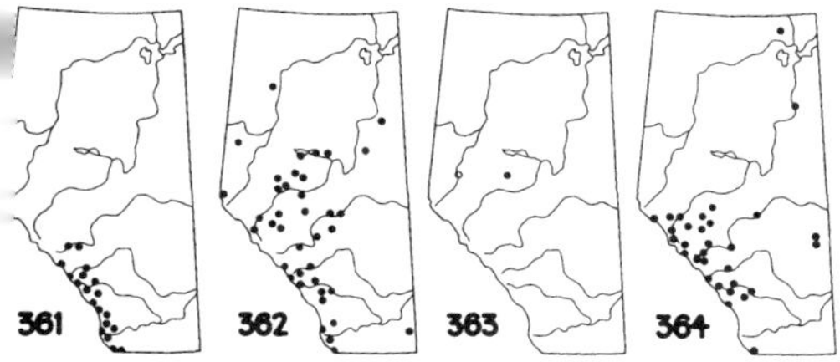

361 *Stenanthium occidentale*
362 *Streptopus amplexifolius*
363 ○ *Streptopus roseus*
● *Streptopus streptopoides*
364 *Tofieldia glutinosa*

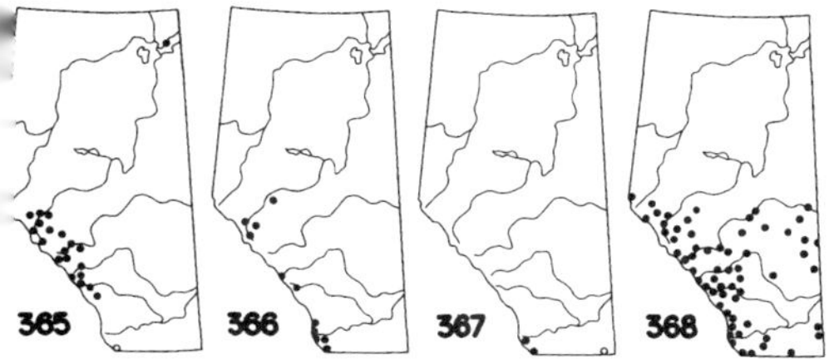

365 ● *Tofieldia pusilla*
○ *Trillium ovatum*
366 *Veratrum eschscholtzii*
367 ● *Xerophyllum tenax*
○ *Yucca glauca*
368 *Zigadenus elegans*

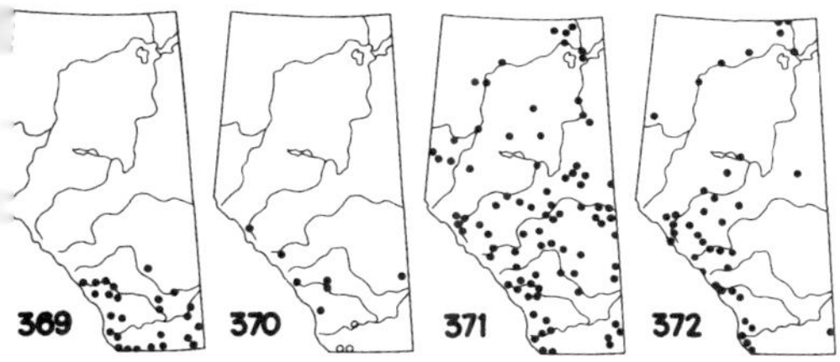

369 *Zigadenus venenosus*
370 ○ *Iris missouriensis*
● *Sisyrinchium septentrionale*
371 *Sisyrinchium montanum*
372 *Calypso bulbosa*

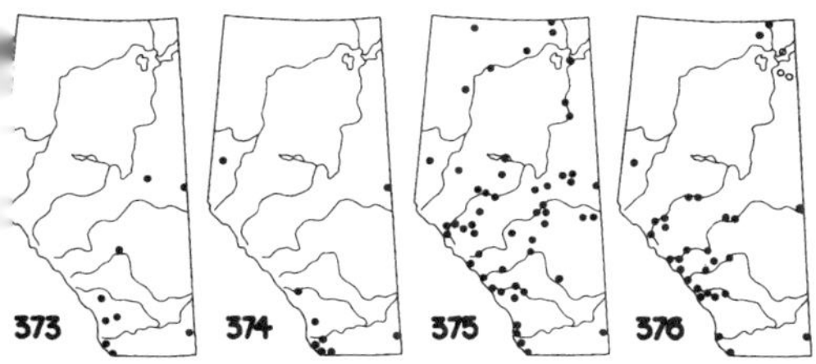

373 *Corallorhiza maculata*
374 *Corallorhiza striata*
375 *Corallorhiza trifida*
376 ○ *Cypripedium acaule*
● *Cypripedium passerinum*

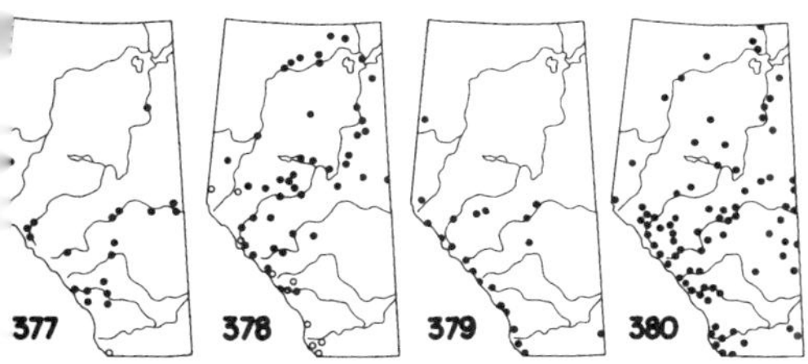

377 ● *Cypripedium calceolus*
○ *Cypripedium montanum*
378 ○ *Goodyera oblongifolia*
● *Goodyera repens*
379 *Habenaria dilatata*
380 *Habenaria hyperborea*

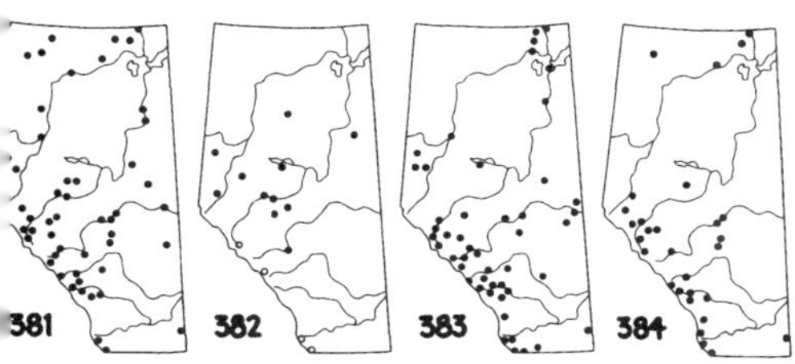

381 *Habenaria obtusata*
382 ● *Habenaria orbiculata*
○ *Habenaria unalascensis*
383 *Habenaria viridis*
384 *Listera borealis*

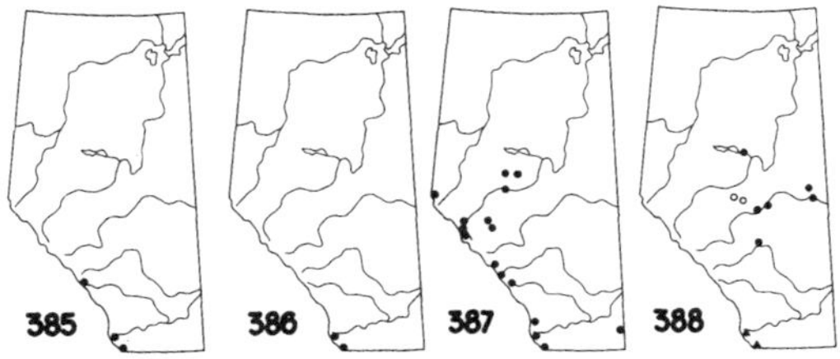

385 *Listera caurina*
386 *Listera convallarioides*
387 *Listera cordata*
388 ● *Malaxis monophylla*
○ *Malaxis paludosa*
▲ *Habenaria saccata*

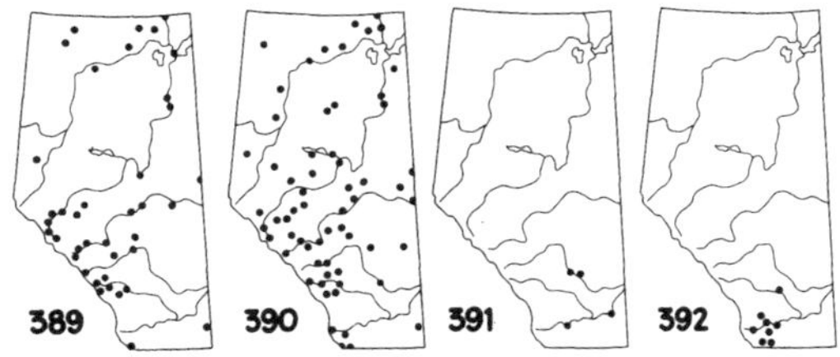

389 *Orchis rotundifolia*
390 *Spiranthes romanzoffiana*
391 *Populus* × *acuminata*
392 *Populus angustifolia*

393 *Populus balsamifera*
394 *Populus deltoides*
395 *Populus tremuloides*
396 ○ *Salix alaxensis*
● *Salix lutea*

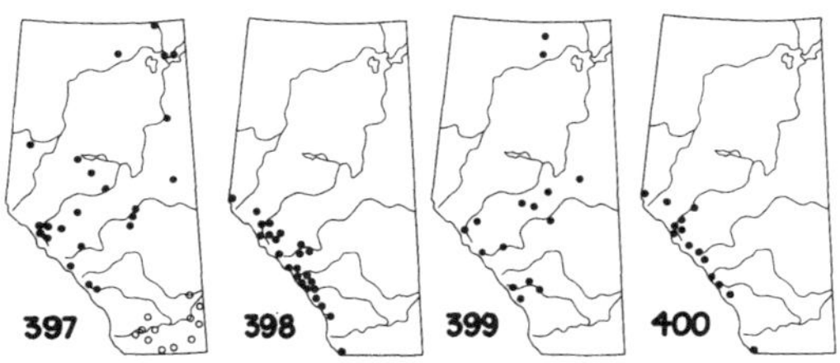

397 ○ *Salix amygdaloides*
● *Salix arbusculoides*
398 *Salix arctica*
399 *Salix athabascensis*
400 *Salix barclayi*

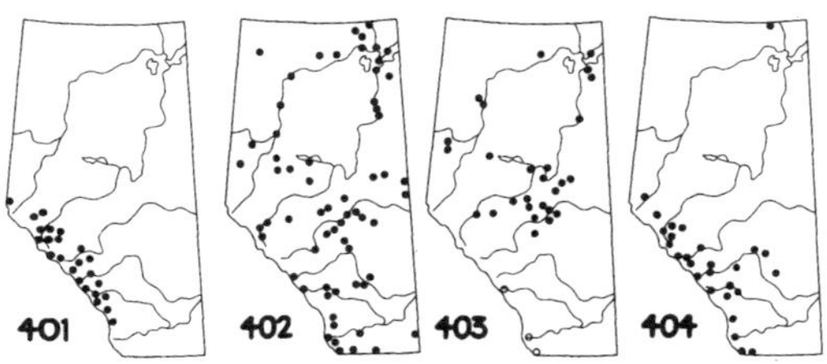

401 *Salix barrattiana*
402 *Salix bebbiana*
403 ○ *Salix boothii*
● *Salix pyrifolia*
404 *Salix brachycarpa*

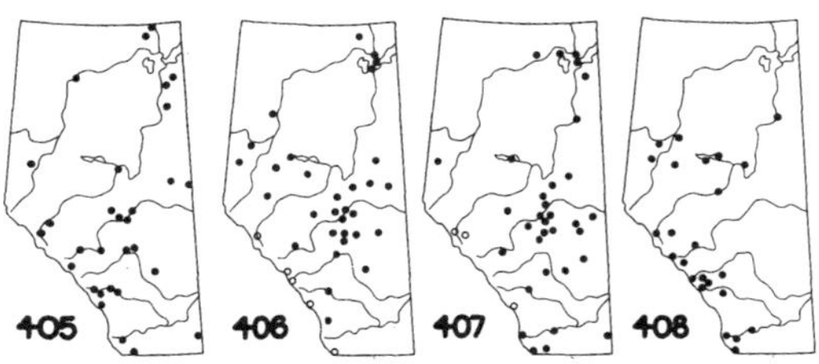

405 *Salix candida*
406 ○ *Salix commutata*
● *Salix petiolaris*
407 ● *Salix discolor*
○ *Salix stolonifera*
408 *Salix drummondiana*

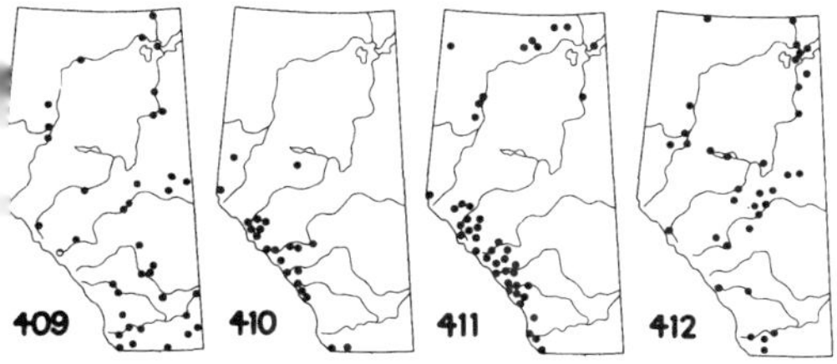

409 ● *Salix exigua*
○ *Salix lanata*
410 *Salix farriae*
411 *Salix glauca*
412 *Salix lucida*

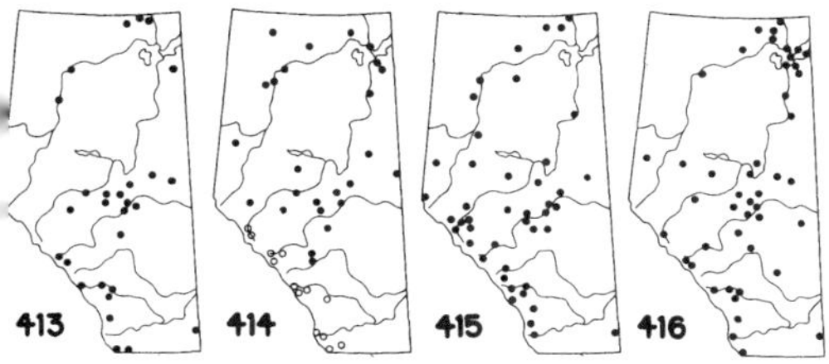

413 *Salix maccalliana*
414 ○ *Salix melanopsis*
● *Salix pedicellaris*
415 *Salix myrtillifolia*
416 *Salix planifolia*

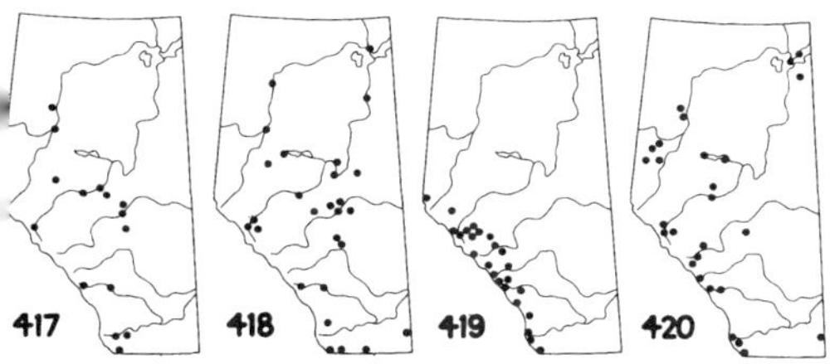

417 *Salix prolixa*
418 *Salix pseudomonticola*
419 *Salix reticulata*
420 *Salix scouleriana*

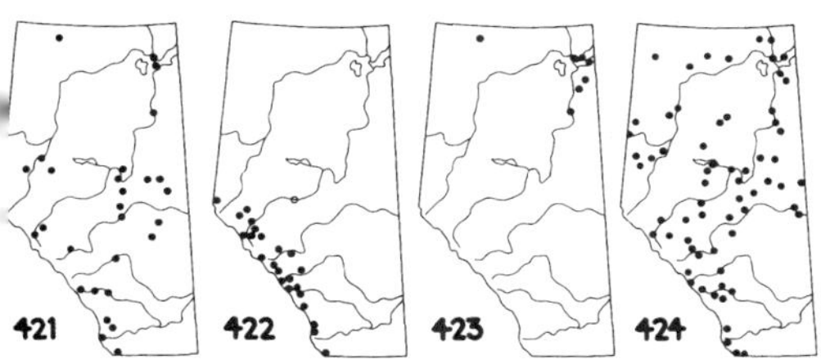

421 *Salix serissima*
422 ○ *Salix sitchensis*
● *Salix vestita*
423 *Myrica gale*
424 *Alnus crispa*

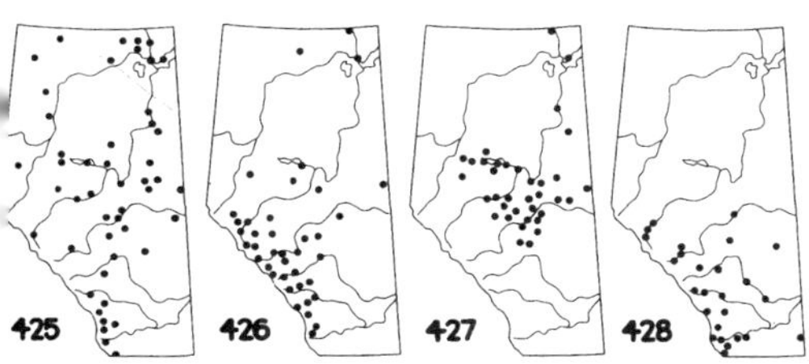

425 *Alnus tenuifolia*
426 *Betula glandulosa*
427 *Betula neoalaskana*
428 *Betula occidentalis*

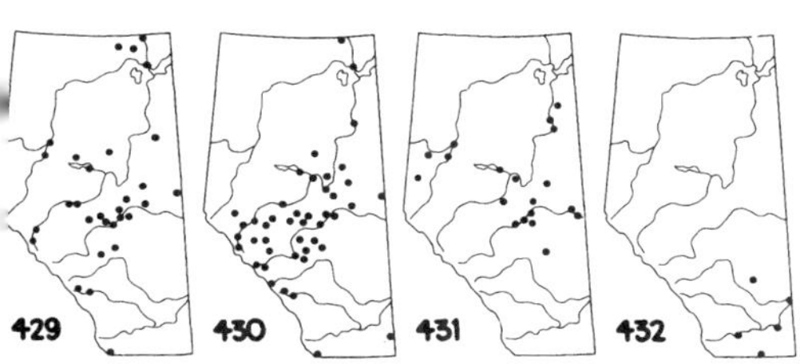

429 *Betula papyrifera*
430 *Betula pumila*
431 *Corylus cornuta*
432 *Parietaria pensylvanica*

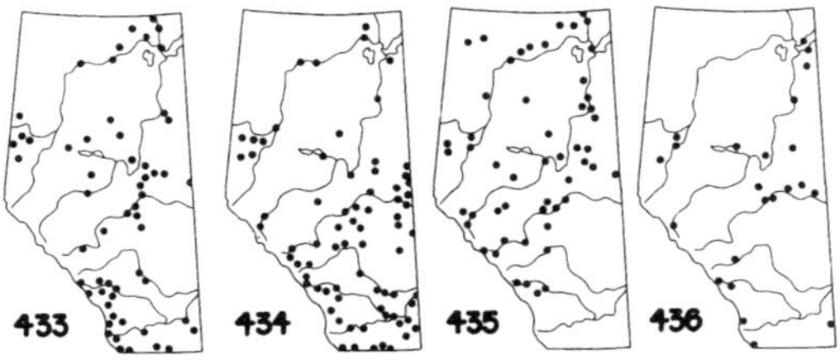

433 *Urtica dioica*
434 *Comandra umbellata*
435 *Geocaulon lividum*
436 *Arceuthobium americanum*

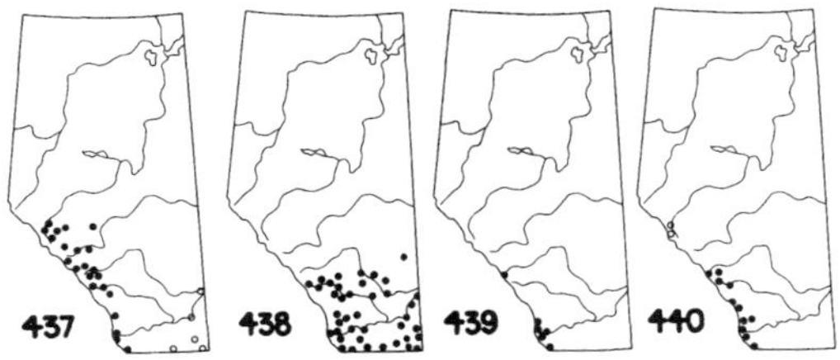

437 ● *Eriogonum androsaceum*
○ *Eriogonum cernuum*
438 *Eriogonum flavum*
439 *Eriogonum ovalifolium*
440 ● *Eriogonum umbellatum*
○ *Koenigia islandica*

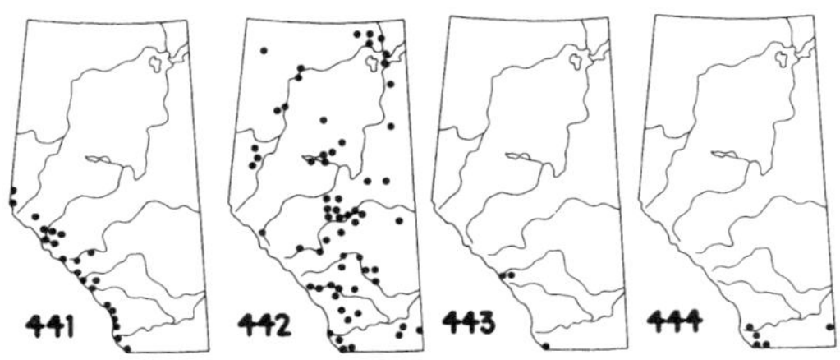

441 *Oxyria digyna*
442 *Polygonum amphibium*
443 *Polygonum austiniae*
444 *Polygonum bistortoides*

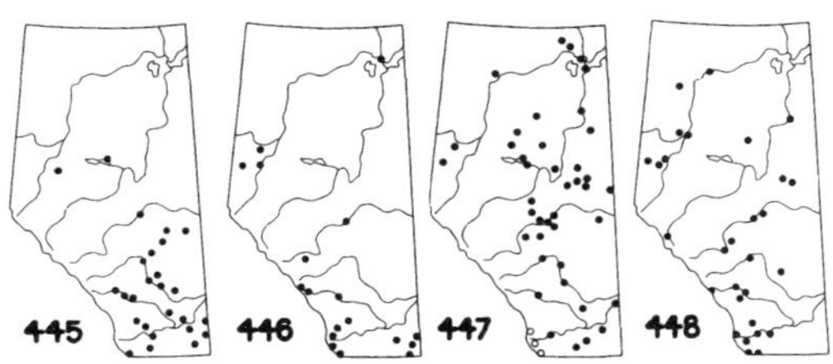

445 *Polygonum coccineum*
446 *Polygonum douglasii*
447 ○ *Polygonum engelmannii*
● *Polygonum lapathifolium*
448 *Polygonum erectum*

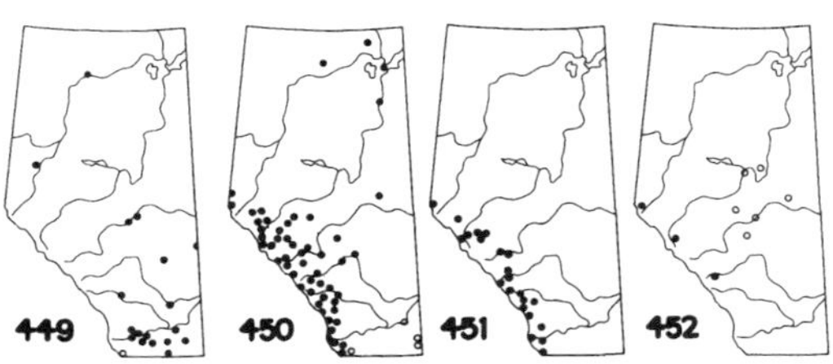

449 ○ *Polygonum minimum*
● *Polygonum ramosissimum*
450 ● *Polygonum viviparum*
○ *Polygonum watsonii*
451 *Rumex acetosa*
452 ○ *Rumex britannica*
● *Rumex paucifolius*

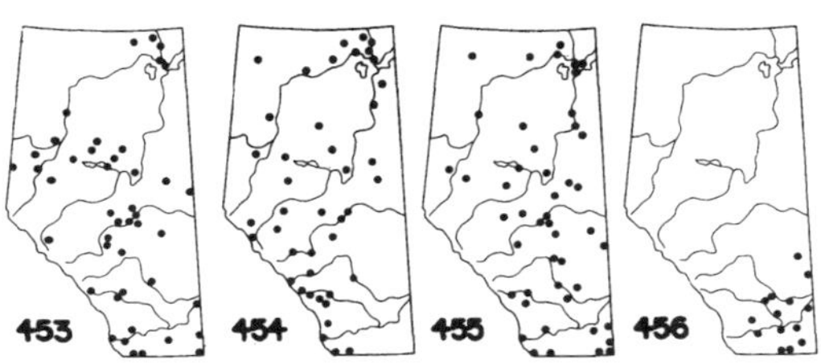

453 *Rumex maritimus*
454 *Rumex occidentalis*
455 *Rumex triangulivalvis*
456 *Rumex venosus*

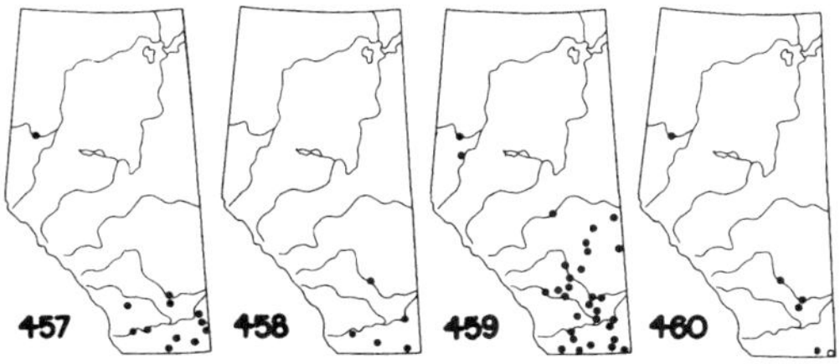

457 *Atriplex argentea*
458 *Atriplex canescens*
459 *Atriplex nuttallii*
460 ● *Atriplex powellii*
○ *Atriplex truncata*

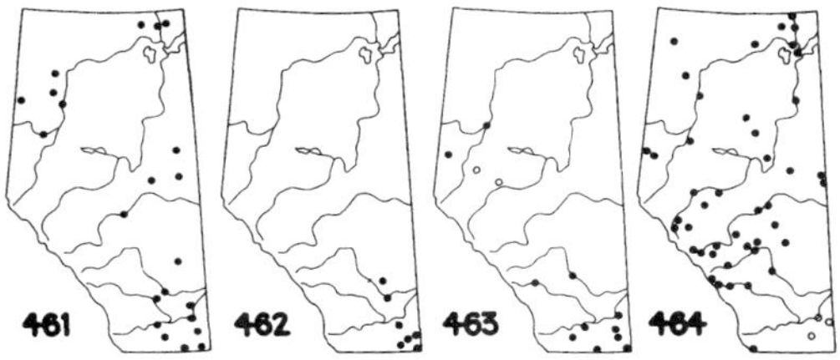

461 *Atriplex subspicata*
462 *Atriplex suckleyi*
463 ○ *Chenopodium atrovirens*
● *Chenopodium berlandieri*
464 ● *Chenopodium capitatum*
○ *Chenopodium desiccatum*

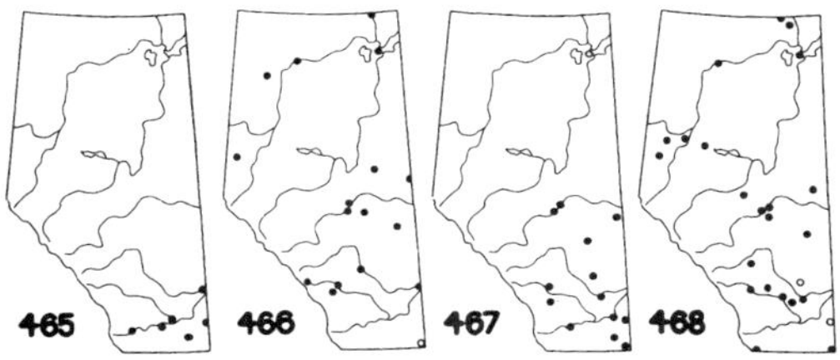

465 *Chenopodium fremontii*
466 ● *Chenopodium gigantospermum*
○ *Chenopodium incanum*
467 ○ *Chenopodium leptophyllum*
● *Chenopodium pratericola*
468 ● *Chenopodium rubrum*
○ *Chenopodium watsonii*

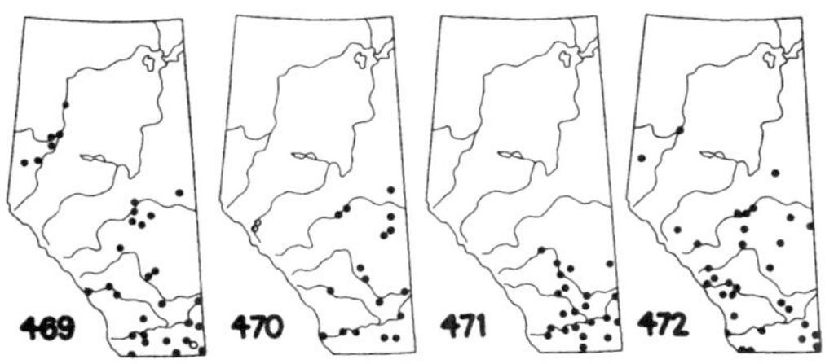

469 ● *Chenopodium salinum*
○ *Chenopodium subglabrum*
470 ● *Corispermum hyssopifolium*
○ *Corispermum nitidum*
471 *Eurotia lanata*
472 *Monolepis nuttalliana*

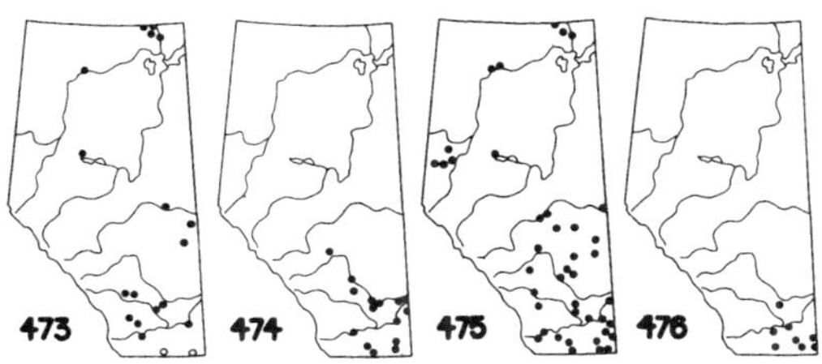

473 ● *Salicornia europaea*
○ *Suaeda moquinii*
474 *Sarcobatus vermiculatus*
475 *Suadea calceoliformis*
476 *Suckleya suckleyana*

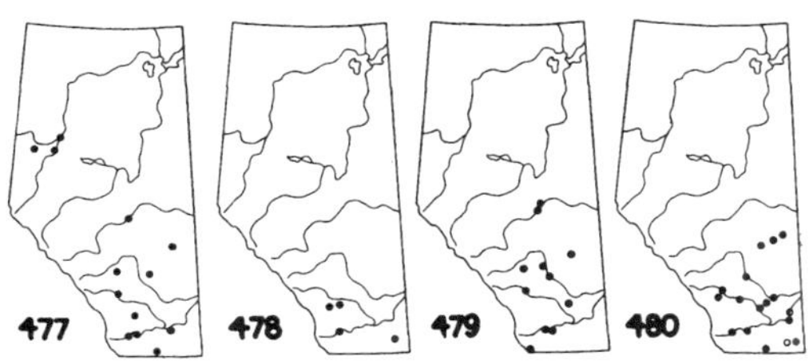

477 *Amaranthus albus*
478 *Amaranthus californicus*
479 *Amaranthus graecizans*
480 ○ *Abronia micrantha*
● *Mirabilis hirsuta*

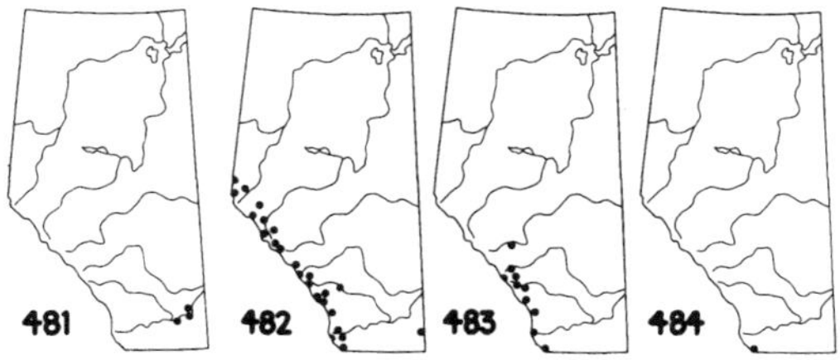

481 *Mirabilis nyctaginea*
482 *Claytonia lanceolata*
483 *Claytonia megarhiza*
484 *Lewisia pygmaea*

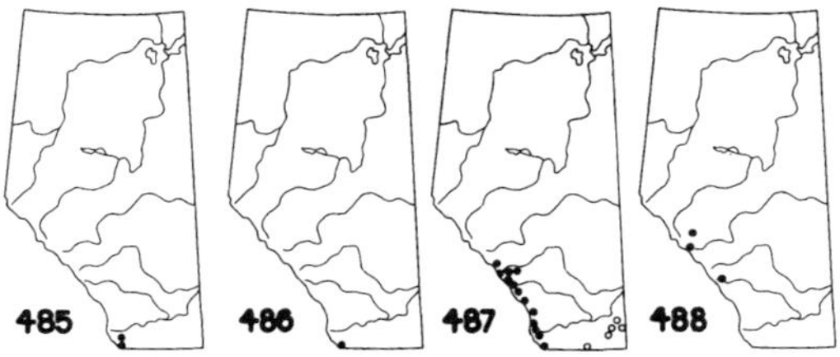

485 *Montia linearis*
486 *Montia parvifolia*
487 ● *Arenaria capillaris*
○ *Arenaria congesta*
488 *Arenaria longipedunculata*

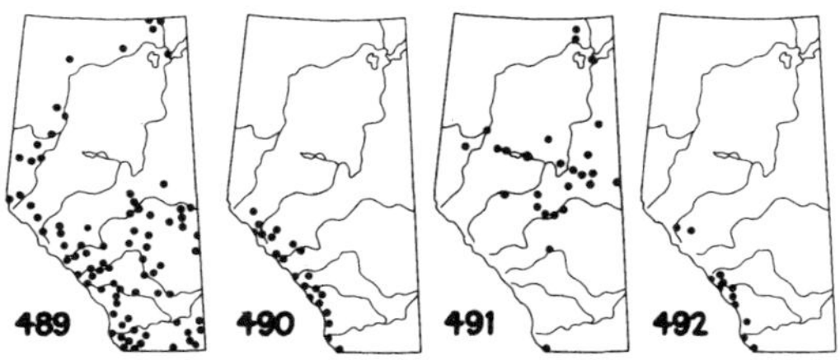

489 *Cerastium arvense*
490 *Cerastium beeringianum*
491 *Cerastium nutans*
492 *Minuartia austromontana*

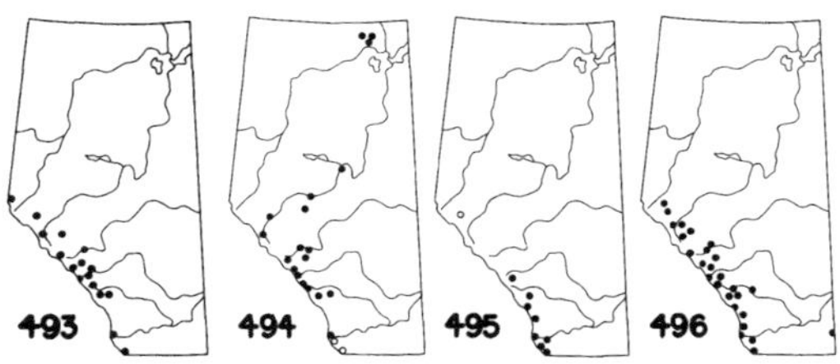

493 *Minuartia biflora*
494 ● *Minuartia dawsonensis*
○ *Minuartia nuttallii*
495 ● *Minuartia obtusiloba*
○ *Minuartia elegans*
496 *Minuartia rubella*

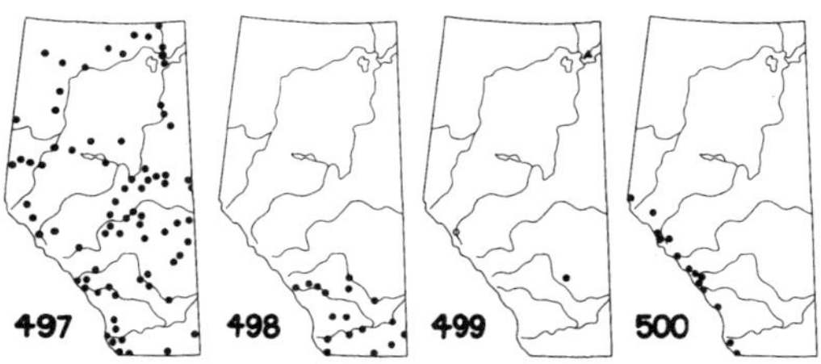

497 *Moehringia lateriflora*
498 *Paronychia sessiliflora*
499 ● *Sagina decumbens*
○ *Sagina nivalis*
▲ *Sagina nodosa*
500 *Sagina saginoides*

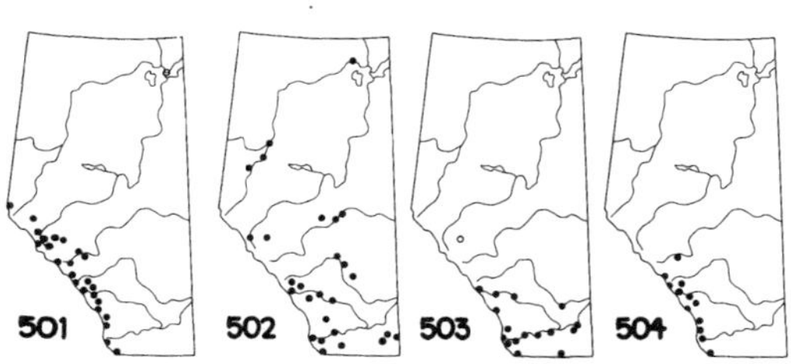

501 ● *Silene acaulis*
○ *Silene antirrhina*
502 *Silene drummondii*
503 ○ *Silene furcata*
● *Silene menziesii*
504 *Silene parryi*

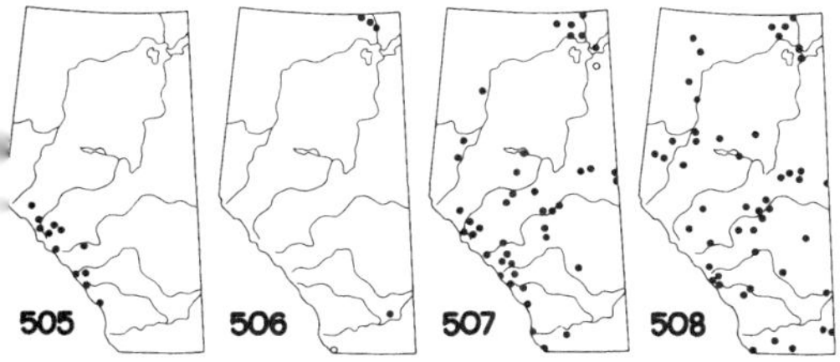

505 *Silene uralensis*
506 ● *Spergularia marina*
○ *Stellaria americana*
507 ○ *Stellaria arenicola*
● *Stellaria calycantha*
508 *Stellaria crassifolia*

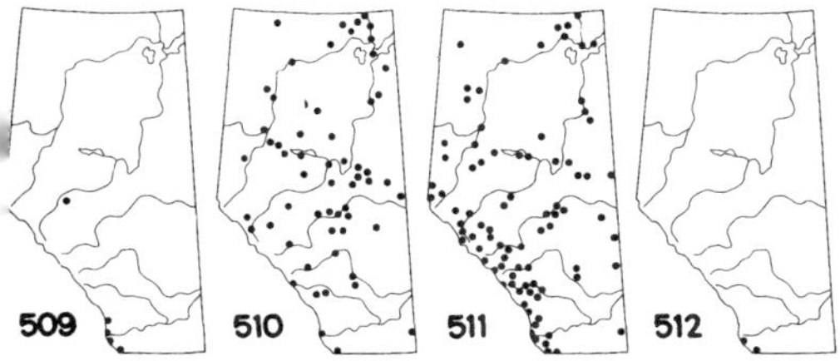

509 *Stellaria crispa*
510 *Stellaria longifolia*
511 *Stellaria longipes*
512 *Stellaria obtusa*

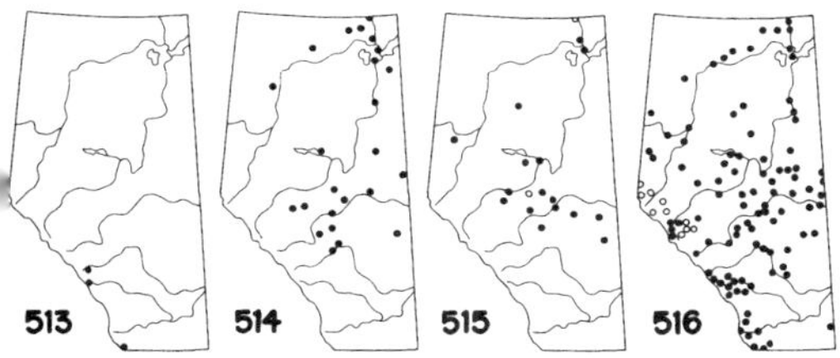

513 *Stellaria umbellata*
514 *Nuphar variegatum*
515 ○ *Nymphaea tetragona*
● *Ceratophyllum demersum*
516 ○ *Aconitum delphinifolium*
● *Actaea rubra*

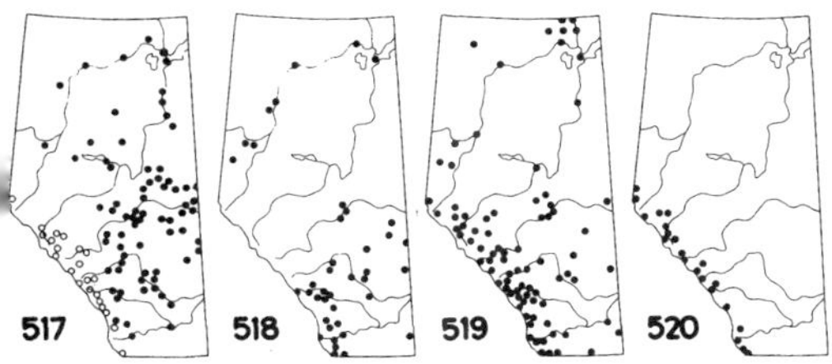

517 ● *Anemone canadensis*
○ *Anemone lithophila*
518 *Anemone cylindrica*
519 *Anemone multifida*
520 *Anemone occidentalis*

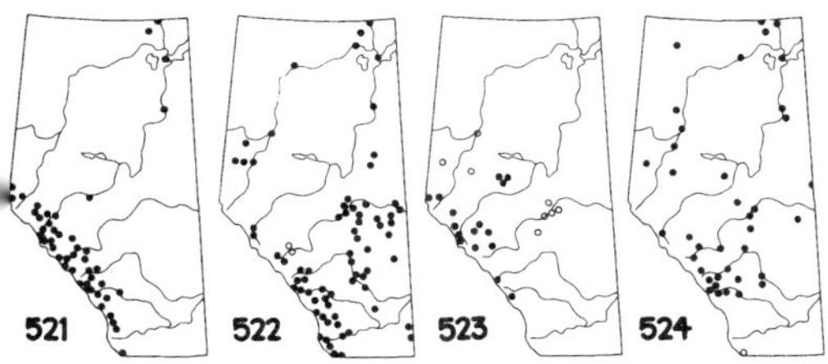

521 *Anemone parviflora*
522 ● *Anemone patens*
○ *Anemone quinquefolia*
523 ● *Anemone richardsonii*
○ *Anemone riparia*
524 ● *Aquilegia brevistyla*
○ *Aquilegia jonesii*

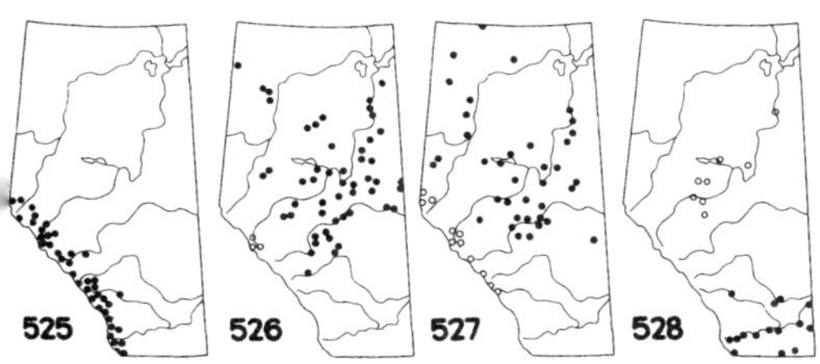

525 *Aquilegia flavescens*
526 ○ *Aquilegia formosa*
● *Caltha palustris*
527 ○ *Caltha leptosepala*
● *Caltha natans*
528 ● *Clematis ligusticifolia*
○ *Coptis trifolia*

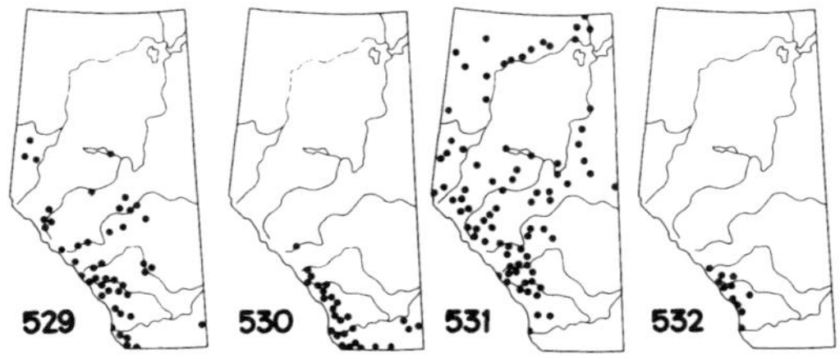

529 Clematis occidentalis
530 Delphinium bicolor
531 Delphinium glaucum
532 Delphinium nuttallianum

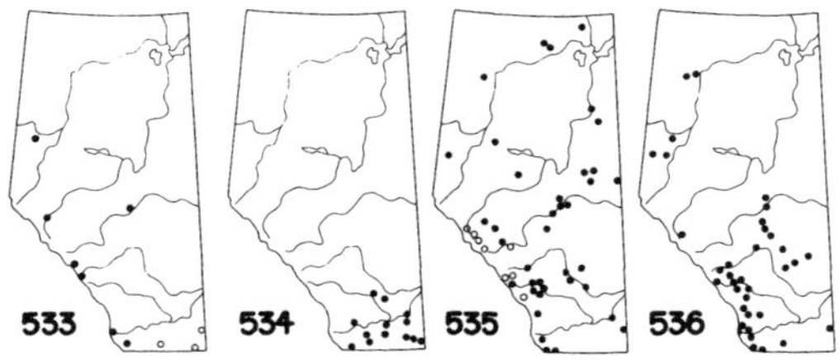

533 ○ Myosurus aristatus
● Ranunculus aquatilis
534 Myosurus minimus
535 ● Ranunculus abortivus
○ Ranunculus grayi
536 Ranunculus cardiophyllus

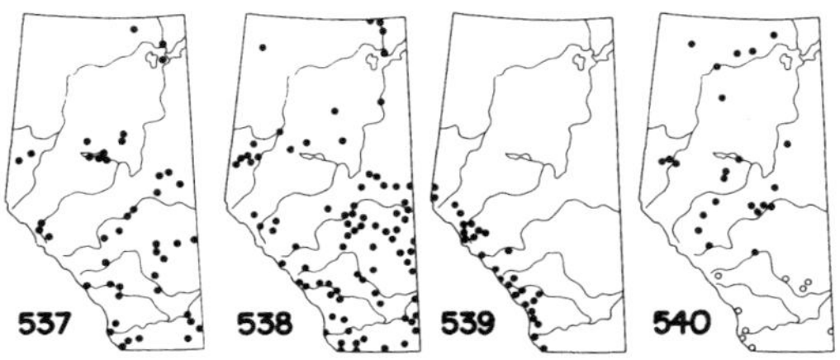

537 Ranunculus circinatus
538 Ranunculus cymbalaria
539 Ranunculus eschscholtzii
540 ○ Ranunculus glaberrimus
● Ranunculus lapponicus

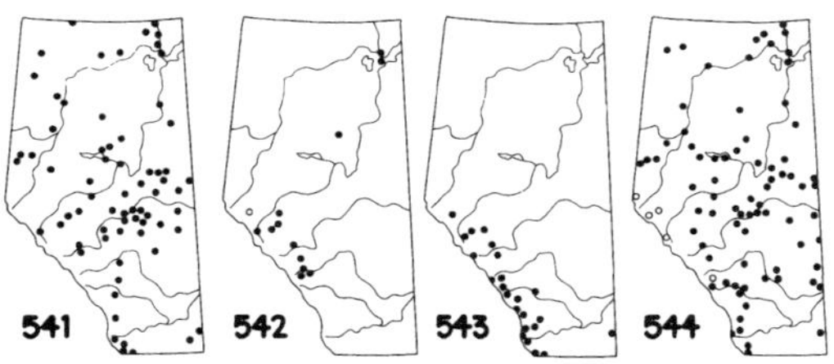

541 Ranunculus gmelinii
542 ● Ranunculus hyperboreus
○ Ranunculus nivalis
543 Ranunculus inamoenus
544 ● Ranunculus macounii
○ Ranunculus occidentalis

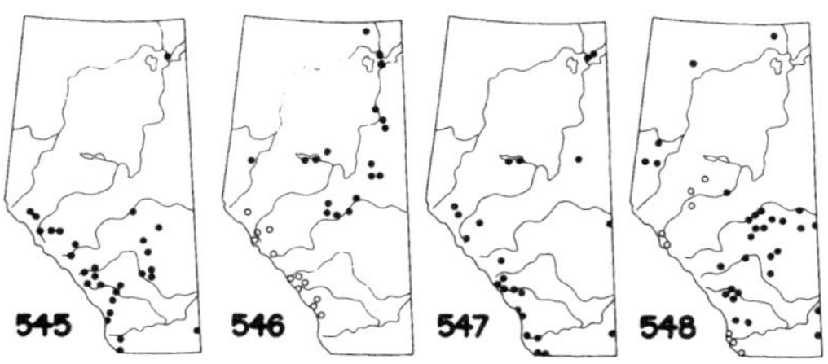

545 Ranunculus pedatifidus
546 ● Ranunculus pensylvanicus
○ Ranunculus pygmaeus
547 Ranunculus reptans
548 ● Ranunculus rhomboideus
○ Ranunculus uncinatus

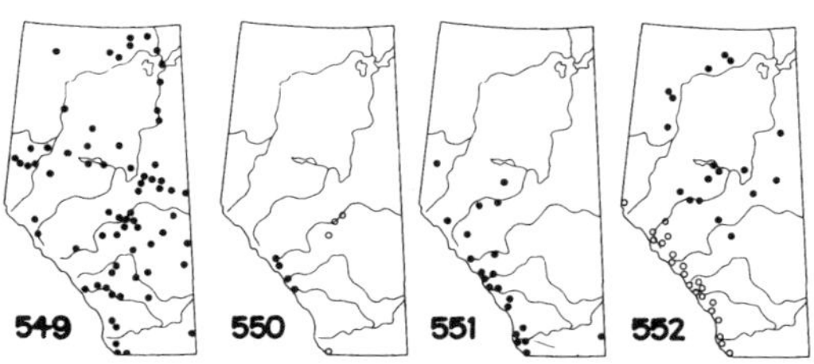

549 Ranunculus sceleratus
550 ● Ranunculus verecundus
○ Thalictrum dasycarpum
551 Thalictrum occidentale
552 ● Thalictrum sparsiflorum
○ Trollius albiflorus

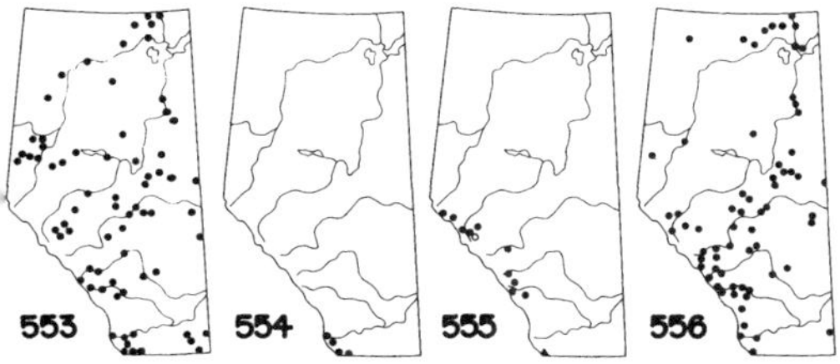

553 *Thalictrum venulosum*
554 *Berberis repens*
555 ○ *Papaver freedmanianum*
● *Papaver kluanensis*
▲ *Papaver pygmaeum*
556 *Corydalis aurea*

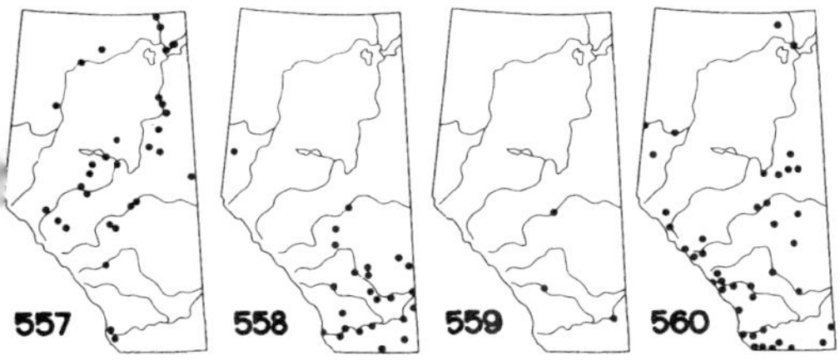

557 *Corydalis sempervirens*
558 *Cleome serrulata*
559 *Polanisia dodecandra*
560 *Arabis divaricarpa*

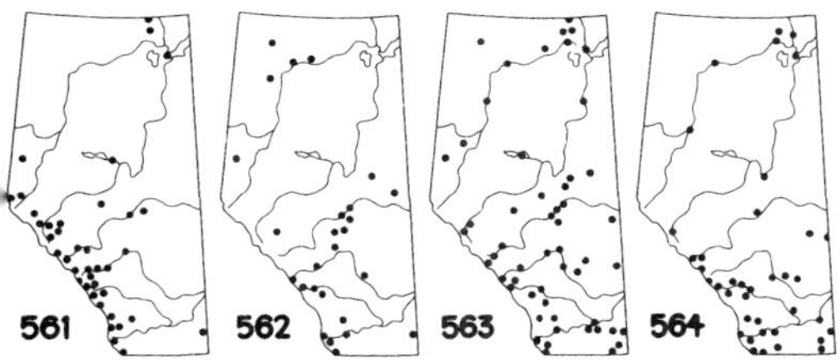

561 *Arabis drummondii*
562 *Arabis glabra*
563 *Arabis hirsuta*
564 *Arabis holboellii*

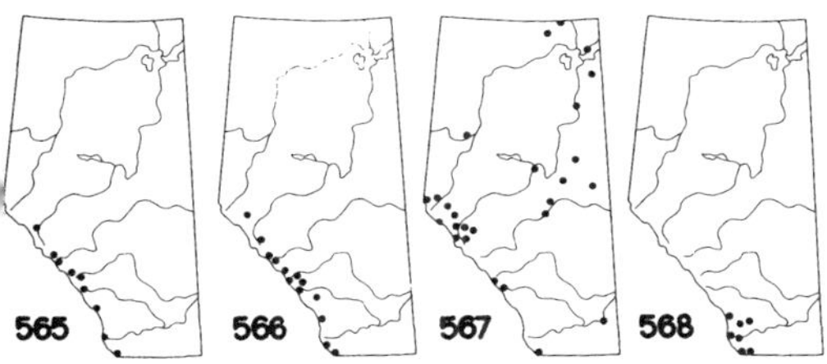

565 *Arabis lemmonii*
566 *Arabis lyallii*
567 *Arabis lyrata*
568 *Arabis nuttallii*

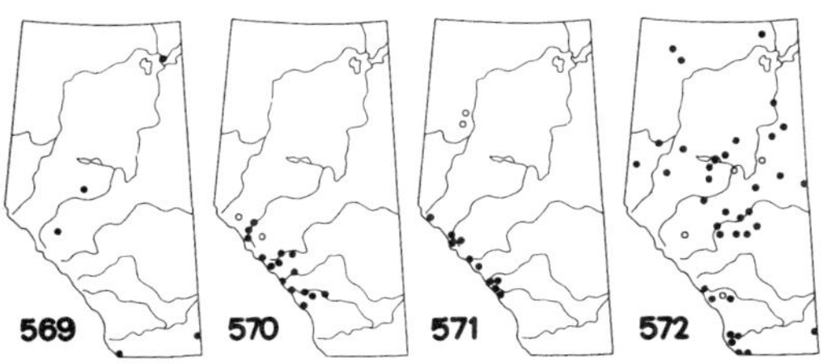

569 *Barbarea orthoceras*
570 ● *Braya humilis*
○ *Braya purpurascens*
571 ● *Cardamine bellidifolia*
○ *Cardamine parviflora*
572 ● *Cardamine pensylvanica*
○ *Cardamine pratensis*

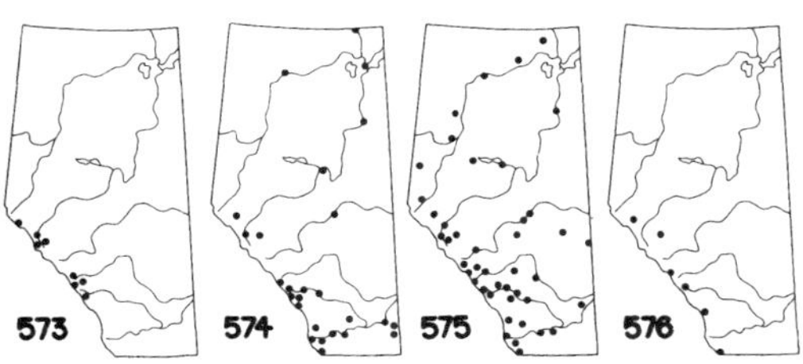

573 *Cardamine umbellata*
574 *Descurainia pinnata*
575 *Descurainia richardsonii*
576 *Draba albertina*

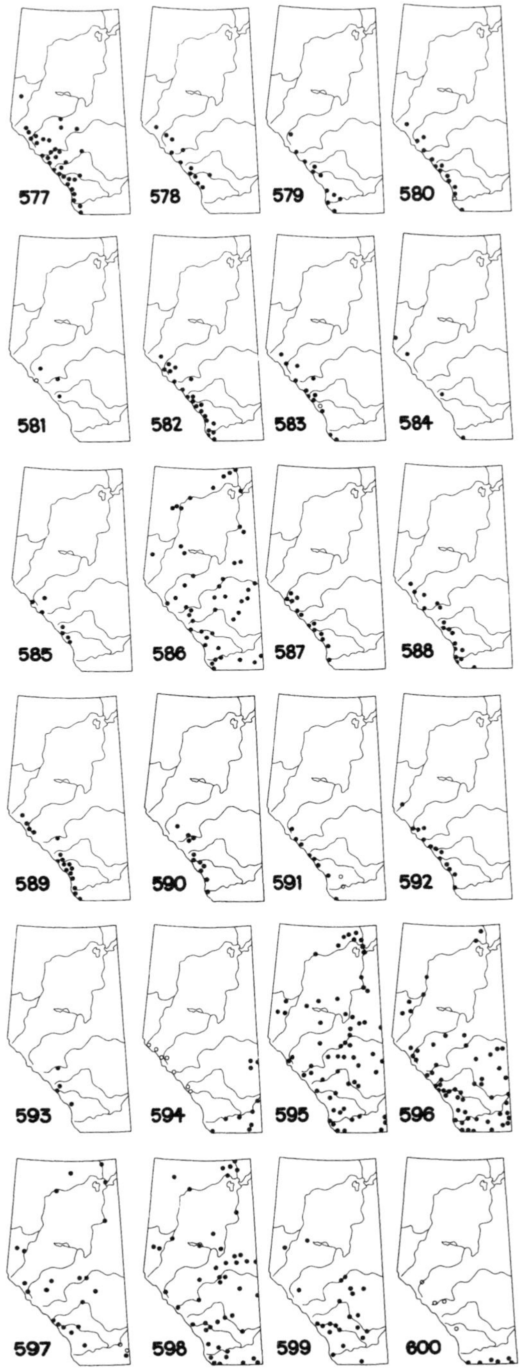

577 *Draba aurea*
578 *Draba borealis*
579 *Draba cana*
580 ● *Draba crassifolia*
○ *Draba densifolia*

581 ● *Draba fladnizensis*
○ *Draba glabella*
582 *Draba incerta*
583 ● *Draba lonchocarpa*
○ *Draba kananaskis*
584 *Draba longipes*

585 *Draba macounii*
586 *Draba nemorosa*
587 *Draba nivalis*
588 *Draba oligosperma*

589 *Draba paysonii*
590 *Draba porsildii*
591 ● *Draba praealta*
○ *Draba reptans*
592 *Draba stenoloba*

593 *Draba ventosa*
594 ● *Erysimum asperum*
○ *Erysimum pallasii*
595 *Erysimum cheiranthoides*
596 *Erysimum inconspicuum*

597 ○ *Halimolobos virgata*
● *Lepidium bourgeauanum*
598 *Lepidium densiflorum*
599 *Lepidium ramosissimum*
600 ● *Lesquerella alpina*
○ *Lesquerella arctica*

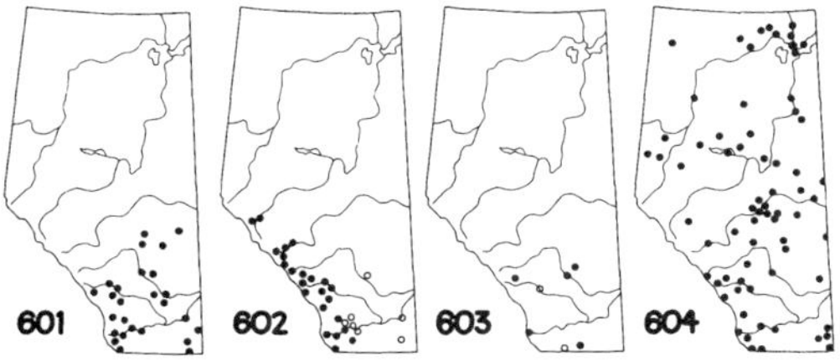

601 *Lesquerella arenosa*
602 ● *Physaria didymocarpa*
○ *Rorippa sinuata*
603 ● *Rorippa curvipes*
○ *Rorippa tenerrima*
604 *Rorippa palustris*

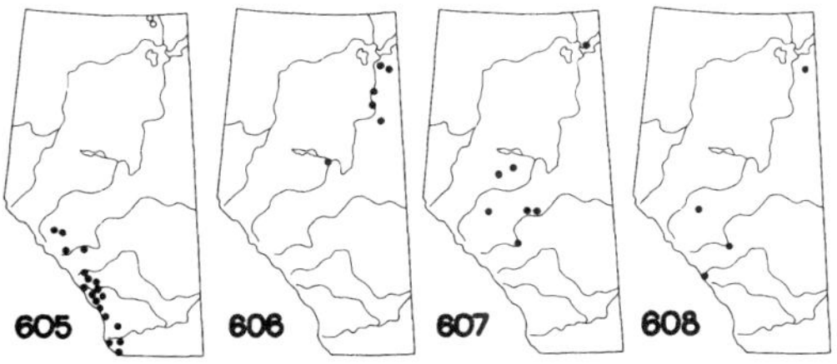

605 ● *Smelowskia calycina*
○ *Thellungiella salsuginea*
606 *Sarracenia purpurea*
607 *Drosera anglica*
608 *Drosera linearis*

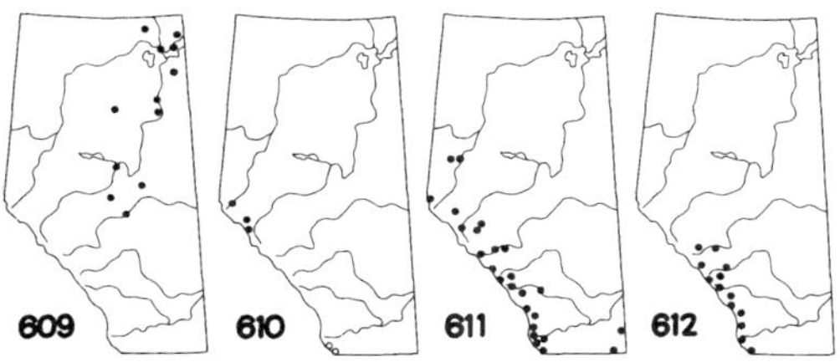

609 *Drosera rotundifolia*
610 ● *Sedum divergens*
○ *Sedum stenopetalum*
611 *Sedum lanceolatum*
612 *Tolmachevia integrifolia*

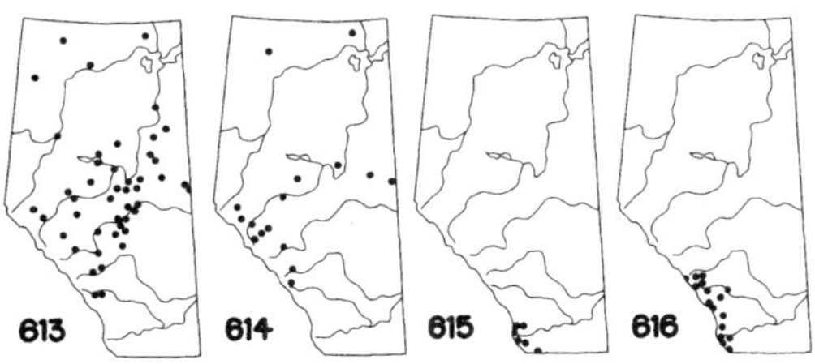

613 *Chrysosplenium iowense*
614 *Chrysosplenium tetrandrum*
615 *Conimitella williamsii*
616 *Heuchera cylindrica*

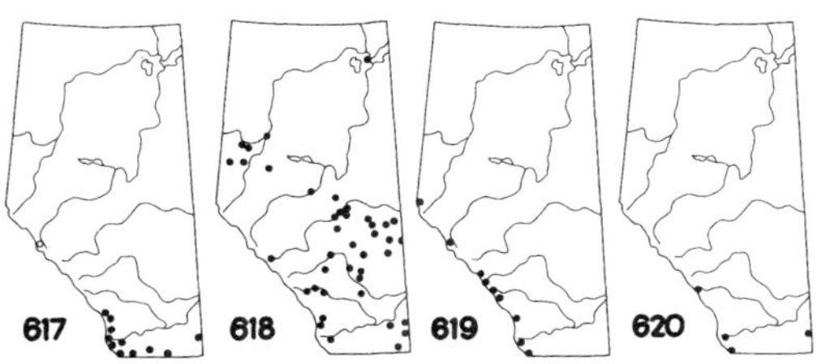

617 ○ *Heuchera glabra*
● *Heuchera parvifolia*
618 *Heuchera richardsonii*
619 *Leptarrhena pyrolifolia*
620 *Lithophragma glabrum*

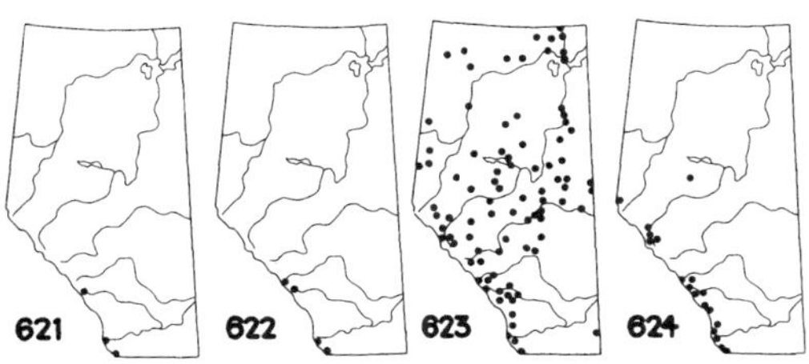

621 *Lithophragma parviflorum*
622 *Mitella breweri*
623 *Mitella nuda*
624 *Mitella pentandra*

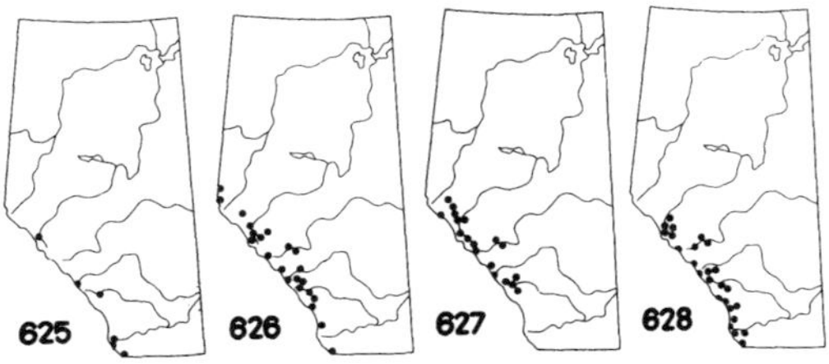

625 *Mitella trifida*
626 *Saxifraga adscendens*
627 *Saxifraga aizoides*
628 *Saxifraga bronchialis*

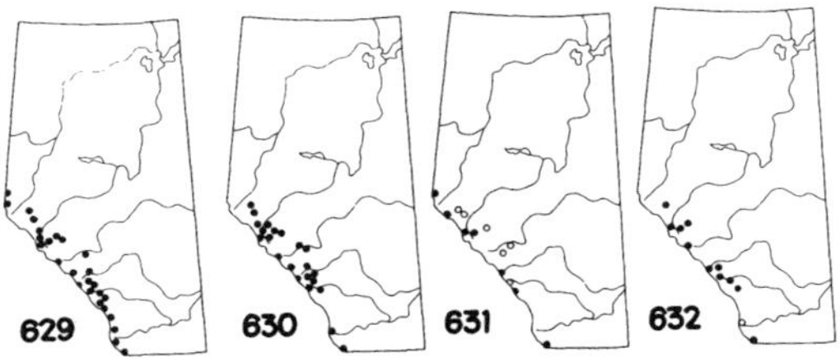

629 *Saxifraga caespitosa*
630 *Saxifraga cernua*
631 ● *Saxifraga ferruginea*
○ *Saxifraga flagellaris*
632 ● *Saxifraga hyperborea*
○ *Saxifraga oregana*

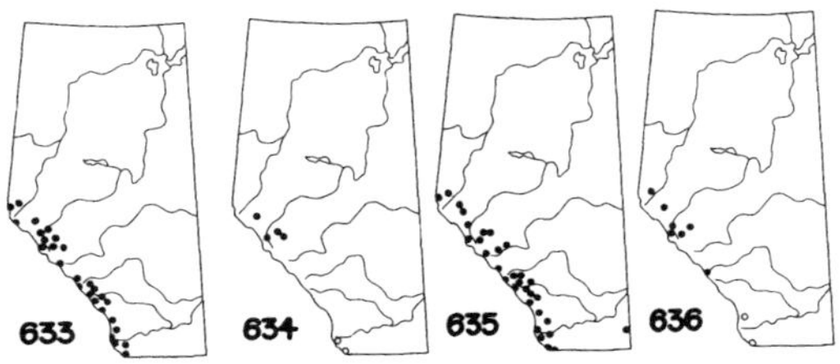

633 *Saxifraga lyallii*
634 ○ *Saxifraga mertensiana*
● *Saxifraga nivalis*
635 *Saxifraga occidentalis*
636 ○ *Saxifraga odontoloma*
● *Saxifraga nelsoniana*

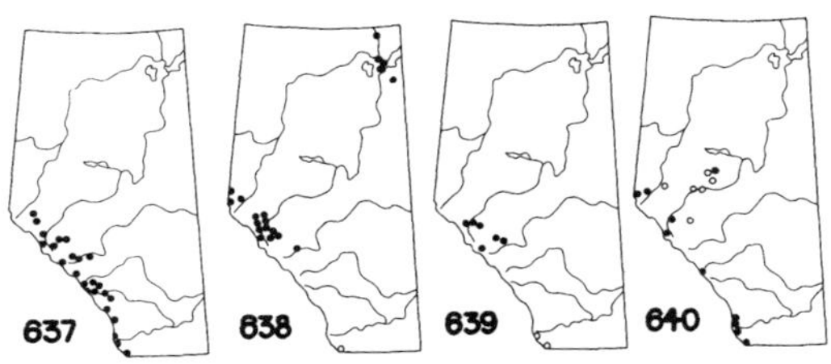

637 *Saxifraga oppositifolia*
638 ● *Saxifraga tricuspidata*
○ *Suksdorfia ranunculifolia*
639 ○ *Suksdorfia violacea*
● *Telesonix heucheriformis*
640 ○ *Tiarella trifoliata*
● *Tiarella unifoliata*

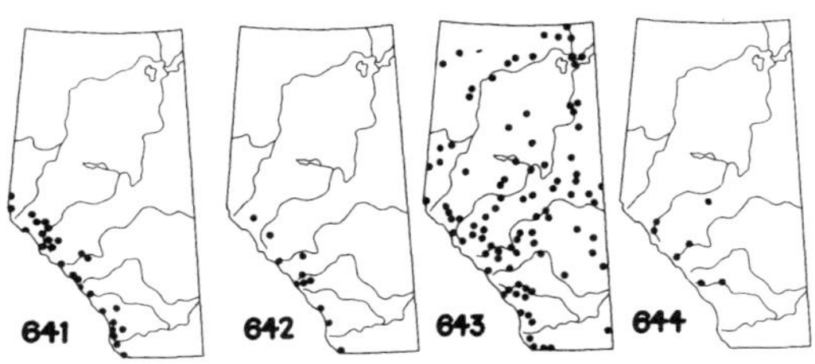

641 *Parnassia fimbriata*
642 *Parnassia kotzebuei*
643 *Parnassia palustris*
644 *Parnassia parviflora*

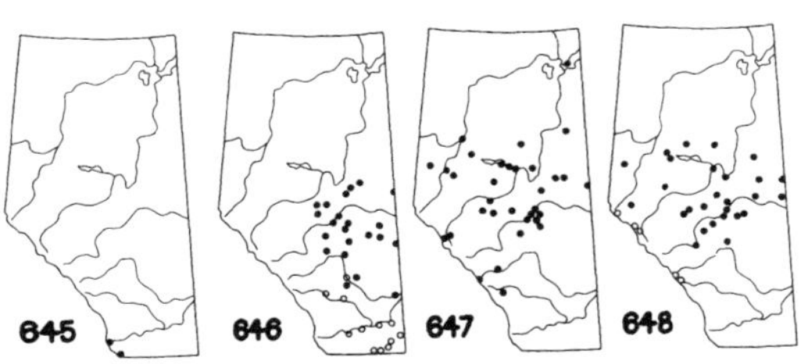

645 *Philadelphus lewisii*
646 ● *Ribes americanum*
○ *Ribes aureum*
647 *Ribes glandulosum*
648 ● *Ribes hirtellum*
○ *Ribes laxiflorum*

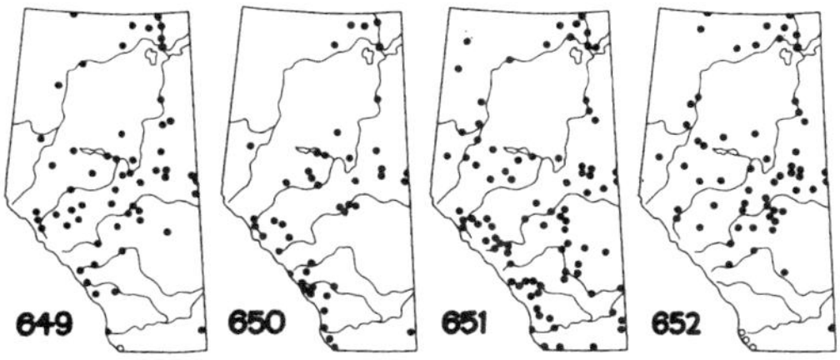

649 ● *Ribes hudsonianum*
○ *Ribes inerme*
650 *Ribes lacustre*
651 *Ribes oxyacanthoides*
652 ● *Ribes triste*
○ *Ribes viscosissimum*

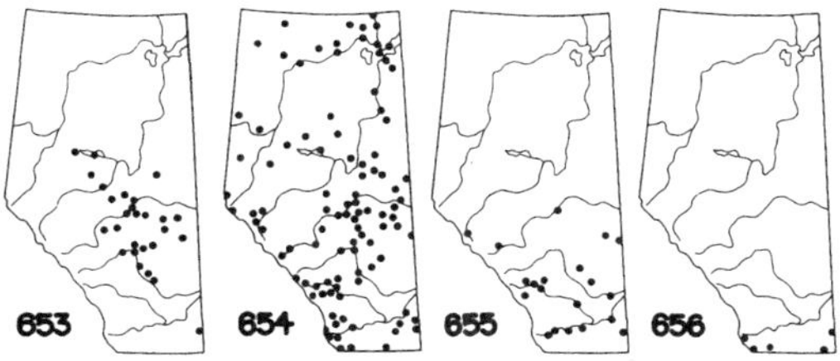

653 *Agrimonia striata*
654 *Amelanchier alnifolia*
655 *Chamaerhodos erecta*
656 *Crataegus douglasii*

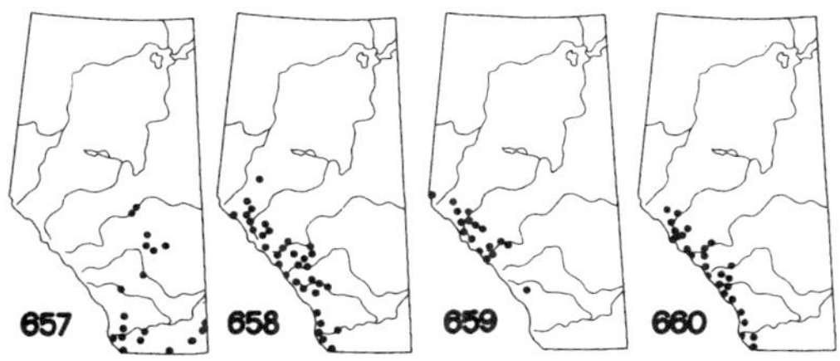

657 *Crataegus rotundifolia*
658 *Dryas drummondii*
659 *Dryas integrifolia*
660 *Dryas octopetala*

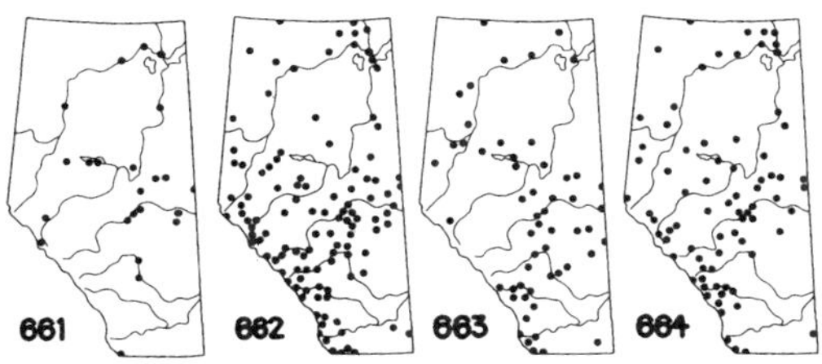

661 *Fragaria vesca*
662 *Fragaria virginiana*
663 *Geum aleppicum*
664 *Geum macrophyllum*

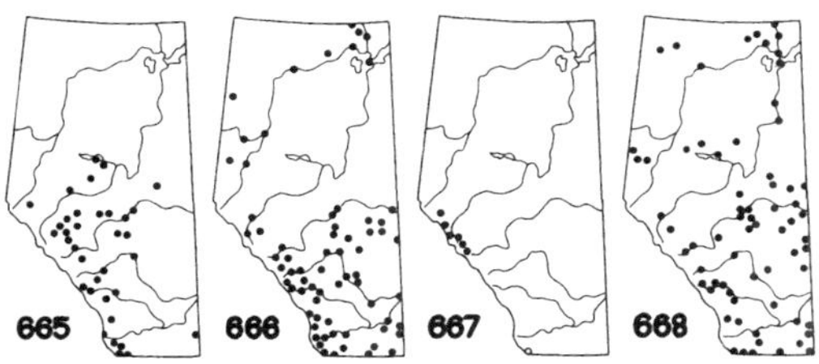

665 *Geum rivale*
666 *Geum triflorum*
667 ● *Luetkea pectinata*
○ *Physocarpus malvaceus*
668 *Potentilla anserina*

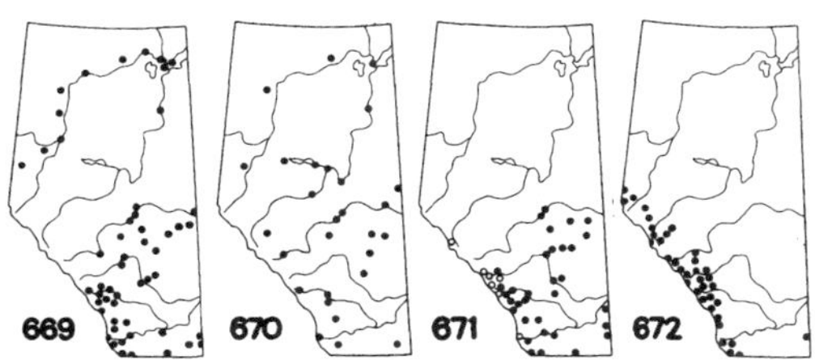

669 *Potentilla arguta*
670 *Potentilla bipinnatifida*
671 ● *Potentilla concinna*
○ *Potentilla drummondii*
672 *Potentilla diversifolia*

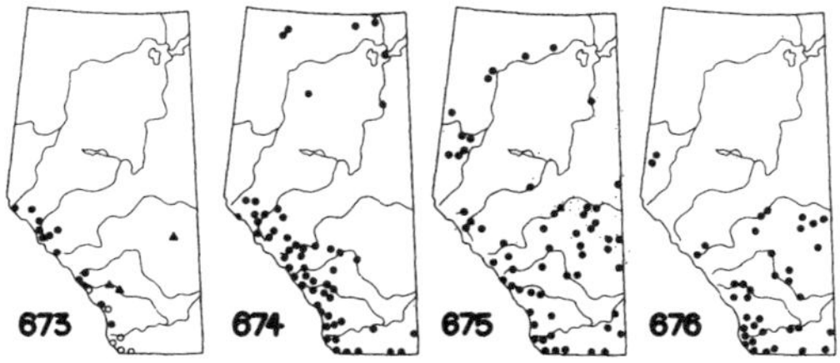

673 ▲ *Potentilla finitima*
○ *Potentilla glandulosa*
● *Potentilla hyparctica*
674 *Potentilla fruticosa*
675 *Potentilla gracilis*
676 *Potentilla hippiana*

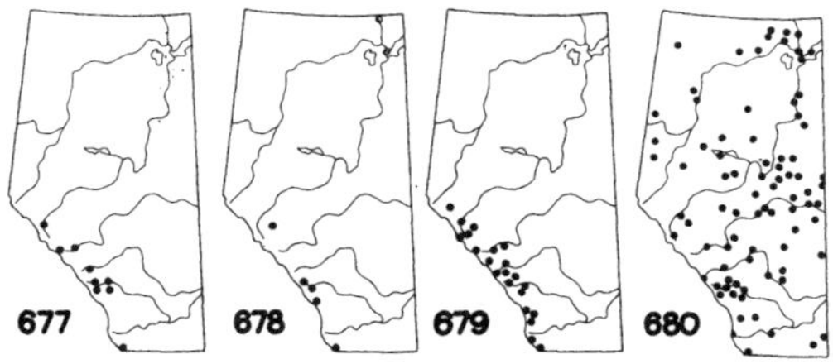

677 *Potentilla hookeriana*
678 ○ *Potentilla multifida*
● *Potentilla multisecta*
679 *Potentilla nivea*
680 *Potentilla norvegica*

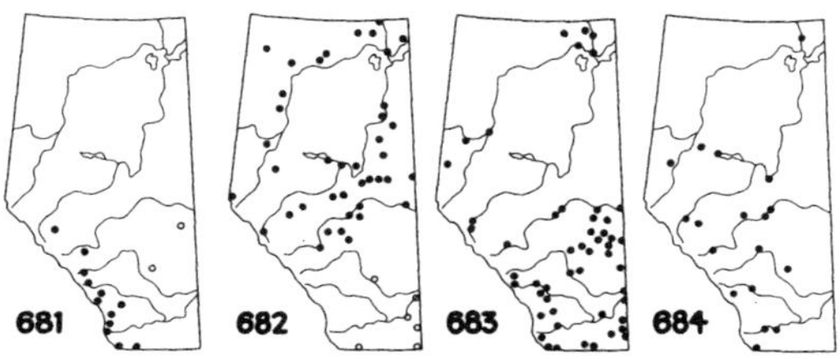

681 ● *Potentilla ovina*
○ *Potentilla plattensis*
682 ● *Potentilla palustris*
○ *Potentilla paradoxa*
683 *Potentilla pensylvanica*
684 *Potentilla rivalis*

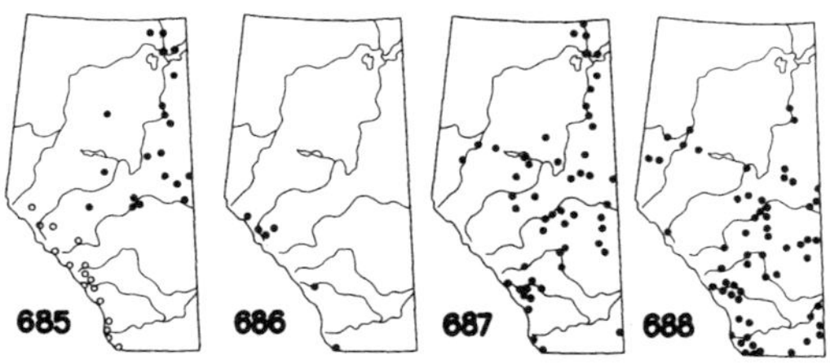

685 ● *Potentilla tridentata*
○ *Potentilla uniflora*
686 *Potentilla villosa*
687 *Prunus pensylvanica*
688 *Prunus virginiana*

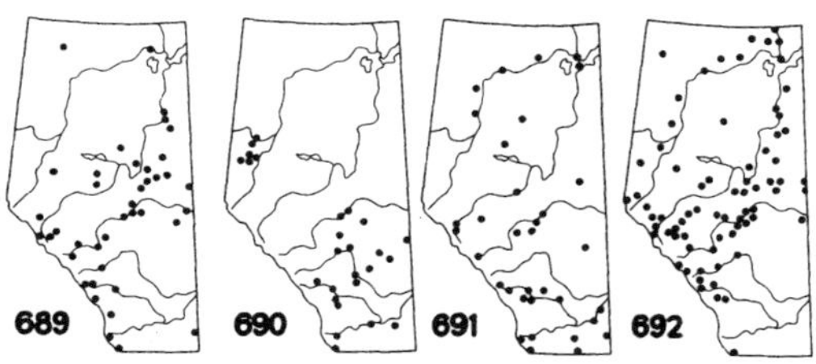

689 *Rosa acicularis*
690 *Rosa arkansana*
691 *Rosa woodsii*
692 *Rubus arcticus*

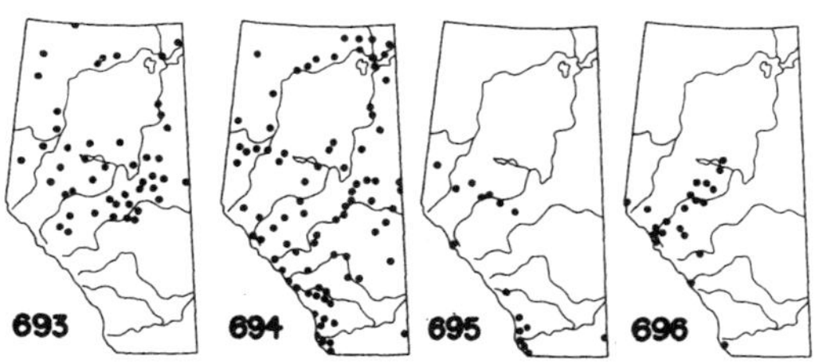

693 *Rubus chamaemorus*
694 *Rubus idaeus*
695 *Rubus parviflorus*
696 *Rubus pedatus*

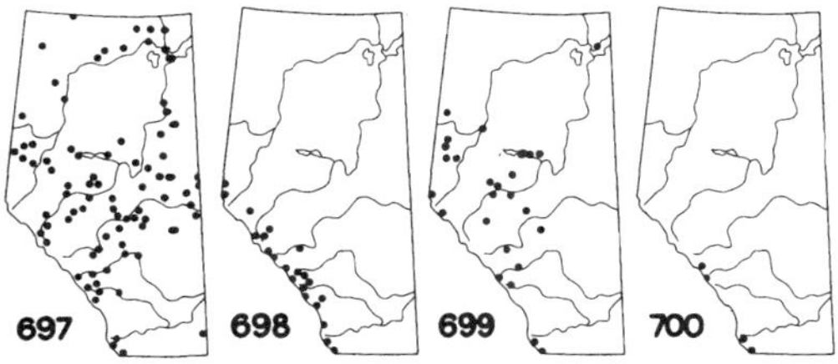

697 *Rubus pubescens*
698 *Sibbaldia procumbens*
699 *Sorbus scopulina*
700 *Sorbus sitchensis*

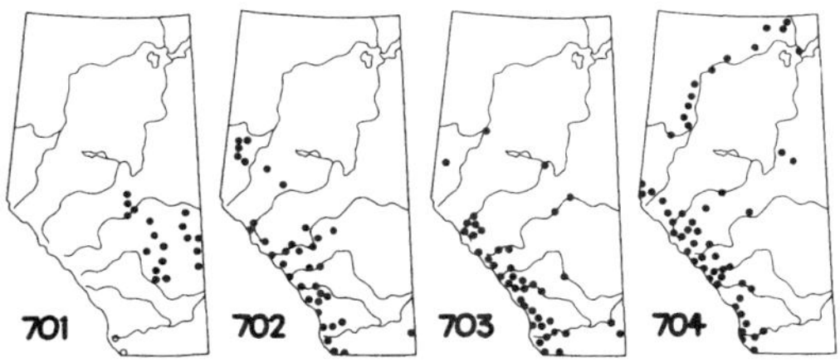

701 ● *Spiraea alba*
○ *Spiraea densiflora*
702 *Spiraea betulifolia*
703 *Astragalus aboriginum*
704 *Astragalus alpinus*

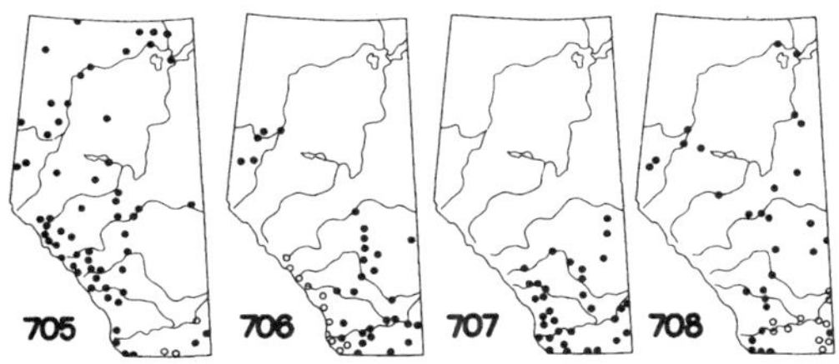

705 ● *Astragalus americanus*
○ *Astragalus kentrophyta*
706 ● *Astragalus bisulcatus*
○ *Astragalus bourgovii*
707 *Astragalus crassicarpus*
708 ● *Astragalus canadensis*
○ *Astragalus spatulatus*

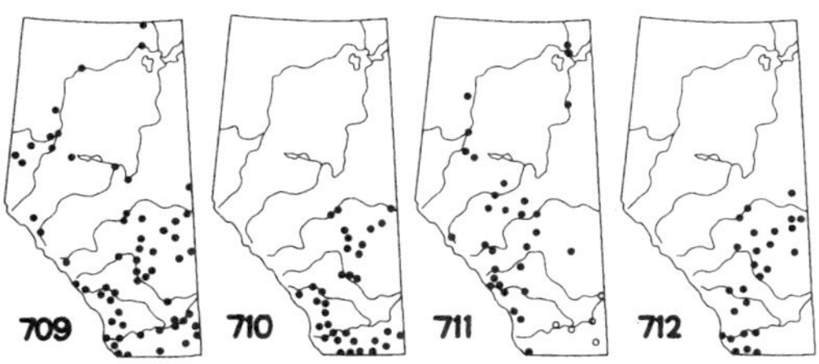

709 *Astragalus dasyglottis*
710 *Astragalus drummondii*
711 ● *Astragalus eucosmus*
○ *Astragalus lotiflorus*
712 *Astragalus flexuosus*

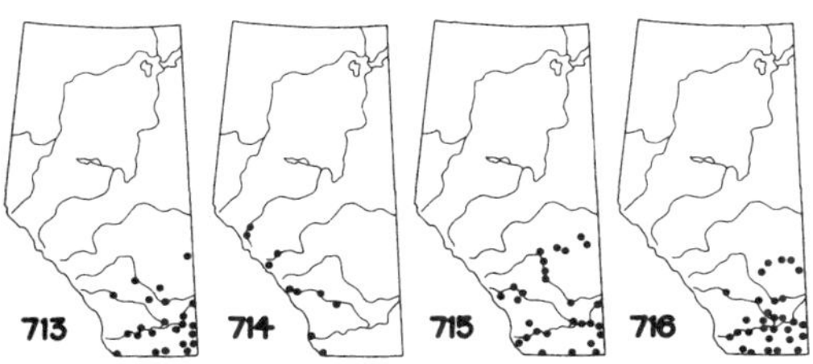

713 *Astragalus gilviflorus*
714 *Astragalus miser*
715 *Astragalus missouriensis*
716 *Astragalus pectinatus*

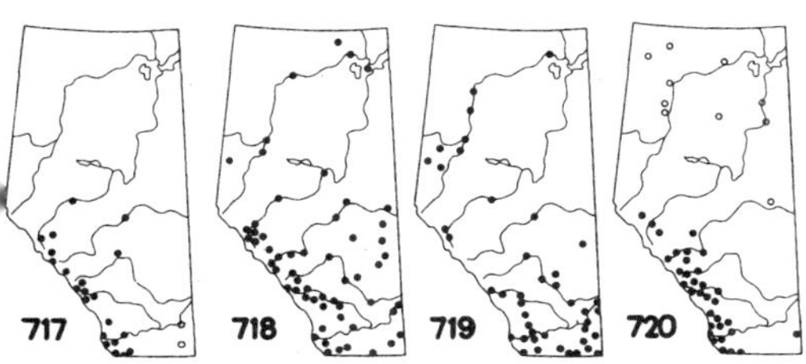

717 ○ *Astragalus purshii*
● *Astragalus robbinsii*
718 *Astragalus striatus*
719 *Astragalus tenellus*
720 ● *Astragalus vexilliflexus*
○ *Astragalus bodinii*

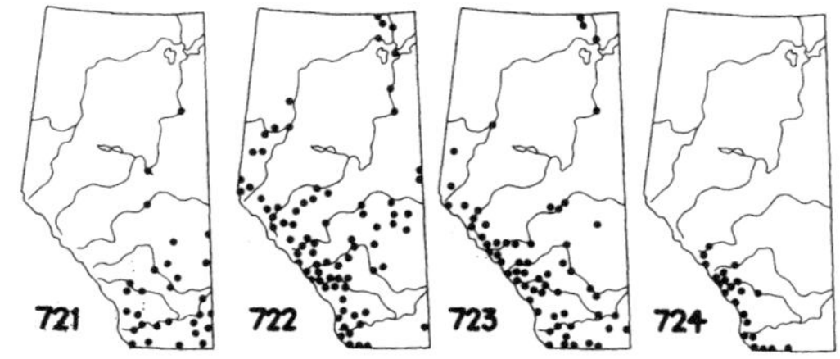

721 *Glycyrrhiza lepidota*
722 *Hedysarum alpinum*
723 *Hedysarum boreale*
724 *Hedysarum sulphurescens*

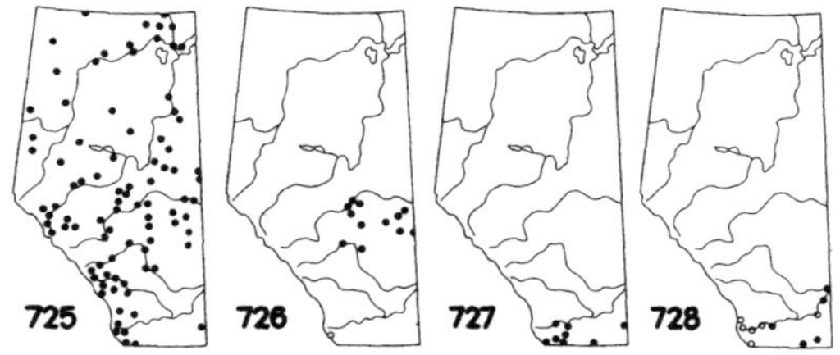

725 *Lathyrus ochroleucus*
726 ● *Lathyrus venosus*
○ *Lupinus polyphyllus*
727 *Lupinus argenteus*
728 ○ *Lupinus lepidus*
● *Lupinus pusillus*

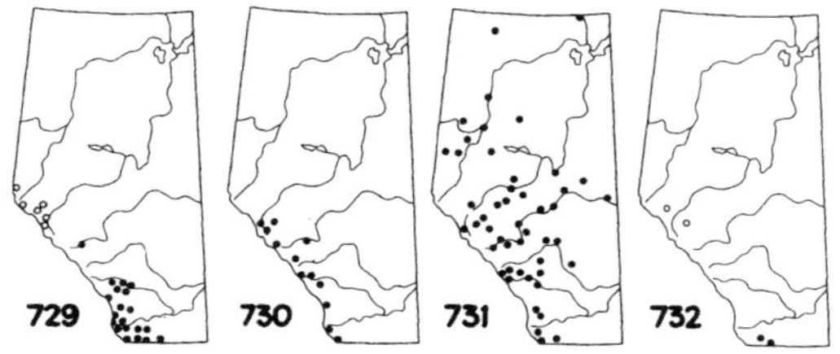

729 ○ *Lupinus nootkatensis*
● *Lupinus sericeus*
730 *Oxytropis cusickii*
731 *Oxytropis deflexa*
732 ● *Oxytropis lagopus*
○ *Oxytropis jordalii*

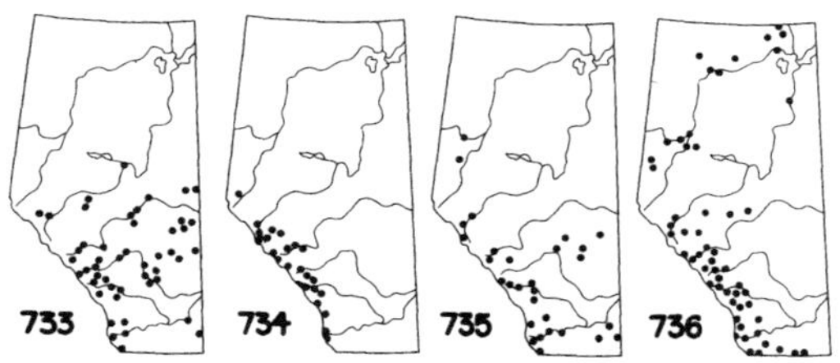

733 *Oxytropis monticola*
734 *Oxytropis podocarpa*
735 *Oxytropis sericea*
736 *Oxytropis splendens*

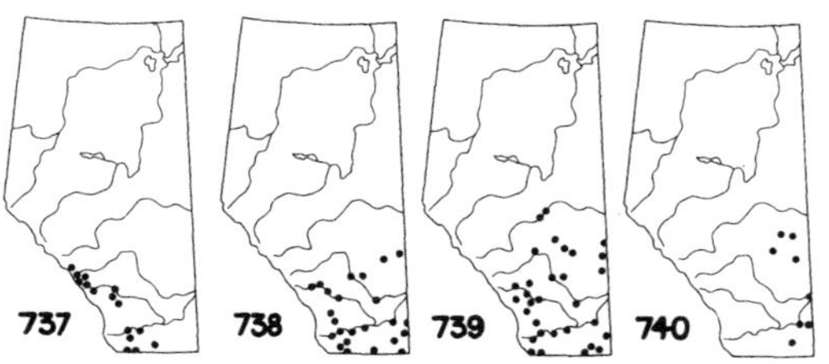

737 *Oxytropis viscida*
738 *Petalostemon candidum*
739 *Petalostemon purpureum*
740 *Psoralea agrophylla*

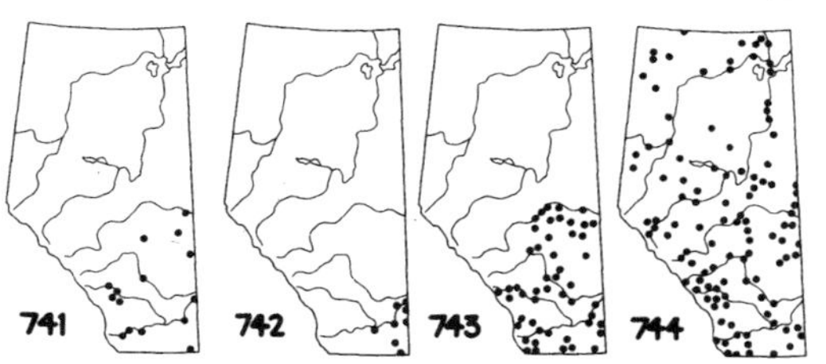

741 *Psoralea esculenta*
742 *Psoralea lanceolata*
743 *Thermopsis rhombifolia*
744 *Vicia americana*

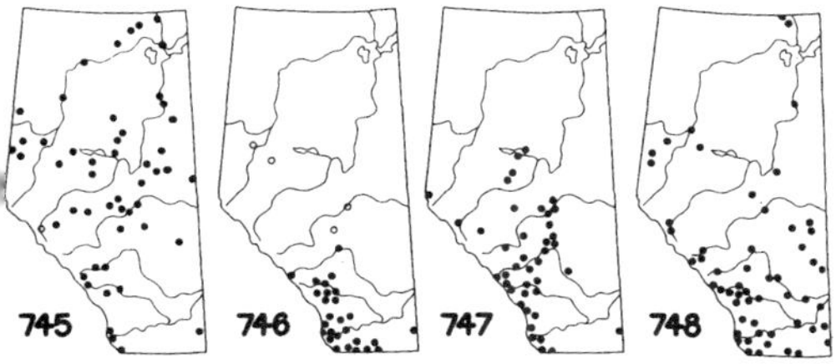

745 ● *Geranium bicknellii*
○ *Geranium erianthum*
746 ○ *Geranium carolinianum*
● *Geranium viscosissimum*
747 *Geranium richardsonii*
748 *Linum lewisii*

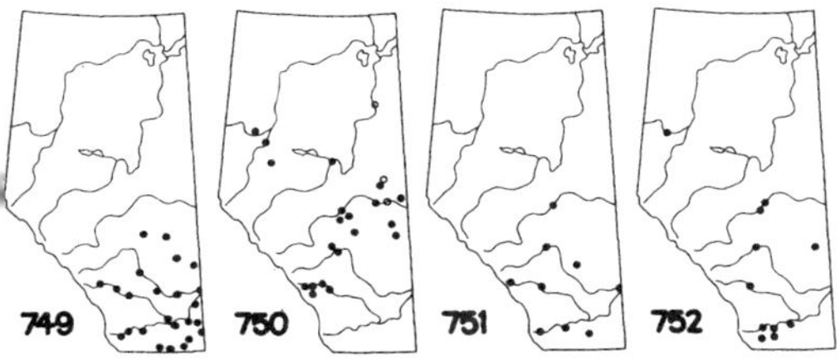

749 *Linum rigidum*
750 ○ *Polygala paucifolia*
● *Polygala senega*
751 *Euphorbia glyptosperma*
752 *Euphorbia serpyllifolia*

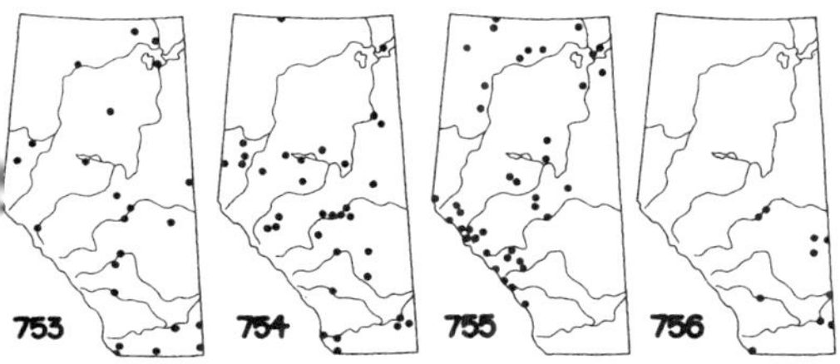

753 *Callitriche hermaphroditica*
754 *Callitriche verna*
755 *Empetrum nigrum*
756 *Rhus radicans*

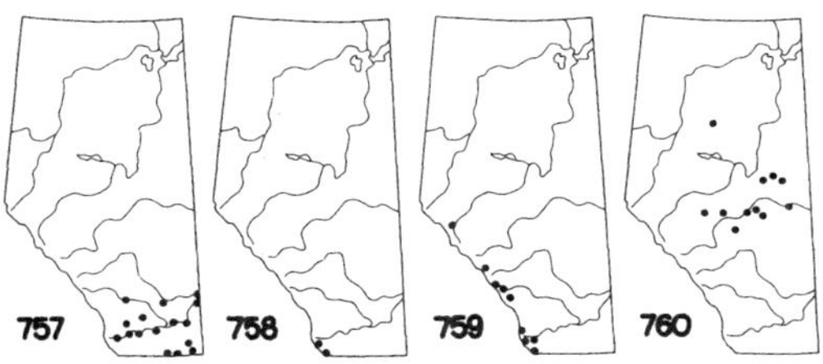

757 *Rhus trilobata*
758 *Pachistima myrsinites*
759 *Acer glabrum*
760 *Impatiens capensis*

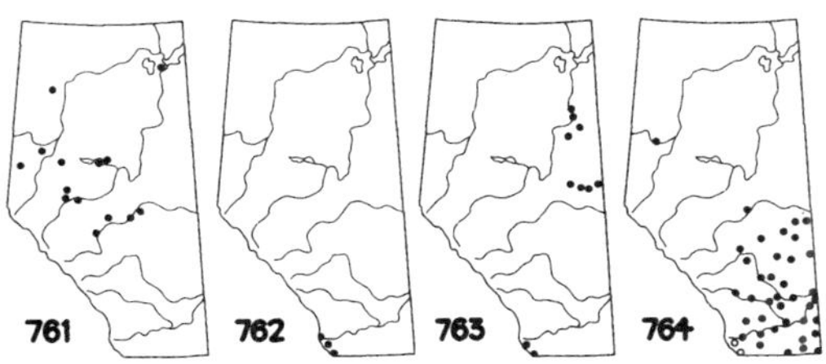

761 *Impatiens noli-tangere*
762 *Ceanothus velutinus*
763 *Rhamnus alnifolia*
764 ● *Sphaeralcea coccinea*
○ *Iliamna rivularis*

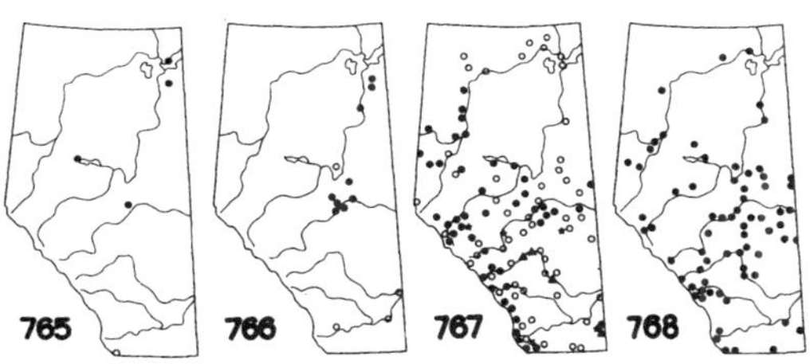

765 ○ *Hypericum formosum*
● *Hypericum majus*
766 ○ *Elatine triandra*
● *Hudsonia tomentosa*
767 *Viola adunca*
768 *Viola canadensis*

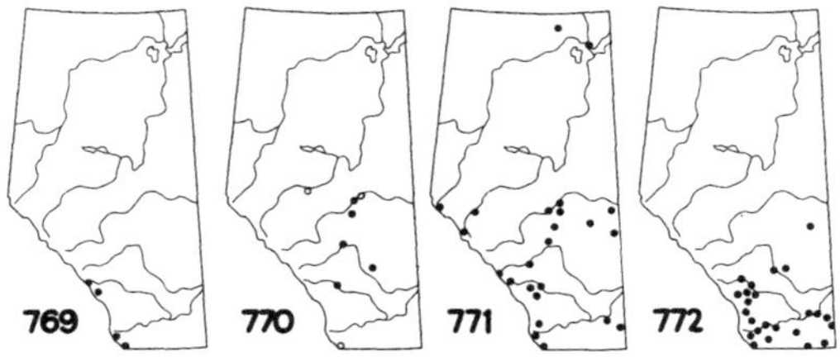

769 *Viola glabella*
770 ○ *Viola macloskeyi*
● *Viola pedatifida*
771 *Viola nephrophylla*
772 *Viola nuttallii*

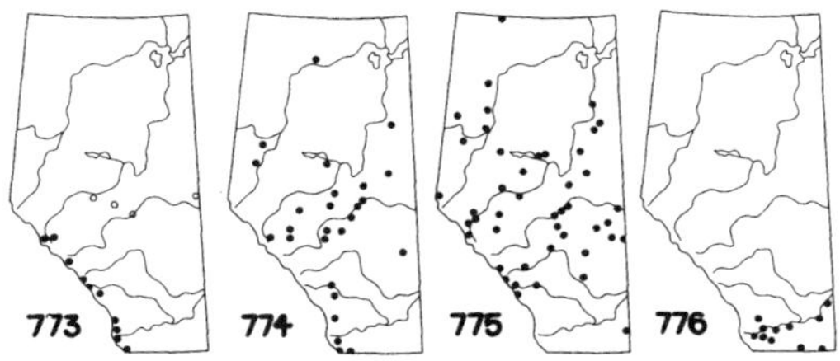

773 ● *Viola orbiculata*
○ *Viola selkirkii*
774 *Viola palustris*
775 *Viola renifolia*
776 *Mentzelia decapetala*

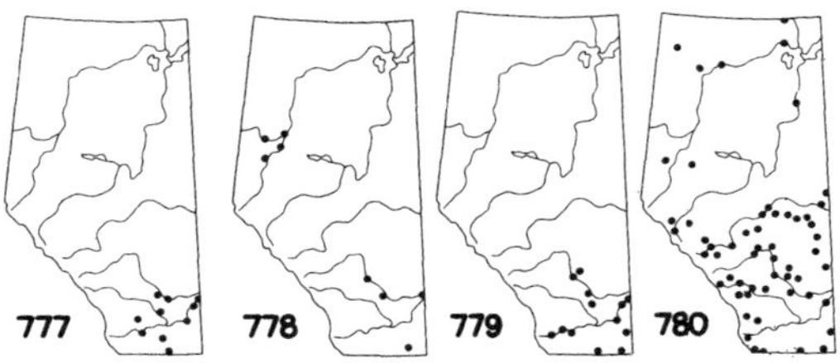

777 *Coryphantha vivipara*
778 *Opuntia fragilis*
779 *Opuntia polyacantha*
780 *Elaeagnus commutata*

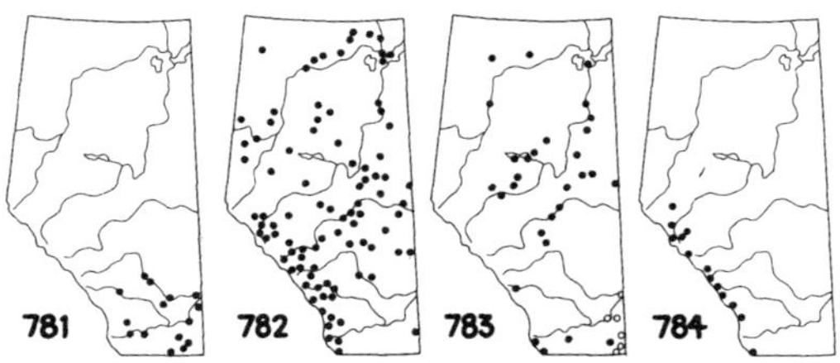

781 *Shepherdia argentea*
782 *Shepherdia canadensis*
783 ○ *Boisduvalia glabella*
● *Circaea alpina*
784 *Epilobium anagallidifolium*

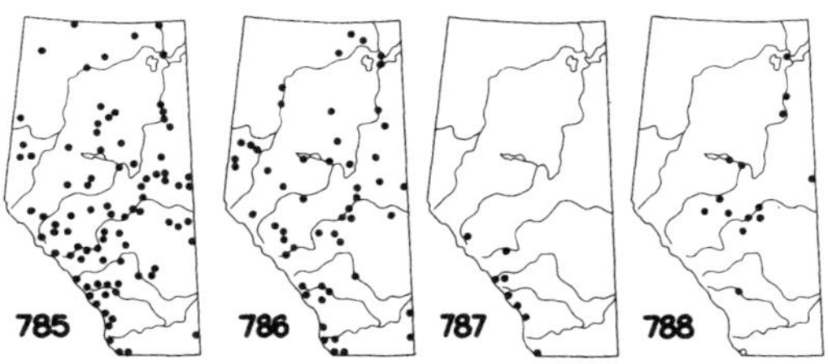

785 *Epilobium angustifolium*
786 *Epilobium ciliatum*
787 *Epilobium clavatum*
788 ○ *Epilobium glaberrimum*
● *Epilobium leptophyllum*

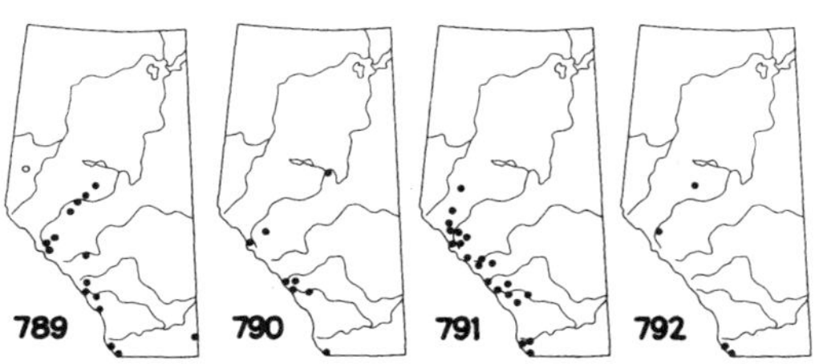

789 ○ *Epilobium halleanum*
● *Epilobium hornemannii*
790 *Epilobium lactiflorum*
791 *Epilobium latifolium*
792 *Epilobium leptocarpum*

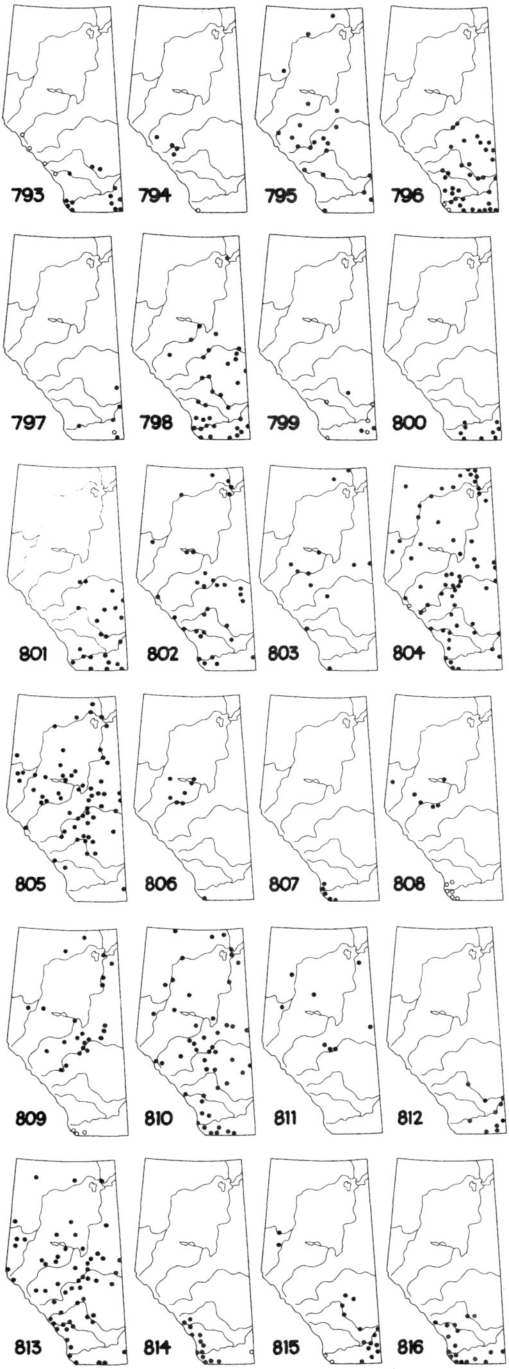

793 ○ *Epilobium luteum*
● *Epilobium paniculatum*
794 ○ *Epilobium mirabile*
● *Epilobium saximontanum*
795 *Epilobium palustre*
796 ● *Gaura coccinea*
○ *Gayophytum racemosum*

797 ○ *Oenothera andina*
● *Oenothera serrulata*
798 *Oenothera biennis*
799 ● *Oenothera breviflora*
○ *Oenothera flava*
800 *Oenothera caespitosa*

801 *Oenothera nuttallii*
802 *Myriophyllum exalbescens*
803 *Myriophyllum verticillatum*
804 ○ *Hippuris montana*
● *Hippuris vulgaris*

805 *Aralia nudicaulis*
806 *Oplopanax horridum*
807 *Angelica arguta*
808 ○ *Angelica dawsonii*
● *Angelica genuflexa*

809 ○ *Bupleurum americanum*
● *Cicuta bulbifera*
810 *Cicuta maculata*
811 *Cicuta virosa*
812 *Cymopteris acaulis*

813 *Heracleum lanatum*
814 ○ *Lomatium cous*
● *Lomatium dissectum*
815 ● *Lomatium foeniculaceum*
○ *Lomatium sandbergii*
816 *Lomatium macrocarpum*

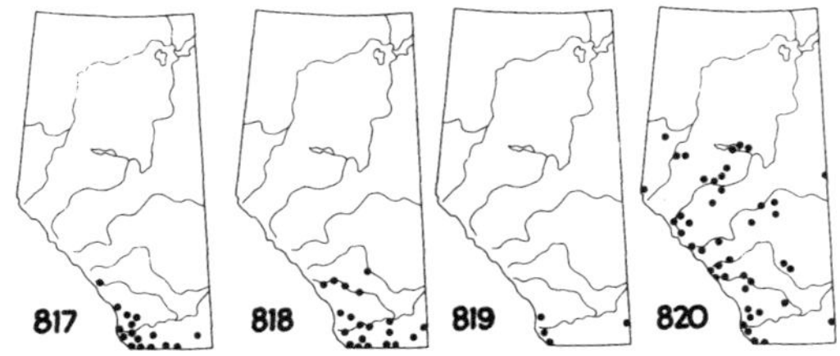

817 *Lomatium triternatum*
818 *Musineon divaricatum*
819 *Osmorhiza chilensis*
820 *Osmorhiza depauperata*

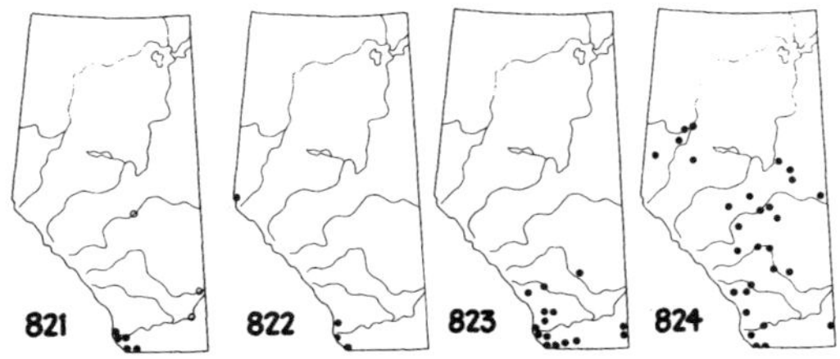

821 ○ *Osmorhiza longistylis*
● *Osmorhiza occidentalis*
822 *Osmorhiza purpurea*
823 *Perideridia gairdneri*
824 *Sanicula marilandica*

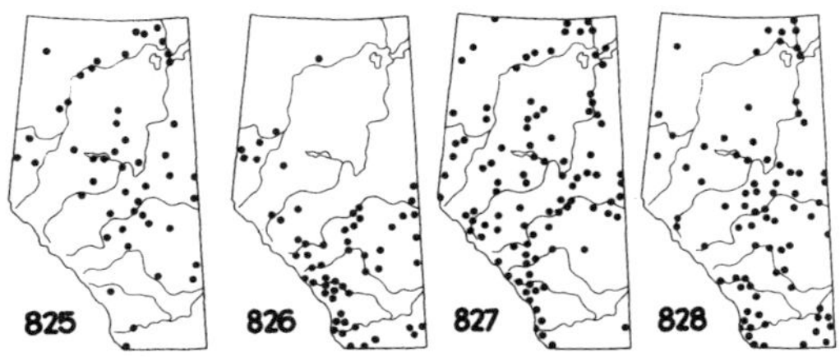

825 *Sium suave*
826 *Zizia aptera*
827 *Cornus canadensis*
828 *Cornus stolonifera*

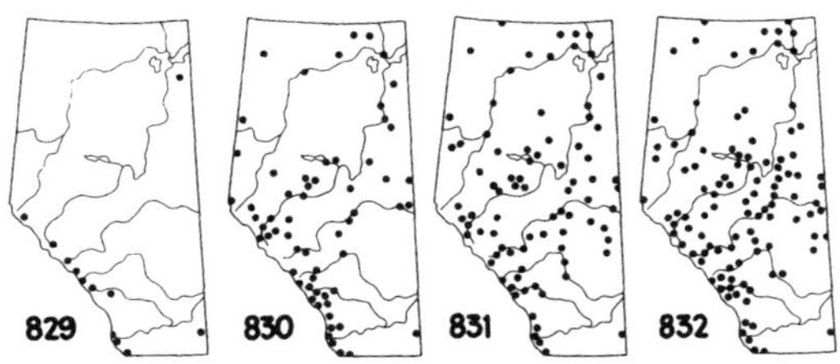

829 *Chimaphila umbellata*
830 *Moneses uniflora*
831 *Orthilia secunda*
832 *Pyrola asarifolia*

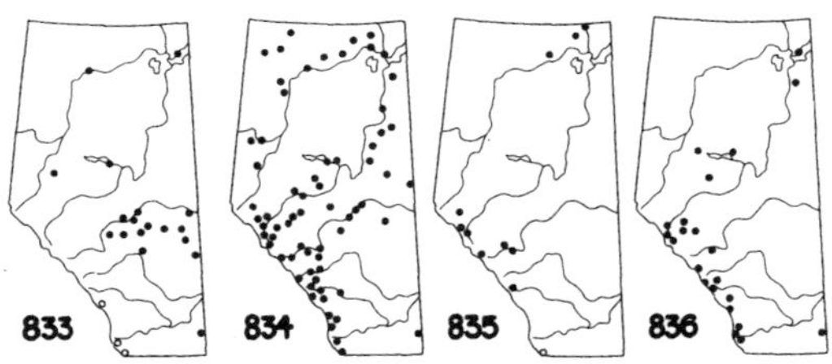

833 ○ *Pyrola bracteata*
● *Pyrola elliptica*
834 *Pyrola chlorantha*
835 ● *Pyrola grandiflora*
○ *Pyrola picta*
836 *Pyrola minor*

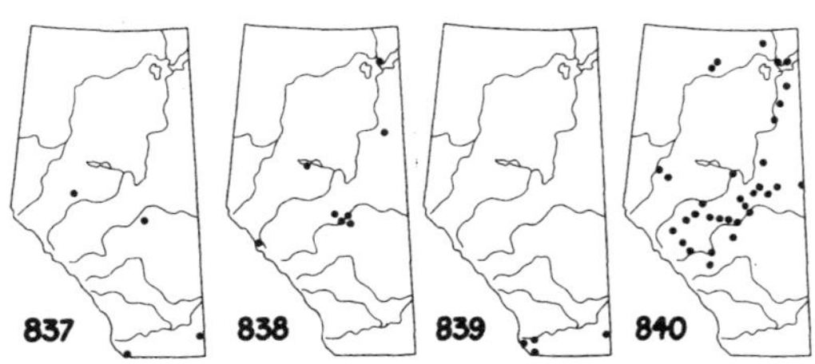

837 *Monotropa hypopitys*
838 *Monotropa uniflora*
839 *Pterospora andromedea*
840 *Andromeda polifolia*

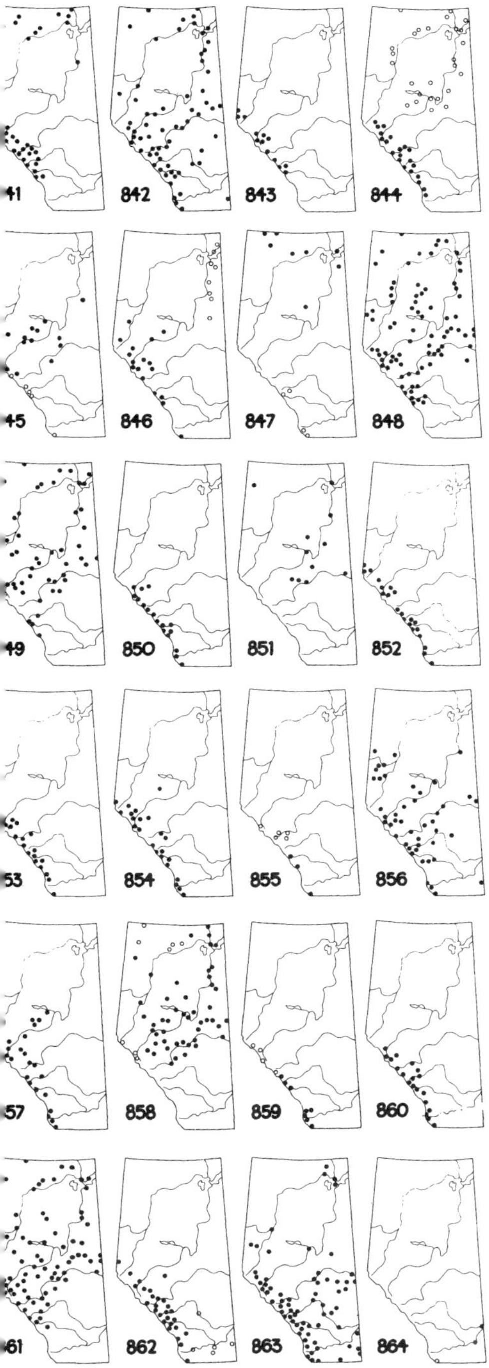

841 *Arctostaphylos rubra*
842 *Arctostaphylos uva-ursi*
843 *Cassiope mertensiana*
844 ● *Cassiope tetragona*
○ *Chamaedaphne calyculata*

845 ● *Gaultheria hispidula*
○ *Gaultheria humifusa*
846 ● *Kalmia microphylla*
○ *Kalmia polifolia*
847 ○ *Ledum glandulosum*
● *Ledum palustre*
848 *Ledum groenlandicum*

849 ○ *Loiseleuria procumbens*
● *Oxycoccus microcarpus*
850 *Menziesia ferruginea*
851 *Oxycoccus quadripetalus*
852 *Phyllodoce empetriformis*

853 *Phyllodoce glanduliflora*
854 *Rhododendron albiflorum*
855 ○ *Rhododendron lapponicum*
● *Phyllodoce × intermedia*
856 *Vaccinium caespitosum*

857 *Vaccinium membranaceum*
858 ● *Vaccinium myrtilloides*
○ *Vaccinium uliginosum*
859 ● *Vaccinium myrtillus*
○ *Vaccinium ovalifolium*
860 *Vaccinium scoparium*

861 *Vaccinium vitis-idaea*
862 ● *Androsace chamaejasme*
○ *Androsace occidentalis*
863 *Androsace septentrionalis*
864 ● *Centunculus minimus*
○ *Douglasia montana*

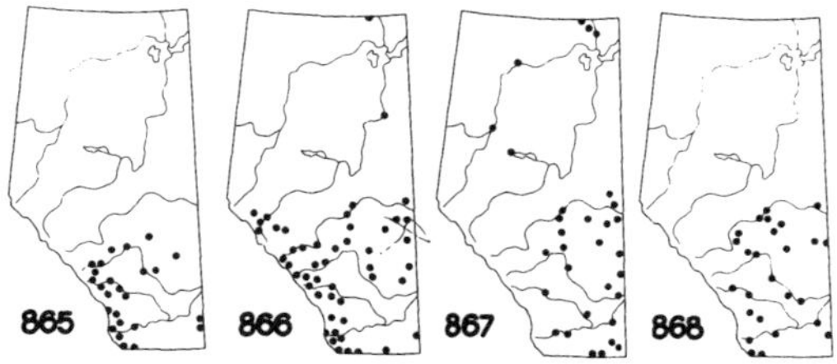

865 *Dodecatheon conjugens*
866 *Dodecatheon pulchellum*
867 *Glaux maritima*
868 *Lysimachia ciliata*

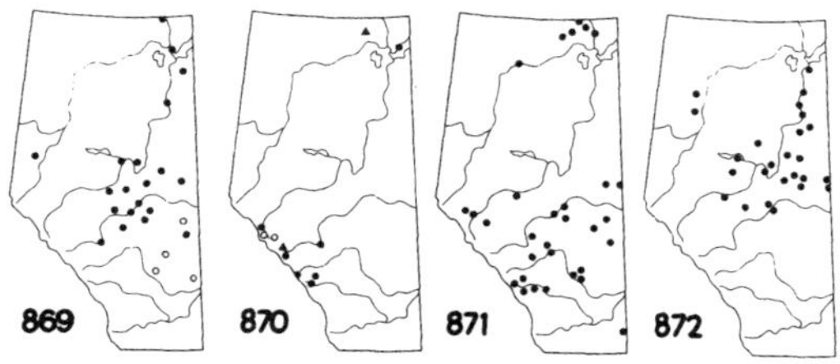

869 ○ *Lysimachia lanceolata*
● *Lysimachia thyrsiflora*
870 ○ *Primula egaliksensis*
● *Primula mistassinica*
▲ *Primula stricta*
871 *Primula incana*
872 *Trientalis borealis*

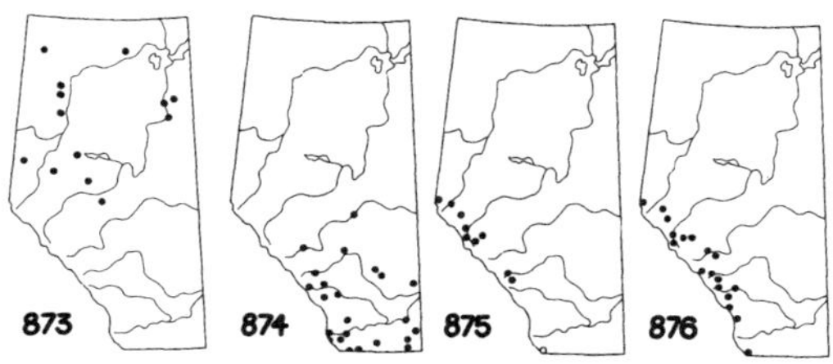

873 *Trientalis europaea*
874 *Gentiana affinis*
875 ○ *Gentiana calycosa*
● *Gentiana glauca*
876 *Gentiana prostrata*

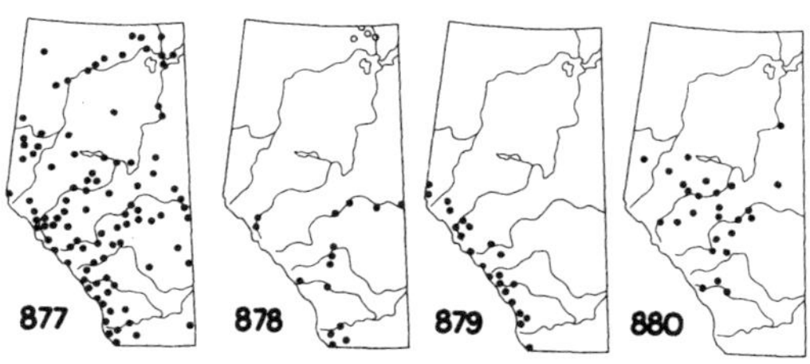

877 *Gentianella amarella*
878 ● *Gentianella crinita*
○ *Gentianella detonsa*
879 *Gentianella propinqua*
880 *Halenia deflexa*

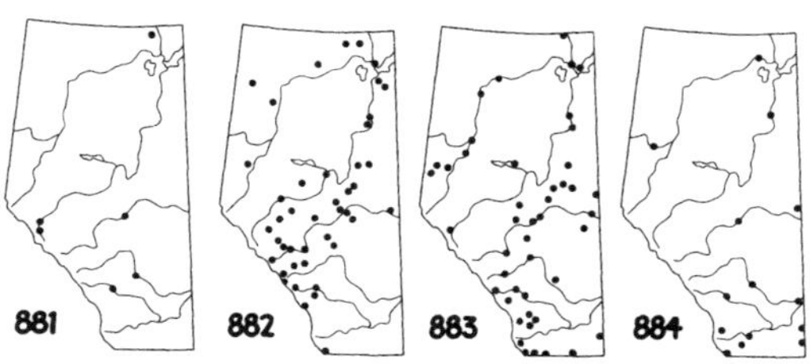

881 *Lomatagonium rotatum*
882 *Menyanthes trifoliata*
883 *Apocynum androsaemifolium*
884 *Apocynum cannabinum*

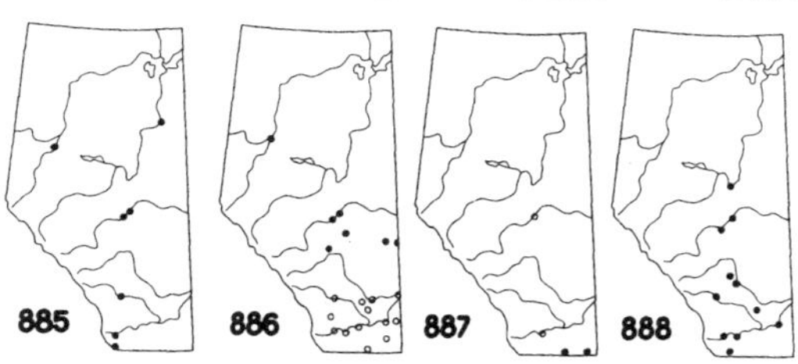

885 *Apocynum × medium*
886 ● *Asclepias ovalifolia*
○ *Asclepias speciosa*
887 ● *Asclepias viridiflora*
○ *Cuscuta gronovii*
888 *Convolvulus sepium*

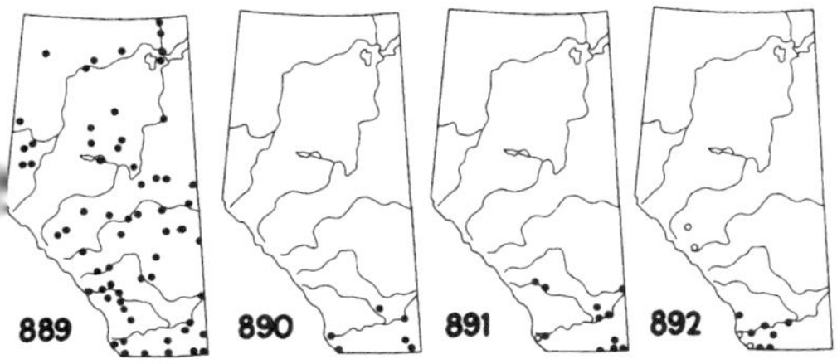

889 *Collomia linearis*
890 *Linanthus septentrionalis*
891 ○ *Microsteris gracilis*
● *Navarretia minima*
892 ● *Phlox alyssifolia*
○ *Polemonium viscosum*

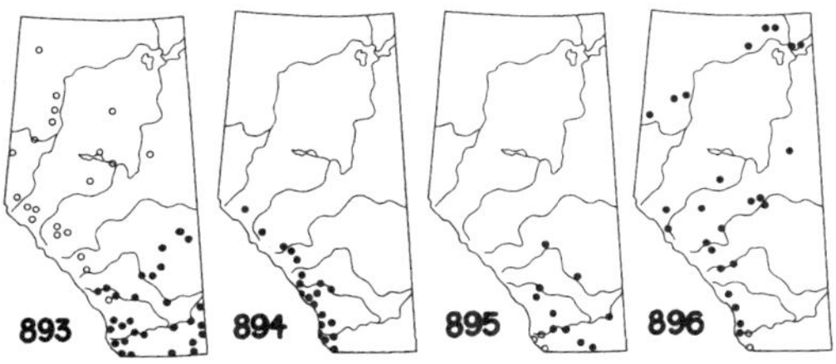

893 ● *Phlox hoodii*
○ *Polemonium acutiflorum*
894 *Polemonium pulcherrimum*
895 ● *Ellisia nyctelea*
○ *Hydrophyllum capitatum*
896 ○ *Nemophila breviflora*
● *Phacelia franklinii*

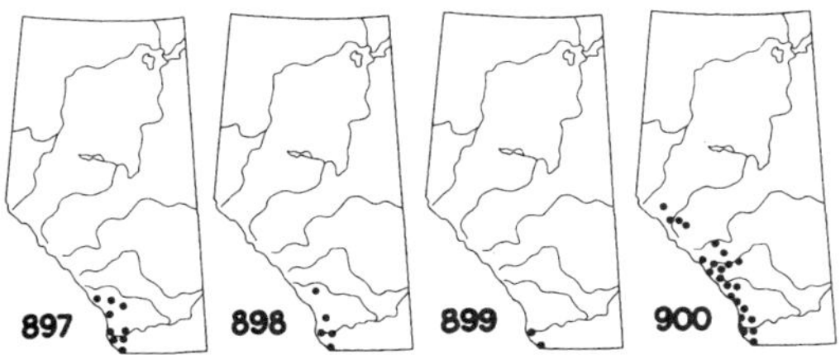

897 *Phacelia hastata*
898 *Phacelia linearis*
899 *Phacelia lyallii*
900 *Phacelia sericea*

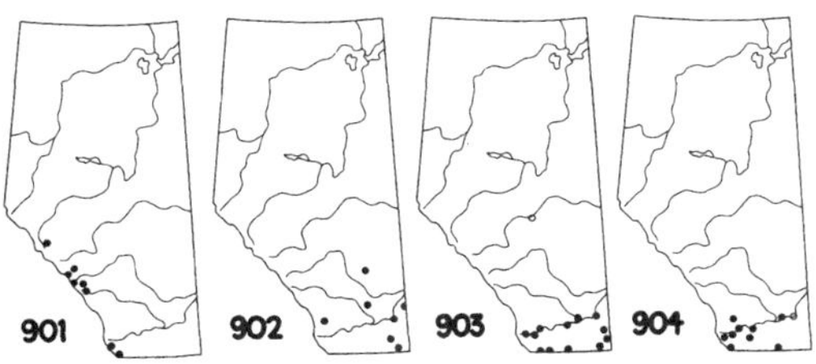

901 *Romanzoffia sitchensis*
902 *Cryptantha fendleri*
903 ● *Cryptantha macounii*
○ *Cynoglossum boreale*
904 ● *Cryptantha nubigena*
○ *Cryptantha minima*

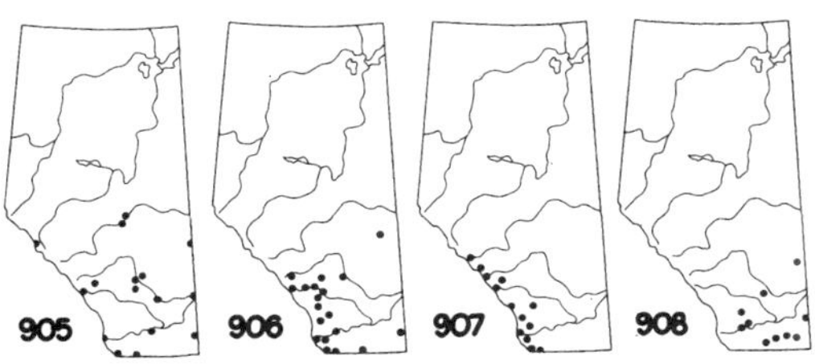

905 *Hackelia americana*
906 *Hackelia floribunda*
907 *Hackelia jessicae*
908 *Heliotropium curassavicum*

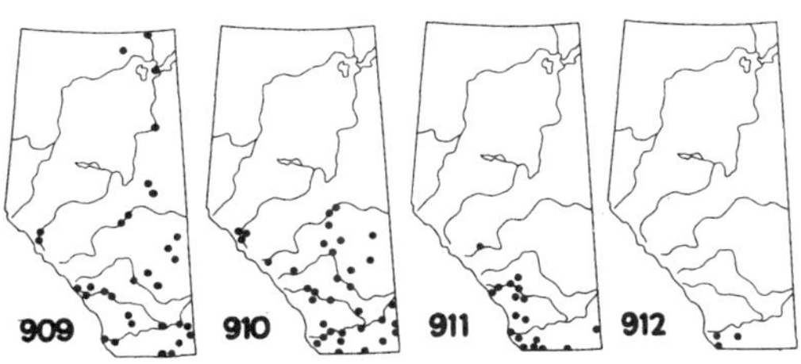

909 *Lappula occidentalis*
910 *Lithospermum incisum*
911 *Lithospermum ruderale*
912 *Mertensia lanceolata*

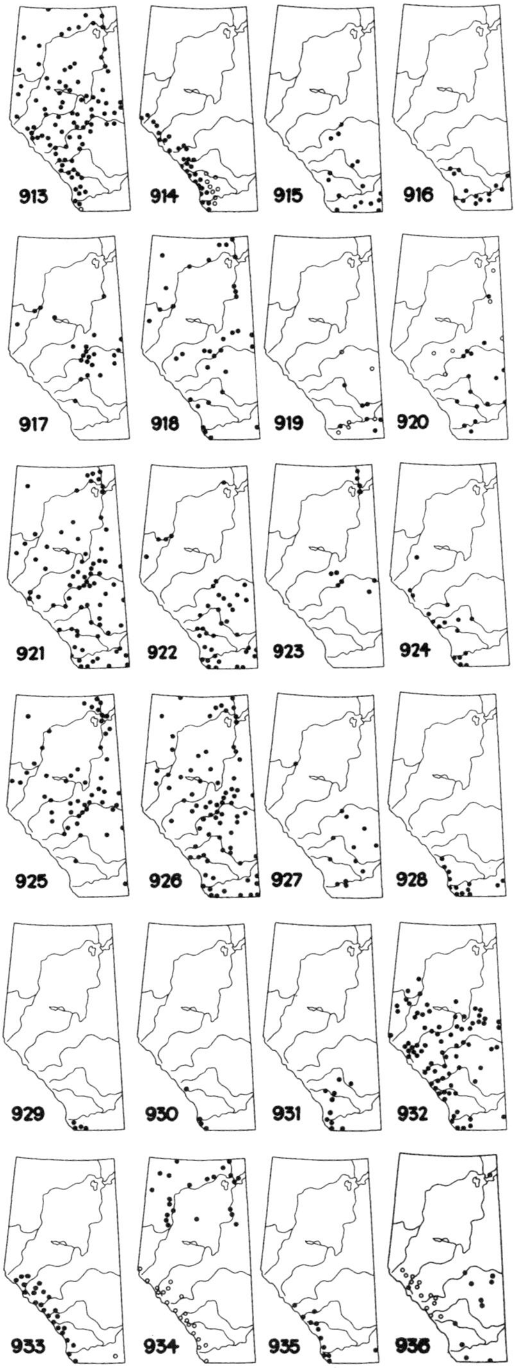

913 ○ *Mertensia longiflora*
● *Mertensia paniculata*
914 ● *Myosotis alpestris*
○ *Onosmodium molle*
915 *Plagiobothrys scouleri*
916 *Verbena bracteata*

917 *Agastache foeniculum*
918 *Dracocephalum parviflorum*
919 ● *Hedeoma hispidum*
○ *Lycopus americanus*
920 ● *Lycopus asper*
○ *Lycopus uniflorus*

921 *Mentha arvensis*
922 *Monarda fistulosa*
923 *Physostegia parviflora*
924 *Prunella vulgaris*

925 *Scutellaria galericulata*
926 *Stachys palustris*
927 *Solanum triflorum*
928 *Besseya wyomingensis*

929 *Castilleja cusickii*
930 *Castilleja hispida*
931 *Castilleja lutescens*
932 *Castilleja miniata*

933 ● *Castilleja occidentalis*
○ *Castilleja sessiliflora*
934 ● *Castilleja raupii*
○ *Castilleja rhexifolia*
935 *Collinsia parviflora*
936 ○ *Euphrasia arctica*
● *Gratiola neglecta*

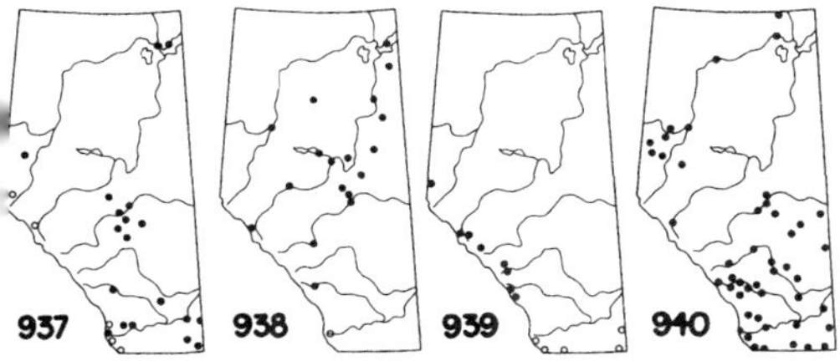

937 ● *Limosella aquatica*
○ *Mimulus lewisii*
938 ● *Melampyrum lineare*
○ *Mimulus floribundus*
939 ○ *Mimulus guttatus*
● *Pedicularis arctica*
940 *Orthocarpus luteus*

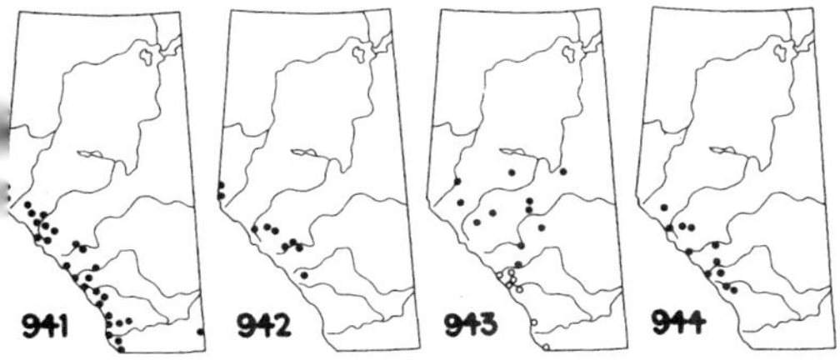

941 *Pedicularis bracteosa*
942 *Pedicularis capitata*
943 ○ *Pedicularis contorta*
● *Pedicularis parviflora*
944 *Pedicularis flammea*

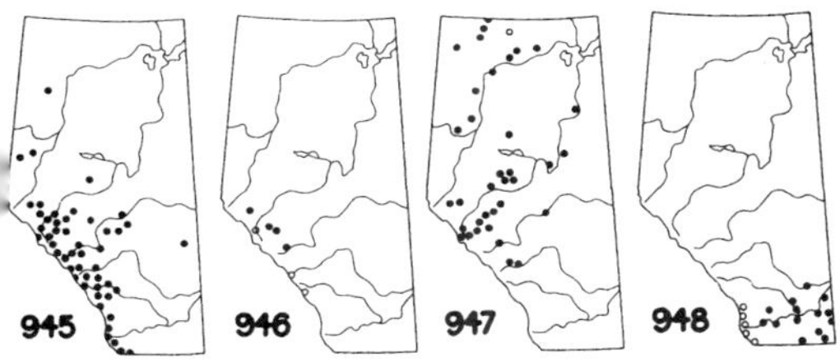

945 *Pedicularis groenlandica*
946 ● *Pedicularis lanata*
○ *Pedicularis racemosa*
947 ● *Pedicularis labradorica*
○ *Pedicularis sudetica*
948 ● *Penstemon albidus*
○ *Penstemon albertinus*

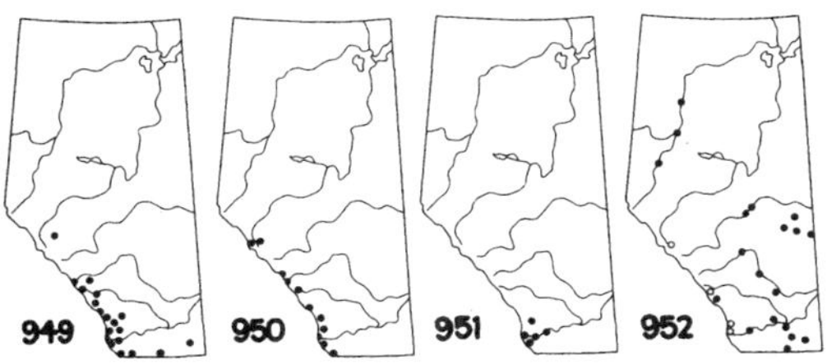

949 *Penstemon confertus*
950 *Penstemon ellipticus*
951 *Penstemon eriantherus*
952 ○ *Penstemon fruticosus*
● *Penstemon gracilis*

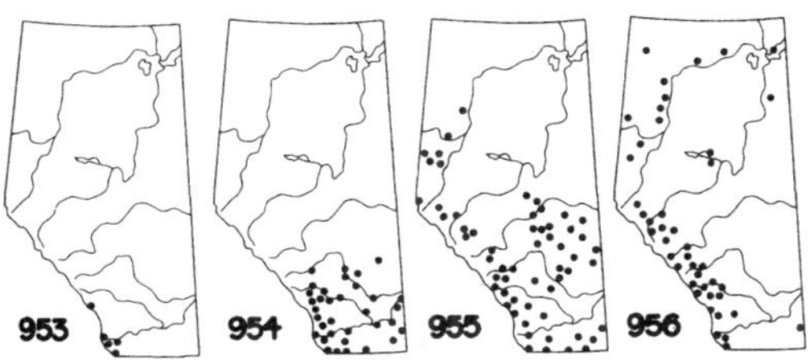

953 *Penstemon lyallii*
954 *Penstemon nitidus*
955 *Penstemon procerus*
956 *Rhinanthus minor*

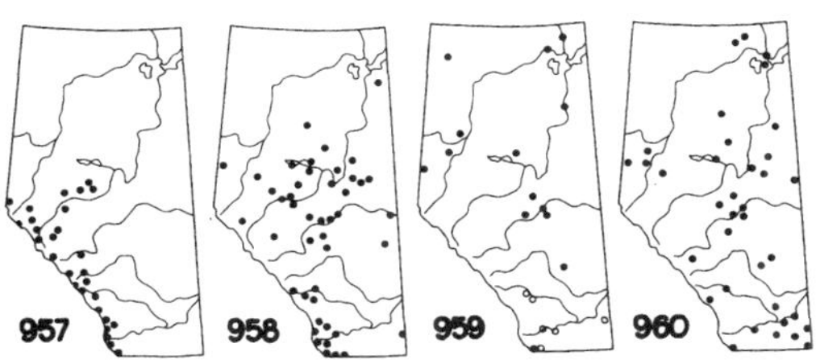

957 *Veronica alpina*
958 *Veronica americana*
959 ○ *Veronica catenata*
● *Veronica scutellata*
960 *Veronica peregrina*

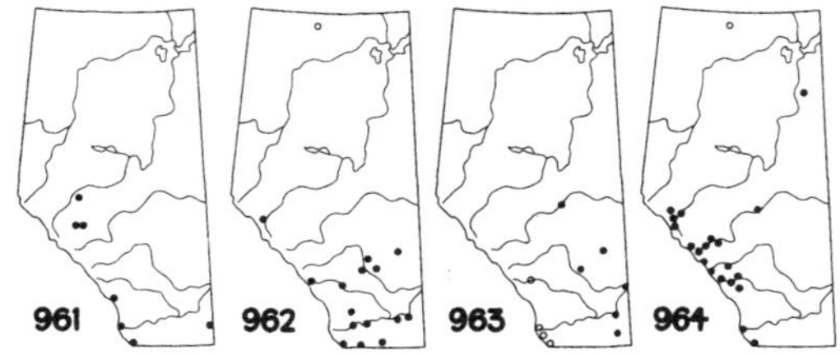

961 *Veronica serpyllifolia*
962 ○ *Boschniakia rossica*
● *Orobanche fasciculata*
963 ● *Orobanche ludoviciana*
○ *Orobanche uniflora*
964 ○ *Pinguicula villosa*
● *Pinguicula vulgaris*

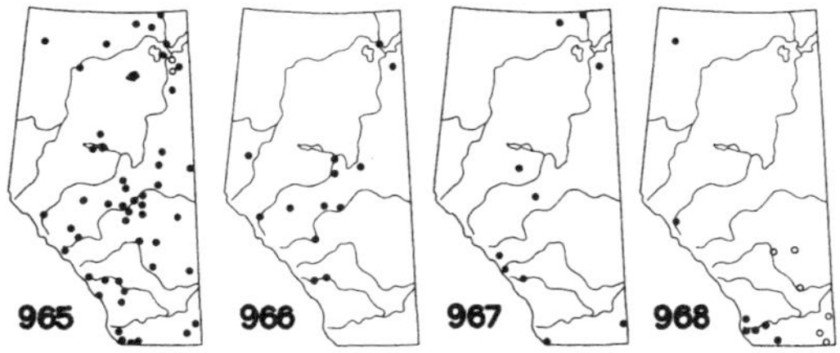

965 ○ *Utricularia cornuta*
● *Utricularia vulgaris*
966 *Utricularia intermedia*
967 *Utricularia minor*
968 ● *Plantago canescens*
○ *Plantago elongata*

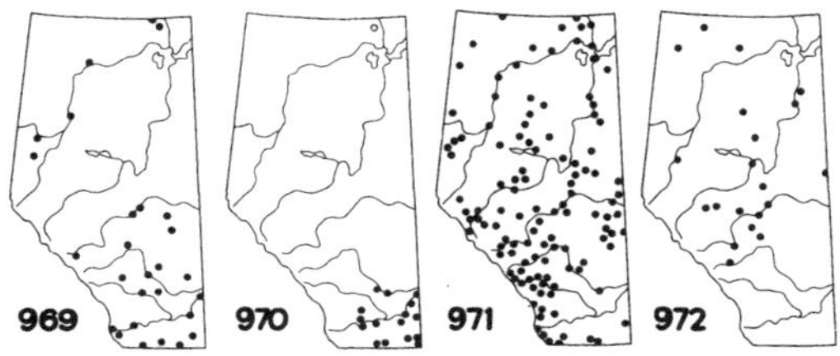

969 *Plantago eriopoda*
970 ○ *Plantago maritima*
● *Plantago patagonica*
971 *Galium boreale*
972 *Galium labradoricum*

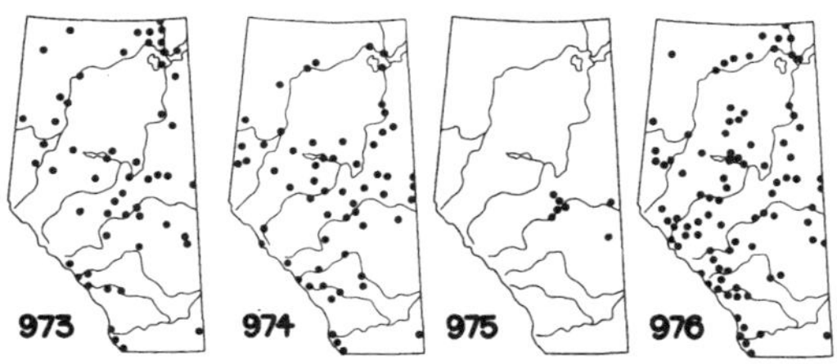

973 *Galium trifidum*
974 *Galium triflorum*
975 *Houstonia longifolia*
976 *Linnaea borealis*

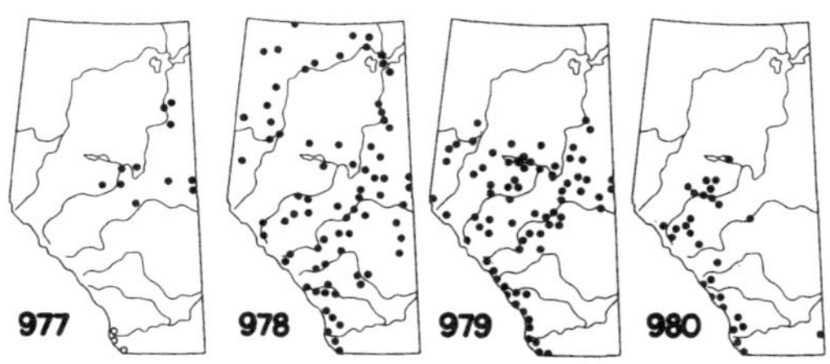

977 ● *Lonicera caerulea*
○ *Lonicera utahensis*
978 *Lonicera dioica*
979 *Lonicera involucrata*
980 *Sambucus racemosa*

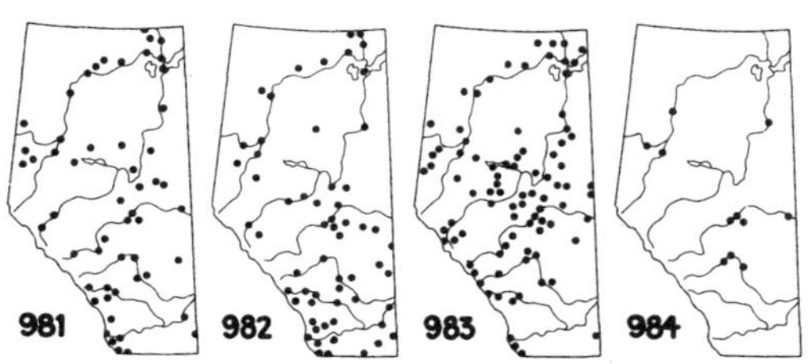

981 *Symphoricarpos albus*
982 *Symphoricarpos occidentalis*
983 *Viburnum edule*
984 *Viburnum opulus*

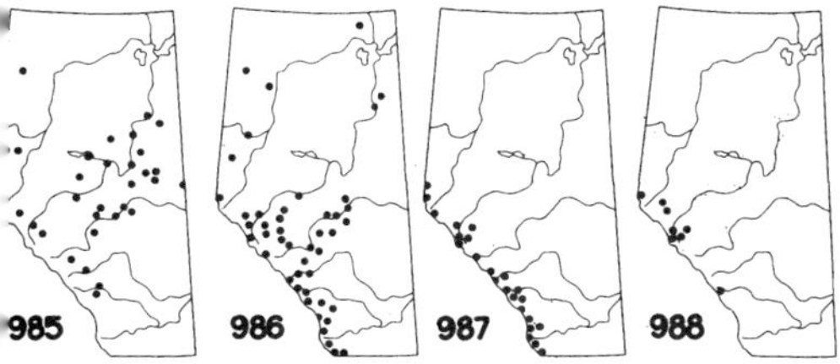

985 *Adoxa moschatellina*
986 *Valeriana dioica*
987 *Valeriana sitchensis*
988 *Campanula lasiocarpa*

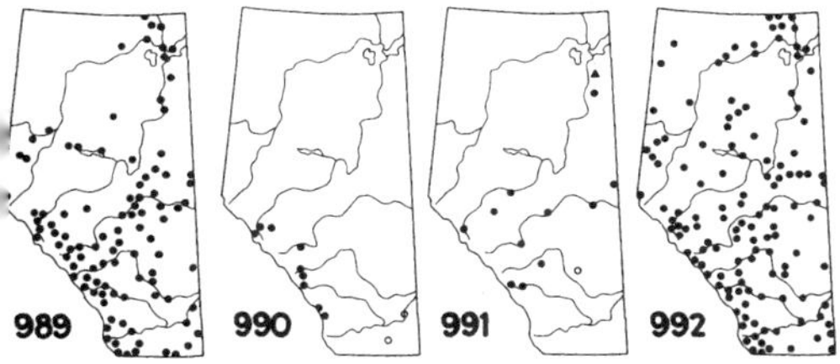

989 *Campanula rotundifolia*
990 ● *Campanula uniflora*
○ *Downingia laeta*
991 ▲ *Lobelia dortmanna*
● *Lobelia kalmii*
○ *Lobelia spicata*
992 *Achillea millefolium*

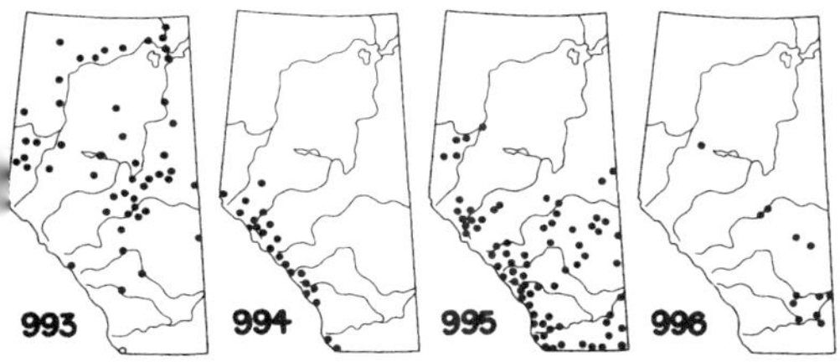

993 ● *Achillea sibirica*
○ *Adenocaulon bicolor*
994 *Agoseris aurantiaca*
995 *Agoseris glauca*
996 *Ambrosia artemisiifolia*

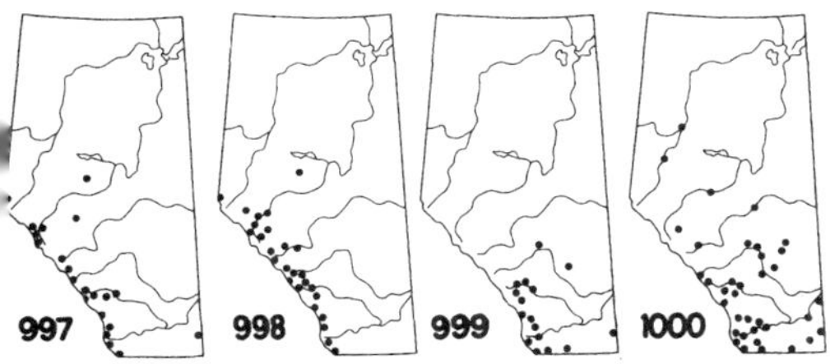

997 *Anaphalis margaritacea*
998 *Antennaria alpina*
999 *Antennaria anaphaloides*
1000 *Antennaria aprica*

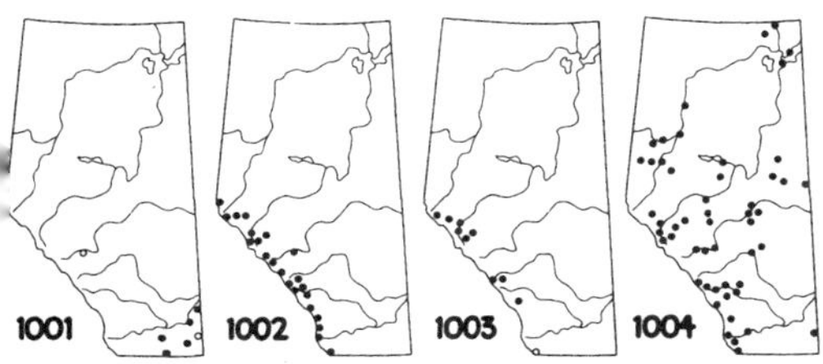

1001 ○ *Antennaria corymbosa*
● *Antennaria dimorpha*
1002 *Antennaria lanata*
1003 ○ *Antennaria luzuloides*
● *Antennaria monocephala*
1004 *Antennaria neglecta*

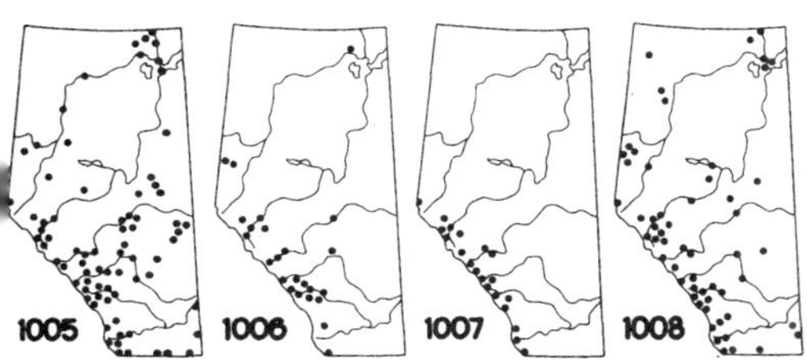

1005 *Antennaria parvifolia*
1006 *Antennaria pulcherrima*
1007 *Antennaria racemosa*
1008 *Antennaria rosea*

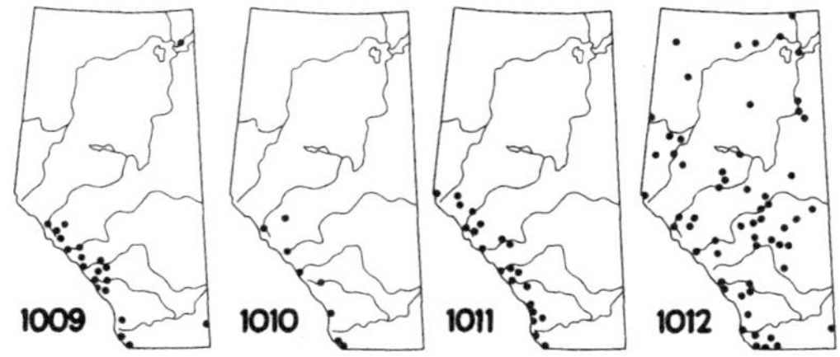

1009 *Antennaria umbrinella*
1010 *Arnica amplexicaulis*
1011 *Arnica angustifolia*
1012 *Arnica chamissonis*

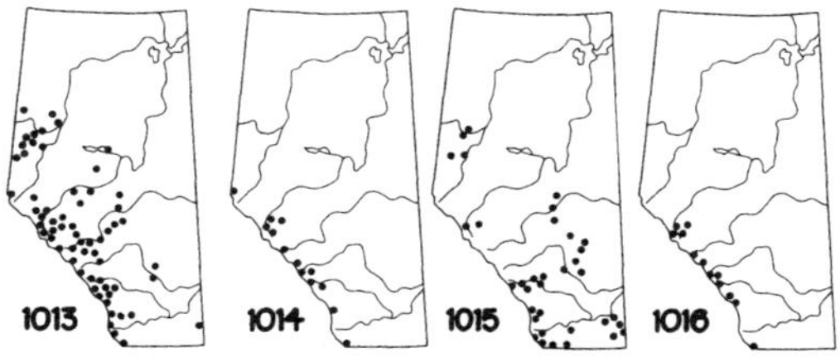

1013 *Arnica cordifolia*
1014 *Arnica diversifolia*
1015 *Arnica fulgens*
1016 *Arnica gracilis*

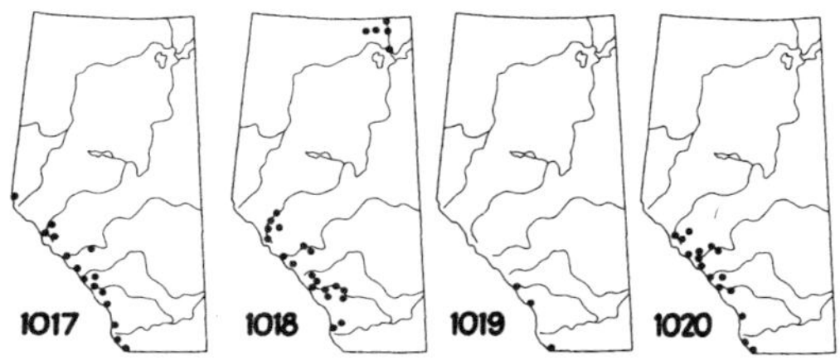

1017 *Arnica latifolia*
1018 *Arnica lonchophylla*
1019 *Arnica longifolia*
1020 *Arnica louiseana*

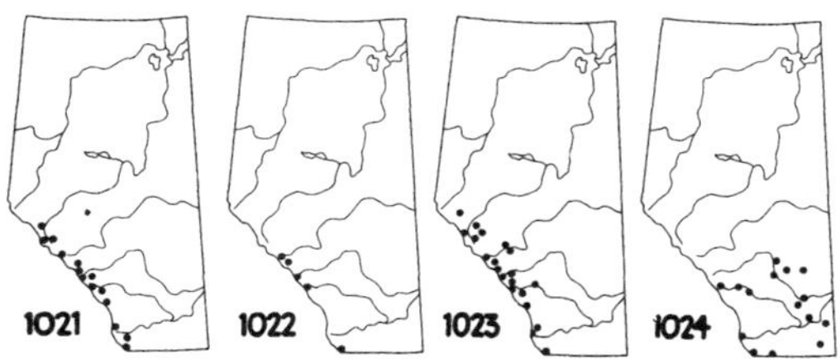

1021 *Arnica mollis*
1022 *Arnica parryi*
1023 *Arnica rydbergii*
1024 *Arnica sororia*

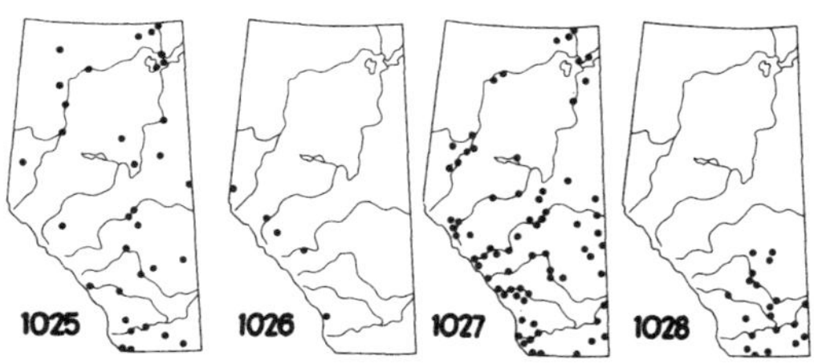

1025 *Artemisia biennis*
1026 *Artemisia borealis*
1027 *Artemisia campestris*
1028 *Artemisia cana*

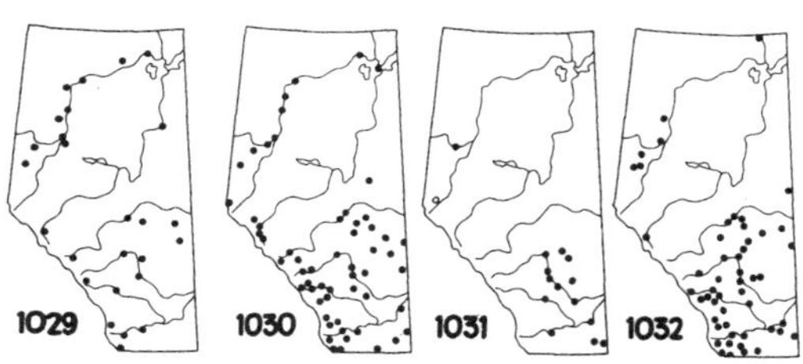

1029 *Artemisia dracunculus*
1030 *Artemisia frigida*
1031 ● *Artemisia longifolia*
○ *Artemisia furcata*
1032 *Artemisia ludoviciana*

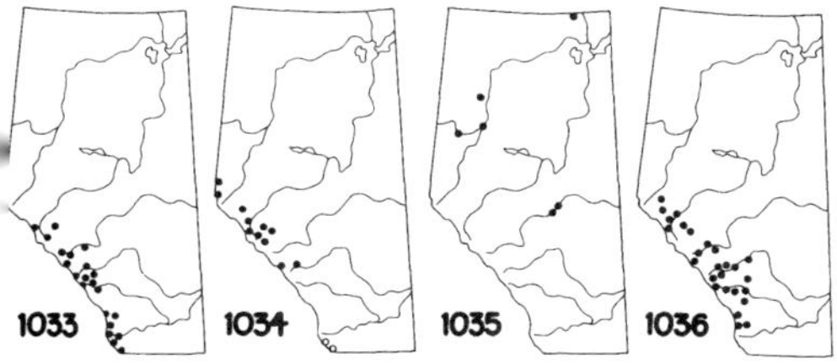

1033 *Artemisia michauxiana*
1034 ● *Artemisia norvegica*
○ *Artemisia tridentata*
1035 *Artemisia tilesii*
1036 *Aster alpinus*

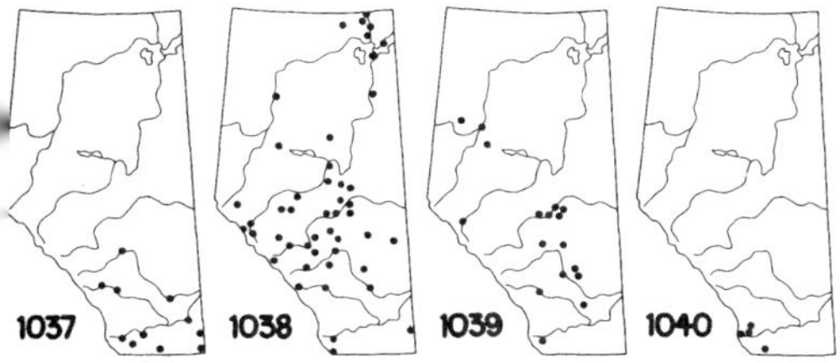

1037 *Aster ascendens*
1038 *Aster borealis*
1039 *Aster brachyactis*
1040 *Aster campestris*

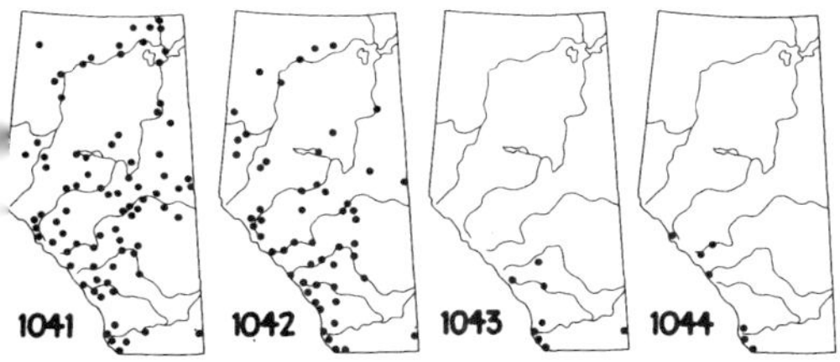

1041 *Aster ciliolatus*
1042 *Aster conspicuus*
1043 *Aster eatonii*
1044 *Aster engelmannii*

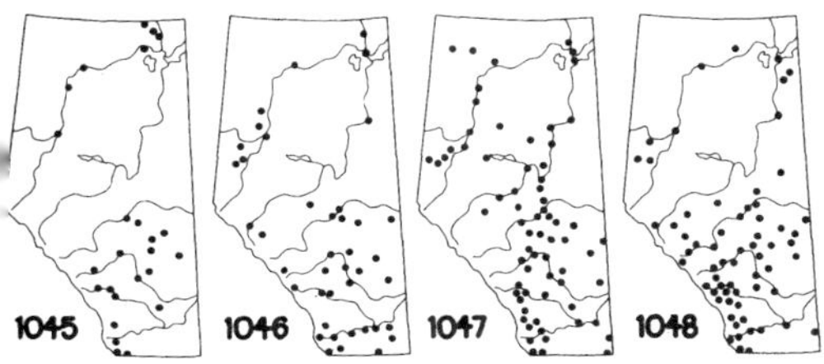

1045 *Aster ericoides*
1046 *Aster falcatus*
1047 *Aster hesperius*
1048 *Aster laevis*

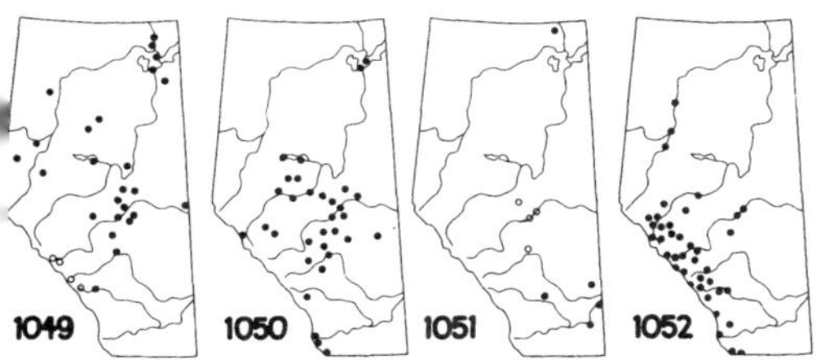

1049 ○ *Aster* × *maccallae*
● *Aster puniceus*
1050 *Aster modestus*
1051 ● *Aster pauciflorus*
○ *Aster umbellatus*
1052 *Aster sibircus*

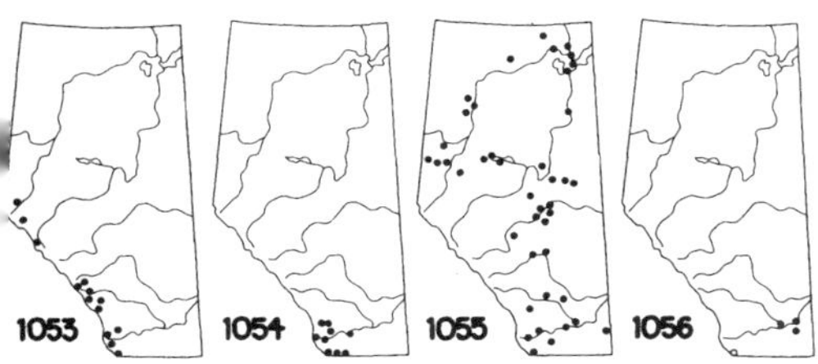

1053 *Aster subspicatus*
1054 *Balsamorhiza sagittata*
1055 *Bidens cernua*
1056 ● *Bidens frondosa*
○ *Brickellia grandiflora*

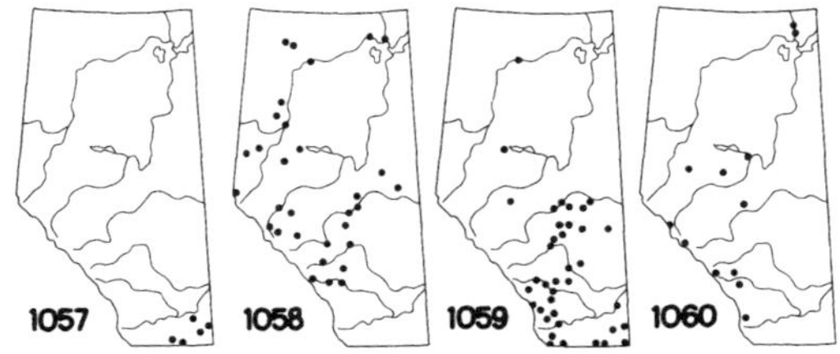

1057 *Chrysothamnus nauseosus*
1058 *Cirsium drummondii*
1059 *Cirsium flodmanii*
1060 *Cirsium foliosum*

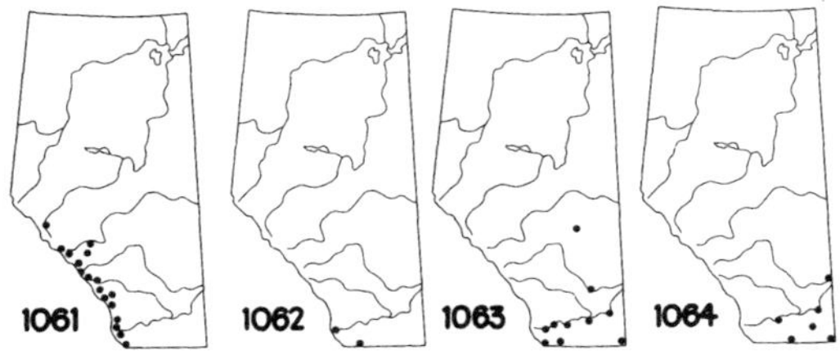

1061 *Cirsium hookerianum*
1062 *Cirsium scariosum*
1063 *Cirsium undulatum*
1064 *Coreopsis tinctoria*

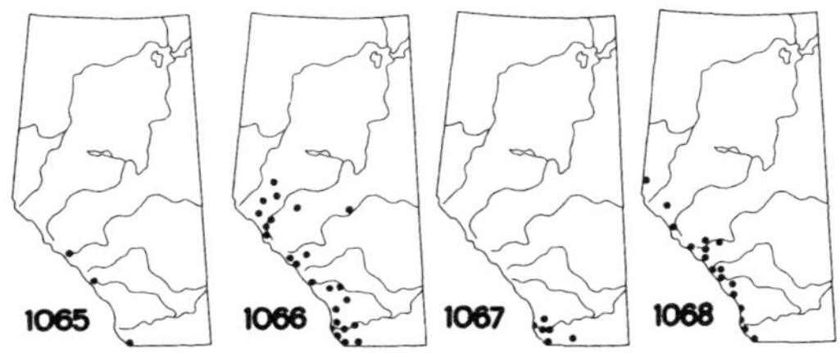

1065 *Crepis atrabarba*
1066 *Crepis elegans*
1067 *Crepis intermedia*
1068 *Crepis nana*

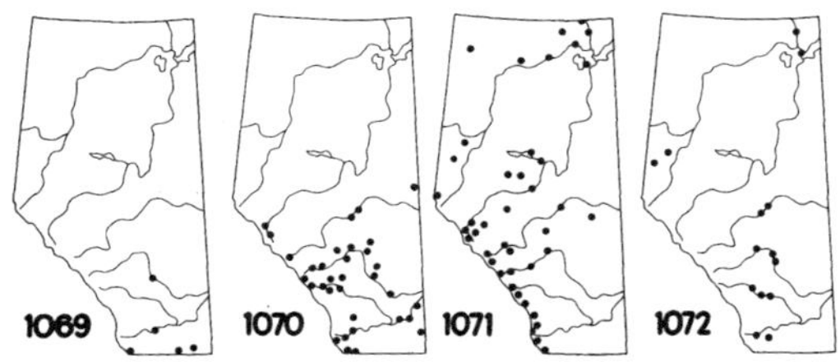

1069 *Crepis occidentalis*
1070 *Crepis runcinata*
1071 *Erigeron acris*
1072 *Erigeron annuus*

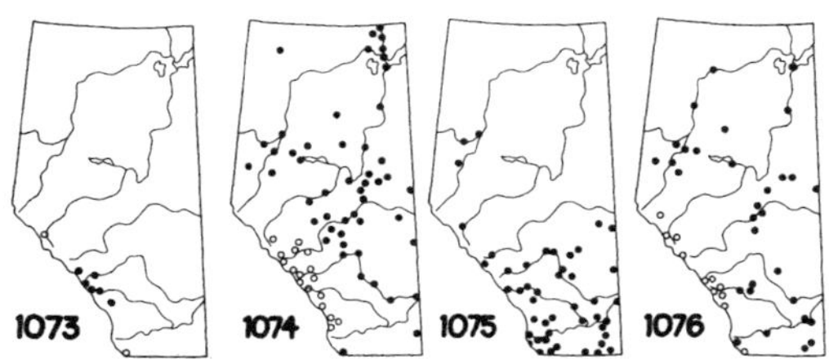

1073 ● *Erigeron* × *arthurii*
○ *Erigeron flagellaris*
1074 ○ *Erigeron aureus*
● *Erigeron philadelphicus*
1075 *Erigeron caespitosus*
1076 ● *Erigeron canadensis*
○ *Erigeron lanatus*

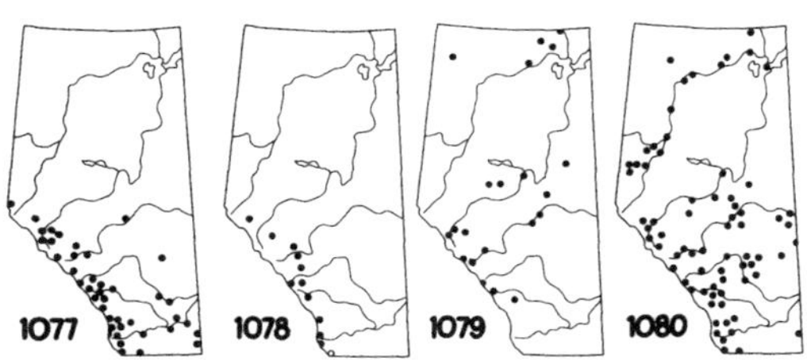

1077 *Erigeron compositus*
1078 ○ *Erigeron divergens*
● *Erigeron grandiflorus*
1079 *Erigeron elatus*
1080 *Erigeron glabellus*

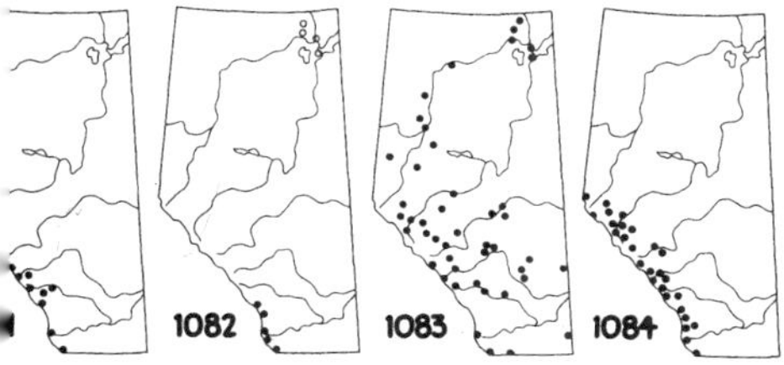

1081 *Erigeron humilis*
1082 ○ *Erigeron hyssopifolius*
● *Erigeron ochroleucus*
1083 *Erigeron lonchophyllus*
1084 *Erigeron peregrinus*

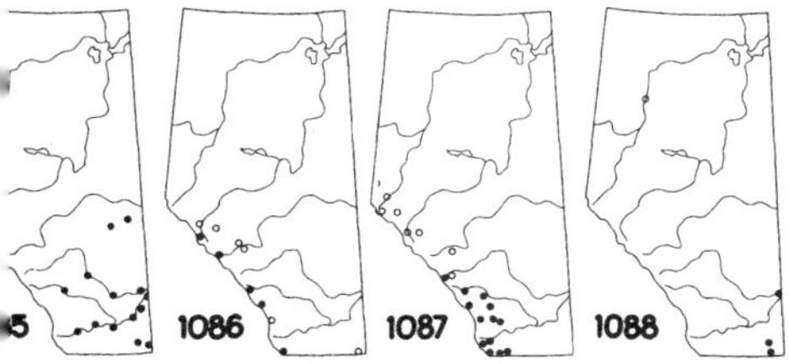

1085 *Erigeron pumilus*
1086 ● *Erigeron purpuratus*
○ *Erigeron radicatus*
1087 ● *Erigeron speciosus*
○ *Erigeron trifidus*
1088 ○ *Eupatorium maculatum*
● *Franseria acanthicarpa*

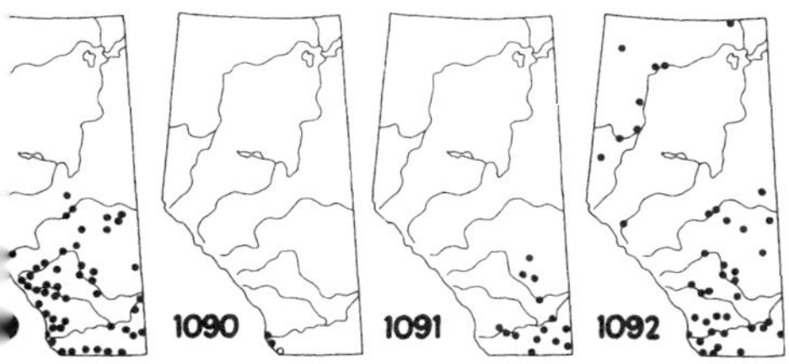

1089 *Gaillardia aristata*
1090 ○ *Gnaphalium microcephalum*
● *Gnaphalium viscosum*
1091 *Gnaphalium palustre*
1092 *Grindelia squarrosa*

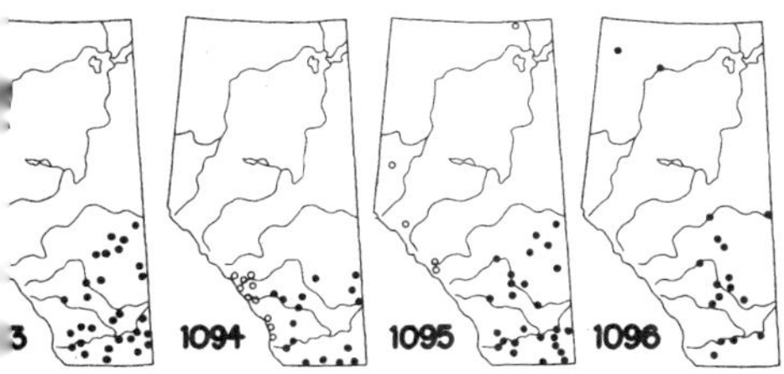

1093 *Gutierrezia sarothrae*
1094 ● *Haplopappus lanceolatus*
○ *Haplopappus lyallii*
1095 ● *Haplopappus spinulosus*
○ *Haplopappus uniflorus*
1096 *Helenium autumnale*

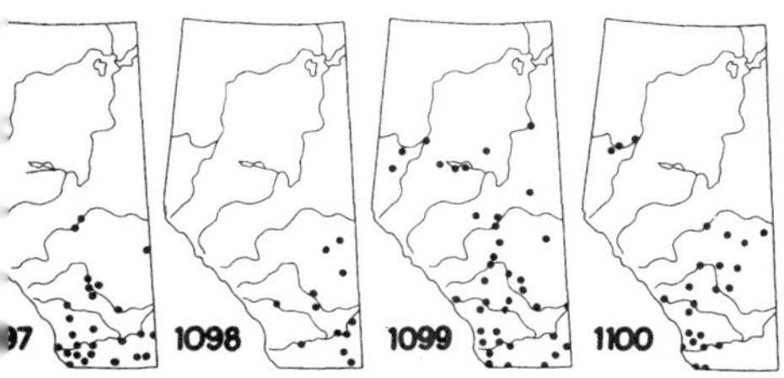

1097 *Helianthus annuus*
1098 *Helianthus couplandii*
1099 *Helianthus nuttallii*
1100 *Helianthus subrhomboideus*

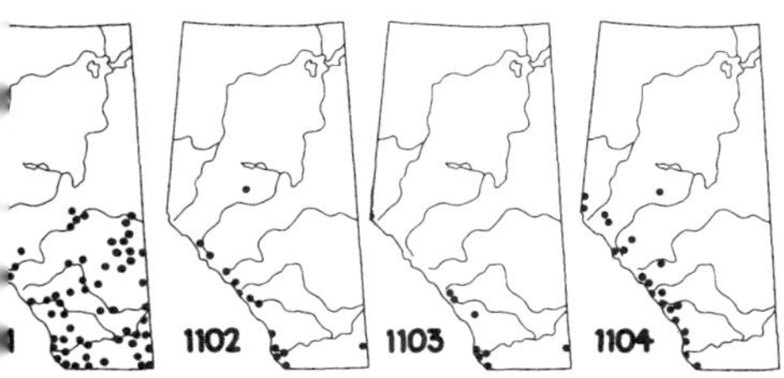

1101 *Heterotheca villosa*
1102 *Hieracium albiflorum*
1103 *Hieracium cynoglossoides*
1104 *Hieracium triste*

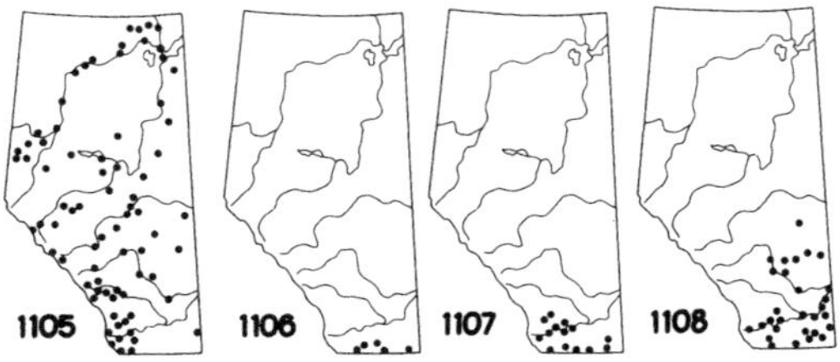

1105 *Hieracium umbellatum*
1106 *Hymenopappus filifolius*
1107 *Hymenoxys acaulis*
1108 *Hymenoxys richardsonii*

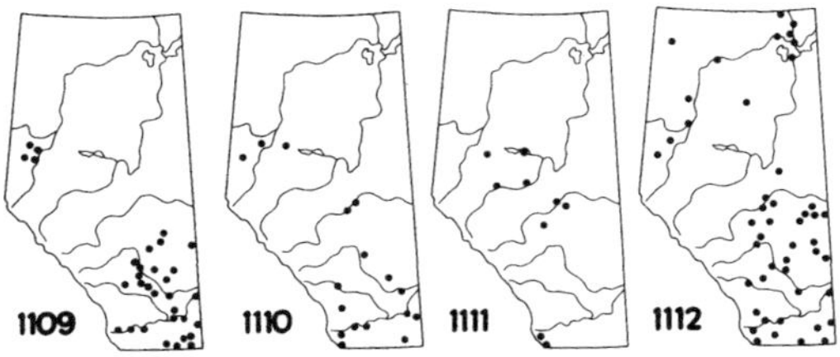

1109 *Iva axillaris*
1110 *Iva xanthifolia*
1111 *Lactuca biennis*
1112 *Lactuca pulchella*

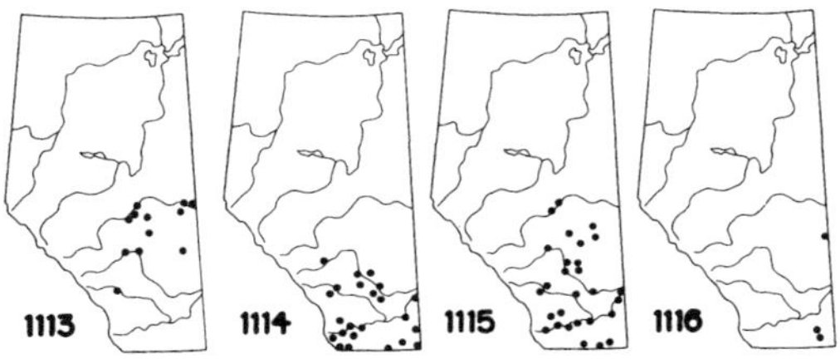

1113 *Liatris ligulistylis*
1114 *Liatris punctata*
1115 *Lygodesmia juncea*
1116 *Lygodesmia rostrata*

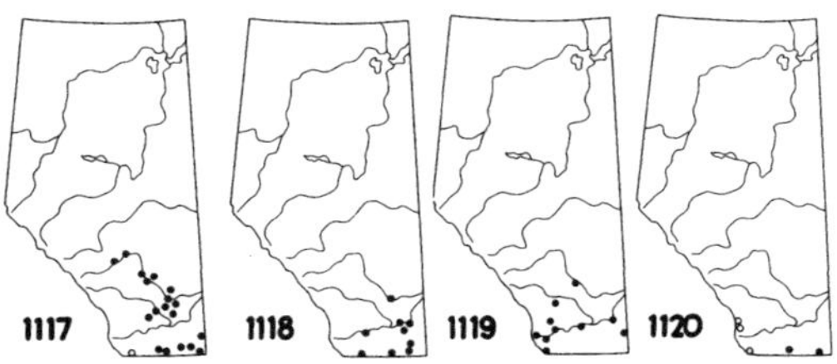

1117 ● *Machaeranthera canescens*
○ *Machaeranthera tanacetifoli*
1118 *Machaeranthera grindelioid*
1119 *Madia glomerata*
1120 ○ *Microseris nutans*
● *Nothocalais cuspidata*

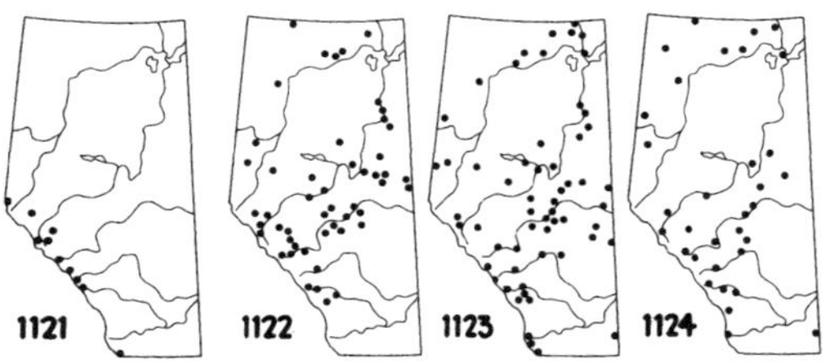

1121 *Petasites nivalis*
1122 *Petasites palmatus*
1123 *Petasites sagittatus*
1124 *Petasites vitifolius*

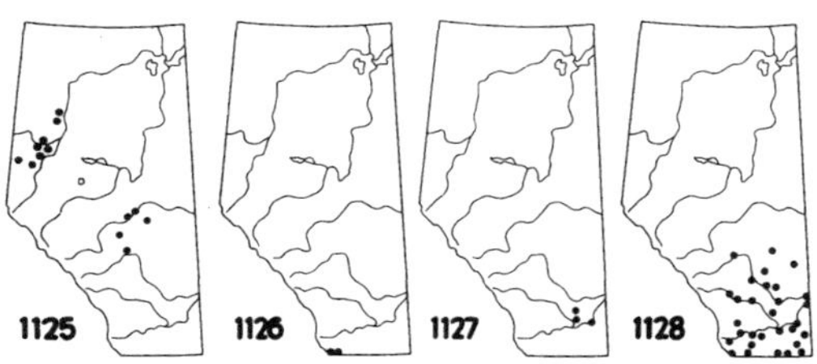

1125 ○ *Prenanthes alata*
● *Prenanthes racemosa*
1126 *Prenanthes sagittata*
1127 *Psilocarphus elatior*
1128 *Ratibida columnifera*

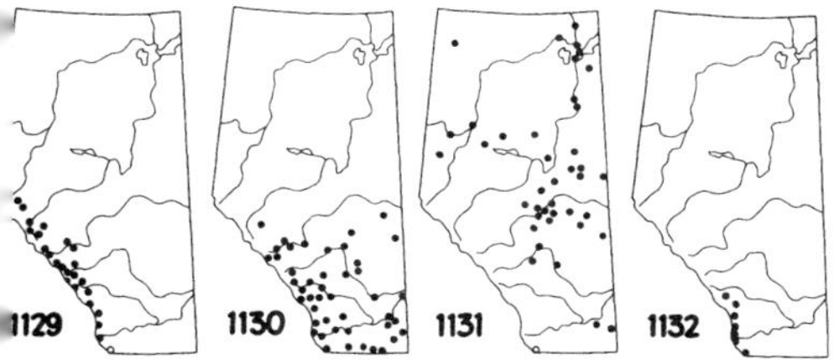

1129 ○ *Saussurea americana*
● *Saussurea nuda*
1130 *Senecio canus*
1131 ● *Senecio congestus*
○ *Senecio cymbalarioides*
1132 *Senecio conterminus*

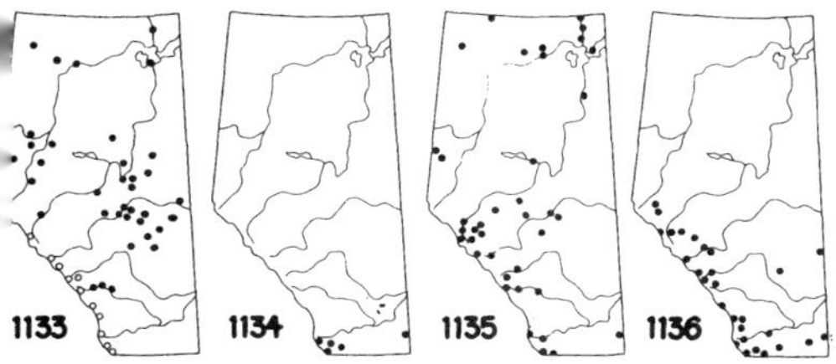

1133 ● *Senecio eremophilus*
○ *Senecio fremontii*
1134 *Senecio foetidus*
1135 *Senecio indecorus*
1136 *Senecio integerrimus*

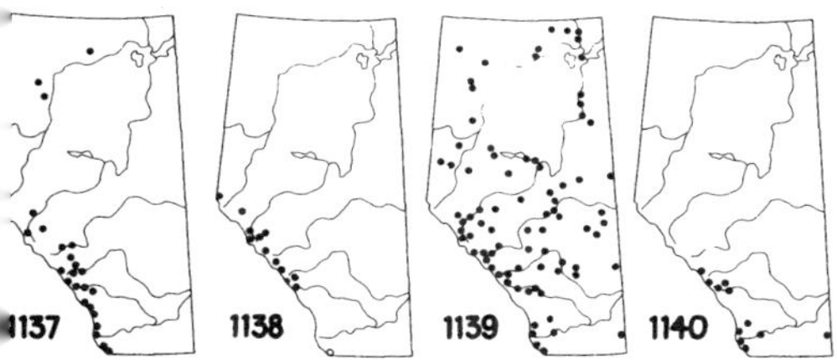

1137 *Senecio lugens*
1138 ○ *Senecio megacephalus*
● *Senecio pauciflorus*
1139 *Senecio pauperculus*
1140 *Senecio pseudaureus*

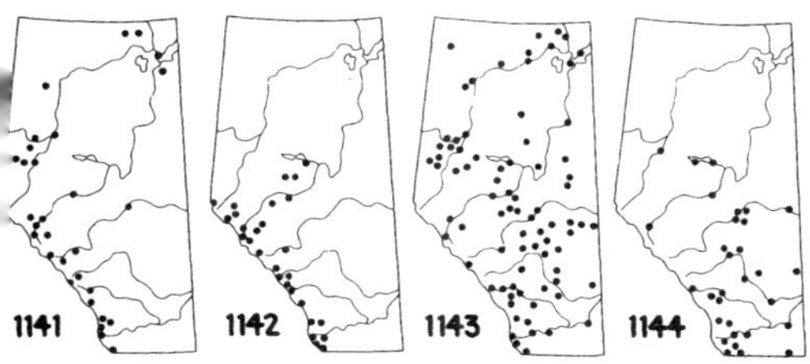

1141 *Senecio streptanthifolius*
1142 *Senecio triangularis*
1143 *Solidago canadensis*
1144 *Solidago gigantea*

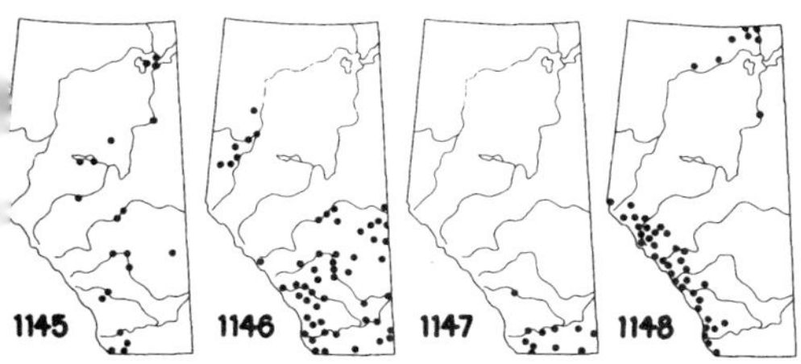

1145 *Solidago graminifolia*
1146 *Solidago missouriensis*
1147 *Solidago mollis*
1148 *Solidago multiradiata*

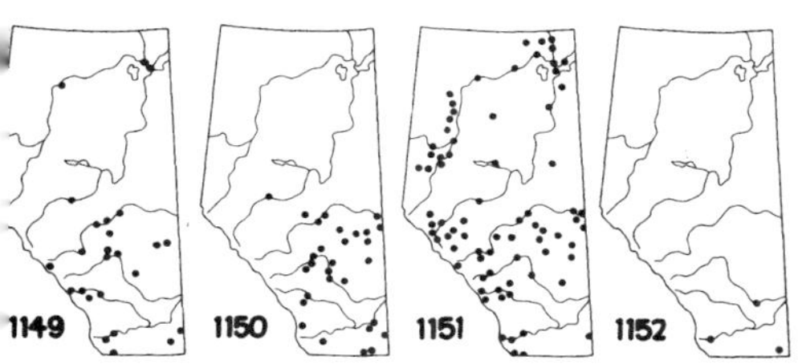

1149 *Solidago nemoralis*
1150 *Solidago rigida*
1151 *Solidago spathulata*
1152 *Stephanomeria runcinata*

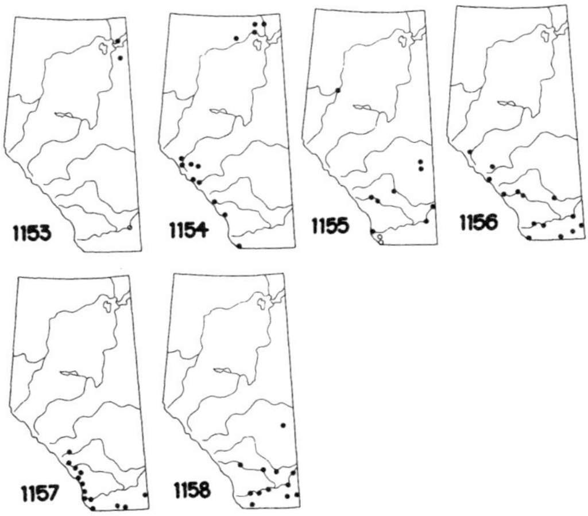

1153 ● *Tanacetum huronense*
○ *Thelesperma marginatum*
1154 *Taraxacum ceratophorum*
1155 ○ *Townsendia condensata*
● *Townsendia exscapa*
1156 *Townsendia hookeri*

1157 *Townsendia parryi*
1158 *Xanthium strumarium*

Index

Abies 46
  balsamea 46
  lasiocarpa 47
Abronia micrantha 249
Absinthe 541
Abutilon theophrasti 402
Acer 399
  douglasii 399
  glabrum 399
  negundo 399
Aceraceae 399
Achillea 526
  lanulosa 526
  millefolium 526
  nigrescens 527
  ptarmica 527
  sibirica 527
Aconite 273
Aconitum
  delphinifolium 273
Acorus 166
  americanus 166
  calamus 167
Actaea 273
  arguta 273
  rubra 273
Adder's-mouth 195
Adder's-tongue Family 32
Adenocaulon bicolor 527
Adiantum pedatum 35
Adoxa moschatellina 515
Adoxaceae 515
Agastache foeniculum 479
Agoseris 527
  aurantiaca 528
  glauca 528
Agrimonia striata 347
Agrimony 347
× Agroelymus 67
  bowdenii 67
  hirtiflorus 67
  mossii 68
  turneri 68
× Agrohordeum
  macounii 68
Agropyron 68
  albicans 69
  × brevifolium 70
  cristatum 70
  dasystachyum 68, 69, 88
  desertorum 70
  griffithsii 69
  inerme 71
  latiglume 71
  pectiniforme 70
  × pseudorepens 69
  repens 70
  riparium 69
  scribneri 70, 71
  sibiricum 70
  smithii 67, 70, 88
  spicatum 69, 71
  subsecundum 71
  trachycaulum 67, 68, 69, 71, 88
  triticeum 89
  violaceum 71
Agrostis 71
  alba 73
  borealis 72
  exarata 72
  geminata 73
  humilis 72
  mertensii 72
  palustris 73
  scabra 73
  stolonifera 73
  thurberiana 73
  variabilis 73
Alder 215
Alfalfa 383
Alisma 60
  brevipes 61
  geyeri 61
  gramineum 61
  plantago-aquatica 61
  triviale 61
Alismataceae 60
Alkali grass 109
Allionia
  hirsuta 250
  pilosa 250
Allium 178
  cernuum 178
  geyeri 178
  recurvatum 179
  rubrum 179
  rydbergii 179
  schoenoprasum 179
  sibiricum 179
  textile 179
Alnus 215
  crispa 216
  incana 216
  rugosa 216
  tenuifolia 216
Alopecurus 73
  aequalis 74
  alpinus 74
  occidentalis 74
  pratensis 74
Alpine Fir 47
Alum-root 331
Alyssum 297
  alyssoides 297
  desortorum 298
  murale 298
Amaranth 248, 249
Amaranthaceae 248
Amaranthus 248

albus 248
californicus 248
graecizans 249
powellii 249
retroflexus 249
Ambrosia 528
artemisiifolia 529
psilostachya 529
trifida 529
Amelanchier alnifolia 347
Amerorchis rotundifolia 196
Amsinckia menziesii 469
Anacardiaceae 398
Anaphalis margaritacea 529
Andromeda polifolia 440
Andropogon scoparius 112
Androsace 450
chamaejasme 450
occidentalis 450
septentrionalis 450
Anemone 273
canadensis 274
cylindrica 274
drummondii 275
lithophila 275
multifida 275
nemorosa 276
occidentalis 275
parviflora 275
patens 275
quinquefolia 276
richardsonii 276
riparia 276
Anethum graveolens 426
Angelica 426
arguta 427
dawsonii 427
genuflexa 427
Anogra nuttallii 422
Antennaria 530
aizoides 533
alpina 531
anaphaloides 531
angustata 532
aprica 531
campestris 532
canescens 531
concinna 533
corymbosa 531
dimorpha 531
glabrata 532
howellii 532
isolepis 533
lanata 532
luzuloides 532
media 531
microphylla 532
monocephala 532
neglecta 532
neodioica 532
nitida 532
oxyphylla 533
parvifolia 531, 532
pulcherrima 533
racemosa 533
rosea 533
sansonii 533
sedoides 533
umbrinella 533
ungavensis 531
Anthemis 533
cotula 534
tinctoria 534
Anticlea elegans 186
Apocynaceae 459
Apocynum 459
androsaemifolium 459
cannabinum 460
× medium 460
sibiricum 460
Aquilegia 276
brevistyla 277
flavescens 277
formosa 277
jonesii 277
Arabidopsis salsuginea 325
Arabis 298
divaricarpa 299
drummondii 299
glabra 299
hirsuta 300
holboellii 300
lemmonii 300
lyallii 300
lyrata 301
nuttallii 301
Araceae 166
Aralia nudicaulis 424
Araliaceae 424
Arceuthobium americanum 222
Arctagrostis 74
arundinacea 74
latifolia 74
Arctium 534
lappa 534
minus 534
tomentosum 535
Arctostaphylos 440
rubra 441
uva-ursi 441
Arctous rubra 441
Arenaria 254
capillaris 254
congesta 254
dawsonensis 259
humifusa 255
lateriflora 260
longipedunculata 255
nuttallii 260
obtusiloba 260
rossii 259
rubella 260
sajanensis 259
serpyllifolia 255
Aristida longiseta 74
Armoracia rusticana 301
Arnica 535
alpina 536
amplexicaulis 536
angustifolia 536
chamissonis 537
cordifolia 537
diversifolia 537
fulgens 537
gracilis 537
griscomii 538
latifolia 538
lonchophylla 538
longifolia 538
louiseana 538
mollis 539
parryi 539
rydbergii 539
sororia 539
Arrow-grass 59
Arrow-head 61
Artemisia 539
abrotanum 540
absinthium 541
biennis 541
borealis 541
campestris 541

camporum 541
cana 541
canadensis 541
caudata 541
dracunculus 542
forwardii 541
frigida 542
furcata 542
herriotii 543
longifolia 542
ludoviciana 542
michauxiana 543
norvegica 543
pabularis 543
pacifica 541
tilesii 543
tridentata 543
vulgaris 542, 543
Arum Family 166
Asclepiadaceae 460
Asclepias 460
ovalifolia 460
speciosa 460
viridiflora 461
Asparagus 179
Asparagus officinalis 179
Aspen 198
Asperugo procumbens 469
Asperula arvensis 509
Asplenium viride 35
Aster 543
alpinus 545
ascendens 545
borealis 546
brachyactis 546
campestris 546
canescens 581
ciliolatus 546, 548
conspicuus 546
eatonii 547
engelmannii 547
ericoides 547
falcatus 547
foliaceus 549
hesperius 548
junciformis 546
laevis 548
× maccallae 548
modestus 548
pansus 547
pauciflorus 548
puniceus 549
sibiricus 549
subspicatus 548, 549
umbellatus 549
Astragalus 371
aboriginum 373
agrestis 375
alpinus 373
americanus 374
bisulcatus 374
bodinii 374
bourgovii 374
caespitosus 377
canadensis 374
cicer 375
crassicarpus 375
dasyglottis 375
decumbens 376
drummondii 375
eucosmus 375
flexuosus 376
frigidus 374
gilviflorus 376
kentrophyta 376
lotiflorus 376
miser 376
missouriensis 376
occidentalis 377
pectinatus 377
purshii 377
robbinsii 377
spatulatus 377
striatus 377
tenellus 378
triphyllus 376
vexilliflexus 378
yukonis 374
Athyrium 35
alpestre 36
distentifolium 36
felix-femina 36
Atriplex 236
argentea 237
canescens 237
heterosperma 237
hortensis 237
nuttallii 237
oblongifolia 238
patula 238
powellii 238
prostrata 238
rosea 238
subspicata 238
suckleyi 239
truncata 239
Avena 75
fatua 75
hookeri 93
sativa 75
Avenochloa hookeri 94
Avens 351
Axyris amaranthoides 239

Baby-blue-eyes 466
Baby's-breath 258
Bachelor's Button 552
Bahia oppositifolia 550
Baked-apple Berry 366
Ball Mustard 321
Balsam Apple 517
Balsam Fir 46
Balsaminaceae 400
Balsamorhiza sagittata 550
Balsam-root 550
Baneberry 273
Barbarea 301
americana 302
orthoceras 301
vulgaris 302
Barberry Family 290
Barley 94
Barnyard Grass 86
Bassia hyssopifolia 240
Bastard Toad-flax 222
Bear Grass 185
Bearberry 440
Beard-tongue 497
Beckmannia syzigachne 75
Bedstraw 509
Bee Plant 293
Beggar-ticks 473, 550
Bent Grass 71
Berberidaceae 290
Berberis repens 290
Bergamot 483
Berteroa incana 302
Besseya 488
cinerea 488
wyomingensis 488
Betula 216
× arbuscula 216
× eastwoodae 216

glandulosa 217
nana 217
neoalaskana 217
occidentalis 218
papyrifera 218
× piperi 217
pumila 219
× sandbergii 217
× sargentii 217
× uliginosa 217
× winteri 217
Betulaceae 215
Bidens 550
cernua 550
comosa 551
frondosa 551
tripartita 551
vulgata 551
Bilberry 447, 448
Bilderdykia
convolvulus 228
scandens 230
Bindweed 228, 461
Birch 216
Bird's-eye 503
Bishop's-cap 333
Bistort 228
Bitter Cress 305
Bitter-root 251
Bittersweet Family 399
Black-eyed Susan 586
Bladder Fern 37
Bladder-pod 320, 321
Bladderwort 505, 506
Blazing Star 579
Blue Buttons 516
Blue Devil 472
Bluebell 474, 517
Blueberry 448
Blue-bur 473
Blue-eyed Grass 187
Blue-eyed Mary 491
Bluegrass 103
Bluestem 112
Bluets 511
Bog Orchid 191
Bog Rosemary 440
Boisduvalia glabella 413
Borage 468, 470
Boraginaceae 468
Borago officinalis 470
Boschniakia rossica 504
Botrychium 32
boreale 32
dusenii 33
lanceolatum 33
lunaria 33
multifidum 33
silaifolium 33
simplex 33
virginianum 33
Botrypus virginianum 33
Bouncing Bet 262
Bouteloua gracilis 75
Bracken 42
Bracyactis angusta 546
Bramble 365
Brassica 302
campestris 302
hirta 303
juncea 303
kaber 303
napus 303
Braya 303
glabella 303
humilis 303
purpurascens 303
richardsonii 303
Brickellia grandiflora 551
Bristly Ox-tongue 584
Brome Grass 76
Bromus 76
altissimus 77
anomalus 77
carinatus 77
ciliatus 77
commutatus 77
inermis 78
japonicus 78
latiglumis 77
marginatus 77
mollis 78
porteri 77
pumpellianus 78
purgans 77
richardsonii 77
secalinus 78
squarrosus 78
tectorum 79
vulgaris 79
Bronze-bells 183
Brook Grass 82
Brooklime 502
Broomcorn Millet 101
Broom-rape Family 504
Broomweed 570
Buck-bean 459
Buckbrush 514
Buckthorn 400, 401
Buckwheat 223, 225, 228
Buffalo Bean 375
Buffalo Grass 99
Buffalo-berry 411
Bugloss 472, 474
Bugseed 244
Bulrush 162
Bunchberry 434
Bupleurum 427
americanum 427
triradiatum 427
Burdock 534
Burning Bush 245
Bur-reed 51
Butte Marigold 577
Butter-and-eggs 493
Buttercup 281
Butterfly-weed 420
Butterwort 506

Cactaceae 409
Cactus 409
Calamagrostis 79
canadensis 80
inexpansa 80
lapponica 80
montanensis 80
neglecta 81
purpurascens 81
rubescens 81
stricta 81
Calamovilfa longifolia 81
Calamus 166
Calla palustris 167
Callitrichaceae 397
Callitriche 397
hermaphroditica 397
palustris 397
verna 397
Calochortus apiculatus 179
Caltha 278
leptosepala 278
natans 278
palustris 278

Calylophus serrulatus 422
Calypso bulbosa 188
Calystegia sepium 461
Camas 180, 186
Camassia quamash 180
Camelina 304
  microcarpa 304
  parodii 304
  sativa 305
Camissonia
  andina 421
  breviflora 422
Campanula 517
  lasiocarpa 518
  rapunculoides 518
  rotundifolia 518
  uniflora 518
Campanulaceae 517
Campion 263
Canary Grass 102
Cancer-root 505
Caper Family 293
Capparidaceae 293
Caprifoliaceae 511
Capsella bursa-pastoris 305
Cannabinaceae 219
Cannabis sativa 219
Caragana arborescens 378
Caraway 427
Cardamine 305
  bellidifolia 305
  parviflora 306
  pensylvanica 306
  pratensis 306
  umbellata 306
Cardaria 306
  chalepensis 307
  draba 307
  pubescens 307
Carduus nutans 551
Carex 120
  abbreviata 155
  adusta 131
  aenea 131
  albo-nigra 131
  altior 132
  aperta 131
  aquatilis 131
  arcta 132
  athabascensis 152
  atherodes 132
  athrostachya 132
  atrata 133
  atratiformis 150
  atrosquama 132
  aurea 133
  backii 133
  bebbii 133
  bipartita 133
  brevior 134
  brunnescens 134
  buxbaumii 134
  canescens 136
  capillaris 134
  capitata 135
  chordorrhiza 135
  concinna 135
  concinnoides 135
  crawei 135
  crawfordii 136
  curta 136
  deflexa 136
  deweyana 136
  diandra 137, 148
  dioica 139
  disperma 137
  douglasii 137
  drummondiana 151
  eburnea 137
  eleocharis 153
  enanderi 138
  epapillosa 138
  eurystachya 138
  festivella 144
  filifolia 138
  flava 138
  foenea 152
  franklinii 139
  garberi 133
  geyeri 139
  glacialis 139
  gymnoclada 152
  gynocrates 139
  haydeniana 139
  helionastes 140
  heliophila 147
  hoodii 140
  hookerana 140
  houghtoniana 140
  hystricina 141
  illota 141
  incondita 143
  incurviformis 143
  interior 141
  kelloggii 141
  lachenallii 133
  lacustris 141
  lagopina 133
  lanuginosa 142
  lasiocarpa 142
  lenticularis 141, 142
  leptalea 142
  leptopoda 136
  limnophila 144
  limosa 142
  livida 143
  loliacea 143
  macloviana 140, 143
  magellanica 146
  maritima 143
  media 145
  mertensii 144
  microglochin 144
  microptera 144
  misandra 144
  montanensis 148
  nardina 144
  nebraskensis 145
  nigricans 145
  nigromarginata 147
  norvegica 145
  obtusata 145
  oligosperma 145
  pachystachya 146
  parryana 146
  pauciflora 146
  paupercula 146
  paysonis 146
  peckii 147
  pensylvanica 147
  petasata 147
  petricosa 147
  phaeocephala 147
  physocarpa 151
  platylepis 148
  podocarpa 148
  praegracilis 148
  prairea 148
  praticola 148
  preslii 149
  pseudo-cyperus 149
  pyrenaica 149
  raymondii 149

raynoldsii 150
retrorsa 150
richardsonii 150
rossii 150
rostrata 151
rupestris 151
sartwellii 151
saxatilis 151
saximontana 133
scirpiformis 152
scirpoidea 152
scoparia 152
scopulorum 152
siccata 152
simulata 153
spectabilis 153
sprengellii 153
stans 132
stenochlaena 152
stenophylla 153
stipitata 153
sychnocephala 154
tenera 154
tenuiflora 154
tincta 154
tolmiei 147
torreyi 154
trichocarpa 132
trisperma 155
umbellata 155
vaginata 155
vesicaria 155
viridula 156
vulpinoidea 156
xerantica 156
Carrot Family 425
Carthamus tinctorius 552
Carum carvi 427
Caryophyllaceae 253
Cassiope 441
mertensiana 442
tetragona 442
Castilleja 488
cusickii 489
elegans 490
hispida 489
lutescens 490
miniata 490
occidentalis 490
raupii 490
rhexifolia 491
sepentrionalis 489, 490
sessiliflora 491
sulphurea 490
Catabrosa aquatica 82
Catchfly 263
Catnip 483
Cattail 51
Ceanothus velutinus 401
Cedar 45
Celestraceae 399
Centaurea 552
cyanus 552
diffusa 553
maculosa 553
repens 553
solstitialis 553
Centunculus minimus 451
Cerastium 255
arvense 255
beeringianum 256
nutans 256
vulgatum 256
Ceratophyllaceae 272
Ceratophyllum demersum 272
Chaenorrhinum minus 491
Chaffweed 451
Chamaedaphne calyculata 442
Chamaepericlymenum canadense 435
Chamaerhodos erecta 348
Chamaesaracha grandiflora 485
Chamomile 533, 582
Chamomilla
recutita 582
suaveolens 582
Charlock 303, 322
Cheilanthes 36
feei 36
gracillima 36
Chenopodiaceae 234
Chenopodium 240
album 241
atrovirens 241
berlandieri 242
bonus-henricus 242
capitatum 242
desiccatum 242
fremontii 242
gigantospermum 243
glaucum 244
hybridum 243
incanum 243
leptophyllum 243
pratericola 243
rubrum 244
salinum 244
subglabrum 244
watsonii 244
Chickweed 268
Chicory 554
Chimaphila umbellata 435
Chiogenes hispidula 442
Chives 179
Choke Cherry 364
Chrysanthemum 553
bipinnatum 597
leucanthemum 553
Chrysopsis villosa 574
Chrysosplenium 329
iowense 329
tetrandrum 330
Chrysothamnus nauseosus 554
Cichorium 554
endivia 554
intybus 554
Cicuta 428
bulbifera 428
douglasii 428
mackenzieana 428
maculata 428
virosa 428
Cinna latifolia 82
Cinquefoil 353
Circaea 413
alpina 413
pacifica 413
Cirsium 555
arvense 555
drummondii 555
flodmanii 556
foliosum 556
hookerianum 556
scariosum 556
undulatum 557
vulgare 557
Cistaceae 404
Clammyweed 293
Claytonia 251

lanceolata 251
linearis 252
megarhiza 251
parvifolia 252
Cleavers 509
Clematis 278
ligusticifolia 279
occidentalis 279
tangutica 279
verticellaris 279
Cleome serrulata 293
Clintonia uniflora 180
Cloudberry 366
Clover 389
Club-Moss 24
Cocklebur 600
Cockspur Grass 86
Coeloglossum viride 193
Collinsia parviflora 491
Collomia linearis 462
Colorado Rubber-plant 577
Coltsfoot 583
Columbine 276
Colza 302
Comandra 222
livida 222
pallida 222
umbellata 222
Comarum palustre 361
Comfrey 477
Compositae 519
Composite Family 519
Cone-flower 585
Conimitella williamsii 330
Conium maculatum 429
Conringia orientalis 307
Convolvulaceae 461
Convolvulus 461
arvensis 461
sepium 461
Coptis 279
groenlandica 279
trifolia 279
Corallorhiza 189
innata 190
maculata 189
multiflora 189
striata 189
trifida 189
Coral-root 189
Cord Grass 113
Coreopsis tinctoria 557
Corispermum 244
hyssopifolium 245
nitidum 245
orientale 245
Corn Spurry 267
Cornaceae 434
Cornflower 552
Cornus 434
canadensis 434
stolonifera 435
Coronilla varia 378
Corydalis 292
aurea 292
sempervirens 293
Corylus 219
cornuta 219
rostrata 219
Coryphantha vivipara 409
Cotton Grass 159
Cottonwood 196
Couch Grass 70
Cow Parsnip 429
Cowberry 449
Cow-cockle 271
Cow-wheat 493
Crab Grass 85
Cranberry 445, 449, 515
Crane's-bill 392
Crassulaceae 327
Crataegus 348
chrysocarpa 349
douglasii 348
rotundifolia 349
Creeping Mahonia 290
Creeping Wintergreen 442
Crepis 557
atrabarba 558
capillaris 558
elegans 559
intermedia 559
nana 559
occidentalis 559
runcinata 559
tectorum 560
Crowberry 397
Crowfoot 272, 281
Cruciferae 294
Cryptantha 470
bradburyana 471
fendleri 470
macounii 470
minima 471
nubigena 470, 471
Cryptogramma 36
acrostichoides 37
crispa 37
stelleri 37
Cucurbitaceae 517
Cudweed 569
Cupressaceae 44
Currant 342
Cuscuta 462
gronovii 462
umbrosa 462
Cymopteris acaulis 429
Cynoglossum 471
boreale 471
officinale 471
Cyperaceae 120
Cyperus 156
schweinitzii 157
squarrosus 157
Cypripedium 190
acaule 190
calceolus 190
montanum 190
parviflorum 190
passerinum 191
Cystopteris 37
fragilis 37
montana 38
Cytherea bulbosa 189

Dactylis glomerata 82
Daisy 553, 560
Dame's Rocket 318
Dandelion 597
Danthonia 82
californica 83
intermedia 83
parryi 83
spicata 83
unispicata 83
Darnel 96
Dead Nettle 481
Delphinium 280
bicolor 280
glaucum 280
nuttallianum 280
Deschampsia 84
atropurpurea 119

cespitosa 84
elongata 84
Descurainia 307
pinnata 308
richardsonii 308
sophia 308
Devil's Paint-brush 575
Devil's-club 424
Dewberry 367
Dianthus 256
barbatus 257
deltoides 257
plumarius 257
seguieri 257
Dicanthelium 84
acuminatum 85
leibergii 85
oligosanthes 85
Digitaria 85
ischaemum 85
sanguinalis 86
Dill 426
Diphasiastrum
alpinum 25
complanatum 25
sitchense 26
Diplotaxis muralis 308
Dipsacaceae 516
Disporum 180
hookeri 180
oreganum 180
trachycarpum 180
Distichlis stricta 86
Dock 230
Dodder 462
Dodecatheon 451
conjugens 451
pulchellum 451
radicatum 452
Dog Mustard 316
Dogbane 459
Dog-fennel 534
Dog-tooth Violet 181
Dogwood 434, 435
Douglas Fir 50
Douglasia montana 452
Downingia laeta 518
Downy Chess 79
Draba 308
albertina 310
aurea 311
borealis 311
cana 311
crassifolia 312
densifolia 312
fladnizensis 312
glabella 312
incerta 312
kananaskis 313
lanceolata 311
lonchocarpa 313
longipes 313
macounii 313
nemorosa 313
nivalis 313, 314
oligosperma 314
paysonii 314
porsildii 314
praealta 314
reptans 315
stenoloba 315
ventosa 315
verna 315
Dracocephalum 479
nuttallii 483
parviflorum 479
thymiflorum 480
Dragonhead 479, 480, 483
Dragonwort 542
Dropseed 115
Drosera 326
anglica 326
linearis 327
rotundifolia 327
Droseraceae 326
Dryad 349
Dryas 349
drummondii 349
hookeriana 350
integrifolia 350
octopetala 350
Drymocallis
arguta 357
glandulosa 358
Dryopteris 38
assimilis 38
carthusiana 38
cristata 39
dilatata 38
disjuncta 40
felix-mas 39
fragrans 39
spinulosa 39
× uliginosa 39
Duckweed 167
Dune Grass 88

Echinochloa 86
crusgalli 86
pungens 86
Echinocystis lobata 517
Echium vulgare 472
Elaeagnaceae 410
Elaeagnus 410
angustifolia 410
commutata 410
Elatinaceae 404
Elatine triandra 404
Elderberry 514
Eleocharis 157
acicularis 158
compressa 158
elliptica 158
engelmannii 158
ovata 158
palustris 158
pauciflora 158
quinquefolia 158
uniglumis 158
Elephant Head 496
Ellisia nyctelea 465
Elodea 62
canadensis 62
longivaginata 62
Elymus 86
canadensis 68, 87
cinereus 88
glaucus 88
hirtiflorus 67
innovatus 67, 68, 70, 88
junceus 88
macounii 68
mollis 88
piperi 88
virginicus 89
Empetraceae 397
Empetrum nigrum 397
Enchanter's Nightshade 413
Endive 554
Endolepis suckleyi 239
Epilolobium 413
alpinum 415

anagallidifolium 415
angustifolium 415
ciliatum 415, 419
clavatum 416
glaberrimum 416
halleanum 416
hornemannii 417, 419
lactiflorum 417
latifolium 417
leptocarpum 418
leptophyllum 418
luteum 418
mirabile 419
palustre 419
paniculatum 419
platyphyllum 416
saximontanum 420
× treleasianum 419
Equisetaceae 28
Equisetum 28
arvense 29
× ferrissii 30
fluviatile 30
hyemale 30, 31
kansanum 30
laevigatum 30, 31
limosum 30
× nelsonii 30
palustre 30
pratense 31
prealtum 30
scirpoides 31
sylvaticum 31
×trachydon 31
variegatum 31
Eragrostis 89
cilianensis 89
megastachya 89
Eremopyrum triticeum 89
Ericaceae 439
Erigeron 560
acris 562, 563, 564
annuus 562
× arthurii 563
asper 564
aureus 563
caespitosus 563
canadensis 563
compositus 563, 568
divergens 564
elatus 564
flagellaris 564
formosissimus 564
glabellus 564
grandiflorus 565
humilus 565
hyssopifolius 565
lanatus 565, 568
lonchophyllus 566
ochroleucus 566
pallens 567
peregrinus 566
philadelphicus 566
pumilus 567
purpuratus 567
radicatus 567
scotteri 562
speciosus 567
strigosus 562
subtrinervis 567
trifidus 568
Eriogonum 223
androsaceum 224
cernuum 224
flavum 224
ovalifolium 224
piperi 224
subalpinum 224
umbellatum 224
Eriophorum 159
alpinum 164
angustifolium 160
brachyantherum 159
callitrix 160
chamissonis 160
gracile 160
opacum 159
polystachion 160
scheuchzeri 160
vaginatum 161
viride-carinatum 161
Erodium cicutarium 391
Erophila verna 315
Eruca sativa 316
Erucastrum
gallicum 316
Eryngium planum 429
Erysimum 316
asperum 316
cheiranthoides 317
inconspicuum 317
pallasii 317
Erythronium
grandiflorum 181
Eupatorium maculatum 568
Euphorbia 395
esula 396
glyptosperma 396
helioscopia 396
peplus 396
serpyllifolia 396
Euphorbiaceae 395
Euphrasia arctica 492
Eurotia lanata 245
Evening Primrose 412, 421
Evening Star 408
Everlasting 530
Eyebright 492

Fagopyrum 225
esculentum 225
sagittatum 225
tartaricum 225
Fairy Candelabra 450
Fairy-bells 180
False Asphodel 184
False Buckwheat 230
False Dandelion 527
False Flax 304
False Gromwell 476
False Heather 404
False Hellebore 185
False Melic 111
False Mitrewort 340
False Ragweed 577
Felwort 457, 458
Fern Family 34
Fescue 89, 119
Festuca 89
altaica 92
baffinensis 90
brachyphylla 90, 91
elatior 91
idahoensis 90
occidentalis 91
octoflora 119
ovina 90, 91
pratensis 91
rubra 91
saximontana 91
scabrella 92
subulata 92
Figwort Family 487

Fir 46
Fireweed 415
Flag 187
Flax 393
Fleabane 560
Flowering-quillwort 59
Fluminea festucaea 112
Forget-me-not 475
Foxtail 73, 112
Fragaria 350
  vesca 351
  virginiana 351
Franseria acanthicarpa 568
Fritillaria pudica 181
Fumariaceae 292
Fumitory Family 292

Gaillardia aristata 568
Galeopsis 480
  speciosa 480
  tetrahit 480
Galinsoga ciliata 569
Galium 509
  aparine 510
  boreale 510
  labradoricum 510
  septentrionale 510
  spurium 510
  trifidum 510
  triflorum 511
  verum 511
Garden Cress 320
Garden Rocket 316
Gaultheria 442
  hispidula 442
  humifusa 443
Gaura coccinea 420
Gayophytum
    racemosum 420
Gentian Family 455
Gentiana 456
  affinis 456
  amarella 457
  arctophila 458
  calycosa 456
  crinita 457
  detonsa 458
  fremontii 457
  glauca 456
  propinqua 458
  prostrata 457
Gentianaceae 455
Gentianella 457
  amarella 457
  crinita 457
  detonsa 458
  propinqua 458
Geocaulon lividum 222
Geraniaceae 391
Geranium 392
  bicknellii 392
  carolinianum 392
  erianthum 393
  nervosum 393
  richardsonii 393
  sphaerospermum 392
  viscosissimum 393
Geranium Family 391
Geum 351
  aleppicum 351, 352
  × aurantiacum 352
  macrophyllum 352
  × pervale 352
  rivale 352
  triflorum 352
Giant Hyssop 479
Gill-over-the-ground 480
Ginseng Family 424
Glacier Lily 181
Glaux maritima 452
Glecoma hederacea 480
Globe-flower 290
Glyceria 92
  borealis 92
  elata 93
  grandis 93
  maxima 93
  pauciflora 111
  pulchella 93
  striata 93
Glycyrrhiza lepidota 378
Gnaphalium 569
  microcephalum 569
  palustre 570
  uliginosum 570
  viscosum 570
Goat's-beard 600
Golden Aster 574
Golden Bean 389
Goldenrod 592
Goldthread 279
Goodyera 191
  oblongifolia 191
  repens 191
Gooseberry 342
Goosefoot 234, 240, 246
Gourd Family 517
Grama Grass 75
Gramineae 62
Grape Fern 32
Grass Family 62
Grass-of-Parnassus 341
Gratiola neglecta 492
Greasewood 246
Green-thread 598
Grindelia squarrosa 570
Grossulariaceae 342
Ground Cedar 25
Ground Ivy 480
Ground Pine 26
Ground Plum 375
Groundsel 587
Grouse-berry 448
Gumweed 570
Gutierrezia sarothrae 570
Gymnocarpium 39
  dryopteris 40
  × intermedium 40
  jessoense 40
  robertianum 40
Gypsophila 257
  elegans 258
  paniculata 258
  scorzonerifolia 258

Habenaria 191
  dilatata 192
  hyperborea 192
  obtusata 192
  orbiculata 193
  saccata 193
  unalascensis 193
  viridis 193
Hackelia 472
  americana 472
  deflexa 472
  floribunda 472
  jessicae 473
Hair Grass 73, 84
Halenia deflexa 458
Halimolobos virgata 317
Haloragaceae 423
Hammarbya paludosa 195

Haplopappus 571
  lanceolatus 571
  lyallii 571
  nuttallii 581
  spinulosus 571
  uniflorus 572
Harebell 518
Hare's-ear Mustard 307
Hawksbeard 557
Hawkweed 574
Hawthorn 348
Hazelnut 219
Heal-all 484
Heath Family 439
Heather 446
Hedeoma hispidum 481
Hedge Mustard 325
Hedge Nettle 484
Hedysarum 379
  alpinum 379
  boreale 379
  mackenzii 379
  sulphurescens 380
Helenium autumnale 572
Helianthus 572
  annuus 573
  couplandii 573
  laetiflorus 574
  maximilianii 573
  petiolaris 573
  nuttallii 573
  subrhomboideus 574
Helictotrichon hookeri 93
Heliotrope 473
Heliotropium
    curassavicum 473
Hemieva ranunculifolia 339
Hemlock 429
Hemp 219
Hemp Nettle 480
Henbane 485
Henbit 481
Heracleum lanatum 429
Hesperis matronalis 318
Heterotheca villosa 574
Heuchera 331
  cylindrica 331
  flabellifolia 332
  glabra 331
  parvifolia 331
  richardsonii 332
Hierachloe 94
  alpina 94
  odorata 94
Hieracium 574
  albertinum 575
  albiflorum 575
  aurantiacum 575
  canadense 576
  cynoglossoides 575
  gracile 576
  scabriusculum 576
  triste 575
  umbellatum 576
Hippochaete
  hyemalis 30
  laevigata 30
  scirpoides 31
  variegata 31
Hippophae rhamnoides 411
Hippuridaceae 423
Hippuris 423
  montana 424
  vulgaris 424
Hoary Alyssum 302
Hoary Cress 306
Holly Fern 42
Hollyhock 402
Honeysuckle 511, 512
Hops 220
Hordeum 94
  brachyantherum 95
  distichon 96
  glaucum 95
  jubatum 68, 71, 95
  pusillum 95
  vulgare 95
Horned Pondweed 58
Hornwort 272
Horse Mint 483
Horse-radish 301
Horsetail 28
Horseweed 563
Houstonia longifolia 511
Hudsonia tomentosa 404
Humulus lupulus 220
Huperzia selago 26
Hydrangeaceae 342
Hydrocharitaceae 62
Hydrophyllaceae 465
Hydrophyllum
    capitatum 466
Hymenopappus
    filifolius 576
Hymenoxys 576
  acaulis 577
  richardsonii 577
Hyoscyamus niger 485
Hypericaceae 403
Hypericum 403
  formosum 404
  majus 404
Hypopitys monotropa 439

Iliamna rivularis 402
Impatiens 400
  capensis 400
  noli-tangere 400
Indian Bread-root 388
Indian Hemp 459
Indian Paint-brush 488
Indian Pipe 438, 439
Iridaceae 186
Iris missouriensis 187
Isoetaceae 27
Isoetes 27
  bolanderi 28
  braunii 28
  echinospora 28
  muricata 28
Iva 577
  axillaris 577
  xanthifolia 577

Jacob's-ladder 464
Japanese Chess 78
Jewelweed 400
Johnny-jump-up 408
Juncaceae 168
Juncaginaceae 59
Juncus 168
  albescens 170
  alpinoarticulatus 170
  alpinus 170
  arcticus 171
  balticus 170
  biglumis 171
  brevicaudatus 171
  bufonius 171
  castaneus 171
  confusus 172
  drummondii 172

dudleyi 174
ensifolius 172
filiformis 172
litorum 171
longistylis 172
macer 174
mertensianus 173
nevadensis 173
nodosus 173
parryi 173
saximontanus 172
sphaerocarpus 171
stygius 173
tenuis 173
torreyi 174
tracyi 174
triglumis 170
vallicola 171
vaseyi 174
June Grass 96
June-berry 347
Juniper 44
Juniperus 44
communis 45
× fassettii 45
horizontalis 45
scopulorum 45
sibirica 45

Kalmia 443
microphylla 443
polifolia 443
Kinnikinnick 435, 441
Kitten-tails 488
Knapweed 553
Knautia arvensis 516
Knawel 263
Knotweed 226
Kobresia 161
bellardii 161
myosuroides 161
simpliciuscula 162
Kochia scoparia 245
Koeleria 96
cristata 96
macrantha 96
nitida 96
Koenigia islandica 225
Krascheninnikovia lanata 245

Labiatae 478
Labrador Tea 443
Lactuca 578
biennis 578
pulchella 578
sativa 578
serriola 579
tartarica 578
Ladies'-tresses 196
Lady Fern 36
Lady's-slipper 190
Lamb's quarters 241
Lamium amplexicaule 481
Lapland Rose-bay 446
Lappula 473
echinata 473
occidentalis 473
redowskii 473
squarrosa 473
Lapsana communis 579
Larch 47
Larix 47
laricina 47
lyallii 47
occidentalis 47
Larkspur 280
Lathyrus 380
ochroleucus 380
venosus 380
Lavauxia flava 422
Leather-leaf 442
Ledum 443
glandulosum 444
groenlandicum 444
palustre 444
Leguminosae 369
Lemna 167
minor 167
trisulca 168
Lemnaceae 167
Lentibulariaceae 505
Leonurus cardiaca 481
Lepidium 318
bourgeauanum 318
campestre 319
densiflorum 319
latifolium 319
perfoliatum 319
ramosissimum 320
sativum 320
Leptarrhena pyrolifolia 332
Lesquerella 320
alpina 320
arctica 320
arenosa 321
Lettuce 578
Lewisia pygmaea 251
Liatris 579
ligulistylis 579
punctata 580
Lilaea 59
scilloides 59
subulata 60
Lilaeaceae 59
Liliaceae 176
Lilium philadelphicum 181
Lily Family 176
Lily-of-the-valley 181
Limosella aquatica 492
Linaceae 393
Linanthus 463
harknessii 463
septentrionalis 463
Linaria 492
dalmatica 493
vulgaris 493
Linnaea borealis 512
Linum 393
lewisii 394
perenne 394
rigidum 394
usitatissimum 394
Listera 194
borealis 194
caurina 194
convallarioides 194
cordata 195
Lithophragma 332
bulbifera 333
glabrum 332
parviflorum 333
Lithospermum 473
incisum 474
ruderale 474
Little Club-moss 26
Little Red Elephant 496
Loasa Family 408
Loasaceae 408
Lobelia 519
dortmanna 519
kalmii 519
spicata 519
Lobeliaceae 518

Loco-weed 384
Loiseleuria
procumbens 444
Lolium 96
multiflorum 96
perenne 96
persicum 97
temulentum 97
Lomatium 430
cous 430
dissectum 430
foeniculaceum 430
macrocarpum 431
sandbergii 431
simplex 431
triternatum 431
Lomatogonium
rotatum 458
Lonicera 512
caerulea 513
dioica 513
ebraculata 513
involucrata 513
tartarica 513
utahensis 513
villosa 513
Loosestrife 412, 452
Loranthaceae 222
Lotus corniculatus 380
Lousewort 495
Lucerne 383
Luetkea pectinata 352
Lungwort 474
Lupine 381
Lupinus 381
argenteus 381
lepidus 381
minimus 381
nootkatensis 382
polyphyllus 382
pusillus 382
sericeus 382
Luzula 174
acuminata 175
arcuata 175
campestris 176
glabrata 175
hitchcockii 175
multiflora 175
parviflora 176
piperi 176
spicata 176
wahlenbergii 176
Lychnis 263
alba 266
apetala 267
drummondii 265
Lycium halimifolium 486
Lycopodiaceae 24
Lycopodiella inundata 26
Lycopodium 24
alpinum 25
annotinum 25
clavatum 25
complanatum 25
inundatum 25
obscurum 26
sabinaefolium 26
selago 26
sitchense 26
Lycopsis arvensis 474
Lycopus 481
americanus 482
asper 482
uniflorus 482
Lygodesmia 580
juncea 580
rostrata 580
Lysimachia 452
ciliata 453
lanceolata 453
thyrsiflora 453
Lythraceae 412
Lythrum salicaria 412

Machaeranthera 580
canescens 581
grindelioides 581
tanacetifolia 581
Madder Family 509
Madia glomerata 581
Madwort 469
Mahonia repens 290
Maianthemum
canadense 181
Maidenhair Fern 35
Malaxis 195
brachypoda 195
monophylla 195
paludosa 195
Male Fern 39
Mallow 401, 402
Malva 402
rotundifolia 403
sylvestris 403
verticillata 403
Malvaceae 401
Mamillaria vivipara 409
Manna Grass 92
Maple 399
Mare's-tail Family 423
Marijuana 219
Mariposa Lily 179
Marsh Elder 577
Marsh Marigold 278
Marsilea 44
mucronata 44
vestita 44
Marsileaceae 44
Matricaria 582
chamomilla 582
maritima 582
matricarioides 582
perforata 582
recutita 582
Matrimony Vine 486
Matteuccia
nodulosa 40
struthiopteris 40
Mayweed 534
Meadow Parsnip 434
Meadow Rue 289
Meadowsweet 368
Medicago 382
falcata 383
lupulina 383
sativa 383
Medick 382
Melampyrum lineare 493
Melic Grass 97
Melica 97
smithii 97
spectabilis 97
subulata 98
Melilotus 383
alba 383
officinalis 384
Mentha 482
arvensis 482
× gracilis 482
spicata 483
Mentzelia decapetala 408
Menyanthaceae 459

Menyanthes trifoliata 459
Menziesia 444
  ferruginea 444
  glabella 445
Meriolix serrulata 422
Mertensia 474
  lanceolata 475
  longiflora 475
  paniculata 475
Microseris 582
  cuspidata 583
  nutans 582
Microsteris gracilis 463
Microstylis monophyllos 195
Milfoil 526
Milk Thistle 592
Milk Vetch 371
Milkweed 460
Milkwort Family 394
Mimulus 493
  floribundus 494
  guttatus 494
  lewisii 494
Mint 478, 482
Minuartia 258
  austromontana 259
  biflora 259
  dawsonensis 259
  elegans 259
  nuttallii 260
  obtusiloba 260
  rubella 260
Mirabilis 250
  hirsuta 250
  nyctaginea 250
Mistletoe 222
Mitella 333
  breweri 333
  nuda 334
  pentandra 334
  trifida 334
Mitrewort 333
Moccasin-flower 190
Mock Orange 342
Moehringia lateriflora 260
Moldavica
  parviflora 479
  thymiflora 480
Monarda fistulosa 483
Moneses uniflora 436
Monkey-flower 493
Monkshood 273
Monolepis nuttalliana 246
Monotropa 438
  hypopitys 439
  uniflora 439
Monotropaceae 438
Montia 252
  linearis 252
  parvifolia 252
Montiastrum lineare 252
Moonwort 33
Mooseberry 515
Morning-glory 461
Moschatel 515
Motherwort 481
Mountain Ash 367
Mountain Avens 349
Mountain Heather 441
Mountain Lover 399
Mountain Sorrel 225
Mouse-ear Chickweed 255
Mousetail 281
Mudwort 492
Muhlenbergia 98
  asperifolia 98
  cuspidata 98
  glomerata 99
  racemosa 99
  richardsonis 99
Muhly Grass 98
Mullein 501
Munroa squarrosa 99
Musineon divaricatum 431
Mustard 294, 302
Myosotis 475
  alpestris 476
  arvensis 476
  laxa 476
  micrantha 476
Myosurus 281
  aristatus 281
  minimus 281
Myrica gale 215
Myricaceae 215
Myriophyllum 423
  exalbescens 423
  spicatum 423
  verticillatum 423

Naiad 53
Naiocrene parvifolia 252
Najadaceae 53
Najas flexilis 53
Nasturtium 321
  microphyllum 321
  officinale 321
Navarretia minima 463
Needle Grass 115
Nemophila breviflora 466
Nepeta cataria 483
Neslia paniculata 321
Nettle 220
Nightshade 485, 486
Ninebark 353
Nipplewort 579
Nonea vesicaria 476
Nothocalais cuspidata 583
Nuphar variegatum 271
Nyctaginaceae 249
Nymphaea tetragona 271
Nymphaeaceae 271

Oak Fern 40
Oat 75
Oat Grass 82
Ochrocodon pudicus 181
Odontites serotina 494
Oenothera 421
  andina 421
  biennis 421
  breviflora 422
  caespitosa 422
  flava 422
  nuttallii 422
  serrulata 422
Old Man's Whiskers 352
Oleaster 410
Onagraceae 412
Onion 178
Onobrychis viciifolia 384
Ononis arvensis 384
Onosmodium 476
  molle 476
  occidentale 477
Ophioglossaceae 32
Ophyrs
  borealis 194
  caurina 194
  convallarioides 194
  cordata 195
Oplopanax horridum 424
Opuntia 409

fragilis 409
polyacantha 409
Orache 237
Orchard Grass 82
Orchid Family 188
Orchidaceae 188
Orchis rotundifolia 196
Orobanchaceae 504
Orobanche 504
fasciculata 505
ludoviciana 505
uniflora 505
Orthilia secunda 436
Orthocarpus luteus 494
Oryzopsis 99
asperifolia 100
canadensis 100
exigua 100
hymenoides 100, 117
micrantha 101
pungens 101
Osmorhiza 431
chilensis 432
depauperata 432
longistylis 432
occidentalis 432
purpurea 433
Ostrich Fern 40
Owl-clover 494
Oxybaphus
hirsutus 250
nyctagineus 250
Oxycoccus 445
microcarpus 445
quadripetalus 445
Oxyria digyna 225
Oxytrope 384
Oxytropis 384
alpicola 385
campestris 385, 386
cusickii 385
deflexa 385
jordalii 386
lagopus 386
monticola 386
podocarpa 386
sericea 387
splendens 387
viscida 387

Pachistima myrsinites 399
Pachylophis caespitosa 422
Packera
cana 589
indecora 590
pauciflora 591
paupercula 591
pseudaurea 591
streptanthifolia 592
Panic Grass 101
Panicum 84, 101
ferventicola 85
capillare 101
miliaceum 101
thermale 85
wilcoxianum 85
Pansy 406, 408
Papaver 291
freedmanianum 291
kluanensis 291
nudicaule 291
pygmaeum 292
radicatum 291
somniferum 292
Papaveraceae 291
Parietaria pensylvanica 220
Parnassia 341
fimbriata 341
kotzebuei 341
montanensis 342
palustris 342
parviflora 342
Parnassiaceae 341
Paronychia sessiliflora 261
Parsnip 433
Partridge-foot 352
Pasque Flower 275
Pastinaca sativa 433
Pathfinder 527
Pea Family 369
Pea Vine 380
Pearlwort 262
Pearly Everlasting 529
Pedicularis 495
albertae 496
arctica 495
bracteosa 496
capitata 496
contorta 496
flammea 496
groenlandica 496
labradorica 497
lanata 497
langsdorfii 495
macrodonta 497
parviflora 497
racemosa 497
sudetica 497
Pellaea 40
atropurpurea 41
glabella 41
occidentalis 41
suksdorfiana 41
Pellitory 220
Pembina 515
Pennycress 325
Pennyroyal 481
Penstemon 497
albertinus 498
albidus 498
confertus 499
ellipticus 499
eriantherus 499
fruticosus 499
gracilis 499
linearifolius 499
lyallii 499
nitidus 500
procerus 500
virens 498
Pentaphylloides
floribunda 358
Peppergrass 318
Pepperwort 319, 320
Peramium
menziesii 191
repens 191
Perideridia gairdneri 433
Petalostemon 387
candidum 388
purpureum 388
Petasites 583
frigidus 584
hyperboreus 583
nivalis 583
palmatus 584
sagittatus 584
vitifolius 584
Phacelia 466
franklinii 467
hastata 467
heterophylla 467
linearis 467

lyallii 467
sericea 467
Phalaris 102
arundinacea 102
canariensis 102
Phegopteris
connectilis 41
dryopteris 40
Philadelphus lewisii 342
Phleum 102
alpinum 103
commutatum 103
pratense 103
Phlox 462, 463
alyssifolia 464
gracilis 463
hoodii 464
Phragmites 103
australis 103
communis 103
Phyllodoce 445
empetriformis 446
glanduliflora 446
× intermedia 446
Physalis ixocarpa 486
Physaria didymocarpa 321
Physocarpus
malvaceus 353
Physostegia parviflora 483
Picea 48
engelmannii 48
glauca 48
mariana 48
Picradeniopsis
oppositifolia 550
Picris echioides 584
Pigweed 240, 249
Pin Cherry 363
Pinaceae 46
Pin-cushion Plant 463
Pine 49
Pine Family 46
Pineapple-weed 582
Pine-drops 439
Pine-sap 439
Pinguicula 505
villosa 505
vulgaris 506
Pink 256
Pink Family 253
Pinus 49
albicaulis 49
banksiana 49, 50
contorta 49
flexilis 50
monticola 50
Piperia unalaschensis 193
Pipsissewa 435
Pitcher-plant 326
Plagiobothrys 477
scopulorum 477
scouleri 477
Plantaginaceae 507
Plantago 507
canescens 507
elongata 508
eriopoda 508
major 508
maritima 508
patagonica 508
purshii 509
septata 508
spinulosa 509
Plantain 507
Platanthera
dilatata 192
hyperborea 192
orbiculata 193
saccata 193
unalaschensis 193
Poa 103
agassizensis 109
alpigena 109
alpina 105
ampla 107
annua 105
arctica 106
arida 106
canbyi 106
compressa 106
cusickii 106
epilis 107
glauca 107
gracillima 107
interior 107
juncifolia 107
leptocoma 107
lettermanii 108
longipila 106
nemoralis 107, 108
nervosa 108
nevadensis 108
palustris 108
pattersonii 108
paucispicula 108
pratensis 109
rupicola 107
sandbergii 109
secunda 109
stenantha 109
Podagrostis
humilis 72
thurberiana 73
Poison Ivy 398
Polanisia 293
dodecandra 293
trachysperma 293
Polemoniaceae 462
Polemonium 464
acutiflorum 464
caeruleum 464
pulcherrimum 464
viscosum 465
Polygala 394
paucifolia 395
senega 395
Polygalaceae 394
Polygonaceae 223
Polygonum 226
achoreum 229
amphibium 227
arenastrum 227
austiniae 227
aviculare 227, 229
bistortoides 228
coccineum 228
convolvulus 228
douglasii 228
dumetorum 230
engelmannii 228
erectum 229
hartwrightii 227
lapathifolium 229
minimum 229
monspeliense 229
muhlenbergii 228
natans 227
persicaria 229
prolificum 230
ramosissimum 229
rigidulum 228
scabrum 229
scandens 230

viviparum 230
watsonii 230
Polypodiaceae 34
Polypodium 41
hesperium 41
virginianum 42
vulgare 42
Polypogon monspeliensis 109
Polystichum lonchitis 42
Pond-lily 271
Pondweed 53
Poplar 196
Poppy 291
Populus 196
× acuminata 197
× andrewsii 197
angustifolia 197
balsamifera 197
× brayshawii 197
detoides 197, 198
× dutillyi 197
× jackii 197
sargentii 198
× sennii 197
tacamahacca 198
tremuloides 197, 198
trichocarpa 198
Porcupine Grass 117
Portulaca oleracea 252
Portulacaceae 250
Potamogeton 53
alpinus 55, 56
berchtoldii 57
crispus 55
filiformis 55
foliosus 55
friesii 55
gramineus 55, 56
× griffithii 55
natans 56
× nericus 55
obtusifolius 56
panormitanus 57
pectinatus 55, 56
perfoliatus 57
praelongus 55, 56
pusillus 57
richardsonii 57
robbinsii 57
× sparganifolius 56
strictifolius 57
× suecicus 55
vaginatus 57
zosterifolius 58
zosteriformis 58
Potamogetonaceae 53
Potentilla 353
anserina 356
argentea 356
arguta 356
bipinnatifida 357
concinna 357
diversifolia 357
drummondii 357
effusa 359
finitima 358
flabelliformis 359
fruticosa 358
furcata 359
glandulosa 358
gracilis 358
hippiana 359
hookeriana 359
hyparctica 359
ledebouriana 363
multifida 360
multisecta 360
nivea 359, 360
norvegica 360
nuttallii 359
ovina 361
palustris 361
paradoxa 361
pensylvanica 361
plattensis 362
pulcherrima 359
quinquefolia 359
recta 362
rivalis 362
tridentata 362
uniflora 362
villosa 363
Poverty-weed 577
Prairie Clover 387
Prairie Crocus 275
Prairie Parsley 430
Prairie Pink 580
Prairie Rocket 316
Prairie Smoke 352
Prenanthes 584
alata 585
racemosa 585
sagittata 585
Prickly Pear 409
Primrose 449, 453
Primula 453
egaliksensis 454
incana 454
mistassinica 454
stricta 454
Primulaceae 449
Prince's Pine 435
Prunella vulgaris 484
Prunus 363
pensylvanica 363
virginiana 364
Pseudotsuga 50
menziesii 50
taxifolia 50
Psilocarphus elatior 585
Psoralea 388
argophylla 388
esculenta 388
lanceolata 389
Pteretis
nodulosa 40
pensylvanica 40
Pteridium aquilinum 42
Pterospora andromedea 439
Puccinellia 109
airoides 111
cusickii 110
distans 110
hauptiana 110
nuttalliana 110
pauciflora 111
Puccoon 473
Pulsatilla
occidentalis 275
patens 276
Purslane 250, 252
Pussy-toes 530
Pyrola 436
asarifolia 437
bracteata 437
chlorantha 437
elliptica 437
grandiflora 437
minor 438
picta 438
secunda 436

uniflora 436
virens 437
Pyrolaceae 435

Quack Grass 70
Quillwort 27
Quinsywort 509

Rabbit-brush 554
Rabbitfoot Grass 109
Radish 322
Ragweed 528, 568
Ragwort 587
Ranunculaceae 272
Ranunculus 281
abortivus 283
acris 283
affinis 287
aquatilis 284
bongardii 289
cardiophyllus 284
circinatus 284
cymbalaria 284
eschscholtzii 284
flammula 288
gelidus 285
glaberrimus 285
gmelinii 285
grayi 285
hyperboreus 285
inamoenus 286
lapponicus 286
limosus 285
macounii 286
natans 286
nivalis 286
occidentalis 287
pedatifidus 284, 287
pensylvanicus 287
purshii 285
pygmaeus 287
repens 287
reptans 288
rhomboideus 288
sceleratus 288
trichophyllus 284
uncinatus 288
verecundus 289
Rape 302
Raphanus
raphanistrum 322

Raspberry 365
Ratibida columnifera 585
Rattlesnake plantain 191
Red Osier 435
Redtop 72, 73
Reed 103
Reed Grass 79
Rhamnaceae 400
Rhamnus 401
alnifolia 401
catharticus 401
Rhinanthus 500
borealis 500
crista-galli 500
groenlandicus 500
minor 500
Rhododendron 446
albiflorum 446
lapponicum 446
Rhus 398
aromatica 398
radicans 398
trilobata 398
Rhynchospora
capillacea 162
Ribes 342
americanum 343
aureum 344
glandulosum 344
hirtellum 344
hudsonianum 344
inerme 344
lacustre 345
laxiflorum 345
oxyacanthoides 345
setosum 345
triste 345
viscosissimum 345
Rice Grass 99
Rock Brake 36
Rock Cress 298
Rockrose Family 404
Romanzoffia sitchensis 468
Rorippa 322
austriaca 322
curvipes 323
islandica 323
obtusa 323
nasturtium-aquaticum 321
palustris 323
sinuata 323

sylvestris 323
tenerrima 323
Rosa 364
acicularis 364
alcea 365
arkansana 365
bourgeauiana 365
fendleri 365
macounii 365
suffulta 365
ultramontana 365
woodsii 365
Rosaceae 346
Rose 346, 364
Round-leaved Orchid 196
Rowan Tree 368
Rubiaceae 509
Rubus 365
acaulis 366
arcticus 366
chamaemorus 366
idaeus 366
melanolasius 366
× paracaulis 366
parviflorus 366
pedatus 367
pubescens 367
strigosus 366
Rudbeckia 586
hirta 586
serotina 586
Rumex 230
acetosa 231
acetosella 232
angiocarpus 232
aquaticus 233
britannica 232
confertus 232
crispus 232
dentatus 233
fennicus 234
longifolius 233
maritimus 233
mexicanus 234
occidentalis 233
paucifolius 233
pseudonatronatus 234
stenophyllus 234
triangulivalis 234
venosus 234
Ruppia 58

maritima 58
occidentalis 58
Ruppiaceae 58
Rush 168
Russian Olive 410
Russian Pigweed 239
Russian Thistle 246
Rutabaga 303
Rye 112
Ryegrass 96

Sabina
horizontalis 45
scopulorum 45
Safflower 552
Sagebrush 539
Sagewort 541, 542
Sagina 261
decumbens 261
intermedia 262
nivalis 262
nodosa 262
saginoides 262
Sagittaria 61
cuneata 61
latifolia 61
sagittifolia 61
Sainfoin 384
Salicaceae 196
Salicornia 246
europaea 246
rubra 246
Salix 198
alaxensis 202
alba 198
amygdaloides 202
arbusculoides 203
arctica 203, 208
athabascensis 203
barclayi 204, 205, 206
barrattiana 204, 206, 210
bebbiana 204
boothii 205, 210
brachycarpa 204, 205, 208
candida 205, 212
caudata 209
commutata 204, 206
daphnoides 198
discolor 206, 211
drummondiana 206
exigua 207
farriae 204, 207
fragilis 198
glauca 204, 205, 208
humilis 213
interior 207
lanata 208
lasiandra 209
lucida 208
lutea 209
maccalliana 209
mackenzieana 211
melanopsis 209
myrtillifolia 204, 205, 210
nivalis 213
pedicellaris 210
pentandra 198
petiolaris 205, 210
planifolia 206, 211
prolixa 211
pseudocordata 205
pseudomonticola 206, 212
pyrifolia 212
reticulata 212
scouleriana 211, 213
serissima 213
sitchensis 213
stolonifera 214
vestita 214
Salmon Berry 366
Salsola kali 246
Salt Grass 86
Salt Sage 237
Saltbush 237
Salvia nemorosa 484
Sambucus 514
melanocarpa 514
pubens 514
racemosa 514
Samphire 246
Sand Grass 81
Sand Heather 404
Sand Lily 408
Sand Rocket 308
Sand Spurry 267
Sand Verbena 249
Sandalwood Family 221
Sandwort 254
Sanicula marilandica 433
Santalaceae 221
Saponaria officinalis 262
Sarcobatus vermiculatus 246
Sarracenia purpurea 326
Sarraceniaceae 326
Saskatoon 347
Saussurea 586
americana 586
densa 587
glomerata 586
nuda 587
Saw-wort 586
Saxifraga 334
adscendens 335
aizoides 336
arguta 338
bronchialis 336
caespitosa 336
cernua 336
debilis 337
ferruginea 336
flagellaris 337
hyperborea 337
lyallii 337
mertensiana 337
nelsoniana 337
nivalis 338
occidentalis 338
odontoloma 337, 338
oppositifolia 338
oregana 339
rivularis 337
tricuspidata 339
Saxifragaceae 328
Saxifrage 328, 334
Scarlet Mallow 403
Schedonnardus paniculatus 111
Scheuchzeria palustris 60
Scheuchzeriaceae 60
Schizachne purpurascens 111
Schizachyrium scoparium 112
Schoenocrambe linifolia 324
Schoenoplectus
acutus 163
validus 166
Scholochloa festucaea 112
Scirpus 162
acutus 163

americanus 166
atrocinctus 164
atrovirens 165
cespitosus 163
clintonii 164
cyperinus 164
fluviatilis 164
hudsonianus 164
lacustris 163, 166
maritimus 165
microcarpus 165
nevadensis 165
pallidus 165
paludosus 165
pumilus 165
pungens 166
rollandii 166
rubrotinctus 165
validus 166
Scleranthus annuus 263
Scorpion Weed 466
Scouring Rush 28
Scrophularia lanceolata 500
Scrophulariaceae 487
Scurf Pea 389
Scutellaria galericulata 484
Sea Buckthorn 411
Sea Milkwort 452
Secale cereale 112
Sedge 120
Sedge Family 120
Sedum 327
divergens 327
douglasii 328
lanceolatum 328
rosea 328
stenopetalum 328
Selaginella 26
densa 27
rupestris 27
scopulorum 27
selaginoides 27
standleyi 27
wallacei 27
Selaginellaceae 26
Seneca-root 395
Senecio 587
canus 588
congestus 589
conterminus 589
cymbalarioides 589
eremophilus 589
foetidus 589
fremontii 590
hydrophylloides 590
indecorus 590
integerrimus 590
lugens 590
megacephalus 591
pauciflorus 591
pauperculus 591
pseudaureus 591
resedifolius 589
streptanthifolius 591
subnudus 589
triangularis 592
viscosus 592
vulgaris 592
Service-berry 347
Setaria 112
glauca 113
viridis 113
Shepherdia 411
argentea 411
canadensis 411
Shepherd's-purse 305
Shield Fern 38
Shooting Star 451
Sibbaldia procumbens 367
Silene 263
acaulis 264
antirrhina 264
conoidea 265
cserei 265
cucubalis 265
drummondii 265
furcata 265
menziesii 265
noctiflora 266
parryi 266
pratensis 266
uralensis 267
vulgaris 265
williamsii 266
Silver-berry 410
Silybum marianum 592
Sisymbrium 324
altissimum 324
linifolium 324
loeselii 324
officinale 325
Sisyrinchium 187
montanum 187
sarmentosum 187
septentrionale 187
Sitanion hystrix 113
Sium suave 434
Skeleton-weed 580
Skullcap 484
Skunk Grass 89
Skunk-bush 398
Skunkweed 465
Slender Lip Fern 36
Slough Grass 75
Smartweed 226
Smelowskia calycina 325
Smilacina 182
racemosa 182
stellata 182
trifolia 182
Snake-root 433
Snapdragon 491
Sneezeweed 527, 572
Snowberry 442, 514
Soapweed 185
Soapwort 262
Solanaceae 485
Solanum 486
nigrum 486, 487
rostratum 486
sarrachoides 487
triflorum 487
Solidago 592
canadensis 593
decumbens 595
gigantea 594
graminifolia 594
lepida 593
missouriensis 594
mollis 594
multiradiata 594
nemoralis 595
parvirigida 595
pruinosa 593
rigida 595
shinnersii 594
spathulata 595
Solomon's-seal 182
Sonchus 595
arvensis 596
asper 596
oleraceus 596
uliginosus 596

Sorbus 367
aucuparia 368
scopulina 368
sitchensis 368
Sorrel 230
Southernwood 540
Sow Thistle 595
Spanish Bayonet 185
Sparganiaceae 51
Sparganium 51
angustifolium 52
chlorocarpum 52
emersum 52
eurycarpum 52
fluctuans 52
glomeratum 52
hyperboreum 53
minimum 53
multipedunculatum 52
Spartina 113
gracilis 114
pectinata 114
Spear Grass 116
Spearmint 483
Spearwort 288
Speedwell 501
Spergula arvensis 267
Spergularia 267
diandra 267
marina 267
Sphaeralcea coccinea 403
Sphaerostigma andinum 421
Sphenopholis 114
intermedia 114
obtusata 114
Spike Rush 157
Spiraea 368
alba 369
betulifolia 369
densiflora 369
lucida 369
Spiranthes
romanzoffiana 196
Spirodela polyrhiza 168
Spleenwort 35
Sporobolus 115
cryptandrus 115
neglectus 115
Spotted Joe-Pye Weed 568
Spring Beauty 251
Spruce 48
Spurge 395
Squaw-root 433
Squirreltail 113
Stachys palustris 484
Star Thistle 553
Star-flower 455
Starwort 268
Steironema
ciliatum 453
lanceolatum 453
Stellaria 268
alpestris 269
americana 269
arenicola 269
borealis 269
calycantha 269
crassifolia 269
crispa 269
gonomischa 270
laeta 270
longifolia 269
longipes 270
media 270
monantha 270
obtusa 270
sitchana 269
umbellata 270
Stenanthella occidentalis 183
Stenanthium
occidentale 183
Stephanomeria 596
runcinata 596
tenuifolia 596
Stick-seed 472
Stick-tights 473
Stinkweed 325
Stipa 115
columbiana 116
comata 116
curtiseta 116
nelsonii 116
occidentalis 116
richardsonii 117
spartea 117
viridula 117
× Stiporyzopsis
caduca 101
St John's-wort 403
Stonecrop 327
Stork's-bill 391
Strawberry 350
Streptopus 183
amplexifolius 183, 184
× oreopolus 184
roseus 183
streptopoides 184
Suaeda 247
calceoliformis 247
depressa 247
intermedia 247
moquinii 247
Suckleya suckleyana 247
Suksdorfia 339
ranunculifolia 339
violacea 340
Sumach Family 398
Summer Cypress 245
Sundew 326
Sunflower 572
Swede 303
Sweet Cicely 431
Sweet Clover 383
Sweet Flag 166
Sweet Gale 215
Sweet Grass 94
Symphoricarpos 514
albus 514
occidentalis 514
Symphytum officinale 477
Syringa 342

Tamarack 47
Tanacetum 597
huronense 597
vulgare 597
Tansy 597
Tansy Mustard 307
Taraxacum 597
ceratophorum 598
eriophorum 598
laevigatum 598
lyratum 598
officinale 598
Taraxia breviflora 422
Tarweed 581
Taxaceae 51
Taxus brevifolia 51
Teasel Family 516
Telesonix
heucheriformis 340
Thalictrum 289
dasycarpum 289

occidentale 289
sparsiflorum 290
turneri 290
venulosum 290
Thelesperma
marginatum 598
Thellungiella
salsuginea 325
Thelypteris dryopteris 40
Thermopsis
rhombifolia 389
Thimble Berry 366
Thistle 551, 555
Thlaspi arvense 325
Three-awn 74
Thuja plicata 45
Tiarella 340
trifoliata 340
unifoliata 341
Tickle Grass 73
Tickseed 557
Timothy 102
Tiniaria convolvulus 228
Toad-flax 492
Tofieldia 184
glutinosa 184
pusilla 184
Tolmachevia
integrifolia 328
Tomatillo 486
Torreyochloa pauciflora 111
Touch-me-not 400
Tower Mustard 299
Townsendia 598
condensata 599
exscapa 599
hookeri 599
parryi 599
sericea 599
Toxicodendron rydbergii 398
Toxicoscordion venenosus 186
Tragopogon 599
dubius 600
porrifolius 600
pratensis 600
Trail-plant 527
Trichophorum
alpinum 164
cespitosum 164
pumilum 166
Trientalis 455
borealis 455
europaea 455
Trifolium 389
agrarium 390
aureum 390
hybridum 390
pratense 390
repens 390
Triglochin 59
maritima 59
palustris 59
Trigonella caerulea 390
Trillium ovatum 185
Trisetum 117
canescens 118
cernuum 118
flavescens 118
montanum 118
spicatum 118
wolfii 119
Triticum aestivum 119
Trollius albiflorus 290
Tsuga heterophylla 50
Tumble Grass 111
Tumbleweed 248
Tumbling Mustard 324
Turritis glabra 300
Twayblade 194
Twin-flower 512
Twisted-stalk 183
Typha latifolia 51
Typhaceae 51

Umbelliferae 425
Umbrella-plant 224
Umbrellawort 250
Urtica 220
dioica 221
gracilis 221
lyallii 221
urens 221
Urticaceae 220
Utricularia 506
cornuta 506
intermedia 506
minor 506
vulgaris 507

Vaccaria 271
pyramidata 271
segetalis 271
Vaccinium 447
caespitosum 447
membranaceum 448
myrtilloides 448
myrtillus 448
ovalifolium 448
scoparium 448
uliginosum 448
vitis-idaea 449
Vahlodea atropurpurea 11
Valerian Family 516
Valeriana 516
dioica 516
scouleri 516
septentrionalis 516
sitchensis 516
Valerianaceae 516
Velvet-leaf 402
Venus'-slipper 188
Veratrum 185
eschscholtzii 185
viride 185
Verbascum 500
nigrum 501
phlomoides 501
thapsus 501
Verbena bracteata 477
Verbenaceae 477
Veronica 501
agrestis 502
alpina 502
americana 502
catenata 503
chamaedrys 503
longifolia 503
peregrina 503
persica 503
salina 503
scutellata 503
serpyllifolia 504
Vervain 477
Vetch 390
Vetchling 380
Viburnum 515
edule 515
opulus 515
trilobum 515
Vicia 390
americana 391
cracca 391
sparsifolia 391

Viola 405
  adunca 406
  aduncoides 406
  arvensis 406
  canadensis 406
  glabella 406
  macloskeyi 407
  nephrophylla 407
  nuttallii 407
  orbiculata 407
  palustris 407
  pedatifida 408
  renifolia 408
  rugulosa 406
  selkirkii 408
  tricolor 408
  vallicola 407
Violaceae 405
Violet 405
Virgin's-bower 278
Virgulus
  campestris 546
  ericoides 547
  falcatus 547
Vulpia octoflora 119

Wake-robin 185
Wapato 61
Wartweed 396
Water Arum 167
Water Horehound 481
Water Hemlock 428
Water Parsnip 434
Waterleaf 465, 466
Water-lily 271
Water-milfoil 423
Water-plantain 60
Water-starwort 397
Waterweed 62
Waterwort 404
Wedge Grass 114
Western Hemlock 50
Wheat 119
Wheat Grass 68
White Cockle 266
White Lettuce 584
Whiteman's-foot 508
Whitlow-wort 261
Widgeon Grass 58
Wild Calla 167
Wild Cucumber 517
Wild Licorice 378
Wild Rice 120
Wild Rye 86
Wild Sarsaparilla 424
Wild Tomato 487
Willow 198
Willow-herb 413
Winter Cress 301
Winter Fat 245
Wintergreen 435, 436
Witch Grass 101
Wolf Willow 410
Wolfberry 514
Wood Lily 181
Wood Reed 82
Wood Rush 174
Wood Sage 484
Woodruff 509
Woodsia 42
  glabella 43
  ilvensis 43
  × maxonii 43
  oregana 43
  scopulina 44
Woolly-heads 585
Wormseed Mustard 317
Wormwood 539

Xanthium strumarium 600
Xerophyllum tenax 185

Yampa 433
Yarrow 526
Yellow-bell 181
Yellow Cress 322
Yellow Rattle 500
Yellow Rocket 302
Yew 51
Yucca glauca 185

Zannichellia palustris 58
Zannichelliaceae 58
Zigadenus 186
  elegans 186
  gramineus 186
  venenosus 186
Zizania aquatica 120
Zizia aptera 434